HIGHWAY SUBCOMMITTEE ON DESIGN

Chairman: Byron C. Blaschke, Texas
Vice Chairman: Kenneth C. Afferton, New Jersey
Secretary: Thomas Willett, FHWA

Alabama, Don Arkle, Ray D. Bass, J.F. Caraway
Alaska, Rodney R. Platzke, Timothy Mitchell, Boyd Brownfield
Arizona, Robert P. Mickelson, Dallis B. Saxton, John L. Louis
Arkansas, Bob Walters, Paul DeBusk
California, Walter P. Smith
Colorado, James E. Siebels
Connecticut, Earle R. Munroe, Bradley J. Smith, James F. Byrnes, Jr.
Delaware, Michael A. Angelo, Chao H. Hu
D.C., Charles F. Williams, Sanford H. Vinick
Florida, Bill Deyo, Ray Reissener
Georgia, Walker Scott, Hoyt J. Lively, Roland Hinners
Hawaii, Kenneth W.G. Wong, Albert Yamaguchi
Idaho, Richard Sorensen, Jeff R. Miles
Illinois, Ken Lazar, Dennis Pescitelli
Indiana, Gregory L. Henneke
Iowa, George F. Sisson, Donald L. East, Dave Little
Kansas, Bert Stratmann, James Brewer, Richard G. Adams
Kentucky, Charles S. Raymer, John Sacksteder, Steve Williams
Louisiana, Charles M. Higgins, William Hickey, Nick Kalivado
Maine, Charles Smith, Walter Henrickson
Maryland, Steve Drumm, Robert D. Douglass
Massachusetts, Sherman Eidelman, Frederick J. Nohelty, Jr.
Michigan, Charles J. Arnold
Minnesota, Roger M. Hill
Mississippi, Irving Harris, Wendel T. Ruff, Glenn Calloway
Missouri, Frank Carroll, Bob Sfreddo
Montana, David S. Johnson, Ronald E. Williams, Carl S. Peil
Nebraska, Gerald Grauer, Marvin J. Volf, Eldon D. Poppe
Nevada, Michael W. McFall, Steve R. Oxoby
New Hampshire, Gilbert S. Rogers
New Jersey, Kenneth Afferton, Walter W. Caddell, Charles A. Goessel
New Mexico, Joseph Pacheco, Charles V.P. Trujillo
New York, J. Robert Lambert, Philip J. Clark, Robert A. Dennison
North Carolina, D.R. (Don) Morton, G.T. (Tom) Rearin, J.T. Peacock, Jr.
North Dakota, David K.O. Leer, Ken Birst
Ohio, Donald K. Huhman, George L. Butzer
Oklahoma, Bruce E. Taylor, Richard B. Hankins, C. Wayne Philliber
Oregon, Tom Lulay, Wayne F. Cobine
Pennsylvania, Fred W. Bowser, John J. Faiella, Jr., Dean Schreiber
Puerto Rico, Jose E. Hernandez, Maria M. Casse, Eugenio Davila
Rhode Island, J. Michael Bennett
South Carolina, Robert L. White, William M. DuBose
South Dakota, Lawrence L. Weiss, Larry Engbrecht, Monte Schneider
Tennessee, Paul Morrison, Clellon Loveall, Jerry D. Hughes
Texas, Frank D. Holzmann, William A. Lancaster, Mark Marek
U.S. DOT, Robert Bates (FAA), Thomas O. Willett (FHWA)
Utah, Dyke LeFevre, P.K. Mohanty, Heber Vlam
Vermont, Robert M. Murphy, Donald H. Lathrop, John L. Armstrong
Virginia, E.C. Cochran, Jr., R.E. Atherton, K.F. Phillips
Washington, E.R. (Skip) Burch
West Virginia, Norman Roush, Randolph Epperly
Wisconsin, Joseph W. Dresser, Robert Pfeiffer
Wyoming, Donald A. Carlson

AFFILIATE MEMBERS

Alberta, P.F. (Peter) Tajcnar
Hong Kong, S.K. Kwei
Manitoba, A. Boychuk
Mariana Islands, Nick C. Sablan
New Brunswick, C. Herbert Page
Newfoundland, Terry McCarthy
Northwest Territories, Peter Vician
Nova Scotia, Donald W. MacIntosh
Ontario, Gerry McMillan
Saskatchewan, Ray Gerbrandt

ASSOCIATE MEMBERS—STATE

Mass. Metro. Dist. Comm., E. Leo Lydon
N.J. Turnpike Authority, Arthur A. Linfante, Jr.
Port Auth. of NY & NJ, Harry Schmerl

ASSOCIATE MEMBERS—FEDERAL

Bureau of Indian Affairs—Division of Transportation, Kimo Natewa
U.S. Department of Agriculture—Forest Service, Tom Pettigrew

JOINT TASK FORCE ON PAVEMENTS

Region 1	**Members**	**Representatives**
Connecticut	Charles Dougan	
New York	Wes Yang	
Pennsylvania	Dennis Morian	
Port Authority of NY & NJ	Harry Schmerl	
FHWA	Louis M. Papet (Secretary)	

Region 2

Arkansas	Robert L. Walters (Vice Chairman)	
Florida	William N. Lofroos	
Louisiana	J.B. Esnard, Jr.	
North Carolina	Ken Creech	Tom Hearne

Region 3

Illinois	John Ebers	
Iowa	George Sisson	Brain McWaters
Missouri	Frank L. Carroll	Danny Davidson
Ohio	Aric Morse	

Region 4

California	Bob Doty	
Oregon	Ira J. Huddleston	
Texas	James L. Brown (Chairman)	
Utah	Les Jester	Wade Betenson
Washington	Newt Jackson	
Wyoming	Tom Atkinson	Don Carlson

Representing

Transportation Research Board	Daniel W. Dearsaugh, Jr., Senior Program Officer
Standing Committee on Planning	Fred Van Kirk, West Virginia
Subcommittee on Construction	Dean M. Testa, Kansas
Subcommittee on Maintenance	Robert W. Moseley, Mississippi
Subcommittee on Materials	Larry Epley, Kentucky
Standing Committee on Aviation	Robert Bates, FAA; Roger H. Barcus, Illinois; Craig Smith, South Dakota

AASHTO® Guide for Design of Pavement Structures 1993

Published by the
American Association of State Highway
and Transportation Officials

444 N. Capitol Street, N.W., Suite 249
Washington, D.C. 20001

© Copyright, 1986, 1993 by the American Association of State Highway and Transportation Officials. *All Rights Reserved.* Printed in the United States of America. This book, or parts thereof, may not be reproduced in any form without written permission of the publishers.

ISBN 1-56051-055-2

SPECIAL NOTICE

The *Guide for Design of Pavement Structures*, when it was published in 1986, was published as two volumes. Volume 1 was written as a basic design guide and provided all of the information required to understand and apply the "Guide" to pavement design. Volume 2 was a series of appendices prepared to provide documentation or further explanations for information contained in Volume 1. Volume 2 is not required for design.

This 1993 edition of the "Guide" contains only one Volume. This Volume replaces the 1986 "Guide" Volume 1 and serves the same purpose. The major changes included in the 1993 "Guide" are changes to the overlay design procedure and the accompanying appendices L, M, and N. There are other minor changes and some of an editorial nature throughout the new Volume 1.

Volume 2 of the 1986 "Guide" is still applicable to most sections of Volume 1 of the 1993 "Guide" and is available through AASHTO, 444 N. Capitol Street, N.W., Suite 249, Washington, D.C. 20001; 202-624-5800. Request book code "GDPS3-V2." A copy of the Table of Contents from Volume 2 of the 1986 "Guide" follows.

VOLUME 2 APPENDICES

AA.	Guidelines for the Design of Highway Internal Drainage Systems
BB.	Position Paper on Pavement Management
CC.	Remaining Life Considerations in Overlay Design
DD.	Development of Coefficients for Treatment of Drainage
EE.	Development of Reliability
FF.	Relationship Between Resilient Modulus and Soil Support
GG.	Relationships Between Resilient Modulus and Layer Coefficients
HH.	Development of Effective Roadbed Soil Moduli
II.	Survey of Current Levels of Reliability
JJ.	Development of Design Nomographs
KK.	Determination of J-Factor for Undowelled Pavements
LL.	Development of Models for Effects of Subbase and Loss of Support
MM.	Extension of Equivalency Factor Tables
NN.	Recommendations for the Selection of an AASHTO Overlay Method Using NDT Within the AASHTO Performance Model Framework
OO.	Pavement Recycling Fundamentals
PP.	Development of NDT Structural Capacity Relationships

PREFACE

When construction, maintenance, and rehabilitation costs are considered, the single most costly element of our nation's highway system is the pavement structure. In an effort to reduce this cost, the state highway and transportation departments and the Federal Government have sponsored a continuous program of research on pavements. One output of that research effort was the *Interim Guide for the Design of Pavement Structures* published in 1972 and revised in 1981. It was based largely upon the findings at the AASHO Road Test.

Because this is such an important topic, the Joint Task Force on Pavements—composed of members from the Subcommittee on Design, one member each from the Materials, Construction, and Maintenance Subcommittees, and one from the Planning Committee of AASHTO—was assigned the task of rewriting the Interim Guide incorporating new developments and specifically addressing pavement rehabilitation.

Because many states were found to be using at least portions of the Interim Guide and because no other generally accepted procedures could be identified, it was decided that this Guide would retain the basic algorithms developed from the AASHO Road Test as used in the Interim Guide. Because the Road Test was very limited in scope, i.e. a few materials, one subgrade, non-mixed traffic, one environment, etc., the original Interim Guide contained many additional models to expand the framework so designers could consider other conditions. The new Guide has been further expanded with the following 14 major new considerations:

(1) Reliability
(2) Resilient Modulus for Soil Support
(3) Resilient Modulus for Flexible Pavement Layer Coefficients
(4) Drainage
(5) Improved Environment Considerations
(6) Tied Concrete Shoulders or Widened Lanes
(7) Subbase Erosion for Rigid Pavements
(8) Life Cycle Cost Considerations
(9) Rehabilitation
(10) Pavement Management
(11) Extension of Load Equivalency Values
(12) Improved Traffic Data
(13) Design of Pavements for Low Volume Roads
(14) State of the Knowledge on Mechanistic-Empirical Design Concepts

The Task Force recognizes that a considerable body of information exists to design pavements utilizing so-called mechanistic models. It further believes that significant improvements in pavement design will occur as these mechanistic models are calibrated to in-service performance, and are incorporated in everyday design usage. Part IV of this document summarizes the mechanistic/empirical status.

In order to provide state-of-the-art approaches without lengthy research, values and concepts are shown that have limited support in research or experience. Each user should consider this to be a reference document and carefully evaluate his or her need of each concept and what initial values to use. To most effectively use the Guide it is suggested that the user adopt a process similar to the following:

(1) Conduct a sensitivity study to determine which inputs have a significant effect on pavement design answers for its range of conditions.

(2) For those inputs that are insignificant or inappropriate, no additional effort is required.

(3) For those that are significant and the state has sufficient data or methods to estimate design values with adequate accuracy, no additional effort is required.

(4) Finally, for those sensitive inputs for which the state has no data of methodology to develop the inputs, research will be necessary. Because of the complexity of pavement design and the large expansion of this Guide, it is anticipated that some additional research will be cost-effective for each and every user agency in order to optimally utilize the Guide.

One significant event, the pavement performance research effort being undertaken in the Strategic Highway Research Program (SHRP), should aid greatly in improving this document.

The Task Force believes that pavement design is gradually, but steadily transitioning from an art to a science. However, when one considers the nebulous nature of such difficult, but important inputs to design considerations such as traffic forecasting, weather forecasting, construction control, maintenance practices, etc.; successful pavement design will always depend largely upon the good judgment of the designer. Finally, the national trend toward developing and implementing pavement management systems, PMS, appears to the Task Force to be extremely important in developing the good judgment needed by pavement designers as well as providing many other elements needed for good design, i.e. information to support adequate funding and fund allocation.

The AASHTO Joint Task Force on Pavements

EXECUTIVE SUMMARY

One of the major objectives of the AASHO Road Test was to provide information that could be used to develop pavement design criteria and pavement design procedures. Accordingly, following completion of the Road Test, the AASHO Design Committee (currently the AASHTO Design Committee), through its Subcommittee on Pavement Design Practices, developed and circulated in 1961 the "AASHO Interim Guide for the Design of Rigid and Flexible Pavements." The Guide was based on the results of the AASHO Road Test supplemented by existing design procedures and, in the case of rigid pavements, available theory.

After the Guide had been used for several years, the AASHTO Design Committee prepared and AASHTO published the "AASHTO Interim Guide for Design of Pavement Structures—1972." Revisions were made in 1981 to Chapter III of the Guide relative to design criteria for Portland Cement Concrete pavements. Evaluation of the Guide by the AASHTO Design Committee in 1983 led to the conclusion that some revisions and additions were required. Representations from government, industry, consultants, and academia led to the conclusion that the Guide should be strengthened to incorporate information developed since 1972 and that a new section on rehabilitation should be added. It is also pertinent to note that, based on responses to a questionnaire sent to the States, there was an indication that the Guide was serving its main objectives and no serious problems were indicated. In other words, the States were generally satisfied with the Guide but acknowledged that some improvements could be made.

Based on the overall evaluation of input from user agencies and the status of research, it was determined by the AASHTO Joint Task Force on Pavements that the revisions to the Guide would retain the AASHO Road Test performance prediction equations, as modified for use in the 1972 Guide, as the basic model to be used for pavement design. This determination also established the present serviceability index (PSI) as the performance variable upon which design would be based.

The major changes which have been included in the revised Guide include the following considerations:

1. *Reliability.* The procedure for design of both rigid and flexible pavements provides a common method for incorporating a reliability factor into the design based on a shift in the design traffic.
2. *Soil support value.* AASHTO test method T 274 (resilient modulus of roadbed soils) is recommended as the definitive test for characterizing soil support. The soil property is recommended for use with both flexible and rigid pavement design.
3. *Layer coefficients (flexible pavements).* The resilient modulus test has been recommended as the procedure to be used in assigning layer coefficients to both stabilized and unstabilized material.

    [NOTE: Guidelines for relating resilient modulus to soil support value and layer coefficients are provided in the Guide; however, user agencies are encouraged to obtain equipment and to train personnel in order to measure the resilient modulus directly.]
4. *Drainage.* Provision has been made in the Guide to provide guidance in the design of subsurface drainage systems and for modifying the design equations to take advantage of improvements in performance to good drainage.
5. *Environment.* Improvements in the Guide have been made in order to adjust designs as a function of environment, e.g., frost heave, swelling soils, and thaw-weakening. Major emphasis is given to thaw-weakening and the effect that seasonal variations have on performance.
6. *Tied shoulders and widened lanes (rigid pavements).* A procedure is provided for the design of rigid pavements with tied shoulders or widened outside lanes.
7. *Subbase erosion.* A method for adjusting the design equations to represent possible soil erosion under rigid pavements is provided.
8. *Life-cycle costs.* Information has been added relative to economic analysis and economic comparisons of alternate designs based on life-

cycle costs. Present worth and/or equivalent uniform annual cost evaluations during a specified analysis period are recommended for making economic analyses.

9. *Rehabilitation.* A major addition to the Guide is the inclusion of a section on rehabilitation. Information is provided for rehabilitation with or without overlays.

10. *Pavement management.* Background information is provided regarding pavement management and the role of the Guide in the overall scheme of pavement management.

11. *Load equivalency values.* Load equivalency values have been extended to include heavier loads, more axles, and terminal serviceability levels of up to 3.0.

12. *Traffic.* Extensive information concerning methods for calculating equivalent single axle loads and specific problems related to obtaining reliable estimates of traffic loading are provided.

13. *Low-volume roads.* A special category for design of pavements subjected to a relative small number of heavy loads is provided in the design section.

14. *Mechanistic-Empirical design procedure.* The state of the knowledge concerning mechanistic-empirical design concepts is provided in the Guide. While these procedures have not, as yet, been incorporated into the Guides, extensive information is provided as to how such methods could be used in the future when enough documentation can be provided.

TABLE OF CONTENTS

	Page
Preface	vii
Executive Summary	ix

PART I PAVEMENT DESIGN AND MANAGEMENT PRINCIPLES

Chapter 1 Introduction and Background I-3

- 1.1 Scope of the Guide I-3
- 1.2 Design Considerations I-5
- 1.3 Pavement Performance I-7
- 1.4 Traffic I-10
  - 1.4.1 Evaluation of Traffic I-10
  - 1.4.2 Limitations I-12
  - 1.4.3 Special Cases I-13
- 1.5 Roadbed Soil I-13
- 1.6 Materials of Construction I-15
  - 1.6.1 Flexible Pavements I-16
  - 1.6.2 Rigid Pavements I-21
  - 1.6.3 Shoulders I-22
- 1.7 Environment I-22
- 1.8 Drainage I-27
  - 1.8.1 General Design Considerations I-28
  - 1.8.2 Design of Pavement Subsurface Drainage I-28
  - 1.8.3 Incorporation of Drainage Into Guide I-28
- 1.9 Shoulder Design I-29

Chapter 2 Design Related Project Level Pavement Management I-31

- 2.1 Relationship of Design to Pavement Management I-31
- 2.2 The Guide as Structural Subsystem for a State Project-Level PMS I-34
- 2.3 Pavement Type Selection I-39
- 2.4 Network Level Pavement Management I-39

Chapter 3 Economic Evaluation of Alternative Pavement Design Strategies I-41

- 3.1 Introduction I-41
- 3.2 Life-Cycle Costs I-41
- 3.3 Basic Concepts I-41
- 3.4 Definitions Related to Economic Analysis I-42
  - 3.4.1 Transport Improvement Costs I-42
  - 3.4.2 User Benefits I-42
- 3.5 Factors Involved in Pavement Cost and Benefits I-44
- 3.6 Initial Capital Costs (Investment Costs) I-44
  - 3.6.1 Maintenance Cost I-44
  - 3.6.2 Rehabilitation and Resurfacing Cost I-44

Contents—Continued

	3.6.3 Salvage or Residual Value	I-45
	3.6.4 User Cost	I-45
	3.6.5 Traffic Delay Cost to User	I-46
	3.6.6 Identification of Pavement Benefit	I-46
	3.6.7 Analysis Period	I-46
3.7	Methods of Economic Evaluation	I-47
3.8	Discussion of Interest Rates, Inflation Factors and Discount Rates	I-47
	3.8.1 Discounting and the Opportunity Cost of Capital	I-47
	3.8.2 Inflation	I-48
3.9	Equations for Economic Analysis	I-49
	3.9.1 Equivalent Uniform Annual Cost Method	I-49
	3.9.2 Present Worth Method	I-49
	3.9.3 Summary	I-51

Chapter 4 Reliability I-53

4.1	Definitions	I-53
	4.1.1 General Definition of Reliability	I-53
	4.1.2 Definition of Design Pavement Section	I-53
	4.1.3 Definition of Pavement Condition, Accumulated Axle Loads, and Pavement Performance Variables	I-54
4.2	Variance Components and Reliability Design Factor	I-56
	4.2.1 Components of Pavement Design-Performance Variability	I-56
	4.2.2 Probability Distribution of Basic Deviations	I-57
	4.2.3 Formal Definition of Reliability Level and Reliability Design Factor	I-60
4.3	Criteria for Selection of Overall Standard Deviation	I-62
4.4	Criteria for Selection of Reliability Level	I-62
4.5	Reliability and Stage Construction Alternatives	I-63

Chapter 5 Summary I-65

References for Part I I-67

PART II PAVEMENT DESIGN PROCEDURES FOR NEW CONSTRUCTION OR RECONSTRUCTION

Chapter 1 Introduction II-3

1.1	Background	II-3
1.2	Scope	II-3
1.3	Limitations	II-4
1.4	Organizations	II-4

Chapter 2 Design Requirements II-5

2.1	Design Variables	II-5
	2.1.1 Time Constraints	II-5
	2.1.2 Traffic	II-6
	2.1.3 Reliability	II-9
	2.1.4 Environmental Effects	II-10
2.2	Performance Criteria	II-10
	2.2.1 Serviceability	II-10
	2.2.2 Allowable Rutting	II-12
	2.2.3 Aggregate Loss	II-12

Contents—Continued

2.3	Material Properties for Structural Design	II-12
	2.3.1 Effective Roadbed Soil Resilient Modulus	II-12
	2.3.2 Effective Modulus of Subgrade Reaction	II-16
	2.3.3 Pavement Layer Materials Characterization	II-16
	2.3.4 PCC Modulus of Rupture	II-16
	2.3.5 Layer Coefficients	II-17
2.4	Pavements Structural Characteristics	II-22
	2.4.1 Drainage	II-22
	2.4.2 Load Transfer	II-25
	2.4.3 Loss of Support	II-27
2.5	Reinforcement Variables	II-27
	2.5.1 Jointed Reinforced Concrete Pavements	II-27
	2.5.2 Continuously Reinforced Concrete Pavements	II-28

Chapter 3 Highway Pavement Structural Design II-31

3.1	Flexible Pavement Design	II-31
	3.1.1 Determine Required Structural Number	II-31
	3.1.2 Stage Construction	II-33
	3.1.3 Roadbed Swelling and Frost Heave	II-33
	3.1.4 Selection of Layer Thickness	II-35
	3.1.5 Layered Design Analysis	II-35
3.2	Rigid Pavement Design	II-37
	3.2.1 Develop Effective Modulus of Subgrade Reaction	II-37
	3.2.2 Determine Required Slab Thickness	II-44
	3.2.3 Stage Construction	II-44
	3.2.4 Roadbed Swelling and Frost Heave	II-47
3.3	Rigid Pavement Joint Design	II-48
	3.3.1 Joint Types	II-48
	3.3.2 Joint Geometry	II-49
	3.3.3 Joint Sealant Dimensions	II-50
3.4	Rigid Pavement Reinforcement Design	II-51
	3.4.1 Jointed Reinforced Concrete Pavements	II-51
	3.4.2 Continuously Reinforced Concrete Pavements	II-51
	3.4.3 Transverse Reinforcement	II-62
3.5	Prestressed Concrete Pavement	II-65
	3.5.1 Subbase	II-65
	3.5.2 Slab Length	II-65
	3.5.3 Magnitude of Prestress	II-66
	3.5.4 Tendon Spacing	II-66
	3.5.5 Fatigue	II-66
	3.5.6 PCP Structural Design	II-66

Chapter 4 Low-Volume Road Design II-69

4.1	Design Chart Procedures	II-69
	4.1.1 Flexible and Rigid Pavements	II-69
	4.1.2 Aggregate-Surfaced Roads	II-69
4.2	Design Catalog	II-77
	4.2.1 Flexible Pavement Design Catalog	II-77
	4.2.2 Rigid Pavement Design Catalog	II-81
	4.2.3 Aggregate-Surfaced Road Design Catalog	II-81

References for Part II II-87

Contents—Continued

PART III PAVEMENT DESIGN PROCEDURES FOR REHABILITATION OF EXISTING PAVEMENTS

Chapter 1 Introduction III-3

1.1 Background III-3
1.2 Scope III-3
1.3 Assumptions/Limitations III-4
1.4 Organization III-4

Chapter 2 Rehabilitation Concepts III-7

2.1 Background III-7
2.2 Rehabilitation Factors III-7
    2.2.1 Major Categories III-7
    2.2.2 Recycling Concepts III-7
    2.2.3 Construction Considerations III-7
    2.2.4 Summary of Major Rehabilitation Factors III-7
2.3 Selection of Alternative Rehabilitation Methods III-8
    2.3.1 Overview III-8
    2.3.2 Problem Definition III-9
    2.3.3 Potential Problem Solutions III-12
    2.3.4 Selection of Preferred Solution III-15
    2.3.5 Summary III-16

Chapter 3 Guides for Field Data Collection III-19

3.1 Overview III-19
3.2 The Fundamental Analysis Unit III-19
    3.2.1 General Background III-19
    3.2.2 Methods of Unit Delineation III-19
3.3 Drainage Survey for Rehabilitation III-21
    3.3.1 Role of Drainage in Rehabilitation III-21
    3.3.2 Assessing Need for Drainage Evaluation III-25
    3.3.3 Pavement History, Topography, and Geometry III-25
    3.3.4 Properties of Materials III-25
    3.3.5 Climatic Zones III-26
    3.3.6 Summary III-28
3.4 Condition (Distress) Survey III-28
    3.4.1 General Background III-28
    3.4.2 Minimum Information Needs III-28
    3.4.3 Utilization of Information III-28
3.5 NDT Deflection Measurement III-30
    3.5.1 Overview III-30
    3.5.2 Uses of NDT Deflection Results III-32
    3.5.3 Evaluating the Effective Structural Capacity III-35
    3.5.4 Joint Load Transfer Analysis III-38
    3.5.5 Use in Slab-Void Detection III-44
3.6 Field Sampling and Testing Programs III-45
    3.6.1 Test Types III-45
    3.6.2 Major Parameters III-45
    3.6.3 Necessity for Destructive Testing III-49
    3.6.4 Selecting the Required Number of Tests III-49

Contents—Continued

Chapter 4 Rehabilitation Methods Other Than Overlay **III-59**

4.1 Evaluation of Pavement Condition III-59
    4.1.1 Surface Distress III-59
    4.1.2 Structural Condition III-59
    4.1.3 Functional Condition III-60
4.2 Development of Feasible Alternatives and Strategies III-60
4.3 Major Nonoverlay Methods III-62
    4.3.1 Full-Depth Repair III-62
    4.3.2 Partial-Depth Pavement Repair III-64
    4.3.3 Joint and Crack Sealing III-65
    4.3.4 Subsealing of Concrete Pavements III-66
    4.3.5 Diamond Grinding of Concrete Surfaces and Cold Milling of
           Asphalt Surfaces III-67
    4.3.6 Subdrainage Design III-68
    4.3.7 Pressure Relief Joints III-69
    4.3.8 Restoration of Joint Load Transfer in Jointed Concrete Pavements .. III-70
    4.3.9 Surface Treatments III-71
    4.3.10 Prediction of Life of Rehabilitation Techniques Without Overlay ... III-73

Chapter 5 Rehabilitation Methods With Overlays **III-79**

5.1 Overlay Type Feasibility III-79
5.2 Important Considerations in Overlay Design III-80
    5.2.1 Pre-overlay Repair III-80
    5.2.2 Reflection Crack Control III-80
    5.2.3 Traffic Loadings III-80
    5.2.4 Subdrainage III-81
    5.2.5 Rutting in AC Pavements III-81
    5.2.6 Milling AC Surface III-81
    5.2.7 Recycling the Existing Pavement III-81
    5.2.8 Structural versus Functional Overlays III-81
    5.2.9 Overlay Materials III-81
    5.2.10 Shoulders III-81
    5.2.11 Existing PCC Slab Durability III-82
    5.2.12 PCC Overlay Joints III-82
    5.2.13 PCC Overlay Reinforcement III-82
    5.2.14 PCC Overlay Bonding/Separation Layers III-82
    5.2.15 Overlay Design Reliability Level and Overall Standard Deviation ... III-82
    5.2.16 Pavement Widening III-82
    5.2.17 Potential Errors and Possible Adjustments to Thickness
           Design Procedure III-83
    5.2.18 Example Designs and Documentation III-83
5.3 Pavement Evaluation for Overlay Design III-83
    5.3.1 Design of Overlay Along Project III-84
    5.3.2 Functional Evaluation of Existing Pavement III-84
    5.3.3 Structural Evaluation of Existing Pavement III-85
    5.3.4 Determination of Design M_R III-91
5.4 AC Overlay of AC Pavement III-94
    5.4.1 Feasibility III-94
    5.4.2 Pre-overlay Repair III-94
    5.4.3 Reflection Crack Control III-95
    5.4.4 Subdrainage III-95
    5.4.5 Thickness Design III-95

Contents—Continued

	5.4.6 Surface Milling	III-105
	5.4.7 Shoulders	III-105
	5.4.8 Widening	III-106
5.5	AC Overlay of Fractured PCC Slab Pavement	III-106
	5.5.1 Feasibility	III-107
	5.5.2 Pre-overlay Repair	III-108
	5.5.3 Reflection Crack Control	III-108
	5.5.4 Subdrainage	III-108
	5.5.5 Thickness Design	III-108
	5.5.6 Shoulders	III-111
	5.5.7 Widening	III-111
5.6	AC Overlay of JPCP, JRCP, and CRCP	III-113
	5.6.1 Feasibility	III-113
	5.6.2 Pre-overlay Repair	III-113
	5.6.3 Reflection Crack Control	III-114
	5.6.4 Subdrainage	III-115
	5.6.5 Thickness Design	III-115
	5.6.6 Shoulders	III-125
	5.6.7 Widening	III-125
5.7	AC Overlay of AC/JPCP, AC/JRCP, and AC/CRCP	III-125
	5.7.1 Feasibility	III-125
	5.7.2 Pre-overlay Repair	III-127
	5.7.3 Reflection Crack Control	III-127
	5.7.4 Subdrainage	III-128
	5.7.5 Thickness Design	III-128
	5.7.6 Surface Milling	III-135
	5.7.7 Shoulders	III-135
	5.7.8 Widening	III-136
5.8	Bonded Concrete Overlay of JPCP, JRCP, and CRCP	III-136
	5.8.1 Feasibility	III-136
	5.8.2 Pre-overlay Repair	III-137
	5.8.3 Reflection Crack Control	III-137
	5.8.4 Subdrainage	III-137
	5.8.5 Thickness Design	III-137
	5.8.6 Shoulders	III-143
	5.8.7 Joints	III-143
	5.8.8 Bonding Procedures and Material	III-145
	5.8.9 Widening	III-145
5.9	Unbonded JPCP, JRCP, or CRCP Overlay of JPCP, JRCP, CRCP, or AC/PCC	III-145
	5.9.1 Feasibility	III-145
	5.9.2 Pre-overlay Repair	III-145
	5.9.3 Reflection Crack Control	III-145
	5.9.4 Subdrainage	III-146
	5.9.5 Thickness Design	III-146
	5.9.6 Shoulders	III-151
	5.9.7 Joints	III-151
	5.9.8 Reinforcement	III-153
	5.9.9 Separation Interlayers	III-153
	5.9.10 Widening	III-153
5.10	JPCP, JRCP, and CRCP Overlay of AC Pavement	III-153
	5.10.1 Feasibility	III-153

Contents—Continued

5.10.2	Pre-overlay Repair	III-153
5.10.3	Reflection Crack Control	III-153
5.10.4	Subdrainage	III-154
5.10.5	Thickness Design	III-154
5.10.6	Shoulders	III-155
5.10.7	Joints	III-155
5.10.8	Reinforcement	III-155
5.10.10	Widening	III-155

References for Chapter 5 .... III-157

PART IV MECHANISTIC-EMPIRICAL DESIGN PROCEDURES

1.1	Introduction	IV-3
1.2	Benefits	IV-4
1.3	Framework for Development and Application	IV-4
1.4	Implementation	IV-7
	1.4.1 Design Considerations	IV-8
	1.4.2 Input Data	IV-8
	1.4.3 Equipment Acquisition	IV-9
	1.4.4 Computer Hardware and Software	IV-9
	1.4.5 Training Personnel	IV-9
	1.4.6 Field Testing and Calibration	IV-10
	1.4.7 Testing	IV-10
1.5	Summary	IV-10

References for Part IV IV-11

APPENDICES

A.	Glossary of Terms	A1
B.	Pavement Type Selection Guidelines	B1
C.	Alternate Methods of Design for Pavement Structures	C1
D.	Conversion of Mixed Traffic to Equivalent Single Axle Loads for Pavement Design	D1
E.	Position Paper on Shoulder Design	E1
F.	List of Test Procedures	F1
G.	Treatment of Roadbed Swelling and/or Frost Heave in Design	G1
H.	Flexible Pavement Design Example	H1
I.	Rigid Pavement Design Example	I1
J.	Analysis Unit Delineation by Cumulative Differences	J1
K.	Typical Pavement Distress Type-Severity Descriptions	K1
L.	Documentation of Design Procedures	L1
M.	An Examination of the AASHTO Remaining Life Factor	M1
N.	Overlay Design Examples	N1

Index 1

PART I
PAVEMENT DESIGN AND MANAGEMENT PRINCIPLES

CHAPTER 1
INTRODUCTION AND BACKGROUND

1.1 SCOPE OF THE GUIDE

This *Guide for the Design of Pavement Structures* provides a comprehensive set of procedures which can be used for the design and rehabilitation of pavements; both rigid (portland cement concrete surface) and flexible (asphalt concrete surface) and aggregate surfaced for low-volume roads. The Guide has been developed to provide recommendations regarding the determination of the pavement structure as shown in Figure 1.1. These recommendations will include the determination of total thickness of the pavement structure as well as the thickness of the individual structural components. The procedures for design provide for the determination of alternate structures using a variety of materials and construction procedures.

A glossary of terms, as used in this Guide, is provided in Appendix A. It is recognized that some of the terms used herein may differ from those used in your local practice; however, it is necessary to establish standard terminology in order to facilitate preparation of the Guide for nationwide use. Insofar as is possible, AASHTO definitions have been used herein.

It should be remembered that the total set of considerations required to assure reliable performance of a pavement structure will include many factors other than the determination of layer thicknesses of the structural components. For example, material requirements, construction requirements, and quality control will significantly influence the ability of the pavement structure to perform according to design expectations. In other words, "pavement design" involves more than choosing thicknesses. Information concerning material and construction requirements will be briefly described in this Guide; however, a good pavement designer must be familiar with relevant publications of AASHTO and ASTM, as well as the local agencies, i.e., state agencies or counties, for whom the design is being prepared. It is extremely important that the designer prepare special provisions to the standard specifications when circumstances indicate that nonstandard conditions exist for a specific project. Examples of such a condition could involve a roadbed soil which is known to be expansive or nonstandard materials which are to be stabilized for use in the pavement structure or prepared roadbed.

Part I of this Guide has been prepared as general background material to assist the user in the proper interpretation of the design procedures and to provide an understanding of the concepts used in the development of the Guide. Detailed information related directly to a number of design considerations, e.g., reliability, drainage, life-cycle costs, traffic, and pavement type selection, will be found in the Appendices. References used in the preparation of the Guide can be found following each of the four major Parts.

Part I, Chapter 3 of the Guide provides information concerning economic evaluation of alternate pavement design strategies. It should not be concluded that the selection of a pavement design should be based on economics alone. There are a number of considerations involved in the final design selection. Appendix B of the Guide on pavement type selection provides an extensive list of guidelines which should be used in comparing alternate design strategies.

Part II of this Guide provides a detailed method for the design of new pavements or for reconstruction of existing pavements on the existing alignment with new or recycled materials.

Part III of this Guide provides alternative methods for pavement rehabilitation with or without the addition of an overlay. The methodology used in this part of the Guide represents the state of the knowledge regarding the deterioration of a pavement structure before and after an overlay has been applied. It is recognized that there are alternate methods for the determination of overlay requirements; a number of these methods are cited in Appendix C. The method included in Part III is somewhat more basic in concept than other existing methods and has the capability for broader application to different types of overlays, e.g., flexible on rigid, flexible on flexible, rigid on rigid, and rigid on flexible type pavements. The method is also compatible with the performance and design concepts used in Part II. In this way, consideration of such factors as drainage, reliability, and traffic is the same for both new and rehabilitated (overlayed) pavement structures.

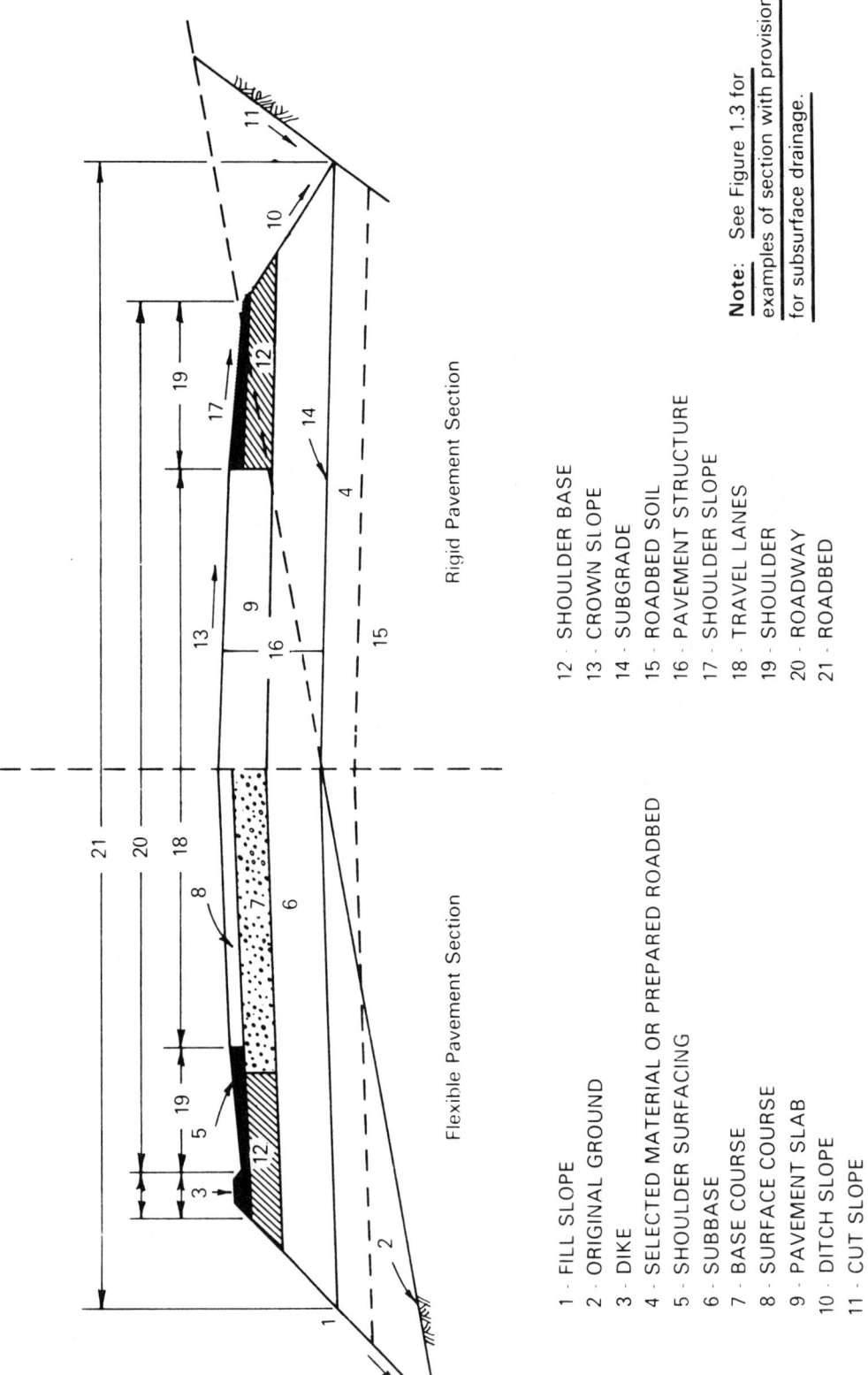

Figure 1.1. Typical Section for Rigid or Flexible Pavement Structure

Introduction and Background

State of the art procedures for rehabilitation of pavement structures without overlay, including drainage and the use of recycled material, are emphasized in Part III. These techniques represent an alternative to overlays which can reduce long-term costs and satisfy design constraints associated with specific design situations.

As an adjunct to pavement rehabilitation it is important to first determine what is wrong with the existing pavement structure. Details of the method for interpretation of the information are contained in Part III. A procedure for measuring or evaluating the condition of a pavement is given in Appendix K and Reference 1. It is beyond the scope of this Guide to discuss further the merits of different methods and equipment which can be used to evaluate the condition of a pavement. However, it is considered essential that a detailed condition survey be made before a set of plans and specifications are developed for a specific project. If at all possible, the designer should participate in the condition survey. In this way, it will be possible to determine if special treatments or methods may be appropriate for site conditions, specifically, if conditions warrant consideration of detailed investigations pertinent to the need for added drainage features.

Part IV of this Guide provides a framework for future developments for the design of pavement structures using mechanistic design procedures. The benefits associated with the development of these methods are discussed; a summary of existing procedures and a framework for development are the major concerns of that portion of the Guide.

1.2 DESIGN CONSIDERATIONS

The method of design provided in this Guide includes consideration of the following items:

(1) pavement performance,
(2) traffic,
(3) roadbed soil,
(4) materials of construction,
(5) environment,
(6) drainage,
(7) reliability,
(8) life-cycle costs, and
(9) shoulder design.

Each of these factors is discussed in Part I. Parts II, III, and IV carry these concepts and procedures forward and incorporate each into a pavement structure design methodology.

It is worth noting again that while the Guide describes and provides a specific method which can be used for the determination of alternate design or rehabilitation recommendations for the pavement structure, there are a number of considerations which are left to the user for final determination, e.g., drainage coefficients, environmental factors, and terminal serviceability.

The Guide by its very nature cannot possibly include all of the site specific conditions that occur in each region of the United States. It is therefore necessary for the user to adapt local experience to the use of the Guide. For example, local materials and environment can vary over an extremely wide range within a state and between states.

The Guide attempts to provide procedures for evaluating materials and environment; however, in the case where the Guide is at variance with proven and documented local experience, the proven experience should prevail. *The designer will need to concentrate on some aspects of design which are not always covered in detail in the Guide.* For example, material requirements and construction specifications are not detailed in this Guide and yet they are an important consideration in the overall design of a pavement structure. The specifics of joint design and joint spacing will need careful consideration. The effect of seasonal variations on material properties and careful evaluation of traffic for the designed project are details which the designer should investigate thoroughly.

The basic design equations used for flexible and rigid pavements in this Guide are as follows:

Flexible Pavements

$$\log_{10}(W_{18}) = Z_R \times S_o + 9.36 \times \log_{10}(SN + 1) - 0.20 + \frac{\log_{10}\left[\frac{\Delta PSI}{4.2 - 1.5}\right]}{0.40 + \frac{1094}{(SN + 1)^{5.19}}} + 2.32 \times \log_{10}(M_R) - 8.07 \quad (1.2.1)$$

where

W_{18} = predicted number of 18-kip equivalent single axle load applications,
Z_R = standard normal deviate,
S_o = combined standard error of the traffic prediction and performance prediction,

ΔPSI = difference between the initial design serviceability index, p_o, and the design terminal serviceability index, p_t, and

M_R = resilient modulus (psi).

SN is equal to the structural number indicative of the total pavement thickness required:

$$SN = a_1D_1 + a_2D_2m_2 + a_3D_3m_3$$

where

a_i = i^{th} layer coefficient,
D_i = i^{th} layer thickness (inches), and
m_i = i^{th} layer drainage coefficient.

Rigid Pavements

$$\log_{10}(W_{18}) = Z_R \times S_o + 7.35 \times \log_{10}(D + 1)$$
$$- 0.06 + \frac{\log_{10}\left[\frac{\Delta PSI}{4.5 - 1.5}\right]}{1 + \frac{1.624 \times 10^7}{(D + 1)^{8.46}}}$$
$$+ (4.22 - 0.32 \times p_t)$$
$$\times \log_{10}\left[\frac{S'_c \times C_d \times (D^{0.75} - 1.132)}{215.63 \times J\left[D^{0.75} - \frac{18.42}{(E_c/k)^{0.25}}\right]}\right]$$
(1.2.2)

where

W_{18} = predicted number of 18-kip equivalent single axle load applications,
Z_R = standard normal deviate,
S_o = combined standard error of the traffic prediction and performance prediction,
D = thickness (inches) of pavement slab,
ΔPSI = difference between the initial design serviceability index, p_o, and the design terminal serviceability index, p_t,
S'_c = modulus of rupture (psi) for portland cement concrete used on a specific project,
J = load transfer coefficient used to adjust for load transfer characteristics of a specific design,
C_d = drainage coefficient,
E_c = modulus of elasticity (psi) for portland cement concrete, and
k = modulus of subgrade reaction (pci).

The design nomographs presented in Part II solve these equations for the structural number (SN) for flexible pavements and thickness of the pavement slab for rigid pavements.

The structural number is an abstract number expressing the structural strength of a pavement required for given combinations of soil support (M_R), total traffic expressed in equivalent 18-kip single axle loads, terminal serviceability, and environment. The required SN must be converted to actual thickness of surfacing, base and subbase, by means of appropriate layer coefficients representing the relative strength of the construction materials. Average values of layer coefficients for materials used in the AASHO Road Test are as follows:

Asphaltic concrete surface course—.44
Crushed stone base course —.14
Sandy gravel subbase —.11

The layer coefficients given in Part II are based on extensive analyses summarized in NCHRP Report 128, "Evaluation of AASHTO Guide for Design of Pavement Structures," (1972). In effect, the layer coefficients are based on the elastic moduli M_R and have been determined based on stress and strain calculations in a multilayered pavement system. Using these concepts, the layer coefficient may be adjusted, increased, or decreased in order to maintain a constant value of stress or strain required to provide comparable performance.

Part II details how each of the design considerations are to be treated in selecting the SN value and how to decompose SN into layers according to material properties and function, i.e., surface, base, subbase, and so forth. The pavement slab thickness, in inches, is provided directly from the design nomographs.

It is important to recognize that equations (1.2.1) and (1.2.2) were derived from empirical information obtained at the AASHO Road Test. As such, these equations represent a best fit to observations at the Road Test. The solution represents the mean value of traffic which can be carried given specific inputs. In other words, there would be a 50-percent chance that the actual traffic to terminal serviceability could be more or less than predicted. In order to decrease the risk of premature deterioration below acceptable levels of serviceability, a reliability factor is included

Introduction and Background

in the design process. An explanation of the reliability factor is given in Chapter 4 of Part I. In order to properly apply the reliability factor, the inputs to the design equation should be the mean value without adjustment. This will be discussed further in Chapter 4 of Part I and in sections of Part II. *The designer must remember to use mean values for such factors as soil support, traffic, layer coefficients, drainage coefficients, etc.* Increased reliability will be obtained by adjustments which are based on uncertainty in *each* of the design variables as well as traffic.

Each of the terms used in the design equations is discussed as necessary in Parts I and II of this Guide. It is pertinent to note that a few changes have been made in the design equations when compared with the 1972 Interim Guide (*2*). The soil support value has been replaced with M_R (flexible) and a drainage coefficient has been added to the rigid equation. For the flexible equation, the structural number (SN) has been modified by the addition of drainage coefficients and the regional factor (R) has been deleted. Lastly, both the rigid and flexible equations have been modified to consider both total serviceability loss ($p_o - p_t$), and terminal serviceability.

There are two important factors to consider concerning these equations: (1) the equations are predictors of the amount of traffic that can be sustained before deteriorating to some selected terminal level of serviceability and (2) the basic prediction equations were developed empirically from field observations at the AASHO Road Test with modifications considered necessary to improve the Guide based on research completed during the past 20 years.

There are a number of alternate procedures which can be used for the design of pavement structures. In fact, all 50 states have adopted their own design procedures, many of which are based on past AASHTO Guide methods. A list of other suitable pavement design procedures is presented in Appendix C.

1.3 PAVEMENT PERFORMANCE

Current concepts of pavement performance include some consideration of functional performance, structural performance, and safety. This Guide is primarily concerned with functional and structural performance. Information pertinent to safety can be found in appropriate publications of NCHRP, FHWA, and AASHTO. One important aspect of safety is the frictional resistance provided at the pavement/tire interface. AASHTO has issued a publication, *Guidelines for Skid Resistant Pavement Design*, which can be referred to for information on this subject.

The structural performance of a pavement relates to its physical condition; i.e., occurrence of cracking, faulting, raveling, or other conditions which would adversely affect the load-carrying capability of the pavement structure or would require maintenance.

The functional performance of a pavement concerns how well the pavement serves the user. In this context, riding comfort or ride quality is the dominant characteristic. In order to quantify riding comfort, the "serviceability-performance" concept was developed by the AASHO Road Test staff in 1957 (*3, 4*). Since the serviceability-performance concept is used as the measure of performance for the design equations in this Guide, an explanation of the concept herein seems worthwhile.

The serviceability-performance concept is based on five fundamental assumptions, summarized as follows (*5*):

(1) Highways are for the comfort and convenience of the traveling public (User).
(2) Comfort, or riding quality, is a matter of subjective response or the opinion of the User.
(3) Serviceability can be expressed by the mean of the ratings given by all highway Users and is termed the serviceability rating.
(4) There are physical characteristics of a pavement which can be measured objectively and which can be related to subjective evaluations. This procedure produces an objective serviceability index.
(5) Performance can be represented by the serviceability history of a pavement.

The serviceability of a pavement is expressed in terms of the present serviceability index (PSI). The PSI is obtained from measurements of roughness and distress, e.g., cracking, patching and rut depth (flexible), at a particular time during the service life of the pavement. Roughness is the dominant factor in estimating the PSI of a pavement. Thus, a reliable method for measuring roughness is important in monitoring the performance history of pavements.

The specific equations developed at the Road Test to calculate the present serviceability index have been modified by most users of the AASHTO Guide. These changes reflect local experience and are assumed to represent results from the Road Test; i.e., the PSI values continue to represent ride quality as evaluated at the Road Test. Because of the relatively small contribution to PSI made by physical distress, and the difficulty in obtaining the information, many agencies rely only on roughness to estimate ride quality. It is acknowledged that physical distress is likely to influence a decision to initiate maintenance or rehabilitation. For purposes of this Guide, it is assumed that the amount of distress associated with the terminal PSI is acceptable.

Because roughness is such an important consideration for the design of pavements, the change in roughness will control the life cycle of pavements. In this regard, the quality of construction will influence performance and the life cycle of the designed pavement. The initial pavement smoothness is an important design consideration. For example, the life cycle of a pavement initially constructed with a smoothness or PSI of 4.5 will have a significantly longer life cycle than one constructed to a PSI of 4.0. Thus, quality control in the construction of a pavement can have a beneficial impact on performance (life cycle).

The scale for PSI ranges from 0 through 5, with a value of 5 representing the highest index of serviceability. For design it is necessary to select both an initial and terminal serviceability index.

The initial serviceability index (p_i) is an estimate by the user of what the PSI will be immediately after construction. Values of p_i established for AASHO Road Test conditions were 4.2 for flexible pavements and 4.5 for rigid pavements. Because of the variation of construction methods and standards, it is recommended that more reliable levels be established by each agency based on its own conditions.

The terminal serviceability index (p_t) is the lowest acceptable level before resurfacing or reconstruction becomes necessary for the particular class of highway. An index of 2.5 or 3.0 is often suggested for use in the design of major highways, and 2.0 for highways with a lower classification. For relatively minor highways, where economic considerations dictate that initial expenditures be kept low, at p_t of 1.5 may be used. Expenditures may also be minimized by reducing the performance period. Such a low value of p_t should only be used in special cases on selected classes of highways.

The major factors influencing the loss of serviceability of a pavement are traffic, age, and environment. Each of these factors has been considered in formulating the design requirements included in this Guide. However, it should be recognized that the separate or the interacting effects of these components are not clearly defined at the present time, especially with regard to age. It is known that the properties of materials used for pavement construction change with time. These changes may be advantageous to performance; however, in most cases, age (time) is a net negative factor and works to reduce serviceability.

An effort has been made in the Guide to account for the effects of environment on pavement performance in situations where swelling clay or frost heave are encountered. Thus, the total change in PSI at any time can be obtained by summing the damaging effects of traffic, swelling clay, and/or frost heave, as shown in Equation 1.3.1 and illustrated in Figure 1.2.

$$\Delta PSI = \Delta PSI_{Traffic} + \Delta PSI_{Swell/Frost\ Heave} \quad (1.3.1)$$

where

ΔPSI = total loss of serviceability,
$\Delta PSI_{Traffic}$ = serviceability loss due to traffic (ESAL's), and
$\Delta PSI_{Swell/Frost\ Heave}$ = serviceability loss due to swelling and/or frost heave of roadbed soil.

It can be noted in Figure 1.2 that the effect of swelling soils or frost heave is to reduce the predicted service life of the pavement. The Guide does not recommend increasing pavement structural thickness to offset the serviceability loss due to swelling soils; but it is feasible, however, to control frost heave by increasing the thickness of non-frost-susceptible material.

In many swelling situations, it may be possible to reduce to acceptable limits the effect of swelling soil by stabilization of the expansive soil or by replacement of these soils with nonexpansive material. When experience indicates this is a viable procedure, it is not necessary to estimate the effect of swelling soil on the life cycle.

The predicted effect of frost heave is based on a limited amount of information available in the literature. If agency design procedures include provisions to mitigate the detrimental effects of frost, the serviceability loss due to frost heave should be ignored, i.e., assumed to be zero. The most accepted procedure to minimize the effect of frost heave is to replace the frost-susceptible material with non-frost-susceptible material to a depth of one-half or more of the frost depth.

A further discussion of the influence of environment will be found in Section 1.7 of this chapter.

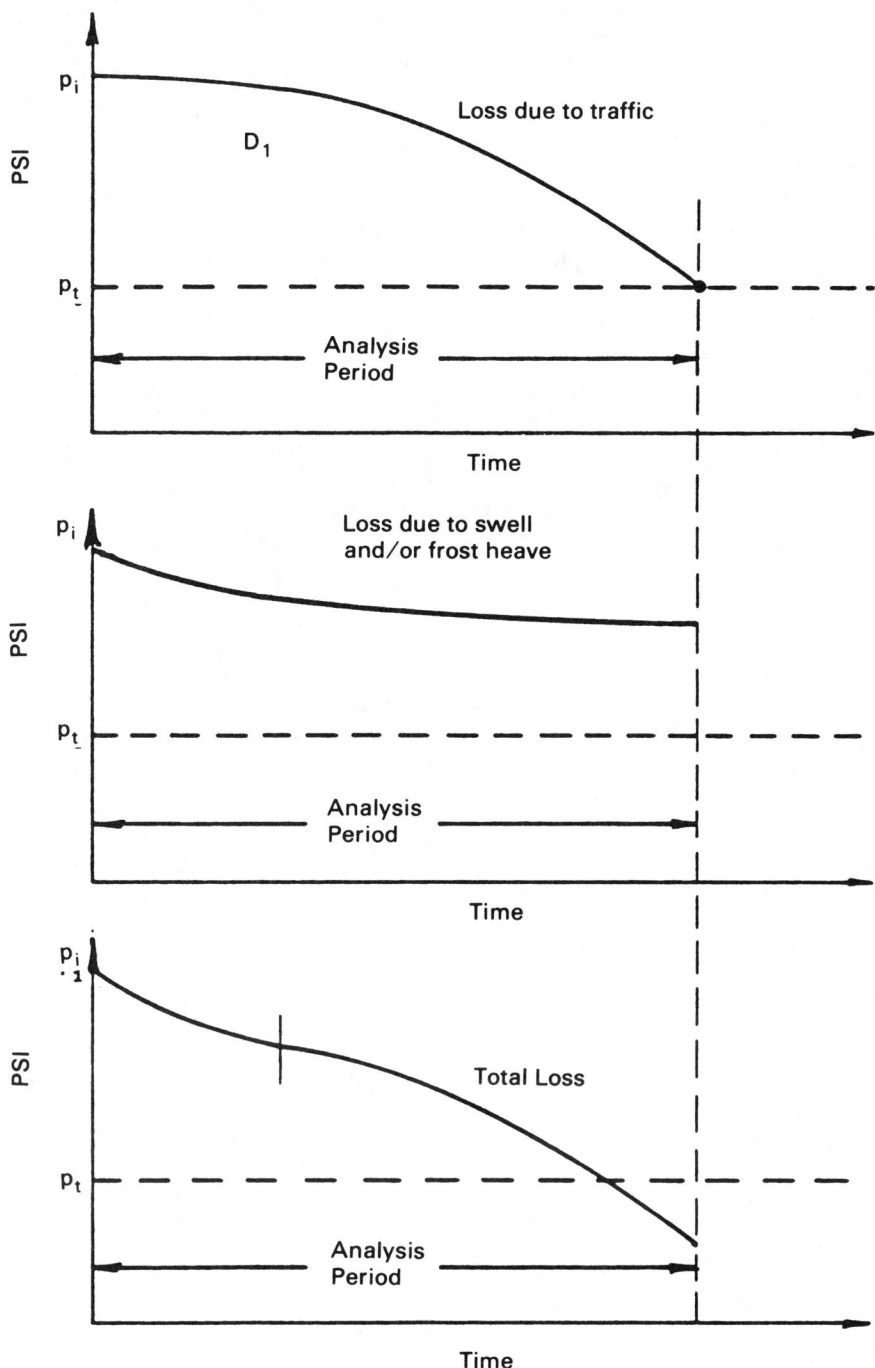

Figure 1.2. Pavement Performance Trends

1.4 TRAFFIC

Traffic information required by the design equations used in this Guide includes axle loads, axle configuration, and number of applications.

The results of the AASHO Road Test have shown that the damaging effect of the passage of an axle of any mass (commonly called load) can be represented by a number of 18-kip equivalent single axle loads or ESAL's. For example, one application of a 12-kip single axle was found to cause damage equal to approximately 0.23 applications of an 18-kip single axle load, and four applications of a 12-kip single axle were required to cause the same damage (or reduction in serviceability) as one application of an 18-kip single axle. This concept has been applied to the design equations and nomographs in Part II. The determination of design ESAL's is a very important consideration for the design of pavement structures using this Guide, as it is in previous versions of the Guide.

1.4.1 Evaluation of Traffic

The procedure used in this Guide to convert a mixed traffic stream of different axle loads and axle configurations into a design traffic number is to convert each expected axle load into an equivalent number of 18-kip single axle loads and to sum these over the design period. The procedure for converting mixed traffic to ESAL's is discussed in Appendix D.

There are four key considerations which influence the accuracy of traffic estimates and which can significantly influence the life cycle of a pavement: (1) the correctness of the load equivalency values used to estimate the relative damage induced by axle loads of different mass and configurations, (2) the accuracy of traffic volume and weight information used to represent the actual loading projections, (3) the prediction of ESAL's over the design period, and (4) the interaction of age and traffic as it affects changes in PSI.

The available load equivalency factors are considered the best available at the present time, representing information derived from the AASHO Road Test. The empirical observations on the Road Test covered a range of axle loads from 2 to 30 kips on single axles and 24 to 48 kips on tandem axles. No tridem axles were included in the Road Test experiment; load equivalency values for tridem axles are included in Appendix D, but they are the result of research carried out since completion of the Road Test. Load equivalency values for single and tandem axles which exceed the loads given above are also extrapolations of the basic data from the Road Test.

It should be noted that load equivalency factors are, to a minor degree, functions of pavement type (rigid or flexible), thickness, and terminal serviceability (p_t) used for design. For designing composite pavements (rigid base with flexible wearing surface), the use of load equivalency values for rigid pavements is recommended.

State DOT's accumulate traffic information in the format of the Federal Highway Administration W-4 truck weight tables, which are tabulations of the number of axles observed within a series of load groups with each load group covering a 2-kip interval. Traffic information relative to truck type, i.e., axle configuration, is provided in W-2 tabulations (distribution of vehicles counted and weighed). As illustrated in Appendix D, these tabulations can be used to estimate the number of equivalent single axle loads associated with mixed traffic at the particular reporting loadometer station. From this information it is possible to obtain average load equivalency factors for all trucks or for trucks by configuration, i.e., the averages for singles, tandems, or tridems.

Most states have taken the information from the W-4 tables and converted it into relatively simple multipliers (truck equivalency factors) which represent each truck type in the traffic stream. These multipliers can be used to convert mixed streams of traffic to ESAL's. It must be realized that such conversions represent estimates when applied to highways other than those from which the data were obtained. Weigh station information represents only a sample of the total traffic stream with weighings at a limited number of locations and for limited periods of time. Such information must be carefully interpreted when applied to specific projects. Results from different weigh stations in one state have been reported to produce truck factors which vary by a factor of 6. Thus, one source of error in ESAL predictions is the use of estimated truck equivalency factors for various classes of highways based on a relatively small sample. Increased sampling of this type of information is necessary in order to reduce the error of the estimate due to insufficient information on a specific project. Users of this Guide are urged to gather the best possible traffic data for each design project.

Since pavements, new or rehabilitated, are usually designed for periods ranging from 10 years to 20 years or more, it is necessary to predict the ESAL's for this period of time, i.e., the performance period. The performance period, often referred to as the design period, is defined as the period of time that an initial

Introduction and Background

(or rehabilitated structure) will last before reaching its terminal serviceability. Any performance period may be used with the Guide since design is based on the total number of equivalent single axle loads; however, experience may indicate a practical upper limit based on considerations other than traffic. The ESAL's for the performance period represent the cumulative number from the time the roadway is opened to traffic to the time when the serviceability is reduced to a terminal value (e.g., p_t equal 2.5 or 2.0). If the traffic is underestimated, the actual time to p_t will probably be less than the predicted performance period, thereby resulting in increased maintenance and rehabilitation costs.

The maximum performance period to be used in designing for a particular pavement type, i.e., flexible, rigid, or composite, should reflect agency experience.

The performance period and corresponding design traffic should reflect real-life experience. The performance period should not be confused with pavement life. The pavement life may be extended by periodic rehabilitation of the surface or pavement structure.

The equivalent loads derived from many traffic prediction procedures represent the totals for all lanes for both directions of travel. This traffic must be distributed by direction and by lanes for design purposes. Directional distribution is usually made by assigning 50 percent of the traffic to each direction, unless available measured traffic data warrant some other distribution. In regard to lane distribution, 100 percent of the traffic in one direction is often assigned to each of the lanes in that direction for purposes of structural design if measured distributions are not available. Some states have developed lane distribution factors for facilities with more than one lane in a given direction. These factors vary from 60 to 100 percent of the one-directional traffic, depending on the total number of lanes in the facility. Part II and Appendix D provide more details pertinent to this lane distribution factor.

Traffic information is often provided to the designer by a Planning or Traffic group. The designers should work closely with traffic personnel to be sure the proper information is provided and that the consequences of poor estimates of present and future traffic are understood by all personnel involved.

Predictions of future traffic are often based on past traffic history. Several factors can influence such predictions.

For purposes of pavement structure design, it is necessary to estimate the cumulative number of 18-kip equivalent single axle loads (ESAL's) for the design (performance) period. The number of ESAL's may or may not be proportional to the average daily traffic. Truck traffic is the essential information required to calculate ESAL's; it is therefore very important to correctly estimate future truck traffic for the facility during the design period.

Traffic may remain constant or increase according to a straight line or at an accelerating (exponential) rate. In most cases, highways classified as principal arterial or interstate will have exponential growth (comparable to compound interest on investments). Traffic on some minor arterial or collector-type highways may increase along a straight line, while traffic on some residential streets may not change because the use remains constant. Thus, the designer must make provision for growth in traffic from the time of the last traffic count or weighing through the performance period selected for the project under consideration. Appendix D provides appropriate information for estimating future traffic growth based on an assumed exponential compounded growth rate. If zero or negative growth in traffic is anticipated, a zero or negative growth factor can be used. In most cases, appropriate growth factors can be selected from the table in Appendix D. For major arterials and interstate highways, the growth rate should be applied by truck class rather than to the total traffic since growth in truck traffic may differ from the total traffic stream.

The percent trucks for the design period is often assumed to be constant; yet on some sections of the interstate system, the truck traffic in rural areas has been reported to increase from an estimated 6 percent to 25 to 30 percent over a 10- to 20-year period.

The load equivalency factor increases approximately as a function of the ratio of any given axle load to the standard 18-kip single axle load raised to the fourth power. For example, the load equivalency of a 12-kip single axle is given as 0.19 (Appendix D), while the load equivalency for 20-kip single axle is 1.51. Thus, the 20-kip load is 8 times as damaging as the 12-kip load, i.e., $(20/12)^4$. This relationship will vary depending on the structural number and terminal serviceability; however, it is generally indicative of load effects. Thus, it is especially important to obtain reliable truck weight information for each truck class and especially for the multi-axle trucks since these vehicles will constitute a high percentage of the total ESAL's on most projects.

Calculation of future ESAL's is often based on truck factors by truck class. For example, based on truck weight information for five-axle tractor and trailer units, it is possible to develop an average multiplier for each five-axle truck. Thus, if the designer can estimate the number of five-axle trucks over the design period, it is possible to calculate the cumulative ESAL's due to this particular truck class. A similar procedure is described in Appendix D for most of the truck classes on the highways at the present time.

In regard to the use of truck factors, it will be important to use truck weight information representative of the truck traffic on the designed facility. Some truck weight data indicate that truck weights can vary by a factor of six or more between weigh stations. Thus, it is very important to obtain information as nearly site specific as possible when estimating ESAL's per truck for each truck classification.

Procedures described in Appendix D may be applied to stage-construction design, i.e., where the initial design (performance) period is varied in order to consider alternative designs for economic comparisons.

It should be clear from this discussion that the estimate or prediction of future traffic (ESAL's) is not a trivial problem. Poor estimates of traffic can produce pavement performance significantly different than that expected and cause a major increase in the cost of the specific project. This increased cost, when applied to all sections being designed by an agency, will adversely affect the overall programming of highway projects and reduce the work which can be done.

Future deregulation or relaxation of truck loads could also result in changes in the load distributions by truck class, possibly resulting in an increased percentage of five-axle (or more) vehicles being used. Also, inflation pressures used in truck tires are increasing as tire manufacturers improve their technology and the truck industry evaluates the potential advantage of using higher inflation tires. It is not known exactly what the net effect of higher tire inflation is; however, pavement engineers and designers need to keep apprised of possible changes which can influence pavement performance.

In summary, reliable information concerning cumulative ESAL's is important for the determination of pavement structure requirements for both new construction and for rehabilitation. Continuous monitoring of traffic on selected routes to compare predicted and actual traffic loadings is an important and vital set of information needed to produce reliable designs.

The reliability factor included in the Guide (Part I, Chapter 4 and Volume 2, Appendix EE) has been developed to provide consideration of uncertainties in both traffic predictions and performance predictions. Investigations by several states and industry have provided some information concerning the uncertainties in traffic predictions, i.e., comparison of predicted ESAL's and actual ESAL's. The standard deviation of the relationship between predicted and actual traffic has been reported (27) to be on the order of 0.2. In effect, the actual traffic may be 1.6 (one standard deviation) to 4.0 times (three standard deviations) as much as predicted. It should be clear that improvements in traffic loading information and predictions will contribute significantly to the precision which can be achieved in thickness design.

Detailed information and procedures for calculating ESAL's are given in Appendix D. Designs in Part II take into consideration the uncertainty in traffic estimates. The designer must use the best estimate for traffic without any adjustment based on his or her interpretation of the accuracy of such information. Provision has been made in the treatment of reliability in Part II to accommodate the overall effect of variances in the cumulative axle load predictions and other design- and performance-related factors.

1.4.2 Limitations

It is pertinent to note that the load equivalency factors used in this Guide are based on observations at the AASHO Road Test in Ottawa, Illinois. In this regard some limitations should be recognized, such as (1) limited pavement types, (2) loads and load applications, (3) age, and (4) environment.

The pavement types at the AASHO Road Test, from which load equivalency values were derived, included conventional flexible construction, i.e., surface, base and subbase, and rigid pavements with and without reinforcement but always with load transfer devices (dowels). The same load equivalency factors are being applied in this Guide to (1) flexible pavements with stabilized base and subbase, (2) rigid pavements without dowels in the transverse joints, and (3) continuously reinforced concrete pavements. Modifications to the load equivalency values can only come through controlled experiments. The values used in this Guide are considered the best available at the present time.

The experimental design at the AASHO Road Test included a wide range of loads as previously discussed (Section 1.4.1); however, the applied loads were limited to a maximum of 1,114,000 axle applications for those sections which survived the full trafficking period. Thus, the maximum number of 18-kip equivalent single axle loads (ESAL's) applied to any test section was approximately one million. However, by applying the concept of equivalent loads to test sections subjected to only 30-kip single axle loads, for example, it

Introduction and Background

is possible to extend the findings to 8×10^6 ESAL's. Use of any design ESAL's above 8×10^6 requires extrapolation beyond the equations developed from the Road Test results. Such extrapolations have, however, provided reasonable results, based on application of the Guide since 1972.

The AASHO Road Test, from which the basic design equations were derived, was completed after 2 years of traffic testing. The prediction models represented by equations (1.2.1) and (1.2.2) do not include a term for age, i.e., an interactive term for age and traffic. For the present state of knowledge there is very little information available to quantify the effect of aging on performance as expressed in terms of PSI or axle load applications. There is a need for more information regarding the combined effect of traffic and aging on performance. If a user agency has such information it may be possible to modify the performance model accordingly. However, this Guide makes no direct evaluation of aging effects. Evaluation of aging factors along with traffic (ESAL's) should be a high priority for long-term monitoring of pavement performance.

Only one set of materials and one roadbed soil were included in the AASHO Road Test for each pavement type. A small experiment also included performance observations of stabilized base materials under asphaltic surfaces. Use of alternate construction materials represents an extrapolation of the basic data. However, as previously indicated, such extrapolations are based on investigations using analytical techniques and are considered reasonable pending results from field investigations.

The weather at the Road Test in Ottawa, Illinois, is representative of a large portion of the United States, subject to freezing temperatures during the winter and medium to high rainfall throughout the year. An effort has been made in Part II of this Guide to provide a procedure for estimating the effects of seasonal conditions and modifying these for site specific locations. More information on environment is provided in a later section of Part I as well as in Part II of the Guide.

A number of new concepts have been included in these Guides, e.g., reliability, drainage coefficients, use of resilient modulus to estimate layer coefficients, remaining life estimates for overlays, and NDT methods to estimate in situ resilient modulus. These concepts have limited documentation based on actual field observations; however, they are based on an extensive evaluation of the present state of the knowledge. To the extent possible, explanations are provided in the Guide in either this volume or Volume 2. It is hoped that these concepts will find sufficient usage in order to evaluate and eventually modify and improve the design procedures and effectiveness of using the Guide.

1.4.3 Special Cases

This Guide is based on performance equations from the AASHO Road Test which may not apply directly to some urban streets, county roads, parkways, or parking lots. For city streets, the major traffic loads will be generated by service vehicles, buses, and delivery trucks. Load equivalency values for such vehicles are not generally well-estimated by truck load equivalency factors from truck weighing stations. If the Guide is used for design of urban streets, an effort should be made to obtain information on actual axle loads and frequencies typical of vehicles operating on those streets. If this is done, the Guide can be used at a selected level of reliability.

For parkways, i.e., highways which limit the use of heavy trucks, it may be necessary to adjust the design based on a combination of traffic factors, environmental factors, and experience. Use of load equivalency factors as given in Appendix D may result in an underdesigned pavement and premature deterioration.

1.5 ROADBED SOIL

The definitive material property used to characterize roadbed soil for pavement design in this Guide is the resilient modulus (M_R). The procedure for determination of M_R is given in AASHTO Test Method T 274.

The resilient modulus is a measure of the elastic property of soil recognizing certain nonlinear characteristics. The resilient modulus can be used directly for the design of flexible pavements but must be converted to a modulus of subgrade reaction (k-value) for the design of rigid or composite pavements. Direct measurement of subgrade reaction can be made if such procedures are considered preferable to the design agency.

The resilient modulus was selected to replace the soil support value used in previous editions of the Design Guide for the following reasons:

(1) It indicates a basic material property which can be used in mechanistic analysis of multi-layered systems for predicting roughness, cracking, rutting, faulting, etc.

(2) Methods for the determination of M_R are described in AASHTO Test Method T 274.
(3) It has been recognized internationally as a method for characterizing materials for use in pavement design and evaluation.
(4) Techniques are available for estimating the M_R properties of various materials in-place from nondestructive tests.

It is recognized that many agencies do not have equipment for performing the resilient modulus test. Therefore, suitable factors are reported which can be used to estimate M_R from standard CBR, R-value, and soil index test results or values. The development of these factors is based on state of the knowledge correlations. It is strongly recommended that user agencies acquire the necessary equipment to measure M_R. In any case, a well-planned experiment design is essential in order to obtain reliable correlations. A range of soil types, saturation, and densities should be included in the testing program to identify the main effects. Guidelines for converting CBR and R-value to M_R are discussed in this chapter. These correlations are used in Part II of this Guide pending the establishment of agency values.

Heukelom and Klomp (6) have reported correlations between the Corps of Engineers CBR value, using dynamic compaction, and the in situ modulus of soil. The correlation is given by the following relationship:

$$M_R(\text{psi}) = 1,500 \times \text{CBR} \quad (1.5.1)$$

The data from which this correlation was developed ranged from 750 to 3,000 times CBR. This relationship has been used extensively by design agencies and researchers and is considered reasonable for fine-grained soil with a soaked CBR of 10 or less. Methods for testing are given in Appendix F. The CBR should correspond to the expected field density.

Similar relationships have also been developed by the Asphalt Institute (7) which relate R-value to M_R as follows:

$$M_R(\text{psi}) = A + B \times (\text{R-value}) \quad (1.5.2)$$

where

A = 772 to 1,155 and
B = 369 to 555.

For the purposes of this Guide, the following correlation may be used for fine-grained soils (R-value less than or equal to 20) until designers develop their own capabilities:

$$M_R = 1,000 + 555 \times (\text{R-value}) \quad (1.5.3)$$

This discussion summarizes estimates for converting CBR and R-values to a resilient modulus for roadbed soil. Similar information is provided for granular materials in Section 1.6, Materials of Construction.

Placement of roadbed soil is an important consideration in regard to the performance of pavements. In order to improve the general reliability of the design, it is necessary to consider compaction requirements. For average conditions it is not necessary to specify special provisions for compaction. However, there are some situations for which the designer should request modifications in the specifications.

(1) The basic criteria for compaction of roadbed soils should include an appropriate density requirement. Inspection procedures must be adequate to assure that the specified density is attained during construction. If, for any reason, the basic compaction requirements cannot be met, the designer should adjust the design M_R value accordingly.
(2) Soils that are excessively expansive or resilient should receive special consideration. One solution is to cover these soils with a sufficient depth of selected material to modify the detrimental effects of expansion or resilience. Expansive soils may often be improved by compaction at water contents of 1 or 2 percent above the optimum. In some cases it may be more economical to treat expansive or resilient soils by stabilizing with a suitable admixture, such as lime or cement, or to encase a substantial thickness in a waterproof membrane to stabilize the water content. Information concerning expansive soil is covered in Reference 8. Methods for evaluating the potential consequences of expansive roadbed soils are provided in Appendix G.
(3) In areas subject to frost, frost-susceptible soils may be removed and replaced with selected, nonsusceptible material. Where such soils are too extensive for economical removal, they may be covered with a sufficient depth of suitable material to modify the detrimental effects of freezing and thawing. Methods for evaluat-

ing the consequences of frost heave are provided in Appendix G and have been reviewed previously in this chapter. Methods for compensating for seasonal thaw-weakening are provided in Part II.

(4) Problems with highly organic soils are related to their extremely compressible nature and are accentuated when deposits are nonuniform in properties or depth. Local deposits, or those of relatively shallow depth, are often most economically excavated and replaced with suitable select material. Problems associated with deeper and more extensive deposits have been alleviated by placing surcharge embankments for preconsolidation, sometimes with special provisions for rapid removal of water to hasten consolidation.

(5) Special provisions for unusually variable soil types and conditions may include: scarifying and recompacting; treatment of an upper layer of roadbed soils with a suitable admixture; using appreciable depths of more suitable roadbed soils (select or borrow); over-excavation of cut sections and placing a uniform layer of selected material in both cut-and-fill areas; or adjustment in the thickness of subbase at transitions from one soil type to another.

(6) Although the design procedure is based on the assumption that provisions will be made for surface and subsurface drainage, some situations may require that special attention be given to design and construction of drainage systems. Drainage is particularly important where heavy flows of water are encountered (i.e., springs or seeps), where detrimental frost conditions are present, or where soils are particularly susceptible to expansion or loss of strength with increase in water content. Special subsurface drainage may include provision of additional layers of permeable material beneath the pavement for interception and collection of water, and pipe drains for collection and transmission of water. Special surface drainage may require such facilities as dikes, paved ditches, and catch-basins.

(7) Certain roadbed soils pose difficult problems in construction. These are primarily the cohesionless soils, which are readily displaced under equipment used to construct the pavement, and wet clay soils, which cannot be compacted at high water contents because of displacement under rolling equipment and which require long periods of time to dry to a suitable water content. Measures used to alleviate such construction problems include: (1) blending with granular materials, (2) adding suitable admixtures to sands to provide cohesion, (3) adding suitable admixtures to clays to hasten drying or increase shear strength, and (4) covering with a layer of more suitable selected material to act as a working platform for construction of the pavement.

Resilient Modulus (M_R) values for pavement structure design should normally be based on the properties of the compact layer of the roadbed soil. It may, in some cases, be necessary to include consideration of the uncompacted foundation if these in situ materials are especially weak. It is important to note that the design of the pavement structure by this Guide is based on the *average* M_R value. Although reliability considers the variation of many factors associated with design, it is treated by adjusting the design traffic. (See Chapter 4.) The design traffic is the expected value of 18-kip ESAL's during the design period. The designer *must not* select a design M_R value based on some minimum or conservative criteria as this will introduce increased conservatism in design beyond that provided by the reliability factor.

1.6 MATERIALS OF CONSTRUCTION

Materials used for construction of the pavement structure can be divided into two general classes; (1) those for flexible pavements and (2) those for rigid pavements. Materials used for composite pavements include those for roadbed preparation, for a subbase, and for a portland cement concrete slab with an asphalt concrete wearing surface. An asphalt concrete overlay on a rigid pavement is considered a composite pavement.

In order to complete the design requirements for flexible pavements, it may be necessary to convert CBR or R-value information to resilient modulus, M_R. In the absence of agency correlations, the following correlations are provided for unbound granular materials (base and subbase):

θ (psi)	M_R (psi)
100	740 × CBR or 1,000 + 780 × R
30	440 × CBR or 1,000 + 450 × R
20	340 × CBR or 1,000 + 350 × R
10	250 × CBR or 1,000 + 250 × R

where θ = sum of the principal stresses, $\sigma_1 + \sigma_2 + \sigma_3$; referring to AASHTO T 274, this corresponds to $\sigma_d + 3\sigma_3$ when $\sigma_d = \sigma_1 - \sigma_3$.

The strength of the granular base or subbase is related to the stress state which will occur under operating conditions. The sum of the principal stresses, θ, is a measure of the stress state, which is a function of pavement thickness, load, and the resilient modulus of each layer. As an agency becomes increasingly familiar with these parameters, it will be possible to determine the stress state from a layered system analysis following procedures given in Part IV of the Guide. However, if such information is not available, estimates of resilient modulus values provided in Part II of this Guide may be used.

1.6.1 Flexible Pavements

As shown in Figure 1.1, flexible pavements generally consist of a prepared roadbed underlying layers of subbase, base, and surface courses. In some cases the subbase and/or base will be stabilized to maximize the use of local materials. The engineering literature contains a good deal of information relative to soil and aggregate stabilization (*9, 10*).

References 9 and 10 provide a state of the knowledge description of procedures for selecting the stabilizing agents appropriate to various soil types and construction methods. Pavement design examples in Reference 9 refer to the 1972 Interim Guide; however, the examples can still be used to illustrate design concepts appropriate for use with stabilized materials.

Prepared Roadbed. The prepared roadbed is a layer of compacted roadbed soil or select borrow material which has been compacted to a specified density.

Subbase Course. The subbase course is the portion of the flexible pavement structure between the roadbed soil and the base course. It usually consists of a compacted layer of granular material, either treated or untreated, or of a layer of soil treated with a suitable admixture. In addition to its position in the pavement, it is usually distinguished from the base course material by less stringent specification requirements for strength, plasticity, and gradation. The subbase material should be of significantly better quality than the roadbed soil. For reasons of economy, the subbase is often omitted if roadbed soils are of high quality.

When roadbed soils are of relatively poor quality and the design procedure indicates that a substantial thickness of pavement is required, several alternate designs should be prepared for structural sections with and without subbase. The selection of an alternate may then be made on the basis of availability and relative costs of materials suitable for base and subbase. Because lower quality materials may be used in the lower layers of a flexible pavement structure, the use of a subbase course is often the most economical solution for construction of pavements over poor roadbed soils.

Although no specific quality requirements for subbase material are presented in this Guide, the *AASHTO Construction Manual for Highway Construction* can be used as a guide. Many different materials have been used successfully for subbase. Local experience can be used as the basis for selection. For use in this design procedure, subbase material, if present, requires the use of a layer coefficient (a_3), in order to convert its actual thickness to a structural number (SN). Special consideration must be given to determining the minimum thickness of base and surfacing required over a given subbase material. Procedures that may be used for this purpose are given in Part II. Procedures for assigning appropriate layer coefficients based on expected M_R are given in Part II.

Untreated aggregate subbase should be compacted to 95 percent of maximum laboratory density, or higher, based on AASHTO Test T 180, Method D, or the equivalent. In addition to the major function as a structural portion of the pavement, subbase courses may have additional secondary functions, such as:

(1) Preventing the intrusion of fine-grained roadbed soils into base courses—relatively dense-graded materials must be specified if the subbase is intended to serve this purpose.

(2) Minimize the damaging effects of frost action—materials not susceptible to detrimental frost action must be specified if the subbase is intended for this purpose.

(3) Preventing the accumulation of free water within or below the pavement structure—a relatively free-draining material may be specified for the subbase if this is the intention. Provisions must also be made for collecting and removing the accumulated water from the subbase if this layer is to be included as part of the drainage system. If the subbase is to be designed as a drainage layer, it will be necessary to limit the fraction passing the No. 8 sieve to a very small percent.

(4) Providing a working platform for construction equipment—important when roadbed soil cannot provide the necessary support.

Base Course. The base course is the portion of the pavement structure immediately beneath the surface course. It is constructed on the subbase course, or, if no subbase is used, directly on the roadbed soil. Its major function in the pavement is structural support. It usually consists of aggregates such as crushed stone, crushed slag, crushed gravel and sand, or combinations of these materials. It may be used untreated or treated with suitable stabilizing admixtures, such as portland cement, asphalt, lime, cement-flyash and lime-flyash, i.e., pozzolonic stabilized bases. Specifications for base course materials are generally considerably more stringent than for subbase materials in requirements for strength, plasticity, and gradation. Guidelines for stabilization can be found in References 9 and 10.

When utilizing pozzolonic stabilized bases under a relatively thin asphaltic wearing surface, it can usually be expected that uncontrolled transverse reflection cracks will occur in the surface in a relatively short period of time, e.g., 1 to 3 years. Sawed and sealed joints (through the asphalt concrete into the base) may be utilized to minimize the adverse effects on appearance and to provide for better future sealing operations. Joint spacing may vary from 20 to 40 feet depending on local experience with past uncontrolled crack-spacing problems.

Although no specific quality requirements for base courses are presented in this Guide, the specifications included in AASHTO's *Manual for Highway Construction* or in ASTM Specification D 2940, "Graded Aggregate Material for Bases or Subbase for Highways and Airports," are often used. Materials varying in gradation and quality from these specifications have been used in certain areas and have provided satisfactory performance. Additional requirements for quality of base materials, based on test procedures used by the constructing agency, may also be included in materials or construction specifications.

Untreated aggregate base should be compacted to at least 95 percent of maximum laboratory density based on AASHTO Test T 180, Method D, or the equivalent. A wide variety of materials unsuitable for use as untreated base course have given satisfactory performance when improved by addition of a stabilizing admixture, such as portland cement, asphalt, or lime. Consideration should be given to the use of such treated materials for base courses whenever they are economically feasible, particularly when suitable untreated materials are in short supply. Economic advantages may result not only from the use of low-cost aggregates but also from possible reduction in the total thickness of the pavement structure that may result from the use of treated materials. Careful study is required in the selection of the type and amount of admixture to be used for optimum performance and economy.

For use in this design procedure, base material must be represented by a layer coefficient (a_2) in order that its actual thickness may be converted to a structural number. Procedures for the determination of layer coefficients based on M_R are given in Part II.

Drainage Layer. A number of agencies are now considering or constructing pavements with a drainage course, or layer, as shown in Figure 1.3 (*11*). Figure 1.3 illustrates one configuration; alternate designs are shown in Appendix AA of Volume 2 and in References 12 and 13.

The cross section shown in Figure 1.3 is illustrative only. The location of the longitudinal drain with respect to the traveled way can vary depending on designer preference and local experience. Also, this figure does not show the collector systems and outlet requirements for a total drainage design. Reference should be made to Appendix AA of Volume 2 and References 11, 12, 13, 22, and 23 for additional information regarding the design of drainage systems.

The designer should give some consideration to the preferred construction sequence when specifying a drainage system, e.g., excavation and installation after the travel lane paving has been completed. Local practice should be followed; however, the designer should be aware that special provisions to the specifications may be necessary. Additional information concerning the design of the drainage layer is provided in Section 1.8 of Part I and in Appendix AA of Volume 2.

Tables 1.1, 1.2, and 1.3 provide some background information for estimating the permeability of various types of material.

Table 1.1 provides general relationships between coarse-graded unstabilized materials and their coefficients of permeability (*11*).

Table 1.2 provides guidelines for the gradation of asphalt-treated permeable material (*11*). At least one state agency has reported the same gradation for porous concrete used as a drainage layer.

Table 1.3 summarizes information relative to the permeability of graded aggregates as a function of the percent passing the No. 200 mesh sieve. Additional information concerning materials to be used for the drainage course is provided in Reference 12.

A. Base is used as the drainage layer.

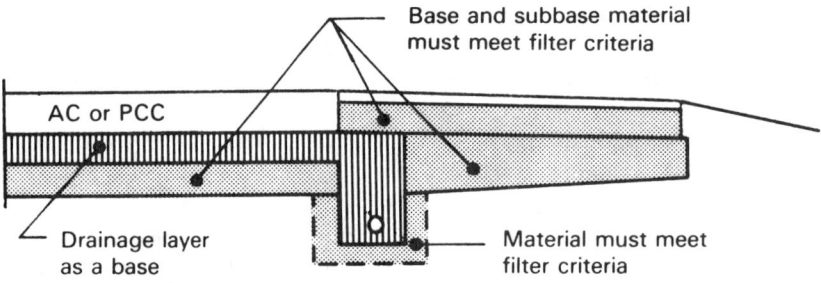

B. Drainage layer is part of or below the subbase.

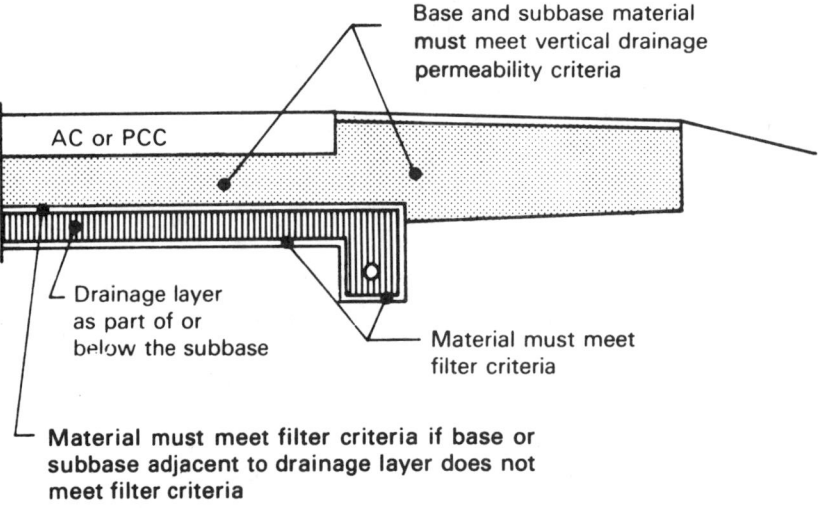

Note: Filter fabrics may be used in lieu of filter material, soil, or aggregate, depending on economic considerations.

Figure 1.3. Example of Drainage Layer in Pavement Structure (*11*)

Table 1.1. Permeability of Graded Aggregates (11)

Percent Passing	Sample Number					
	1	2	3	4	5	6
3/4-inch sieve	100	100	100	100	100	100
1/2-inch sieve	85	84	83	81.5	79.5	75
3/8-inch sieve	77.5	76	74	72.5	69.5	63
No. 4 sieve	58.5	56	52.5	49	43.5	32
No. 8 sieve	42.5	39	34	29.5	22	5.8
No. 10 sieve	39	35	30	25	17	0
No. 20 sieve	26.5	22	15.5	9.8	0	0
No. 40 sieve	18.5	13.3	6.3	0	0	0
No. 60 sieve	13.0	7.5	0	0	0	0
No. 140 sieve	6.0	0	0	0	0	0
No. 200 sieve	0	0	0	0	0	0
Dry density (pcf)	121	117	115	111	104	101
Coefficient of permeability (ft. per day)	10	110	320	1,000	2,600	3,000

NOTE: Subsurface drainage systems should be capable of removing.

The approximate coefficient of permeability of the asphalt-treated permeable material is 3,000 feet or more per day when treated with 2-percent asphalt and 8,000 feet per day with no asphalt.

Table 1.2. Gradation for Asphalt Treated Permeable Layer (11)

Sieve Size	Percent Passing
1"	100
3/4"	90–100
3/8"	30–50
No. 4	0–5
No. 8	0–2

Table 1.3. Effect of Percentage Passing 200 Mesh Sieve on Coefficient of Permeability of Dense Graded Aggregate, Feet Per Day (11)

Types of Fines	Percent Passing No. 200 Sieve			
	0	5	10	15
Silica or limestone	10	0.07	0.08	0.03
Silt	10	0.08	0.001	0.0002
Clay	10	0.01	0.0005	0.00009

Specifications, for both design and construction, of drainage courses are under development; hence, material requirements should be referenced to the latest guide specifications of AASHTO, ASTM, or the appropriate state agency responsible for developing statewide criteria and requirements. Information in Tables 1.1, 1.2, and 1.3 provides some guidelines for estimating permeability.

The N.J. Department of Transportation has developed specifications for bituminous stabilized and non-stabilized open-graded mixes for drainage layers. The gradation requirements used by the NJDOT are:

Sieve Size	Percent Passing
1.5 in.	100
1.0 in.	95–100
0.5 in.	60–80
No. 4	40–55
No. 8	5–25
No. 16	0–8
No. 50	0–5

This material can be made with a 50/50 blend of No. 57 and No. 9 stone of a crushed stone. The target permeability suggested by NJDOT is 1,000–3,000 ft. per day. Laboratory testing for permeability is recommended prior to approval of the porous layer material.

A "cookbook" approach to the internal drainage problem is given by G.S. Kozloo in Transportation Record 993.

The measurement of subsurface drainage is generally based on the time required for 50-percent of the unbound water to be removed from the layer to be drained. The Casagrande flow equation for estimating the 50-percent drainage time is expressed as:

$$t_{50} = (\eta_e \times L^2)/[2 \times K \times (H + L \times \tan \alpha)] \quad (1.6.1)$$

where

t_{50} = time for 50 percent of unbound water to drain (days),
η_e = effective porosity (80 percent of absolute porosity),
L = length of flow path (feet),
K = permeability constant (ft./day), and
$\tan \alpha$ = slope of the base layer.

Filter Material. A detailed description of filter layers is contained in Appendix AA, Volume 2. Ridgeway (*11*) provides the following general comments:

> The drainage layer and the collector system must be prevented from clogging if the system is to remain functioning for a long period of time. This is accomplished by means of a filter between the drain and the adjacent material. The filter material, which is made from select aggregates or fabrics, must meet three general requirements: (1) it must prevent finer material, usually the subgrade, from piping or migrating into the drainage layer and clogging it; (2) it must be permeable enough to carry water without any resistance; and (3) it must be strong enough to carry the loads applied and, for aggregate, to distribute live loads to the subgrade.

Surface Course. The surface course of a flexible structure consists of a mixture of mineral aggregates and bituminous materials placed as the upper course and usually constructed on a base course. In addition to its major function as a structural portion of the pavement, it must also be designed to resist the abrasive forces of traffic, to reduce the amount of surface water penetrating the pavement, to provide a skid-resistance surface, and to provide a smooth and uniform riding surface.

The success of a surface course depends to a degree on obtaining a mixture with the optimum gradation of aggregate and percent of bituminous binder to be durable and to resist fracture and raveling without becoming unstable under expected traffic and climatic conditions. The use of a laboratory design procedure is essential to ensure that a mixture will be satisfactory.

Although dense-graded aggregates with a maximum size of about 1 inch are most commonly specified for surface courses for highways, a wide variety of other gradations, from sands to coarse, open-graded mixtures, have been used and have provided satisfactory performance for specific conditions. Surface courses are usually prepared by hot plant mixing with an asphalt cement, but satisfactory performance has also been obtained by cold plant mixing, or even mixing, in-place, with liquid asphalts or asphalt emulsions. Hot plant mixes, e.g., asphalt concrete, are recommended for use on all moderate to heavily trafficked highways.

Construction specifications usually require that a bituminous material be applied on untreated aggregate base courses as a prime coat, and on treated base courses and between layers of the surface course to serve as a tack coat.

No specific quality requirements for surface courses are presented in this Guide. It is recognized that each agency will prepare specifications that are based on performance, local construction practices, and the most economical use of local materials. ASTM Specification D 3515 provides some guidelines for designing asphalt concrete paving mixes.

It is particularly important that surface courses be properly compacted during construction. Improperly compacted surface courses are more likely to exhibit a variety of types of distress that tend to reduce the life and overall level of performance of the pavement. Types of distress that are often related to insufficient compaction during construction include rutting resulting from further densification under traffic, structural failure resulting from excess infiltration of surface water through the surface course, and cracking or raveling of the surface course resulting from embrittlement of the bituminous binder by exposure to air and water in the mixture. Specific criteria for compaction must be established by each highway agency based on local experience. Theoretical maximum densities of 92 percent or more are sometimes specified for dense-graded mixes.

1.6.2 Rigid Pavements

As shown in Figure 1.1, rigid pavements generally consist of a prepared roadbed underlying a layer of subbase and a pavement slab. The subbase may be stabilized or unstabilized. In cases of low volume road design where truck traffic is low, a subbase layer may not be necessary between the prepared roadbed and the pavement slab.

A drainage layer can be included in rigid pavements in much the same manner described for flexible pavements as shown in Figure 1.3. Alternate drainage designs are shown in Appendix AA, Volume 2.

Subbase. The subbase of a rigid pavement structure consists of one or more compacted layers of granular or stabilized material placed between the subgrade and the rigid slab for the following purposes:

(1) to provide uniform, stable, and permanent support,
(2) to increase the modulus of subgrade reaction (k),
(3) to minimize the damaging effects of frost action,
(4) to prevent pumping of fine-grained soils at joints, cracks, and edges of the rigid slab, and
(5) to provide a working platform for construction equipment.

If the roadbed soils are of a quality equal to that of a subbase, or in cases where design traffic is less than 1,000,000 18-kip ESAL's, an additional subbase layer may not be needed.

A number of different types of subbases have been used successfully. These include graded granular materials and materials stabilized with suitable admixtures. Local experience may also provide useful criteria for the selection of subbase type. The prevention of water accumulations on or in roadbed soils or subbases is essential if satisfactory performance of the pavement structure is to be attained. It is recommended that the subbase layer be carried 1 to 3 feet beyond the paved roadway width or to the inslope if required for drainage.

Problems with the erosion of subbase material under the pavement slab at joints and at the pavement edge have led some designers to use a lean concrete or porous layers for subbase. While the use of a porous layer is encouraged it should be noted that design criteria for such materials are still in the development stage and the designer should review the literature or contact agency personnel familiar with current requirements.

Pavement Slab. The basic materials in the pavement slab are portland cement concrete, reinforcing steel, load transfer devices, and joint sealing materials. Quality control on the project to ensure that the materials conform to AASHTO or the agency specifications will minimize distress resulting from distortion or disintegration.

Portland Cement Concrete. The mix design and material specifications for the concrete should be in accordance with, or equivalent to, the requirements of the AASHTO *Guide Specifications for Highway Construction* and the *Standard Specifications for Transportation Materials*. Under the given conditions of a specific project, the minimum cement factor should be determined on the basis of laboratory tests and prior experience of strength and durability.

Air-entrained concrete should be used whenever it is necessary to provide resistance to surface deterioration from freezing and thawing or from salt or to improve the workability of the mix.

Reinforcing Steel. The reinforcing steel used in the slab should have surface deformations adequate to bond and develop the working stresses in the steel. For smooth wire mesh, this bond is developed through the welded cross wires. For deformed wire fabric, the bond is developed by deformations on the wire and at the welded intersections.

Joint Sealing Materials. Three basic types of sealants are presently used for sealing joints:

(1) *Liquid sealants.* These include a wide variety of materials including: asphalt, hot-poured rubber, elastomeric compounds, silicone, and polymers. The materials are placed in the joint in a liquid form and allowed to set. When using liquid sealants, care should be taken to provide the proper shape factor for the movement expected.
(2) *Preformed elastomeric seals.* These are extruded neoprene seals having internal webs that exert an outward force against the joint face. The size and installation width depend on the amount of movement expected at the joint.
(3) *Cork expansion joint filler.* There are two types of cork fillers: (a) standard expansion joint filler, and (b) self-expanding (SE) type.

Longitudinal Joints. Longitudinal joints are needed to form cracks at the desired location so that they may be adequately sealed. They may be keyed, butted, or tied joints, or combinations thereof. Longitudinal joints should be sawed or formed to a minimum depth of one-fourth of the slab thickness. Timing of the sawcutting is critical to the crack formation at the desired location. The maximum recommended longitudinal joint spacing is 16 feet.

Load-Transfer Devices. Mechanical load-transfer devices for *transverse joints* should possess the following attributes:

(1) They should be simple in design, be practical to install, and permit complete encasement by the concrete.
(2) They should properly distribute the load stresses without overstressing the concrete at its contact with the device.
(3) They should offer little restraint to longitudinal movement of the joint at any time.
(4) They should be mechanically stable under the wheel load weights and frequencies that will prevail in practice.
(5) They should be resistant to corrosion when used in those geographic locations where corrosive elements are a problem. (Various types of coatings are often used to minimize corrosion.)

A commonly used load-transfer device is the plain, round steel dowel conforming to AASHTO Designation M 31-Grade 60 or higher. Specific design requirements for these relative to diameter, length, and spacing are provided in Part II. Although round dowels are the most commonly used, other mechanical devices that have proven satisfactory in field installations may also be used.

Consideration may also be given to omitting load transfer devices from transverse weakened plane joints in plain jointed concrete pavement when supported on a treated permeable base.

Tie Bars. Tie bars, either deformed steel bars or connectors, are designed to hold the faces of abutting slabs in firm contact. Tie bars are designed to withstand the maximum tensile forces required to overcome subgrade drag. They are not designed to act as load-transfer devices.

Deformed bars should be fabricated from billet or axle steel of Grade 40 conforming to AASHTO M 31 or M 53. Specific recommendations on bar sizes, lengths, and spacings for different pavement conditions are presented in Part II.

Other approved connectors may also be used. The tensile strength of such connectors should be equal to that of the deformed bar that would be required. The spacing of these connectors should conform to the same requirements given for deformed tie bars in Part II.

Consideration should be given to the use of corrosion-resistant materials or coatings for both tie bars and dowels where salts are to be applied to the surface of the pavement.

1.6.3 Shoulders

Shoulders have often in the past been constructed of a flexible base with an asphalt surfacing or of a stabilized base with an asphalt surfacing. The combination of a dissimilarity between the outside lane and shoulder and the encroachment of heavy wheel loads onto the shoulder have sometimes resulted in joint problems between the travel lanes and the shoulder. Research has shown that strengthening of the shoulder and adding special sealants have helped to alleviate this problem. The use of tied concrete shoulders or 3-foot monolithic widening of the outside PCC lane has also proven beneficial (1.5-foot monolithic widening is acceptable if a rumble strip is provided as a deterrent to edge encroachment). Thickening the outside edge of the travel lane or using a monolithic curb (where appropriate) also strengthens the pavement edge and reduces the shoulder-joint problem. Provision for slab design which incorporates tied shoulders and widened outside lanes is provided in Part II of this Guide.

Additional information pertinent to shoulder design is given in Section 1.9.

1.7 ENVIRONMENT

Two main environmental factors are considered with regard to pavement performance and pavement structure design in this Guide; specifically, these are temperature and rainfall.

Temperature will affect (1) the creep properties of asphalt concrete, (2) thermal-induced stresses in asphalt concrete, (3) contraction and expansion of portland cement concrete, and (4) freezing and thawing of the roadbed soil. Temperature and moisture differential between the top and bottom of concrete slabs in jointed concrete pavements creates an upward curling

Introduction and Background

and warping of the slab ends which can result in pumping and structural deterioration of undrained sections.

Rainfall, if allowed to penetrate the pavement structure or roadbed soil, will influence the properties of those materials. This section of the Guide covers problems associated with temperature. Section 1.8 covers drainage requirements as related to rainfall.

Freezing and thawing of roadbed soil has traditionally been a major concern of pavement designers. The major effect is with regard to the thaw-weakening which can occur during the spring thaw period. Figure 1.4 illustrates the seasonal effects which can occur in many regions of the United States. A second effect of freezing is the occurrence of frost heaving, causing a reduction in the serviceability of the pavement.

Procedures for calculating the damage during various seasons of the year as a function of thaw-weakening and frost heaving are given in Part II. It is beyond the scope of the Guide to describe in detail the mechanism related to frost susceptibility, thaw-weakening, and frost heaving. The user is referred to Reference 14 for more information on this subject. A few of the more pertinent considerations from Reference 14 which relate to pavement structure design in frost areas are reproduced in this section of the Guide.

Frost heaving of soil within or beneath a pavement is caused by the accumulation of ice within the larger soil voids and, usually, a subsequent expansion to form continuous ice lenses, layers, veins, or other ice masses. The growth of such distinct bodies of ice is termed ice segregation. A lens grows in thickness in the direction of heat transfer until the water supply is depleted, as by formation of a new lens at a lower level, or until freezing conditions at the freezing interface will no longer support further crystallization. Investigations (*12, 13, 16*) have shown that ice segregation occurs only in soils containing fine particles. Such soils are said to be frost susceptible; clean sands and gravels are nonfrost-susceptible soils. The degree of frost susceptibility is principally a function of the percentage of fine particles and, to a lesser degree, of particle shape, distribution of grain sizes, and mineral composition.

The following three conditions of soil, temperature, and water must be present simultaneously in order for ice segregation to occur in the subsurface materials:

(1) *Soil.* The soil must be frost susceptible.
(2) *Temperature.* Freezing temperatures must penetrate the soil. In general, the thickness of a particular layer or lens of ice is inversely proportional to the rate of penetration of freezing temperature into the soil.
(3) *Water.* A source of water must be available from the underlying groundwater table, infiltration or gravitational flow, an aquifer, or the water held within the voids of fine-grained soil.

Periods of thawing are among the most critical phases in the annual cycle of environmental changes affecting pavements in seasonal frost areas. Such thawing cycles are in many cases very disruptive, depending on the rapidity of the thaw and the drainage capabilities of the pavement system. During thaw periods considerable melting of snow may occur, with melt water filling the ditches and infiltrating into the pavement from the shoulders and through surface cracks in the pavement itself. During thawing periods, the bearing capacity of the roadbed soil may be severely reduced, and frost heaving frequently is more severe after midwinter thaw periods. In areas of deep frost penetration, the period of complete thawing of thicker pavement structures in the spring is usually the most damaging type of thaw period because it affects the roadbed as well as subbase and base layers. The severity of the adverse effect on the supporting capacity of a given roadbed is largely dependent on the temperature distribution in the ground during the thawing period.

Thawing can proceed from the top downward, from the bottom upward, or both. The manner of thawing depends on the pavement surface temperature. During a sudden spring thaw, melting will proceed almost entirely from the surface downward. This type of thawing leads to extremely adverse drainage conditions. The still-frozen soil beneath the thawed layer traps the water released by the melting ice lenses so that lateral and surface drainage are the only means of egress. In granular soils, lateral drainage may be restricted by still-frozen shoulders resulting from the insulating effect of snow and/or different thermal conductivity and surface reflectivity characteristics. If air temperatures in the spring remain cool and frosty at night, upward conduction of heat stored in the ground from the previous summer and of heat from the interior of the earth will produce thawing, principally from the bottom upward. Such thawing permits soil moisture from melted ice lenses to drain downward while the material above it remains frozen.

The climatic factors of air temperature, solar radiation received at the surface, wind, and precipitation are major parameters that effect the severity of frost effects in a given geographical area. The first three

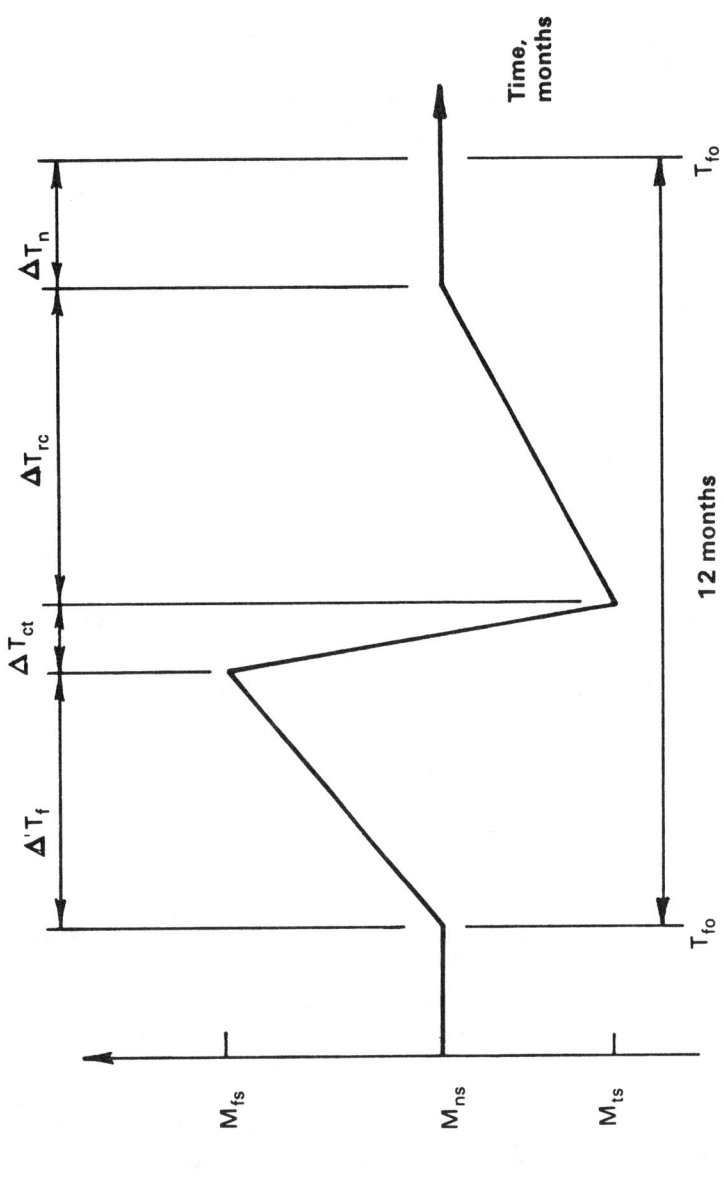

Figure 1.4. Representation of Roadbed Modulus Variations throughout Year

mainly affect the temperature regime in the pavement structure, including the important parameters of depth of frost penetration, number of freeze-thaw cycles, and duration of the freezing and thawing periods. Precipitation affects mainly the moisture regime but causes changes in the thermal properties of the soil and interacts with the other climatic variables determining ground temperatures as well.

Investigators who have endeavored to calculate the depth of frost penetration have found it convenient to make use of a freezing index (15), which expresses the cumulative effect of intensity and duration of subfreezing air temperatures. The freezing index is expressed in degree days and represents the difference between the highest and lowest points on a curve of cumulative degree days versus time for one freezing season. The degree days for any one day equals the difference between the average daily air temperature and 32°F. Degree days are plus when the average daily temperature is below 32°F (freezing degree days) and minus when above 32°F (thawing degree days). Thus, an average daily temperature of 31°F is equal to one degree day, 33°F is equal to minus one degree day, and 22°F is equal to 10 degree days.

The freezing index for a given year and site location can be calculated from average daily air temperature records, which should be obtained from a station situated close to the construction site. This is necessary because differences in elevation and topography, and nearness to centers of population or bodies of water (rivers, lakes, seacoast) and other sources of heat, are likely to cause considerable variations in the value of the freezing index over short distances. Such variations may be of sufficient magnitude to affect a pavement design based on depth of frost penetration, particularly in areas where the freezing index used in the calculation is more than about 100 degree days. Table 1.4 provides an indication of the depth of frost based on the penetration of the 32°F (0°C) isotherm below the surface of 12 inches of portland cement concrete. Variations due to pavement type, soil type, duration of low temperature, and water content may affect the actual frost penetration; however, it is clear that frost penetration can extend well into the roadbed soils during sustained periods of freezing temperatures.

Most studies have shown that a soil is frost susceptible only if it contains fine particles. Soils free of material passing the 200 mesh sieve generally do not develop significant ice segregation or frost heave.

A reliable method for recognizing a frost susceptible material for site specific conditions has not, as yet, been identified. Some guidelines are available in the literature and are described by Johnson, et al. (14). The U.S. Corps of Engineers have reported that most inorganic soils containing 3 percent (by weight) or more of grains finer than 0.02 mm in diameter are considered frost susceptible for pavement design purposes (16).

In summary, frost action due to freezing temperatures in soil, can cause both heaving and thaw-weakening. However, thaw-weakening is not necessarily directly proportional to heaving since field experience shows that thaw-weakened but well-drained sandy or gravelly materials recover bearing strength quite rapidly, whereas clayey soils may show little heave but recover their stability very slowly (14). The design procedure in Part II of this Guide provides for both frost heave and thaw-weakening.

The period of thaw-weakening can be estimated from deflection measurements, as shown in Figure 1.5. These data were obtained at the AASHO Road Test and indicate that the thaw-weakening period can range from a few weeks to a few months, with varying degrees of reduction in structural capacity. Further guidelines relative to thaw-weakening periods are given in Part II of this Guide; however, user agencies are encouraged to develop these relationships based on site specific measurements within their areas and to compare such experience with other agencies nationally.

Laboratory tests and field evaluations indicate that the retained modulus during the thaw-weakening period may be 20 to 50 percent of the normal modulus obtained during the summer and fall periods.

It should be noted that the resilient modulus for roadbed soils may also vary by season even when no thaw-weakening period is experienced. For example, during the heavy rainy periods it might be expected that some seasonal variation in bearing capacity will occur. There may be other situations in which no seasonal variations occur and a constant modulus can be used for the roadbed soil. Note that the modulus is

Table 1.4. Frost Penetration under Portland Cement Concrete Pavement (11)

Air-Freezing Index (degree days)	Frost Penetration (feet)
200	1.8
400	3.0
600	4.0
800	5.0
1,000	6.0

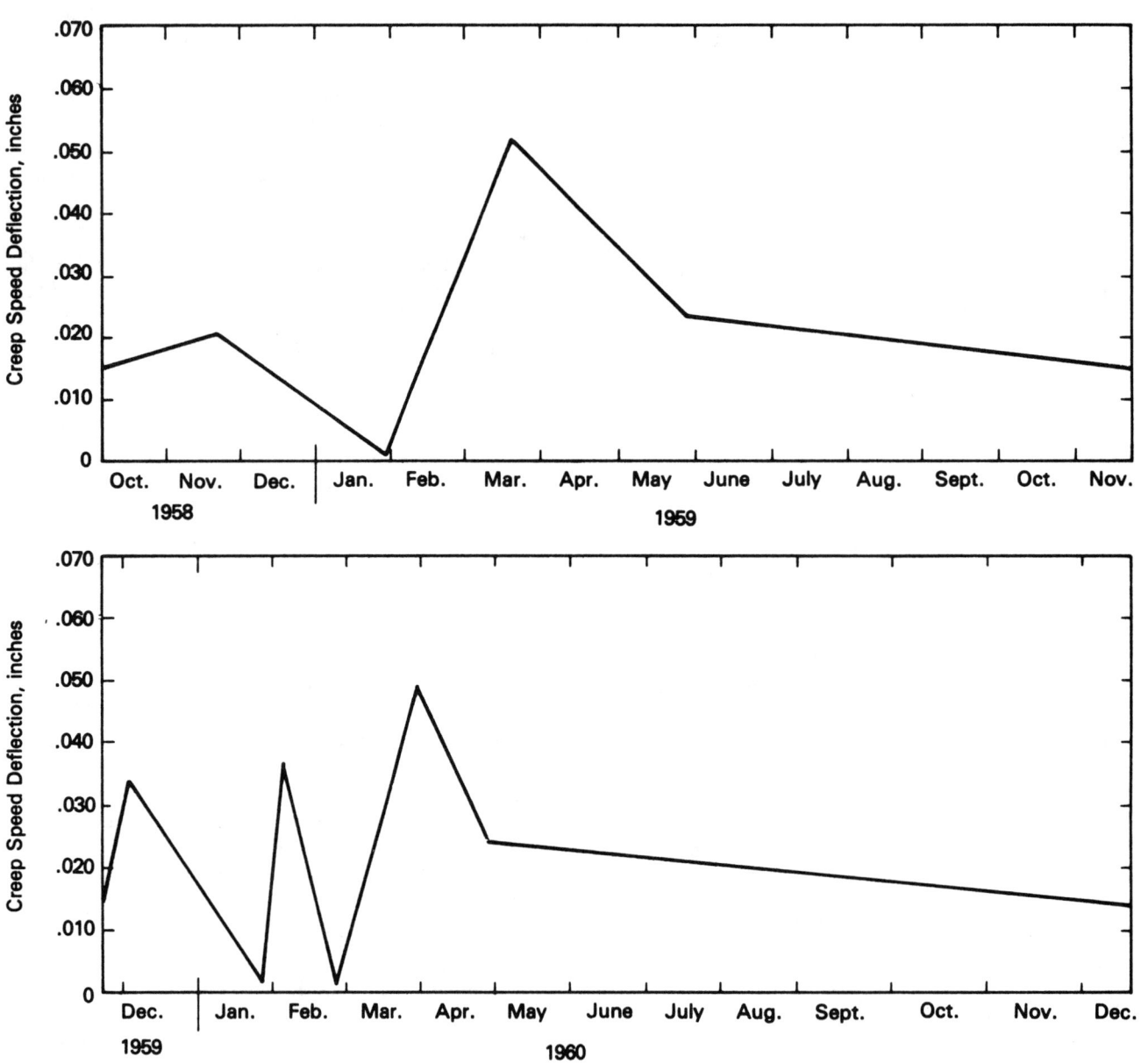

Figure 1.5. Seasonal Deflection on Nontraffic Loop, 6-kip Single Axle Load (17)

related to CBR, R-value, or plate bearing value and, hence, experience with these types of strength tests can be used to infer the seasonal effects on the modulus.

While information is generally lacking with regard to the effects of thaw-weakening or water saturation on untreated aggregate base and subbase, some research does suggest that a reduced modulus can occur during certain critical seasons. Reference 18, based on results of spring deflection measurements at the AASHO Road Test, indicates that the retained modulus ranged from 80 to 85 percent of the normal modulus obtained during the summer and fall. If these layers had been well drained no reduction in modulus would be anticipated.

In order to compensate for thaw-weakening effects on pavement performance, provision has been made in Part II to calculate an effective annual roadbed soil resilient modulus. The effective modulus used for a full 12 months will produce the same change in PSI as would be obtained by calculating the change with the respective seasonal moduli.

The design for frost areas included in this Guide depend to a large extent on the performance of rigid and flexible pavements at the AASHO Road Test. It is recognized that experience in some northern tier states and Alaska may indicate that alternate procedures can be used. For example, some state agencies require a 12- to 24-inch granular layer over frost susceptible roadbed soils. Other agencies require full or partial replacement of frost susceptible materials (*16*). Such requirements could increase the total thickness of the pavement structure when compared with requirements of this Guide. Careful review of the cost and benefit (performance) of such design policies should be considered; however, if field data indicate that life-cycle costs can be reduced by following such a procedure there should be no problem in justifying this type of design and construction.

In addition to the seasonal effect on the subgrade and granular materials, temperature will also influence the characteristics of the asphalt concrete. Performance will be affected in three ways: (1) low temperature cracking, (2) fatigue cracking, and (3) rutting. It is not clear from research studies just how much these factors will influence PSI (*19*). However, low temperature cracking and fatigue cracking will increase maintenance costs, and rutting is a safety consideration related to potential hydroplaning as well as a maintenance problem.

Reference 20 summarizes information concerning premature cracking in asphalt concrete due to low temperature induced stresses and fatigue due to traffic. The recommendations from this study indicate that the softer grades of asphalt, i.e., AC-5 or equivalent, should be used in cold climates (when the mean annual air temperature is less than 45°F); and harder grades, i.e., AC-20 or equivalent, in hot climates (when the mean annual air temperature is greater than 75°F). The specific selection of asphalt grade will be a function of local experience; however, it is recommended that consideration be given to the above guidelines.

For thick, full-depth asphalt concrete, there are indications from research that fatigue cracking can be significantly affected by temperature (*21*). In general, these findings suggest that the harder grades of asphalt will provide improved performance in terms of fatigue cracking. Thus, an AC-40 would be appropriate in warm climates for thick (7 inches or more) pavements.

It should be noted that the selection of the grade of asphalt, per se, will not solve all of the problems of premature cracking. The designer must also give careful attention to *all* of the factors which can influence performance, e.g., structural design, drainage, construction, thaw-weakening, etc.

1.8 DRAINAGE

Drainage of water from pavements has always been an important consideration in road design; however, current methods of design have often resulted in base courses that do not drain well. This excess water combined with increased traffic volumes and loads often leads to early pavement distress in the pavement structure.

Water enters the pavement structure in many ways, such as through cracks, joints, or pavement infiltration, or as groundwater from an interrupted aquifer, high water table, or localized spring. Effects of this water (when trapped within the pavement structure) on pavements include:

(1) reduced strength of unbounded granular materials,
(2) reduced strength of roadbed soils,
(3) pumping of concrete pavements with subsequent faulting, cracking, and general shoulder deterioration, and
(4) pumping of fines in aggregate base under flexible pavements with resulting loss of support.

Less frequently noticed problems due to entrapped water include (but are not limited to):

(1) stripping of asphaltic concrete,

(2) differential heaving over swelling soils, and
(3) frost heave.

Prior editions of the AASHTO *Guide for Design of Pavement Structures* have not treated the effects of drainage on pavement performance. In this Guide, drainage effects are directly considered in terms of the effect of moisture on roadbed soil and base strength (for flexible pavements) and the effect of moisture on subgrade strength and on base erodability (for concrete pavements). Though consideration for stripping of asphalt concrete is not directly considered, the effects of swelling soils and frost heave are.

1.8.1 General Design Considerations

Methods for treating water in pavements have generally consisted of:

(1) preventing water from entering the pavement,
(2) providing drainage to remove excess water quickly, and
(3) building the pavement strong enough to resist the combined effect of load and water.

When all possible sources of water are considered, protection of the pavement structural section from water entry requires interception of groundwater as well as sealing of the pavement surface. Considerable attention has generally been given to intercepting groundwater, whereas less attention has been given to sealing the surface to exclude infiltration from rain and snow melt. As a result, a considerable amount of water often enters the pavement substructure, resulting in a need for some type of drainage.

To obtain adequate pavement drainage, the designer should consider providing three types of drainage systems: (1) surface drainage, (2) groundwater drainage, and (3) structural drainage. Such systems, however, are only effective for "free water." Water held by capillary forces in soils and in fine aggregates cannot be drained. The effects of this "bound" moisture must be considered in the design of pavement structures through its effect on the pavement material properties. Most existing pavements do not include drainage systems capable of quickly removing free water.

Most existing design methods have relied on the practice of building pavements strong enough to resist the combined effects of load and water. However, they do not always account for the potential destructive effects of water within the pavement structure. As a result, increased emphasis is needed to exclude water from the pavement and provide for rapid drainage. While both approaches are extremely difficult, this Guide will emphasize only the latter treatment. However, maintenance policies should recognize the benefits and necessity of maintaining the joint sealant and thus preventing water from leaking into the subbase layer.

1.8.2 Design of Pavement Subsurface Drainage

Two general types of pavement subsurface design criteria have been proposed for use in pavements (*11*). These include:

(1) criterion for the time of drainage of the base or subbase beginning with the flooded condition and continuing to an established acceptable level, and
(2) an inflow-outflow criterion, by which drainage occurs at a rate greater than or equal to the inflow rate, thus avoiding saturation.

Removal of the free water can be accomplished by draining the free water vertically into the subgrade, or laterally through a drainage layer into a system of pipe collectors. Generally, the actual process will be a combination of the two.

1.8.3 Incorporation of Drainage Into Guide

Drainage effects on pavement performance have been considered in this Guide. Drainage is treated by considering the effect of water on the properties of the pavement layers and the consequences to the structural capacity of the pavement. Additional work is needed to document the actual effect of drainage on pavement life.

For new design (Part II), the effect of drainage is considered by modifying the structural layer coefficient (for flexible pavements) and the load transfer coefficient (for rigid pavements) as a function of

(1) the quality of drainage (e.g., the time required for the pavement to drain), and
(2) the percent of time the pavement structure is exposed to moisture levels approaching saturation.

For rehabilitation of existing pavements, additional questions need to be asked. These include (*22*):

(1) Is the original drainage design adequate for the existing road?

Introduction and Background

(2) What changes are necessary to ensure that drainage inadequacies, which may contribute to structural distress, are corrected?

(3) If the original drainage system design was adequate, have environmental or structural changes taken place since it was built that require reconstruction of the system?

(4) Does the present or projected land use in areas adjacent to the road indicate that surface drainage flow patterns have changed or are likely to change, thus rendering existing drainage facilities inadequate?

Details of the design of subsurface drainage systems are important and, therefore, Appendix AA of Volume 2 has been provided to assist the engineer in this effort.

1.9 SHOULDER DESIGN

As defined by AASHTO, a highway shoulder is the "portion of roadways contiguous with the traveled way for accommodation of stopped vehicles for emergency use, and for lateral support of base and subbase courses." The shoulder is also considered by some agencies as a temporary detour to be used during rehabilitation of the usual traveled way.

No specific design criteria are provided in this Guide for the determination of the pavement structure for shoulders. An AASHTO position paper on shoulder design is included herein as Appendix E.

A number of agencies have developed specific design criteria for shoulders. Where such criteria are available within specific governmental jurisdictions it is recommended that such criteria be followed pending the development of more specific recommendations by AASHTO.

If design criteria for shoulders are based on pavement structure requirements similar to those used for the traveled way, the design and rehabilitation procedures included in Parts II and III of this Guide are considered applicable.

The use of tied shoulders or a widened width of paving in the lane adjacent to the shoulder has proven to be beneficial to overall performance of rigid pavements. Provision has been made in both Parts II and III to recognize the benefits to be derived from this type of design.

It is recognized that paved shoulders adjacent to flexible pavements will provide lateral support for the base and surface courses. No provision is made in this Guide to modify the design of flexible pavements as a function of shoulder design. Local practice, experience, and cost analysis should, in all cases, be considered as prime factors in shoulder design. The benefits of a paved shoulder will be enhanced if the traffic is concentrated in the traffic lanes. The use of a contrasting shoulder color or texture (seal coats) will help achieve this objective. Truck encroachment onto the shoulder is a major cause of shoulder distress; hence, any treatment which will minimize operations on the shoulder will benefit the performance of pavements in the traveled way and on the shoulder.

CHAPTER 2
DESIGN-RELATED PROJECT LEVEL PAVEMENT MANAGEMENT

Pavement management in its broadest sense encompasses all the activities involved in the planning, design, construction, maintenance, evaluation, and rehabilitation of the pavement portion of a public works program. A pavement management system (PMS) is a set of tools or methods that assist decision-makers in finding optimum strategies for providing, evaluating, and maintaining pavements in a serviceable condition over a given period of time. The function of a PMS is to improve the efficiency of decision-making, expand its scope, provide feedback on the consequences of decisions, facilitate the coordination of activities within the agency, and ensure the consistency of decisions made at different management levels within the same organization.

In this sense, pavement "design," as covered by this design Guide, and "rehabilitation," as covered in Part III of the Guide, are vital parts of the overall pavement management process. The purpose of this chapter is to show more clearly the interrelations of design and rehabilitation with pavement management and with existing or potential pavement management systems.

The detailed structure of a PMS depends on the organization of the particular agency within which it is implemented. Nevertheless, an overall, generally applicable framework can be defined or established without regard to any particular detailed departmental organization. Other reports outline rather complete, long-term concepts of pavement management, and provide guidelines for immediate application based on existing technology (29), and thus it is not our purpose here to include such guidelines.

It is convenient to describe pavement management in terms of two generalized levels: (1) the network management level, sometimes called the program level, where key administrative decisions that affect programs for road networks are made, and (2) the project management level, where technical management decisions are made for specific projects. Early formal pavement management systems development occurred at the project level. More recently, extensive development in maintenance management and data management methodologies provides opportunities for development of more comprehensive pavement management systems, where more activities can be included and explicitly interfaced with each other at the network level.

Pavement management systems can provide several benefits for highway agencies at both the network and project levels. Foremost among these is the selection of cost-effective alternatives. Whether new construction, rehabilitation, or maintenance is concerned, PMS can help management achieve the best possible value for the public dollar.

At the network level, the management system provides information pertinent to the development of a statewide or agencywide program of new construction, maintenance, or rehabilitation that will optimize the use of available resources. This relationship is illustrated on the left side of Figure 2.1.

Considering the needs of the network as a whole, a total PMS provides a comparison of the benefits and costs for several alternative programs, making it possible to identify that budget or program which will have the least total cost, or greatest benefit, over the selected analysis period. The benefits of using such a system have been proven in practice.

At the project level, detailed consideration is given to alternative design, construction, maintenance, or rehabilitation activities for a particular roadway section or project within the overall program. Here again, by comparing the benefits and costs associated with several alternative activities, an optimum strategy is identified that will provide the desired benefits or service levels at the least total cost over the analysis period.

2.1 RELATIONSHIP OF DESIGN TO PAVEMENT MANAGEMENT

From Figure 2.1 we see that "design" is primarily a project level activity since design is normally not done until budgets are allotted and programs are set. Figure 2.2 illustrates the better known relationships

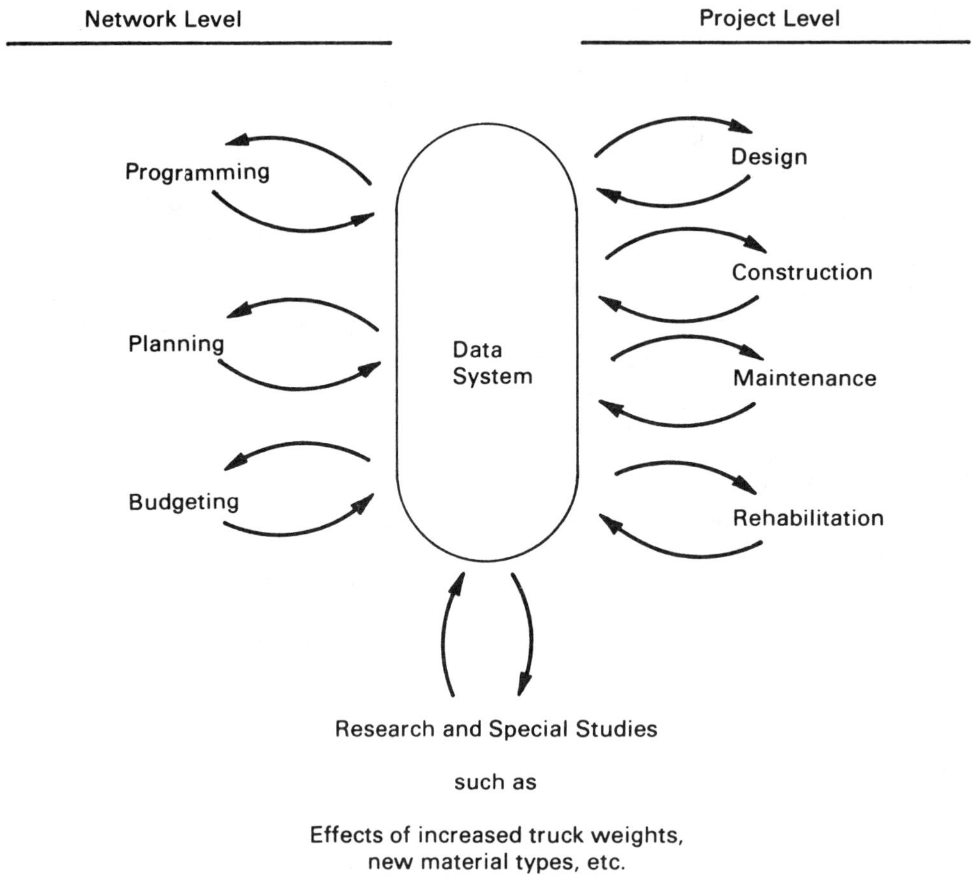

Figure 2.1. Activities of a Pavement Management System (29)

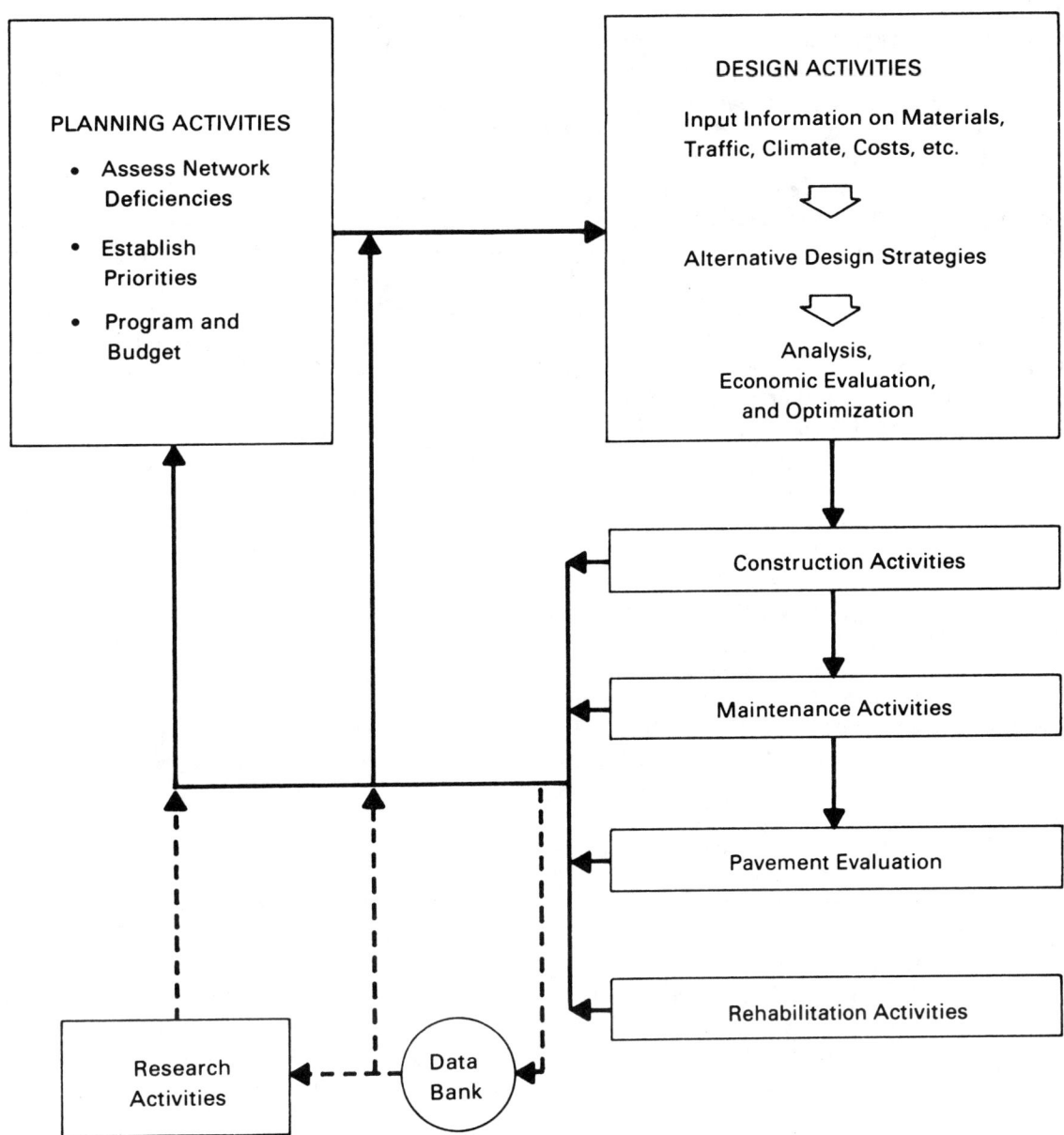

Figure 2.2. Major Classes of Activities in a Pavement Management System

between design and other typical project level activities once a project or roadway section is selected for construction, and design activities begin.

Too often in the past, design alternatives have considered only those structural sections or design strategies which are expected to last the entire predicted service life or selected performance period. It is vital to note that pavement management provides an organized approach to correcting these deficiencies. It is apparent in Figure 2.3 that the life-cycle economics and the interaction of initial construction and subsequent overlay were often not included in past design analyses.

More explicitly, a pavement management system (PMS) provides an organized coordinated way of handling the pavement management process. The amount of data involved and the number of calculations required to check the available alternatives clearly indicate the need to have some type of device to assist the engineer. Normally a computer, either micro or mainframe, fills this need very well.

Currently then the design function as defined covers new design (Part II of the Guide) as well as rehabilitation (Part III of the Guide). Pavement management also provides a straightforward mechanism for comparing the advantages of various pavement types and selecting the best pavement type for a given situation or set of circumstances. It is also essential, of course, that construction provide the as-built pavement as designed. This is noted in Section 4.1.2.

It should be reiterated here that a PMS does not make decisions but provides a method for processing data and making comparisons which then permit the designer or decision-maker to sort out the results and compare alternate possibilities based on practical realistic decision criteria.

How then does the design process as outlined in Parts II and III of this Guide relate to project level pavement management? Simply put, the solution from the Guide methodology for a single fixed set of inputs is only one alternative way of fulfilling the requirement of the design. Figure 2.4 illustrates this aspect of the broader pavement concern.

Given the inputs, which can, of course, be the same as the inputs to be used in the Guide, the Guide equations or nomographs become one of the "models of pavement structure" shown near the top of Figure 2.4. There are several models involved, of course, illustrated by the fact that there are different models for flexible pavements and rigid pavements. Using one of these models will produce an estimate of the design life related to a particular set of inputs tested on a first or second trial for example. This may or may not meet, with sufficient reliability, the performance period or required design period constraints set forth. If a given design trial satisfies these constraints, then it moves on to the economic evaluation block of the process. That means that the particular combination of inputs used for that trial, including the thicknesses and materials used, satisfy the constraints imposed and provide a serviceability history which survives for the entire performance period or design life, as illustrated in Figure 2.5 for Trial B.

Trial A on the other hand is not acceptable as a "total" design since it does not reach the designated design life T_d. Trial A, however, is not dead yet; although unacceptable as a total design, it may be economically acceptable if combined with an adequate overlay applied at or before time T_A. The decision will involve life-cycle costs, including user costs and benefits.

Many possibilities arise from adding overlays; two of these are illustrated by Trials A1 and A2 in Figure 2.5. Thus, Trial A1 is rejected because it still does not meet life and traffic constraints. The design developed in Trial A2 on the other hand is acceptable structurally and now passes on to the economic evaluation subsystem for comparison with the total economics of other acceptable trial designs. Figure 2.6 illustrates the more complete design concept, which allows Trials $A-A_1$ and $A-A_2$ to be tested as overall economic designs. The results will depend on the economic analysis. The details of the economic evaluation or life-cycle costing are presented in Chapter 3.

2.2 THE GUIDE AS STRUCTURAL SUBSYSTEM FOR A STATE PROJECT LEVEL PMS

The contents of this Guide can be used very effectively as the structural model or subsystem for a state project level PMS. It will work most effectively, of course, when the models (equations and nomographs) are properly set up for rapid comparative solutions of subsequent trials, such as on a computer or calculator.

The process can begin for new construction or for reconstruction as rehabilitation as long as the proper relationships and input value requirements are combined into the process. Any state with an existing project level pavement management system would be well advised to examine the modification of its PMS to make use of the new guides. States using a network level PMS but no formal project level system, should consider early development of a PMS addition which uses the Guide models combined with life-cycle cost

Design-Related Project Level Pavement Management

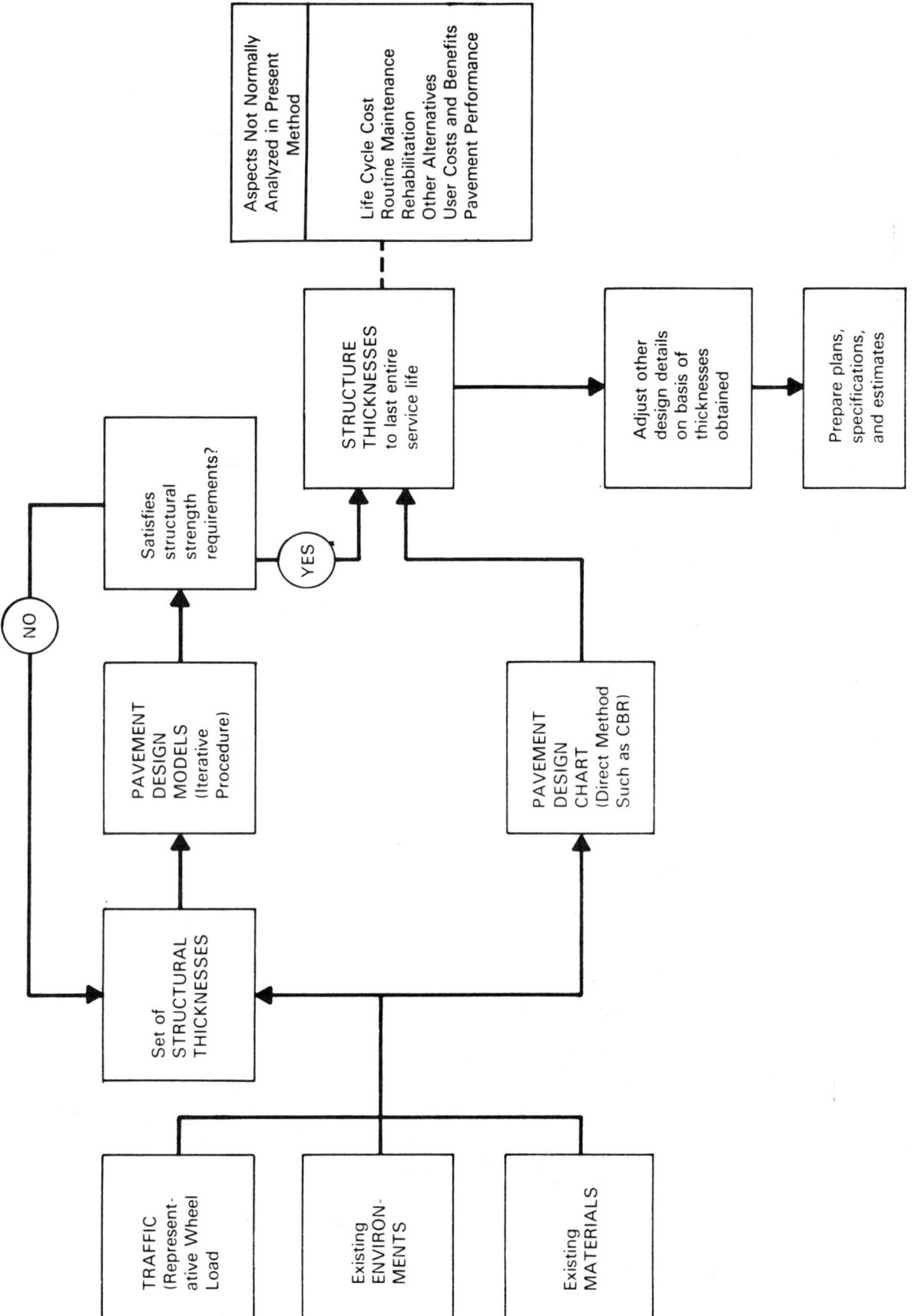

Figure 2.3. Schematic Diagram of Typical Past Pavement Design Practice

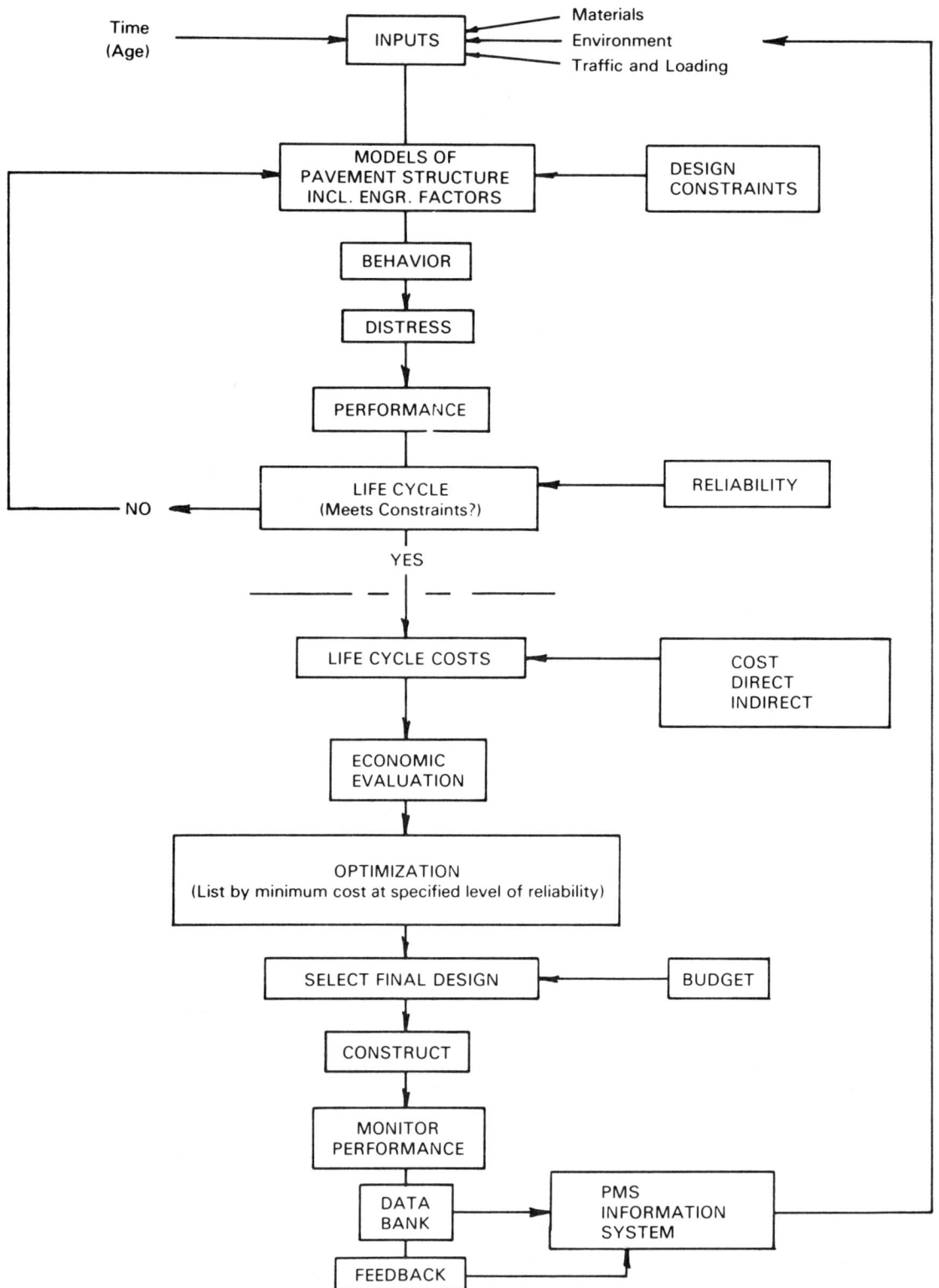

Figure 2.4. Flow Diagram of a Pavement Management System

Design-Related Project Level Pavement Management

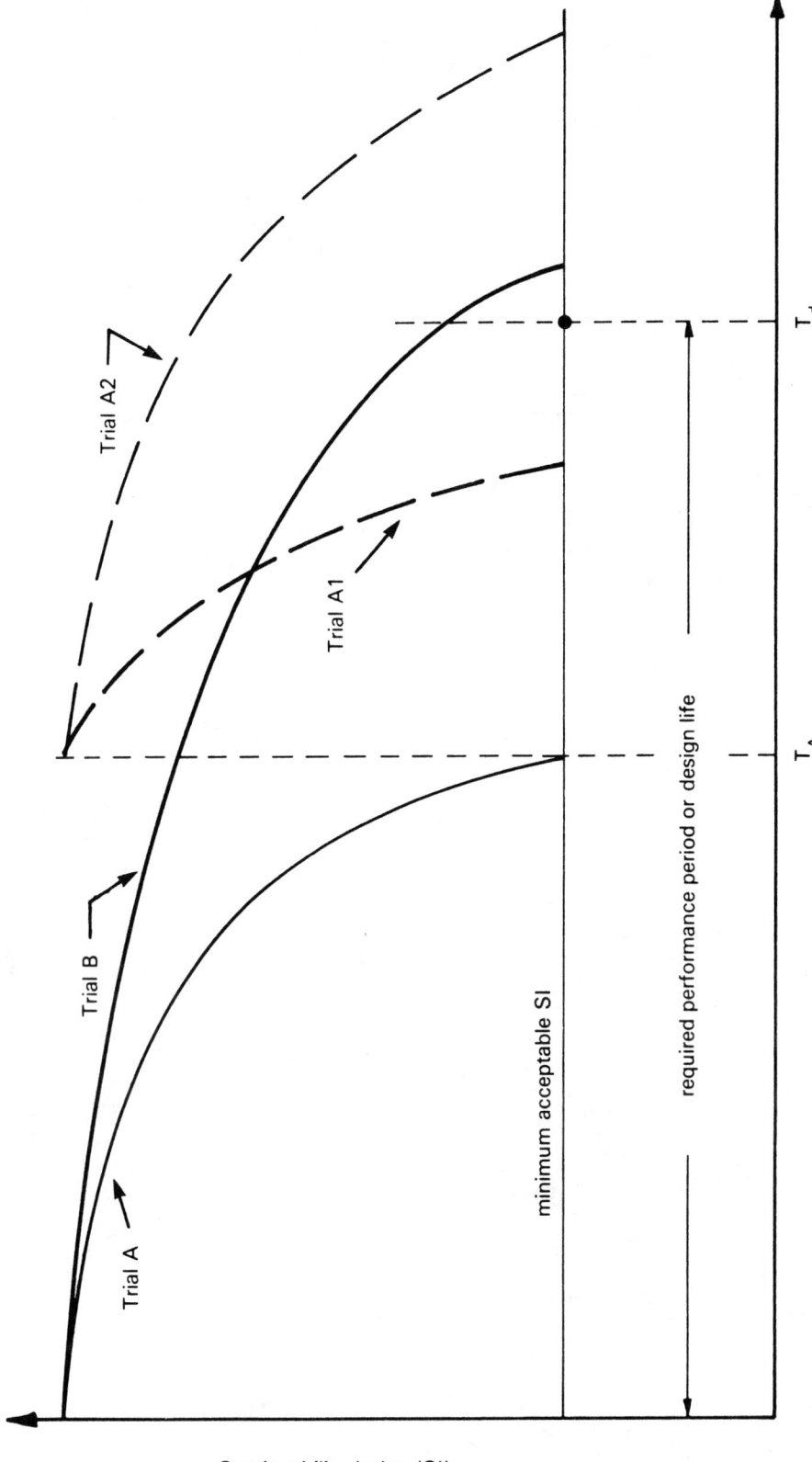

Figure 2.5. Illustrated Service Histories of Several Trial Designs

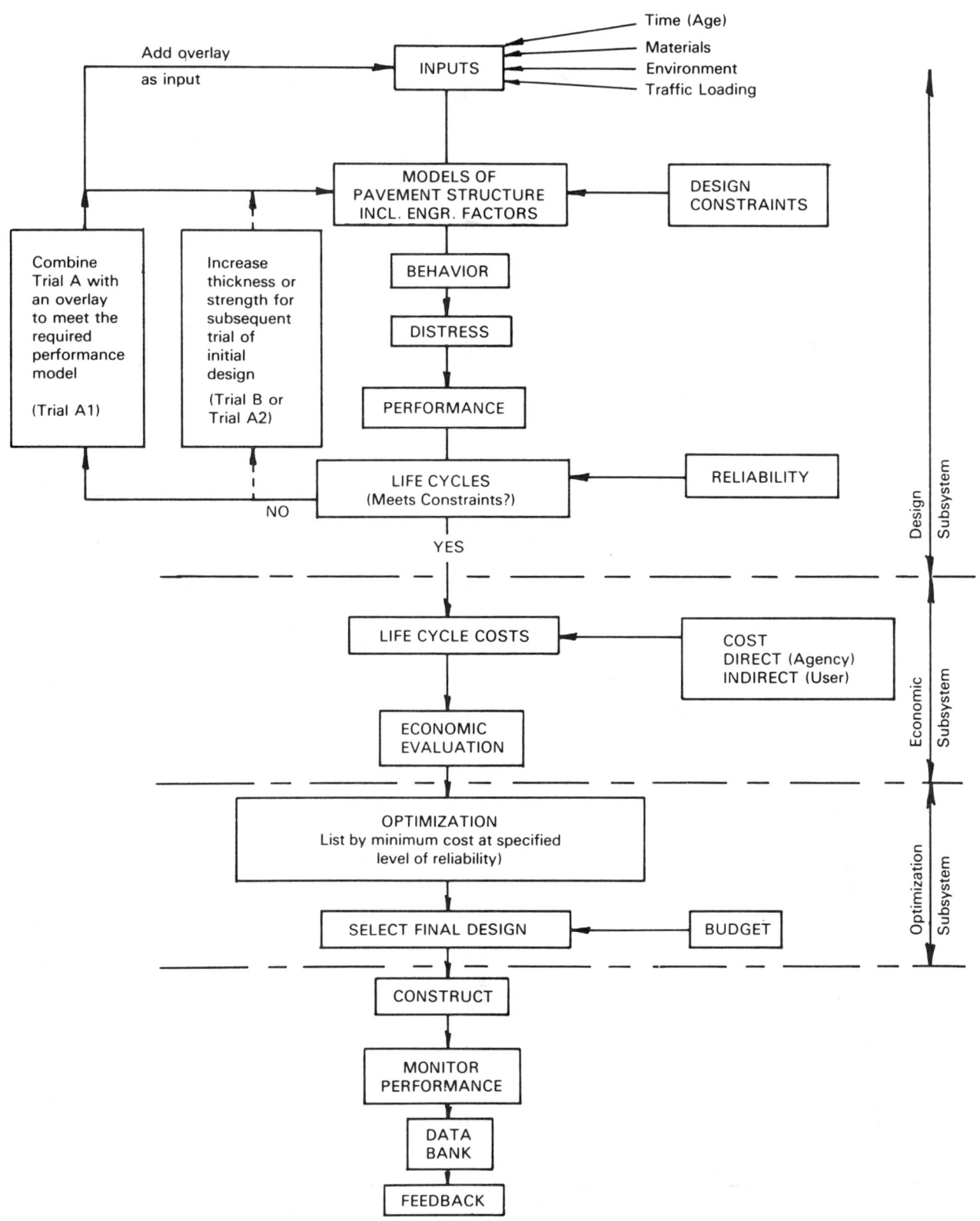

Figure 2.6. Design Process with Possibility of Overlays to Complete Design Life

calculations and optimization routines to provide an ordered set of economical designs from which a "final" design can be selected and implemented. AASHTO has prepared a written guideline on pavement management which is presented in Appendix BB, Volume 2.

2.3 PAVEMENT TYPE SELECTION

The process of selecting the proper pavement type is complex and hard to define. In the final analysis the selection process is an economic decision, although all engineering factors must be properly and carefully considered in such an analysis. If all engineering factors could be properly modeled and all costs properly compared and discounted to present value the ultimate lowest cost pavement of whatever type or design would be the proper pavement type to construct. Or, depending on economy and the models chosen, the pavement type yielding the highest benefit/cost ratio would be the proper choice. Unfortunately, the models used to compare pavement types are often not as good as they should be. Lack of long-term pavement observations has limited our ability to model the performance of various pavement types on a common basis, particularly with respect to long-term environmental effects, and the effect and relative costs of maintenance.

In the face of these imperfections in models, errors can result and be transmitted between the network and project phases of the PMS process. Thus, if the cost estimates used for each pavement section in the network reflects a proper estimate of pavement strength made using the Guide models, the resulting funding allocation to each respective project will more directly meet the actual needs of the final project level designs, also made using the Guide and its models.

Pavement type selection guidelines are reproduced in Appendix B. Currently, the most realistic pavement type selection process can result by obtaining 5 to 10 most nearly optimal cost solutions for each pavement type being considered and examining these options qualitatively in the light of the factors outlined in the selection guidelines.

2.4 NETWORK LEVEL PAVEMENT MANAGEMENT

Pavement management is an important process at the network level, but this Guide is not concerned with pavement management at this level. The relationship is much less direct than for project level PMS. However, any network level PMS must have some estimate of pavement condition and related pavement performance and cost predictions as a function of time and expected traffic. A simplified version of the models and equations presented in this Guide could be used for this purpose. The benefits of such a process would include improved interaction and cost estimates, as outlined above.

CHAPTER 3
ECONOMIC EVALUATION OF ALTERNATIVE PAVEMENT DESIGN STRATEGIES

3.1 INTRODUCTION

The application of principles of engineering economy to pavement projects occurs generally at two levels. First, there are the management decisions required to determine the feasibility and programming of a project; second, there is the requirement to achieve the maximum economy within that project if the project is economically feasible as a whole. The second level might be considered suboptimization with respect to the first level, but it is more important to the designer.

Project feasibility is determined at the network level, by comparison with other potential projects, whereas within-project economy is achieved by considering a variety of alternatives capable of satisfying the overall project requirements.

The major difference in economic evaluation between these two levels of pavement management concerns the amount of detail and information required. Otherwise, the basic principles involved are the same. This chapter considers both these principles and their incorporation into methods of economic evaluation. Such models then become a vital part of the pavement design process.

3.2 LIFE-CYCLE COSTS

It is essential in economic evaluation that all costs occurring during the life of the facility be included. When making economic comparisons this has not always been carefully practiced or even understood by pavement designers because comparisons were often made over a fixed, equal design period. Thus, designers assumed that first-cost comparisons were adequate for economic studies. This is not true, and, in order to emphasize the need for a complete cost analysis, the term "life-cycle costs" was coined about 1970 for use with pavements.

Life-cycle costs refer to all costs (and, in the complete sense, all benefits) which are involved in the provision of a pavement during its complete life cycle. These include, of course, construction costs, maintenance costs, rehabilitation costs, etc. In order to compare the costs and value of two automobiles for purchase we all realize the need to include (1) purchase price, (2) gasoline and operating costs, such as buying tires, (3) repairs (maintenance), (4) trade-in value (salvage), etc. The same kind of comparison should be recognized for pavements.

Also required, of course, is a consideration of the useful life of the car. An inexpensive car may last 4 years while an expensive one, carefully selected, may last 15 years. Since all of these costs do not occur at the same time, it is useful to determine the amount of money which could be invested at a fixed time (usually the beginning) and would earn enough money at a specific interest rate to permit payment of all costs when they occur. Thus, an interest rate or time value of money becomes important in the calculations.

"Life-cycle costs" then is a term coined to call special attention to the fact that a complete and current economic analysis is needed if alternatives are to be truly and correctly compared to each other.

3.3 BASIC CONCEPTS

A great deal has been written on the basic principles of engineering economy and methods of economic evaluation. Those principles that are applicable to pavement design can be summarized as follows:

(1) The level of management at which the evaluation is to be performed should be clearly identified; this can range from the planning or programming (network) level (i.e., project-to-project comparison) to a sublevel of design where one element, such as surface type, is being considered within a project.

(2) Economic analysis provides the basis for decision but does not provide a decision. Criteria for such decisions must be separately formulated before the results of the economic evaluation can be applied. Moreover, the economic

evaluation itself has no relationship to the method or source of financing a project.

(3) An economic evaluation should consider many possible alternatives within the constraints of time and design resources. This includes the need for comparing alternatives, not only with an existing situation, but with each other.

(4) Alternatives should be compared over the same time period. This time period should be chosen so that the factors involved in the comparison can be defined with reasonable accuracy.

(5) The economic evaluation of pavements should include agency costs and user costs and benefits if possible.

Principle 5 is not normally stated for transport projects because it is an accepted requirement. However, in the pavement field, the usual practice has been to consider only capital and maintenance costs, with the implied assumption that user costs do not vary. This approach is inadequate because, as demonstrated by McFarland (30) and by Kher, et al. (31), user costs can vary significantly with these factors. Benefits can then be considered as cost reductions (32).

3.4 DEFINITIONS RELATED TO ECONOMIC ANALYSIS

The definitions that follow include the principal technical terms used in text of the AASHTO economic analysis manual (37). The listing is broken down into two categories: economic analysis concepts or constants, and highway traffic characteristics. The definitions have been simplified in some cases for use with pavement projects.

3.4.1 Transportation Improvement Costs

This refers to the sum of highway investment cost, highway maintenance cost, and highway user cost associated with a given highway improvement. That is, for purposes of economic analysis, only transportation costs that are the direct result of the studied improvement should be considered. The components of transportation improvement costs are defined as follows:

Highway or Facility Investment Cost. Total investment required to prepare a highway improvement for service, including engineering design and supervision, right-of-way acquisition, construction, traffic control devices (e.g., signals and signs), and landscaping.

Highway Maintenance Cost. The cost of keeping a highway and its appurtenances in serviceable condition. Changes in administrative costs that can be allocated to a particular improvement should also be included.

Highway User Costs. The sum of (1) motor vehicle running cost, (2) the value of vehicle user travel time, and (3) traffic accident cost.

Motor Vehicle Running Cost. The mileage-dependent cost of running automobiles, trucks, and other motor vehicles on the highway, including the expense of fuel, tires, engine oil, maintenance, and that portion of vehicle depreciation attributable to highway mileage traveled. Operating and ownership costs that do not vary with mileage are excluded from running cost; e.g., license and parking fees, insurance premiums, the time-dependent portion of depreciation, and any other costs of off-highway use.

Value of Travel Time. The result of vehicle travel time multiplied by the average unit value of time.

Vehicle Travel Time. The total vehicle-hours of time traveled by a specific type of vehicle.

Unit Value of Time. The value attributed to 1 hour of travel time, usually different for passenger cars and trucks.

Traffic Accident Costs. The cost attributable to motor vehicle traffic accidents, usually estimated by multiplying estimated accident rates by the average cost per accident.

User Costs. The sum of highway user costs.

3.4.2 User Benefits

This refers to the advantages, privileges, and/or cost reductions that accrue to highway motor vehicle users (drivers or owners) through the use of a particular transportation facility constructed a particular way as compared with the use of another. For pavement, at the project level, the comparison is between two pavement strategies. Benefits are generally measured in terms of a decrease in user costs.

Incremental Cost. The net change in dollar costs directly attributable to a given decision or proposal compared with some other alternative (which could be the existing situation, or the "do-nothing" alternative). This definition includes cost reductions that result in negative incremental costs or, equivalently, incremental benefits. To illustrate, if the existing, do-nothing situation calls for no capital (investment) expenditures and the particular improvement proposed would require a $1 million capital outlay, the incremental capital cost would be $1 million. If, on the other hand, we are comparing two improvement alternatives, A and B, where A costs $1 million and B costs $3 million, then the incremental cost of proposal B compared to A would be $2 million. As another illustration, if current user costs associated with a given highway facility are $100 per thousand vehicle miles and a highway improvement would result in a unit user cost of $80 per thousand vehicle miles, then the incremental unit user cost would be minus $20 per thousand vehicle miles (equivalent to a $20 per thousand vehicle mile benefit). The only costs that are relevant to a given proposal are incremental future costs, in contrast to sunk costs of the past, which are irrelevant to future decisions.

Present Value (PV). An economic concept that represents the translation of specified amounts of costs or benefits occurring in different time periods into a single amount at a single instant (usually the present). Another name for present value is "present worth." The term "net present value" (NPV) refers to the net cumulative present value of a series of costs and benefits stretching over time. It is derived by applying to each cost or benefit in the series an appropriate discount factor, which converts each cost or benefit to a present value. Two related considerations underlie the need for computing present values: (1) the fact that money has an intrinsic capacity to earn interest over time (known as the time value of money) due to its productiveness and scarcity, and (2) the need in an economic study for comparing or summing incremental outlays or savings of money in different time periods.

Equivalent Uniform Annual Cost (or Benefit). A uniform annual cost (or benefit) that is the equivalent, spread over the entire period of analysis, of all incremental disbursements or costs incurred on (or benefits received from) a project. Equivalent annual cost (or benefit) is an obverse form of present value. That is, the present value of the uniform series of equivalent annual costs equals the present value of all project disbursements.

Discount Rate (Interest Rate, Time Value of Money). A percentage figure—usually expressed as an annual rate—representing the rate of interest money can be assumed to earn over the period of time under analysis. A governmental unit that decides to spend money improving a highway, for example, loses the opportunity to "invest" this money elsewhere. That rate at which money could be invested elsewhere is sometimes known as the "Opportunity Cost of Capital" and is the appropriate discount rate for use in economic studies. Discount factors derived as a function of the discount rate and time period relative to the present can be used to convert periodic benefits and costs for a project into present value or into equivalent uniform annual cost. However, calculating benefits in constant dollars and using market rates of interest is an error because the market rate of return includes an allowance for expected inflation. Hence, if future benefits and cost are calculated in constant dollars, only the real cost of capital should be represented in the discount rate used. The discount rate assumes annual end-of-year compounding, unless otherwise specified. The sum of $100 in cash today is equivalent, at a 10-percent discount rate, to $110 a year from now, $121 at the end of the second year, and $259.37 at the end of the tenth year. Correspondingly, a commitment to spend $259.37 in the tenth year discounted at 10 percent has a present value of $100.

Analysis Period. The length of time (usually the number of years) chosen for consideration and study of incremental benefits and costs in an economic analysis. The final year of construction is usually designated year 0 (zero). Subsequent years are designated year 1, year 2, and so on. Projects entailing stage construction that extends over more than 4 or 5 years should, where possible, be divided into separate projects for separable stages (for which separable benefits can be ascertained). Where such is not possible, the final year of construction for the first major stage should be used as year 0. Prior capital outlays should be compounded to their present equivalent value in year 0.

Residual or Salvage Value. The value of an investment or capital outlay remaining at the end of the study or analysis period.

Project. Any relatively independent component of a proposed highway improvement. By this definition, independent links of a large improvement proposal can be evaluated separately. Where alternative construction improvements are being considered, separate projects can be defined.

Project Alternatives. Any variations to a basic project plan that (1) entail significantly different costs, (2) result in significantly different levels of service or demand, or (3) incorporate different route locations or other distinctive design features such as surfacing type.

3.5 FACTORS INVOLVED IN PAVEMENT COSTS AND BENEFITS

The major initial and recurring costs that should be considered in the economic evaluation of alternative pavement strategies include the following:

(1) Agency costs
 (a) Initial construction costs
 (b) Future construction or rehabilitation costs (overlays, seal coats, reconstruction, etc.)
 (c) Maintenance costs, recurring throughout the design period
 (d) Salvage return or residual value at the end of the design period (which may be a "negative cost")
 (e) Engineering and administration costs
 (f) Traffic control costs if any are involved
(2) User costs
 (a) Travel time
 (b) Vehicle operation
 (c) Accidents
 (d) Discomfort
 (e) Time delay and extra vehicle operating costs during resurfacing or major maintenance

3.6 INITIAL CAPITAL COSTS (INVESTMENT COSTS)

Computing the initial cost of construction involves the calculation of material quantities to be provided in each pavement structure and multiplication by their unit prices. Material quantities are generally direct functions of their thicknesses in the structure. They are also functions of thicknesses of other layers and the width of pavement and shoulders.

The cost of in-place material in a pavement structure is not directly proportional to the volume required. Unit material price is dependent on material quantity to be provided, construction procedure employed, length of project, etc. Therefore, care should be taken to estimate quantities and true expected costs carefully. A 2-inch layer, for example, may not be twice as expensive as a 1-inch layer because the labor involved in each operation is the same. Engineering and administrative costs associated with the design should also be included.

3.6.1 Maintenance Cost

The estimation of all costs which are essential to maintaining pavement investment at a desirable specified level of service, or at a specified rate of deteriorating service, is essential to a proper economic analysis. The level of maintenance, i.e., the type and extent of maintenance operations, determines the rate of loss of riding quality or serviceability index.

There are various maintenance operations which are carried out for a highway. Maintenance of pavement, shoulders, drainage, erosion, vegetation, and structures, plus snow and ice control, are some of the major categories. For pavement economic analysis, only those categories of maintenance which directly affect the performance of a pavement should be considered. This normally includes maintenance of pavement surface, shoulders, and related drainage.

Some agencies refer to a category of "major maintenance"; we have chosen to stay with only two categories, maintenance and rehabilitation, which include all activities carried out subsequent to construction.

3.6.2 Rehabilitation and Resurfacing Cost

Rehabilitation cost includes future overlays and/or upgrading made necessary when the riding quality of a pavement decreases to a certain minimum level of acceptability, for example, a present serviceability index (PSI) of 2.5. For purposes of this report, resurfacing costs are included in the rehabilitation category.

Maintenance. As defined in Section 101 of Title 23, U.S. Code, "The preservation of the entire roadway, including surface, shoulders, roadside, structures, and such traffic-control devices as are necessary for its safe and efficient utilization." Pavement maintenance then involves the preservation of the pavement including shoulders and related drainage.

Pavement Rehabilitation. Work undertaken to extend the service life of an existing facility. This includes placement of additional surfacing material and/or other work necessary to return an existing roadway, including shoulders, to a condition of structural or functional adequacy. This could include the partial removal and replacement of the pavement structure.

Pavement rehabilitation work shall not include normal periodic maintenance activities. Periodic maintenance is interpreted to include such items as resurfacing less than K-inch in thickness or of short length; patching, filling potholes, sealing cracks and joints or repair of minor failures, and undersealing of concrete slabs other than as an essential part of rehabilitation; and other work intended primarily for preservation of the existing roadway.

Pavement rehabilitation projects should substantially increase the service life of a significant length of roadway. The following are a few examples of possible pavement rehabilitation work appropriate for major highway projects:

(1) resurfacing to provide improved structural capacity or serviceability (including in some cases cracking and seating);
(2) replacing or restoring malfunctioning joints;
(3) substantial pavement undersealing when essential for stabilization;
(4) grinding or grooving of pavements to restore smoothness or skid resistance, providing adequate structural thickness remains;
(5) removing and replacing deteriorated materials;
(6) reworking or strengthening of bases or subbases;
(7) recycling of existing materials;
(8) cracking and seating of PCC pavements with AC overlays; and
(9) adding underdrains.

This list is not all-inclusive. There are other items that could be added which satisfy the above definition. However, it is imperative that the definition be applied consistently nationwide.

The common practice of selecting a rehabilitation technique only because it has the lowest initial construction cost is a poor engineering practice and can lead to serious future pavement problems. The consideration of life-cycle costs is recommended in selecting the preferred alternative. The various costs of the pavement rehabilitation alternatives are the major consideration in selecting the preferred alternative. Life-cycle costs include (1) costs to the highway agency of initial design and construction, future maintenance and rehabilitation, and salvage value; and (2) costs to the highway user including travel delays from lane closures and rough pavements, vehicle operation, accidents, and discomfort. Although difficulties exist in estimating these costs, it is believed that this approach will provide the best pavement for the lowest annual cost. While available funding may not always permit the lowest user cost improvement to be constructed, it is a good tool to use in evaluating the feasible alternatives.

3.6.3 Salvage or Residual Value

Salvage or residual value is used by some agencies in economic evaluation. It can be significant in the case of pavements because it involves the value of reusable materials at the end of the design period. With the depletion of resources, such materials can become increasingly important in the future, especially when used in a new pavement by reworking or reprocessing. The practice of recycling pavements provides a dramatic and recognizable illustration of the reasons for using salvage value, as well as a basis for determining it.

Salvage value of a material depends on several factors, such as volume and position of the material, contamination, age or durability, anticipated use at the end of the design period, etc. It can be represented as a percentage of the original cost.

Salvage value can be relatively easy to calculate; however, the choice of values to be assigned will pose a problem for the analyst. For example, what value to assign to a 15-year-old base or a moderately damaged asphalt concrete which is 10 years old. Such questions must be left to each agency until such time as objective methods based on structural analysis are developed.

3.6.4 User Cost

Each alternative pavement strategy is associated with a number of indirect or nonagency (soft) costs which accrue to the road user and must be considered for a rational economic analysis. Such costs cannot be ignored because, similar to pavement costs, user costs are related to the roughness or serviceability history of the pavement. A pavement strategy which provides an overall high level of roughness over a larger time period will result in a higher user cost than a strategy which carries the traffic on a relatively smooth surface for most of the time.

Three major types of user costs associated with a pavement's performance are as follows:

(1) Vehicle operating cost
    (a) Fuel consumption
    (b) Tire wear
    (c) Vehicle maintenance
    (d) Oil consumption
    (e) Vehicle depreciation
    (f) Parts replacement
(2) User travel time cost
(3) Accident cost
    (a) Fatal accidents
    (b) Nonfatal accidents
    (c) Property damage

Each of the costs given above is a function of roughness level as well as vehicle speed resulting from such roughness level. As a pavement becomes rougher, the operating speeds of vehicles are generally reduced (*41*). Lower speeds and rough pavements result in higher travel time, discomfort, and other user costs. This is alleviated to some degree by lower fuel costs at the lower speeds (*42*). Since level of roughness for a pavement strategy depends, among other things, on its initial construction thicknesses and materials provided, the extent and times of rehabilitations, and the extent of major and minor maintenance provided during its service life, user cost is interrelated with all of these factors.

3.6.5 Traffic Delay Cost To User

Major maintenance or overlay placement is generally accompanied by disturbance to normal traffic flow and even lane closure. This results in vehicle speed fluctuations, stops and starts, and time losses. The extra user cost thus incurred can in certain cases become a significant factor in choice of designs and may warrant its inclusion in the economic cost calculations. Though this indirect (nonagency) cost is sometimes considered to be a "soft" cost, (i.e., not a part of the actual spending of an agency), it is certainly borne by the road users and this justifies its inclusion in the economic analysis.

Broadly, traffic delay cost is a function of traffic volume, road geometrics, time and duration of overlay construction, road geometrics in the overlay zone, and the traffic diversion method adopted. Cost is comprised of vehicle operating and user time values for driving slowly, fluctuating speeds, stopping, accelerating, idling, and vehicle accidents.

3.6.6 Identification of Pavement Benefits

Pavement benefits accrue primarily from direct reductions in transportation costs of the user, as listed in the preceding section. It is also possible to consider benefits in terms of additional road user taxes generated by a project, but this has several deficiencies and is not recommended for pavement projects.

In order to measure or calculate pavement benefits, it is necessary to define those pavement characteristics that will affect the previously noted user costs to vehicle operation, travel time, accidents, and discomfort. These could include roughness, level of serviceability, slipperiness, appearance, color, light reflection characteristics, and so on. However, two factors, serviceability (as it affects vehicle operating costs, travel time costs, accident costs, and discomfort costs) and slipperiness (as it affects accident costs) have the major influence.

(1) As serviceability decreases, travel time costs increase because drivers slow down and average travel speed decreases (in a nonlinear manner).
(2) When rehabilitation occurs (i.e., there is major maintenance, resurfacing, or reconstruction), high travel time costs can occur because of traffic delays during the construction.
(3) User benefits are not usually considered in making economic analyses for new construction or comparisons between alternative rehabilitation or treatments of pavements. In most economic analyses, user costs are considered as an added cost to the user as a pavement deteriorates and, thus, are added to maintenance and construction cost. However, when establishing priorities, user benefits may be considered. For example, in evaluating two pavements to determine which pavement to correct, user benefits could be included in the decision criteria for a pavement management system. In effect, a benefit-cost ratio approach could be considered as the basis for prioritizing the expenditure of funds for rehabilitation or reconstruction.

3.6.7 Analysis Period

The analysis period refers to the time for which the economic analysis is to be conducted. The analysis period can include provision for periodic surface renewal or rehabilitation strategies which will extend the overall service life of a pavement structure to 30 or 50 years before complete reconstruction is required.

3.7 METHODS OF ECONOMIC EVALUATION

There are a number of methods of economic analysis that are applicable to the evaluation of alternative pavement design strategies.

(1) Equivalent uniform annual cost method, often simply termed the "annual cost method"
(2) Present worth method for:
  (a) costs,
  (b) benefits, or
  (c) benefits minus costs, usually termed the "net present worth" or "net present value method"
(3) Rate-of-return method
(4) Benefit-cost ratio method
(5) Cost-effectiveness method

A common feature of these methods is the ability to consider future streams of costs (i.e., methods 1, 2a, and 5) or of costs and benefits (i.e., methods 2c, 3, and 4), so that alternative investments may be compared. Differences in the worth of money over time, as reflected in the compound interest equations used, provide the means for such comparisons.

There are several basic considerations in selecting the most appropriate (but not necessarily the best) method for economic evaluation of alternative pavement strategies. It is useful to present these prior to discussing details of the methods themselves and their advantages and limitations. They include the following:

(1) How important is the initial capital expenditure in comparison to future expected expenditures? Often, public officials and private interests (say in the case of paving a large parking lot) are concerned primarily with initial costs. An economic analysis may indicate, for example, that a low capital expenditure today can result in excessive future costs for a particular alternative (of course, the opposite could also occur). Yet the low capital expenditure is perhaps the only consideration of relevance to decision-making officials, especially if they do not know what funds they will have available several years hence. Such situations may not represent good economy to the analyst, but they do often represent reality.

(2) What method of analysis is most understandable to the decision-maker? This consideration again represents reality. For example, consider an agency that has used a benefit-cost ratio method for some years, with a good degree of subjective grasp of the results of the analysis. It may well be that this is not the best overall method for their situation; however, changing to a better method could be quite difficult and lengthy.
Another aspect of this consideration is the level of decision-making involved (i.e., at the network level or the project level). It is possible, for example, that a highway agency could use the rate-of-return method for analyzing its proposed investments over the network, whereas a net present value analysis is used by the pavement designer at the project level.

(3) What method best suits the requirements of the particular DOT involved? Although the net present value method is preferable for providing pavements, an annual cost method might be more suitable for a privately provided pavement (such as a large shopping complex).

(4) Are benefits included in the analysis? Any method that does not consider the differences in benefits between pavement alternatives is basically incomplete for use by a public agency. However, for the previously mentioned private situation, an implicit assumption of equal benefits for various alternatives may be satisfactory.

3.8 DISCUSSION OF INTEREST RATES, INFLATION FACTORS, AND DISCOUNT RATE

Many authors have considered the effects of inflation and interest rates on economic analyses, including Winfrey (32), Grant and Ireson (40), Wohl and Martin (34), and Sandler (38).

Of particular value is the lucid discussion presented by the last listed author (38) in his 1984 Transportation Research Board paper, which is presented here for its applicable insight.

3.8.1 Discounting and the Opportunity Cost of Capital

The concept of life-cycle costing (LCC) should be understood to represent an economic assessment of competing design alternatives, considering all significant costs over the life of each alternative, expressed in equivalent dollars (39). A significant key to LCC is the economic assessment using equivalent dollars. For example, assume one person has $1,000 on hand, another has $1,000 promised 10 years from now, and a third is collecting $100 a year for 10 years. Each has assets of $1,000. However, are the assets equivalent? The answer is not so simple because the assets are spread across different periods of time. To determine whose assets are worth more, a baseline time reference must first be established. All dollar values are then brought back to the baseline, using proper economic procedures to develop an equivalent dollar value. Money invested in any form earns, or has the capacity to earn, interest; so that a dollar today is worth more than the prospect of a dollar at some future time. The same principle applies when comparing the cost of various pavement design alternatives over time. Each alternative may have a different stream of costs which must be transformed into a single equivalent dollar value before a meaningful comparison can be made. The rate at which these alternative cost streams are converted into a single equivalent dollar value is referred to as the discount rate.

The discount rate is used to adjust future expected costs or benefits to present day value. It provides the means to compare alternative uses of funds, but it should not be confused with interest rate which is associated with the costs of actually borrowing money.

The time value of money concept applies far beyond the financial aspects of interest paid on borrowed money. First of all, money is only a medium of exchange which represents ownership of real resources—land, labor, raw materials, plant, and equipment. Second, the most important concept in the use of a discount rate is the opportunity cost of capital (32, 33). Any funds expended for a pavement project would not otherwise stand idle. They are funds collected from the private sector, either by taxation or by borrowing, or from the government itself by diverting funds from other purposes. If left in the private sector, they can be put to use there and earn a return that measures the value society places on the use of the funds. If the funds are diverted to government use, the true cost of the diversion is the return that would otherwise have been earned. That cost is the opportunity cost of capital and is the correct discount rate to use in calculating the LCC of various pavement design alternatives.

3.8.2 Inflation

The issue of how to deal with inflation in LCC studies is important because the procedure adopted for the treatment of inflation can have a decided effect on the results of an analysis. First, one must carefully identify the difference between two types of price changes: general inflation and differential price changes. The former may be defined as an increase in the general level of prices and income throughout the economy. Differential price change means the difference between the price trend of the goods and services being analyzed and the general price trend. During the period of analysis, some prices may decline whereas others remain fairly constant, keep pace with, or exceed the general trend in prices.

Distortions in the analysis caused by general inflation can be avoided by appropriate decisions regarding the discount rate and the treatment of future costs. The discount rate for performing present value calculations on public projects should represent the opportunity cost of capital to the taxpayer as reflected by the average market rate of return. However, the market or nominal rate of interest includes an allowance for expected inflation as well as a return that represents the real cost of capital. For example, a current market rate of interest of 12 percent may well represent a 7-percent opportunity cost component and a 5-percent inflation component. The practice of expressing future costs in constant dollars and then discounting these costs using the market, or nominal, rate of interest is in error and will understate the LCC of an alternative. Similarly, the practice of expressing future costs in inflated, or current dollars and then discounting the costs using the real cost of capital would overstate the LCC of an alternative.

The distortion caused by general inflation may be neutralized in two ways. One is to use the nominal rate of interest (including its inflation premium) for discounting, while all costs are projected in inflated or current dollars. The other is to adjust the nominal rate of interest for inflation, discounting with the real rate component only, while measuring the cost stream in terms of constant dollars.

Because of the uncertainty associated with predicting future rates of inflation and in view of the similar results achieved by following either method, Sandler et al., elected to use a discount rate which represents the real cost of capital while calculating LCC in terms of constant dollars. Because it avoids the need for speculation about inflation in arriving at the economic merit of a project, this is the generally accepted procedure used in the engineering profession and is recommended by the U.S. Office of Management and Budget.

The final choice of discount rate, interest, or inflation and the method of interpretation is left to each analyst or decision-maker. Consultation with agency authorities and familiarity with policy will help provide appropriate values to use. It should be emphasized that the final determination of the discount rate will have a significant impact on the results of the analysis.

Although the distortions caused by general price inflation can be easily neutralized, the issue of incorporating differential, or real, price changes into an economic analysis is an extremely complex matter. Authorities, such as Winfrey (32), and Lee and Grant (33, 40), have recommended the use of differential prices only when there is overwhelming or substantial evidence that certain inputs, such as land costs, are expected to experience significant changes relative to the general price level. Such circumstances seldom relate to pavement costs and thus differential cost analysis should not be used with the Guide.

3.9 EQUATIONS FOR ECONOMIC ANALYSIS

For this report only the annual cost and present worth methods of analysis are presented because of their wide applicability and acceptance. The material has been adapted from Haas and Hudson (5), who also present details of the remaining methods of economic analysis for those who desire to compare methods. The AASHTO *Manual on User Benefit Analysis* also presents comprehensive details for those desiring more information (37).

3.9.1 Equivalent Uniform Annual Cost Method

The equivalent uniform annual cost method combines all initial capital costs and all recurring future expenses into equal annual payments over the analysis period. In equation form, this method may be expressed as (5):

$$AC_{x_1,n} = crf_{x_1}(ICC)_{x_1} + (AAMO)_{x_1} + (AAUC)_{x_1}$$
$$- crf_{i,n}(SV)_{x_1,n} \qquad (3.9.1)$$

where

$AC_{x_1,n}$ = equivalent uniform annual cost for alternative x_1, for a service life or analysis period of n years,

$crf_{i,n}$ = capital recovery factor for interest rate i and n years,
= $i(1 + i)^n/(1 + i)^n - 1$,

$(ICC)_{x_1}$ = initial capital costs of construction (including actual construction costs, materials costs, engineering costs, etc.),

$(AAMO)_{x_1}$ = average annual maintenance plus operation costs for alternative x_1,

$(AAUC)_{x_1}$ = average annual user costs for alternative x_1 (including vehicle operation, travel time, accidents and discomfort if designated), and

$(SV)_{x_1,n}$ = salvage value, if any, for alternative x_1 at the end of n years.

Equation (3.9.1) considers annual maintenance and operating costs, and user costs, on an average basis. This can be satisfactory for many purposes. Where such costs do not increase uniformly, however, an exponential growth factor can easily be applied.

3.9.2 Present Worth Method

The present worth of costs method is directly comparable to the equivalent uniform annual cost method for comparable conditions, e.g., costs, discount rates, and analysis periods. The present worth method can consider either costs alone, benefits alone, or costs and benefits together. It involves the discounting of all future sums to the present, using an appropriate discount rate. The factor (5) for discounting either costs or benefits is:

$$pwf_{i,n} = 1/(1 + i)^n \qquad (3.9.2)$$

where

$pwf_{i,n}$ = present worth factor for a particular i and n,
i = discount rate, and
n = number of years to when the sum will be expended, or saved.

Published tables for pwf, or the crf of equation (3.9.1), are readily available in a wide variety of references, including Winfrey (32).

The present worth method for costs alone can be expressed in terms of the following equation (5):

$$TPWC_{x_1,n} = (ICC)_{x_1} + \sum_{t=0}^{t=1} pwf_{i,t}$$
$$* [(CC)_{x_1,t} + (MO)_{x,t} + (UC)_{x_1,t}]$$
$$- (SV)_{x_1,n} pwf_{i,n} \quad (3.9.3)$$

where

$TPWC_{x_1,n}$ = total present worth of costs for alternative x_1, for an analysis period of n years,

$(ICC)_{x_1}$ = initial capital costs of construction, etc., for alternative x_1,

$(CC)_{x_1,t}$ = capital costs of construction, etc., for alternative x_1, in year t, where t is less than n,

$pwf_{i,t}$ = present worth factor for discount rate, i, for t years,
= $1/(1 + i)^t$,

$(MO)_{x_1,t}$ = maintenance plus operation costs for alternative x_1 in year t,

$(UC)_{x_1,t}$ = user costs (including vehicle operation, travel live, accidents, and discomfort if designated) for alternative x_1, in year t, and

$(SV)_{x_1,n}$ = salvage value, if any, for alternative x_1, at the end of the design period, n years.

Although the present worth of costs method is directly comparable to the equivalent uniform annual cost method, it is only in recent years that it has begun to be applied to the pavement field.

The present worth of costs is used in the equivalent uniform annual cost method when additional capital expenditures occur before the end of the analysis period, i.e., when the service life is less than the analysis period; and future rehabilitation, such as overlays or seal coats, is needed. The equation (5) for this situation, as modified from that suggested by Baldock (35) to include user costs, is:

$$AC_{x_1,n} = crf_{i,n}[(ICC)_{x_1} + R_1 pwf_{i,a_1} + R_2 pwf_{i,a_2}$$
$$+ \cdots + R_j pwf_{i,a_j} + (AAMO)_{x_1}$$
$$+ (AAUC)_{x_1} - crf_{i,n}(SV)_{x_1,n}] \quad (3.9.4)$$

where

$AC_{x_1,n}$ = equivalent uniform annual cost for alternative x_1, for an analysis period of n years,

$R_1, R_2, \ldots, R_j$ = costs of first, second, ..., j^{th} resurfacings, respectively, and

$a_1, a_2, \ldots, a_j$ = ages at which the first, second, ..., j^{th} resurfacings occur, respectively.

All other factors are as previously defined.

The present worth of benefits can be calculated in the same manner as the present worth of costs using the following equation (5):

$$TPWB_{x_1,n} = \sum_{t=0}^{n} pwf_{i,t}$$
$$* [(DUB)_{x_1,t} + (IUB)_{x_1,t} + (NUB)_{x_1,t}] \quad (3.9.5)$$

where

$TPWB_{x_1,n}$ = total present worth of benefits for alternative x_1 for an analysis period of n years,

$(DUB)_{x_1,t}$ = direct user benefits accruing from alternative x_1 in year t,

$(IUB)_{x_1,t}$ = indirect user benefits accruing from alternative x_1 in year t, and

$(NUB)_{x_1,t}$ = non-user benefits accruing from project x_1 in year t.

It is questionable, for pavements, whether or not non-user benefits and indirect user benefits can be measured adequately. Consequently, it is perhaps reasonable to consider only direct user benefits until such time as the state of the art is sufficiently advanced to allow the other factors to be measured.

The net present value method follows from the foregoing methods because it is simply the difference between the present worth of benefits and the present worth of costs. Obviously, benefits must exceed costs if a project is to be justified on economic grounds. The equation (5) for net present value is:

$$NPV_{x_1} = TPWB_{x_1,n} - TPWC_{x_1,n} \quad (3.9.6)$$

where

NPV_{x_1} = net present value of alternative x_1 (and $TPWB_{x_1,n}$ and $TPWC_{x_1,n}$ are as previously defined).

However, for a pavement project alternative, x_1, equation (3.9.6) is not applicable directly to x_1 itself but rather to the difference between it and some other suitable alternative, say x_o. Considering only direct user benefits, these are then calculated as the user savings (resulting from lower vehicle operating costs, lower travel time costs, lower accident costs, and lower discomfort costs) realized by x_1 over x_o.

Thus, the net present value method can be applied to pavements only on the basis of project comparison, where the project alternatives are mutually exclusive. When a project alternative is evaluated, it needs to be compared not only with some standard or base alternative but also with all the other project alternatives. In the case of pavements, the base alternative may be that of no capital expenditures for improvements (where increased maintenance and operation costs are required to keep it in service). The equation form of the net present value method for pavements (5) may then be expressed as:

$$NPV_{x_1} = TPWC_{x_o,n} - TPWC_{x_1,n} \quad (3.9.7)$$

where

NPV_{x_1} = net present value of alternative x_1, and

$TPWC_{x_o,n}$ = total present worth of costs, for alternative x_o (where x_o can be the standard or base alternative, or any other feasible mutually exclusive alternative $x_1, x_2, \ldots, x_k$) for an analysis period of n years, and $TPWC_{x_1,n}$ is as previously defined.

The net present value method is preferred for the transportation field by some writers, such as Wohl and Martin (34). Others, such as Winfrey (32), consider that it has no particular advantage in economic studies of highways. Although there are certain limitations to the method, the advantages outweigh the disadvantages. Thus, it is the preferred approach for evaluating alternative pavement strategies when public investments are involved. Moreover, with increasing use of this approach in the overall transport planning field, its application to pavements will undoubtedly find much greater acceptance in the next decade.

In many cases, and for most agencies, however, only equation (3.9.3) is used, without the user costs term, either because the data are unavailable to relate user costs to pavement factors or because the policy is to consider only agency costs. The comparison between alternatives is conducted in such cases on the basis of least total present worth of costs.

There are a number of advantages inherent in the net present value method that make it perhaps the most feasible for the highway field in comparison to the "traditional" annual cost and benefit-cost methods. These advantages include the following:

(1) The benefits and costs of a project are related and expressed as a single value.
(2) Projects of different service lives, and with stage development, are directly and easily comparable.
(3) All monetary costs and benefits are expressed in present-day terms.
(4) Nonmonetary benefits (or costs) can be evaluated subjectively and handled with a cost-effectiveness evaluation.
(5) The answer is given as a total payoff for the project.
(6) The method is computationally simple and straightforward.

There are several disadvantages to the net present value method, including the following:

(1) The method cannot be applied to single alternatives where the benefits of those single alternatives cannot be estimated. In such cases, each alternative must be considered in comparison to the other alternatives, including the standard or base alternative.
(2) The results, in terms of a lump sum, may not be easily understandable to some people as a rate of return or annual cost. In fact, the summation of costs in this form can tend to act as a deterrent to investment in some cases.

Wohl and Martin (34) have extensively considered these advantages and disadvantages not only for the net present value method, but also for other methods of economic analysis. They conclude that the net present value method is the only one that will always give the correct answer. The other methods may, under certain situations, give incorrect or ambiguous answers.

3.9.3 Summary

Either the net present worth value or the equivalent uniform annual cost may be used to determine life-cycle costs for comparisons of alternate pavement design or rehabilitation strategies. In either case, it is essential that comparisons only be made for analysis periods of equal length.

CHAPTER 4
RELIABILITY

4.1 DEFINITIONS

This section provides general definitions for the concept of pavement design reliability and specific definitions that are required for the evaluation of reliability.

4.1.1 General Definition of Reliability

The following are general definitions that have been selected from the highway research literature;

(1) "Reliability is the probability that serviceability will be maintained at adequate levels from a user's point of view, throughout the design life of the facility" (25).

(2) "Reliability is the probability that the load applications a pavement can withstand in reaching a specified minimum serviceability level is not exceeded by the number of load applications that are actually applied to the pavement" (26).

(3) "Reliability is the probability that the pavement system will perform its intended function over its design life (or time) and under the conditions (or environment) encountered during operation" (27).

Definitions 1, 2, and 3 above are stated in terms of serviceability (PSI). An analogous definition for other measures of pavement condition might be stated as follows:

(4) Reliability is the probability that any particular type of distress (or combination of distress manifestations) will remain below or within the permissible level during the design life.

A final summary description of the reliability concept is given by the following definition:

(5) The reliability of a pavement design-performance process is the probability that a pavement section designed using the process will perform satisfactorily over the traffic and environmental conditions for the design period.*

Evaluation of reliability requires specific definitions for each of the elements of definition 5. The necessary definitions are given in Sections 4.1.2–4.1.3.

[*NOTE: Design period in this chapter, as in other locations in this Guide, refers to the performance period or period of time elapsed as initial or rehabilitated pavement structure deteriorates from its initial to its terminal serviceability.]

4.1.2 Definition of Designed Pavement Section

Design Equation. For the purpose of this discussion, a designed pavement section is defined to be a section that is designed through the use of a specific design equation. The equation is assumed to be an explicit mathematical formula for predicting the number of ESAL that the section can withstand (W_t) before it reaches a specified terminal level of serviceability (p_t). Predictor variables (design factors) in the equation can be put in one or another of four categories:

(1) pavement structure factors (PSF), such as subbase thickness,
(2) roadbed soil factors (RSF) such as roadbed soil resilient modulus,
(3) climate-related factors (CRF) such as drainage coefficients, and
(4) pavement condition factors (PCF), such as terminal PSI.

The design equation may be written in the form:

$$W_t = f(PSF, RSF, CRF, PCF) \quad (4.1.1)$$

wherein every design factor and the mathematical form of the function "f" are completely specified. Such design equations for flexible and rigid pavements are given in Chapter 1, Section 1.2.

Initial Substitutions. Use of the design equation to arrive at a structural design involves the following steps:

(1) insertion of nominal values for the pavement condition factors, (PCF),
(2) use of local climatic data to estimate values for the climate-related factors (CRF) and insertion of these values,
(3) use of on-site roadbed soil data to estimate values for roadbed soil factors (RSF) and insertion of these values,
(4) use of relevant traffic and loadometer data, and specified equivalence factors to predict the total number of ESAL's, w_T, that the section will receive over the design period of T years, and
(5) multiplication of the traffic prediction, w_T, by a reliability design factor, F_R, that is greater than or equal to one, and substitution of $F_R \times w_T$ for W_t in the design equation.

$$W_t = F_R \times w_T \text{ or } F_R = W_t/w_T \quad (4.1.2)$$

Thus, the design equation may be written as follows:

$$F_R \times w_T = f(PSF, RSF, CRF, PCF) \quad (4.1.3)$$

where all italicized factors and variables now have specific numerical values. Further discussion and details for the reliability factor, F_R, are given in Section 4.2.

Selection of Pavement Structure Design. Equation (4.1.3) or its nomograph may now be used to identify one or more combinations of materials and thicknesses (PSF) that will satisfy the reduced design equation. Selection of a final design from the identified alternatives is based on engineering and economic analysis.

Final Specifications for the Designed Pavement Section. It is assumed that fixed values have been specified for all relevant factors, such as shoulder and traffic lane features, that are not accounted for directly by the design equation.

It is also assumed that materials and construction specifications have been prepared for all design factors in the equation and for all supporting factors such as material quality. Use of quality control measures will then produce a degree of compliance between the as-constructed values and the input design values of all controlled factors.

4.1.3 Definition of Pavement Condition, Accumulated Axle Loads, and Pavement Performance Variables

This section defines three types of variables that are essential to the definition of reliability. The variables represent (1) pavement condition, (2) axle load accumulations, and (3) pavement performance. The discussion includes variables that were necessarily introduced in Section 4.1.2 so that the designed pavement section could be completely defined.

Definition of Pavement Condition and Accumulated Axle Load Variables. The only measure of pavement condition that will be considered here is a present serviceability index, denoted by PSI or p, whose value at a particular time depends upon the extent of surface roughness and manifestations of distress such as cracking, rutting, and faulting over the length of the design section. Formulas for flexible and rigid pavement indexes are given in References 3 and 4.

The measure of axle load applications that will be used is the number of 18-kip equivalent single axle loads (ESAL) that have accumulated from the start to some point during the design period. This accumulation is denoted by N.

The serviceability history of a pavement section is represented by the plot of p versus N as shown in Figure 4.1 for two sections, A and B. A design period of T years is also indicated.

For design purposes and reliability calculations, only three points on the (p, N) serviceability curve are of concern:

(1) At the start of the design period:

$p = p_1$ (generally somewhat greater than 4.0)

$N = 0$

(2) When the section's serviceability reaches a terminal or minimum allowable level and must be overlaid or reconstructed:

$p = p_t$, generally assumed to be 2.0 or 2.5 for design,

$N = N_t$

As shown in Figure 4.1, Section A reaches its terminal serviceability (p_t) before (and Section B after) the end of the design period.

Reliability

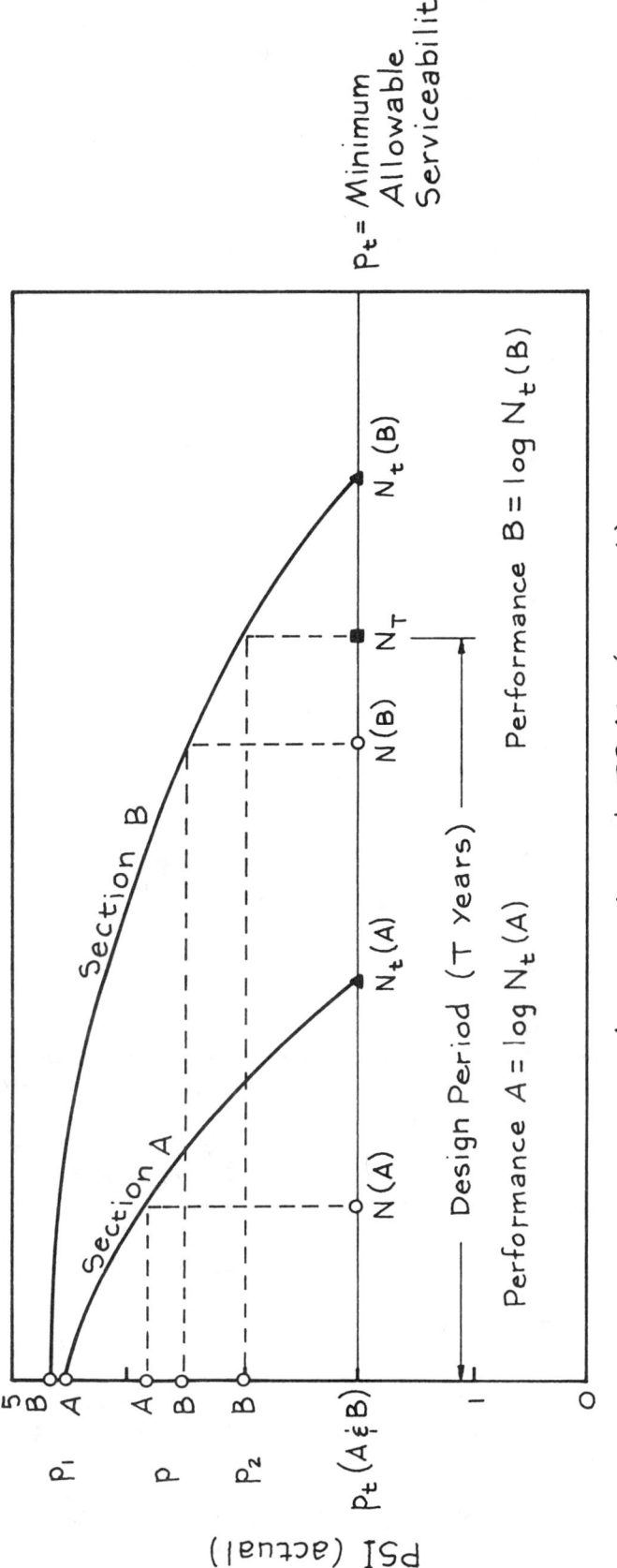

Figure 4.1. Serviceability Index (p) Versus Equivalent Load Applications (N)

(3) At the end of the design period for sections whose serviceability index still exceeds p_t:

$$p = p_2$$
$$N = N_T$$

In this case, $p_1 - p_2$ is the extent of serviceability loss over the design period, T, rather than $p_1 - p_t$.

As was explained in Section 4.1.2, the pavement design process requires a prediction, w_T, of design period ESAL, N_T. Thus,

$$w_T(\text{predicted}) = g \times N_T(\text{actual})$$

where g represents prediction uncertainty which, based on past experience, may range from less than $1/2$ to more than 2 (*28*); i.e., the actual traffic may range from $1/2$ to 2 times the predicted traffic as measured in terms of ESAL.

Definition of Pavement Performance. There are two elements to the definition of pavement performance:

(1) *Actual Performance Relative to Specified Terminal Serviceability.* When PSI (p) is used as a measure of pavement condition, there are at least two indicators that might be used to represent total performance of the pavement section. One would be based on the total area between the serviceability curve and the line $p = p_t$. The other indicator would be based only upon the actual number, N_t, of applications "withstood" by the section before its serviceability reached p_t. All ensuing discussion of reliability will be based on the latter indicator. Specifically, performance relative to a specified terminal serviceability level:

Actual Performance (to PSI = p_t)
$$= \log_{10} N_t \quad (4.1.4)$$

The logarithm is used to induce normality in the probability distributions for the analysis to be discussed in Section 4.2.2.

(2) *Predicted Performance.* The pavement design equation (4.1.1) gives a predicted value, W_t, for N_t when specific values are substituted for all other design factors in the equation. Thus, performance as predicted by the design equation is:

Predicted Performance (to PSI = p_t)
$$= \log W_t = \text{Predicted log } N_t$$

In the design process discussed in Section 4.1.2, W_t is replaced by a multiple (F_R) of w_T, where w_T is a predicted value for N_T, the actual number of design period ESAL. This means that the pavement section is designed to have

Predicted Performance = $\log W_t$
$$= \log (F_R \times w_T) = \log w_T + \log F_R \quad (4.1.5)$$

where

$$F_R \geq 1$$

and

$$\log F_R \geq 0$$

Thus $\log F_R$ is a positive "spacing factor" between $\log w_T$ and $\log W_t$, i.e.,

$$\log F_R = (\log W_t - \log w_T) \geq 0 \quad (4.1.6)$$

4.2 VARIANCE COMPONENTS AND RELIABILITY DESIGN FACTOR

4.2.1 Components of Pavement Design-Performance Variability

As far as reliability is concerned, the pavement design-performance process involves three major steps:

(1) Prediction, w_T, of actual design period ESAL, N_T,
(2) Multiplication of w_T by a selected reliability design factor, $F_R \geq 1$, and
(3) Prediction of actual pavement performance, N_t, by $W_t = w_T \times F_R$ through a design equation that expresses W_t as a function of pavement design factors.

Reliability

The three steps involve four basic points and intervals on ESAL and log ESAL scales as shown in Figure 4.2. The first point is for actual design period traffic (N_T and log N_T); the second is for predicted traffic (w_T and log w_T). The third and fourth points are for pavement performance, predicted (log W_t) and actual (log N_t). The actual performance of a single pavement section is shown at the top of the figure.

The three (log ESAL) intervals formed by the four basic points are shown as basic (level 1) deviations and are as follows:

(1) Prediction error in design period traffic:

$$(\log w_T - \log N_T) = \pm \delta(N_T, w_T)$$

(2) Reliability design factor (log):

$$(\log W_t - \log w_T) = +\log F_R$$

(3) Prediction error in pavement performance:

$$(\log N_t - \log W_t) = \pm \delta(W_t N_t)$$

The fourth basic deviation is the sum of the first three, both geometrically and algebraically:

(4) Overall deviation of actual section performance from actual design period traffic:

$$(\log N_t - \log N_T) = \pm \delta_o$$

At the design stage, the designer has control over log F_R but cannot know either the size or the direction (sign) of the other deviations. For ease of presentation, only positive deviations are shown in Figure 4.2, but each of the remaining (+ or −) combinations are equally likely. For example, it might turn out that all of N_t, w_T, and W_t are to the left of N_T. The only guarantee is that W_t will equal or exceed w_T since F_R is equal to or greater than one by definition. Thus, log F_R is a controlled variation, the remaining deviations are all "chance" variations.

The overall deviation, δ_o, will be positive whenever the actual performance (log N_t) of a pavement section exceeds the corresponding actual design period traffic (log N_T), i.e., for all sections that "survive" the design period traffic by having p greater than p_t at the end of T years. As will be explained, the reliability design factor is used to provide probabilistic assurance that log N_t will exceed log N_T, i.e., that the overall deviation will be positive.

4.2.2 Probability Distributions of Basic Deviations

It is assumed that the set of all possible outcomes for each of the chance deviations would produce a normal probability distribution as shown in Figure 4.3. The distribution for $\delta(N_T w_t)$ is shown at upper left and represents all traffic prediction errors that can be generated by repeated predictions for a given N_T, and for a wide range of N_T values. If the prediction procedure is unbiased, then the set of all possible deviations, $\delta(N_T w_T)$ will have mean value zero and variance S_w^2 (say). Thus S_w is an average (root mean square) or "to be expected" value of $\delta(N_T w_T)$ and is called the standard error of design period traffic prediction.

The probability distribution for $\delta(W_t N_t)$ is shown at upper right and represents all performance prediction errors that can be generated by construction of many pavement sections for a given log $W_t = \log w_T + \log F_R$, and for a wide range of W_t values. Again, if the prediction procedure is unbiased, then the set of all possible deviations, $\delta(W_t N_t)$ will have mean value zero and (root mean square) average value S_N (say). Thus, S_N is the standard error of performance prediction, and S_N^2 is the variance of the distribution of all possible deviations of performance predictions (log W_t) from corresponding actual performances (log N_t) of pavement sections.

The probability distribution for δ_o, shown at the bottom of Figure 4.3, represents the set of all possible overall deviations that arise from corresponding pairs of $\delta(N_T w_T)$ and $\delta(W_t N_t)$. Since $\delta_o = (N_T w_T) + \log F_R + \delta(W_t N_t)$ for every such pair, δ_o is composed one fixed deviation (log F_R) and two chance deviations that are each normally distributed. For this situation, the laws of probability are that δ_o also follows a normal probability distribution whose mean is the sum of the three deviate means and whose variance is the sum of the three deviate variances. Thus,

$$\overline{\delta}_o = \overline{\delta}(N_T w_t) + \log F_R + \overline{\delta}(W_t N_t)$$
$$= 0 + \log F_R + 0 = \log F_R$$

and

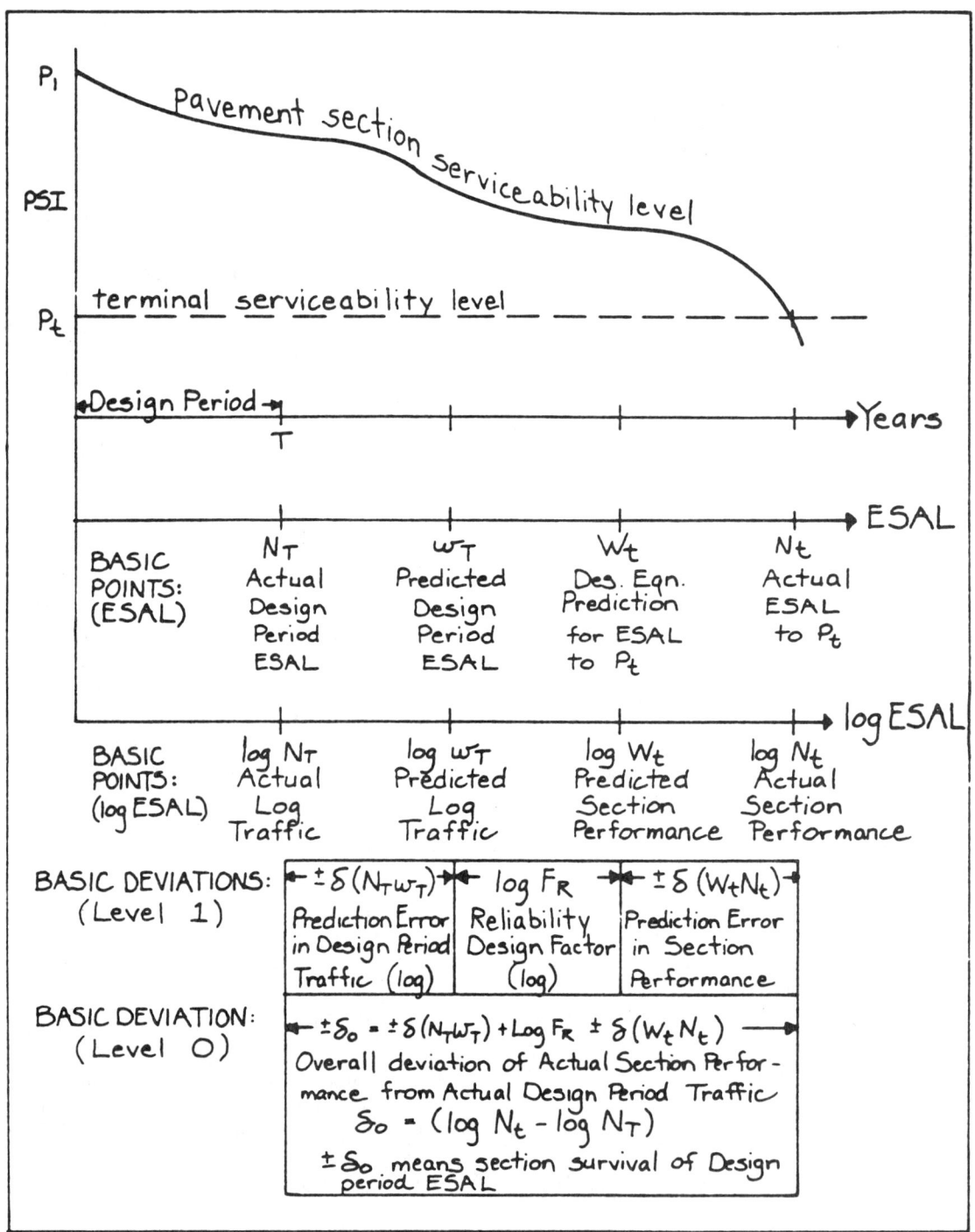

Figure 4.2. Basic Points and Deviations for Design-Performance Reliability

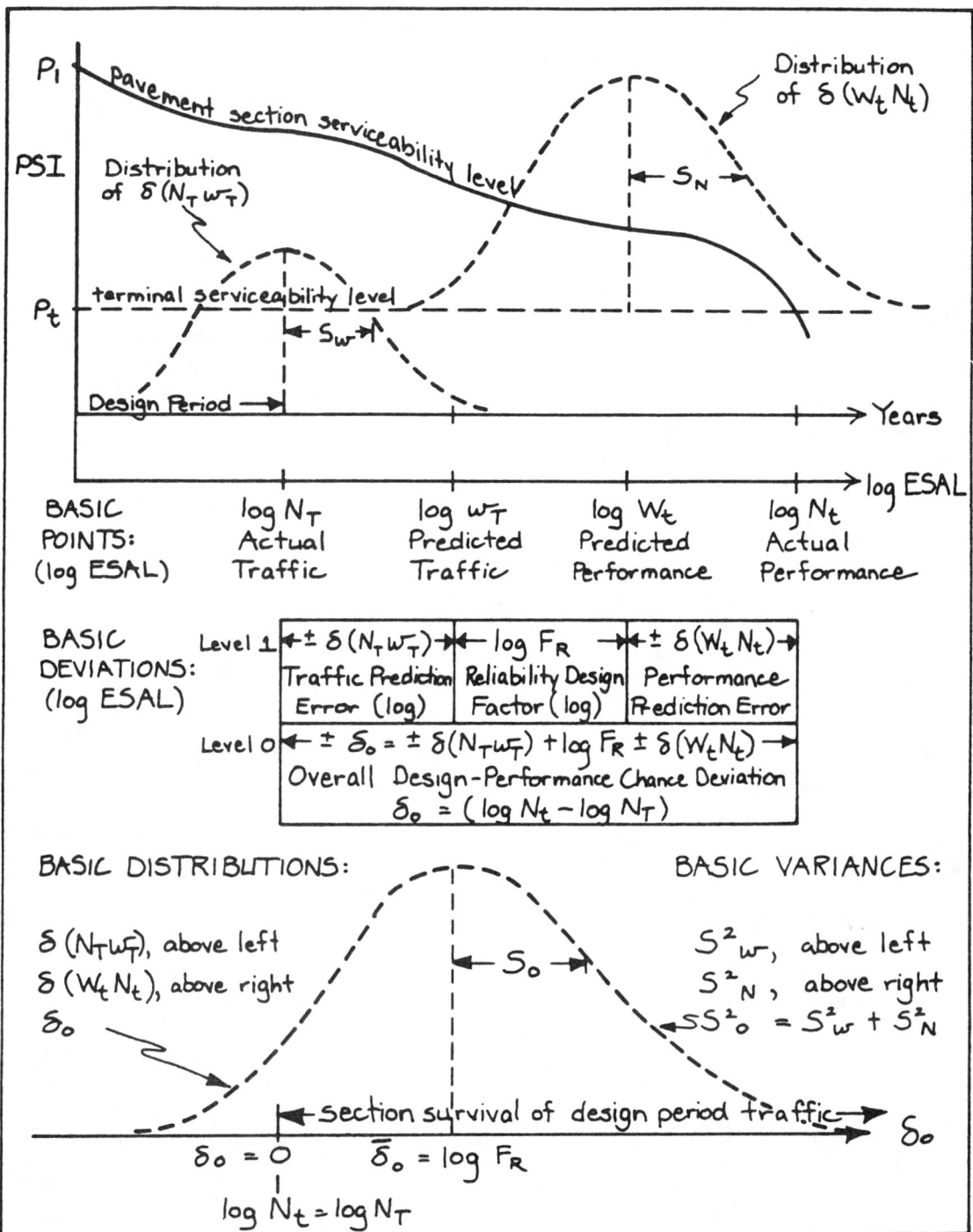

Figure 4.3. Basic Probability Distributions for Design-Performance Reliability

$$S_o^2 = S_w^2 + 0 + S_N^2$$

since log F_R is fixed by design and has no chance variation. Of particular interest is the point where $\delta_o = 0$. Since $\delta_o = (\log N_t - \log N_T)$, this point is where actual performance (log N_t) equals actual design period traffic (log N_T). All points having $\delta_o > 0$ correspond to pavement sections that survive ($p > p_t$) the design period traffic.

4.2.3 Formal Definition of Reliability Level and Reliability Design Factor

The probability distribution for the overall design-performance deviation (δ_o) is repeated in greater detail in Figure 4.4 and is the basis for formal definitions of design-performance reliability and the reliability design factor.

The stippled area above the range $\delta_o \geq 0$ corresponds to the probability that $N_t \geq N_T$, i.e., that a pavement section will survive the design period traffic with $p \geq p_t$. This probability is defined to be the *reliability level*, R/100, of the design-performance process, where R is expressed as a percent. Thus, the formal definition of reliability is given by:

$$R \text{ (percent)} = 100 \times \text{Prob}[N_t \geq N_T]$$
$$= 100 \times \text{Prob}[\delta_o \geq 0] \quad (4.2.1)$$

To calculate R and to evaluate the reliability design factor (log F_R) it is necessary to change the δ_o scale to the corresponding Z-scale for a standard normal deviate by the relationship:

$$Z = (\delta_o - \bar{\delta}_o)/S_o$$
$$= (\delta_o - \log F_R)/S_o \quad (4.2.2)$$

At the point where $\delta_o = 0$, Z becomes Z_R (say) where

$$Z_R = (-\log F_R)/S_o \quad (4.2.3)$$

For a given reliability level, say R equal 90 percent, Z_R can be found in standard normal curve area tables and corresponds to the tabulated tail area from $-\infty$ to $(100 - R)/100$. If R equal 90 percent, the tables show $Z_R = -1.28$ for 10-percent tail area. (For convenience, Table 4.1 is provided here to allow the selection of Z_R values corresponding to specific levels of reliability). Algebraic manipulation of equation (4.2.3) gives:

$$\log F_R = -Z_R \times S_o \quad (4.2.4)$$

or

$$F_R = 10^{-Z_R \times S_o} \quad (4.2.5)$$

either of which may be regarded as an algebraic definition for the reliability design factor. Values for F_R are tabulated in Table EE.9 of Appendix EE, Volume 2 for a wide range of reliability levels (R) and overall variances, S_o^2.

The following summary paragraphs bring out or emphasize salient features of the reliability design process that has been presented.

(1) Some level of reliability is implicit in every pavement design procedure. The methods presented simply make it possible to design at a *predetermined* level of reliability. If, for example, the designer substitutes the traffic prediction (w_T) directly into the design equation for W_t, then $F_R = 1$ and log $F_R = 0$. Figure 4.4 shows that the distribution of δ_o will then be centered over $\bar{\delta}_o = 0$, and that R will then be 50 percent. The designer is thereby taking a 50–50 chance that the designed sections will not survive the design period traffic with $p \geq p_t$.

(2) Log F_R is the positive part of δ_o (see Figure 4.3) that "counteracts" negative errors in both the traffic prediction, $\delta(N_T w_T)$, and performance prediction, $\delta(W_t N_t)$. Geometrically, log F_R is a "spacer" that governs how much of the left tail of the $\delta(W_t N_t)$ distribution will average to extend past N_T. For convenience, F_R is applied as a multiplier of the traffic prediction (w_T), but the value of F_R depends (see equation 4.2.5) both on the reliability level (R) that is selected and the value of S_{o2}, the overall standard deviation. Since $S_o = S_w^2 + S_N^2$, F_R accounts not only for chance variation in the traffic prediction (S_w^2) but also for chance variation in actual performance (S_N^2). Moreover, S_w^2 and S_N^2 by definition account for *all* chance variation in the respective predictions. Thus, S_o^2 and log F_R provide for *all* chance variation

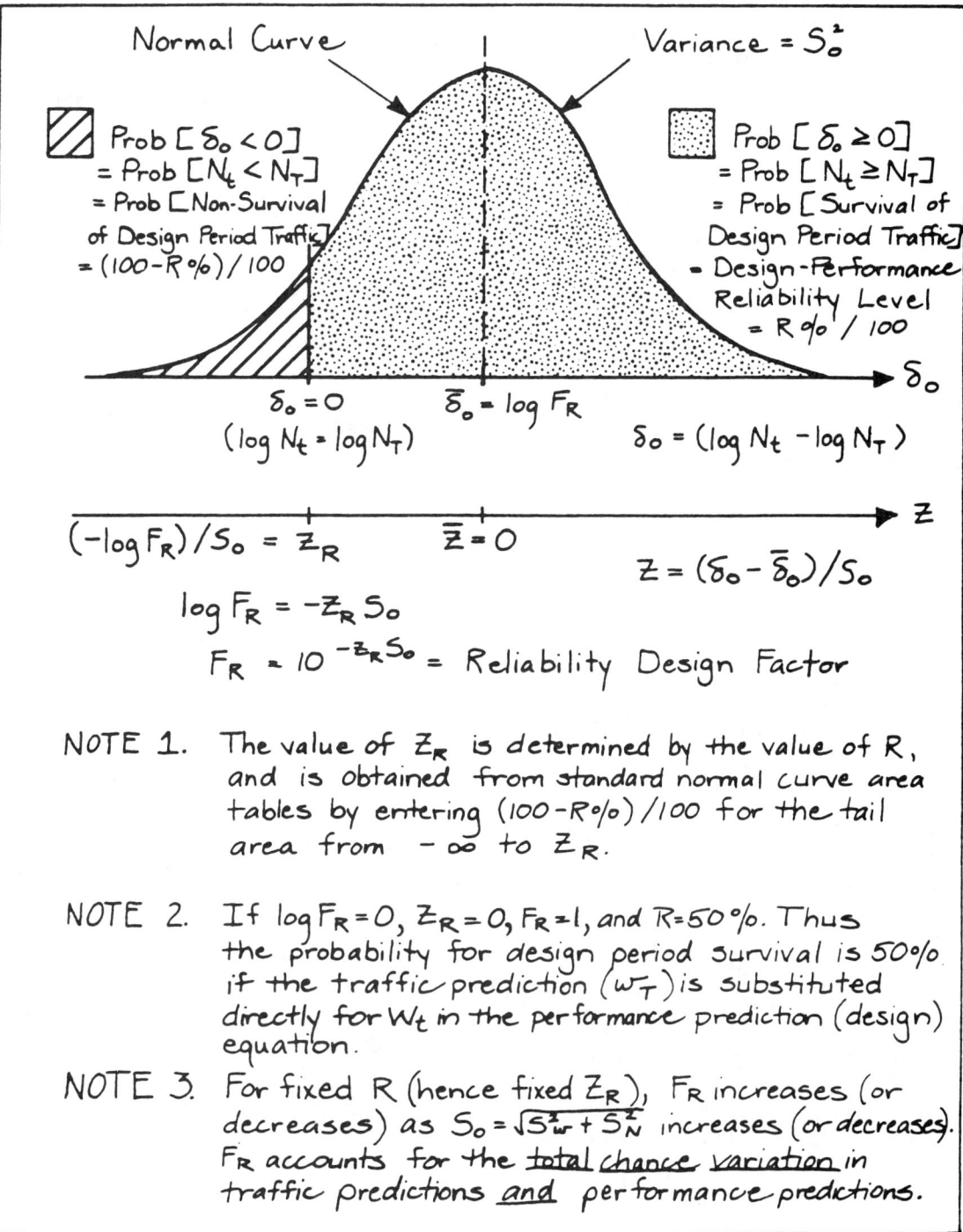

Figure 4.4. Definition of Reliability and Evaluation of Reliability Design Factor

Table 4.1. Standard Normal Deviate (Z_R) Values Corresponding to Selected Levels of Reliability

Reliability, R (percent)	Standard Normal Deviate, Z_R
50	−0.000
60	−0.253
70	−0.524
75	−0.674
80	−0.841
85	−1.037
90	−1.282
91	−1.340
92	−1.405
93	−1.476
94	−1.555
95	−1.645
96	−1.751
97	−1.881
98	−2.054
99	−2.327
99.9	−3.090
99.99	−3.750

in the design-performance process and at a known level of reliability.

Finally, the (level 1) variances S_w^2 and S_N^2 can be decomposed, respectively, into hierarchies of variance components at levels 2, 3, and 4. The decompositions are given in Appendix EE, Volume 2, where numerical estimates are given in Table EE.4 for flexible pavements, in Table EE.5 for rigid pavements, and in Table EE.6 for traffic predictions. For example, level 4 components are measures of chance variation in individual design factors such as surfacing thickness and roadbed soil modulus. The appendix gives guidance for user estimation of each component at each level. Thus, the user may make new estimates for any component and finally arrive at a new estimate for S_o^2 that is applicable to local conditions. Nomographs for the design equations (see Part II, Chapter 3) provide for a range of S_o values at any reliability level, R.

4.3 CRITERIA FOR SELECTION OF OVERALL STANDARD DEVIATION

As just discussed, Appendix EE of Volume 2 provides the guidance necessary for any user to develop levels of overall variance (S_o^2) or overall standard deviation (S_o) suitable to his own specific conditions. In doing so, the appendix identifies variance estimates for each of the individual factors associated with the performance prediction models (including the variance in future traffic projections) and subsequently arrives at overall variance and standard deviation estimates which may be used as interim criteria.

(1) The estimated overall standard deviations for the case where the variance of projected future traffic *is* considered (along with the other variances associated with the revised pavement performance models) are 0.39 for rigid pavements and 0.49 for flexible pavements.

(2) The estimated overall standard deviations for the case when the variance of projected future traffic is *not* considered (and the other variances associated with the revised pavement performance models are 0.34 for rigid pavements and 0.44 for flexible pavements).

(3) The range of S_o values provided in Part II (Section 2.1.3) are based on the values identified above:

0.30–0.40	Rigid Pavements
0.40–0.50	Flexible Pavements

The lower end of each range, however, corresponds roughly to the estimated variances associated with the AASHO Road Test and the original pavement performance models presented in the previous (1972 and 1981) Design Guides.

NOTE: It is useful to recognize that inherent in the S_o values identified in (1) and (2) above is a means for the user to specify an overall standard deviation (S_o) which better represents his ability to project future 18-kip ESAL traffic. If, because of an extensive traffic count and weigh-in-motion program, one state is capable of projecting future traffic better and therefore has a lower traffic variance (than that identified in Appendix EE of Volume 2), then that state might use an S_o-value somewhere between the values identified in (1) and (2). For example, for rigid pavements, where S_o (low) is 0.34 and S_o (high) is 0.39, a value of 0.37 or 0.38 could be used.

4.4 CRITERIA FOR SELECTION OF RELIABILITY LEVEL

The selection of an appropriate level of reliability for the design of a particular facility depends primarily upon the projected level of usage and the consequences (risk) associated with constructing an initially

Reliability

thinner pavement structure. If a facility is heavily trafficked, it may be undesirable to have to close or even restrict its usage at future dates because of the higher levels of distress, maintenance, and rehabilitation associated with an inadequate initial thickness. On the other hand, a thin initial pavement (along with the heavier maintenance and rehabilitation levels) may be acceptable, if the projected level of usage is such that fewer conflicts can be expected.

One means of identifying appropriate design reliability levels is to evaluate the reliability inherent in many of the current pavement design procedures. This approach was used to develop the suggested levels of reliability presented in Part II (Section 2.1). They were derived by surveying the inherent reliability of many current state DOT design procedures considering the functional class of the facility and whether its environment was rural or urban (see Volume 2, Appendix II). Although this approach is sound in that it is based on a considerable amount of past experience, it does not provide a means for selecting a unique level of reliability for a given project. This requires a more detailed consideration of usage and the risk of premature failure.

Figure 4.5 provides a graph illustrating the concept behind this detailed approach to identifying an optimum level of reliability for a particular design project. Three curves are shown in the figure. The first, curve (A), represents the effects of reliability on the cost (expressed in net present value or equivalent uniform annual cost) of the initial pavement structure; as design reliability increases, so does the required initial pavement thickness and its associated cost. The second, curve (B), represents the effects of reliability on the future distress-related costs (maintenance, rehabilitation, user delay, etc.). The third, curve (C), represents the sum total of the first two curves. Since the objective is to minimize the total overall cost, the optimum reliability for a given project corresponds to the minimum value on curve (C).

It should be recognized that this optimum reliability is applicable only to the level of usage and consequences (risk) of failure associated with a particular project. Although other design projects may have the same level of usage, varying soil and environmental conditions may affect the level of risk and, therefore, the optimum reliability.

4.5 RELIABILITY AND STAGE CONSTRUCTION ALTERNATIVES

When considering reliability in stage construction or "planned rehabilitation" design alternatives, it is important to consider the effects of compound reliability. Unless this is recognized, the overall reliability of say a 2-stage strategy (each stage designed for a 90-percent reliability level) would be 0.90 × 0.90 or 81 percent. Such a strategy could not be compared equally with a single-stage strategy designed for 90-percent reliability.

Referring to the formal definition in Section 4.2.3, reliability is basically the probability that a given pavement structure will survive the design (performance) period traffic with $p \geq p_t$. This definition is applicable to the fundamental case where the design period for the initial structure is equivalent to the analysis period. For cases where the initial design period is less than the desired analysis period, stage construction or planned rehabilitation is required (for the design strategy to last the analysis period) and the definition of reliability must be expanded to include the uncertainty associated with the additional stage(s). Assuming that the probability of one stage lasting its design period is independent of that of another stage, the probability or overall reliability that all stages will last their design periods (or that the strategy will last the entire analysis period) is the product of the individual stage reliabilities.

Thus, in order to achieve a certain overall design reliability ($R_{overall}$) in a particular design strategy, the following equation should be applied to establish the individual reliability (R_{stage}) required to design each stage:

$$R_{stage} = (R_{overall})^{1/n} \quad (4.5.1)$$

where n is equal to the number of stages including that of the initial pavement structure.

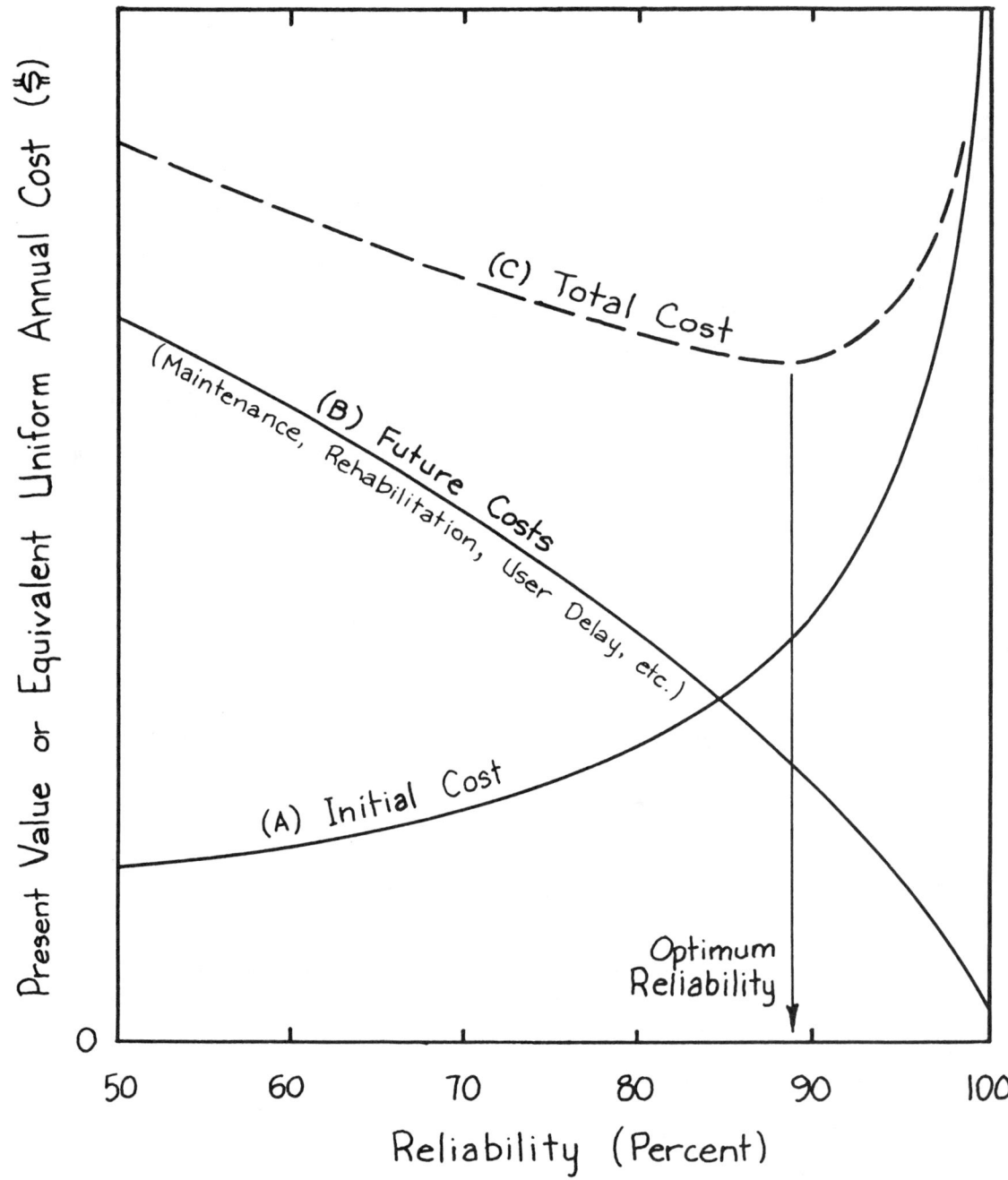

Figure 4.5. Illustration of Approach to Identifying the Optimum Reliability Level for a Given Facility

CHAPTER 5
SUMMARY

This chapter concludes Part I of the Guide, the part which explains general concepts related to pavement design and performance. Every attempt has been made to provide the potential users of the Guide with useful background information related to (1) design inputs, (2) pavement management, (3) economics, and (4) reliability. Of course, it is not possible to give complete details on any of these subjects in this Guide, and the users of the Guide are encouraged to examine the Appendices and to review important references which are cited herein for explicit detail for their specific needs.

This Guide can, and hopefully will, be used by many pavement agencies ranging from the federal level through the states to counties and cities. For this reason, flexibility has been provided to adapt the Guide to your use. However, many new developments and concepts are also presented in the Guide. Please consider carefully these new aspects before you discard them or modify them in favor of existing methods. Change is not easy, but nationwide experience has shown the need to modify this Guide, and its application to your agency probably also deserves some changes or at least serious consideration of change.

Chapter 1 of Part I addresses the detailed design factors and inputs required for using the Guide. The application and determination of final design details will be accomplished by using the methodologies which are presented in Part II for New Design and in Part III for Rehabilitation Design. It is important that you carefully review Chapter 1 and Parts II and III of the Guide *before* you undertake any specific design activities.

Chapter 2 of the Guide describes the relationships between pavement design and pavement management with particular attention to the pavement management system (PMS) at the project level. The users of the Guide should continue to study the relationship of design to pavement management and consider using the Guide's nomographs and equations as the appropriate models for the design subsystem of PMS in their agencies.

Chapter 3 examines the economic aspects of pavement design and rehabilitation. The design activities outlined in Part II and Part III do not include economics per se. After alternative designs are developed with the Guide, they should be compared with a true economic analysis, as outlined in Chapter 3. This, of course, includes the comparisons of life-cycle costs and is best done in the context of a good complete project level PMS methodology, such as SAMP-6 (*36*) and FPS-13 (*5*), to name a few.

Chapter 4 covers the very important area of reliability and its application in pavement design. The users of the Guide should remember that much of the misunderstanding of pavement design, and the resulting pavement failures for the past 20 years, have been associated with uncertainty and the resulting lack of reliability in design. Any design method based on average conditions has only a 50-percent chance of fulfilling its required performance life. The associated appendices present a rational and straightforward approach to this problem. We realize this is complex material, but users of the Guide should try to understand and use this section of the Guide. The reliability methodologies discussed here are used in Parts II and III.

Having completed the reading and studying of Part I, the user will move on to Part II—New Design, and Part III—Rehabilitation in the Guide. Care should also be exercised in the proper review of the related Appendices, which provide additional background material. Good pavement design is not simple. It cannot be done on the back of an envelope. Please realize that a reasonable degree of complexity is involved, but the Guide can be used successfully with study, training, and careful application of engineering expertise.

Part IV of the Guide provides more detail concerning the background of pavement theory and the possible application of such mechanistic methods to future pavement design or to special cases of difficult design requiring more detailed study. Reading and study of Part IV can be very useful to the serious pavement designer. A second volume resulting from the efforts is also being made available. Volume 2 will provide detailed background on how the Guide and the design equations were developed, including the analytical and empirical basis thereof.

REFERENCES FOR PART I

1. Smith, R., Darter, M., and Herrin, M., "Highway Pavement Distress Identification Manual," Federal Highway Administration, Report FHWA-79-66, March 1979.
2. "AASHTO Interim Guide for Design of Pavement Structures—1972," published by American Association of State Highway and Transportation Officials.
3. Carey, W., and Irick, P., "The Pavement Serviceability—Performance Concept," Highway Research Board Special Report 61E, AASHO Road Test, pp. 291–306, 1962.
4. Carey, W., and Irick, P., "The Pavement Serviceability—Performance Concept," Highway Research Board Record 250, 1960.
5. Haas, R., and Hudson, W.R., "Pavement Management Systems," Krieger Publishing Company, Malabor, Florida, 1982.
6. Heukelom, W., and Klomp, A.J.G., "Dynamic Testing as a Means of Controlling Pavements During and After Construction," *Proceedings* of the First International Conference on Structural Design of Asphalt Pavements, University of Michigan, 1962.
7. The Asphalt Institute, "Research and Development of the Asphalt Institute's Thickness Design Manual, Ninth Edition," Research Report No. 82-2, pp. 60–, 1982.
8. Federal Highway Administration, "Technical Guidelines for Expansive Soils in Highway Subgrades," FHWA-RD-79-51, June 1979.
9. Terrel, R.L., Epps, J., Barenberg, E.J., Mitchell, J., and Thompson, M., "Soil Stabilization in Pavement Structures—A User's Manual, Volume 1, Pavement Design and Construction Considerations."
10. Terrel, R.L., Epps, J., Barenberg, E.J., Mitchell, J., and Thompson, M., "Soil Stabilization in Pavement Structures—A User's Manual, Volume 2, Mixture Design Considerations," FHWA Report IP 80-2, October 1979.
11. Ridgeway, H.H., "Pavement Subsurface Drainage Systems," NCHRP Synthesis of Highway Practice, Report 96, November 1982.
12. Moulton, L.K., "Highway Subdrainage Design," Federal Highway Administration, Report No. FHWA-TS-80-224, August 1980 (reprinted 1982).
13. Thompson, D., "Improving Subdrainage and Shoulders of Existing Pavements," Report No. FHWA/RD/077, State of the Art, 1982.
14. Johnson, T.C., Berg, R.L., Carey, K.L., and Kaplan, C.W., "Roadway Design in Seasonal Frost Areas," NCHRP Synthesis in Highway Practice, Report 26, 1974.
15. Yoder, E., and Witczak, M., "Principles of Pavement Design," Second Edition, John Wiley & Co., pp. 179–180.
16. Linell, K.A., Hennion, F.B., and Lobacy, E.F., "Corps of Engineers' Pavement Design in Areas of Seasonal Frost," Highway Research Board Record 33, pp. 76–136, 1963.
17. The AASHO Road Test—Report 5—Pavement Research," Highway Research Board, Special Report 61E, p. 107, 1962.
18. Finn, F.N., Saraf, C.L., Kulkarni, R., Nair, K., Smith, W., and Abdullah, A., "Development of Pavement Structural Subsystems," Final Report NCHRP 1-10, February 1977.
19. Hudson, W.R., Finn, F.N., Pedigo, R.D., and Roberts, F.L., "Relating Pavement Distress to Serviceability and Performance," Report No. FHWA RD 80/098, July 1980.
20. Finn, F.N., Nair, K., and Hilliard, J., "Minimizing Premature Cracking in Asphaltic Concrete Pavement," NCHRP Report 195, 1978.
21. Thompson, M.R., and Cation, K., "Characterization of Temperature Effects for Full Depth AC Pavement Design," Department of Civil Engineering, University of Illinois, Illinois Cooperative Highway Research Program, IRH-510, October 1984.
22. Cedergren, H.R., et al., "Guidelines for Design of Subsurface Drainage Systems for Highway Structural Sections," FHWA-RD-72-30, 1972.
23. Ridgeway, H.H., "Infiltration of Water Through the Pavement Surface," Transportation Research Board Record 616, 1976.

24. The Asphalt Institute, "Asphalt Overlays and Pavement Rehabilitation," Manual Series No. 17, November 1969.
25. Lemer, A.C., and Noavenzadeh, F., "Reliability of Highway Pavements," Highway Research Record No. 362, Highway Research Board, 1971.
26. Kher, R.K., and Darter, M.I., "Probabilistic Concepts and Their Applications to AASHO Interim Guide for Design of Rigid Pavements," Highway Research Record No. 466, Highway Research Board, 1973.
27. Darter, M.I., and Hudson, W.R., "Probabilistic Design Concepts Applied to Flexible Pavement System Design," Research Report 123-18, Center for Transportation Research, University of Texas at Austin, 1973.
28. Deacon, J.A., and Lynch, R.L., "Deterioration of Traffic Parameters for the Prediction, Projection, and Computation of EWL's," Final Report KYHPR-64-21 HPR-1(4), Kentucky Highway Department, 1968.
29. Hudson, W.R., Haas, R., and Pedigo, R.D., "Pavement Management System Development," NCHRP Report 215, November 1979.
30. McFarland, W.F., "Benefit Analysis for Pavement Design Systems," Res. Report 123-13, jointly published by Texas Highway Department, Texas Transportation Institute of Texas at Austin, April 1972.
31. Kher, R., Phang, W.A., and Haas, R.C.G., "Economic Analysis Elements in Pavement Design," Highway Research Board Record 572, 1976.
32. Winfrey, R., "Economic Analysis for Highways," International Textbook Company, 1969.
33. Lee, R.R., and Grant, E.L., "Inflation and Highway Economy Studies," Highway Research Board Record 100, 1965.
34. Wohl, M., and Martin, B., "Evaluation of Mutually Exclusive Design Projects," Highway Research Board Special Report 92, 1967.
35. Baldock, R.H., "The Annual Cost of Highways," Highway Research Board Record 12, 1963.
36. Lytton, R.L., and McFarland, W.F., "Systems Approach to Pavement Design-Implementation Phase," Final Report: Prepared for Highway Research Board, NCHRP, National Academy of Sciences, March 1974.
37. AASHTO, "A Manual on User Benefit Analysis of Highway and Bus-Transit Improvements," 1977.
38. Sandler, R.D., "A Comparative Economic Analysis of Asphalt and Concrete Pavements," Transportation Research Board, January 1984.
39. Dellisola, A.J., and Kirk, S.J., "Life Cycle Costing for Design Professionals," McGraw-Hill Book Co., New York, 1982.
40. Grant E.L., Ireson, W.G., and Leavenworth, R.J., "Principles of Engineering Economy," 6th Edition, McGraw-Hill Book Co., New York, 1976, p. 293.
41. Karan, M.A., Haas, R., and Kher, R., "Effects of Pavement Roughness on Vehicle Speeds," Transportation Research Board Record 602, 1976.

PART II
PAVEMENT DESIGN PROCEDURES FOR NEW CONSTRUCTION OR RECONSTRUCTION

CHAPTER 1
INTRODUCTION

This chapter first discusses the background relative to the development of pavement design procedures for new construction and reconstruction. This is followed by a brief discussion of the scope of Part II. Next, the limitations of the design procedures are discussed followed by the concluding section, which briefly discusses the organization of this Part.

It is assumed in this text that the reader has studied Part I, "Pavement Design and Management Principles" prior to applying the design procedures described herein. The basic principles are contained in Part I.

1.1 BACKGROUND

One of the major objectives of the AASHO Road Test was to provide information that could be used in developing pavement design criteria and pavement design procedures. Accordingly, following completion of the Road Test, the AASHO Design Committee, through its Subcommittee on Pavement Design Practices, developed the AASHO *Interim Guide for the Design of Rigid and Flexible Pavements*. The Guide was based on the results of the AASHO Road Test supplemented by existing design procedures and, in the case of rigid pavements, available theory.

After the Guide was used for a few years by the states, the AASHTO Design Committee, in 1972, issued the AASHTO *Interim Guide for Design of Pavement Structures* that incorporated experience that had accrued since the original issue of the Guide. In 1981, the rigid pavement portion of the Guide (Chapter III) was revised.

This issue of the Guide contains the following modifications to the 1981 version, which were defined by the Subcommittee on Pavement Design Practices:

(1) The following modifications are included in the flexible pavement design procedures:
   (a) The soil support number is replaced by the resilient modulus to provide a rational testing procedure that may be used by an agency to define the material properties.
   (b) The layer coefficients for the various materials are defined in terms of resilient modulus as well as standard methods (CBR and R-value).
   (c) The environmental factors of moisture and temperature are objectively included in the Guide so that environmental considerations could be rationally accounted for in the design procedure. This approach replaced the subjective regional factor term previously used.
   (d) Reliability is introduced to permit the designer to use the concept of risk analysis for various classes of roadways.
   (e) Stage construction (i.e., planned rehabilitation) design procedures are incorporated.
(2) The following modifications are made in the design procedures for rigid pavements:
   (a) Reliability concepts identical to those used for the flexible pavements are introduced.
   (b) The environmental aspects of design are introduced in the same format as for flexible pavements.
   (c) The design procedure is modified to include such factors as tied shoulders, subbase erosion, and lean subbase designs.

The material from the 1972 version is reorganized and presented in a new format, as described in Part I of this Guide. Basically, the approach is to describe the input, present the design equation (nomographs, etc.), and, finally, describe the results of the design process.

1.2 SCOPE

The procedure contained herein is basically an extension of the algorithms originally developed from the AASHO Road Test. The extensions provide the designer with the opportunity to use the latest state of the art techniques. If all the inputs of the AASHO Road Test are entered into the design procedures, the results will be the same as from those equations developed at the AASHO Road Test.

The material contained in this Part deals with the design of a new roadway or reconstruction of an existing one. The concepts of stage construction are also presented to provide the designer with the option of examining numerous alternatives for selection of an optimum pavement design strategy for a facility.

Part II also permits the designer to account for pavement serviceability loss resulting from both traffic loads and environment. The environmental aspects are considered in terms of both their direct and indirect effects on the serviceability index. The direct environmental effects are in terms of swelling and frost heave of the roadbed soil, while the indirect effects are in terms of the seasonal variation of material properties and their impact on traffic load associated serviceability loss. The designer has the option of not considering either of these environmental factors, if so desired.

1.3 LIMITATIONS

The limitations inherent in the original AASHO Road Test equations are still applicable:

(1) specific set of pavement materials and one roadbed soil,
(2) single environment,
(3) an accelerated procedure for accumulating traffic (a 2-year testing period extrapolated to a 10- or 20-year design), and
(4) accumulating traffic on each test section by operating vehicles with identical axle loads and axle configurations, as opposed to mixed traffic.

These basic limitations are reduced to some extent by experiences of various agencies which have been incorporated into this edition of the Guide, as well as into previous editions.

1.4 ORGANIZATION

Basically, the material contained herein is presented in a modular form. First, the procedures for major highways are presented. These are then followed by the design procedures for low-volume roads.

Although this Guide is not intended to be a user's manual for computer application, the material is presented in a format suitable for utilization with the computer. Computer programs are available for solving the basic equations and generating multiple design strategies so that the designer may select an optimum economical solution. These programs are not, however, documented in this Guide. Thus, the designer must refer to other AASHTO documents for user manuals. The version presented in this Part is basically a simplified approach in which nomographs are used to solve the basic equations. If the designer solves an extensive array of problems, he will arrive at the same optimum solution as the computer approach.

In addition to the design chart procedure, a simplified approach is provided for the design of low-volume roads. Basically, it consists of a catalog of designs which requires a minimum of user input. This is intended to be used as a guideline by those agencies with minimal available funds for design. It is not intended to serve as a replacement for a rigorous design procedure.

CHAPTER 2
DESIGN REQUIREMENTS

This chapter discusses the preparation and/or selection of the inputs required for new (or reconstructed) pavement design. Since this chapter addresses the design requirements for several types of pavement structures on both highways and low-volume roads, only certain sets of inputs are required for a given structural design combination. Table 2.1 identifies all possible design input requirements and indicates the specific types of structural designs for which they are required. A one (1) means that a particular design input (or set of inputs) *must be determined* for that structural combination. A two (2) indicates that the design input *should be considered* because of its potential impact on the results. Under the "Flexible" heading, AC refers to asphalt concrete surfaces and ST to surface treatments. Under "Rigid," JCP refers to plain jointed concrete pavement, JRCP to jointed reinforced concrete pavement, CRCP to continuously reinforced concrete pavement, and PCP to prestressed concrete pavements. PCP is not shown as a column in Table 2.1, however, since detailed design input requirements are not available at this time.

For ease of description these inputs are classified under five separate categories:

Design Variables. This category refers to the set of criteria which must be considered for each type of road surface design procedure presented in this Guide.

Performance Criteria. This represents the user-specified set of boundary conditions within which a given pavement design alternative should perform, e.g., serviceability.

Material Properties for Structural Design. This category covers all the pavement and roadbed soil material properties that are required for structural design.

Structural Characteristics. This refers to certain physical characteristics of the pavement structure which have an effect on its performance.

Reinforcement Variables. This category covers all the reinforcement design variables needed for the different types of rigid (PCC) pavements considered.

Important. Because of the treatment of reliability in this Guide (as discussed in Part I and later in this section), *it is strongly recommended* that the designer *use mean (average) values* rather than "conservative estimates" for each of the design inputs required by the procedures. This is important since the equations were developed using mean values and actual variations. Thus, the designer *must use mean values* and *standard deviations* associated with his or her conditions.

2.1 DESIGN VARIABLES

2.1.1 Time Constraints

This section involves the selection of performance and analysis period inputs which affect (or constrain) pavement design from the dimension of time. Consideration of these constraints is required for both highway and low-volume road design. Time constraints permit the designer to select from strategies ranging from the initial structure lasting the entire analysis period (i.e., performance period equals the analysis period) to stage construction with an initial structure and planned overlays.

Performance Period. This refers to the period of time that an initial pavement structure will last before it needs rehabilitation. It also refers to the performance time between rehabilitation operations. In the design procedures presented in this Guide, the performance period is equivalent to the time elapsed as a new, reconstructed, or rehabilitated structure deteriorates from its initial serviceability to its terminal serviceability. For the performance period, the designer must select minimum and maximum bounds that are established by agency experience and policy. It is important to note that in actual practice the performance period can be significantly affected by the type and

Table 2.1. Design Requirements for the Different Initial Pavement Types that can be Considered

Description	Flexible		Rigid		Aggr. Surf.
	AC	ST	JCP/JRCP	CRCP	
2.1 DESIGN VARIABLES					
2.1.1 Time Constraints					
Performance Period	1	1	1	1	1
Analysis Period	1	1	1	1	1
2.1.2 Traffic	1	1	1	1	1
2.1.3 Reliability	1	1	1	1	
2.1.4 Environmental Impacts					
Roadbed Swelling	2	2	2	2	
Frost Heave	2	2	2	2	
2.2 PERFORMANCE CRITERIA					
2.2.1 Serviceability	1	1	1	1	1
2.2.2 Allowable Rutting					1
2.2.3 Aggregate Loss					1
2.3 MATERIAL PROPERTIES FOR STRUCTURAL DESIGN					
2.3.1 Effective Roadbed Soil Resilient Modulus	1	1			1
2.3.2 Effective Modulus of Subgrade Reaction			1	1	
2.3.3 Pavement Layer Materials Characterization	2	2	1	1	1
2.3.4 PCC Modulus of Rupture			1	1	
2.3.5 Layer Coefficients	1	1			
2.4 PAVEMENT STRUCTURAL CHARACTERISTICS					
2.4.1 Drainage					
Flexible Pavements	1	1			
Rigid Pavements		1	1		
2.4.2 Load Transfer					
Jointed Pavements			1		
Continuous Pavements				1	
Tied Shoulders or Widened Outside Lanes			2	2	
2.4.3 Loss of Support			1	1	
2.5 REINFORCEMENT VARIABLES					
2.5.1 Jointed Pavements					
Slab Length			1		
Working Stress			1		
Friction Factor			1		
2.5.2 Continuous Pavements					
Concrete Tensile Strength				1	
Concrete Shrinkage				1	
Concrete Thermal Coefficient				1	
Bar Diameter				1	
Steel Thermal Coefficient				1	
Design Temperature Drop				1	
Friction Factor				1	

AC—Asphalt Concrete
ST—Surface Treatment
JCP—Jointed Concrete Pavement
JRCP—Jointed Reinforced Concrete Pavement
CRCP—Continuously Reinforced Concrete Pavement
1—Design input variable that must be determined
2—Design variable that should be considered

Design Requirements

level of maintenance applied. The predicted performance inherent in this procedure is based on the maintenance practices at the AASHO Road Test.

The *minimum performance period* is the shortest amount of time a given stage should last. For example, it may be desirable that the initial pavement structure last at least 10 years before some major rehabilitation operation is performed. The limit may be controlled by such factors as the public's perception of how long a "new" surface should last, the funds available for initial construction, life-cycle cost, and other engineering considerations.

The *maximum performance period* is the maximum practical amount of time that the user can expect from a given stage. For example, experience has shown in areas that pavements originally designed to last 20 years required some type of rehabilitation or resurfacing within 15 years after initial construction. This limiting time period may be the result of PSI loss due to environmental factors, disintegration of surface, etc. The selection of longer time periods than can be achieved in the field will result in unrealistic designs. Thus, if life-cycle costs are to be considered accurately, it is important to give some consideration to the maximum practical performance period of a given pavement type.

Analysis Period. This refers to the period of time for which the analysis is to be conducted, i.e., the length of time that any design strategy must cover. The analysis period is analogous to the term "design life" used by designers in the past. Because of the consideration of the maximum performance period, it may be necessary to consider and plan for stage construction (i.e., an initial pavement structure followed by one or more rehabilitation operations) to achieve the desired analysis period.

In the past, pavements were typically designed and analyzed for a 20-year performance period since the original Interstate Highway Act in 1956 required that traffic be considered through 1976. It is now recommended that consideration be given to longer analysis periods, since these may be better suited for the evaluation of alternative long-term strategies based on life-cycle costs. Consideration should be given to extending the analysis period to include one rehabilitation. For high-volume urban freeways, longer analysis periods may be considered. Following are general guidelines:

Highway Conditions	Analysis Period (years)
High-volume urban	30–50
High-volume rural	20–50
Low-volume paved	15–25
Low-volume aggregate surface	10–20

2.1.2 Traffic

The design procedures for both highways and low-volume roads are all based on cumulative expected 18-kip equivalent single axle loads (ESAL) during the analysis period ($\hat{w}_{18}$). The procedure for converting mixed traffic into these 18-kip ESAL units is presented in Part I and Appendix D of this Guide. Detailed equivalency values are given in Appendix D. For any design situation in which the initial pavement structure is expected to last, the analysis period without any rehabilitation or resurfacing, all that is required is the total traffic over the analysis period. If, however, stage construction is considered, i.e., rehabilitation or resurfacing is anticipated (due to lack of initial funds, roadbed swelling, frost heave, etc.), then the user must prepare a graph of cumulative 18-kip ESAL traffic versus time, as illustrated in Figure 2.1. This will be used to separate the cumulative traffic into the periods (stages) during which it is encountered.

The predicted traffic furnished by the planning group is generally the cumulative 18-kip ESAL axle applications expected on the highway, whereas the designer requires the axle applications in the design lane. Thus, unless specifically furnished, the designer must factor the design traffic by direction and then by lanes (if more than two). The following equation may be used to determine the traffic (w_{18}) in the design lane:

$$w_{18} = D_D \times D_L \times \hat{w}_{18}$$

where

D_D = a directional distribution factor, expressed as a ratio, that accounts for the distribution of ESAL units by direction, e.g., east-west, north-south, etc.,

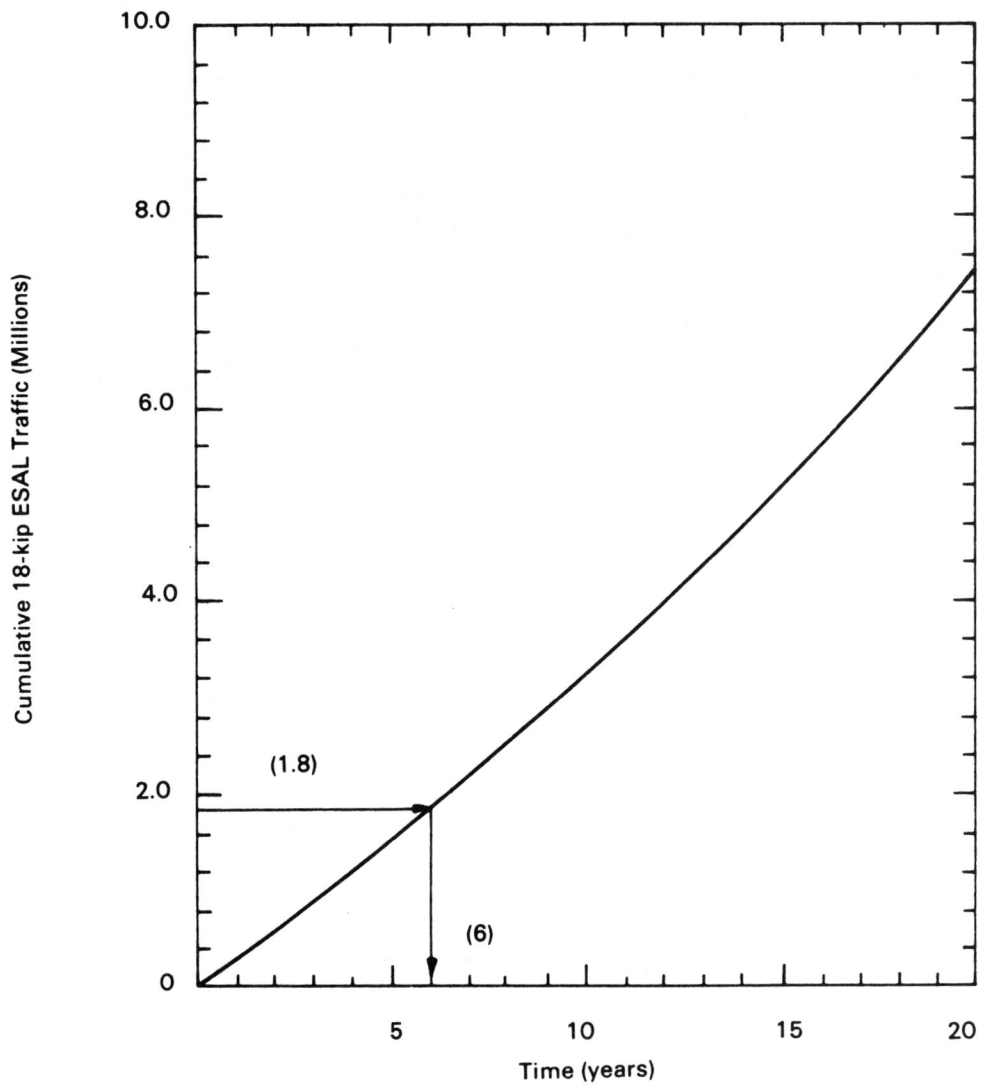

Figure 2.1. Example Plot of Cumulative 18-kip ESAL Traffic Versus Time

Design Requirements

D_L = a lane distribution factor, expressed as a ratio, that accounts for distribution of traffic when two or more lanes are available in one direction, and

$\hat{w}_{18}$ = the cumulative two-directional 18-kip ESAL units predicted for a specific section of highway during the analysis period (from the planning group).

Although the D_D factor is generally 0.5 (50 percent) for most roadways, there are instances where more weight may be moving in one direction than the other. Thus, the side with heavier vehicles should be designed for a greater number of ESAL units. Experience has shown that D_D may vary from 0.3 to 0.7, depending on which direction is "loaded" and which is "unloaded."

For the D_L factor, the following table may be used as a guide:

Number of Lanes in Each Direction	Percent of 18-kip ESAL in Design Lane
1	100
2	80–100
3	60–80
4	50–75

2.1.3 Reliability

Reliability concepts were introduced in Chapter 4 of Part I and are developed fully in Appendix EE of Volume 2. Basically, it is a means of incorporating some degree of certainty into the design process to ensure that the various design alternatives will last the analysis period. The reliability design factor accounts for chance variations in both traffic prediction (w_{18}) and the performance prediction (W_{18}), and therefore provides a predetermined level of assurance (R) that pavement sections will survive the period for which they were designed.

Generally, as the volume of traffic, difficulty of diverting traffic, and public expectation of availability increases, the risk of not performing to expectations must be minimized. This is accomplished by selecting higher levels of reliability. Table 2.2 presents recommended levels of reliability for various functional classifications. Note that the higher levels correspond

Table 2.2. Suggested Levels of Reliability for Various Functional Classifications

Functional Classification	Recommended Level of Reliability	
	Urban	Rural
Interstate and Other Freeways	85–99.9	80–99.9
Principal Arterials	80–99	75–95
Collectors	80–95	75–95
Local	50–80	50–80

NOTE: Results based on a survey of the AASHTO Pavement Design Task Force.

to the facilities which receive the most use, while the lowest level, 50 percent, corresponds to local roads.

As explained in Part I, Chapter 4, design-performance reliability is controlled through the use of a reliability factor (F_R) that is multiplied times the design period traffic prediction (w_{18}) to produce design applications (W_{18}) for the design equation. For a given reliability level (R), the reliability factor is a function of the overall standard deviation (S_o) that accounts for both chance variation in the traffic prediction and normal variation in pavement performance prediction for a given W_{18}.

It is important to note that by treating design uncertainty as a separate factor, the designer should no longer use "conservative" estimates for all the other design input requirements. Rather than conservative values, the designer should use his best estimate of the mean or average value for each input value. The selected level of reliability and overall standard deviation will account for the combined effect of the variation of all the design variables.

Application of the reliability concept requires the following steps:

(1) Define the functional classification of the facility and determine whether a rural or urban condition exists.

(2) Select a reliability level from the range given in Table 2.2. The greater the value of reliability, the more pavement structure required.

(3) A standard deviation (S_o) should be selected that is representative of local conditions. Values of S_o developed at the AASHO Road Test did not include traffic error. However, the performance prediction error developed at the Road Test was .25 for rigid and .35 for flexible pavements. This corresponds to a total stand-

ard deviation for traffic of 0.35 and 0.45 for rigid and flexible pavements, respectively.

2.1.4 Environmental Effects

The environment can affect pavement performance in several ways. Temperature and moisture changes can have an effect on the strength, durability, and load-carrying capacity of the pavement and roadbed materials. Another major environmental impact is the direct effect roadbed swelling, pavement blowups, frost heave, disintegration, etc., can have on loss of riding quality and serviceability. Additional effects, such as aging, drying, and overall material deterioration due to weathering, are considered in this Guide only in terms of their inherent influence on the pavement performance prediction models.

The actual treatment of the effects of seasonal temperature and moisture changes on material properties is discussed in Section 2.3, "Material Properties for Structural Design." This section provides only the criteria necessary for quantifying the input requirements for evaluating roadbed swelling and frost heave. If either of these can lead to a significant loss in serviceability or ride quality during the analysis period, then it (they) should be considered in the design analysis for all pavement structural types, except perhaps aggregate-surfaced roads. As serviceability-based models are developed for such factors as pavement blowups, then they may be added to the design procedure.

The objective of this step is to produce a graph of serviceability loss versus time, such as that illustrated in Figure 2.2. As described in Part I, the serviceability loss due to environment must be added to that resulting from cumulative axle loads. Figure 2.2 indicates that the environmental loss is a result of the summation of losses from both swelling and frost heave. The chart may be used to estimate the serviceability loss at intermediate periods, e.g., at 13 years the loss is 0.73. Obviously, if only swelling or only frost heave is considered, there will be only one curve on the graph. The environmental serviceability loss is evaluated in detail in Appendix G, "Treatment of Roadbed Swelling and/or Frost Heave in Design."

2.2 PERFORMANCE CRITERIA

2.2.1 Serviceability

The serviceability of a pavement is defined as its ability to serve the type of traffic (automobiles and trucks) which use the facility. The primary measure of serviceability is the Present Serviceability Index (PSI), which ranges from 0 (impossible road) to 5 (perfect road). The basic design philosophy of this Guide is the serviceability-performance concept, which provides a means of designing a pavement based on a specific total traffic volume and a minimum level of serviceability desired at the end of the performance period.

Selection of the lowest allowable PSI or *terminal serviceability index* (p_t) is based on the lowest index that will be tolerated before rehabilitation, resurfacing, or reconstruction becomes necessary. An index of 2.5 or higher is suggested for design of major highways and 2.0 for highways with lesser traffic volumes. One criterion for identifying a minimum level of serviceability may be established on the basis of public acceptance. Following are general guidelines for minimum levels of p_t obtained from studies in connection with the AASHO Road Test (*14*):

Terminal Serviceability Level	Percent of People Stating Unacceptable
3.0	12
2.5	55
2.0	85

For relatively minor highways where economics dictate that the initial capital outlay be kept at a minimum, it is suggested that this be accomplished by reducing the design period or the total traffic volume, rather than by designing for a terminal serviceability less than 2.0.

Since the time at which a given pavement structure reaches its terminal serviceability depends on traffic volume and the original or initial serviceability (p_o), some consideration must also be given to the selection of p_o. (It should be recognized that the p_o values observed at the AASHO Road Test were 4.2 for flexible pavements and 4.5 for rigid pavements.)

Once p_o and p_t are established, the following equation should be applied to define the total change in serviceability index:

$$\Delta PSI = p_o - p_t$$

The equation is applicable to flexible, rigid, and aggregate-surfaced roads.

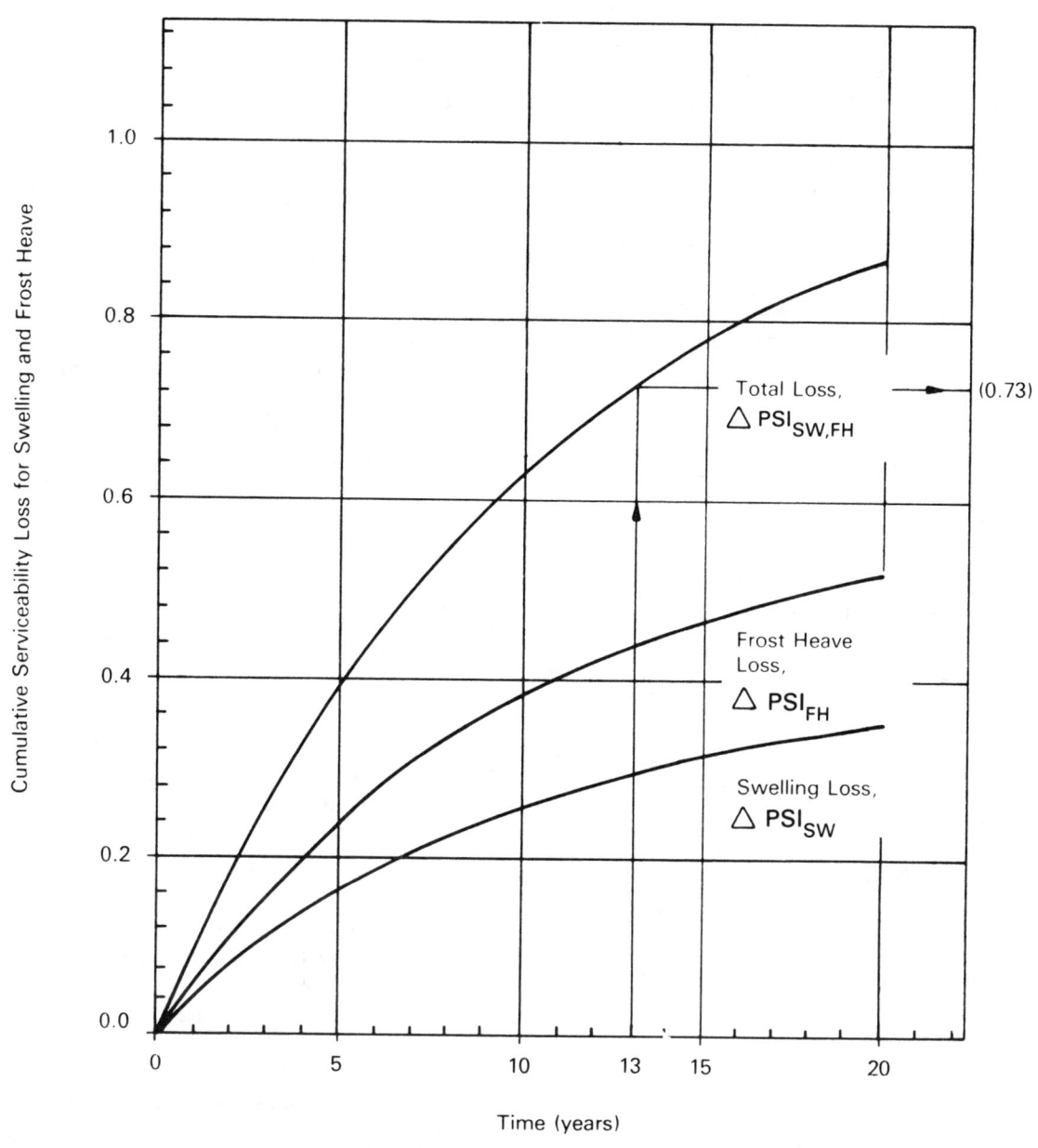

Figure 2.2. A Conceptual Example of the Environmental Serviceability Loss Versus Time Graph that may be Developed for a Specific Location

2.2.2 Allowable Rutting

In this design guide, rutting is considered only as a performance criterion for aggregate-surfaced roads. Although rutting is a problem with asphalt concrete surface pavements, no design model suitable for incorporation into this Guide is available at this time. It is important to note that the rut depth failure predicted by the aggregate-surfaced road model does not refer to simple surface rutting (which can be corrected by normal blading operations), but to serious rutting associated with deformation of the pavement structure and roadbed support. The allowable rut depth for an aggregate-surfaced road is dependent on the average daily traffic. Typically, allowable rut depths range from 1.0 to 2.0 inches for aggregate-surfaced roads.

2.2.3 Aggregate Loss

For aggregate-surfaced roads, an additional concern is the aggregate loss due to traffic and erosion. When aggregate loss occurs, the pavement structure becomes thinner and the load-carrying capacity is reduced. This reduction of the pavement structure thickness increases the rate of surface deterioration.

To treat aggregate loss in the procedure, it is necessary to estimate (1) the total thickness of aggregate that will be lost during the design period, and (2) the minimum thickness of aggregate that is required to keep a maintainable working surface for the pavement structure.

Unfortunately, there is very little information available today to predict the rate of aggregate loss. Below is an example of a prediction equation developed with limited data on sections experiencing greater than 50 percent truck traffic (*15, 16*):

$$GL = 0.12 + 0.1223(LT)$$

where

GL = total aggregate loss in inches, and
LT = number of loaded trucks in thousands.

A second equation, which was developed from a recent study in Brazil on typical rural sections, can be employed by the user to determine the input for gravel loss (*15, 16*):

$$GL = (B/25.4)/(.0045 LADT + 3380.6/R + 0.467G)$$

where

GL = aggregate loss, in inches, during the period of time being considered,
B = number of bladings during the period of time being considered,
LADT = average daily traffic in design lane (for one-lane road use total traffic in both directions),
R = average radius of curves, in feet, and
G = absolute value of grade, in percent.

Another equation, developed through a British study done in Kenya, is more applicable to areas where there is very little truck activity and thus the facility is primarily used by cars. Since this equation (below) is for annual gravel loss, the total gravel loss (GL) would be estimated by multiplying by the number of years in the performance period:

$$AGL = [T^2/(T^2 + 50)] \times f(4.2 + .092T + 0.889R^2 + 1.88VC)$$

where

AGL = annual aggregate loss, in inches,
T = annual traffic volume in both directions, in thousands of vehicles,
R = annual rainfall, in inches,
VC = average percentage gradient of the road, and
f = .037 for lateritic gravels,
 = .043 for quartzitic gravels,
 = .028 for volcanic gravels, and
 = .059 for coral gravels.

It should be noted that there are serious drawbacks with all the equations shown here; therefore, whenever possible, local information about aggregate loss should be used as input to the procedure.

2.3 MATERIAL PROPERTIES FOR STRUCTURAL DESIGN

2.3.1 Effective Roadbed Soil Resilient Modulus

As discussed previously in this Part and Part I, the basis for materials characterization in this Guide is

Design Requirements

elastic or resilient modulus. For roadbed materials, laboratory resilient modulus tests (AASHTO T 274) should be performed on representative samples in stress and moisture conditions simulating those of the primary moisture seasons. Alternatively, the seasonal resilient modulus values may be determined by correlations with soil properties, i.e., clay content, moisture, PI, etc. The purpose of identifying seasonal moduli is to quantify the relative damage a pavement is subjected to during each season of the year and treat it as part of the overall design. An effective roadbed soil resilient modulus is then established which is equivalent to the combined effect of all the seasonal modulus values. (The development of the procedure for generating an effective roadbed soil resilient modulus is presented in Appendix HH of Volume 2 of this Guide.)

The seasonal moisture conditions for which the roadbed soil samples should be tested are those which result in significantly different resilient moduli. For example, in a climate which is not subjected to extended sub-freezing temperatures, it would be important to test for differences between the wet (rainy) and dry seasons. It would probably not be necessary, however, to test for the difference between spring-wet and fall-wet, unless there is significant difference in the average rainfall during spring and fall. If operations make it difficult to test the roadbed soil for spring-thaw or winter-frozen conditions, then, for these extreme cases, practical values of resilient moduli of 20,000 to 50,000 psi may be used for frozen conditions, and for spring-thaw conditions, the retained modulus may be 20 to 30 percent of the normal modulus during the summer and fall periods.

Two different procedures for determining the seasonal variation of the modulus are offered as guidelines. One method is to obtain a laboratory relationship between resilient modulus and moisture content. Then, with an estimate of the in situ moisture content of the soil beneath the pavement, the resilient modulus for each of the seasons may be estimated. An alternate procedure is to back calculate the resilient modulus for different seasons using the procedure described in Part III using deflections measured on in-service pavements. These may be used as adjustment factors to correct the resilient modulus for a reference condition.

Besides defining the seasonal moduli, it is also necessary to separate the year into the various component time intervals during which the different moduli are effective. In making this breakdown, it is not necessary to specify a time interval of less than one-half month for any given season. If it is not possible to adequately estimate the season lengths, the user may refer to Section 4.1.2, which provides criteria suggested for the design of low-volume roads.

At this point, the length of the seasons and the seasonal roadbed resilient moduli are all that is required in terms of roadbed support for the design of rigid pavements and aggregate-surfaced roads. For the design of flexible pavements, however, the seasonal data must be translated into the effective roadbed soil resilient modulus described earlier. This is accomplished with the aid of the chart in Figure 2.3. The effective modulus is a weighted value that gives the equivalent annual damage obtained by treating each season independently in the performance equation and summing the damage. It is important to note, however, that the effective roadbed soil resilient modulus determined from this chart applies only to flexible pavements designed using the serviceability criteria. It is not necessarily applicable to other resilient modulus-based design procedures.

Since a mean value of resilient modulus is used, design sections with coefficient of variations greater than 0.15 (within a season) should be subdivided into smaller sections. For example, if the mean value of resilient modulus is 10,000 psi, then approximately 99 percent of the data should be in a range of 5,500 to 14,500 psi.

The first step of this process is to enter the seasonal moduli in their respective time periods. If the smallest season is one-half month, then all seasons must be defined in terms of half months and each of the boxes must be filled. If the smallest season is one month, then all seasons must be defined in terms of whole months and only one box per month may be filled in.

The next step is to estimate the relative damage (u_f) values corresponding to each seasonal modulus. This is done using the vertical scale or the corresponding equation shown in Figure 2.3. For example, the relative damage corresponding to a roadbed soil resilient modulus of 4,000 psi is 0.51.

Next, the u_f values should all be added together and divided by the number of seasonal increments (12 or 24) to determine the average relative damage. The effective roadbed soil resilient modulus (M_R), then, is the value corresponding to the average relative damage on the M_R u_f scale. Figure 2.4 provides an example of the application of the effective M_R estimation process. Again, it is emphasized that this effective M_R value should be used only for the design of flexible pavements based on serviceability criteria.

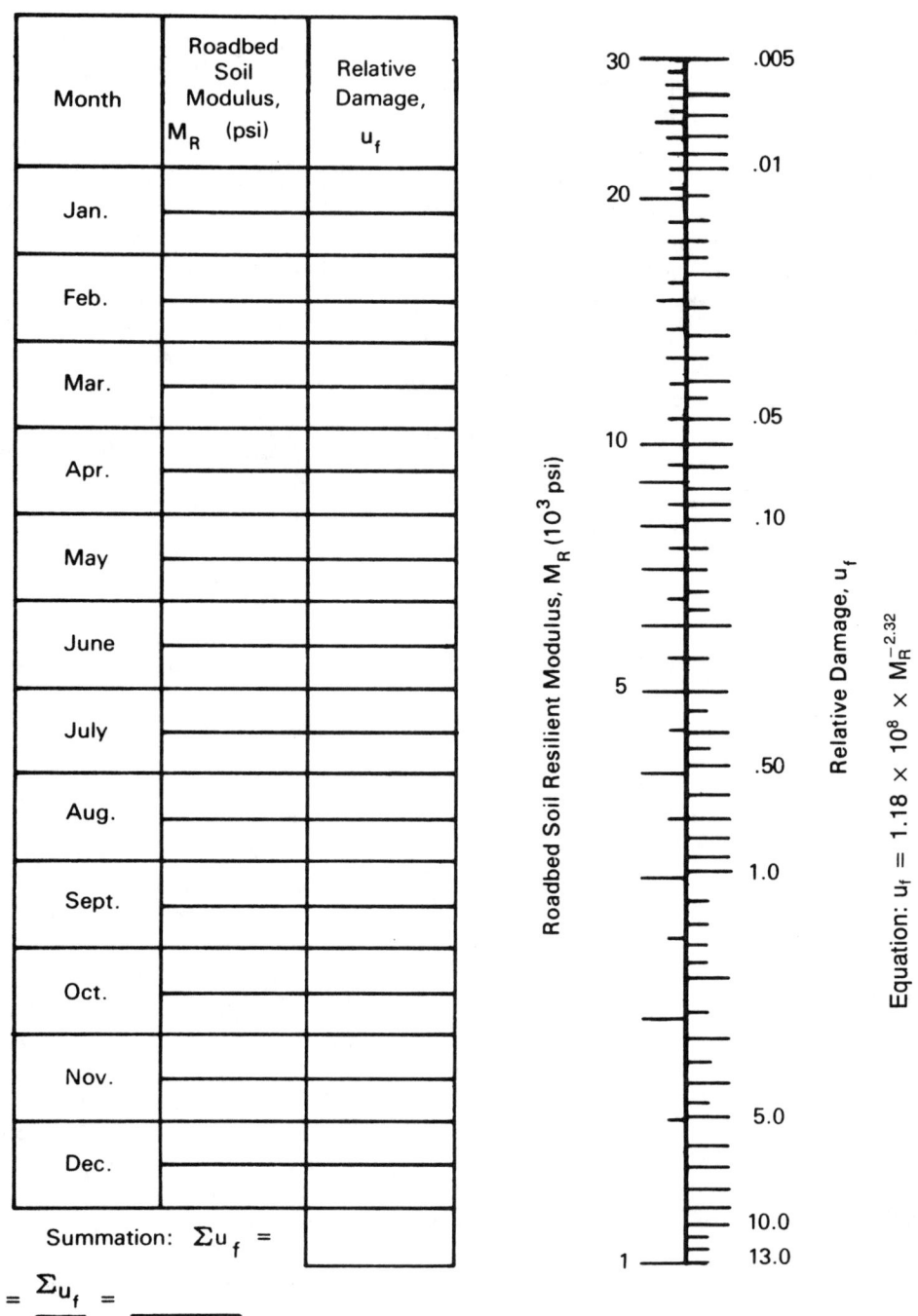

Figure 2.3. Chart for Estimating Effective Roadbed Soil Resilient Modulus for Flexible Pavements Designed Using the Serviceability Criteria

Design Requirements

Month	Roadbed Soil Modulus, M_R (psi)	Relative Damage, u_f
Jan.	20,000	0.01
Feb.	20,000	0.01
Mar.	2,500	1.51
Apr.	4,000	0.51
May	4,000	0.51
June	7,000	0.13
July	7,000	0.13
Aug.	7,000	0.13
Sept.	7,000	0.13
Oct.	7,000	0.13
Nov.	4,000	0.51
Dec.	20,000	0.01
Summation: $\Sigma u_f =$		3.72

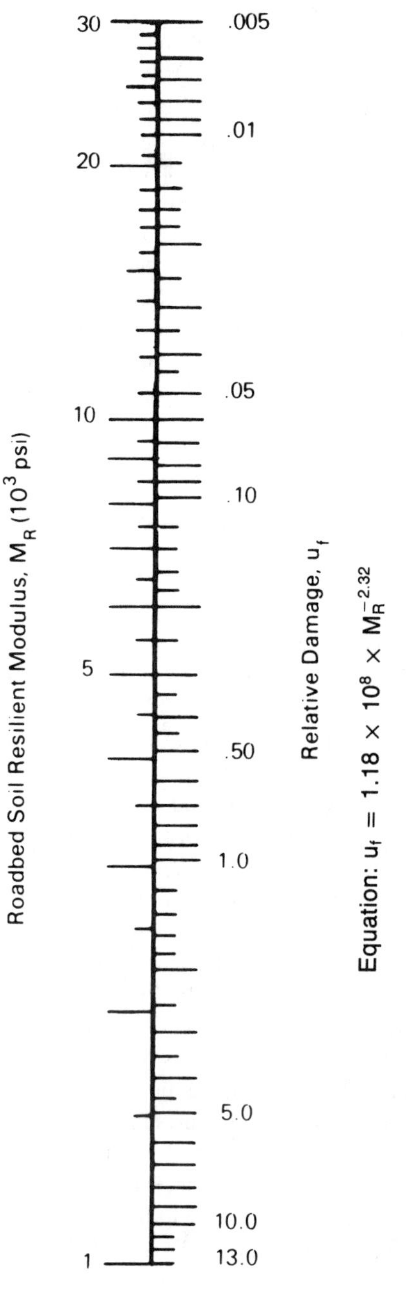

Average: $\bar{u}_f = \dfrac{\Sigma u_f}{n} = \dfrac{3.72}{12} = 0.31$

Equation: $u_f = 1.18 \times 10^8 \times M_R^{-2.32}$

Effective Roadbed Soil Resilient Modulus, M_R (psi) = __5,000__ (corresponds to $\bar{u}_f$)

Figure 2.4. Chart for Estimating Effective Roadbed Soil Resilient Modulus for Flexible Pavements Designed Using the Serviceability Criteria

2.3.2 Effective Modulus of Subgrade Reaction

Like the effective roadbed soil resilient modulus for flexible pavement design, an effective modulus of subgrade reaction (k-value) will be developed for rigid pavement design. Since the k-value is directly proportional to roadbed soil resilient modulus, the season lengths and seasonal moduli developed in the previous section will be used as input to the estimation of an effective design k-value. But, because of the effects of subbase characteristics on the effective design k-value, its determination is included as a step in an iterative design procedure. (See Part II, Chapter 3.) The development of the actual procedure for generating this effective modulus of subgrade reaction is presented in Appendix HH of Volume 2 of this Guide.

2.3.3 Pavement Layer Materials Characterization

Although there are many types of material properties and laboratory test procedures for assessing the strength of pavement structural materials, one has been adopted as a basis for design in this Guide. If, however, the user should have a better understanding of the "layer coefficients" (see Section 2.3.5) that have traditionally been used in the original AASHTO flexible pavement design procedure, it is not essential that the elastic moduli of these materials be characterized. In general, layer coefficients derived from test roads or satellite sections are preferred.

Elastic modulus is a fundamental engineering property of any paving or roadbed material. For those material types which are subject to significant permanent deformation under load, this property may not reflect the material's behavior under load. Thus, resilient modulus refers to the material's stress-strain behavior under normal pavement loading conditions. The strength of the material is important in addition to stiffness, and future mechanistic-based procedures may reflect strength as well as stiffness in the materials characterization procedures. In addition, stabilized base materials may be subject to cracking under certain conditions and the stiffness may not be an indicator for this distress type. It is important to note, that, although resilient modulus can apply to any type of material, the notation M_R as used in this Guide applies only to the roadbed soil. Different notations are used to express the moduli for subbase (E_{SB}), base (E_{BS}), asphalt concrete (E_{AC}), and portland cement concrete (E_C).

The procedure for estimating the resilient modulus of a particular pavement material depends on its type. Relatively low stiffness materials, such as natural soils, unbound granular layers, and even stabilized layers and asphalt concrete, should be tested using the resilient modulus test methods (AASHTO T 274). Although the testing apparatus for each of these types of materials is basically the same, there are some differences, such as the need for triaxial confinement for unbound materials.

Alternatively, the bound or higher stiffness materials, such as stabilized bases and asphalt concrete, may be tested using the repeated-load indirect tensile test (ASTM D 4123). This test still relies on the use of electronic gauges to measure small movements of the sample under load, but is less complex and easier to run than the triaxial resilient modulus test.

Because of the small displacements and brittle nature of the stiffest pavement materials, i.e., portland cement concrete and those base materials stabilized with a high cement content, it is difficult to measure the modulus using the indirect tensile apparatus. Thus, it is recommended that the elastic modulus of such high-stiffness materials be determined according to the procedure described in ASTM C 469.

The elastic modulus for any type of material may also be estimated using correlations developed by the state's department of transportation or by some other reputable agency. The following is a correlation recommended by the American Concrete Institute (4) for normal weight portland cement concrete:

$$E_c = 57,000(f'_c)^{0.5}$$

where

E_c = PCC elastic modulus (in psi), and
f'_c = PCC compressive strength (in psi) as determined using AASHTO T 22, T 140, or ASTM C 39.

2.3.4 PCC Modulus of Rupture

The modulus of rupture (flexural strength) of portland cement concrete is required only for the design of a rigid pavement. The modulus of rupture required by the design procedure is the mean value determined after 28 days using third-point loading (AASHTO T 97, ASTM C 78). If standard agency practice dictates the use of center-point loading, then a correlation should be made between the two tests.

Design Requirements

Because of the treatment of reliability in this Guide, it is strongly recommended that the normal construction specification for modulus of rupture (flexural strength) *not* be used as input, since it represents a value below which only a small percent of the distribution may lie. If it is desirable to use the construction specification, then some adjustment should be applied, based on the standard deviation of modulus of rupture and the percent (PS) of the strength distribution that normally falls below the specification:

$$S'_c(\text{mean}) = S_c + z(SD_s)$$

where

S'_c = estimated mean value for PCC modulus of rupture (psi),
S_c = construction specification on concrete modulus of rupture (psi),
SD_s = estimated standard deviation of concrete modulus of rupture (psi), and
z = standard normal variate:
= 0.841, for PS = 20 percent,*
= 1.037, for PS = 15 percent,
= 1.282, for PS = 10 percent,
= 1.645, for PS = 5 percent, and
= 2.327, for PS = 1 percent.

*NOTE: Permissible number of specimens, expressed as a percentage, that may have strengths less than the specification value.

2.3.5 Layer Coefficients

This section describes a method for estimating the AASHTO structural layer coefficients (a_i values) required for standard flexible pavement structural design. A value for this coefficient is assigned to each layer material in the pavement structure in order to convert actual layer thicknesses into structural number (SN). This layer coefficient expresses the empirical relationship between SN and thickness and is a measure of the relative ability of the material to function as a structural component of the pavement. The following general equation for structural number reflects the relative impact of the layer coefficients (a_i) and thickness (D_i):

$$SN = \sum_{i=1} a_i D_i$$

Although the elastic (resilient) modulus has been adopted as the standard material quality measure, it is still necessary to identify (corresponding) layer coefficients because of their treatment in the structural number design approach. Though there are correlations available to determine the modulus from tests such as the R-value, the procedure recommended is direct measurement using AASHTO Method T 274 (subbase and unbound granular materials) and ASTM D 4123 for asphalt concrete and other stabilized materials. Research and field studies indicate many factors influence the layer coefficients, thus the agency's experience must be included in implementing the results from the procedures presented. For example, the layer coefficient may vary with thickness, underlying support, position in the pavement structure, etc.

It should be noted that laboratory resilient modulus values can be obtained that are significantly different from what may exist for an in situ condition. For example, the presence of a very stiff unbound layer over a low stiffness layer may result in decompaction and a corresponding reduction of stiffness. As a guideline for successive layers of unbound materials, the ratio of resilient modulus of the upper layer to that of the lower layer should not exceed values that result in tensile stresses in unbound granular layers.

The discussion of how these coefficients are estimated is separated into five categories, depending on the type and function of the layer material. These are asphalt concrete, granular base, granular subbase, cement-treated, and bituminous base. Other materials such as lime, lime flyash, and cement flyash are acceptable materials, and each agency should develop charts.

Asphalt Concrete Surface Course. Figure 2.5 provides a chart that may be used to estimate the structural layer coefficient of a dense-graded asphalt concrete surface course based on its elastic (resilient) modulus (E_{AC}) at 68°F. Caution is recommended for modulus values above 450,000 psi. Although higher modulus asphalt concretes are stiffer and more resistant to bending, they are also more susceptible to thermal and fatigue cracking.

Granular Base Layers. Figure 2.6 provides a chart that may be used to estimate a structural layer coefficient, a_2, from one of four different laboratory test results on a granular base material, including base resilient modulus, E_{BS}. The AASHO Road Test basis for these correlations is:

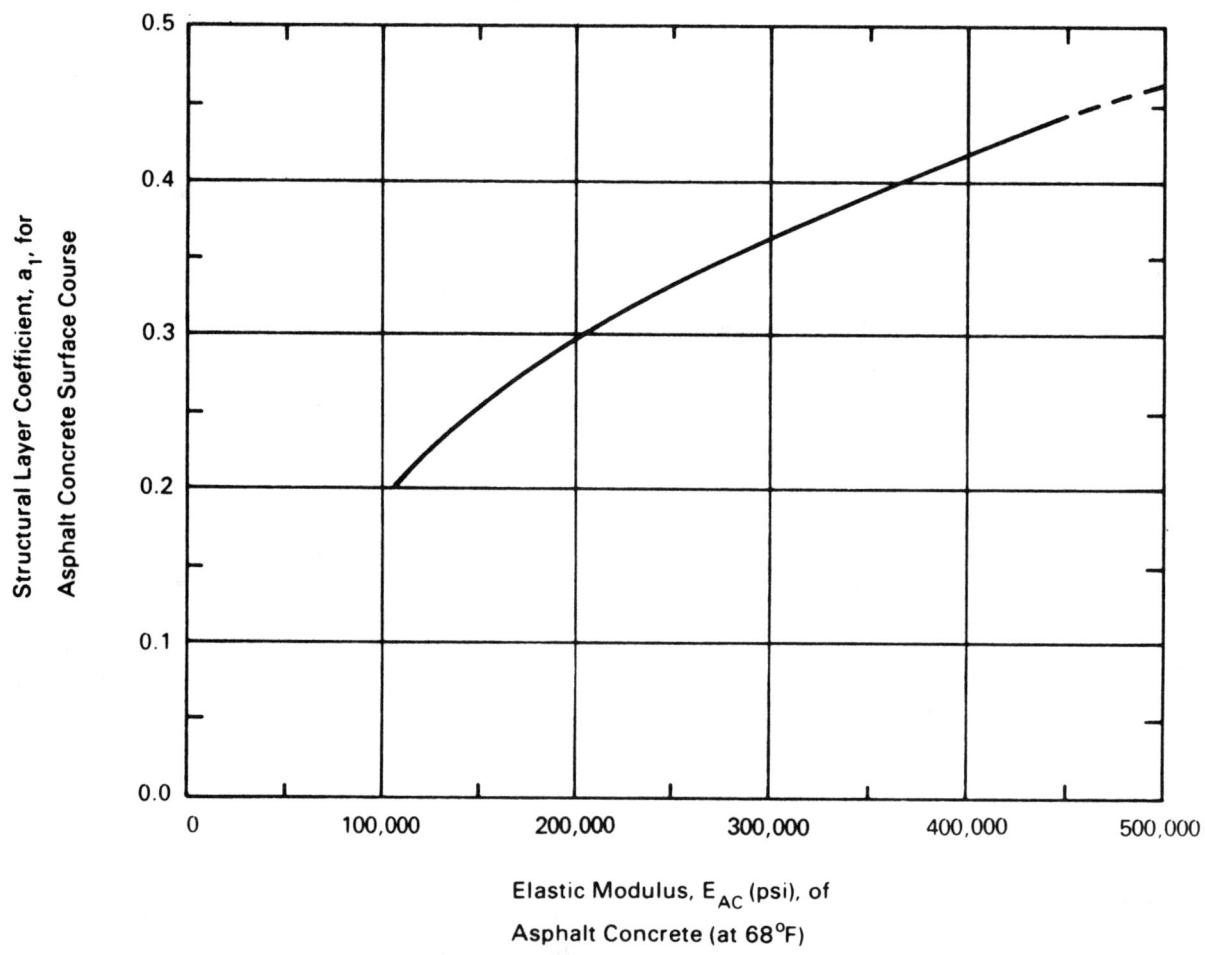

Figure 2.5. Chart for Estimating Structural Layer Coefficient of Dense-Graded Asphalt Concrete Based on the Elastic (Resilient) Modulus (3)

Design Requirements

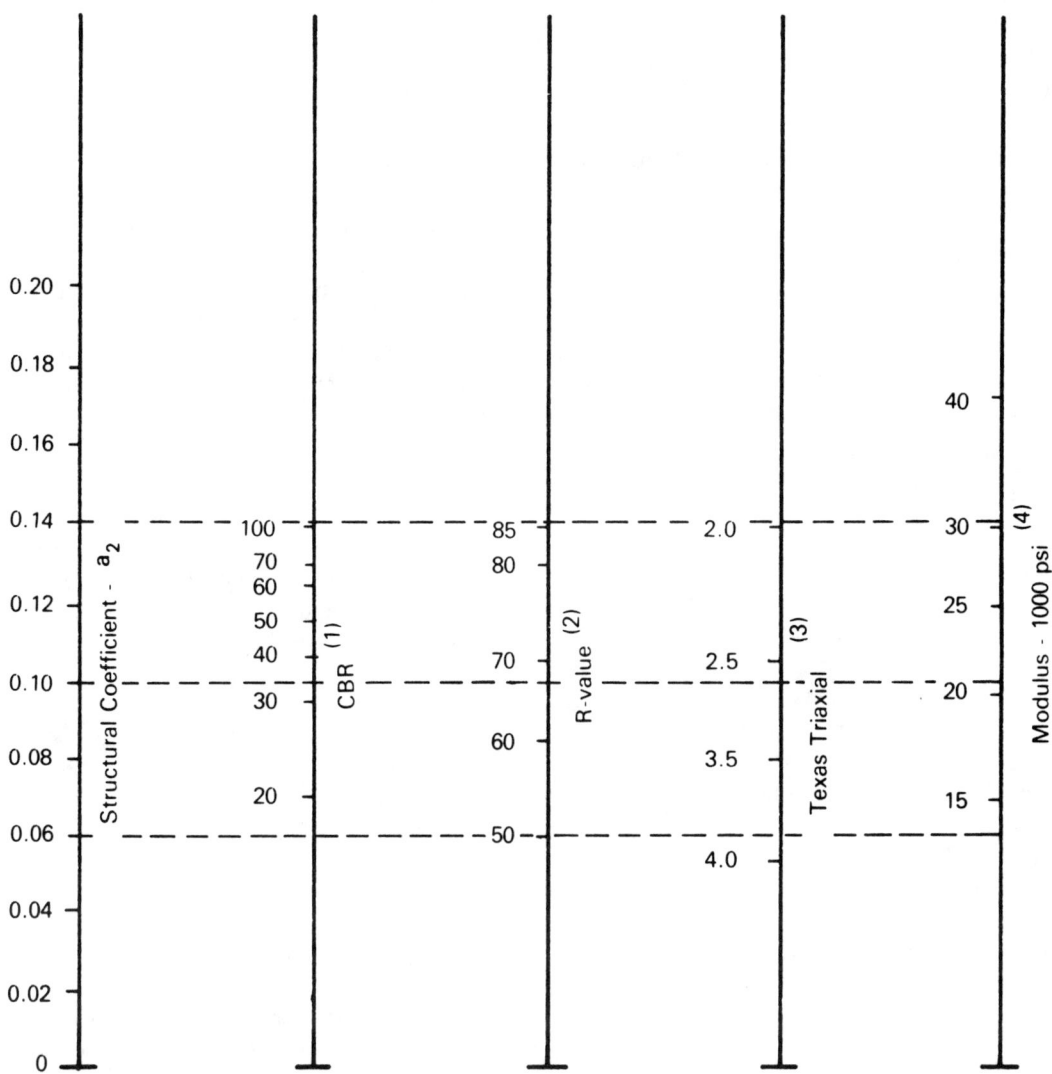

(1) Scale derived by averaging correlations obtained from Illinois.
(2) Scale derived by averaging correlations obtained from California, New Mexico and Wyoming.
(3) Scale derived by averaging correlations obtained from Texas.
(4) Scale derived on NCHRP project *(3)*.

Figure 2.6. Variation in Granular Base Layer Coefficient (a_2) with Various Base Strength Parameters (*3*)

$$a_2 = 0.14$$
$$E_{BS} = 30{,}000 \text{ psi}$$
$$CBR = 100 \text{ (approx.)}$$
$$R\text{-value} = 85 \text{ (approx.)}$$

The following relationship may be used in lieu of Figure 2.6 to estimate the layer coefficient, a_2, for a granular base material from its elastic (resilient) modulus, E_{BS} (5):

$$a_2 = 0.249(\log_{10} E_{BS}) - 0.977$$

For aggregate base layers, E_{BS} is a function of the stress state (θ) within the layer and is normally given by the relation:

$$E_{BS} = k_1 \theta^{k_2}$$

where

θ = stress state or sum of principal stresses $\sigma_1 + \sigma_2 + \sigma_3$ (psi), and

k_1, k_2 = regression constants which are a function of material type.

Typical values for base materials are:

$k_1 = 3{,}000$ to $8{,}000$
$k_2 = 0.5$ to 0.7

At the AASHO Road Test, modulus values (E_{BS} in psi) for the base were as follows:

Moisture State	Equation	Stress State (psi)			
		$\theta = 5$	$\theta = 10$	$\theta = 20$	$\theta = 30$
Dry	$8{,}000\theta^{0.6}$	21,012	31,848	48,273	61,569
Damp	$4{,}000\theta^{0.6}$	10,506	15,924	24,136	30,784
Wet	$3{,}200\theta^{0.6}$	8,404	12,739	19,309	24,627

Note, E_{BS} is a function of not only moisture but also the stress state (θ). Values for the stress state within the base course vary with the subgrade modulus and thickness of the surface layer. Typical values for use in design are:

Asphalt Concrete Thickness (inches)	Roadbed Soil Resilient Modulus (psi)		
	3,000	7,500	15,000
Less than 2	20	25	30
2–4	10	15	20
4–6	5	10	15
Greater than 6	5	5	5

For intermediate values of roadbed soil resilient modulus, interpolation can be used.

Each agency is encouraged to develop relationships for their specific base materials (e.g., $M_R = k_1 \theta^{k_2}$) using AASHTO Method T 274; however, in the absence of this data, values given in Table 2.3 can be used.

Granular Subbase Layers. Figure 2.7 provides a chart that may be used to estimate a structural layer coefficient, a_3, from one of four different laboratory results on a granular subbase material, including subbase resilient modulus, E_{SB}. The AASHO Road Test basis for these correlations is:

$$a_3 = 0.11$$
$$E_{SB} = 15{,}000 \text{ psi}$$
$$CBR = 30 \text{ (approx.)}$$
$$R\text{-value} = 60 \text{ (approx.)}$$

Table 2.3. Typical Values for k_1 and k_2 for Unbound Base and Subbase Materials ($M_R = k_1 \theta^{k_2}$)

Moisture Condition	k_1*	k_2*
(a) Base		
Dry	6,000–10,000	0.5–0.7
Damp	4,000–6,000	0.5–0.7
Wet	2,000–4,000	0.5–0.7
(b) Subbase		
Dry	6,000–8,000	0.4–0.6
Damp	4,000–6,000	0.4–0.6
Wet	1,500–4,000	0.4–0.6

*Range in k_1 and k_2 is a function of the material quality.

Design Requirements

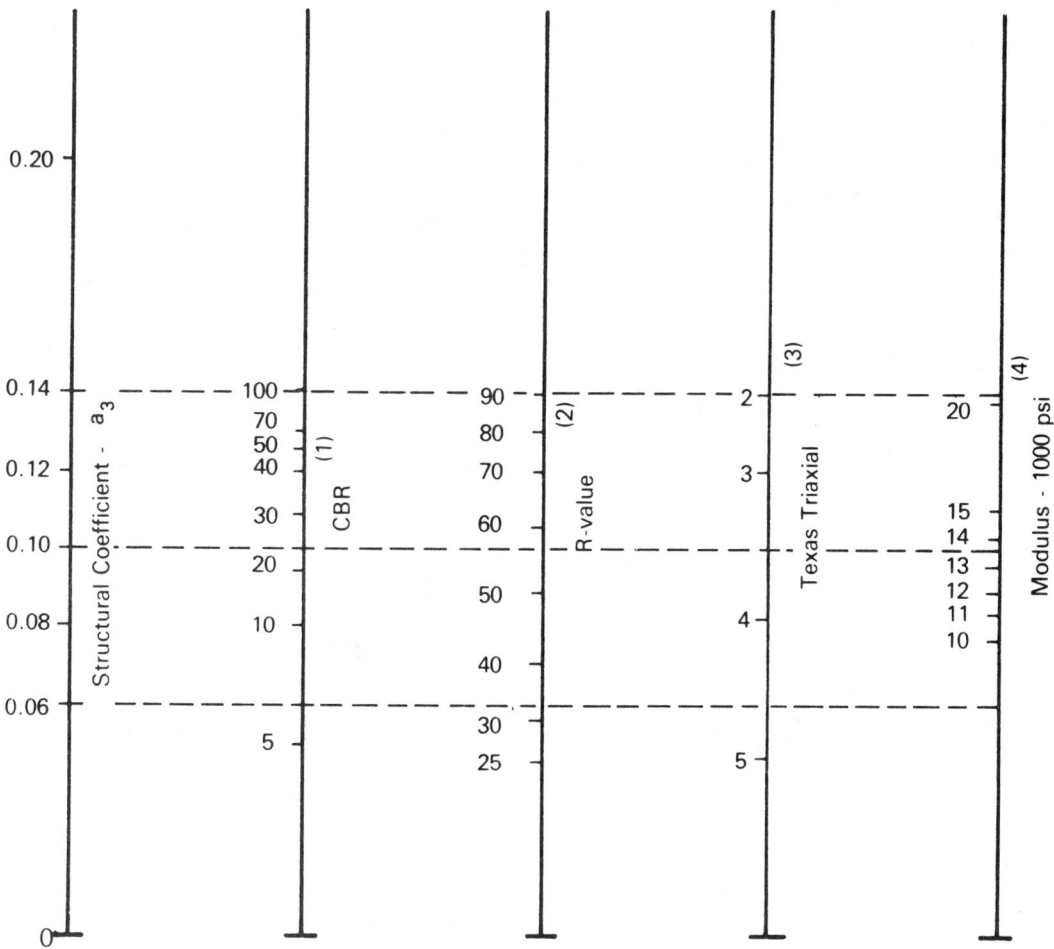

(1) Scale derived from correlations from Illinois.
(2) Scale derived from correlations obtained from The Asphalt Institute, California, New Mexico and Wyoming.
(3) Scale derived from correlations obtained from Texas.
(4) Scale derived on NCHRP project *(3)*.

Figure 2.7. Variation in Granular Subbase Layer Coefficient (a_3) with Various Subbase Strength Parameters (*3*)

The E_{SB} versus a_2 relationship (5) similar to that for granular base materials is as follows:

$$a_3 = 0.227(\log_{10} E_{SB}) - 0.839$$

For aggregate subbase layers, E_{SB} is affected by stress state (θ) in a fashion similar to that for the base layer. Typical values for k_1 range from 1,500 to 6,000, while k_2 varies from 0.4 to 0.6. Values for the AASHO Road Test subbase material were (13):

Moisture State	Developed Relationship	Stress State (psi)		
		$\theta = 5$	$\theta = 7.5$	$\theta = 10$
Damp	$M_R = 5,400\theta^{0.6}$	14,183	18,090	21,497
Wet	$M_R = 4,600\theta^{0.6}$	12,082	15,410	18,312

As with the base layers, each agency is encouraged to develop relationships for their specific materials; however, in lieu of this data, the values in Table 2.3 can be used.

Stress states (θ) which can be used as a guide to select the modulus value for subbase thicknesses between 6 and 12 inches are as follows:

Asphalt Concrete Thickness (inches)	Stress State (psi)
Less than 2	10.0
2–4	7.5
Greater than 4	5.0

Cement-Treated Bases. Figure 2.8 provides a chart that may be used to estimate the structural layer coefficient, a_2, for a cement-treated base material from either its elastic modulus, E_{BS}, or, alternatively, its 7-day unconfined compressive strength (ASTM D 1633).

Bituminous-Treated Bases. Figure 2.9 presents a chart that may be used to estimate the structural layer coefficient, a_2, for a bituminous-treated base material from either its elastic modulus, E_{BS}, or, alternatively, its Marshall stability (AASHTO T 245, ASTM D 1559). This is not shown in Figure 2.9.

2.4 PAVEMENT STRUCTURAL CHARACTERISTICS

2.4.1 Drainage

This section describes the selection of inputs to treat the effects of certain levels of drainage on predicted pavement performance. Guidance is not provided here for any detailed drainage designs or construction methods. Furthermore, criteria on the ability of various drainage methods to remove moisture from the pavement are not provided. It is up to the design engineer to identify what level (or quality) of drainage is achieved under a specific set of drainage conditions. Below are the general definitions corresponding to different drainage levels from the pavement structure:

Quality of Drainage	Water Removed Within
Excellent	2 hours
Good	1 day
Fair	1 week
Poor	1 month
Very poor	(water will not drain)

For comparison purposes, the drainage conditions at the AASHO Road Test are considered to be fair, i.e., free water was removed within 1 week.

Flexible Pavements. The treatment for the expected level of drainage for a flexible pavement is through the use of modified layer coefficients (e.g., a higher effective layer coefficient would be used for improved drainage conditions). The factor for modifying the layer coefficient is referred to as an m_i value and has been integrated into the structural number (SN) equation along with layer coefficient (a_i) and thickness (D_i); thus:

$$SN = a_1 D_1 + a_2 D_2 m_2 + a_3 D_3 m_3$$

(The possible effect of drainage on the asphalt concrete surface course is not considered.) The conversion of the structural number into actual pavement layer thicknesses is discussed in more detail in Part II, Chapter 3.

Table 2.4 presents the recommended m_i values as a function of the quality of drainage and the percent of time during the year the pavement structure would normally be exposed to moisture levels approaching

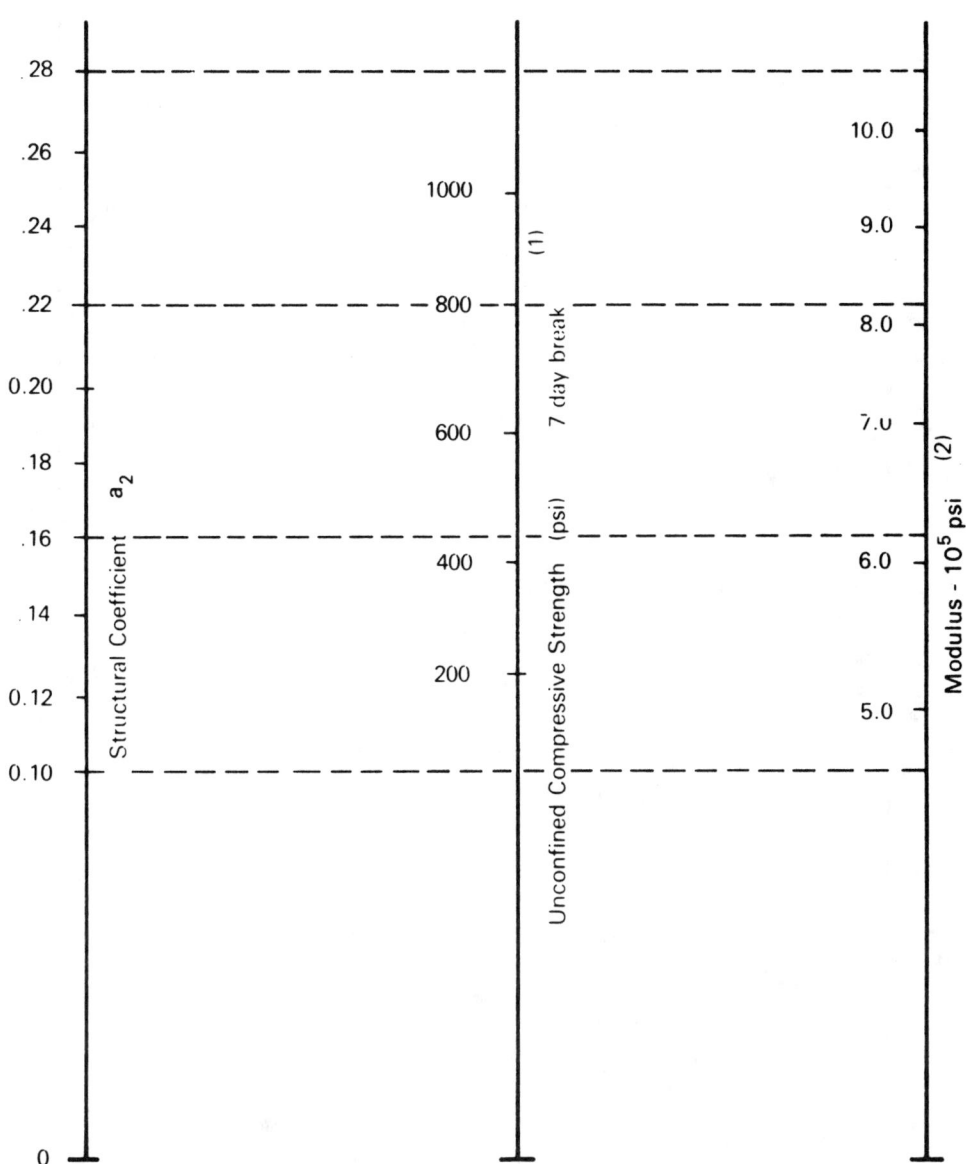

(1) Scale derived by averaging correlations from Illinois, Louisiana and Texas.
(2) Scale derived on NCHRP project (3).

Figure 2.8. Variation in a for Cement-Treated Bases with Base Strength Parameter (3)

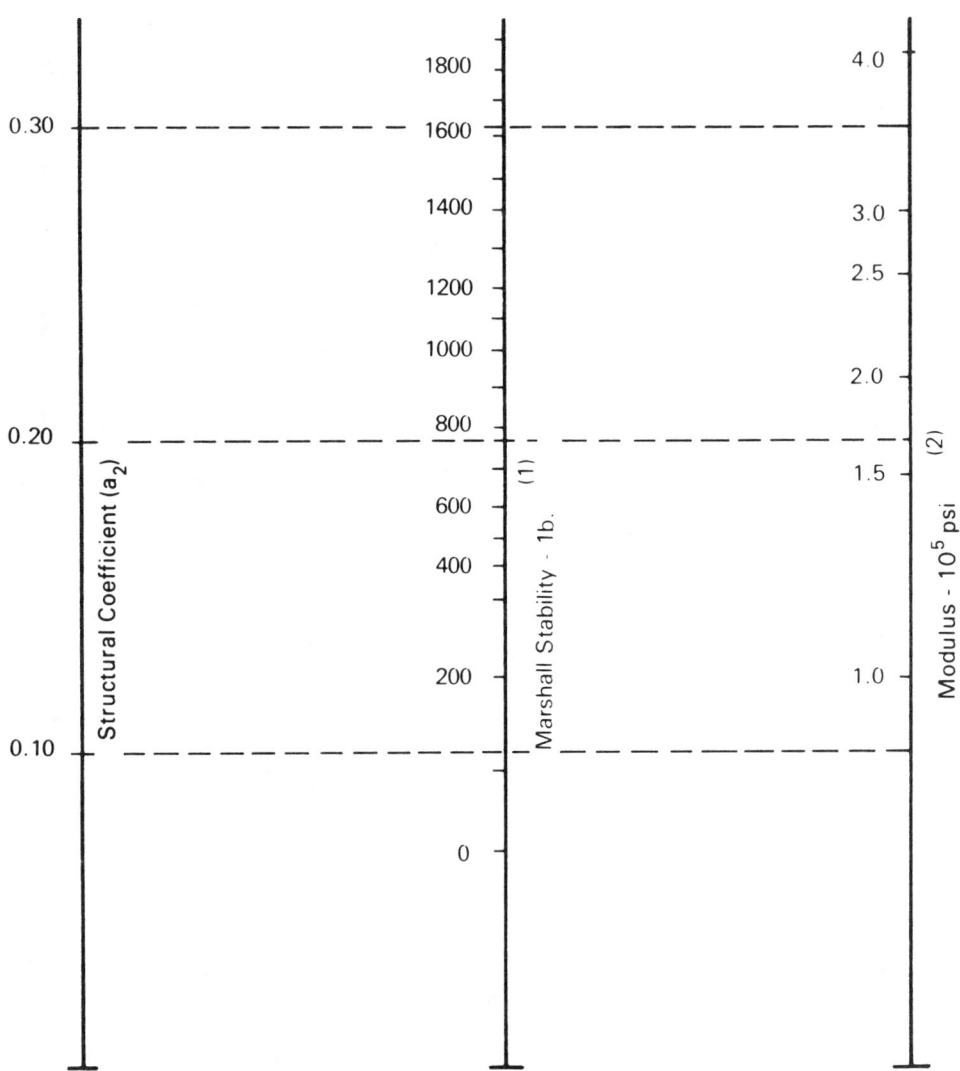

(1) Scale derived by correlation obtained from Illinois.
(2) Scale derived on NCHRP project (3).

Figure 2.9. Variation in a_2 for Bituminous-Treated Bases with Base Strength Parameter (3)

Table 2.4. Recommended m_i Values for Modifying Structural Layer Coefficients of Untreated Base and Subbase Materials in Flexible Pavements

Quality of Drainage	Percent of Time Pavement Structure is Exposed to Moisture Levels Approaching Saturation			
	Less Than 1%	1–5%	5–25%	Greater Than 25%
Excellent	1.40–1.35	1.35–1.30	1.30–1.20	1.20
Good	1.35–1.25	1.25–1.15	1.15–1.00	1.00
Fair	1.25–1.15	1.15–1.05	1.00–0.80	0.80
Poor	1.15–1.05	1.05–0.80	0.80–0.60	0.60
Very poor	1.05–0.95	0.95–0.75	0.75–0.40	0.40

saturation. Obviously, the latter is dependent on the average yearly rainfall and the prevailing drainage conditions. As a basis for comparison, the m_i value for conditions at the AASHO Road Test is 1.0, regardless of the type of material. A discussion of how these recommended m_i values were derived is presented in Appendix DD of Volume 2.

Finally, it is also important to note that these values apply *only* to the effects of drainage on untreated base and subbase layers. Although improved drainage is certainly beneficial to stabilized or treated materials, the effects on performance of flexible pavements are not as profound as those quantified in Table 2.4.

Rigid Pavements. The treatment for the expected level of drainage for a rigid pavement is through the use of a drainage coefficient, C_d, in the performance equation. (It has an effect similar to that of the load transfer coefficient, J.) As a basis for comparison, the value for C_d for conditions at the AASHO Road Test is 1.0.

Table 2.5 provides the recommended C_d values, depending on the quality of drainage and the percent of time during the year the pavement structure would normally be exposed to moisture levels approaching saturation. As before, the latter is dependent on the average yearly rainfall and the prevailing drainage conditions. A discussion of how these recommended C_d values were derived is also presented in Appendix DD of Volume 2.

2.4.2 Load Transfer

The load transfer coefficient, J, is a factor used in rigid pavement design to account for the ability of a concrete pavement structure to transfer (distribute) load across discontinuities, such as joints or cracks. Load transfer devices, aggregate interlock, and the presence of tied concrete shoulders all have an effect on this value. Generally, the J-value for a given set of conditions (e.g., jointed concrete pavement with tied shoulders) increases as traffic loads increase since aggregate load transfer decreases with load repetitions. Table 2.6 establishes ranges of load transfer coefficients for different conditions developed from experience and mechanistic stress analysis. As a general guide for the range of J-values, higher J's should be used with low k-values, high thermal coefficients, and large variations of temperature. (The development of the J-factor terms is provided in Appendix KK of Volume 2.) Each agency should, however, develop criteria for their own aggregates, climatic conditions, etc.

If dowels are used, the size and spacing should be determined by the local agency's procedures and/or experience. As a general guideline, the dowel diameter should be equal to the slab thickness multiplied by ⅛ inch (e.g., for a 10-inch pavement, the diameter is 1¼ inch. The dowel spacing and length are normally 12 inches and 18 inches, respectively.

Jointed Pavements. The value of J recommended for a plain jointed pavement (JCP) or jointed reinforced concrete pavement (JRCP) with some type of load transfer device (such as dowel bars) at the joints is 3.2 ("protected corner" condition at the AASHO Road Test). This value is indicative of the load transfer of jointed pavements without tied concrete shoulders.

For jointed pavements without load transfer devices at the joints, a J-value of 3.8 to 4.4 is recommended. (This basically accounts for the higher bending stresses that develop in undowelled pavements, but also includes some consideration of the increased potential for faulting.) If the concrete has a high thermal

Table 2.5. Recommended Values of Drainage Coefficient, C_d, for Rigid Pavement Design

Quality of Drainage	Percent of Time Pavement Structure is Exposed to Moisture Levels Approaching Saturation			
	Less Than 1%	1–5%	5–25%	Greater Than 25%
Excellent	1.25–1.20	1.20–1.15	1.15–1.10	1.10
Good	1.20–1.15	1.15–1.10	1.10–1.00	1.00
Fair	1.15–1.10	1.10–1.00	1.00–0.90	0.90
Poor	1.10–1.00	1.00–0.90	0.90–0.80	0.80
Very poor	1.00–0.90	0.90–0.80	0.80–0.70	0.70

coefficient, then the value of J should be increased. On the other hand, if few heavy trucks are anticipated such as a low-volume road, the J-value may be lowered since the loss of aggregate interlock will be less. Part I of this Guide provides some other general criteria for the consideration and/or design of expansion joints, contraction joints, longitudinal joints, load transfer devices, and tie bars in jointed pavements.

Continuously Reinforced Pavements. The value of J recommended for continuously reinforced concrete pavements (CRCP) without tied concrete shoulders is between 2.9 to 3.2, depending on the capability of aggregate interlock (at future transverse cracks) to transfer load. In the past, a commonly used J-value for CRCP was 3.2, but with better design for crack width control each agency should develop criteria based on local aggregates and temperature ranges.

Tied Shoulders or Widened Outside Lanes. One of the major advantages of using tied PCC shoulders (or widened outside lanes) is the reduction of slab stress and increased service life they provide. To account for this, significantly lower J-values may be used for the design of both jointed and continuous pavements.

For continuously reinforced concrete pavements with tied concrete shoulders (the minimum bar size and maximum tie bar spacing should be the same as that for tie bars between lanes), the range of J is between 2.3 and 2.9, with a recommended value of 2.6. This value is considerably lower than that for the design of concrete pavements without tied shoulders because of the significantly increased load distribution capability of concrete pavements with tied shoulders.

For jointed concrete pavements with dowels and tied shoulders, the value of J should be between 2.5 and 3.1 based on the agency's experience. The lower J-value for tied shoulders assumes traffic is not permitted to run on the shoulder.

NOTE: Experience has shown that a concrete shoulder of 3 feet or greater may be considered a tied shoulder. Pavements with monolithic or tied curb and gutter that provides additional stiffness and keeps

Table 2.6. Recommended Load Transfer Coefficient for Various Pavement Types and Design Conditions

Shoulder	Asphalt		Tied P.C.C.	
Load Transfer Devices	Yes	No	Yes	No
Pavement Type				
1. Plain jointed and jointed reinforced	3.2	3.8–4.4	2.5–3.1	3.6–4.2
2. CRCP	2.9–3.2	N/A	2.3–2.9	N/A

Design Requirements

traffic away from the edge may be treated as a tied shoulder.

2.4.3 Loss of Support

This factor, LS, is included in the design of rigid pavements to account for the potential loss of support arising from subbase erosion and/or differential vertical soil movements. It is treated in the actual design procedure (discussed in Part II, Chapter 3) by diminishing the effective or composite k-value based on the size of the void that may develop beneath the slab. Table 2.7 provides some suggested ranges of LS depending on the type of material (specifically its stiffness or elastic modulus). Obviously, if various types of base or subbase are to be considered for design, then the corresponding values of LS should be determined for each type. A discussion of how the loss of support factor was derived is present in Appendix LL of Volume 2 of this Guide.

The LS factor should also be considered in terms of differential vertical soil movements that may result in voids beneath the pavement. Thus, even though a non-erosive subbase is used, a void may still develop, thus reducing pavement life. Generally, for active swelling clays or excessive frost heave, LS values of 2.0 to 3.0 may be considered. Each agency's experience in this area should, however, be the key element in the selection of an appropriate LS value. Examination of the effect of LS on reducing the effective k-value of the roadbed soil (see Figure 3.6) may also be helpful in selecting an appropriate value.

2.5 REINFORCEMENT VARIABLES

Because of the difference in the reinforcement design procedures between jointed and continuous pavements, the design requirements for each are separated into two sections. Information is also provided here for the design of prestressed concrete pavement. In addition to dimensions, consideration should be given to corrosion resistance of reinforcement, especially in areas where pavements are exposed to variable moisture contents and salt applications.

2.5.1 Jointed Reinforced Concrete Pavements

There are two types of rigid pavement which fall under the "jointed" category: plain jointed pavement (JCP), which is designed not to have steel reinforcement, and jointed reinforced concrete pavement (JRCP), which is designed to have significant steel reinforcement, in terms of either steel bars or welded steel mats. The steel reinforcement is added if the probability of transverse cracking during pavement life is high due to such factors as soil movement and/or temperature/moisture change stresses.

For the case of plain jointed concrete pavements (JCP), the joint spacing should be selected at values so that temperature and moisture change stresses do not produce intermediate cracking between joints. The maximum joint spacing will vary, depending on local conditions, subbase types, coarse aggregate types, etc. In addition, the maximum joint spacing may be selected to minimize joint movement and, consequently, maximize load transfer. Each agency's experience should be relied on for this selection.

Following are the criteria needed for the design of jointed pavements which are steel reinforced (JRCP). These criteria apply to the design of both longitudinal and transverse steel reinforcement.

Slab Length. This refers to the joint spacing or distance, L (feet), between free (i.e., untied) transverse joints. It is an important design consideration since it has a large impact on the maximum concrete tensile stresses and, consequently, the amount of steel

Table 2.7. Typical Ranges of Loss of Support (LS) Factors for Various Types of Materials (6)

Type of Material	Loss of Support (LS)
Cement Treated Granular Base (E = 1,000,000 to 2,000,000 psi)	0.0 to 1.0
Cement Aggregate Mixtures (E = 500,000 to 1,000,000 psi)	0.0 to 1.0
Asphalt Treated Base (E = 350,000 to 1,000,000 psi)	0.0 to 1.0
Bituminous Stabilized Mixtures (E = 40,000 to 300,000 psi)	0.0 to 1.0
Lime Stabilized (E = 20,000 to 70,000 psi)	1.0 to 3.0
Unbound Granular Materials (E = 15,000 to 45,000 psi)	1.0 to 3.0
Fine Grained or Natural Subgrade Materials (E = 3,000 to 40,000 psi)	2.0 to 3.0

NOTE: E in this table refers to the general symbol for elastic or resilient modulus of the material.

reinforcement required. Because of this effect, slab length (joint spacing) is an important factor that must be considered in the design of any reinforced or unreinforced jointed concrete pavement. The selection of an appropriate value is covered in more detail in Part II, Chapter 3.

Steel Working Stress. This refers to the allowable working stress, f_s (psi), in the steel reinforcement. Typically, a value equivalent to 75 percent of the steel yield strength is used for working stress. For Grade 40 and Grade 60 steel, the allowable working stresses are 30,000 and 45,000 psi, respectively. For Welded Wire Fabric (WWF) and Deformed Wire Fabric (DWF), the steel yield strength is 65,000 psi and the allowable working stress is 48,750 psi. The minimum wire size should be adequate so that potential corrosion does not have a significant impact on the cross-sectional area.

Friction Factor. This factor, F, represents the frictional resistance between the bottom of the slab and the top of the underlying subbase or subgrade layer and is basically equivalent to a coefficient of friction. Recommended values for natural subgrade and a variety of subbase materials are presented in Table 2.8.

2.5.2 Continuously Reinforced Concrete Pavements

The principal reinforcement in continuously reinforced concrete pavements (CRCP) is the longitudinal steel which is essentially "continuous" throughout the length of the pavement. This longitudinal reinforcement is used to control cracks which form in the pavement due to volume change in the concrete. The reinforcement may be either reinforcing bars or deformed wire fabric. It is the restraint of the concrete due to the steel reinforcement and subbase friction which causes the concrete to fracture. A balance between the properties of the concrete and the reinforcement must be achieved for the pavement to perform satisfactorily. The evaluation of this interaction forms the basis for longitudinal reinforcement design.

The purpose of transverse reinforcement in a CRC pavement is to control the width of any longitudinal cracks which may form. Transverse reinforcement may not be required for CRC pavements in which no longitudinal cracking is likely to occur based on observed experience of concrete pavements with same soils, aggregate types, etc. However, if longitudinal cracking does occur, transverse reinforcement will restrain lateral movement and minimize the deleterious effects of a free edge. Transverse reinforcement should be designed based on the same criteria and methodology used for jointed pavements.

The following are the requirements for the design of longitudinal steel reinforcement in CRC pavements.

Concrete Tensile Strength. Two measures of concrete tensile strength are used in separate sections of this design procedure. The modulus of rupture (or flexural strength) derived from a flexural beam test (with third point loading) is used for determination of the required slab thickness (see Section 2.3.4). Steel reinforcement design is based on the tensile strength derived from the indirect tensile test which is covered under AASHTO T 198 and ASTM C 496 test specifications. The strength at 28 days should be used for both of these values. Also, these two strengths should be consistent with each other. For this design procedure, the indirect tensile strength will normally be about 86 percent of concrete modulus of rupture.

Concrete Shrinkage. Drying shrinkage in the concrete from water loss is a significant factor in the reinforcement design. Other factors affecting shrinkage include cement content, chemical admixtures, curing method, aggregates, and curing conditions. The value of shrinkage at 28 days is used for the design shrinkage value.

Both shrinkage and strength of the concrete are strongly dependent upon the water-cement ratio. As more water is added to a mix, the potential for shrinkage will increase and the strength will decrease. Since shrinkage can be considered inversely proportional to strength, Table 2.9 may be used as a guide in selecting a value corresponding to the indirect tensile strength determined in Section 2.5.2.

Table 2.8. Recommended Friction Factors (7)

Type of Material Beneath Slab	Friction Factor (F)
Surface treatment	2.2
Lime stabilization	1.8
Asphalt stabilization	1.8
Cement stabilization	1.8
River gravel	1.5
Crushed stone	1.5
Sandstone	1.2
Natural subgrade	0.9

Design Requirements

Table 2.9. Approximate Relationship Between Shrinkage and Indirect Tensile Strength of Portland Cement Concrete (6)

Indirect Tensile Strength (psi)	Shrinkage (in./in.)
300 (or less)	0.0008
400	0.0006
500	0.00045
600	0.0003
700 (or greater)	0.0002

Concrete Thermal Coefficient. The thermal coefficient of expansion for portland cement concrete varies with such factors as water-cement ratio, concrete age, richness of the mix, relative humidity, and the type of aggregate in the mix. In fact, the type of coarse aggregate exerts the most significant influence. Recommended values of PCC thermal coefficient (as a function of aggregate type) are presented in Table 2.10.

Bar or Wire Diameter. Typically, No. 5 and No. 6 deformed bars are used for longitudinal reinforcement in CRCP. The No. 6 bar is the largest practical size that should be used in CRCP to meet bond requirements and to control crack widths. The design nomographs for reinforcement limit the bar selection to a range of No. 4 to No. 7. The nominal diameter of a reinforcing bar, in inches, is simply the bar number divided by 8. The wire diameter should be large enough so that possible corrosion will not significantly reduce the cross section diameter. Also, the relationship between longitudinal and transverse wire should conform to manufacturers' recommendations.

Steel Thermal Coefficient. Unless specific knowledge of the thermal coefficient of the reinforcing steel is known, a value of 5.0×10^{-6} in./in./°F may be assumed for design purposes.

Design Temperature Drop. The temperature drop used in the reinforcement design is the difference between the average concrete curing temperature and a design minimum temperature. The average concrete curing temperature may be taken as the average daily high temperature during the month the pavement is expected to be constructed. This average accounts for the heat of hydration. The design minimum temperature is defined here as the average daily low temperature for the coldest month during the pavement life. If not available, the needed temperature data may be obtained from U.S. Government weather records. The design temperature drop which is entered in the longitudinal reinforcement design procedure is:

$$DT_D = T_H - T_L$$

where

DT_D = design temperature drop, °F,
T_H = average daily high temperature during the month the pavement is constructed, °F, and
T_L = average daily low temperature during the coldest month of the year, °F.

Friction Factor. The criteria for the selection of a slab-base friction factor for CRC pavements is the same as that for jointed pavements (see Section 2.5.1).

Table 2.10. Recommended Value of the Thermal Coefficient of PCC as a Function of Aggregate Types (8)

Type of Coarse Aggregate	Concrete Thermal Coefficient (10^{-6}/°F)
Quartz	6.6
Sandstone	6.5
Gravel	6.0
Granite	5.3
Basalt	4.8
Limestone	3.8

CHAPTER 3
HIGHWAY PAVEMENT STRUCTURAL DESIGN

This chapter describes the application of design procedures for both flexible and rigid highway pavements. Flexible pavement design includes asphalt concrete (AC) surfaces and surface treatments (ST). Rigid pavement design includes plain jointed (JCP), jointed reinforced (JRCP), and continuously reinforced (CRCP) concrete pavements. General criteria are also provided for the design of prestressed concrete pavements (PCP). Pavements designed using these procedures are expected to carry significant levels of traffic and require a paved surface.

With the exception of prestressed concrete pavements, the design procedures in this chapter are based on the original AASHTO pavement performance equations, which have been modified to include design factors not considered in the previous Interim Design Guide. The design process relies exclusively on the design requirements developed in Part II, Chapter 2 and a series of nomographs which solve the design equations. It should be noted that because of the additional complexity, computer-based design procedures for both rigid and flexible pavements need to be treated in separate design manuals. It should also be noted that the design chart procedures presented here do have some inherent assumptions and simplifications which, in some cases, make their solution somewhat less precise than that provided by the corresponding computer solution.

The design approaches for both flexible and rigid pavements permit both traffic and environmental loss of serviceability to be taken into account. If the designer desires that only the serviceability loss due to traffic be considered, then Sections 3.1.3 and 3.2.4 may be ignored.

The basic concept of design for both flexible and rigid pavements is to first determine the required thickness based on the level of traffic. The associated performance period is then corrected for any environmental-associated losses of serviceability. A stage construction option is provided to allow the designer to consider planned rehabilitation for either environmental or economic reasons. Thus, numerous strategies for original design thickness and subsequent rehabilitation may be developed.

Finally, it is strongly recommended that the life-cycle cost economic analysis method described in Part I be used as a basis to compare the alternate pavement designs generated by this design chart procedure for a given pavement type. Because of certain fundamental differences between flexible and rigid pavements and the potential difference in relative costs, it is recommended that this life-cycle economic analysis be a factor, but not be the sole criteria for pavement type selection.

3.1 FLEXIBLE PAVEMENT DESIGN

This section describes the design for both asphalt concrete (AC) pavements and surface treatments (ST) which carry significant levels of traffic (i.e., greater than 50,000 18-kip ESAL) over the performance period. For both the AC and ST surface types, the design is based on identifying a flexible pavement structural number (SN) to withstand the projected level of axle load traffic. It is up to the designer to determine whether a single or double ST or a paved AC surface is required for the specific conditions. An example of the application of the flexible pavement design procedure is presented in Appendix H.

3.1.1 Determine Required Structural Number

Figure 3.1 presents the nomograph recommended for determining the design structural number (SN) required for specific conditions, including

(1) the estimated future traffic, W_{18} (Section 2.1.2), for the performance period,
(2) the reliability, R (Section 2.1.3), which assumes all input is at average value,
(3) the overall standard deviation, S_o (Section 2.1.3),
(4) the effective resilient modulus of roadbed material, M_R (Section 2.3.1), and
(5) the design serviceability loss, $\Delta PSI = p_o - p_t$ (Section 2.2.1).

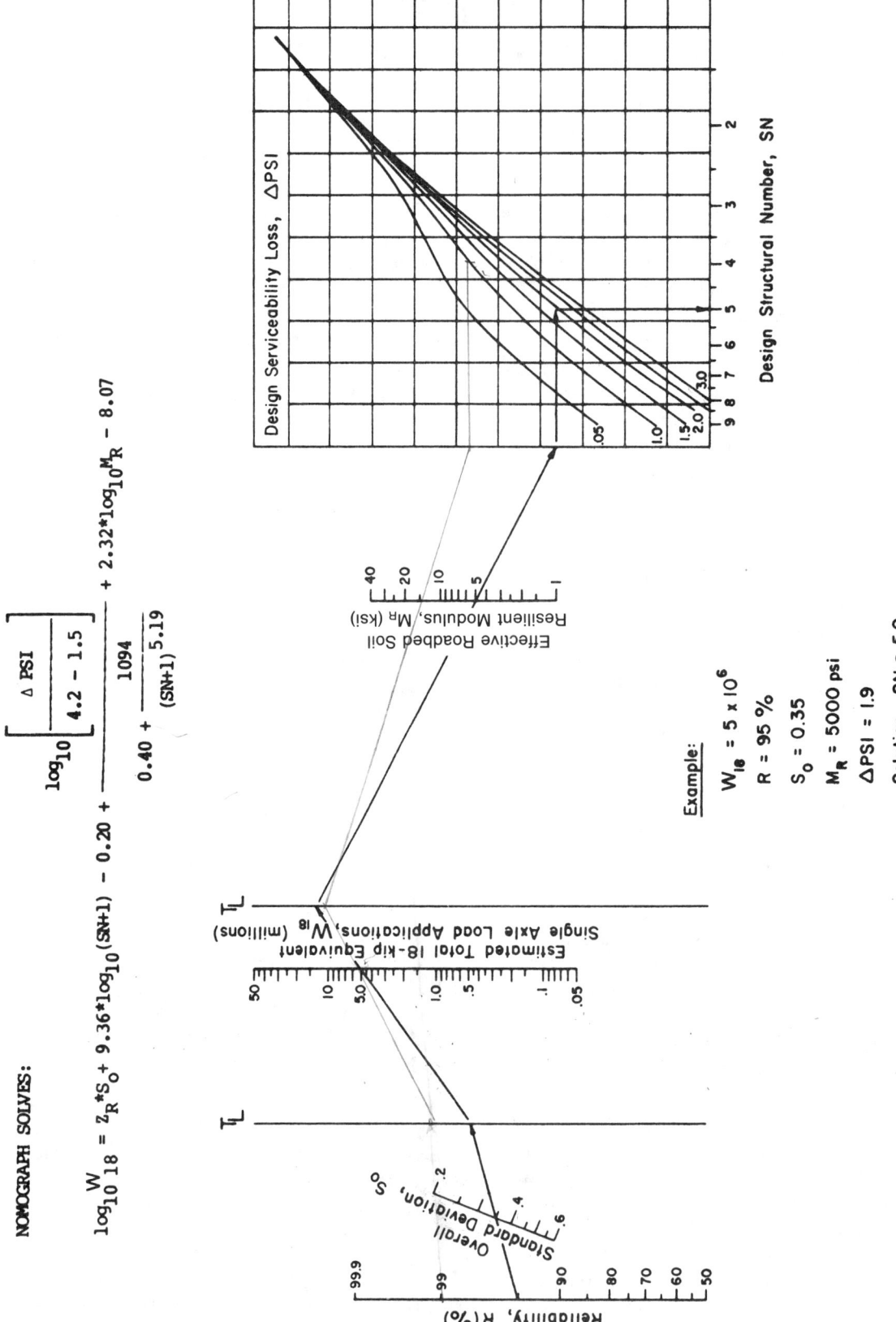

Figure 3.1. Design Chart for Flexible Pavements Based on Using Mean Values for Each Input

Highway Pavement Structural Design

3.1.2 Stage Construction

Experience in some states has shown that regardless of the strength (or load-carrying capacity) of a flexible pavement, there may be a maximum performance period (Section 2.1.1) associated with a given initial structure which is subjected to some significant level of truck traffic. Obviously, if the analysis period (Section 2.1.1) is 20 years (or more) and this practical maximum performance period is less than 20 years, there may be a need to consider stage construction (i.e., planned rehabilitation) in the design analysis. This is especially true if life-cycle economic analyses are to be performed, where the trade-offs between the thickness designs of the initial pavement structure and any subsequent overlays can be evaluated. In such instances, where stage construction alternatives are to be considered, it is important to check the constraint on minimum performance period (Section 2.1.1) within the various candidate strategies. It is also important to recognize the need to compound the reliability for each individual stage of the strategy. For example, if each stage of 3-stage strategy (an initial pavement with two overlays) has a 90-percent reliability, the overall reliability of the design strategy is $0.9 \times 0.9 \times 0.9$ or 72.9 percent. Conversely, if an overall reliability of 95 percent is desired, the individual reliability for each stage must be $(0.95)^{1/3}$ or 98.3 percent. It is important to recognize compounding of reliability may be severe for stage construction, and later opportunities to correct problem areas may be considered.

To evaluate stage construction alternatives, the user should refer to Part III of this Guide which addresses pavement rehabilitation. That Part provides not only a procedure for designing an overlay, but also criteria for the application of other rehabilitation methods that may be used to improve the serviceability and extend the load-carrying capacity of the pavement. The design example in Appendix H provides an illustration of the application of the stage construction approach using a planned future overlay.

3.1.3 Roadbed Swelling and Frost Heave

Roadbed swelling and/or frost heave are both important environmental considerations because of their potential effect on the rate of serviceability loss. Swelling refers to the localized volume changes that occur in expansive roadbed soils as they absorb moisture. A drainage system can be effective in minimizing roadbed swelling if it reduces the availability of moisture for absorption.

Frost heave, as it is treated here, refers to the localized volume changes that occur in the roadbed soil as moisture collects, freezes into ice lenses and produces permanent distortions in the pavement surface. Like swelling, the effects of frost heave can be decreased by providing some type of drainage system. Another effective measure is to provide a layer of nonfrost-susceptible material thick enough to insulate the roadbed from frost penetration. This not only protects against frost heave, but may also significantly reduce or even eliminate the thaw-weakening that occurs in the roadbed soil during early spring.

If either swelling or frost heave are to be considered in terms of their effects on serviceability loss and the need for future overlays, then the following procedure should be applied. It does require the plot of serviceability loss versus time that was developed in Section 2.1.4.

The procedure for considering environmental serviceability loss is similar to the treatment of stage construction strategies because of the planned future need for rehabilitation. In the stage construction approach, the structural number of the initial pavement is selected and its corresponding performance period (service life) determined. An overlay (or series of overlays) which will extend the combined performance periods past the desired analysis period is then identified. The difference in the stage construction approach when swelling and/or frost heave are considered is that an iterative process is required to determine the length of the performance period for each stage of the strategy. The objective of this iterative process is to determine when the combined serviceability loss due to traffic and environment reaches the terminal level. It is described with the aid of Table 3.1.

Step 1. Select an appropriate structural number (SN) for the initial pavement. Because of the relatively small effect the structural number has on minimizing swelling and frost heave, the maximum initial SN recommended is that derived for conditions assuming no swelling or frost heave. For example, if the desired overall reliability is 90 percent (since an overlay is expected, the design reliability for both the initial pavement and overlay is $0.9^{1/2}$ or 95 percent), the effective roadbed soil modulus is 5,000 psi, the initial serviceability expected is 4.4, the design terminal serviceability is 2.5, and a 15-year performance period (along with a corresponding 5 million 18-kip ESAL application) for the initial pavement is as-

Table 3.1. Example of Process Used to Predict the Performance Period of an Initial Pavement Structure Considering Swelling and/or Frost Heave

Initial PSI ___4.4___

Maximum Possible Performance Period (years) ___15___

Design Serviceability Loss, $\Delta PSI = p_o - p_t =$ ___$4.4 - 2.5 = 1.9$___

(1) Iteration No.	(2) Trial Performance Period (years)	(3) Total Serviceability Loss Due to Swelling and Frost Heave $\Delta PSI_{SW,FH}$	(4) Corresponding Serviceability Loss Due to Traffic ΔPSI_{TR}	(5) Allowable Cumulative Traffic (18-kip ESAL)	(6) Corresponding Performance Period (years)
1	13.0	0.73	1.17	2.0×10^6	6.3
2	9.7	0.63	1.27	2.3×10^6	7.2
3	8.5	0.56	1.34	2.6×10^6	8.2

Column No.	Description of Procedures
2	Estimated by the designer (Step 2).
3	Using estimated value from Column 2 with Figure 2.2, the total serviceability loss due to swelling and frost heave is determined (Step 3).
4	Subtract environmental serviceability loss (Column 3) from design total serviceability loss to determine corresponding serviceability loss due to traffic.
5	Determined from Figure 3.1 keeping all inputs constant (except for use of traffic serviceability loss from Column 4) and applying the chart in reverse (Step 5).
6	Using the traffic from Column 5, estimate net performance period from Figure 2.1 (Step 6).

sumed, the maximum structural number (determined from Figure 3.1) that should be considered for swelling/frost heave conditions is 4.4. Anything less than a SN of 4.4 may be appropriate, so long as it does not violate the minimum performance period (Section 2.1.1).

Step 2. Select a trial performance period that might be expected under the swelling/frost heave conditions anticipated and enter in Column 2. This number should be less than the maximum possible performance period corresponding to the selected initial pavement structural number. In general, the greater the environmental loss, the smaller the performance period will be.

Step 3. Using the graph of cumulative environmental serviceability loss versus time developed in Section 2.1.4 (Figure 2.2 is used as an example), estimate the corresponding total serviceability loss due to swelling and frost heave ($\Delta PSI_{SW,FH}$) that can be expected for the trial period from Step 2, and enter in Column 3.

Step 4. Subtract this environmental serviceability loss (Step 3) from the desired total serviceability loss ($4.4 - 2.5 = 1.9$ is used in the example) to establish the corresponding traffic serviceability loss. Enter result in Column 4.

$$\Delta PSI_{TR} = \Delta PSI - \Delta PSI_{SW,FH}$$

Step 5. Use Figure 3.1 to estimate the allowable cumulative 18-kip ESAL traffic corresponding to the traffic serviceability loss determined in Step 4 and enter in Column 5. Note that it is important to use the same levels of reliability, effective roadbed soil resilient modulus, and initial structural number when applying the flexible pavement chart to estimate this allowable traffic.

Step 6. Estimate the corresponding year at which the cumulative 18-kip ESAL traffic (determined in Step 5) will be reached and enter in Column 6. This should be accomplished with the aid of the cumulative traffic versus time plot developed in Section 2.1.2. (Figure 2.1 is used as an example.)

Step 7. Compare the trial performance period with that calculated in Step 6. If the difference is greater than 1 year, calculate the average of the two and use this as the trial value for the start of the next iteration (return to Step 2). If the difference is less than 1 year, convergence is reached and the average is said to be the predicted performance period of the initial pavement structure corresponding to the selected initial SN. In the example, convergence was reached after three iterations and the predicted performance period is about 8 years.

The basis of this iterative process is exactly the same for the estimation of the performance period of any subsequent overlays. The major differences in actual application are that (1) the overlay design methodology presented in Part III is used to estimate the performance period of the overlay and (2) any swelling and/or frost heave losses predicted after overlay should restart and then progress from the point in time when the overlay was placed.

3.1.4 Selection of Layer Thicknesses

Once the design structural number (SN) for an initial pavement structure is determined, it is necessary to identify a set of pavement layer thicknesses which, when combined, will provide the load-carrying capacity corresponding to the design SN. The following equation provides the basis for converting SN into actual thicknesses of surfacing, base and subbase:

$$SN = a_1 D_1 + a_2 D_2 m_2 + a_3 D_3 m_3$$

where

a_1, a_2, a_3 = layer coefficients representative of surface, base, and subbase courses, respectively (see Section 2.3.5),

D_1, D_2, D_3 = actual thicknesses (in inches) of surface, base, and subbase courses, respectively, and

m_2, m_3 = drainage coefficients for base and subbase layers, respectively (see Section 2.4.1).

The SN equation does not have a single unique solution; i.e., there are many combinations of layer thicknesses that are satisfactory solutions. The thickness of the flexible pavement layers should be rounded to the nearest 1/2 inch. When selecting appropriate values for the layer thicknesses, it is necessary to consider their cost effectiveness along with the construction and maintenance constraints in order to avoid the possibility of producing an impractical design. From a cost-effective view, if the ratio of costs for layer 1 to layer 2 is less than the corresponding ratio of layer coefficients times the drainage coefficient, then the optimum economical design is one where the minimum base thickness is used. Since it is generally impractical and uneconomical to place surface, base, or subbase courses of less than some minimum thickness, the following are provided as minimum practical thicknesses for each pavement course:

Minimum Thickness (inches)

Traffic, ESAL's	Asphalt Concrete	Aggregate Base
Less than 50,000	1.0 (or surface treatment)	4
50,001–150,000	2.0	4
150,001–500,000	2.5	4
500,001–2,000,000	3.0	6
2,000,001–7,000,000	3.5	6
Greater than 7,000,000	4.0	6

Because such minimums depend somewhat on local practices and conditions, individual design agencies may find it desirable to modify the above minimum thicknesses for their own use.

Individual agencies should also establish the effective thicknesses and layer coefficients of both single and double surface treatments. The thickness of the surface treatment layer may be neglectible in computing SN, but its effect on the base and subbase properties may be large due to reductions in surface water entry.

3.1.5 Layered Design Analysis

It should be recognized that, for flexible pavements, the structure is a layered system and should be designed accordingly. The structure should be designed in accordance with the principles shown in Figure 3.2. First, the structural number required over the roadbed soil should be computed. In the same way, the structural number required over the subbase layer and the base layer should also be computed, using the applicable strength values for each. By working with differences between the computed structural numbers

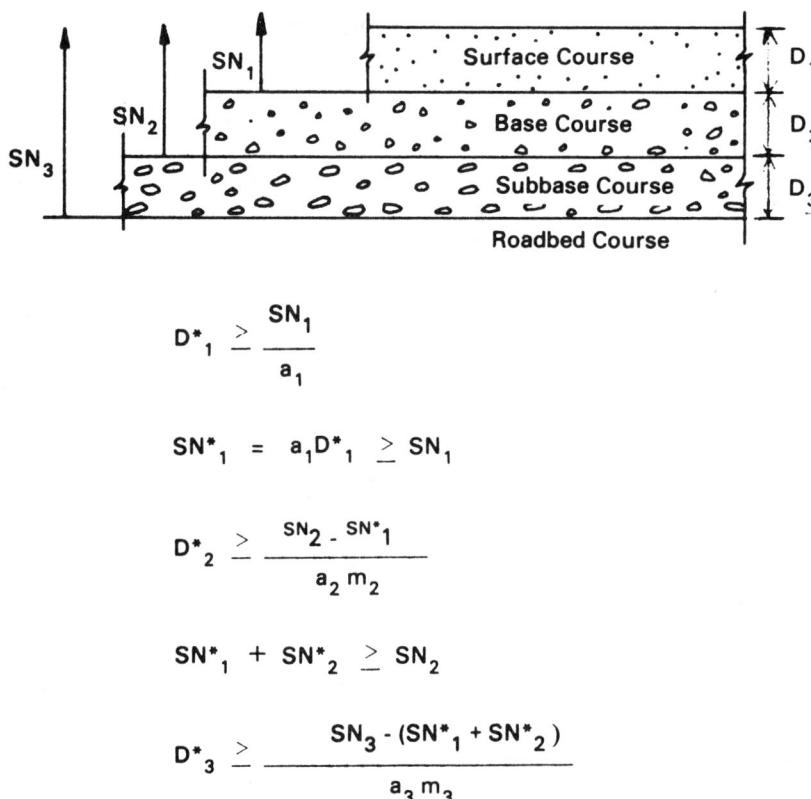

1) a, D, m and SN are as defined in the text and are minimum required values.

2) An asterisk with D or SN indicates that it represents the value actually used, which must be equal to or greater than the required value.

Figure 3.2. Procedure for Determining Thicknesses of Layers Using a Layered Analysis Approach

required over each layer, the maximum allowable thickness of any given layer can be computed. For example, the maximum allowable structural number for the subbase material would be equal to the structural number required over the subbase subtracted from the structural number required over the roadbed soil. In a like manner, the structural numbers of the other layers may be computed. The thicknesses for the respective layers may then be determined as indicated on Figure 3.2.

It should be recognized that this procedure should not be applied to determine the SN required above subbase or base materials having a modulus greater than 40,000 psi. For such cases, layer thicknesses of materials above the "high" modulus layer should be established based on cost effectiveness and minimum practical thickness considerations.

3.2 RIGID PAVEMENT DESIGN

This section describes the design for portland cement concrete pavements, including plain jointed (JCP), jointed reinforced (JRCP), and continuously reinforced (CRCP). As in the design for flexible pavements, it is assumed that these pavements will carry traffic levels in excess of 50,000 18-kip ESAL over the performance period. An example of the application of this rigid pavement design procedure is presented in Appendix L.

The AASHTO design procedure is based on the AASHO Road Test pavement performance algorithm. Inherent in the use of the procedure is the use of dowels at transverse joints. Hence, joint faulting was not a distress manifestation at the Road Test. If the designer wishes to consider nondowelled joints, he may develop an appropriate J-factor (see Section 2.4.2, "Load Transfer") or check his design with another agency's procedure, such as the PCA procedure (9).

3.2.1 Develop Effective Modulus of Subgrade Reaction

Before the design chart for determining design slab thickness can be applied, it is necessary to estimate the possible levels of slab support that can be provided. This is accomplished using Table 3.2 and Figures 3.3, 3.4, 3.5, and 3.6 to develop an effective modulus of subgrade reaction, k. An example of this process is demonstrated in Table 3.3.

Since the effective k-value is dependent upon several different factors besides the roadbed soil resilient modulus, the first step is to identify the combinations (or levels) that are to be considered and enter them in the heading of Table 3.2.

(1) Subbase types—Different types of subbase have different strengths or modulus values. The consideration of a subbase type in estimating an effective k-value provides a basis for evaluating its cost-effectiveness as part of the design process.

(2) Subbase thicknesses (inches)—Potential design thicknesses for each subbase type should also be identified, so that its cost-effectiveness may be considered.

(3) Loss of support, LS—This factor, quantified in Section 2.4.3, is used to correct the effective k-value based on potential erosion of the subbase material.

(4) Depth to rigid foundation (feet)—If bedrock lies within 10 feet of the surface of the subgrade for any significant length along the project, its effect on the overall k-value and the design slab thickness for that segment should be considered.

For each combination of these factors that is to be evaluated, it is necessary to prepare a separate table and develop a corresponding effective modulus of subgrade reaction.

The second step of the process is to identify the seasonal roadbed soil resilient modulus values (from Section 2.3.1) and enter them in Column 2 of each table. As before, if the length of the smallest season is one-half month, then all seasons must be defined in terms of consecutive half-month time intervals in the table. (The same seasonal roadbed soil resilient modulus values used for the example in Section 2.3.1 are used in the example presented in Table 3.3.)

The third step in estimating the effective k-value is to assign subbase elastic (resilient) modulus (E_{SB}) values for each season. These values, which were discussed in Section 2.3.3, should be entered in Column 3 of Table 3.2 and should correspond to those for the seasons used to develop the roadbed soil resilient modulus values. For those types of subbase material which are insensitive to season (e.g., cement-treated material), a constant value of subbase modulus may be assigned for each season. For those unbound materials which are sensitive to season but were not tested for the extreme conditions, values for E_{SB} of 50,000 psi and 15,000 psi may be used for the frozen and spring thaw periods, respectively. For unbound materials, the ratio of the subbase to the roadbed soil resilient

Table 3.2. Table for Estimating Effective Modulus of Subgrade Reaction

Trial Subbase: Type _____ Depth to Rigid Foundation (feet) _____

Thickness (inches) _____ Projected Slab Thickness (inches) _____

Loss of Support, LS _____

(1) Month	(2) Roadbed Modulus, M_R (psi)	(3) Subbase Modulus, E_{SB} (psi)	(4) Composite k-Value (pci) (Fig. 3.3)	(5) k-Value (pci) on Rigid Foundation (Fig. 3.4)	(6) Relative Damage, u_r (Fig. 3.5)
Jan.					
Feb.					
Mar.					
Apr.					
May					
June					
July					
Aug.					
Sept.					
Oct.					
Nov.					
Dec.					

Average: $\bar{u}_r = \dfrac{\Sigma u_r}{n} =$ _____   Summation: $\Sigma u_r =$ _____

Effective Modulus of Subgrade Reaction, k (pci) = _____

Corrected for Loss of Support: k (pci) = _____

Highway Pavement Structural Design

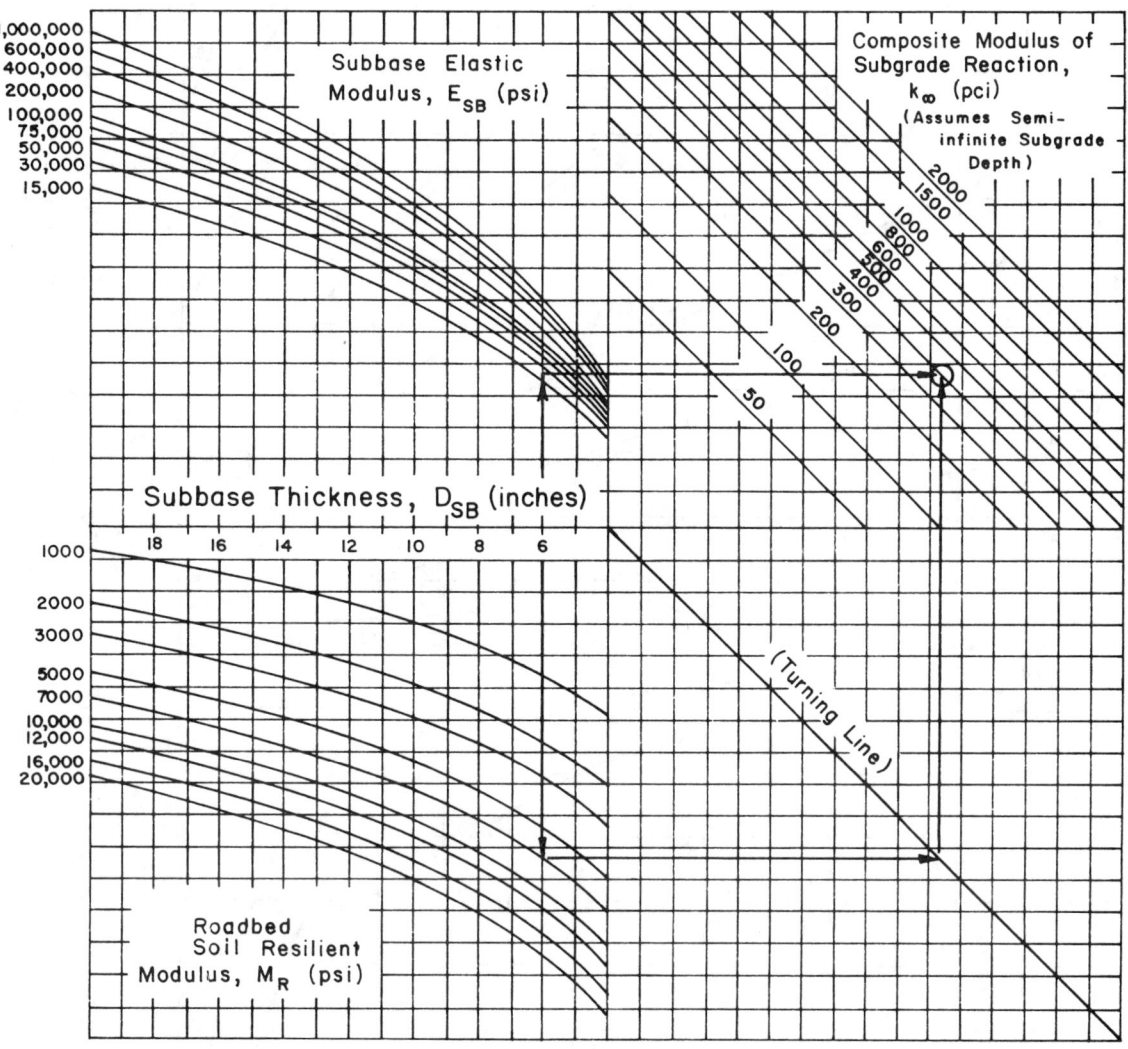

Figure 3.3. Chart for Estimating Composite Modulus of Subgrade Reaction, k_∞, Assuming a Semi-Infinite Subgrade Depth. (For practical purposes, a semi-infinite depth is considered to be greater than 10 feet below the surface of the subgrade.)

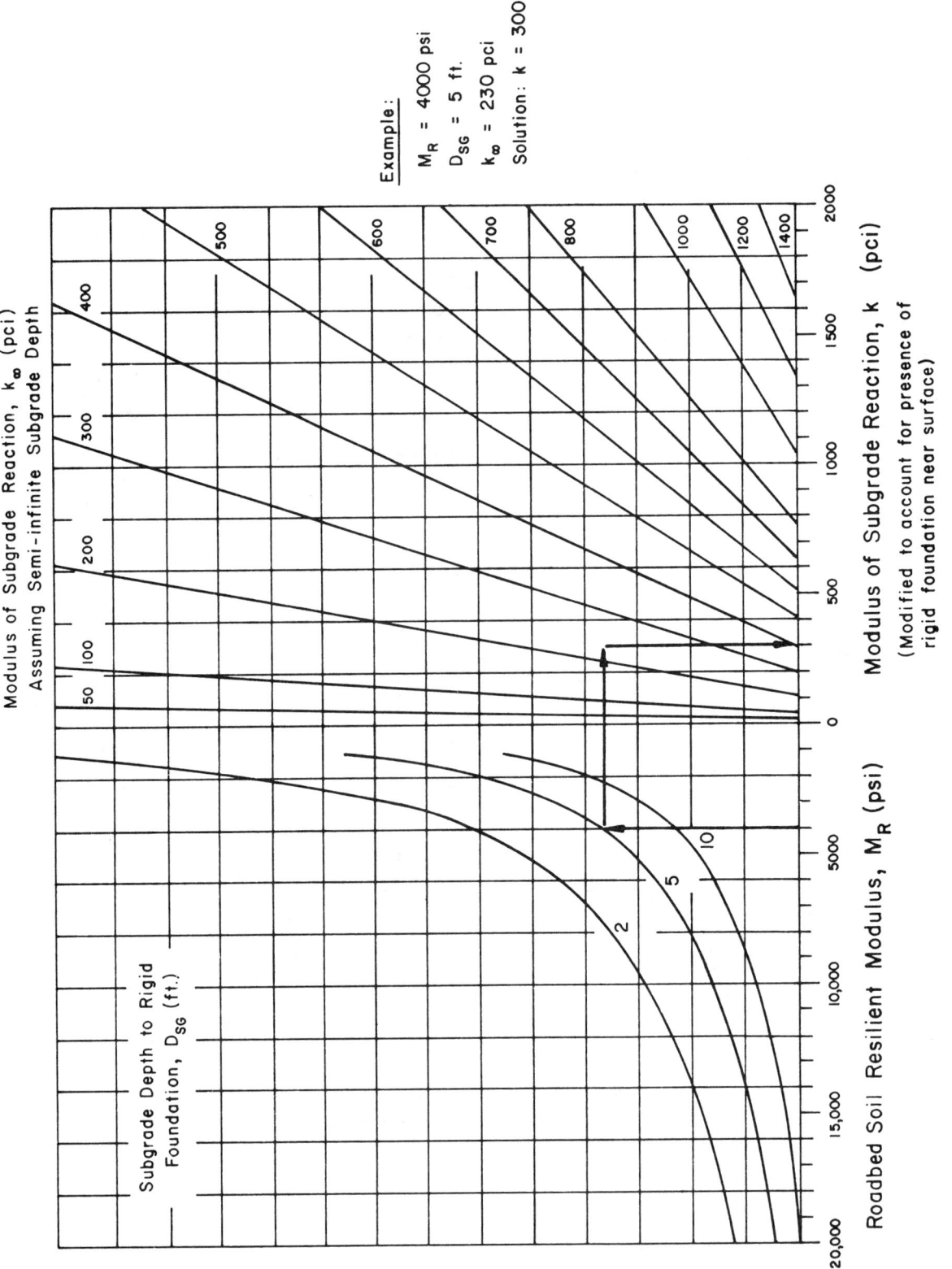

Figure 3.4. Chart to Modify Modulus of Subgrade Reaction to Consider Effects of Rigid Foundation Near Surface (within 10 feet)

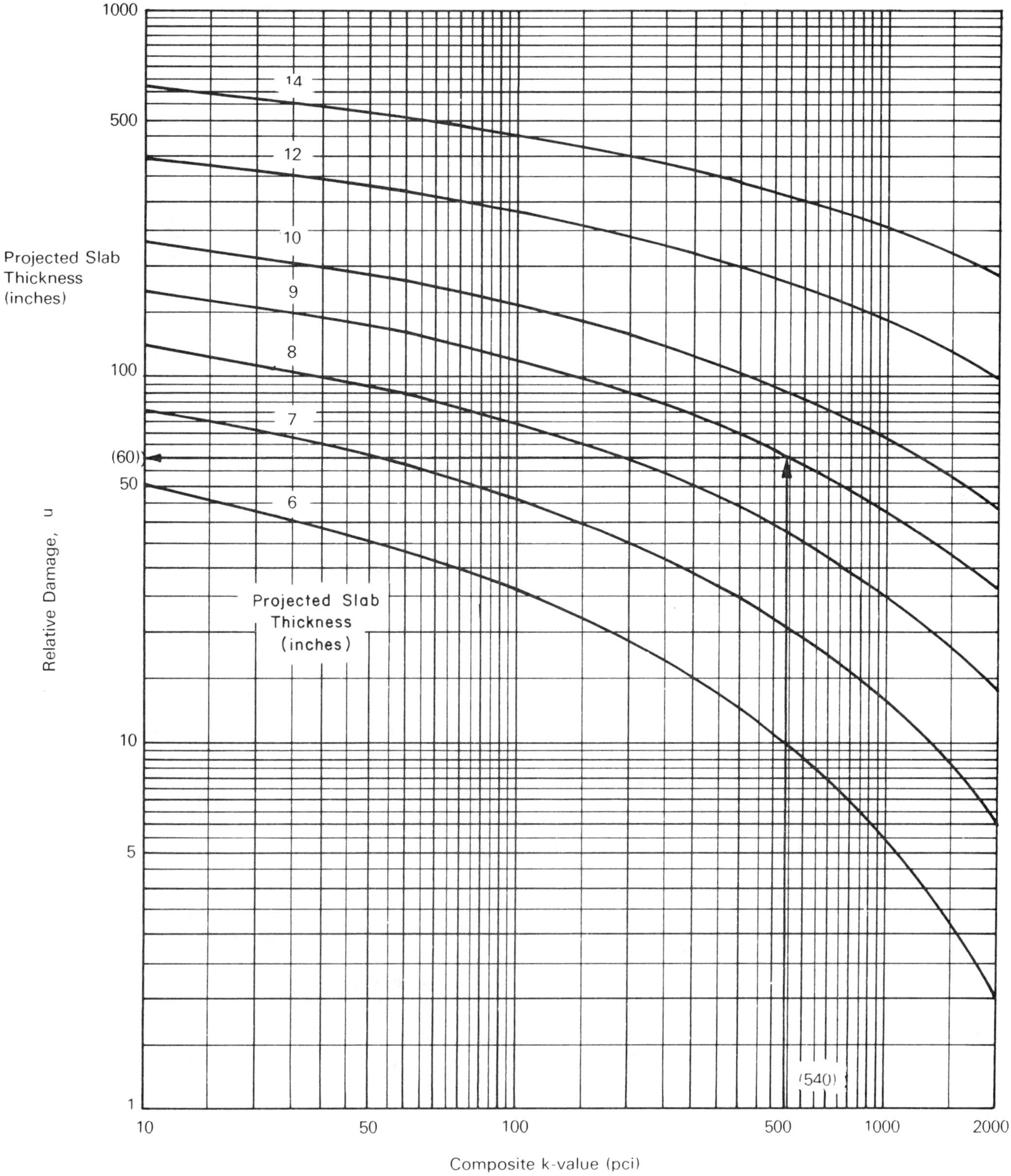

Figure 3.5. Chart for Estimating Relative Damage to Rigid Pavements Based on Slab Thickness and Underlying Support

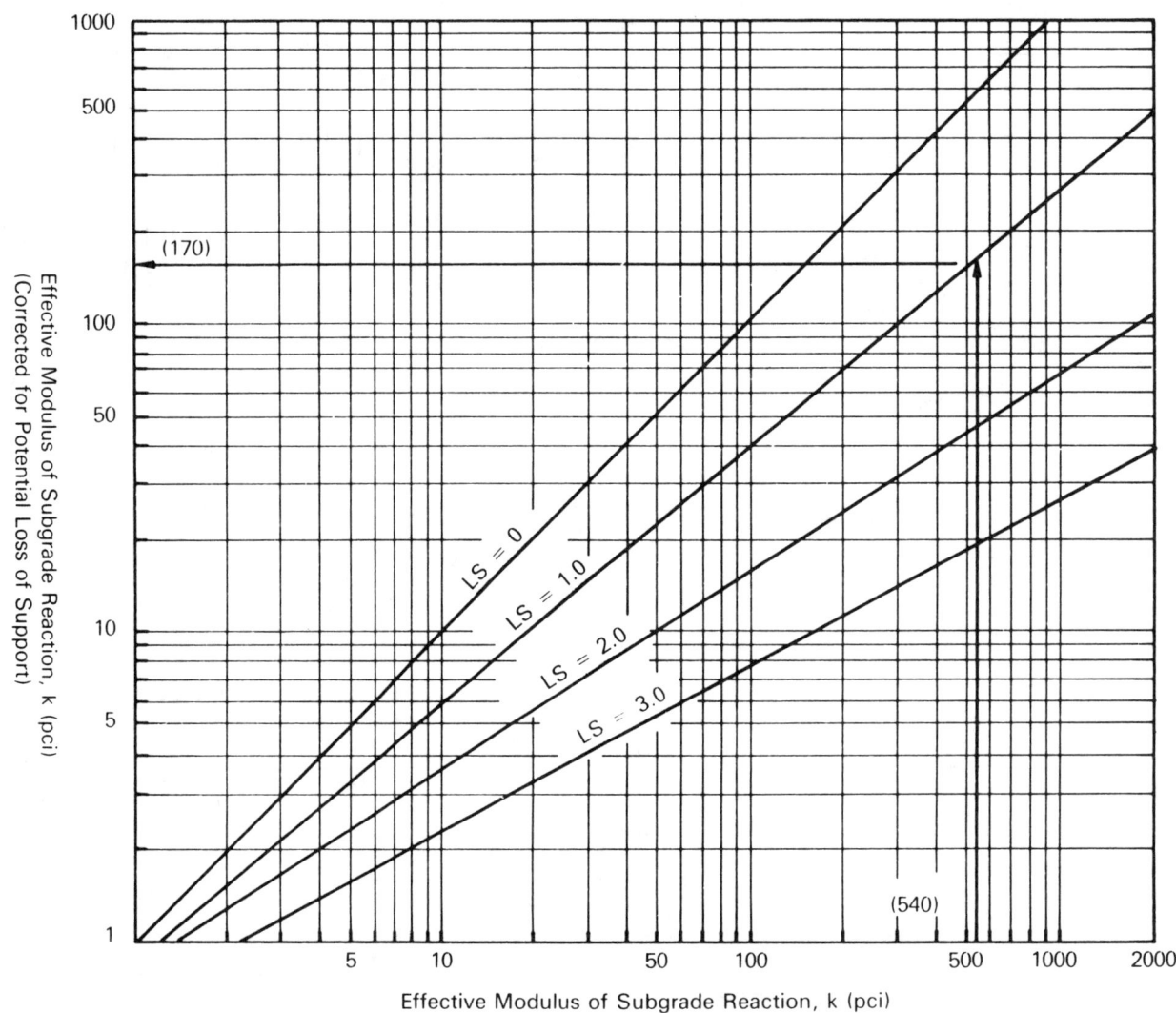

Figure 3.6. Correction of Effective Modulus of Subgrade Reaction for Potential Loss of Subbase Support (6)

Highway Pavement Structural Design

Table 3.3. Example Application of Method for Estimating Effective Modulus of Subgrade Reaction

Trial Subbase: Type ___Granular___   Depth to Rigid Foundation (feet) ___5___
Thickness (inches) ___6___   Projected Slab Thickness (inches) ___9___
Loss of Support, LS ___1.0___

(1)	(2)	(3)	(4)	(5)	(6)
Month	Roadbed Modulus, M_R (psi)	Subbase Modulus, E_{SB} (psi)	Composite k-Value (pci) (Fig. 3.3)	k-Value (pci) on Rigid Foundation (Fig. 3.4)	Relative Damage, u_r (Fig. 3.5)
Jan.	20,000	50,000	1,100	1,350	0.35
Feb.	20,000	50,000	1,100	1,350	0.35
Mar.	2,500	15,000	160	230	0.86
Apr.	4,000	15,000	230	300	0.78
May	4,000	15,000	230	300	0.78
June	7,000	20,000	410	540	0.60
July	7,000	20,000	410	540	0.60
Aug.	7,000	20,000	410	540	0.60
Sept.	7,000	20,000	410	540	0.60
Oct.	7,000	20,000	410	540	0.60
Nov.	4,000	15,000	230	300	0.78
Dec.	20,000	50,000	1,100	1,350	0.35

Average: $\bar{u}_r = \dfrac{\Sigma u_r}{n} = \dfrac{7.25}{12} = 0.60$ Summation: $\Sigma u_r = $ 7.25

Effective Modulus of Subgrade Reaction, k (pci) = ___540___
Corrected for Loss of Support: k (pci) = ___170___

modulus should not exceed 4 to prevent an artificial condition.

The fourth step is to estimate the composite modulus of subgrade reaction for each season, assuming a semi-infinite subgrade depth (i.e., depth to bedrock greater than 10 feet) and enter in Column 4. This is accomplished with the aid of Figure 3.3. Note that the starting point in this chart is subbase thickness, D_{SB}. If the slab is placed directly on the subgrade (i.e., no subbase), the composite modulus of subgrade reaction is defined using the following theoretical relationship between k-values from a plate bearing test and elastic modulus of the roadbed soil:

$$k = M_R/19.4$$

NOTE: The development of this relationship is described as part of Volume 2, Appendix HH.

The fifth step is to develop a k-value which includes the effect of a rigid foundation near the surface. This step should be disregarded if the depth to a rigid foundation is greater than 10 feet. Figure 3.4 provides the chart that may be used to estimate this modified k-value for each season. It considers roadbed soil resilient modulus and composite modulus of subgrade reaction, as well as the depth to the rigid foundation. The values for each modified k-value should subsequently be recorded in Column 5 of Table 3.2.

The sixth step in the process is to estimate the thickness of the slab that will be required, and then use Figure 3.5 to determine the relative damage, u_r, in each season and enter them in Column 6 of Table 3.2.

The seventh step is to add all the u_r values (Column 6) and divide the total by the number of seasonal increments (12 or 24) to determine the average relative damage, u_r. The effective modulus of subgrade reaction, then, is the value corresponding to the average relative damage (and projected slab thickness) in Figure 3.5.

The eighth and final step in the process is to adjust the effective modulus of subgrade reaction to account for the potential loss of support arising from subbase erosion. Figure 3.6 provides the chart for correcting the effective modulus of subgrade reaction based on the loss of support factor, LS, determined in Section 2.4.3. Space is provided in Table 3.2 to record this final design k-value.

3.2.2 Determine Required Slab Thickness

Figure 3.7 (in 2 segments) presents the nomograph used for determining the slab thickness for each effective k-value identified in the previous section. The designer may then select the optimum combination of slab and subbase thicknesses based on economics and other agency policy requirements. Generally, the layer thickness is rounded to the nearest inch, but the use of controlled grade slip form pavers may permit ½-inch increments. In addition to the design k-value, other inputs required by this rigid pavement design nomograph include:

(1) the estimated future traffic, W_{18} (Section 2.1.2), for the performance period,
(2) the reliability, R (Section 2.1.3),
(3) the overall standard deviation, S_o (Section 2.1.3),
(4) design serviceability loss, $\Delta PSI = p_i - p_t$ (Section 2.2.1),
(5) concrete elastic modulus, E_c (Section 2.3.3),
(6) concrete modulus of rupture, S'_c (Section 2.3.4),
(7) load transfer coefficient, J (Section 2.4.2), and
(8) drainage coefficient, C_d (Section 2.4.1).

3.2.3 Stage Construction

Experience in some states has shown that there may be a practical maximum performance period (Section 2.1.1) associated with a given rigid pavement which is subjected to some significant level of truck traffic. To consider analysis periods which are longer than this maximum expected performance period or to more rigorously consider the life-cycle costs of rigid pavement designs which are initially thinner, it is necessary to consider the stage construction (planned rehabilitation) approach in the design process. It is also important to recognize the need to compound the reliability for each individual stage of the strategy. For example, if both stages of a two-stage strategy (an initial PCC pavement with one overlay) have a 90-percent reliability, the overall reliability of the design strategy would be 0.9 × 0.9 or 81 percent. Conversely, if an overall reliability of 95 percent is desired, the individual reliability for each stage must be $(0.95)^{1/2}$ or 97.5 percent.

To evaluate secondary stages of such stage construction alternatives, the user should refer to Part III of this Guide which addresses the design for pavement

Highway Pavement Structural Design

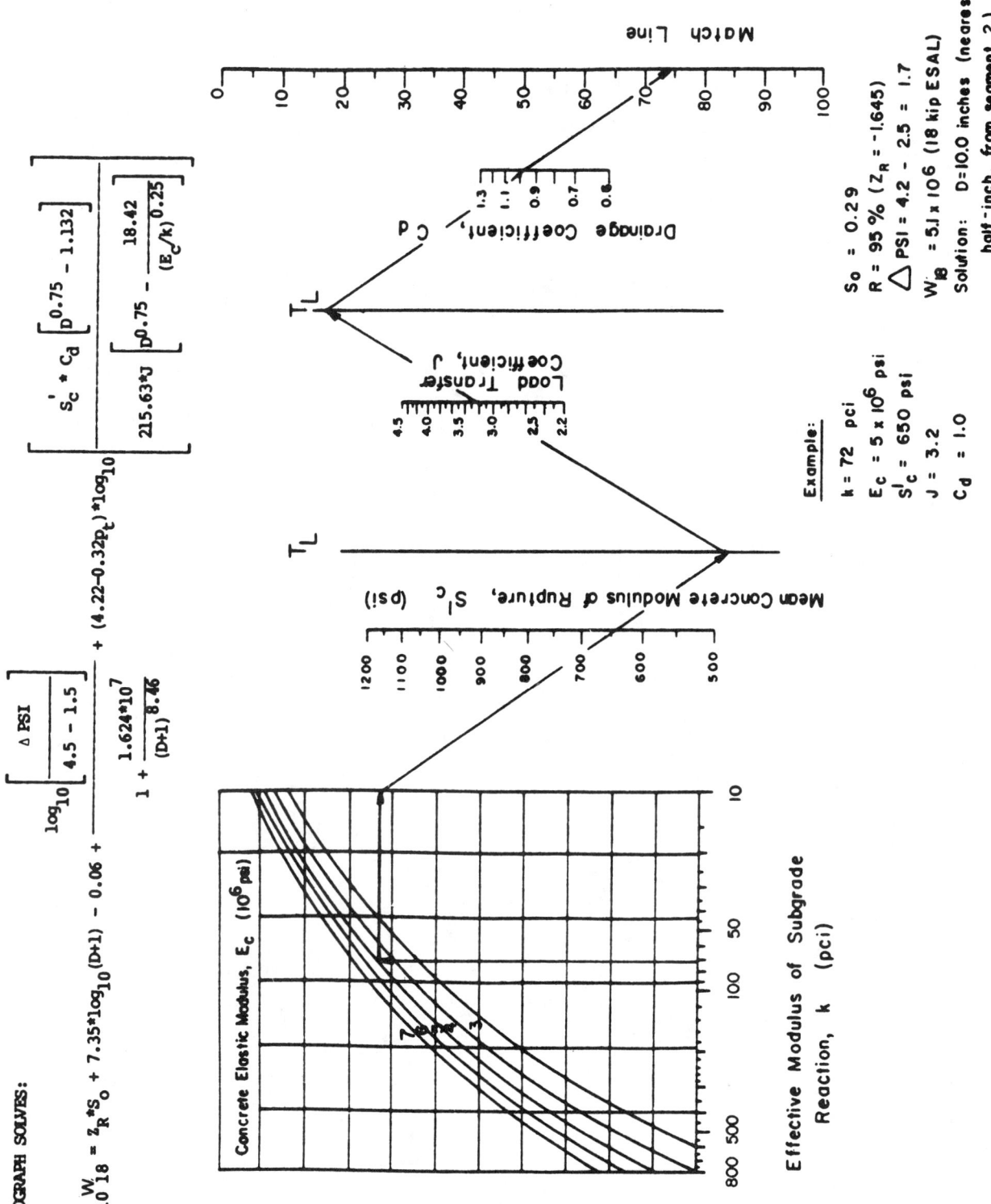

Figure 3.7. Design Chart for Rigid Pavement Based on Using Mean Values for Each Input Variable (Segment 1)

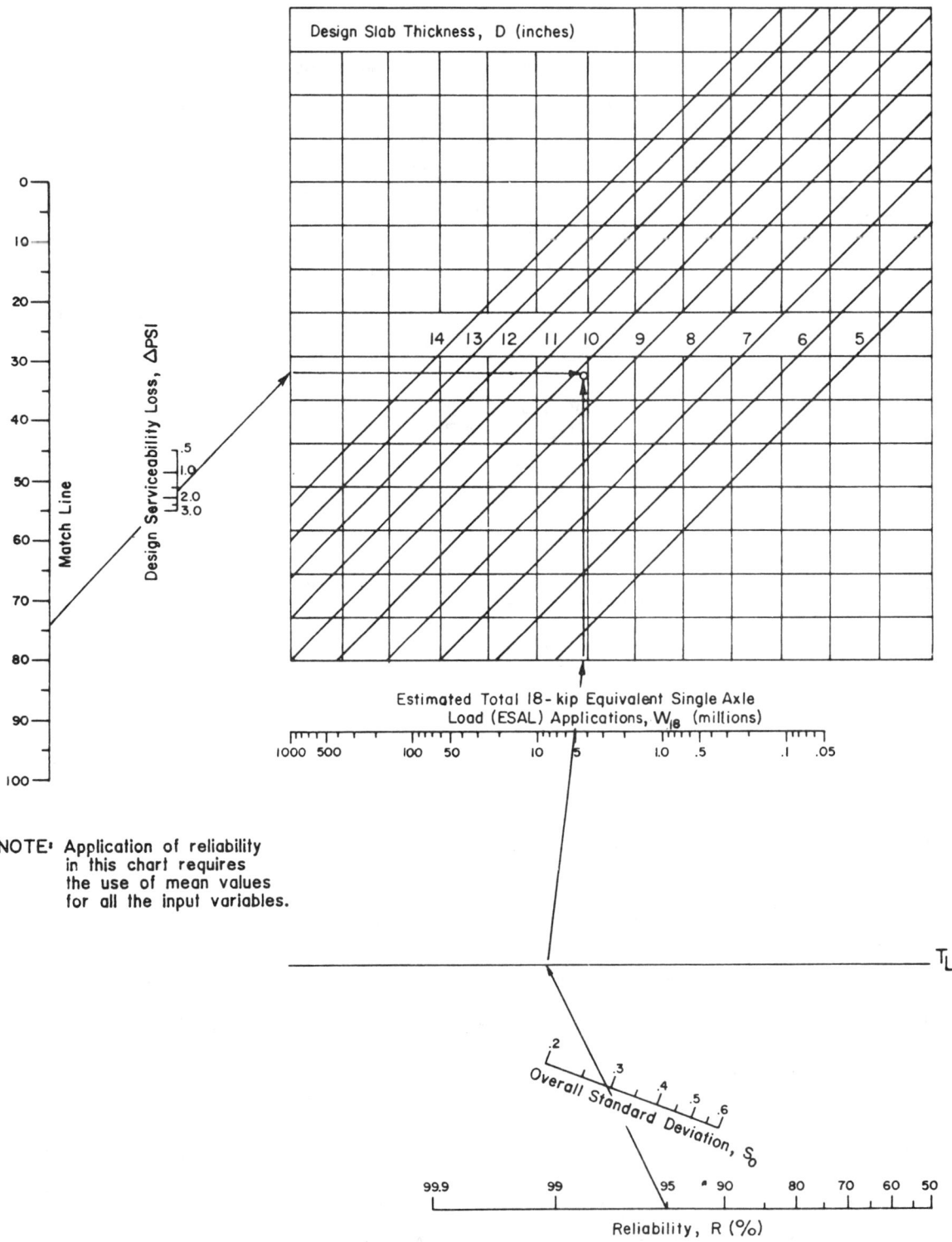

Figure 3.7. Continued—Design Chart for Rigid Pavements Based on Using Mean Values for Each Input Variable (Segment 2)

rehabilitation. That part not only provides a procedure for designing overlays, but also provides criteria for the application of other rehabilitation methods that may be used to improve the serviceability and extend the load-carrying capacity of the pavement. The design example in Appendix I provides an illustration of the application of the stage construction approach using a planned future overlay.

3.2.4 Roadbed Swelling and Frost Heave

The approach to considering the effects of swelling and frost heave in rigid pavement design is almost identical to that for flexible pavements (Section 3.1.3). Thus, some of the discussion is repeated here.

Roadbed swelling and frost heave are both important environmental considerations because of their potential effect on the rate of serviceability loss. Swelling refers to the localized volume changes that occur in expansive roadbed soils as they absorb moisture. A drainage system can be effective in minimizing roadbed swelling if it reduces the availability of moisture for absorption.

Frost heave, as it is treated here, refers to the localized volume changes that occur in the roadbed as moisture collects, freezes into ice lenses, and produces distortions on the pavement surface. Like swelling, the effects of frost heave can be decreased by providing some type of drainage system. Perhaps a more effective measure is to provide a layer of non-frost-susceptible material thick enough to insulate the roadbed soil from frost penetration. This not only protects against frost heave, but also significantly reduces or even eliminates the thaw-weakening that may occur in the roadbed soil during early spring.

If either swelling or frost heave is to be considered in terms of their effects on serviceability loss and the need for future overlays, then the following procedure should be applied. It requires the plot of serviceability loss versus time developed in Section 2.1.4.

The procedure for considering environmental serviceability loss is similar to the treatment of stage construction strategies because of the planned future need for rehabilitation. In the stage construction approach, an initial PCC slab thickness is selected and the corresponding performance period (service life) determined. An overlay (or series of overlays) which will extend the combined performance periods past the desired analysis period is then identified. The difference in the stage construction approach when swelling and/or frost heave are considered is that an iterative process is required to determine the length of the performance period for each stage of the strategy. The objective of this iterative process is to determine when the combined serviceability loss due to traffic and environment reaches the terminal level. This is described with the aid of Table 3.4.

Step 1. Select an appropriate slab thickness for the initial pavement. Because of the relatively small effect slab thickness has on minimizing swelling and frost heave, the maximum initial thickness recommended is that derived for conditions assuming no swelling or frost heave. Referring to the example problem presented in Figure 3.7, the maximum feasible slab thickness is 9.5 inches. Any practical slab thickness less than this value may be appropriate for swelling or frost heave conditions, so long as it does not violate the minimum performance period (Section 2.1.1).

It is important to note here that for this example, an overall reliability of 90 percent is desired. Since it is expected that one overlay will be required to reach the 20-year analysis period, the individual reliability that must be used for the design of both the initial pavement and the overlay is $0.90^{1/2}$ or 95 percent.

Step 2. Select a trial performance period that might be expected under the swelling/frost heave conditions anticipated and enter in Column 2. This number should be less than the maximum possible performance period corresponding to the selected initial slab thickness. In general, the greater the environmental loss, the smaller the performance period will be.

Step 3. Using the graph of cumulative environmental serviceability loss versus time developed in Section 2.1.4 (Figure 2.2 is used as an example), estimate the corresponding total environmental serviceability loss due to swelling and frost heave ($\Delta PSI_{SW,FH}$) that can be expected for the trial period from Step 2 and enter in Column 3.

Step 4. Subtract this environmental serviceability loss (Step 3) from the desired total serviceability loss (4.2 − 2.5 = 1.7 used in the example) to establish the corresponding traffic serviceability loss. Enter in Column 4.

$$\Delta PSI_{TR} = \Delta PSI - \Delta PSI_{SW,FH}$$

Step 5. Use Figure 3.7 to estimate the allowable cumulative 18-kip ESAL traffic corresponding to the

Table 3.4. Example of Process Used to Predict the Performance Period of an Initial Rigid Pavement Structure Considering Swelling and/or Frost Heave

Slab Thickness (inches) ___9.5___

Maximum Possible Performance Period (years) ___20___

Design Serviceability Loss, $\Delta PSI = p_i - p_t =$ ___4.2 − 2.5 = 1.7___

(1) Iteration No.	(2) Trial Performance Period (years)	(3) Total Serviceability Loss Due to Swelling and Frost Heave $\Delta PSI_{SW,FH}$	(4) Corresponding Serviceability Loss Due to Traffic ΔPSI_{TR}	(5) Allowable Cumulative Traffic (18-kip ESAL)	(6) Corresponding Performance Period (years)
1	14.0	0.75	0.95	3.1×10^6	9.6
2	11.8	0.69	1.01	3.3×10^6	10.2
3	11.0	0.67	1.03	3.4×10^6	10.4

Column No.	Description of Procedures
2	Estimated by the designer (Step 2).
3	Using estimated value from Column 2 with Figure 2.2, the total serviceability loss due to swelling and frost heave is determined (Step 3).
4	Subtract environmental serviceability loss (Column 3) from design total serviceability loss to determine corresponding serviceability loss due to traffic.
5	Determined from Figure 3.5 keeping all inputs constant (except for use of traffic serviceability loss from Column 4) and applying the chart in reverse (Step 5).
6	Using the traffic from Column 5, estimate net performance period from Figure 2.1 (Step 6).

traffic serviceability loss determined in Step 4 and enter in Column 5. Note that it is important to use the same levels of reliability, effective modulus of subgrade reaction, etc., when applying the rigid pavement design chart to estimate the allowable traffic.

Step 6. Estimate the corresponding year at which the cumulative 18-kip ESAL traffic (determined in Step 5) will be reached and enter in Column 6. This should be accomplished with the aid of the cumulative traffic versus time plot developed in Section 2.1.2. (Figure 2.1 is used as an example.)

Step 7. Compare the trial performance period with that calculated in Step 6. If the difference is greater than 1 year, calculate the average of the two and use this as the trial value for the start of the next iteration (return to Step 2). If the difference is less than 1 year, convergence is reached and the average is said to be the predicted performance period of the initial pavement structure corresponding to the selected design slab thickness. In the example, convergence was reached after three iterations and the predicted performance period is about 10.5 years.

The basis of this iterative process is exactly the same for the estimation of the performance period of any subsequent overlays. The major differences in actual application are that (1) the overlay design methodology presented in Part III is used to estimate the performance period of the overlay, and (2) any swelling and/or frost heave losses predicted after overlay should restart and then progress from the point when the overlay was placed.

3.3 RIGID PAVEMENT JOINT DESIGN

This section covers the design considerations for the different types of joints in portland cement concrete pavements. This criteria is applicable to the design of joints in both jointed and continuous pavements.

3.3.1 Joint Types

Joints are placed in concrete pavements to permit expansion and contraction of the pavement, thereby relieving stresses due to environmental changes (i.e., temperature and moisture), friction, and to facilitate construction. There are three general types of joints: contraction, expansion, and construction. These joints and their functions are as follows:

(1) Contraction or weakened-plane (dummy) joints are provided to relieve the tensile stresses due to temperature, moisture, and friction, thereby controlling cracking. If contraction joints were not installed, random cracking would occur on the surface of the pavement.

(2) The primary function of an expansion joint is to provide space for the expansion of the pavement, thereby preventing the development of compressive stresses, which can cause the pavement to buckle.

(3) Construction joints are required to facilitate construction. The spacing between longitudinal joints is dictated by the width of the paving machine and by the pavement thickness.

3.3.2 Joint Geometry

The joint geometry is considered in terms of the spacing and general layout.

Joint Spacing. In general, the spacing of both transverse and longitudinal contraction joints depends on local conditions of materials and environment, whereas expansion and construction joints are primarily dependent on layout and construction capabilities. For contraction joints, the spacing to prevent intermediate cracking decreases as the thermal coefficient, temperature change, or subbase frictional resistance increases; and the spacing increases as the concrete tensile strength increases. The spacing also is related to the slab thickness and the joint sealant capabilities. At the present time, the local service records are the best guide for establishing a joint spacing that will control cracking. Local experience must be tempered since a change in coarse aggregate type may have a significant impact on the concrete thermal coefficient and consequently, the acceptable joint spacing. As a rough guide, the joint spacing (in feet) for plain concrete pavements should not greatly exceed twice the slab thickness (in inches). For example, the maximum joint spacing for an 8-inch slab is 16 feet. Also, as a general guideline, the ratio of slab width to length should not exceed 1.25.

The use of expansion joints is generally minimized on a project due to cost, complexity, and performance problems. They are used at structures where pavement types change (e.g., CRCP to jointed), with prestressed pavements, and at intersections.

The spacing between construction joints is generally dictated by field placement and equipment capabilities. Longitudinal construction joints should be placed at lane edges to maximize pavement smoothness and minimize load transfer problems. Transverse construction joints occur at the end of a day's placement or in connection with equipment breakdowns.

Joint Layout. Skewing and randomization of joints minimize the effect of joint roughness, thereby improving the pavement riding quality.

Skewed transverse joints will improve joint performance and extend the life or rigid pavements, i.e., plain or reinforced, doweled, or undoweled. The joint is skewed sufficiently so that wheel loads of each axle cross the joint one at a time. The obtuse angle at the outside pavement edge should be ahead of the joint in the direction of traffic since that corner receives the greatest impact from the sudden application of wheel loads. Skewed joints have these advantages:

(1) reduced deflection and stress at joints, thereby increasing the load-carrying capacity of the slab and extending pavement life, and

(2) less impact reaction in vehicles as they cross the joints, and hence a smoother ride if the joints have some roughness.

A further refinement for improving performance of plain pavements is to use skewed joints at randomized or irregular spacings. Randomized spacing patterns prevent rhythmic or resonant responses in vehicles moving at normal rural expressway speeds. Research at a motor vehicle proving ground indicated that slab spacing patterns of 7.5 feet should be avoided.

Joint Dimensions. The width of the joint is controlled by the joint sealant extension and is covered in Section 2.4.6, "Joint Sealant Dimensions." The depth of contraction joints should be adequate enough to ensure that cracking occurs at the desired location rather than in a random pattern. Normally, the depth of transverse contraction joints should be $1/4$ of the slab thickness, and longitudinal joints $1/3$ of the thickness. These joints may be developed by sawing, inserts, or forming. Time of sawing is critical to prevent uncontrolled cracking, and joints should be sawed consecutively to ensure all commence working together. The length of time from concrete placement to

3.3.3 Joint Sealant Dimensions

The joint sealant dimension guidelines are discussed for each joint type in the following sections.

Contraction Joints. Joint movement and the capabilities of the sealant material must be optimized. In general, the quality of the joint sealant material should increase as the expected joint movement increases. Increased joint movement can be the result of longer slab length, higher temperature change, and/or higher concrete thermal coefficient.

Joint movement in pavements is influenced by factors such as slab length volume change characteristics of the concrete, slab temperature range, and friction between the slab and subbase (or subgrade). Note that because of subgrade friction and end restraints, changes in joint width are less than what would be predicted by simple thermal contraction and expansion.

In order to maintain an effective field-molded seal, the sealant reservoir must have the proper shape factor (depth-to-width ratio). Within the practical limitations of minimum joint depth, the reservoir should be as nearly square as possible and recessed below the surface a minimum of 1/8 inch. This means that a sealant reservoir normally must be formed by increasing the width and reducing the depth of the top portion of the joint to hold the sealant. For narrow joints with close joint spacing, the reservoir can be formed by inserting a cord or other material to a predetermined depth to define the reservoir. This method minimizes the amount of joint sealant required. In general, the depth to width of sealant ratio should be within a range of 1 to 1 1/2, with a minimum depth of 3/8 and 1/2 inch for longitudinal and transverse joints, respectively.

The joint width is defined as the maximum value that occurs at the minimum temperature. Thus, the maximum value includes the anticipated horizontal movement plus residual width due to sealant properties. The horizontal movement can be calculated by considering the seasonal openings and closings caused by temperature cycles plus concrete shrinkage. The amount of opening and closing depends on temperature and moisture change, spacing between working joints or cracks, friction between the slab and base, the condition of the joint load transfer devices, etc.

For design purposes, the mean transverse joint opening over a time interval can be computed approximately. The joint width must account for the movement plus the allowable residual strain in the joint sealant, and may be computed by the following:

$$\Delta L = \frac{CL(\alpha_c \times DT_D + Z)}{S} \times 100$$

where

ΔL = the joint opening caused by temperature changes and drying shrinkage of the PCC, in.,

S = allowable strain of joint sealant material. Most current sealants are designed to withstand strains of 25 to 35 percent, thus 25 percent may be used as a conservative value,

α_c = the thermal coefficient of contraction of portland cement concrete, °F,

Z = the drying shrinkage coefficient of the PCC slab, which can be neglected for a resealing project, in./in.,

L = joint spacing, in.,

DT_D = the temperature range, °F, and

C = the adjustment factor due to subbase/slab friction restraint. Use 0.65 for stabilized subbase, 0.80 for granular base.

For premolded sealants, the material and the movement must be optimized. The manufacturers generally publish aids for selecting dimensions to suit their product. The sealant should be compressed between 20 to 50 percent of its nominal width. The sealant should be placed 1/8 to 1/2 inch below the surface of the pavement.

Expansion Joints. The movement at expansion joints should be based on the agency's experience. The sealant reservoir dimensions should be optimized based on movement and material capabilities. In general, the dimensions will be much larger than for contraction joints.

Construction Joints. The discussion pertaining to transverse contraction joints is also applicable to construction and other longitudinal joints.

3.4 RIGID PAVEMENT REINFORCEMENT DESIGN

The purpose of distributed steel reinforcement in reinforced concrete pavement is not to prevent cracking, but to hold tightly closed any cracks that may form, thus maintaining the pavement as an integral structural unit. The physical mechanism through which cracks develop is affected by (1) temperature and/or moisture-related slab contractions, and (2) frictional resistance from the underlying material. As temperature drops or moisture content decreases, the slab tends to contract. This contraction is resisted by the underlying material through friction and shear between it and the slab. The restraint of slab contraction results in tensile stresses which reach a maximum at midslab. If these tensile stresses exceed the tensile strength of the concrete, a crack will develop and all the stresses are transferred to the steel reinforcement. Thus, the reinforcement must be designed to carry these stresses without any appreciable elongation that would result in excessive crack width.

Because the longitudinal steel reinforcement requirements between jointed reinforced (JRCP) and continuously reinforced concrete pavement (CRCP) are significantly different, the reinforcement designs are treated separately. It should be recognized, however, that the design for transverse steel in CRCP is exactly the same as the design for longitudinal and transverse steel reinforcement in JRCP. In all cases, the amount of reinforcement required is specified as a percentage of the concrete cross-sectional area.

3.4.1 Jointed Reinforced Concrete Pavements

The nomograph for estimating the percent of steel reinforcement required in a jointed reinforced concrete pavement is presented in Figure 3.8. The inputs required include:

(1) slab length, L (Section 2.5.1),
(2) steel working stress, f_s (Section 2.5.1), and
(3) friction factor, F (Section 2.5.1).

This chart applies to the design of transverse steel reinforcement (Section 3.3.3) in both jointed and continuously reinforced concrete pavements, as well as to the design of longitudinal steel reinforcement in JRCP. Normally for joint spacing, less than 15 feet transverse cracking is not anticipated; thus steel reinforcement would not be required.

3.4.2 Continuously Reinforced Concrete Pavements

This section is for the design of longitudinal reinforcing steel in continuously reinforced concrete pavements. The design procedure presented here may be systematically performed using the worksheet in Table 3.5. In this table, space is provided for entering the appropriate design inputs, intermediate results and calculations for determining the required longitudinal steel percentage. A separate worksheet, presented in Table 3.6, is provided for design revisions. Although the examples use reinforcing bars, the use of deformed wire fabric (DWF) is also an acceptable alternative.

The design inputs required by this procedure are as follows:

(1) concrete indirect tensile strength, f_t (Section 2.5.2),
(2) concrete shrinkage at 28 days, Z (Section 2.5.2),
(3) concrete thermal coefficient, α_c (Section 2.5.2),
(4) reinforcing bar or wire diameter, ϕ/ (Section 2.5.2),
(5) steel thermal coefficient, α_s (Section 2.5.2), and
(6) design temperature drop, DT_D (Section 2.5.2).

These data should be recorded in the space provided in the top portion of Table 3.5.

An additional input required by the procedure is the wheel load tensile stress developed during initial loading of the constructed pavement by either construction equipment or truck traffic. Figure 3.9 may be used to estimate this wheel load stress based on the design slab thickness, the magnitude of the wheel load, and the effective modulus of subgrade reaction. This value should also be recorded in the space provided in Table 3.5.

Limiting Criteria. In addition to the inputs required for the design of longitudinal reinforcing steel, there are three limiting criteria which must be considered: crack spacing, crack width, and steel stress. Acceptable limits of these are established below to ensure that the pavement will respond satisfactorily under the anticipated environmental and vehicular loading conditions.

(1) The limits on *crack spacing* are derived from consideration of spalling and punchouts. To minimize the incidence of crack spalling, the

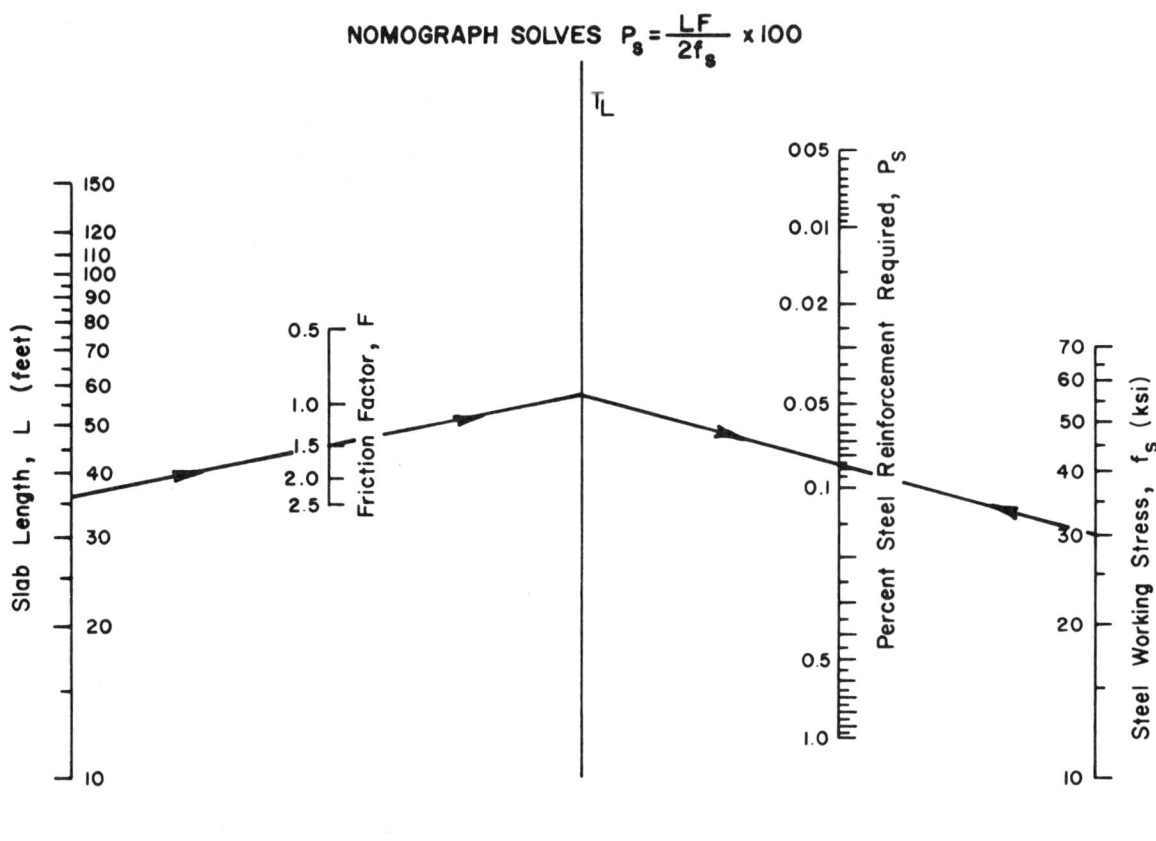

Figure 3.8. Reinforcement Design Chart for Jointed Reinforced Concrete Pavements

Table 3.5. Worksheet for Longitudinal Reinforcement Design

DESIGN INPUTS

Input Variable	Value	Input Variable	Value
Reinforcing Bar/Wire Diameter, ϕ (inches)		Thermal Coefficient Ratio, α_s/α_c (in./in.)	
Concrete Shrinkage, Z (in./in.)		Design Temperature Drop, DT_D (°F)	
Concrete Tensile Strength, f_t (psi)		Wheel Load Stress, σ_w (psi)	

DESIGN CRITERIA AND REQUIRED STEEL PERCENTAGE

	Crack Spacing, $\bar{x}$ (feet)	Allowable Crack Width, CW_{max} (inches)	Allowable Steel Stress, $(\sigma_s)_{max}$ (ksi)	
Value of Limiting Criteria	Max. 8.0 Min. 3.5			
Minimum Required Steel Percentage				(P_{min})*
Maximum Allowable Steel Percentage				P_{max}

*Enter the largest percentage across line.
**If $P_{max} < P_{min}$, then reinforcement criteria are in conflict, design not feasible.

---

maximum spacing between consecutive cracks should be no more than 8 feet. To minimize the potential for the development of punchouts, the minimum desirable crack spacing that should be used for design is 3.5 feet. These limits are already recorded in Table 3.5.

(2) The limiting criterion on *crack width* is based on a consideration of spalling and water penetration. The allowable crack width should not exceed 0.04 inch. In final determination of the longitudinal steel percentage, the predicted crack width should be reduced as much as possible through the selection of a higher steel percentage or smaller diameter reinforcing bars.

(3) Limiting criteria placed on *steel stress* are to guard against steel fracture and excessive permanent deformation. To guard against steel fracture, a limiting stress of 75 percent of the *ultimate* tensile strength is set. The conventional limit on Figure 3.9 steel stress is 75 percent of the yield point so that the steel does not undergo any plastic deformation. Based on past experience, many miles of CRC pavements have performed satisfactorily even though the steel stress was predicted to be above the yield point. This led to reconsideration of this criteria and allowance for a small amount of permanent deformation (*10*).

Values of allowable mean steel working stress for use in this design procedure are listed in Table 3.7 as a function of reinforcing bar size and concrete strength. The indirect tensile strength should be that determined in Section 2.5.2. The limiting steel working stresses in Table 3.7 are for the Grade 60 steel (meeting ASTM A 615 specifications) recommended for longitudinal reinforcement in CRC pavements (guidance for determination of allowable steel stress for other types of steel provided in Reference 10). Once the allowable steel working stress is determined, it should be entered in the space provided in Table 3.5.

Design Procedure. The following procedure may be used to determine the amount of longitudinal reinforcement required:

Table 3.6. Worksheet for Revised Longitudinal Reinforcement Design

Parameter	Change in Value from Previous Trial				
	Trial 2	Trial 3	Trial 4	Trial 5	Trial 6
[2]Reinforcing Bar/Wire Diameter, ϕ (inches)					
Concrete Shrinkage, Z (in./in.)					
[2]Concrete Tensile Strength, f_t (psi)					
Wheel Load Stress, σ_w (psi)					
[1]Design Temperature Drop, DT_D (°F)					
Thermal Coefficient Ratio, α_s/α_c					
Allowable Crack Width Criterion, CW_{max} (inches)					
Allowable Steel Stress Criterion, $(\sigma_s)_{max}$ (ksi)					
Required Steel % for Crack Spacing — min.					
Required Steel % for Crack Spacing — max.					
Minimum Required Steel % for Crack Width					
Minimum Required Steel % for Steel Stress					
Minimum % Reinforcement, P_{min}					
Maximum % Reinforcement, P_{max}					

[1]Change in this parameter will affect crack width criterion.
[2]Change in this parameter will affect steel stress criterion.

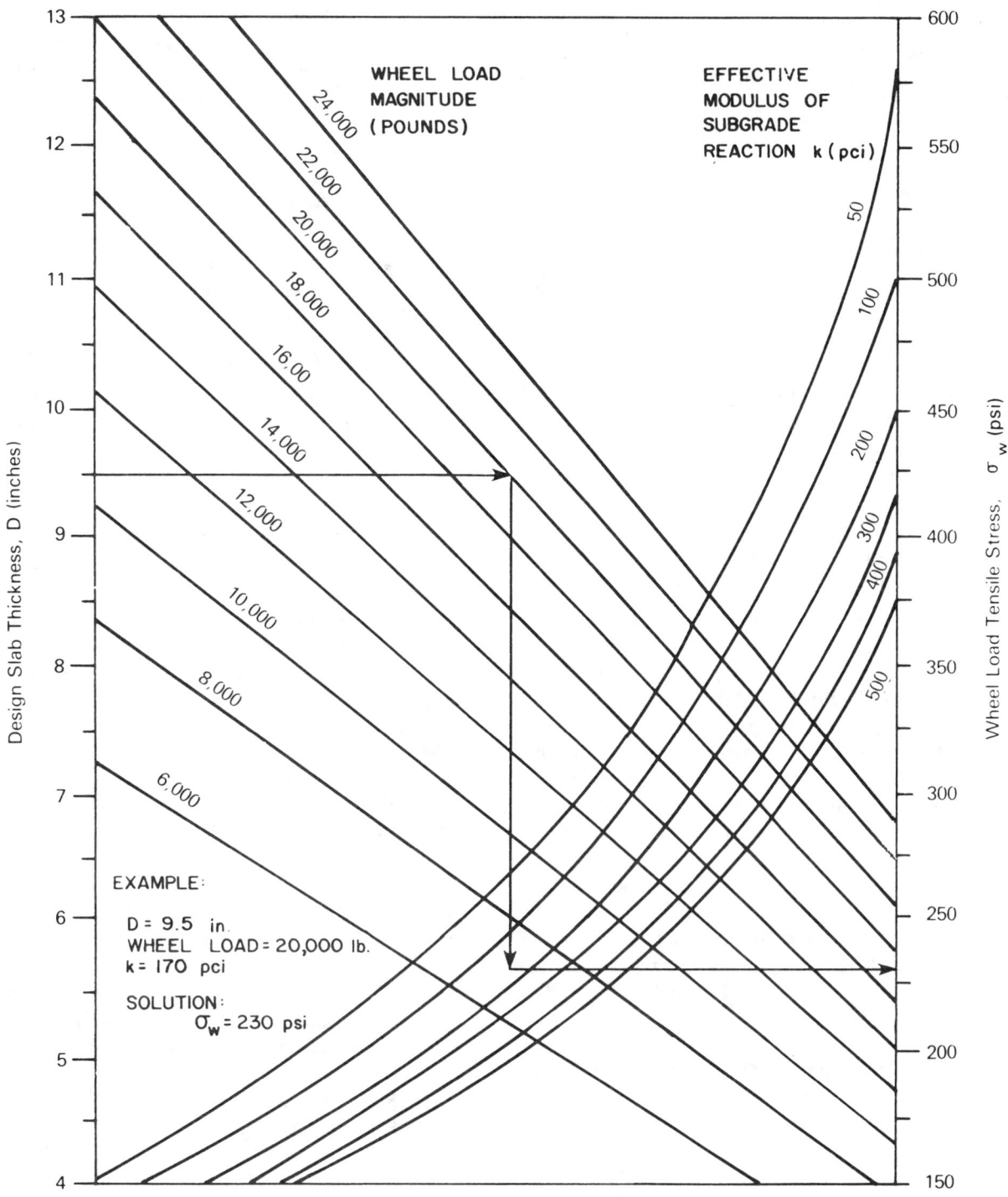

Figure 3.9. Chart for Estimating Wheel Load Tensile Stress

Table 3.7. Allowable Steel Working Stress, ksi (10)

Indirect Tensile Strength of Concrete at 28 days, psi	Reinforcing Bar Size*		
	No. 4	No. 5	No. 6
300 (or less)	65	57	54
400	67	60	55
500	67	61	56
600	67	63	58
700	67	65	59
800 (or greater)	67	67	60

*For DWF proportional adjustments may be made using the wire diameter to bar diameter.

Step 1. Solve for the required amount of steel reinforcement to satisfy each limiting criterion using the design charts in Figures 3.10, 3.11, and 3.12. Record the resulting steel percentages in the spaces provided in the worksheet in Table 3.5.

Step 2. If P_{max} is greater than or equal to P_{min}, go to Step 3. If P_{max} is less than P_{min}, then

(1) Review the design inputs and decide which input to revise.
(2) Indicate the revised design inputs in the worksheet in Table 3.6. Make any corresponding change in the limiting criteria as influenced by the change in design parameter and record this in Table 3.6. Check to see if the revised inputs affect the subbase and slab thickness design. It may be necessary to reevaluate the subbase and slab thickness design.
(3) Rework the design nomographs and enter the resulting steel percentages in Table 3.6.
(4) If P_{max} is greater than or equal to P_{min}, go to Step 3. If P_{max} is less than P_{min}, repeat this step using the space provided in Table 3.6 for additional trials.

Step 3. Determine the range in the number of reinforcing bars or wires required:

$$N_{min} = 0.01273 \times P_{min} \times W_S \times D/\phi/^2, \text{ and}$$
$$N_{max} = 0.01273 \times P_{max} \times W_S \times D/\phi/^2$$

where

N_{min} = minimum required number of reinforcing bars or wires,
N_{max} = maximum required number of reinforcing bars or wires,
P_{min} = minimum required percent steel,
P_{max} = maximum required percent steel,
W_S = total width of pavement section (inches),
D = thickness of concrete layer (inches), and
ϕ = reinforcing bar or wire diameter (inches), which may be increased if loss of cross section is anticipated due to corrosion.

Step 4. Determine the final steel design by selecting the total number of reinforcing bars or wires in the final design section, N_{Design}, such that N_{Design} is a whole integer number between N_{min} and N_{max}. The appropriateness of these final design alternatives may be checked by converting the whole integer number of bars or wires to percent steel and working backward through the design charts to estimate crack spacing, crack width, and steel stress.

Design Example. The following example is provided to demonstrate the CRCP longitudinal reinforcement design procedure. Two trial designs are evaluated; the first considers ⅝-inch (No. 5) reinforcing bars and the second trial design examines ¾-inch (No. 6) bars. Below are the input requirements selected for this example. These values are also recorded for both of the trial designs in the example worksheets presented in Tables 3.8 and 3.9.

(1) Concrete tensile strength, f_t: 550 psi. (This is approximately 86 percent of the modulus of rupture used in the slab thickness design example, see Figure 3.7.)
(2) Concrete shrinkage, Z: 0.0004 in./in. (This corresponds to the concrete tensile strength; see Table 2.7.)
(3) Wheel load stress, σ_w: 230 psi. (This is based on the earlier slab thickness design example, 9.5-inch slab with a modulus of subgrade reaction equal to 170 pci; see Figure 3.9.)
(4) Ratio of steel thermal coefficient to that of Portland Cement Concrete, α_s/α_c: 1.32 (For steel, the thermal coefficient is 5×10^{-6} in./in./15°F. (See Section 2.5.2). Assume limestone coarse aggregate in concrete, therefore, the thermal coefficient is 3.8×10^{-6} in./in./°F. (See Table 2.9.)
(5) Design temperature drop, DT_D: 55°F. (Assume high temperature is 75°F and low is 20°F.)

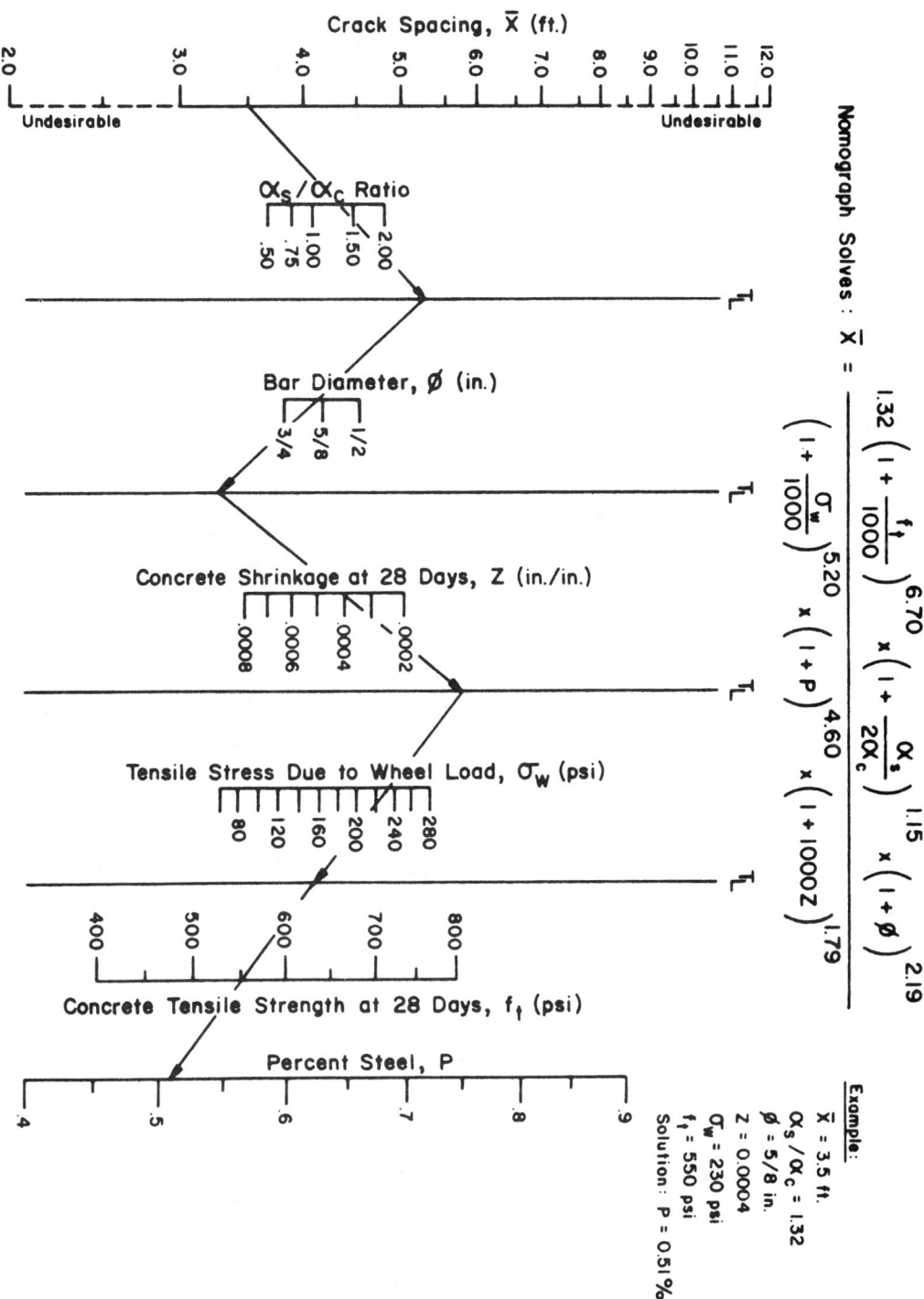

Figure 3.10. Percent of Longitudinal Reinforcement to Satisfy Crack Spacing Criteria

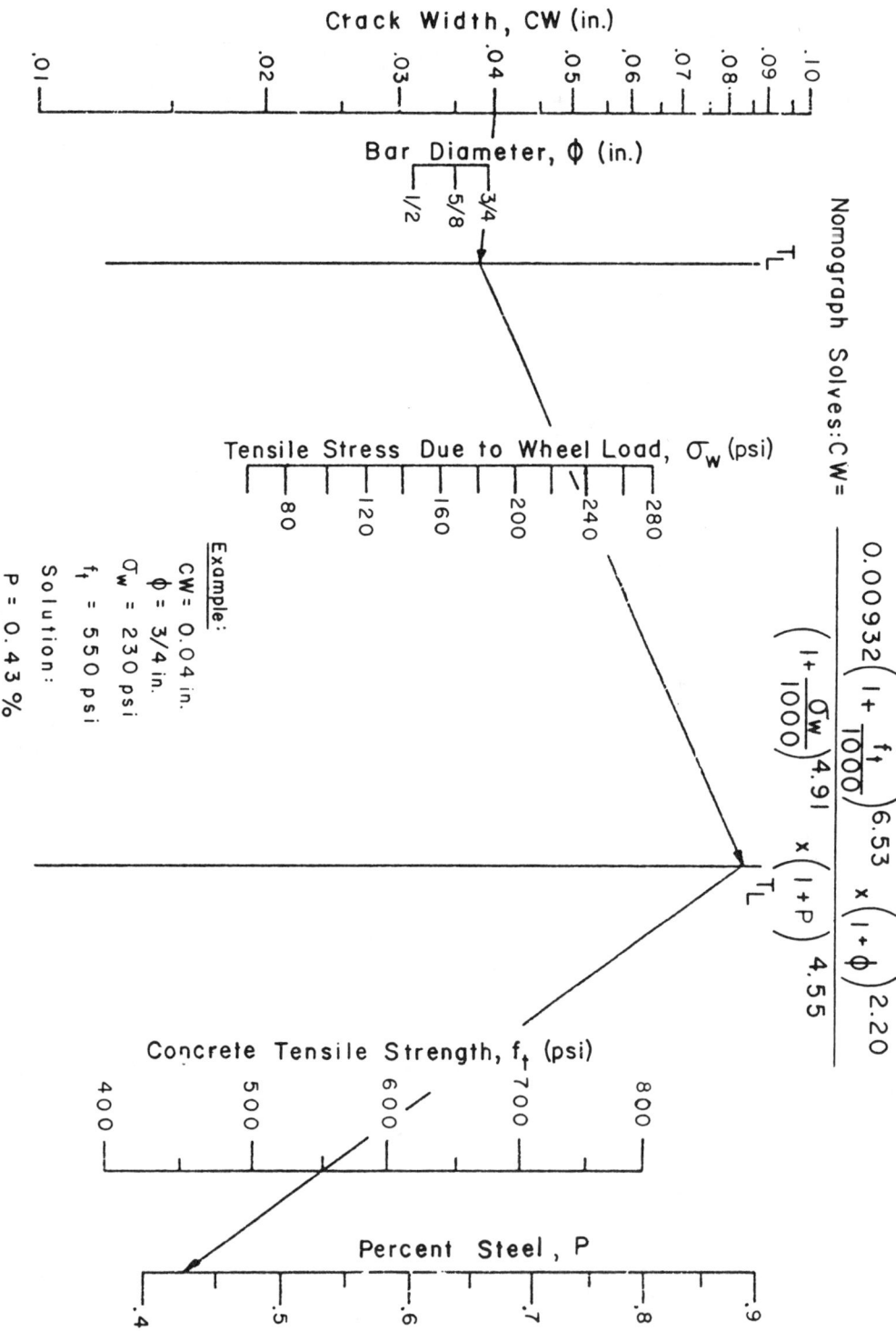

Figure 3.11. Minimum Percent Longitudinal Reinforcement to Satisfy Crack Width Criterion

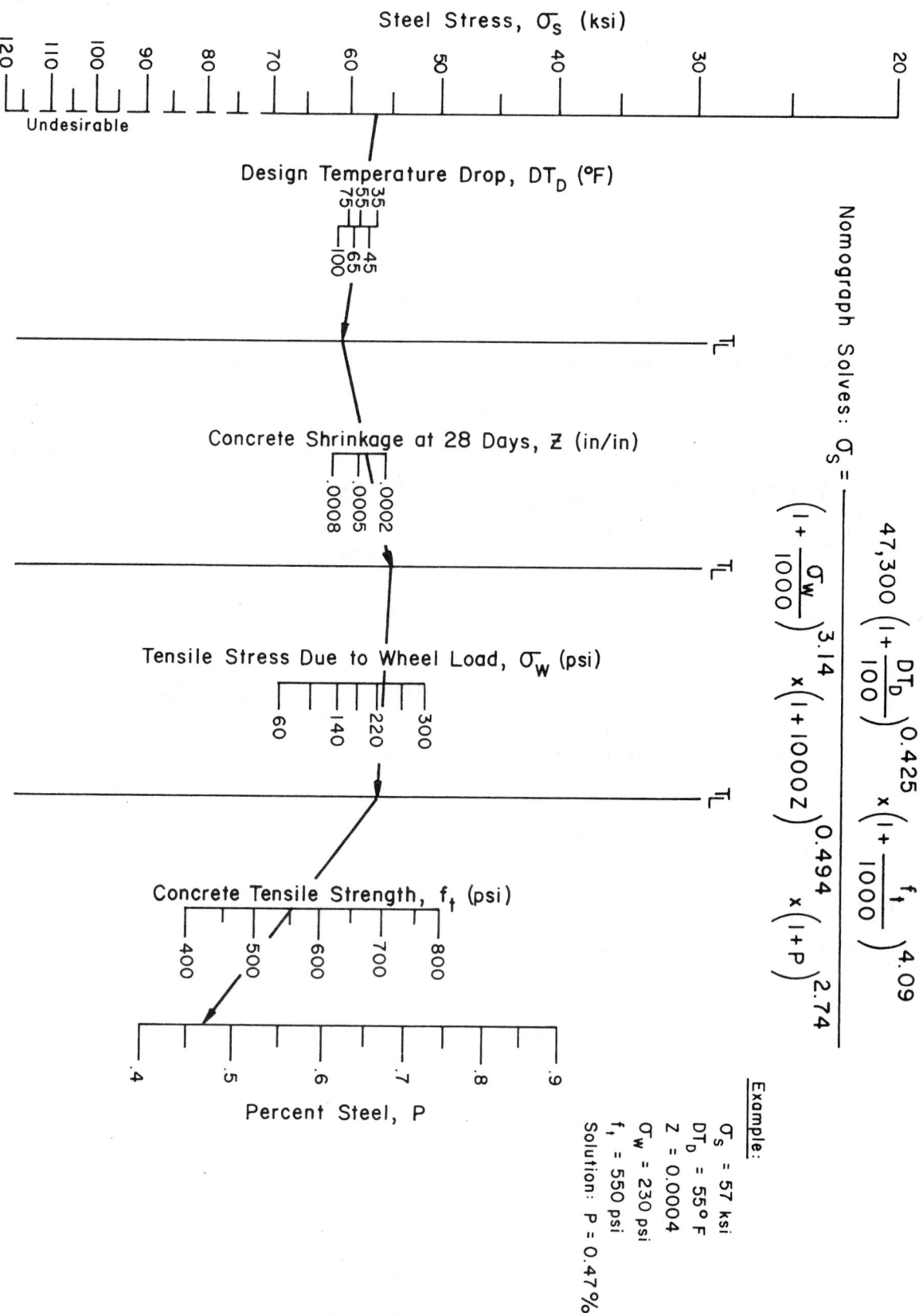

Figure 3.12. Minimum Percent Longitudinal Reinforcement to Satisfy Steel Stress Criteria

Table 3.8. Example Application of Worksheet for Longitudinal Reinforcement Design

DESIGN INPUTS

Input Variable	Value	Input Variable	Value
Reinforcing Bar/Wire Diameter, ϕ (inches)	5/8 (No. 5)	Thermal Coefficient Ratio, α_s/α_c (in./in.)	1.32
Concrete Shrinkage, Z (in./in.)	0.0004	Design Temperature Drop, DT_D (°F)	55
Concrete Tensile Strength, f_t (psi)	550	Wheel Load Stress, σ_w (psi)	230

DESIGN CRITERIA AND REQUIRED STEEL PERCENTAGE

	Crack Spacing, $\bar{x}$ (feet)	Allowable Crack Width, CW_{max} (inches)	Allowable Steel Stress, $(\sigma_s)_{max}$ (ksi)	Design Steel Range**
Value of Limiting Criteria	Max. 8.0 Min. 3.5	0.04	62	
Minimum Required Steel Percentage	<0.40%	<0.40%	0.43%	0.43% (P_{min})*
Maximum Allowable Steel Percentage	0.51%			0.51% P_{max}

*Enter the largest percentage across line.

**If $P_{max} < P_{min}$, then reinforcement criteria are in conflict, design not feasible.

Table 3.9. Example Application of Worksheet for Revised Longitudinal Reinforcement Design

Parameter		Change in Value from Previous Trial				
		Trial 2	Trial 3	Trial 4	Trial 5	Trial 6
[2]Reinforcing Bar/Wire Diameter, ϕ (inches)		3/4 (No. 6)				
Concrete Shrinkage, Z (in./in.)		0.0004				
[2]Concrete Tensile Strength, f_t (psi)		550				
Wheel Load Stress, σ_w (psi)		230				
[1]Design Temperature Drop, DT_D (°F)		550				
Thermal Coefficient Ratio, α_s/α_c		1.32				
Allowable Crack Width Criterion, CW_{max} (inches)		0.04				
Allowable Steel Stress Criterion, $(\sigma_s)_{max}$ (ksi)		57				
Required Steel % for Crack Spacing	min.	<0.04%				
	max.	0.57%				
Minimum Required Steel % for Crack Width		0.43%				
Minimum Required Steel % for Steel Stress		0.47%				
Minimum % Reinforcement, P_{min}		0.47%				
Maximum % Reinforcement, P_{max}		0.57%				

[1]Change in this parameter will affect crack width criterion.
[2]Change in this parameter will affect steel stress criterion.

The limiting criteria corresponding to these design conditions are as follows:

(1) Allowable crack width, CW: 0.04 inch for both trial designs. (See Section 3.3.2, "Continuously Reinforced Concrete Pavements; Limiting Criteria.")
(2) Allowable steel stress, σ_s: 62 ksi for ⅝-inch bars (Trial 1) and 57 ksi for ¾-inch bars. (See Table 3.7 using tensile strength of 550 psi.)

Application of the design nomographs in Figures 3.10, 3.11, and 3.12 yields the following limits on steel percentage for the two trial designs:

Trial Design 1: $P_{min} = 0.43\%$, $P_{max} = 0.51\%$

Trial Design 2: $P_{min} = 0.47\%$, $P_{max} = 0.57\%$

The range (N_{min} to N_{max}) of the number of reinforcing bars requires (assuming a 12-foot-wide lane) for each trial design is

Trial Design 1 (No. 5 bars): $N_{min} = 19.2$,

$N_{max} = 22.7$

Trial Design 2 (No. 6 bars): $N_{min} = 14.6$,

$N_{max} = 17.6$

Using twenty No. 5 bars for Trial 1 (P = 0.45%) and fifteen No. 6 bars for Trial 2 (P = 0.48%), the longitudinal reinforcing bar spacings would be 7.2 and 9.6 inches, respectively. The predicted crack spacing, crack width, and steel stress for these two trial designs are:

Predicted Response	Trial Design 1 (20 No. 5 Bars, P = 0.45%)	Trial Design 2 (15 No. 6 Bars, P = 0.48%)
Crack Spacing, x (feet)	4.3	4.6
Crack Width, CW (inches)	0.031	0.032
Steel Stress, σ_s (ksi)	60	55

Inspection of these results indicates that there is no significant difference in the predicted response of these two designs such that one should be selected over the other. Thus, in this case, the selection should be based on economics and/or ease of construction.

3.4.3 Transverse Reinforcement

Transverse steel is included in either jointed or continuous pavements for conditions where soil volume changes (due to changes in either temperature or moisture) can result in longitudinal cracking. Steel reinforcement will prevent the longitudinal cracks from opening excessively, thereby maintaining maximum load transfer and minimizing water entry.

If transverse reinforcement and/or tie bars are desired, then the information collected under Section 2.5.1, "Reinforcement Variables for Jointed Reinforced Concrete Pavements," is applicable. In this case, the "slab length" should be considered as the distance between *free* longitudinal edges. If tie bars are placed within a longitudinal joint, then that joint is not a free edge.

For normal transverse reinforcement, Figure 3.8 may be used to determine the percent transverse steel. The percent transverse steel may be converted to spacing between reinforcing bars as follows:

$$Y = \frac{A_s}{P_t D} \times 100$$

where

Y = transverse steel spacing (inches),
A_s = cross-sectional area of transverse reinforcing steel (in.^{2}),
P_t = percent transverse steel, and
D = slab thickness (inches).

Figures 3.13 and 3.14 may be used to determine the tie bar spacing for ½- and ⅝-diameter deformed bars, respectively. The designer enters the figure on the horizontal with the distance to the closest free edge axis and proceeds vertically to the pavement thickness obtained from Section 3.2.2, "Determine Required Slab Thickness." From the pavement thickness, move horizontally and read the tie bar spacing from the vertical scale. These nomographs are based on Grade 40 steel and a subgrade friction factor of 1.5.

Note that since steel stress decreases from a maximum near the center of the slab (between the free edges) to zero at the free edges, the required minimum tie bar spacing increases. Thus, in order to design the

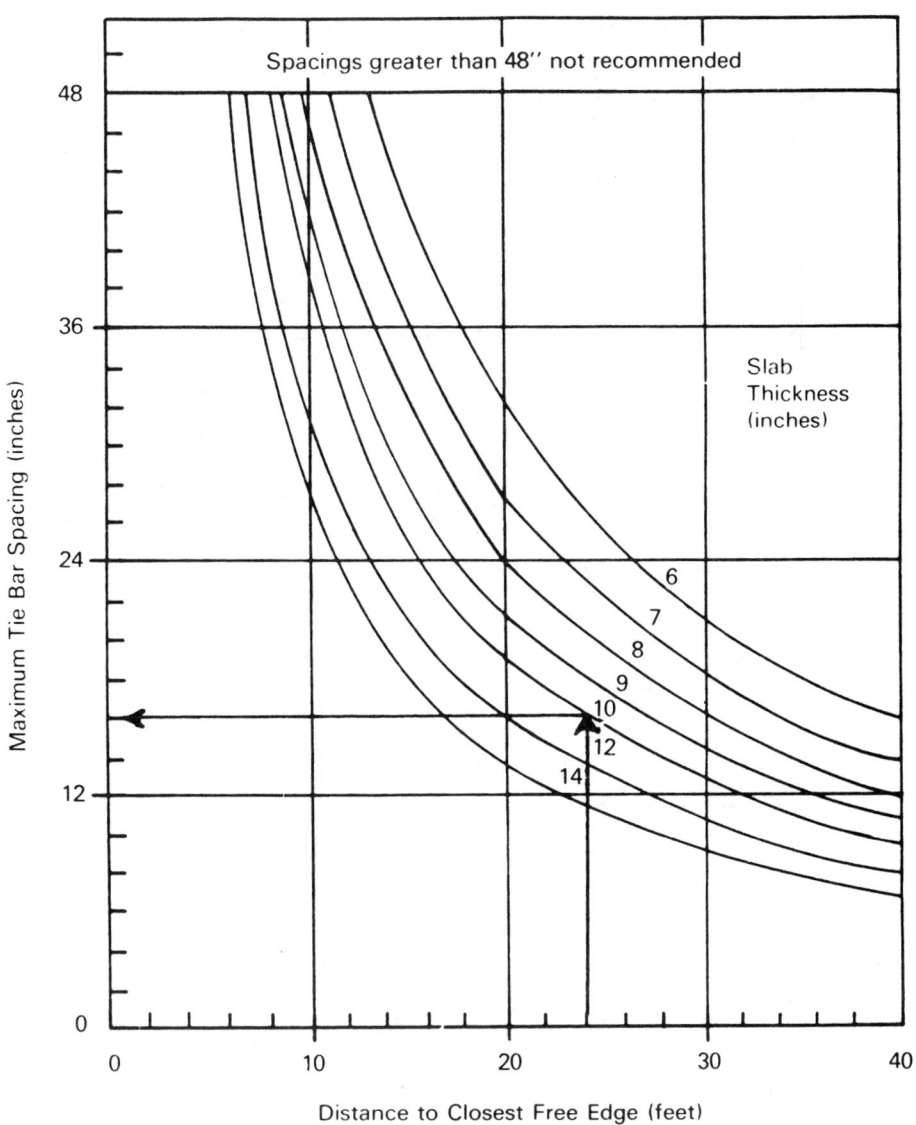

Example: Distance from free edge = 24 ft.
D = 10 in.

Answer: Spacing = 16 in.

Figure 3.13. Recommended Maximum Tie Bar Spacings for PCC Pavements Assuming 1/2-inch Diameter Tie Bars, Grade 40 Steel, and Subgrade Friction Factor of 1.5

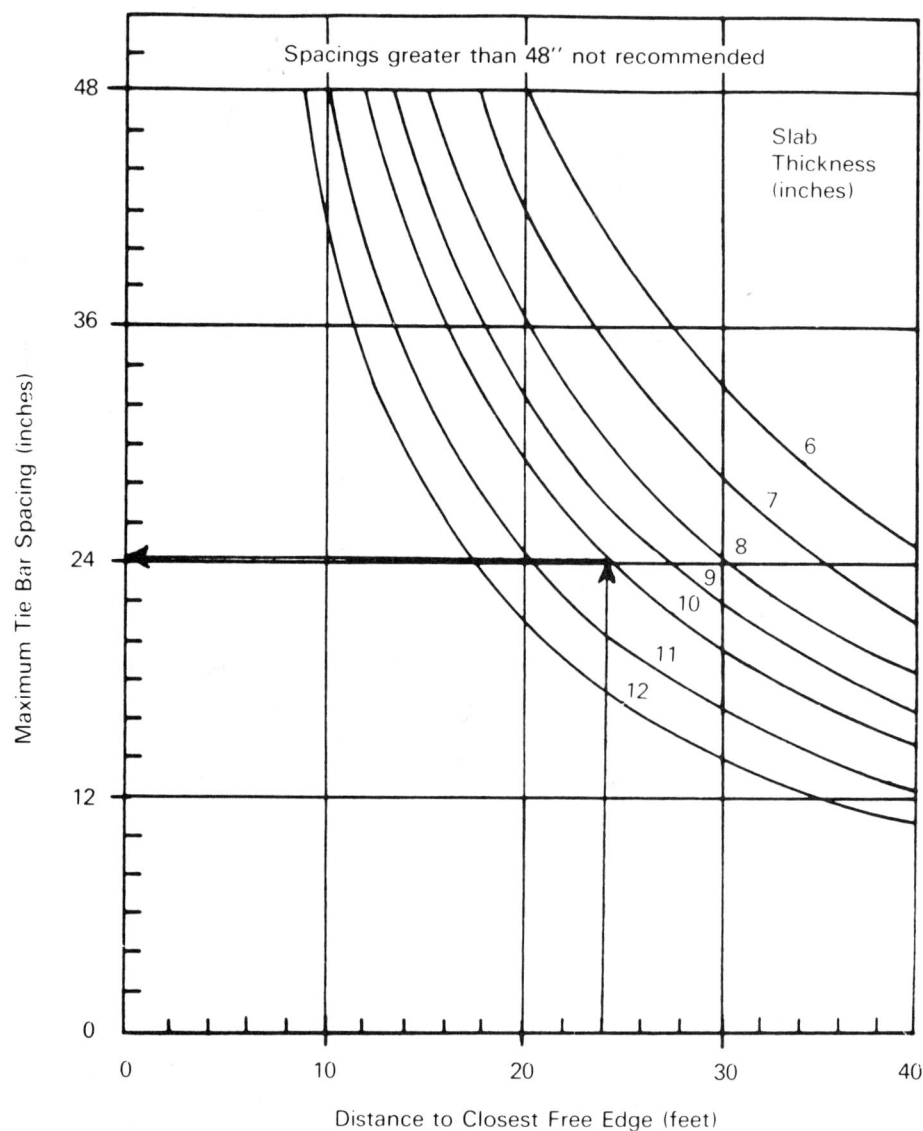

Example: Distance from free edge   24 ft.
D = 10 in.

Answer: Spacing = 24 in.

Figure 3.14. Recommended Maximum Tie Bar Spacings for PCC Pavements Assuming 5/8-inch Diameter Tie Bars, Grade 40 Steel, and Subgrade Friction of 1.5

tie bars efficiently, the designer should first select the layout of the longitudinal construction joints.

Finally, if bending of the tie bars is to be permitted during construction, then to prevent steel failures, the use of brittle (high carbon content) steels should be avoided and an appropriate steel working stress level selected.

3.5 PRESTRESSED CONCRETE PAVEMENT

This section is provided to give the user some general guidelines on the design of prestressed concrete pavement. No specific design procedure can be provided at this time.

A prestressed concrete pavement (PCP) is one in which a permanent and essentially horizontal compressive stress has been introduced prior to the application of any wheel loads. Past experience has indicated the potential of PCP in at least two significant respects:

(1) more efficient use of construction materials; and
(2) fewer required joints and less probability of cracking, resulting in less required maintenance and longer pavement life.

In conventional concrete pavement design, stresses due to wheel loads are restricted to the elastic range of the concrete. Thus, the pavement thickness is determined such that the extreme fiber tensile stress due to applied loads does not exceed the flexural strength or modulus of rupture of the concrete. In this conventional design approach, the concrete between the extreme top and bottom fibers of the slab is not fully utilized to resist stresses due to applied loads, resulting in an inefficient use of construction materials.

With PCP, the effective flexural strength of the concrete is increased by the induced compressive stress and is no longer limited in load-carrying capacity by the modulus of rupture of concrete. Consequently, the required pavement thickness for a given load is significantly less than that required for a conventional concrete pavement.

On most of the previously constructed PCP's, one of the following prestressed orientations was employed:

(1) Prestress is only applied parallel to the longitudinal axis of the pavement. The pavement may be either unreinforced or reinforced in the transverse direction.

(2) Prestress is applied both parallel and perpendicular to the longitudinal axis of the pavement.
(3) Prestress is applied diagonally at an angle to the longitudinal axis of the pavement. Desired prestress levels both parallel and perpendicular to the longitudinal axis of the pavement can be obtained by merely adjusting the angle at which the prestress is applied.

The particular prestress orientation that the designer wants to employ on a given project may have a significant influence on the prestressing method that is used.

The following factors have a direct influence on the performance of a PCP and must be considered in any rational PCP design approach: subbase support, slab length, magnitude of prestress, tendon spacing, and concrete fatigue. Each is discussed in the following sections.

3.5.1 Subbase

Although it has been demonstrated that acceptable performance of PCP can be obtained with low-strength support if provisions are taken to prevent pumping, virtually all previous subbases for PCP have been fairly high-strength (usually 200 psi, or higher, modulus of subgrade reaction). This is due primarily to an unwillingness of the designers to risk failure of the pavement if it is constructed on a low-strength subbase. Although, soil cement and bituminous concrete bases have been used to increase the strength of support, the most common method has been the use of a layer of compacted granular material. The thickness of the layer has generally been on the order of 6 to 12 inches, but as little as 4 inches and as much as 18 inches has been used.

3.5.2 Slab Length

Slab length refers to the distance between active transverse joints and not to the distance between intermediate inactive construction joints. There are two main factors which must be considered when selecting the optimum slab length for PCP. These are: (1) The prestress force required to overcome the frictional restraint between the subgrade and the slab and to provide the desired minimum compressive stress at the midlength of the slab so that it is proportional to the slab length. The cost associated with providing the prestress force is, in turn, proportional to the magnitude of the required force. (2) The number of, and the total cost for, transverse joints is inversely proportional to the slab length. Since transverse joints are probably the largest maintenance item for a pavement, total cost for transverse joints should not be based only on the initial cost, but should also include an estimate of the maintenance cost over the life of the facility. Generally, a compromise must be sought between these two factors. Based on PCP projects built to date, a pavement length on the order of 400 feet appears to strike a reasonable balance between these two constraints. Slabs as long as 760 feet in length have been built in the United States and some over 1,000 feet in length have been built in Europe; however, these are exceptions.

3.5.3 Magnitude of Prestress

The magnitude of the longitudinal and transverse prestress must be great enough to provide sufficient compressive stress at the midlength and possible midwidth of the pavement slab during a period of contraction to sustain the stresses occurring during the passage of a load. Many factors must be taken into account to assure that the desired prestress level is obtained including the magnitude of the frictional restraint between the slab and the subgrade, the slab thickness, the slab length, and the maximum temperature differential anticipated during the life of the pavement.

On some of the early PCP projects, relatively high prestress levels were used so that sufficient prestress was assured. However, it has been shown by means of small-scale laboratory tests and full-scale field tests that structural benefits do not increase in proportion to increases in the prestress level. Therefore, more recent projects have used prestress levels ranging from 100 to 300 psi longitudinally and from 0 to 200 psi transversely.

3.5.4 Tendon Spacing

The main factors governing tendon spacing are tendon size, magnitude of design prestress, allowable concrete bearing stress at the tendon anchorages, and permissible tendon anchoring stress. Although bar and stranded cable tendon spacings have varied from a minimum of two to a maximum of eight times the slab thickness, more typically, spacings of two to four times and three to six times the slab thickness have been utilized for the longitudinal and transverse tendons, respectively. The allowable stress in the tendon is set at 0.8 yield stress, and generally 0.6-inch strands are used.

3.5.5 Fatigue

Since very little data exists for the relationship between number of load repetitions and design requirements, it is recommended the designer use conservative load repetition factors at the present time. This is supported by the observation that little advance warning accompanies the load failure of PCP, i.e., a PCP may require only a few additional load repetitions to go from a few initial signs of distress to complete failure.

3.5.6 PCP Structural Design

At this time, the design of PCP is primarily the application of experience and engineering judgment. The designer should recognize the basic principle that the greater the prestress level, the thinner the pavement; however, full potential cannot be recognized since adequate thickness must be maintained to prevent excessive deflection and the resulting problems. The basic steps to PCP design are as follows:

(1) Select a pavement thickness using the criteria in the following section, and a practical magnitude of prestress to be achieved at the center of slab.
(2) Using the selected joint spacing and subbase friction, compute the loss due to subgrade restraint as outlined in a following section.
(3) Estimate the loss of prestress as described in a following section.
(4) Add the desired magnitude of prestress from Step 1 to the losses from Steps 2 and 3 to obtain the prestress level that must be applied at the slab end.
(5) The spacing of the tendons may be obtained by the following formula:

$$Y_t = \frac{f_t \times A_f}{\sigma_p \times D}$$

where

Y_t = spacing of tendons (in.),

f_t = allowable working stress in tendon (psi),
A_f = cross-sectional area of tendon (in.^{2}),
D = selected pavement thickness (in.), and
σ_p = prestress level at end from Step 4.

Pavement Thickness. Many factors of roadbed strength, concrete strength, magnitude of prestress, and expected traffic loads should be taken into account when determining the required thickness of PCP. In the past, highway PCP pavement thickness has generally been determined more on the basis of providing the minimum allowable concrete cover on the prestressing tendons than on the basis of load-carrying considerations. This procedure has resulted in PCP thicknesses on the order of 40 to 50 percent of equivalent conventional concrete pavement. On previous projects, highway pavement thicknesses have usually been on the order of 4 to 6 inches.

Subgrade Restraint. Differential movement of PCP relative to the subbase occurs as a result of the elastic shortening of the pavement at the time of stressing, moisture/thermal changes in the pavement and creep of the pavement. This movement is resisted by the friction between the pavement and the subgrade which induces restraint stresses in the pavement. These restraint stresses are additive to the design prestress during periods when the pavement is increasing in length and subtractive from the design prestress when the pavement is decreasing in length.

The magnitude of the restraint stresses is a function of the coefficient of subgrade friction and the dimensions of the slab, and is at maximum at the midlength and midwidth of the slab. The maximum value of this stress, from concrete having a unit weight of 144 pcf, is given by the following equation:

$$f_{SR} = \mu L/2$$

where

f_{SR} = maximum subgrade restraint stress (psi),
μ = coefficient of subgrade friction, and
L = length of slab (feet).

PCP's have generally been constructed on some type of friction-reducing layer such as sand and building paper, or sand and polyethylene sheeting. When a friction-reducing layer is provided, the coefficients of subgrade friction usually range from 0.4 to 1.0.

Prestress Losses. Factors contributing to loss of prestress include: (1) elastic shortening of the concrete; (2) creep of the concrete; (3) shrinkage of the concrete; (4) relaxation of the stressing tendons; (5) slippage of the stressing tendons in the anchorage devices; (6) friction between the stressing tendons and the enclosing conduits; and (7) hydrothermal contraction of the pavement.

Due to the above factors, prestress losses of approximately 15 to 20 percent of the applied prestress force should be expected for a carefully constructed pretensioned or post-tensioned PCP. For a post-stressed PCP, all of the prestress may be lost unless proper provision is made. These losses must be accounted for in the design of a PCP in order to ensure that the required prestress level is maintained over the service life of the pavement.

Prestress losses for pretensioned and post-tensioned PCP are generally expressed as a stress loss in the tendons. Therefore, the prestress applied to the pavement by means of the tendons must be increased to counter the stress losses resulting from natural adjustments in the materials during and after construction.

CHAPTER 4
LOW-VOLUME ROAD DESIGN

Pavement structural design for low-volume roads is divided into three categories:

(1) flexible pavements,
(2) rigid pavements, and
(3) aggregate-surfaced roads.

This chapter covers the design of low-volume roads for these three surface types using procedures based on design charts (nomographs) and design catalogs. These two procedures are covered in Sections 4.1 and 4.2, respectively. For surface treatment or chip seal pavement structures, the procedures for flexible pavements may be used.

Because the primary basis for all rational pavement performance prediction methods is cumulative heavy axle load applications, it is necessary in this Guide to use the 18-kip equivalent single axle load (ESAL) design approach for low-volume roads, regardless of how low the traffic level is or what the distribution is between automobiles and trucks.

Since many city streets and county roads that fall under the low-volume category may still carry significant levels of truck traffic, the maximum number of 18-kip ESAL applications considered for flexible and rigid pavement design is 700,000 to 1 million. The practical minimum traffic level that can be considered for any flexible or rigid pavement during a given performance period is about 50,000 18-kip ESAL applications. For the aggregate-surfaced (gravel) roads used for many county and forest roads, the maximum traffic level considered is 100,000 18-kip ESAL applications, while the practical minimum level (during a single performance period) is 10,000.

4.1 DESIGN CHART PROCEDURES

4.1.1 Flexible and Rigid Pavements

The low-volume road design chart procedures for flexible and rigid pavements are basically the same as those for highway pavement design. The low-volume road procedure basically relies on the set of design requirements (developed in Chapter 2) as well as the basic step-by-step procedures described in Chapter 3. The primary difference in the design for low-volume roads is the level of reliability that may be used. Because of their relative low usage and the associated low level of risk, the level of reliability recommended for low-volume road design is 50 percent. The user may, however, design for higher levels of 60 to 80 percent, depending on the actual projected level of traffic and the feasibility of rehabilitation, importance of corridor, etc.

If, in estimating an effective resilient modulus of the roadbed material (M_R) or an effective modulus of subgrade reaction (k), it is not possible to determine the lengths of the seasons or even the seasonal roadbed soil resilient moduli, the following suggestions should be considered.

Season Lengths. Figure 4.1 provides a map showing six different climatic regions of the United States and the environmental characteristics associated with each. Based on these regional characteristics, Table 4.1 may be used to define the season lengths needed for determining the effective roadbed soil resilient modulus (Section 2.3.1) for flexible pavement design or the effective modulus of subgrade reaction (Section 3.2.1) for rigid pavement design.

Seasonal Roadbed Soil Resilient Moduli. Table 4.2 provides roadbed soil resilient modulus values that may be used for low-volume road design if the user can classify the general quality of the roadbed material as a foundation for the pavement structure. If the suggested values in this table are combined with the suggested season lengths identified in the previous section, effective roadbed soil resilient modulus values (for flexible pavement design only) can be generated for each of the six U.S. climatic regions. These M_R values are presented in Table 4.3.

4.1.2 Aggregate-Surfaced Roads

The basis for treating the effects of seasonal moisture changes on roadbed soil resilient modulus, M_R, is

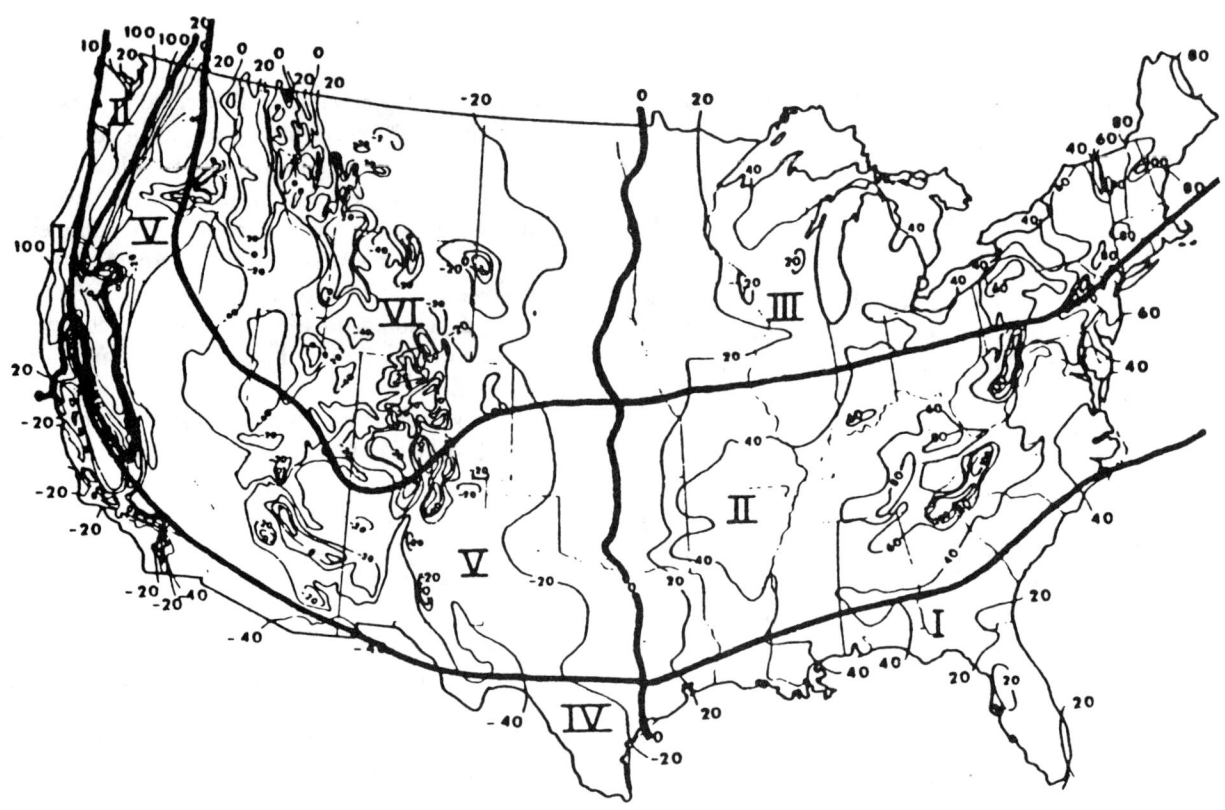

REGION	CHARACTERISTICS
I	Wet, no freeze
II	Wet, freeze-thaw cycling
III	Wet, hard-freeze, spring thaw
IV	Dry, no freeze
V	Dry, freeze-thaw cycling
VI	Dry, hard freeze, spring thaw

Figure 4.1. The Six Climatic Regions in the United States (*12*)

Low-Volume Road Design

Table 4.1. Suggested Seasons Length (Months) for the Six U.S. Climatic Regions

U.S. Climatic Region	Season (Roadbed Soil Moisture Condition)			
	Winter (Roadbed Frozen)	Spring-Thaw (Roadbed Saturated)	Spring/Fall (Roadbed Wet)	Summer (Roadbed Dry)
I	0.0*	0.0	7.5	4.5
II	1.0	0.5	7.0	3.5
III	2.5	1.5	4.0	4.0
IV	0.0	0.0	4.0	8.0
V	1.0	0.5	3.0	7.5
VI	3.0	1.5	3.0	4.5

*Number of months for the season.

Table 4.2. Suggested Seasonal Roadbed Soil Resilient Moduli, M_R (psi), as a Function of the Relative Quality of the Roadbed Material

Relative Quality of Roadbed Soil	Season (Roadbed Soil Moisture Condition)			
	Winter (Roadbed Frozen)	Spring-Thaw (Roadbed Saturated)	Spring/Fall (Roadbed Wet)	Summer (Roadbed Dry)
Very good	20,000*	2,500	8,000	20,000
Good	20,000	2,000	6,000	10,000
Fair	20,000	2,000	4,500	6,500
Poor	20,000	1,500	3,300	4,900
Very poor	20,000	1,500	2,500	4,000

*Values shown are Resilient Modulus in psi.

Table 4.3. Effective Roadbed Soil Resilient Modulus Values, M_R (psi), That May be Used in the Design of Flexible Pavements for Low-Volume Roads. Suggested values depend on the U.S. climatic region and the relative quality of the roadbed soil.

U.S. Climatic Region	Relative Quality of Roadbed Soil				
	Very Poor	Poor	Fair	Good	Very Good
I	2,800*	3,700	5,000	6,800	9,500
II	2,700	3,400	4,500	5,500	7,300
III	2,700	3,000	4,000	4,400	5,700
IV	3,200	4,100	5,600	7,900	11,700
V	3,100	3,700	5,000	6,000	8,200
VI	2,800	3,100	4,100	4,500	5,700

*Effective Resilient Modulus in psi.

the same for aggregate-surfaced road design as it is for flexible or rigid pavement design. Unlike the flexible or rigid design procedures, however, the design chart-based procedure for aggregate-surfaced roads requires a graphical solution. It is important to note that the effective modulus of the roadbed soil developed for flexible pavement design should *not* be used in lieu of the procedure described here.

The primary design requirements for aggregate-surfaced roads (*17*) include:

(1) the predicted future traffic, w_{18} (Section 2.1.2), for the period,
(2) the lengths of the seasons (Section 2.3.1; or criteria in Section 4.1.1 may be used if better information is not available),
(3) seasonal resilient moduli of the roadbed soil (Section 2.3.1 or general criteria in Section 4.1.1 may be used if better information is not available),
(4) elastic modulus, E_{BS} (psi), of aggregate base layer (Section 2.3.3),
(5) elastic modulus, E_{SB} (psi), of aggregate sub-base layer (Section 2.3.3),
(6) design serviceability loss, ΔPSI (Section 2.2.1),
(7) allowable rutting, RD (inches), in surface layer (Section 2.2.2), and
(8) aggregate loss, GL (inches), of surface layer (Section 2.2.3).

These design requirements are used in conjunction with the computational chart in Table 4.4 and the design nomographs for serviceability (Figure 4.2) and rutting (Figure 4.3). An example of the application of certain steps of this procedure is presented in Table 4.5.

Step 1. Select four levels of aggregate base thickness, D_{BS}, which should bound the probable solution. For this, four separate tables, identical to Table 4.4, should be prepared. Enter each of the four trial base thickness, D_{BS}, in the upper left-hand corner of each of the four tables (D_{BS} = 8 inches is used in the example).

Step 2. Enter the design serviceability loss as well as the allowable rutting in the appropriate boxes of each of the four tables.

Step 3. Enter the appropriate seasonal resilient (elastic) moduli of the roadbed (M_R) and the aggregate base material, E_{BS} (psi), in Columns 2 and 3, respectively, of Table 4.4. The base modulus values may be proportional to the resilient modulus of the roadbed soil during a given season. A constant value of 30,000 psi was used in the example, however, since a portion of the aggregate base material will be converted into an equivalent thickness of subbase material (which will provide some shield against the environmental moisture effects).

Step 4. Enter the seasonal 18-kip ESAL traffic in Column 4 of Table 4.4. Assuming that truck traffic is distributed evenly throughout the year, the lengths of the seasons should be used to proportion the total projected 18-kip ESAL traffic to each season. If the road is load-zoned (restricted) during certain critical periods, the total traffic may be distributed only among those seasons when truck traffic is allowed. (Total traffic of 21,000 18-kip ESAL applications and a seasonal pattern corresponding to U.S. Climatic Region III was used in the example in Table 4.5.)

Step 5. Within each of the four tables, estimate the allowable 18-kip ESAL traffic for each of the four seasons using the serviceability-based nomograph in Figure 4.2, and enter in Column 5. If the resilient modulus of the roadbed soil (during the frozen season) is such that the allowable traffic exceeds the upper limit of the nomograph, assume a practical value of 500,000 18-kip ESAL.

Step 6. Within each of the four tables, estimate the allowable 18-kip ESAL traffic for each of the four seasons using the rutting-based nomograph in Figure 4.3, and enter in Column 7. Again, if the resilient modulus of the roadbed soil is such that the allowable traffic exceeds the upper limit of the nomograph, assume a practical value of 500,000 18-kip ESAL.

Step 7. Compute the seasonal damage values in each of the four tables for the serviceability criteria by dividing the projected seasonal traffic (Column 4) by the allowable traffic in that season (Column 5). Enter these seasonal damage values in Column 6 of Table 4.4 corresponding to serviceability criteria. Next, follow these same instructions for rutting criteria, i.e., divide Column 4 by Column 7 and enter in Column 8.

Step 8. Compute the total damage for both the serviceability and rutting criteria by adding the seasonal damages. When this is accomplished for all four tables (corresponding to the four trial base thicknesses), a graph of total damage versus base layer thickness should be prepared. The average base layer thickness, $\overline{D}_{BS}$, required is determined by interpolat-

Low-Volume Road Design

Table 4.4. Chart for Computing Total Pavement Damage (for both Serviceability and Rutting Criteria) Based on a Trial Aggregate Base Thickness

	TRIAL BASE THICKNESS, D_{BS} (inches) _____			Serviceability Criteria, $\Delta PSI =$ _____		Rutting Criteria, RD (inches) = _____	
(1) Season (Roadbed Moisture Condition)	(2) Roadbed Resilient Modulus, M_R (psi)	(3) Base Elastic Modulus, E_{BS} (psi)	(4) Projected 18-kip ESAL Traffic, w_{18}	(5) Allowable 18-kip ESAL Traffic, $(W_{18})_{PSI}$	(6) Seasonal Damage, $\dfrac{w_{18}}{(W_{18})_{PSI}}$	(7) Allowable 18-kip ESAL Traffic, $(W_{18})_{RUT}$	(8) Seasonal Damage, $\dfrac{w_{18}}{(W_{18})_{RUT}}$
Winter (Frozen)							
Spring/Thaw (Saturated)							
Spring/Fall (Wet)							
Summer (Dry)							
			Total Traffic =	Total Damage =		Total Damage =	

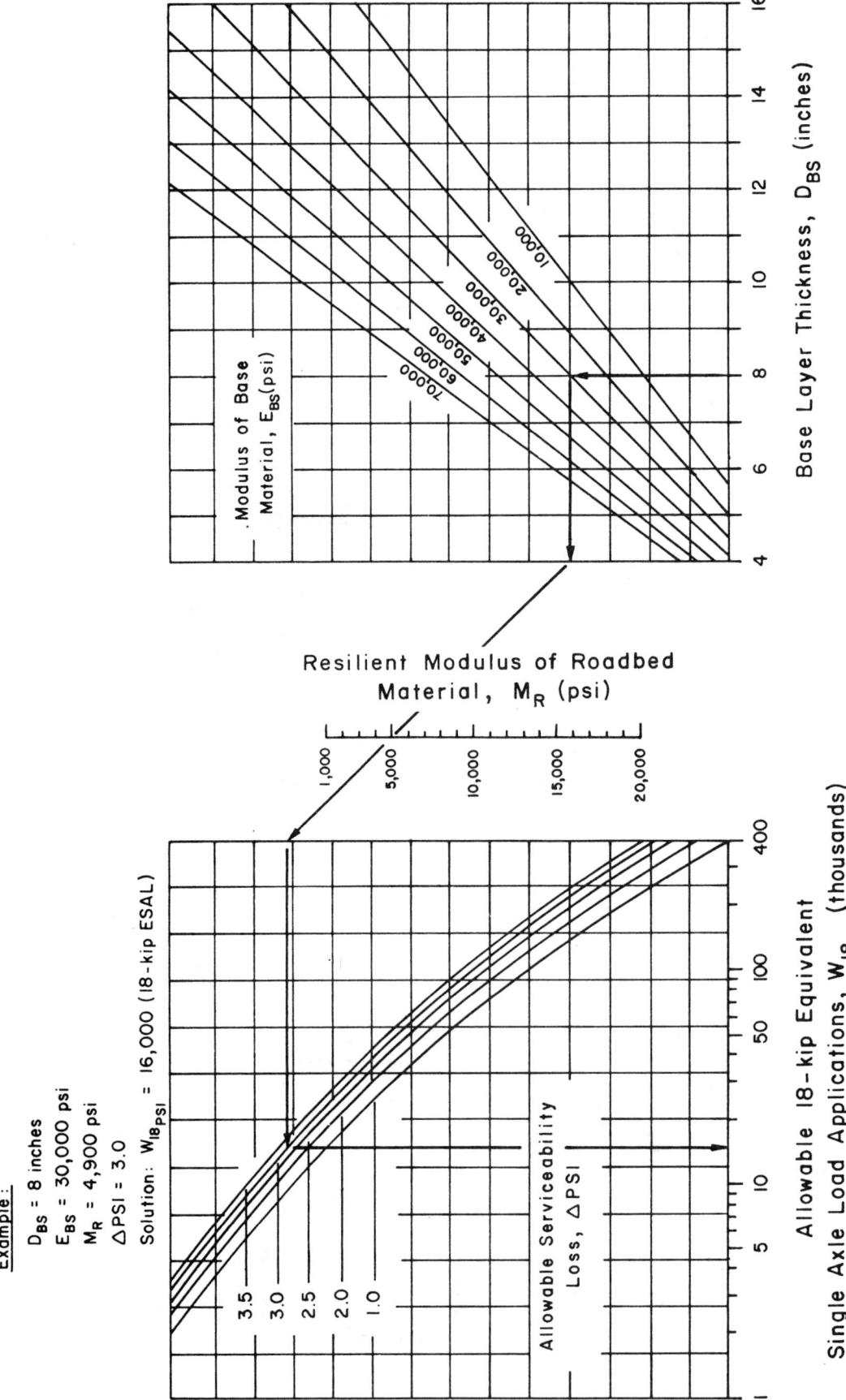

Figure 4.2. Design Chart for Aggregate-Surfaced Roads Considering Allowable Serviceability Loss

Low-Volume Road Design

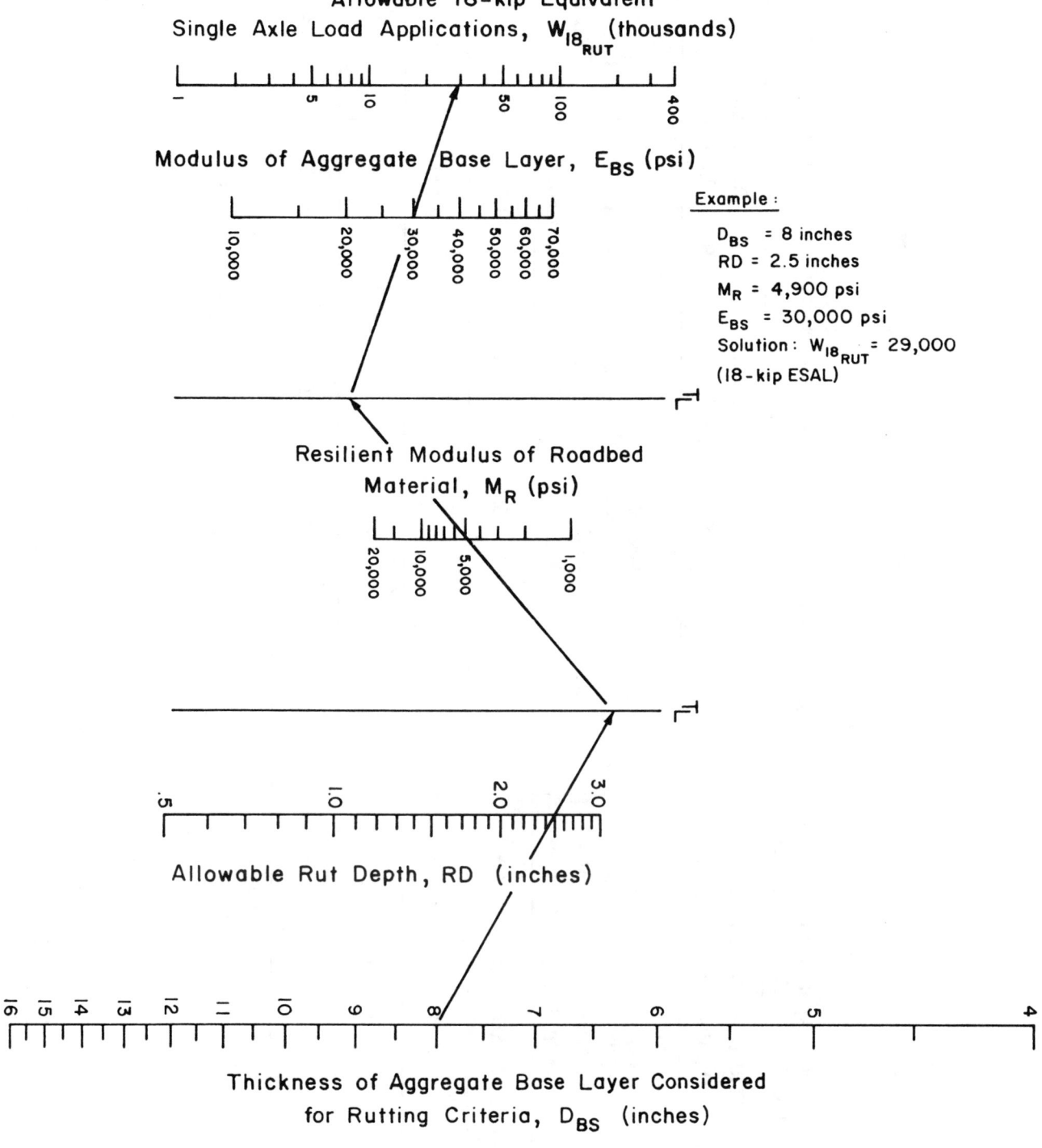

Figure 4.3. Design Chart for Aggregate-Surfaced Roads Considering Allowable Rutting

Table 4.5. Example Application of Chart for Computing Total Pavement Damage (for both Serviceability and Rutting Criteria) Based on a Trial Aggregate Base Thickness

TRIAL BASE THICKNESS, D_{BS} (inches) ___8___

(1) Season (Roadbed Moisture Condition)	(2) Roadbed Resilient Modulus, M_R (psi)	(3) Base Elastic Modulus, E_{BS} (psi)	(4) Projected 18-kip ESAL Traffic, w_{18}	Serviceability Criteria, $\Delta PSI = $ ___3.0___		Rutting Criteria, RD (inches) = ___2.5___	
				(5) Allowable 18-kip ESAL Traffic, $(W_{18})_{PSI}$	(6) Seasonal Damage, $\dfrac{w_{18}}{(W_{18})_{PSI}}$	(7) Allowable 18-kip ESAL Traffic, $(W_{18})_{RUT}$	(8) Seasonal Damage, $\dfrac{w_{18}}{(W_{18})_{RUT}}$
Winter (Frozen)	20,000	30,000	4,400	400,000	0.01	130,000	0.03
Spring/Thaw (Saturated)	1,500	30,000	2,600	4,900	0.53	8,400	0.31
Spring/Fall (Wet)	3,300	30,000	7,000	8,400	0.83	20,000	0.35
Summer (Dry)	4,900	30,000	7,000	16,000	0.44	29,000	0.24
			Total Traffic = 21,000		Total Damage = 1.81		Total Damage = 0.93

Low-Volume Road Design

ing in this graph for a total damage equal to 1.0. Figure 4.4 provides an example in which the design is controlled by the serviceability criteria: $\overline{D}_{BS}$ is equal to 10 inches.

Step 9. The base layer thickness determined in the last step should be used for design if the effects of aggregate loss are negligible. If, however, aggregate loss is significant, then the design thickness is determined using the following equation:

$$D_{BS} = \overline{D}_{BS} + (0.5 \times GL)$$

where

GL = total estimated aggregate (gravel) loss (in inches) over the performance period.

If, for example, the total estimated gravel loss was 2 inches and the average base thickness required was 10 inches, the design thickness of the aggregate base layer would be

$$D_{BS} = 10 + (0.5 \times 2) = 11 \text{ inches}$$

Step 10. The final step of the design chart procedure for aggregate-surfaced roads is to convert a portion of the aggregate base layer thickness to an equivalent thickness of subbase material. This is accomplished with the aid of Figure 4.5. Select the final base thickness desired, D_{BS_f} (6 inches is used in the example). Draw a line to the estimated modulus of the subbase material, E_{SB} (15,000 psi is used in the example). Go across and through the scale corresponding to the reduction in base thickness, $D_{BS_i} - D_{BS_f}$ (11 minus 6 equal to 5 inches is used in the example). Then, for the known modulus of the base material, E_{BS} (30,000 psi in the example), determine the required subbase thickness, D_{SB} (8 inches).

4.2 DESIGN CATALOG

The purpose of this Section is to provide the user with a means for identifying reasonable pavement structural designs suitable for low-volume roads. The catalog of designs presented here covers aggregate-surfaced roads as well as both flexible and rigid pavements. It is important to note, however, that although the structural designs presented represent precise solutions using the design procedure described in the previous section, they are based on a unique set of assumptions relative to design requirements and environmental conditions. The following specific assumptions apply to all three types of structural designs considered:

(1) All designs are based on the structural requirement for one performance period, regardless of the time interval. The range of traffic levels for the flexible and rigid pavement designs is between 50,000 and 1,000,000 18-kip ESAL applications. The allowable range of relative traffic for aggregate-surfaced road design is between 10,000 and 100,000 18-kip ESAL applications.

(2) All designs presented are based on either a 50- or 75-percent level of reliability.

(3) The designs are for environmental conditions corresponding to all six of the U.S. climatic regions. (See map in Figure 4.1.)

(4) The designs are for five qualitative levels of roadbed soil strength or support capability: Very Good, Good, Fair, Poor, and Very Poor. Table 4.2 indicates the levels of roadbed soil resilient modulus that were used for each soil classification. Table 4.1 indicates the actual lengths of the seasons used to quantify the effects of each of the six climatic regions on pavement performance.

(5) The terminal serviceability for the flexible and rigid pavement designs is 1.5 and the overall design serviceability loss used for aggregate-surfaced roads is 3.0. (Thus, if the initial serviceability of an aggregate-surfaced road was 3.5, the corresponding terminal serviceability inherent in the design solution is 0.5.)

4.2.1 Flexible Pavement Design Catalog

Tables 4.6 and 4.7 present a catalog of flexible pavement SN values (structural numbers) that may be used for the design of low-volume roads when the more detailed design approach is not possible. Table 4.6 is based on the 50-percent reliability level and Table 4.7 is based on a 75-percent level. The range of SN values shown for each condition is based on a specific range of 18-kip ESAL applications at each traffic level:

High	700,000 to 1,000,000
Medium	400,000 to 600,000
Low	50,000 to 300,000

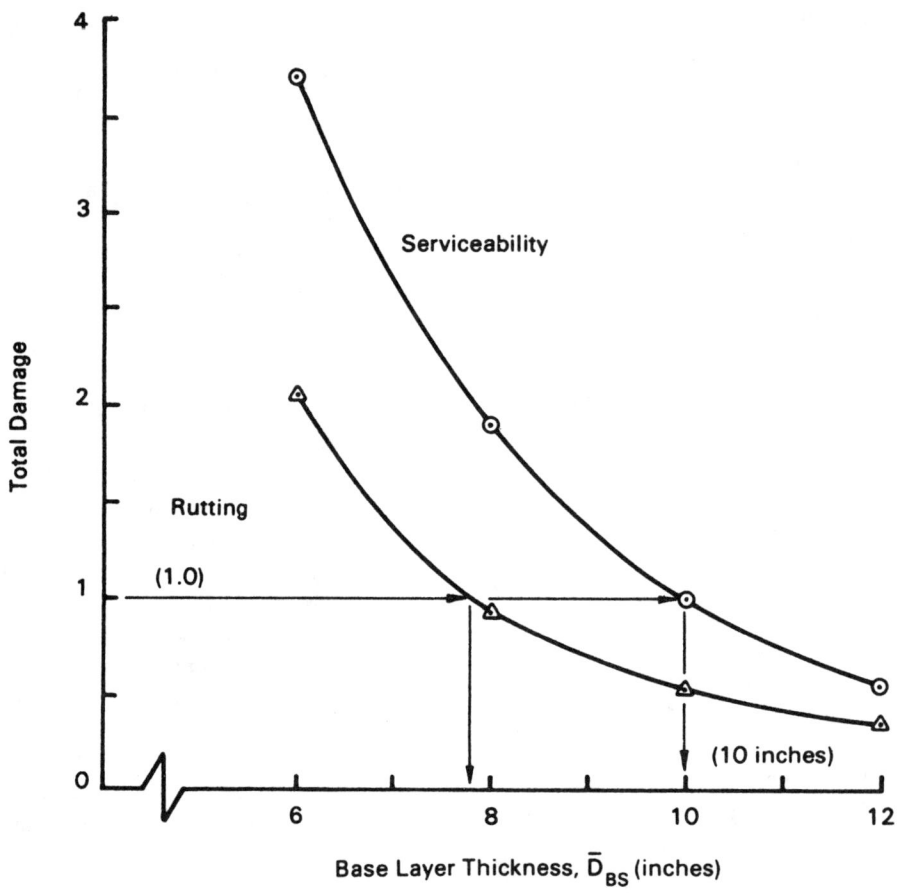

Figure 4.4. Example Growth of Total Damage Versus Base Layer Thickness for Both Serviceability and Rutting Criteria

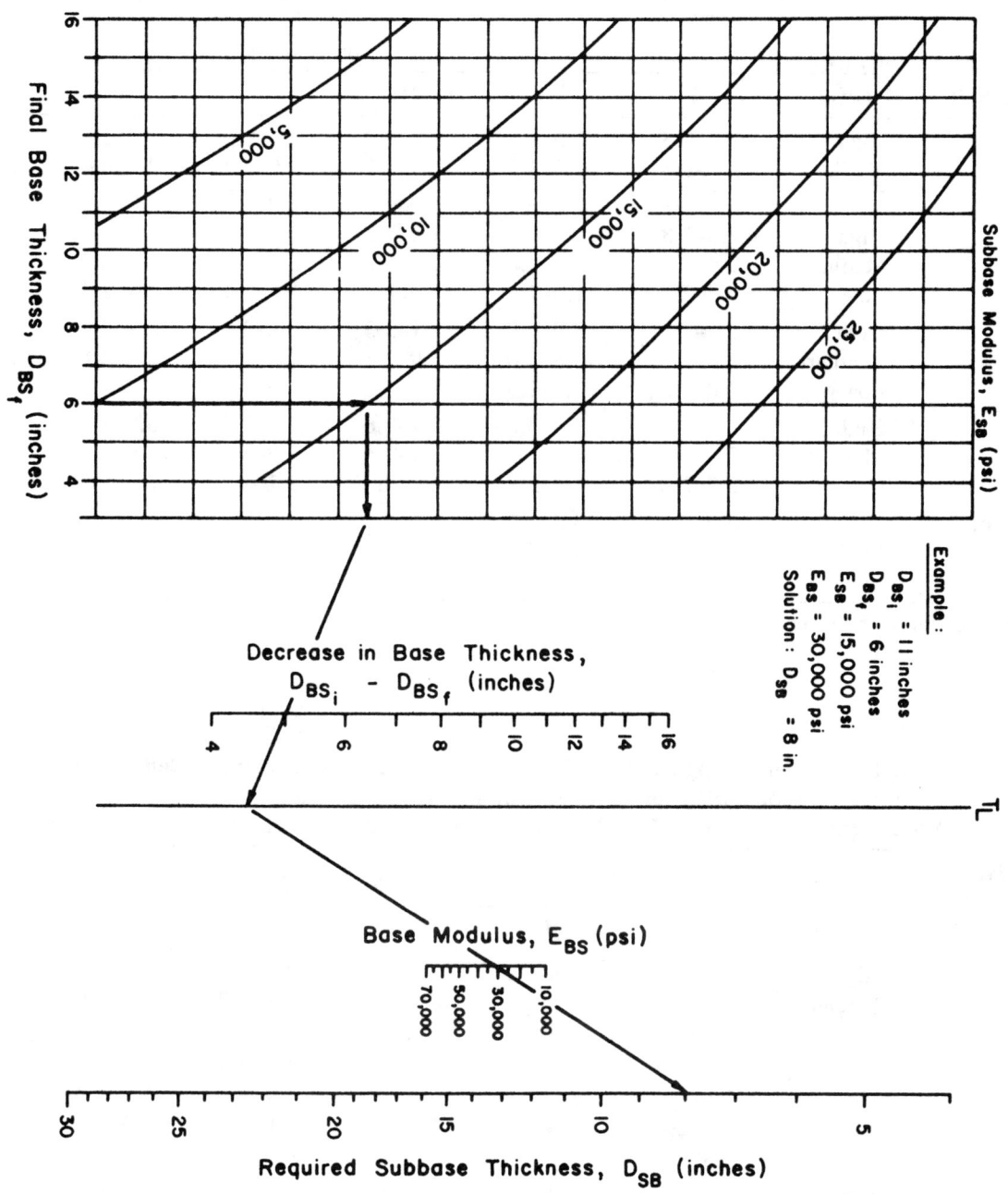

Figure 4.5. Chart to Convert a Portion of the Aggregate Base Layer Thickness To an Equivalent Thickness of Subbase

Table 4.6. Flexible Pavement Design Catalog for Low-Volume Roads: Recommended Ranges of Structural Number (SN) for the Six U.S. Climatic Regions, Three Levels of Axle Load Traffic and Five Levels of Roadbed Soil Quality—Inherent Reliability: 50 percent

Relative Quality of Roadbed Soil	Traffic Level	U.S. Climatic Region					
		I	II	III	IV	V	VI
Very good	High	2.3–2.5*	2.5–2.7	2.8–3.0	2.1–2.3	2.4–2.6	2.8–3.0
	Medium	2.1–2.3	2.3–2.5	2.5–2.7	1.9–2.1	2.2–2.4	2.5–2.7
	Low	1.5–2.0	1.7–2.2	1.9–2.4	1.4–1.8	1.6–2.1	1.9–2.4
Good	High	2.6–2.8	2.8–3.0	3.0–3.2	2.5–2.7	2.7–2.9	3.0–3.2
	Medium	2.4–2.6	2.6–2.8	2.8–3.0	2.2–2.4	2.5–2.7	2.7–2.9
	Low	1.7–2.3	1.9–2.4	2.0–2.7	1.6–2.1	1.8–2.4	2.0–2.6
Fair	High	2.9–3.1	3.0–3.2	3.1–3.3	2.8–3.0	2.9–3.1	3.1–3.3
	Medium	2.6–2.8	2.8–3.0	2.9–3.1	2.5–2.7	2.6–2.8	2.8–3.0
	Low	2.0–2.6	2.0–2.6	2.1–2.8	1.9–2.4	1.9–2.5	2.1–2.7
Poor	High	3.2–3.4	3.3–3.5	3.4–3.6	3.1–3.3	3.2–3.4	3.4–3.6
	Medium	3.0–3.2	3.0–3.2	3.1–3.4	2.8–3.0	2.9–3.2	3.1–3.3
	Low	2.2–2.8	2.2–2.9	2.3–3.0	2.1–2.7	2.2–2.8	2.3–3.0
Very poor	High	3.5–3.7	3.5–3.7	3.5–3.7	3.3–3.5	3.4–3.6	3.5–3.7
	Medium	3.2–3.4	3.3–3.5	3.3–3.5	3.1–3.3	3.1–3.3	3.2–3.4
	Low	2.4–3.1	2.4–3.1	2.4–3.1	2.3–3.0	2.3–3.0	2.4–3.1

*Recommended range of structural number (SN).

Table 4.7. Flexible Pavement Design Catalog for Low-Volume Roads: Recommended Ranges of Structural Number (SN) for Six U.S. Climatic Regions, Three Levels of Axle Load Traffic and Five Levels of Roadbed Soil Quality— Inherent Reliability: 75 percent

Relative Quality of Roadbed Soil	Traffic Level	U.S. Climatic Region					
		I	II	III	IV	V	VI
Very good	High	2.6–2.7*	2.8–2.9	3.0–3.2	2.4–2.5	2.7–2.8	3.0–3.2
	Medium	2.3–2.5	2.5–2.7	2.7–3.0	2.1–2.3	2.4–2.6	2.7–3.0
	Low	1.6–2.1	1.8–2.3	2.0–2.6	1.5–2.0	1.7–2.2	2.0–2.6
Good	High	2.9–3.0	3.0–3.2	3.3–3.4	2.7–2.8	3.0–3.1	3.3–3.4
	Medium	2.6–2.8	2.7–3.0	3.0–3.2	2.4–2.6	2.6–2.9	2.9–3.2
	Low	1.9–2.4	2.0–2.6	2.2–2.8	1.8–2.3	2.0–2.5	2.2–2.8
Fair	High	3.2–3.3	3.3–3.4	3.4–3.5	3.0–3.2	3.2–3.3	3.4–3.5
	Medium	2.8–3.1	2.9–3.2	2.7–3.3	2.7–3.0	2.8–3.1	3.0–3.3
	Low	2.1–2.7	2.2–2.8	2.3–2.9	2.0–2.6	2.1–2.7	2.3–2.9
Poor	High	3.5–3.6	3.6–3.7	3.7–3.9	3.4–3.5	3.5–3.6	3.7–3.8
	Medium	3.1–3.4	3.2–3.5	3.4–3.6	3.0–3.3	3.1–3.4	3.3–3.6
	Low	2.4–3.0	2.4–3.0	2.5–3.2	2.3–2.8	2.3–2.9	2.5–3.2
Very poor	High	3.8–3.9	3.8–4.0	3.8–4.0	3.6–3.8	3.7–3.8	3.8–4.0
	Medium	3.4–3.7	3.5–3.8	3.5–3.7	3.3–3.6	3.3–3.6	3.4–3.7
	Low	2.6–3.2	2.5–3.3	2.6–3.3	2.5–3.1	2.5–3.1	2.6–3.3

*Recommended range of structural number (SN).

Low-Volume Road Design

Once a design structural number is selected, it is up to the user to identify an appropriate combination of flexible pavement layer thicknesses which will provide the desired load-carrying capacity. This may be accomplished using the criteria for layer coefficients (a_i-values) presented in Section 2.3.5 and the general equation for structural number:

$$SN = a_1D_1 + a_2D_2 + a_3D_3$$

where

a_1, a_2, a_3 = layer coefficient for surface, base, and subbase course materials, respectively, and

D_1, D_2, D_3 = thickness (in inches) of surface, base, and subbase course, respectively.

4.2.2 Rigid Pavement Design Catalog

Tables 4.8a, 4.8b, 4.9a, and 4.9b present the catalog of portland cement pavement slab thicknesses that may be used for the design of low-volume roads when the more detailed design approach is not possible. Tables 4.8a and 4.8b are based on a 50-percent reliability level, without granular subbase and with granular subbase, respectively. Tables 4.9a and 4.9b are based on a 75-percent level, without granular subbase and with granular subbase, respectively. The assumptions inherent in these design catalogs are as follows:

(1) Slab thickness design recommendations apply to all six U.S. climatic regions.
(2) If the option to use a subbase is chosen, it consists of 4 to 6 inches of high quality granular material.
(3) Mean PCC modulus of rupture (S'_c) is 600 or 700 psi.
(4) Mean PCC elastic modulus (E_c) is 5,000,000 psi.
(5) Drainage (moisture) conditions are fair ($C_d = 1.0$).
(6) The 18-kip ESAL traffic levels are:

High	700,000 to 1,000,000
Medium	400,000 to 600,000
Low	50,000 to 300,000

(7) The levels of roadbed soil quality and corresponding ranges of effective modulus of subgrade reaction (k-value) are:

Very Good	Greater than 550 pci
Good	400 to 550 pci
Fair	250 to 350 pci
Poor	150 to 250 pci
Very Poor	Less than 150 pci

4.2.3 Aggregate-Surfaced Road Design Catalog

Table 4.10 presents a catalog of aggregate base layer thicknesses that may be used for the design of low-volume roads when the more detailed design approach is not possible. The thicknesses shown are based on specific ranges of 18-kip ESAL applications at traffic levels:

High	60,000 to 100,000
Medium	30,000 to 60,000
Low	10,000 to 30,000

One other assumption inherent in these base thickness recommendations is that the effective resilient modulus of the aggregate base material is 30,000 psi, regardless of the quality of the roadbed soil. This value should be used as input to the nomograph in Figure 4.5 to convert a portion of the aggregate base thickness to an equivalent thickness of subbase material with an intermediate modulus value between the base and roadbed soil.

Table 4.8(a). Rigid Design Catalog for Low-Volume Roads: Recommended Minimum PCC Slab Thickness (Inches) for Three Levels of Axle Load Traffic and Five Levels of Roadbed Soil Quality

Inherent reliability: 50 percent.
Without Granular Subbase

Load Transfer Devices	No				Yes			
Edge Support	No		Yes		No		Yes	
S'_c (psi)	600	700	600	700	600	700	600	700
Relative Quality of Roadbed Soil				Low Traffic				
Very good & good	5.5	5	5	5	5.25	5	5	5
Fair	5.5	5	5.25	5	5.25	5	5	5
Poor & very poor	5.5	5.25	5.25	5	5.5	5	5	5
				Medium Traffic				
Very good & good	6.25	5.75	5.75	5.25	6	5.5	5.5	5
Fair	6.25	5.75	5.75	5.25	6	5.5	5.5	5
Poor & very poor	6.25	5.75	5.75	5.25	6	5.5	5.5	5
				High Traffic				
Very good & good	7	6.25	6.25	5.25	6.5	6	5.75	5.25
Fair	7	6.25	6.25	5.75	6.5	6	6	5.5
Poor & very poor	7	6.5	6.5	6	6.5	6	6	5.5

Table 4.8(b). Rigid Design Catalog for Low-Volume Roads: Recommended Minimum PCC Slab Thickness (Inches) For Three Levels of Axle Load Traffic and Five Levels of Roadbed Soil Quality

Inherent reliability: 50 percent.
With Granular Subbase

Load Transfer Devices	No				Yes			
Edge Support	No		Yes		No		Yes	
S'_c (psi)	600	700	600	700	600	700	600	700
Relative Quality of Roadbed Soil				Low Traffic				
Very good & good	5	5	5	5	5	5	5	5
Fair	5.25	5	5	5	5	5	5	5
Poor & very poor	5.25	5	5	5	5	5	5	5
				Medium Traffic				
Very good & good	5.75	5.25	5.25	5	5.5	5	5	5
Fair	5.75	5.25	5.5	5	5.5	5	5	5
Poor & very poor	6	5.5	5.5	5	5.75	5.25	5	5
				High Traffic				
Very good & good	6.5	6	6	5.5	6	5.5	5.25	5
Fair	6.5	6	6	5.5	6	5.5	5.5	5
Poor & very poor	6.75	6	6	5.5	6.25	5.75	5.5	5

Table 4.9(a). Rigid Design Catalog for Low-Volume Roads: Recommended Minimum PCC Slab Thickness (Inches) for Three Levels of Axle Load Traffic and Five Levels of Roadbed Soil Quality

Inherent reliability: 75 percent.
Without Granular Subbase

Load Transfer Devices	No				Yes			
Edge Support	No		Yes		No		Yes	
S'_c (psi)	600	700	600	700	600	700	600	700
Relative Quality of Roadbed Soil	\multicolumn{8}{c}{Low Traffic}							
Very good & good	6	5.5	5.5	5	5.75	5.25	5.25	5
Fair	6	5.5	5.75	5.25	5.75	5.25	5.25	5
Poor & very poor	6	5.5	5.75	5.25	6	5.5	5.25	5
	\multicolumn{8}{c}{Medium Traffic}							
Very good & good	6.75	6.25	6.25	5.75	6.5	6	6	5.5
Fair	6.75	6.25	6.25	5.75	6.5	6	6	5.5
Poor & very poor	6.75	6.25	6.25	5.75	6.5	6	6	5.5
	\multicolumn{8}{c}{High Traffic}							
Very good & good	7.5	7	7	6.25	7	6.5	6.5	6
Fair	7.5	7	7	6.25	7	6.5	6.5	6
Poor & very poor	7.5	7	7	6.5	7.25	6.5	6.5	6

Table 4.9(b). Rigid Design Catalog for Low-Volume Roads: Recommended Minimum PCC Slab Thickness (Inches) for Three Levels of Axle Load Traffic and Five Levels of Roadbed Soil Quality

Inherent reliability: 75 percent.
With Granular Subbase

Load Transfer Devices	No				Yes			
Edge Support	No		Yes		No		Yes	
S'_c (psi)	600	700	600	700	600	700	600	700
Relative Quality of Roadbed Soil								
Low Traffic								
Very good & good	5.5	5	5	5	5	5	5	5
Fair	5.75	5.25	5	5	5	5	5	5
Poor & very poor	5.75	5.25	5	5	5	5	5	5
Medium Traffic								
Very good & good	6.25	5.75	5.75	5.25	6	5.5	5.5	5
Fair	6.5	5.75	6	5.5	6.25	5.5	5.5	5
Poor & very poor	6.5	6	6	5.5	6.25	5.75	5.5	5.25
High Traffic								
Very good & good	7.25	6.5	6.5	6	6.75	6	6	5.5
Fair	7.25	6.5	6.5	6	6.75	6	6	5.5
Poor & very poor	7.25	6.75	6.75	6	6.75	6.25	6.25	5.5

Table 4.10. Aggregate Surfaced Road Design Catalog: Recommended Aggregate Base Thickness (in Inches) for the Six U.S. Climatic Regions, Five Relative Qualities of Roadbed Soil and Three Levels of Traffic

Relative Quality of Roadbed Soil	Traffic Level	U.S. Climatic Region					
		I	II	III	IV	V	VI
Very good	High	8*	10	15	7	9	15
	Medium	6	8	11	5	7	11
	Low	4	4	6	4	4	6
Good	High	11	12	17	10	11	17
	Medium	8	9	12	7	9	12
	Low	4	5	7	4	5	7
Fair	High	13	14	17	12	13	17
	Medium	11	11	12	10	10	12
	Low	6	6	7	5	5	7
Poor	High	**	**	**	**	**	**
	Medium	**	**	**	15	15	**
	Low	9	10	9	8	8	9
Very poor	High	**	**	**	**	**	**
	Medium	**	**	**	**	**	**
	Low	11	11	10	8	8	9

*Thickness of aggregate base required (in inches).

**Higher type pavement design recommended.

REFERENCES FOR PART II

1. "Flexible Pavement Designer's Manual—Part I," Texas State Department of Highways and Public Transportation, Highway Division, 1972.
2. "Design Manual for Controlled Access Highways," Texas Highway Department, January 1960.
3. Van Til, C.J., McCullough, B.F., Vallerga, B.A., and Hicks, R.G., "Evaluation of AASHO Interim Guides for Design of Pavement Structures," NCHRP Report 128, 1972.
4. American Concrete Institute, "Building Code Requirements For Reinforced Concrete," (ACI 318-77).
5. Rada, G., and Witczak, M.W., "A Comprehensive Evaluation of Laboratory Resilient Moduli Results for Granular Material," TRB Papers, 1981.
6. McCullough, B.F., and Elkins, G.E., "CRC Pavement Design Manual," Austin Research Engineers, Inc., October 1979.
7. McCullough, B.F., "An Evaluation of Terminal Anchorage Installations on Rigid Pavements," Research Report No. 39-4F, Texas Highway Department, September 1966.
8. "Mass Concrete for Dams and Other Massive Structures," *Proceedings*, Journal of the American Concrete Institute, Vol. 67, April 1970.
9. Portland Cement Association, "Thickness Design for Concrete Highway and Street Pavements," 1984.
10. Majidzadeh, K., "Observations of Field Performance of Continuously Reinforced Concrete Pavements in Ohio," Report No. Ohio-DOT-12-77, Ohio Department of Transportation, September 1978.
11. Kaplar, C.W., "A Laboratory Freezing Test to Determine the Relative Frost Susceptibility of Soils," *Technical Report TR 250*, Cold Regions Research and Engineering Laboratory (CRREL), U.S. Army Corps of Engineers, 1974.
12. Lister, N.W., "Deflection Criteria for Flexible Pavements and Design of Overlays," *Proceedings*, Third International Conference on Structural Design of Asphalt Pavements, Ann Arbor, 1972.
13. Finn, F.N., and Saraf, C.L., "Development of Pavement Structural Subsystems," NCHRP Project No. 1-10B, Woodward-Clyde Consultants, February 1977.
14. Carey, W., and Irick, P., "The Pavement Serviceability Performance Concept," Highway Research Board Record 250, 1980.
15. Roberts, F.L., McCullough, B.F., Williamson, H.J., and Wallin, W.R., "A Pavement Design and Management System for Forest Service Roads: A Working Model—Phase II," Research Report 43, Council for Advanced Transportation Studies, University of Texas at Austin, February 1977.
16. McCullough, B.F., and Luhr, D.R., "A Pavement Design and Management System for Forest Service Roads: Implementation—Phase III," Research Report 60, Council for Advanced Transportation Studies, University of Texas at Austin, January 1979.
17. McCullough, B.F., and Luhr, D.R., "The New Chapter 50" Revisions to the Transportation Engineering Handbook and New Pavement Design and Management System; Draft Report Project FSH 7709.11, submitted by the Center for Transportation Research to Forest Service, June 1982.

PART III
PAVEMENT DESIGN PROCEDURES FOR REHABILITATION OF EXISTING PAVEMENTS

CHAPTER 1
INTRODUCTION

This chapter provides an overview to Part III of the Design Guide which examines the rehabilitation of existing pavement systems. A brief background relative to the analysis procedures for rehabilitation is first presented, followed by a discussion of the scope of Part III. Assumptions and limitations associated with this material are discussed, as well as the general organization and objectives of the chapters comprising Part III.

1.1 BACKGROUND

The 1981 edition of the Design Guide contained a specific chapter dealing with overlay design procedures, but no unique AASHTO overlay method was introduced. The Guide simply presented a brief summary overview of various overlay approaches and noted that, "state highway agencies are encouraged to develop procedures applicable to their specific conditions and requirements."

In recent years, the emphasis of highway construction has gradually shifted from new design and construction activities to maintenance and rehabilitation of the existing network. This critical change in project emphasis clearly necessitates the development of guidelines for specific major rehabilitation procedures and their engineering consequences. Thus, Part III has been developed to expand the previous treatment of rehabilitation in the AASHTO *Design Guide for Pavement Structures*.

The Guide methodologies presented in Part II (Design of New/Reconstructed Pavements), coupled with the methodologies of Part III (Rehabilitation), afford the engineer with the means to develop a comprehensive approach to pavement performance analysis on a project level management system framework. When Parts II and III are used collectively, pavement performance may be assessed within an analysis period that may encompass one or more rehabilitation cycles. In addition, both of these Parts are flexible in that they may be used independently to provide detailed guidance relative to either new designs or major rehabilitation.

1.2 SCOPE

The major objective of Part III is to present the comprehensive framework of a method for selecting the best major rehabilitation strategy (or strategies) for use on a specific project. It is important to recognize that major rehabilitation activities discussed in Part III encompass not only structural overlay procedures, but other major rehabilitation methods as well. Of equal importance is the fact that no guidance is presented in Part III for the use of overlays as a tool to improve the skid-resistant qualities of a pavement surface. Guidance on skid resistance is contained in the 1976 AASHTO publication *Guidelines for Skid Resistant Pavement Design*.

The overall philosophy of the rehabilitation approach is based upon the AASHTO design-serviceability-performance concepts used in Part II for new pavement designs. This performance-based framework allows for a combined design-rehabilitation strategy to be analyzed over a predefined analysis period. This, in turn, allows for a comprehensive framework to be developed in order to estimate the probable life-cycle costs of any given strategy within the analysis period. Such an approach is necessary if economic principles are to be applied as one of the decision criteria for the eventual selection of the preferred rehabilitation strategy from several possible (and technically feasible) solutions.

While Part III is intended to serve as a self-contained solution method, the user will quickly discover the need to make direct use of the methodology presented in Part II. This is necessary because the structural overlay procedure presented here requires new structural designs, found in Part II, as an integral part of the rehabilitation analysis. Also noteworthy is that the approach presented for the structural overlay analysis of pavement systems lends itself to developing input for use with the more mechanistic overlay approaches discussed in Part IV of the Guide.

The structural overlay analysis presented in Part III is based, in part, on two relatively new concepts. First, the role of nondestructive dynamic deflection testing is emphasized as the key tool in evaluating characteristics of the existing pavement. In addition, the concept of remaining pavement life is directly incorporated into the overlay methodology.

The rehabilitation methodology of Part III is applicable to all major types of existing pavement systems. Similarly, methods for both flexible and rigid overlays are presented for any type of existing pavement system. Also discussed within the overall approach is the use of either new (virgin) or recycled material as the sole source of material.

Finally, while Part III examines a comprehensive approach to the rehabilitation of pavements, the user will note that the philosophy of methodology is broader in scope than the more well-defined, methodical solution of Part II. The major reason for this is that significant differences exist between the current new design-performance relationships and rehabilitation performance knowledge. While analytical solutions to portions of the rehabilitation methodology are presented, the engineer must recognize that it may be impossible to accurately determine the optimal rehabilitation solution from a rigorous analytical model. However, the user should not be discouraged from employing this approach but rather feel encouraged to use every available tool at his/her disposal to determine the problem cause, identify potentially sound and economic solution alternatives, and then select the most preferred rehabilitation strategy from sound engineering experience.

1.3 ASSUMPTIONS/LIMITATIONS

Because the structural overlay method is based, in part, on the AASHTO design-performance concepts of Part II, the limitations and assumptions associated with the new pavement design methodology are applicable to the overlay portion of Part III. The fundamental approach used for all overlay-existing pavement combinations is based on the "Thickness Deficiency" overlay approach (i.e., the existing thickness is inadequate for anticipated future traffic). This requires evaluation of the existing pavement system, principally through the use of nondestructive testing (NDT), to determine the effective structural capacity of the existing pavement prior to overlay.

While the Thickness Deficiency approach has been used in practice for many years, it lacks some degree of field verification for design-performance prediction when compared to the procedures for new pavement designs. In addition, while the state of the art associated with the use/analysis of NDT deflection data is considered good, changes and advancements in NDT technology are constantly improving the accuracy of this methodology in practice. While the recognition of possible future improvements should be a consideration, the fundamental approach presented in Part III can serve as the basic framework for structural overlay evaluation for the foreseeable future.

Part III also incorporates the use of major rehabilitation methods other than overlays. In general, one of the least understood areas of state of the art rehabilitation concerns the ability to confidently and accurately predict probable performance (e.g., serviceability-traffic loading/time) for nonoverlay rehabilitation solutions. This is one of the most significant limitations of the rehabilitation guidelines, and user agencies are strongly encouraged to build a continuous and accurate performance data base to increase the overall accuracy and confidence level of performance predictions. In addition, while major nonoverlay rehabilitation methods are presented in Part III, the user must not view these as being all-inclusive. As the state of the art increases, future revisions of Part III will incorporate additional nonoverlay rehabilitation methods that have been successfully used in practice.

The overlay design procedure for flexible pavement presented by these guidelines is considered to represent the state of the art with respect to the rehabilitation of pavements with structural sections deficient in strength and/or thickness for the traffic loadings which have been applied, as evidenced by permanent deformation. For those pavements in which the primary distress mechanism is fatigue cracking without permanent deformation, other empirical or mechanistic-empirical design procedures based on nondestructive testing may be more appropriate.

With respect to rigid pavements, the following procedures are considered applicable and appropriate for those situations in which, based on visual observation and the results of nondestructive tests, there exists a structural section deficiency. In those cases where the distress mechanism is due to causes other than a deficiency in structural section thickness and/or strength, avoidance of reflective cracking will control the design of the rehabilitation.

1.4 ORGANIZATION

Part III is organized into three major sections. Chapter 2 presents the general fundamentals associated with pavement rehabilitation, rehabilitation types, approaches to use, and the decision process for selecting preferred rehabilitation treatment. Chapter 3 details guidelines for collecting information from both

Introduction

the field and historic records for use in the rehabilitation process. This information then forms the basis for the rehabilitation methodology presented. Chapters 4 and 5 discuss the specific rehabilitation methods. In Chapter 4, rehabilitation approaches other than overlays are examined, while Chapter 5 details the structural overlay method for all pavement types. Examples are presented in both chapters to illustrate and clarify procedure specifics.

CHAPTER 2
REHABILITATION CONCEPTS

2.1 BACKGROUND

The main objective of Part III is to provide guidance for major rehabilitation activities. In this Guide, the term "rehabilitation" encompasses the activities described in the 4R program—resurfacing, restoration, rehabilitation, and reconstruction. In short, major rehabilitation activities will be viewed as any work that is undertaken to significantly extend the service life of an existing pavement through the principles of resurfacing, restoration, and/or reconstruction.

Major rehabilitation activities differ markedly from periodic maintenance activities (sometimes called normal, routine and/or preventive maintenance) in that the primary function of the latter activity is to preserve the existing pavement so that it may achieve its applied loading, while rehabilitation is undertaken to significantly increase the functional life. While periodic maintenance is a vital part of the overall performance cycle of any highway, this topic is not discussed within Part III. Therefore, no guidance is presented relative to the use of thin asphaltic overlays (generally less than $3/4$ inch), overlays of short (spot) length, pavement patching, pothole repairs, routine sealing of cracks and joints, miscellaneous repair of minor pavement failures, slab sealing (other than as an essential part of major rehabilitation), or any other work designed to preserve the existing pavement system.

2.2 REHABILITATION FACTORS

2.2.1 Major Categories

As noted, Part III of the Design Guide specifically addresses major rehabilitation pavement activities. For simplicity, major rehabilitation is subdivided into two major categories:

(1) Rehabilitation Methods Other Than Overlay
(2) Rehabilitation Methods With Overlays

These categories will be discussed in Chapters 4 and 5, respectively. It should be recognized that some methods discussed in Chapter 4 (Rehabilitation Methods Other Than Overlay) may be used/required as pre-overlay treatments in major rehabilitation work.

2.2.2 Recycling Concepts

The broad category of material source is a primary factor in the rehabilitation process for the engineer to consider. Materials used in rehabilitation can be obtained from new (virgin) sources (i.e., aggregates and binders), from recycled sources, or from a combination of the two. Cost should be the primary factor used in deciding to use recycling.

Recycling of existing pavement materials for rehabilitation purposes offers promise as a partial solution by offering the following benefits: conservation of aggregates, binders, and energy; preservation of the environment and existing pavement geometrics and the benefits associated with a potential reduction in project cost. Appendix OO contains a more detailed discussion of recycling in terms of definitions, types of recycling, and design material properties for recycled materials.

2.2.3 Construction Considerations

Another important factor in the major rehabilitation process is the choice of construction method. The engineer should view the full-depth reconstruction of a pavement as the extreme opposite of a full overlay. Obviously, a wide range of construction choices between these two limits is feasible (e.g., partial-depth reconstruction with or without the application of an overlay).

2.2.4 Summary of Major Rehabilitation Factors

The previous sections clearly indicate that major rehabilitation strategies should be viewed in a broad context with reference to three major factors. They are:

(1) the selection of a major rehabilitation category that may or may not involve an overlay (resurfacing).
(2) the decision to use new (virgin) materials, recycled materials, or a combination of both (it should be noted that recycled materials need not be those obtained from the specific pavement project being rehabilitated, but may be obtained from a variety of other recycled material stockpiles).
(3) the decision to employ full reconstruction (i.e., complete removal/replacement), partial reconstruction, a direct (full) overlay, or some combination of reconstruction and overlay.

Since the major factors listed above may act in combination with each other, the engineer quickly realizes that a complex combination of rehabilitation alternatives exists for a single project. For example, rehabilitation of a structurally failed (cracked) PCC pavement requires the analysis of several potential rehabilitation strategies before the optimum or preferred strategy can be selected. The optimum solution will be obtained by a life-cycle cost analysis.

Many of the rehabilitation methods available are presently being tried on an experimental basis and lack full verification. Part III deals only with major rehabilitation methods. Table 2.1 summarizes these methods and cites their chapter location. This list simply serves as a reminder of the potentially large number of initial strategies that may be investigated to arrive at a final rehabilitation recommendation. It should be noted that two major rehabilitation concepts, recycling and break/seat approach for asphalt overlays over existing rigid pavements, are directly integrated into discussions/methodologies that deal with the structural analysis of overlay systems.

2.3 SELECTION OF ALTERNATIVE REHABILITATION METHODS

2.3.1 Overview

This section provides overall guidance for the selection of pavement rehabilitation methods. Pavement rehabilitation is as much an art as a science. With the exception of the various overlay models presented in Chapter 5, there are no definitive equations, guides, or step-by-step procedures that one can use to "cookbook" a proper rehabilitation design. Therefore, a considerable amount of both analysis and engineering judgment must be applied to each project. Due to state of the art limitations relative to the entire rehabilitation process, a definite need exists for continuous feedback from agencies on the performance of various rehabilitation methods.

Table 2.1. Major Rehabilitation Concepts in Guide

Description/Factor	Guide Location
Rehabilitation methods other than overlay	Chapter 4
1. Full depth pavement repair	
2. Partial depth pavement repair	
3. Joint and crack sealing	
4. Subsealing of concrete pavements	
5. Grinding/milling of pavements	
6. Subdrainage design	
7. Pressure relief	
8. Restoration of joint load transfer	
9. Surface treatments	
Rehabilitation methods with overlay	Chapter 5
1. Flexible overlay/flexible existing	
2. Flexible overlay/rigid existing	
3. Rigid overlay/flexible existing	
4. Rigid overlay/rigid existing	
Special rehabilitation	
1. Recycling	Appendix OO; Chapter 5
2. Break/seat	Chapter 5

Despite incomplete knowledge, the engineer must make rehabilitation decisions based on the most adequate information available. There are no "right" and "wrong" solutions to pavement rehabilitation problems, but rather "better" or "optimum" solutions. The truly "optimum" solution, which maximizes benefits while minimizing costs, is often not attainable due to constraints imposed (i.e., limited funding). However, there will be a "preferred" solution which is cost-effective, has other desirable characteristics, and meets the existing constraints. The engineer has a responsibility to determine, to the best of his or her ability, the most "preferred" rehabilitation method given particular conditions and limitations.

While selection of the preferred solution is a very complex engineering problem, rehabilitation analysis is made easier by using a logical step-by-step approach. The fundamentals of the approach are based on the necessity to: (1) determine cause of the distress(es) or pavement problems, (2) develop a candidate list of solutions that will properly address (cure and prevent future occurrences) the problem, and (3) select the preferred rehabilitation method given economic and other project constraints. The principal steps in this selection process are illustrated in Figure 2.1 and are described in detail in this section.

2.3.2 Problem Definition

Phase I of the pavement rehabilitation selection process is problem definition. To avoid making an inaccurate problem definition, the engineer must collect and evaluate enough information about the pavement to adequately comprehend the situation. The premature failure of many rehabilitated pavements can be traced to inadequate evaluation. In summary, the first step is to identify/establish the condition of the pavement.

Data Collection. Pavement evaluation requires substantial data collection, which can be divided into the following major categories:

> pavement condition
> shoulder condition
> pavement design
> geometric design
> materials and soils properties
> traffic volumes and loadings
> climate conditions
> drainage conditions
> safety considerations

Specific data collection items depend in part on the type of rehabilitation being considered. For example, if grinding of a concrete pavement is being considered, the pavement design, hardness of the large aggregate in the concrete, traffic level, surface profile, traffic control options, and magnitude of faulting must be known. Figure 2.2 summarizes the data required for specific rehabilitation alternatives.

Each agency should develop guidelines to determine what data to collect, as well as standard procedures for collection.

Data Evaluation. During the data collection and evaluation process, the engineer should acquire adequate information to thoroughly define the problem. Because limited time and funds are allotted to this portion of the rehabilitation process, each agency should develop a standard data collection/evaluation procedure that best suits its information, personnel, and equipment resources. A sample procedure is outlined below.

Step 1. Office Data Collection—includes information such as location of the project, year constructed, year and type of major maintenance, pavement design, materials and soils properties, traffic, climate conditions, and any available performance data.

Step 2. First Field Survey—includes items such as distress, drainage conditions, subjective roughness, traffic control options, and safety considerations. Detailed procedures for collecting pavement condition data are given in Part III, Chapter 3.

Step 3. First Data Evaluation and the Determination of Additional Data Needs—based on this first evaluation, a list of candidate rehabilitation alternatives may be developed to aid in assessing additional data needs.

Step 4. Second Field Survey—detailed measuring and testing; includes such items as coring and sampling, roughness measurement, deflection testing, skid resistance, drainage tests, and vertical clearances.

Step 5. Laboratory Testing of Samples—includes tests such as material strength, resilient modulus, permeability, moisture content, composition, density, and gradations (if felt to be necessary).

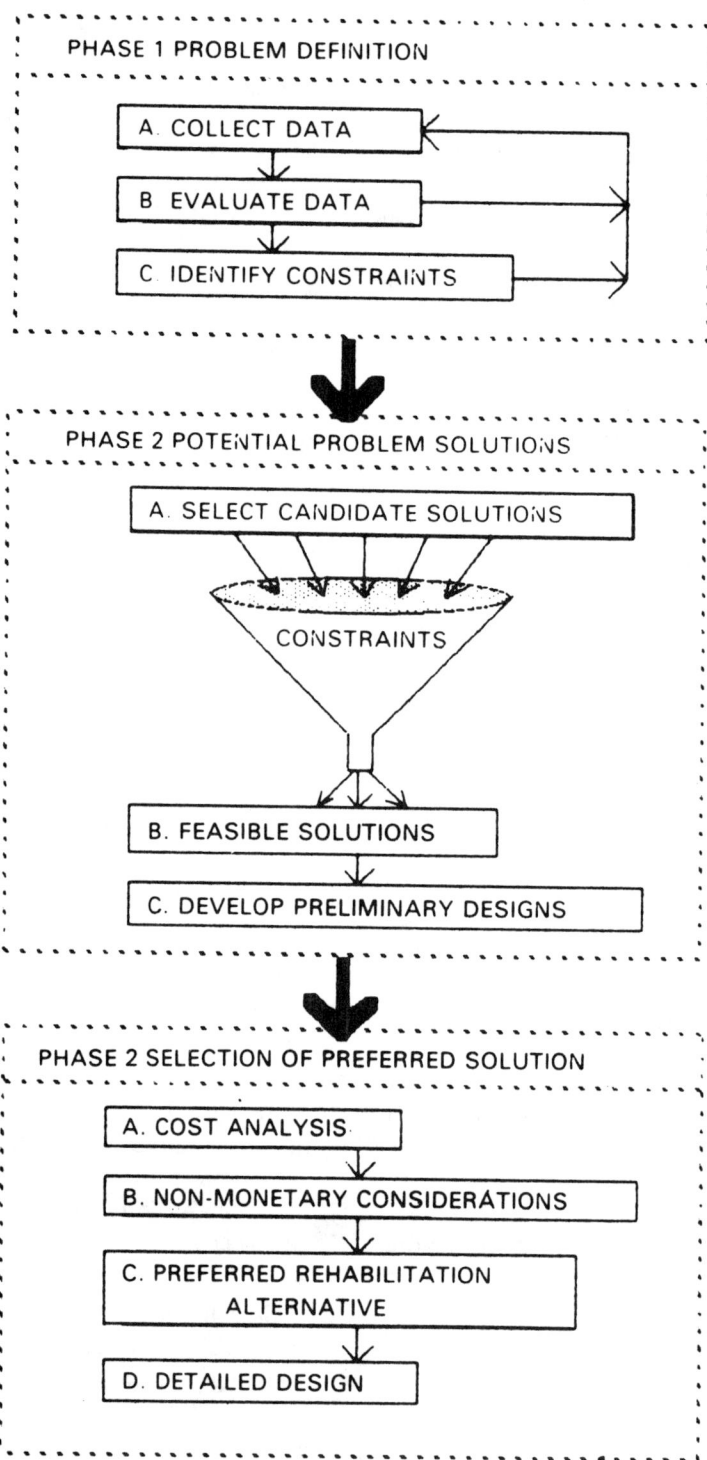

Figure 2.1. The Pavement Rehabilitation Selection Process

Rehabilitation Concepts

DATA REQUIRED	FULL-DEPTH REPAIR	PARTIAL DEPTH PATCHING	GRINDING	RECYCLING	UNDERSEALING	SLAB JACKING	SUBDRAINS	JOINT RESEALING	PRESSURE REBEL JOINTS	LOAD TRANSFER RESTORATION	SURFACE TREATMENT	OVERLAY
PAVEMENT DESIGN	⊙	⊙	⊙	⊙	⊙	⊙	⊙	⊙	⊙	⊙	⊙	⊙
ORIGINAL CONSTRUCTION DATA	•	⊙	•	•	•	•	•	•	•	•	•	•
AGE	•	•	•	•	•	•	•	•				•
MATERIALS PROPERTIES	⊙	•	•	⊙	•		•					⊙
SUBGRADE				⊙		•	⊙	•				⊙
CLIMATE				•	•		•		•		⊙	•
TRAFFIC LOADINGS AND VOLUMES	⊙		⊙	⊙	•		⊙	•	•	⊙	⊙	⊙
DISTRESS	⊙	⊙	⊙	⊙	⊙	⊙	⊙	⊙	⊙	⊙	⊙	⊙
NDT	•		•	•	⊙					⊙	•	⊙
DESTRUCTIVE TESTING & SAMPLING	⊙	⊙	•	⊙	•					•		⊙
ROUGHNESS			•	•		•						
SURFACE PROFILE			⊙			⊙						
DRAINAGE	⊙		•	⊙	•		⊙	⊙			•	⊙
PREVIOUS MAINTENANCE	⊙		•	•	•		•	•	•		•	•
BRIDGE PUSHING									⊙			
UTILITIES	⊙			⊙	•		•					⊙
TRAFFIC CONTROL OPTIONS	⊙	⊙	⊙	⊙	⊙	⊙	⊙	⊙	⊙	⊙	⊙	⊙
VERTICAL CLEARANCE				⊙							•	•
GEOMETRICS				•								•

KEY: ⊙ DEFINITELY NEEDED   • DESIRABLE

Figure 2.2. Data Required for Various Rehabilitation Techniques

Step 6. Second Data Evaluation—includes structural evaluation, functional evaluation, and determination of additional data requirements, if any.

Step 7. Final Field and Office Data Compilation—preparation of a final evaluation report.

To some extent, project size dictates the amount of time and money that may justifiably be spent on pavement evaluation. Major highways and high traffic volume roads certainly require a more thorough and comprehensive evaluation than do low-volume roads.

The collected data must be carefully evaluated and summarized in a systematic fashion. Figure 2.3 presents a comprehensive list of factors to examine in an adequate pavement evaluation. Each agency should adapt this list according to their own particular needs. It is vital that the agency then develop procedures and guidelines for consistently answering the questions on their list. Many items can be obtained for evaluation from existing data routinely collected. Agencies having substantial pavement management systems will already have a large block of information in their data banks. Other items will require direct field testing for current or detailed information.

Identify Constraints. Constraints placed on a pavement rehabilitation project should be identified during the problem definition phase since they frequently affect the choice of rehabilitation alternative. Some constraints which may restrict alternative selection are:

> limited project funding
> traffic control problems (lane closure
>   availability)
> minimum desirable life of rehabilitation
> geometric design problems
> utilities
> clearances
> right-of-way
> available materials and equipment
> contractor expertise and manpower
> agency policies

A particularly difficult constraint to deal with involves network considerations. When evaluating the problems of a particular pavement and the possible rehabilitation alternatives, an agency must consider the needs and priorities of the entire network for which it is responsible. The best rehabilitation approach for an individual project may not be in the best interest of the network as a whole.

Project constraints often limit the number of rehabilitation alternatives available. Where possible, careful planning should be used to circumvent these constraints; the more they are permitted to affect a project, the less likelihood there is of obtaining the best available solution.

2.3.3 Potential Problem Solutions

Phase 2 of the pavement rehabilitation selection process, as outlined in Figure 2.1, is the identification of potential problem solutions. The first step in this phase is the identification of candidate solutions that appear to be technically feasible in solving a pavement deterioration problem. Next, candidate solutions are subjected to the project constraints, and those that meet the constraints are considered feasible rehabilitation solutions.

Select Candidate Solutions. After completion of Phase 1, Problem Definition, the design engineer should be able to suggest several candidate rehabilitation solutions. Candidate solutions are those which address the causes of the deterioration and are effective in both repairing the existing distress and preventing, as much as possible, recurrence. After selecting candidate solutions, the engineer must determine the quantity of work required by each alternative, since this will have a bearing on cost.

It is very easy, and very unwise, to perform a "quick fix," or worse yet, a cosmetic treatment, on a deteriorated pavement. Funds spent on such superficial repairs are funds wasted. If mechanisms which cause distress are not treated, the distress will continue to appear and increase in severity. The short-lived benefits achieved from superficial repairs never justify the costs. The quick fix treatments are not inherently bad; they are simply uneconomical.

In general, rehabilitation is considered only for significantly damaged portions of a pavement. For instance, if one mile of a three-mile pavement section is badly distressed, usually only that one mile receives rehabilitation. This does not mean that only high-severity distress merits rehabilitation work. It may be economically justifiable to spend additional funds repairing some lower-severity distress at the same time adjacent high-severity distress is being corrected. The additional cost must be weighed against the benefit obtained by "intercepting" distress at an earlier stage in its development. Also, in terms of convenience, it may be beneficial to carry out simultaneous repairs on both high- and low-severity distress on a high-volume road if major rehabilitation work creates significant traffic-handling problems.

STRUCTURAL EVALUATION
  Existing distress:
    Little or not load-associated distress
    Moderate load-associated distress
    Major load-associated distress
  Structural Load-Carrying Capacity Deficiency:
    Yes, No

FUNCTIONAL EVALUATION
  Roughness:
    Very Good, Good, Fair, Poor, Very Poor
    Measurement: _____
  Present Serviceability Index/Rating: _____
  Skid Resistance:
    Satisfactory, Questionable, Unsatisfactory
  Rutting Severity:
    Low, Medium, High

VARIATION OF CONDITION EVALUATION
  Systematic variation along project:
    Yes, No
  Systematic variation between lanes:
    Yes, No
  Localized variation (very bad areas) along project:
    Yes, No

CLIMATIC EFFECTS EVALUATION
  Climatic Zone
    Moisture Region: I   Moisture throughout year
                               II Seasonal moisture
                               III Very little moisture
    Temperature Region: A   Severe frost penetration
                               B   Freeze-thaw cycles
                               C   No frost problems
  Severity of moisture-accelerated damage:
    Low, Medium, High
    Describe (asphalt stripping, pumping, _____)
  Subsurface drainage capability-BASE:
    Satisfactory, Marginal, Unacceptable
  Subsurface drainage capability-SUBGRADE:
    Satisfactory, Marginal, Unacceptable
  Surface drainage capability:
    Acceptable, Needs Improvement
    Describe: _____

PAVEMENT MATERIALS EVALUATION
  Surface-Sound condition, Deteriorated
    Describe: _____
  Base-Sound condition, Deteriorated
    Describe: _____
  Subbase-Sound condition, Deteriorated
    Describe: _____

Figure 2.3. Overall Pavement Evaluation Summary and Checklist

SUBGRADE EVALUATION
　Structural support:
　　Low, Medium, High
　Moisture softening potential:
　　Low, Medium, High
　Temperature problems:
　　None, Frost Heaving, Freeze-Thaw Softening
　Swelling Potential:
　　Yes, No

PREVIOUS MAINTENANCE PERFORMED EVALUATION
　Minor, Normal, Major
　Has lack of maintenance contributed to deterioration?
　　Yes, No
　　Describe: _____

RATE OF DETERIORATION EVALUATION
　Long Term:
　　Low, Normal, High
　Short Term:
　　Low, Normal, High

TRAFFIC CONTROL DURING CONSTRUCTION
　Are detours available so that facility can be closed?
　　Yes, No
　Must construction be accomplished under traffic?
　　Yes, No
　Could construction be done at off-peak hours?
　　Describe _____

GEOMETRIC AND SAFETY FACTORS
　Current Capacity:
　　Adequate, Inadequate
　Future Capacity:
　　Adequate, Inadequate
　Widening Required Now:
　　Yes, No
　List high-accident locations: _____
　Bridge clearance problems: _____
　Lateral obstruction problems: _____
　Utilities problems: _____
　Bridge pushing problems: _____

TRAFFIC LOADINGS
　ADT(two-way): _____
　AADT(two-way): _____
　Accumulated 18-kip ESAL/year: _____
　Current 18-kip ESAL/year: _____

SHOULDERS
　Pavement Condition:
　　Good, Fair, Poor
　Localized Deteriorated Areas:
　　Yes, No

Figure 2.3. Continued—Overall Pavement Evaluation Summary and Checklist

Feasible Rehabilitation Solutions. As stated, feasible rehabilitation solutions for a particular case of pavement distress are obtained by weighing candidate solutions against project constraints. A feasible alternative is defined as one that addresses the cause of the distress and is effective in both repairing the existing deterioration and preventing its recurrence, while satisfying all the imposed constraints.

A feasible rehabilitation alternative may encompass more than one repair technique. Combined rehabilitation techniques may be necessary to repair either single- or multiple-distress types for a particular project. It is the engineer's responsibility, based on project evaluation results, to determine the techniques or combination of techniques to be considered as feasible rehabilitation alternatives for a particular pavement.

Development of Preliminary Designs. After all feasible alternatives have been selected, preliminary designs should be prepared. Preliminary design, including such things as approximate overlay thickness selection, requires only approximate cost estimates. Design rehabilitation projects require as much technical expertise as new pavement design.

2.3.4 Selection of Preferred Solution

Phase 3 of the pavement rehabilitation selection process, as illustrated in Figure 2.1, is the selection of a preferred solution. There is no infallible method for selecting the most "preferred" rehabilitation alternative for a given project. Rather, the selection process requires considerable engineering judgment, creativity, and flexibility. Each agency should develop a procedure to select preferred solutions for their projects using both monetary and nonmonetary considerations.

Cost Analysis. Cost of rehabilitation alternatives is generally considered the most important decision criteria when choosing the preferred solution. The various types of costs incurred over the life of a pavement are discussed in Part I of this Guide. Presented here are a few important points about life-cycle cost analysis as it pertains to the selection of a rehabilitation method.

Life-cycle cost analysis requires inputs of both cost and time. Unfortunately, both of these elements are subject to a large degree of uncertainty. For instance, the effective life of a rehabilitation technique is subject to the following influences:

the skill and care with which the work is performed
the quality of the materials used
environmental conditions prevalent in the region where the pavement exists
the traffic which uses the pavement
other rehabilitation and maintenance work being performed concurrently

Even the engineer familiar with the performance of various rehabilitation methods in his or her local area can appreciate the difficulty of selecting appropriate inputs for use in the life-cycle cost analysis. To eliminate as much uncertainty as possible, it is essential to begin collecting rehabilitation performance data in the pavement management data bank. This is crucial to life-cycle cost analysis.

Another important consideration regarding life-cycle cost analysis is that the same rehabilitation techniques, when applied to different pavements, may have variant effects. Furthermore, some methods may keep a pavement at a consistently high-condition level, while others may allow the condition of the same pavement to fluctuate. Thus, discrepancy is often not revealed by the cost analysis if user costs are not included in the calculations. It is therefore important to include user costs in a cost analysis.

Nonmonetary Considerations. Several nonmonetary factors should be considered when determining the preferred rehabilitation method. Some of these factors are:

service life
duration of construction
traffic control problems
reliability (proven design in region)
constructibility
maintainability

As with monetary considerations, the service life of a rehabilitation method is an important factor. This is particularly significant to agencies responsible for high-volume roads, for which lane closures and traffic delays pose considerable difficulties. The important time parameter is years of pavement life extension achieved by the rehabilitation methods and should be a factor in almost any decision criterion used by the agency.

Preferred Rehabilitation Alternative. The preferred rehabilitation alternative for a project is selected using, first, monetary and then nonmonetary factors. Whenever the cost analysis does not indicate a clear advantage for one of the feasible alternatives, the nonmonetary factors may be used to aid in the selection process. A method for measuring several rehabilitation alternatives against criterion that cannot be expressed in monetary units is depicted in Figure 2.4. First, the relative importance of each criterion is assigned by the design team. Next, the alternatives are rated according to their anticipated performance in the criterion areas. Then, an alternative's rating in an area is multiplied by the assigned weight of that factor to achieve a "score." Finally, all of the scores for an alternative are summed, and the alternative with the highest score is the preferred solution. This procedure has been used successfully on projects to select the preferred pavement rehabilitation alternative.

Detailed Design. Once the preferred rehabilitation method has been selected, detailed design plans, specifications, and estimates are prepared. If a major difference in design, cost, or condition occurs during this phase, it may be necessary to reinvestigate whether this alternative is still a cost-effective solution.

2.3.5 Summary

A logical procedure for selecting the preferred rehabilitation method is presented in Figure 2.1. It provides the engineer with guidance in organizing and evaluating the information available about the pavement, identifying needs for further information and evaluation, developing feasible rehabilitation alternatives, and selecting the preferred alternative from among these using sound engineering principles.

This step-by-step procedure can help the engineer conserve time and money in selecting the rehabilitation method which best meets the pavement's needs, satisfies all the project constraints, and reflects the agency's priorities concerning use of available funds, performance demanded of the rehabilitation work, and needs of the agency's pavement network. If the procedure is well-documented and tempered by good engineering judgement, the selection of a particular rehabilitation method for a project will be justifiable to management and the public. Perhaps most important, a systematic procedure for selecting rehabilitation methods can move an agency away from the traditional "standard fix" approach of rehabilitating its pavements, toward a policy of custom designing rehabilitation to truly meet the pavements needs.

Rehabilitation Concepts

	CRITERIA								
	INITIAL COST	DURATION OF CONSTRUCTION	SERVICE LIFE	REPAIRABILITY & MAINTENANCE EFFORT	RIDEABILITY & TRAFFIC ORIENTATION	PROVEN DESIGN IN STATE CLIMATE		TOTAL COST	RANK
RELATIVE IMPORTANCE	20%	20%	25%	15%	5%	15%		100	
ALTERNATIVE 1	60 / 12	60 / 12	100 / 25	80 / 12	90 / 4.5	100 / 15		80.5	1
ALTERNATIVE 1A	60 / 12	60 / 12	100 / 25	80 / 12	90 / 4.5	100 / 15		80.5	1
ALTERNATIVE 2	60 / 12	60 / 12	70 / 17.5	50 / 7.5	60 / 3	40 / 6		58	5
ALTERNATIVE 2A	60 / 12	60 / 12	70 / 17.5	50 / 7.5	60 / 3	40 / 6		58	5
ALTERNATIVE 3	60 / 12	40 / 8	100 / 25	80 / 12	100 / 5	90 / 13.5		75.5	2
ALTERNATIVE 4	60 / 12	80 / 6	40 / 10	20 / 3	40 / 2	20 / 3		44	8
ALTERNATIVE 5	40 / 8	60 / 12	40 / 10	50 / 7.5	50 / 2.5	30 / 4.5		44.5	7
ALTERNATIVE 6	70 / 14	80 / 16	60 / 12.5	50 / 7.5	80 / 4	40 / 6		60	4
ALTERNATIVE 7	100 / 20	100 / 20	20 / 5	20 / 3	40 / 2	40 / 6		56	6
ALTERNATIVE 8	30 / 6	60 / 12	100 / 25	100 / 15	100 / 5	30 / 4.5		67.5	3

Figure 2.4. Illustrative Method of Selecting Rehabilitation Alternatives

CHAPTER 3
GUIDES FOR FIELD DATA COLLECTION

3.1 OVERVIEW

This chapter provides guidance and background information relative to field data collection surveys and measurements used in the rehabilitation process. Of particular importance are:

(1) the interpretative techniques used with continuously measured pavement variables along a highway, such as deflection, serviceability index, skid number, etc., and the associated methodologies that can be used to define the boundary limits of relatively uniform analysis units;
(2) the development and utilization of pavement condition surveys;
(3) the development and utilization of drainage surveys;
(4) the general considerations associated with NDT (Nondestructive Testing) deflections (types of equipment, use and interpretation of deflection results); and
(5) the use of destructive sampling and testing programs to augment field NDT.

3.2 THE FUNDAMENTAL ANALYSIS UNIT

3.2.1 General Background

When considering a major rehabilitation project, pavement monitoring activities are undertaken to obtain measurements, either continuous or discontinuous/point, which assess pavement response variables. Examples of pavement response variables are deflection, serviceability index, friction number, pavement condition indices, or even individual distress severities such as percent cracking, rut depth, etc.

Figure 3.1 illustrates the typical plot of a response variable as a function of distance along the highway segment. Measurement of a response variable indicates change from one location to another, with some points experiencing changes of major magnitude. At these points of significant change, the overall response of the pavement segments on either side will be noticeably different, as indicated in the figure.

The existence of deviation when measuring a pavement can be traced to two major sources. The first source of variation is termed "between unit variability" and reflects the fact that statistically homogeneous units may exist within a given rehabilitation project. The ability to delineate the general boundary locations of these units is critical in rehabilitation because these units form the basis for the specific analysis to be conducted. For instance, for the variable response depicted in Figure 3.1, four separate rehabilitation studies may be warranted (i.e., four separate overlay design thicknesses).

The other major source of variability is the inherent diversity of the response variable within each unit, thus called "within unit variability." Within unit variability is important because it relates to the eventual rehabilitation design reliability obtained for a given project.

Proper consideration of both between unit and within unit variability has a positive impact on rehabilitation design which cannot be overemphasized. If care is not exercised in the delineation of units and their internal variation, gross inefficiencies in the rehabilitation strategy will occur; every unit will either be underdesigned (i.e., premature failure) or overdesigned (uneconomical use of materials).

3.2.2 Methods of Unit Delineation

Idealized Approach. In order to delineate a pavement length, the engineer should isolate each unique factor influencing potential pavement performance. These factors are:

> pavement type
> construction history (including rehabilitation and major maintenance)
> pavement cross section (layer material type/thickness)
> subgrade (foundation)
> traffic
> pavement condition

III-19

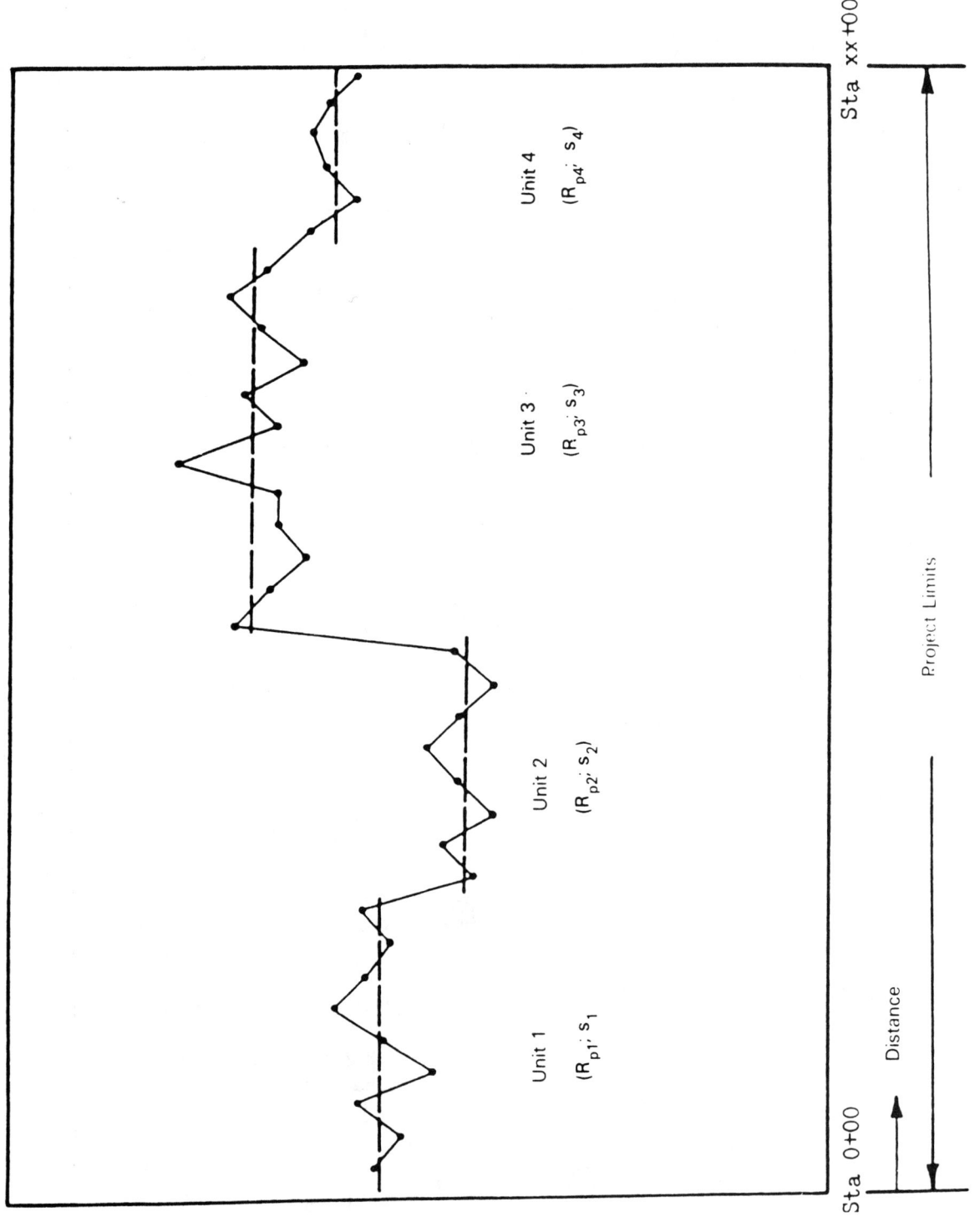

Figure 3.1. Typical Pavement Response Variable Versus Distance Plot for Given Project

R_p Pavement Response Variable

Under ideal circumstances, the engineer will use a historic pavement data base to evaluate these factors. Figure 3.2 illustrates how this information can be used to determine analysis units that are characterized by a unique combination of pavement performance factors.

The validity of the final units is directly related to the accuracy of the historic pavement information available. If accurate records have been kept, this historical data approach has more merit in delineating unique units than a procedure which relies on current observations of condition or performance indicators. The reason for this is that changes in one or more design factors (which indicate points of delineation) are not always evident through observation.

When delineating pavement analysis units, the most difficult factor to assess (without measurement) is the subgrade (foundation) factor. While records may indicate a uniform soil subgrade, the realities of cut-and-fill earthwork operations, variable compactive effort drainage, topographic positions, and groundwater table positions, often alter the in situ response of subgrades even along a "uniform soil type."

Measured Pavement Response Approach. Frequently, the engineer cannot accurately determine the practical extent of the performance factors noted and must rely upon the analysis of a measured pavement response variable (e.g., deflection) for unit delineation. The designer should develop a plot of the measured response variable as a function of the distance along the project. This can be done manually or through computerized data analysis-graphic systems.

To illustrate this approach, Figure 3.3 is a plot of friction number results, FN(40), versus station number along an actual highway system. While this example uses deflection as the pavement response variable, the procedure is identical for any type of pavement response variable selected (i.e., pavement condition, serviceability, rut depth, etc.).

Once the plot of a pavement response variable has been generated, it may be used to delineate units through several methods. The simplest of these is visual examination to subjectively determine where relatively unique units occur. In addition, several analytical methods are available to help delineate units, with the recommended procedure being the "cumulative difference." This analytical procedure, readily adaptable to computerized evaluation, relies on the simple mathematical fact that when the variable Z_c (defined as the difference between the area under the response curve at any distance and the total area developed from the overall project average response at the same distance) is plotted as a function of distance along the project, unit boundaries occur at the location where the slopes (Z_c vs. X) change sign. Figure 3.4 is a plot of the cumulative difference variable (Z_c) for the data shown in Figure 3.3. For this example, 11 preliminary analysis units are defined. The engineer must then evaluate the resulting length of each unit to determine whether two or more units should be combined for practical construction considerations and economic reasons. The combination of units should be done relative to the sensitivity of the mean response values for each unit upon performance of future rehabilitation designs.

Appendix J describes the mathematical background and development of the cumulative difference approach and uses the data presented in Figure 3.4 as an example.

3.3 DRAINAGE SURVEY FOR REHABILITATION

3.3.1 Role of Drainage in Rehabilitation

Distress in both rigid and flexible pavements is often either caused or accelerated by the presence of moisture in the pavement structure. When designing pavement rehabilitation, the engineer must investigate the role of drainage improvements in correcting declining pavement performance. It is also important to recognize when a pavement's distresses are not moisture-related and, therefore, cannot be remedied by drainage improvements.

The condition survey, an essential part of any rehabilitation project evaluation, will often reveal moisture-related distresses. Distress types in flexible pavement which may be caused by or accelerated by moisture in the pavement structure include stripping, rutting, depressions, fatigue cracking, and potholes. Moisture-related distresses in rigid pavements include pumping, "D" cracking, joint deterioration, faulting, and corner breaks.

Further, the condition survey may also show that a pavement has suffered damage due to freezing and subsequent thawing. Differential frost heave and spring breakup (evidence of loss of support) both indicate that the pavement structure retains excess moisture in the winter months. In areas of the country where cycling above and below freezing occurs throughout the winter, pavements will often exhibit distresses related to weakening of the support layers.

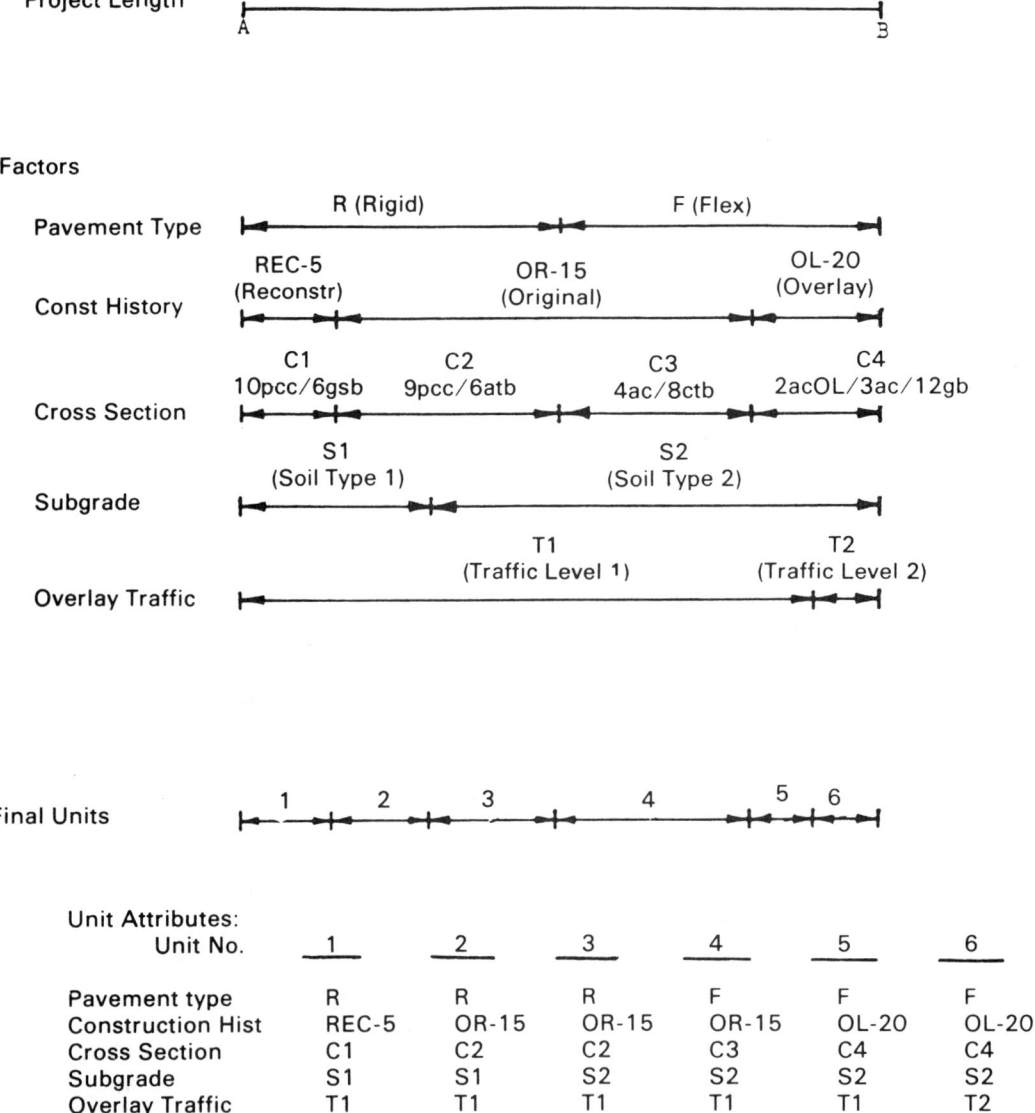

Figure 3.2. Idealized Method for Analysis Unit Delineation

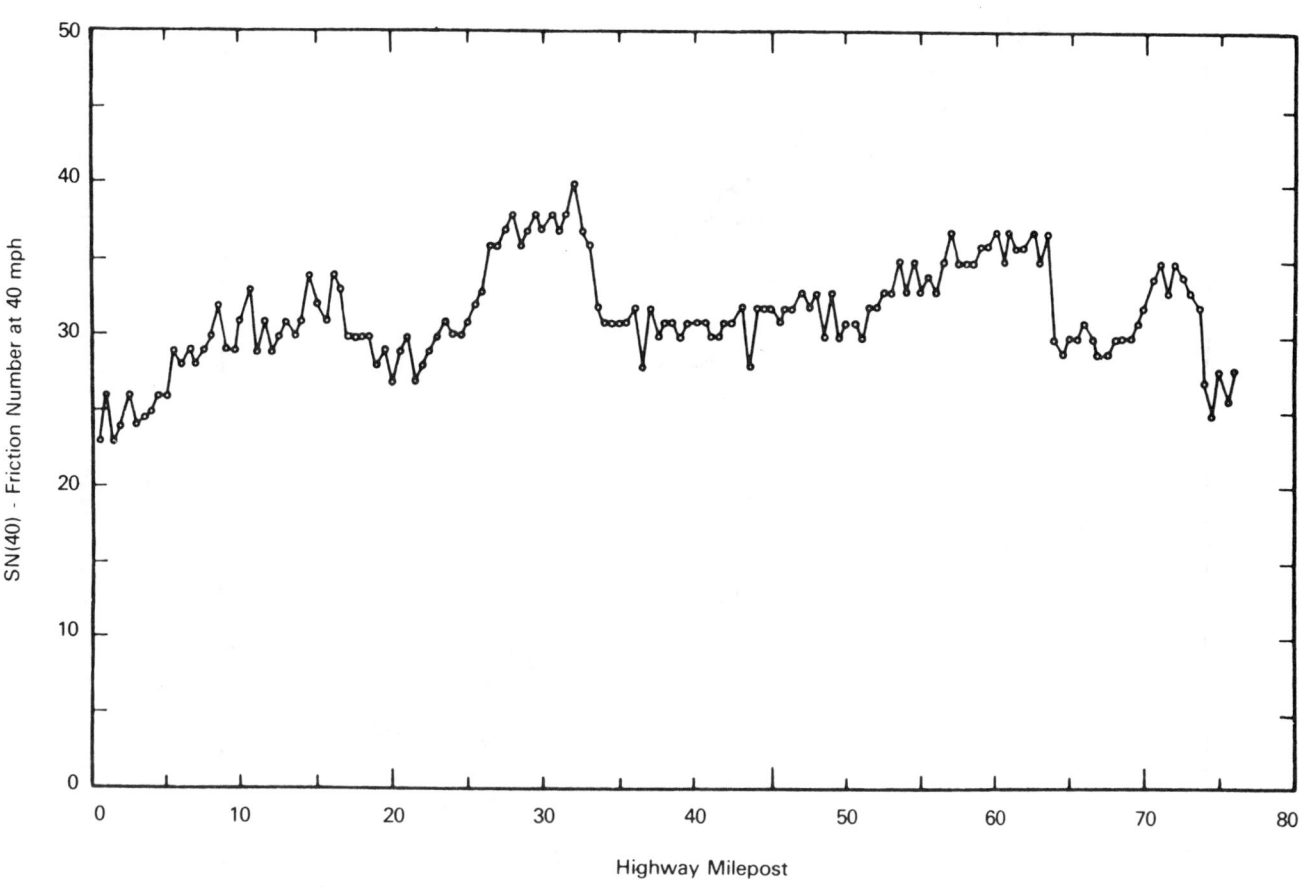

Figure 3.3. FN(40) Results Versus Distance along Project

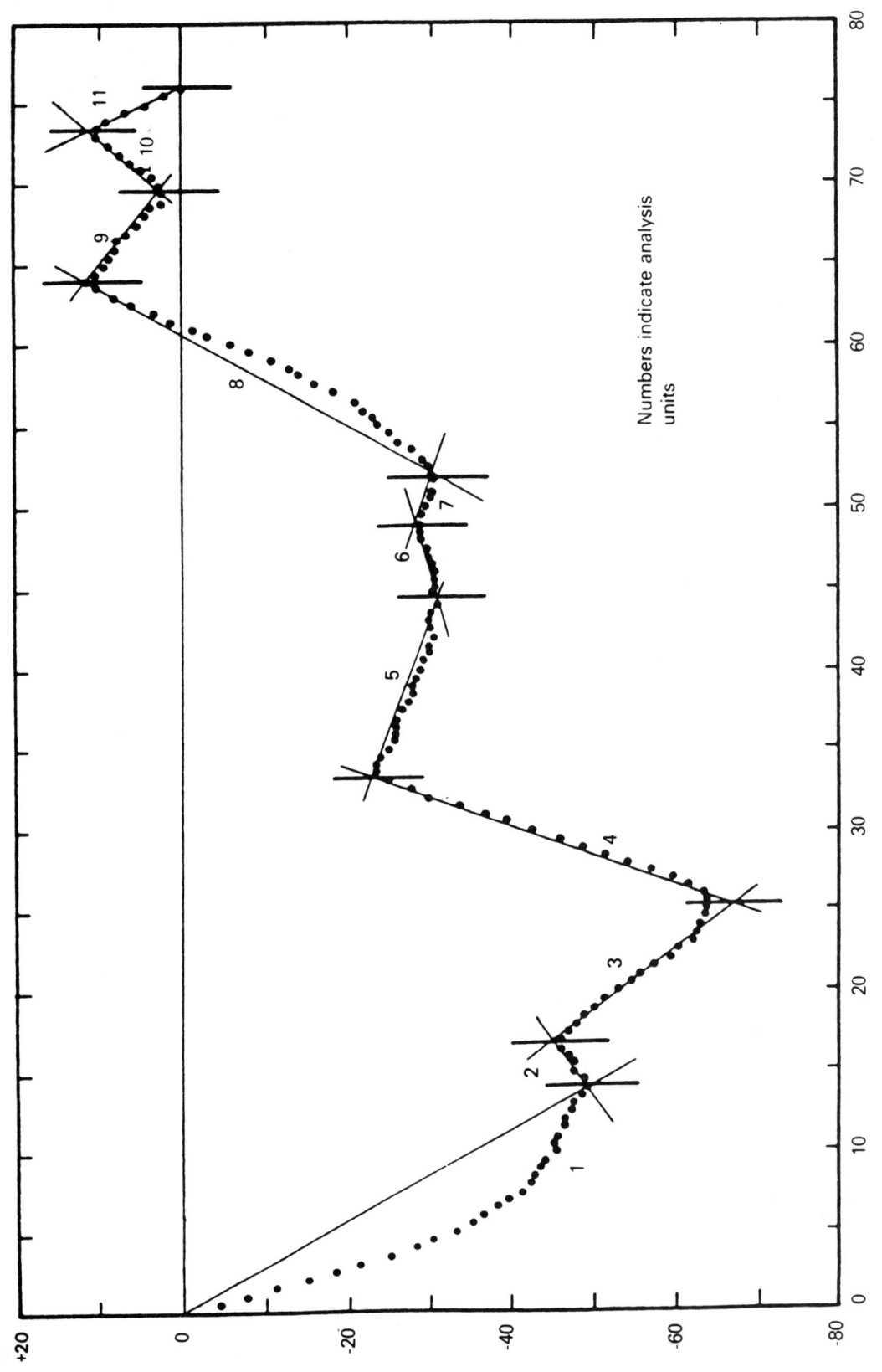

Figure 3.4. Delineating Analysis Units by Cumulative Difference Approach

3.3.2 Assessing Need for Drainage Evaluation

The extent of moisture-related damage in a pavement, as revealed by the condition survey, determines the commitment of time and funds to drainage evaluation. However, the absence of moisture-related distress does not necessarily mean that a pavement is without moisture-related problems; the potential for such distress may exist. During the site investigation, and even during the performance of the rehabilitation work, the engineer should look for deficiencies in the pavement's drainage system which might allow moisture to damage the pavement structure. Maintenance personnel are a good source for this type of information. In order to select a rehabilitation approach that both repairs and prevents moisture-related distress, the engineer must understand the mechanisms by which moisture causes or accelerates distress in the pavement.

3.3.3 Pavement History, Topography, and Geometry

The first step in drainage evaluation is the examination of a pavement's construction records. For instance, what provisions were made for drainage in the original design? Further, the drainage data previously collected should be examined, as well as pavement cross sections and profiles for the following:

- longitudinal grades
- transverse grades
- widths of pavement layers
- layer thicknesses
- cut-and-fill depths
- slopes and dimensions of surface drainage features (ditches, culverts, etc.)
- in-place subsurface drainage

If the pavement has developed moisture-related distress, it is obvious that the original system is inadequate to meet the pavement's present needs. The drainage evaluation will reveal to the engineer whether the existing drainage system only needs to be repaired and maintained, or whether it needs to be augmented with additional drainage features.

The next step in drainage evaluation is the examination of a topographic map for features influencing the surface and subsurface movement of water in the project area. Has the pavement been built in a "bathtub," with no lower ground for the water to drain to? Are there any lakes, streams, or seasonally wet areas above the elevation of the pavement? In addition, regional soil maps should be examined as a further source of information on the movement of surface and subsurface water in the pavement area. They can also provide information about types of soil present.

Drainage evaluation also requires investigation of the problem site, preferably during a wet weather period. Following is a partial list of questions to ask during the site investigation:

- Where and how does water move across the pavement surface?
- Where does water collect on and near the pavement?
- How high is the water level in the ditches?
- Do the joints and cracks contain any water?
- Does water pond on the shoulder?
- Does water-loving vegetation flourish along the roadside?
- Are deposits of fines or other evidence of pumping (blowholes) visible at the pavement's edge?
- Do the inlets contain debris or sediment buildup?
- Are the joints and cracks sealed well?

Site investigation should also include an inspection to determine if drainage features planned in the original design were actually constructed. Make no assumptions in this regard since plans are subject to change. Also, look for evidence of in-place drain maintenance, and inquire about scheduled clean-out procedures.

3.3.4 Properties of Materials

The determination of which material properties to investigate depends on two factors: the type of moisture-related distress present in the pavement, and the pavement layer(s) in which the distress appears. Table 3.1 lists some of the material properties which might be investigated for each of three layers—subgrade, granular, and surface. Many county maps are available that provide information about the engineering properties of soils and should be used as a source of data.

When possible, the collection of materials data for both drainage evaluation and overall project evaluation should be coordinated. For example, if coring must be performed to determine layer thicknesses, samples of subgrade soil can be taken at the same time for drainage-related testing. In this way, drainage evaluation expenses will be minimized.

Table 3.1. Material Properties Associated with Drainage Problems in Pavements

Subgrade	
General categorization:	gradation
	classification
Weight-volume relationships:	optimum lab dry density
	optimum lab moisture content
	in situ dry density
	in situ moisture content
Other drainage-related characteristics:	permeability
	effective porosity
	frost susceptibility
	capillarity
Granular Layers	
General categorization:	gradation
	percent fines
	Atterberg limits
	classification
	optimum lab moisture content
	in situ dry density
	in situ moisture content
Other drainage-related characteristics:	permeability
	effective porosity
	frost susceptibility
	capillarity
Surface	
Aggregate:	"D" cracking susceptibility
	freeze-thaw susceptibility
	stripping
	Aggregate reaction

3.3.5 Climatic Zones

The United States can be divided into nine regional climatic zones which are formed by the intersection of three moisture regions and three temperature regions. Figure 3.5 illustrates the nine climatic zones. The three moisture regions are:

Region I—High potential for moisture presence in the entire pavement structure throughout the year.

Region II—Seasonal variability of moisture in the pavement structure.

Region III—Very little moisture in the pavement structure during the year.

The three temperature regions are:

Region A—Severe winters with a high potential for frost penetration to appreciable depths into the subgrade.

Region B—Freeze-thaw cycles in the surface and base. Severe winters may produce frozen subgrades, but long-term freezing problems are minor.

Region C—Low temperatures are not a problem. Stability at high temperature should be considered.

Pavements within a given climatic zone typically exhibit similarities in performance, moisture-related distress, and drainage-related rehabilitation work required.

Guides for Field Data Collection

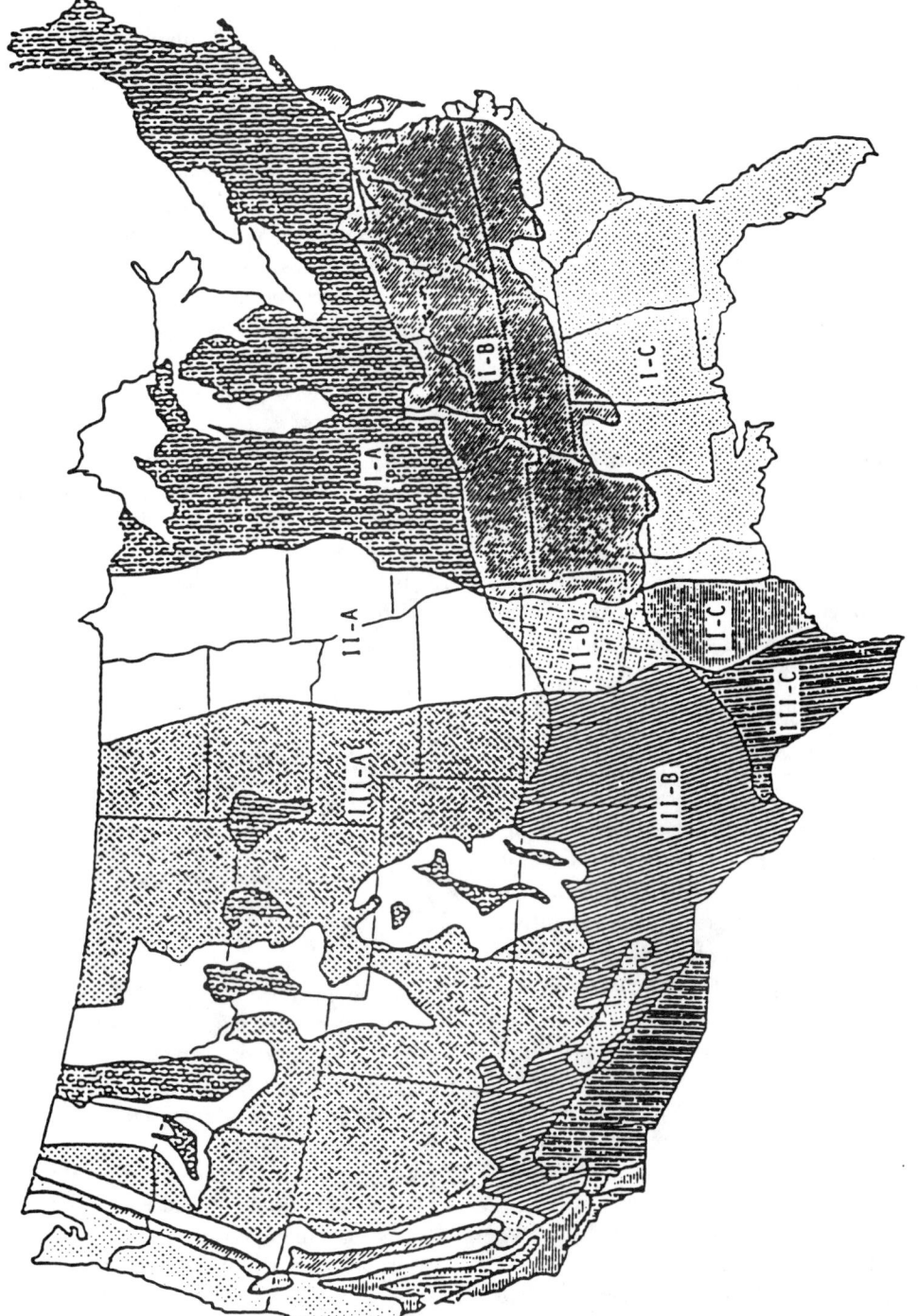

Figure 3.5. Climatic Zones Based on Thornthwaite Potential Evaporation and Moisture Index and Their Interaction with Performance, with Similar Performance Expected in Similar Climatic Regions

3.3.6 Summary

Only when the engineer recognizes a pavement's moisture-related problems and understands how they developed can he or she design rehabilitation alternatives which address the problems and prevent their recurrence. To increase the economics of the drainage survey process, every effort should be made to develop drainage rehabilitation alternatives which are compatible with alternatives being considered for the correction of other pavement distresses present.

3.4 CONDITION (DISTRESS) SURVEY

3.4.1 General Background

Accurate condition surveys which assess a pavement's physical distress are vital to a successful rehabilitation effort. Condition survey results, together with serviceability (roughness), drainage, and structural evaluation surveys, provide the engineer with the necessary information to develop a sound rehabilitation strategy. Thus, an intensive survey is mandatory before any rehabilitation designs are attempted.

In addition, it is important that condition surveys be conducted after new construction or rehabilitation work. Such monitoring is a tool for network assessment and provides information regarding the rate of distress buildup. These survey results are a major input when determining whether to undertake a major rehabilitation project. However, when a rehabilitation project is planned, the use of these periodic condition survey results are insufficient to properly evaluate the necessary rehabilitation steps and the intensive survey, as mentioned above, is vital.

While engineers accept the necessity for condition or distress surveys in broad terms, specific methodologies for such surveys vary from agency to agency. Each agency must develop a survey approach consistent with its use of the data generated, as well as its available manpower and financial resources.

Several agencies have expanded the condition survey concept and combined all of the recorded distress information into a single "condition index" which measures overall pavement condition and probable required maintenance. This approach is encouraged because it provides an additional engineering tool that greatly aids in the overall rehabilitation planning effort at both project and network levels.

3.4.2 Minimum Information Needs

When pavement condition surveys are conducted, there is a minimum information requirement necessary if the engineer is to make knowledgeable decisions regarding rehabilitation needs and strategies. These information requirements are:

(1) *Distress Type*—Identify types of physical distress existing in the pavement. The distress types should be placed in categories according to their casual mechanisms.
(2) *Distress Severity*—Note level of severity for each distress type present to assess degree of deterioration.
(3) *Distress Amount*—Denote relative area (percentage of the project) affected by each combination of distress type and severity.

A technically sound engineering condition survey must address each one of these needs, although the parameters of each category may vary from agency to agency. Appendix C provides example distress-type descriptions and associated severity groups that may be used as a guide for developing or modifying condition (distress) surveys for an agency.

3.4.3 Utilization of Information

A thorough condition survey is an invaluable tool in the rehabilitation process. If properly conducted, the condition survey identifies distress types present which, in turn, assists the engineer in defining probable causes of the distress. Only with the proper identification of probable cause(s) is it possible to select the rehabilitation strategy (overlay or nonoverlay) that will both repair and prevent the problem. As previously noted, not all pavement distress is traceable to structural mechanisms; factors such as climate, construction quality, etc., may also interact in a complex way to cause pavement distress. Furthermore, many observed distresses may be a function of several mechanisms. Tables 3.2, 3.3, and 3.4 categorize pavement distress relative to probable cause for flexible (asphalt), jointed concrete, and continuously reinforced pavement systems.

In addition to identifying probable causes of distress, a properly conducted condition survey will document the location and severity of the distress types. This then indicates the necessity for restoration, if any. Furthermore, the condition survey provides a permanent record of the pavement condition at the time of the survey. From this, significant deviations in

Table 3.2. General Categorization of Asphalt Pavement Distress

Distress Type	Primarily Traffic Load Caused	Primarily Climate/Materials Caused
1. Alligator or fatigue cracking	X	
2. Bleeding		X
3. Block cracking		X
4. Corrugation		X
5. Depression		X
6. Joint reflection cracking from PCC slab		X
7. Lane/shoulder dropoff or heave		X
8. Lane/shoulder separation		X
9. Longitudinal and transverse cracking		X
10. Patch deterioration	X	
11. Polished aggregate	X	
12. Potholes	X	
13. Pumping and water bleeding	X(M,H)	X(L)
14. Raveling and weathering		X
15. Rutting	X	
16. Slippage cracking		X
17. Swell		X

Table 3.3. General Categorization of Jointed Concrete Pavement Distress

Distress Type	Primarily Traffic Load Caused	Primarily Climate/Materials Caused
1. Blow-up		X
2. Corner break	X	
3. Depression		X
4. Durability "D" cracking		X
5. Faulting of transverse joints and cracks	X	
6. Joint load transfer associated distress	X	X
7. Joint seal damage of transverse joints		X
8. Lane/shoulder dropoff or heave		X
9. Lane/shoulder joint separation		X
10. Longitudinal cracks		X
11. Longitudinal joint faulting	X	X
12. Patch deterioration	X(M,H)	X(L)
13. Patch adjacent slab deterioration	X	X
14. Popouts		X
15. Pumping and water bleeding	X(M,H)	X(L)
16. Reactive aggregate durability distress		X
17. Scaling, map cracking and crazing		X
18. Spalling (transverse and longitudinal joints)	X(M,H)	X(L,M,H)
19. Spalling (corner)		X
20. Swell		X
21. Transverse and diagonal cracks	X(L,M,H)	X(L)

Table 3.4. General Categorization of Continuously Reinforced Concrete Pavement Distress

Distress Type	Primarily Traffic Load Caused	Primarily Climate/Materials Caused
1. Asphalt patch deterioration	X	
2. Blow-up		X
3. Concrete patch deterioration	X(M,H)	X(L)
4. Construction joint distress		X
5. Depression		X
6. Durability "D" cracking		X
7. Edge punchout	X	
8. Lane/shoulder dropoff or heave		X
9. Lane/shoulder joint separation		X
10. Localized distress		X
11. Longitudinal cracking		X
12. Longitudinal joint faulting	X	X
13. Patch adjacent slab deterioration	X	X
14. Popouts		X
15. Pumping and water bleeding	X(M,H)	X(L)
16. Reactive aggregate distress		X
17. Scaling, map cracking and crazing		X
18. Spalling	X	X
19. Swell		X
20. Transverse cracking	X(M,H)	X(L,M)

condition can be easily assessed along the entire project length. Also, differences between lanes of a multi-lane pavement facility will be revealed. The presence of distress, as indicated by a condition survey, indicates a decline in pavement serviceability, and more detailed field evaluations should be considered.

In summary, it is again emphasized that periodic condition surveys provide the engineer with the capability to assess impending distress and estimate the probable rate of future pavement deterioration. Thus, recognition of the initial stages of rigid pavement pumping, for example, may allow nonoverlay rehabilitation approaches to be used as compared to the costly rehabilitation of slab fracture, faulting, and joint damage. A more extensive condition survey is essential when embarking on a major rehabilitation project so that the best and most economic solution may be achieved.

3.5 NDT DEFLECTION MEASUREMENT

3.5.1 Overview

Deflection Interpretation. The use of nondestructive deflection testing has been an integral part of the structural evaluation and rehabilitation process for many decades. In its earliest applications, the total measured pavement deflection under a particular load arrangement was used as a direct indicator of structural capacity. Several agencies developed failure criteria, particularly for flexible pavements, that related the maximum measured deflection to the number of allowable load repetitions.

Such criteria have been, and still are, used for the design of both new pavement systems and structural overlay systems. As experience with this approach grew, more accurate performance relationships were obtained by using only the "rebound," "recoverable," or "elastic" portion of the deflection as the key indicator of performance, rather than the total deflection under load. A typical deflection criterion (using a Benkleman Beam) is shown in Figure 3.6. For many years, the Asphalt Institute has used this criterion as the basis for the structural overlay analysis of flexible overlays over existing flexible pavements.

While deflection criteria similar to that shown in Figure 3.6 are in common use, recent technical advances indicate that maximum rebound (elastic) deflection, by itself, is not the most accurate nor applicable parameter for the variety of pavement

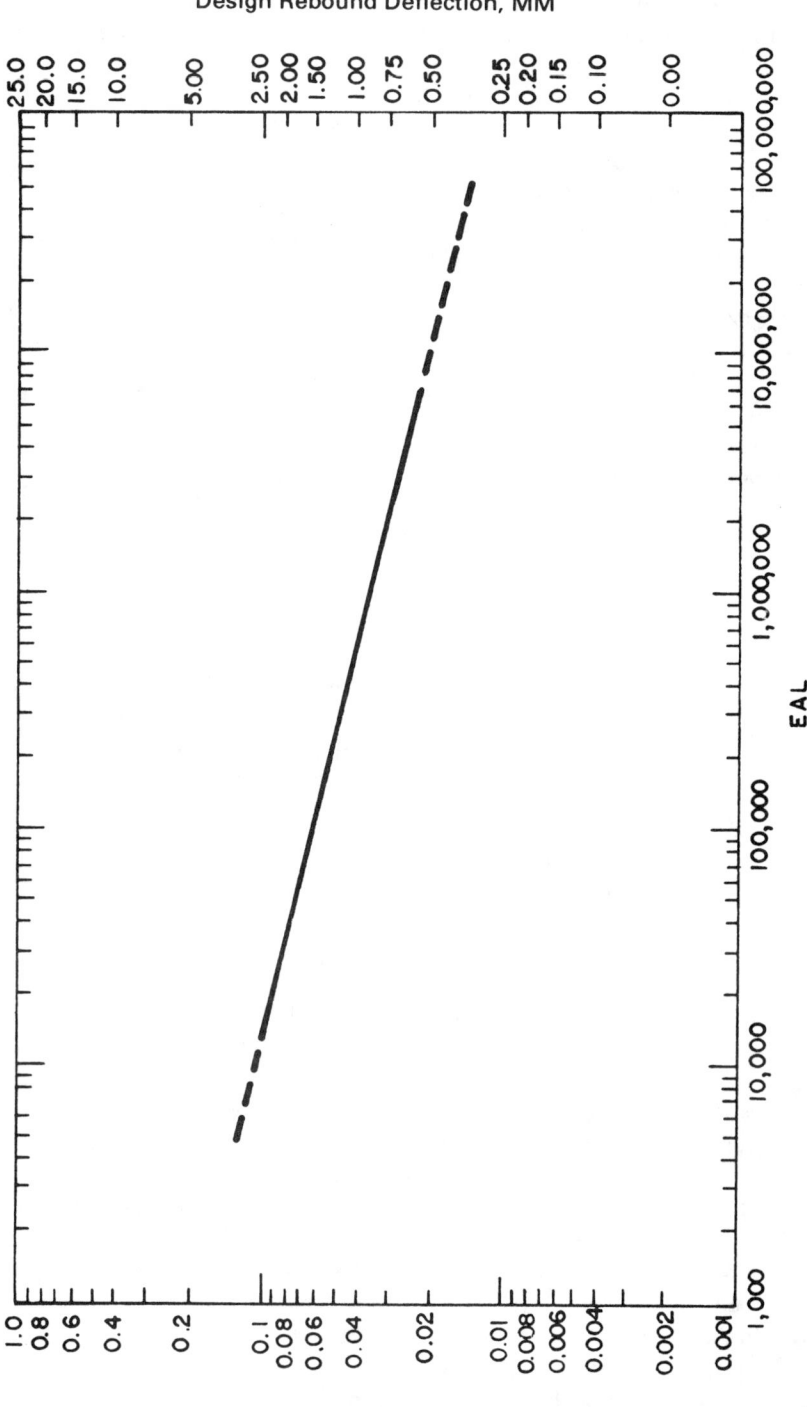

Figure 3.6. Design Rebound Deflection Chart

structures encountered in practice. In reality, all pavements may be structurally distressed by either excessive deformations and/or load-associated fracture of a particular stabilized layer. While the maximum elastic deflection may be more indicative of the pavement's ability to resist repetitive shear displacements leading to rutting, the curvature radius of the pavement under load is more indicative of overall resistance to repeated load fracture of stabilized pavement layers.

As a result, several agencies have refined deflection-repetition-performance criteria to account for this very important concept. Figures 3.7 and 3.8 indicate two such criteria from leading agencies which clearly demonstrate that no single deflection criterion is applicable for all flexible or semi-rigid pavement systems. In essence, these figures illustrate the following fundamental principles of deflection testing: (1) multiple structural distress types (deformation and fracture) must be logically accounted for in the interpretation of deflection testing results; and (2) pavement layer material type (quality) and layer thickness also must be considered if deflection-repetition-performance curves are to be used.

In summary, the most accurate assessment of pavement performance is achieved through the use of maximum elastic deflection in combination with an indicator of the radius of curvature of the pavement under load. In this Guide, the NDT deflection pavement structural capacity method requires the use of deflection basin measurements under load, rather than maximum deflections alone. The details, use, and interpretation of this fundamental approach are presented in this section.

Environmental Adjustments. When deflection measurements are taken on an asphalt pavement, the results must be corrected (standardized) to a particular type of loading system (vehicle or NDT device) and normalized to an arbitrarily defined set of environmental conditions. In general, measured deflections must be adjusted to a reference pavement temperature (usually 70°F) to account for the effect of this variable upon asphaltic-stabilized material modulus. This factor significantly affects the interpretation of flexible pavement deflections. Because deflection testing is generally conducted at a particular time of year, the engineer must make a deflection adjustment to ensure that the most critical moisture regime, within a typical year, is used in the analysis. In areas subject to frost penetration into the pavement, this time is always associated with the spring thaw. For pavements not experiencing frost, the critical deflection period is a direct function of when the pavement is weakened due to larger than average mixture conditions in the unbound pavement layers. This is illustrated by Figure 3.9.

Procedures for use in adjusting NDT deflection measurements are uniquely dependent upon the specific NDT deflection methodology introduced later in this section. Detailed steps to adjust for environmental conditions are presented for each of two recommended approaches in Chapter 5.

Deflection Measuring Systems. Several NDT deflection measuring systems are available for use in pavement evaluation work. In general, systems can be categorized into five major groups:

Static-Creep Deflection Methods
Automated Deflection Beams
Steady State (Sinusoidal) Deflection Devices
Impulse Devices
Wave Propagation Devices

The latter three measuring systems use "dynamic" deflection equipment to exert loads (stress forms) of short duration and, thus, simulate to variable degrees the dynamic stress conditions caused by moving wheel loads. At present, wave propagation approaches are primarily experimental and are not considered as current "production"-oriented NDT field devices.

3.5.2 Uses of NDT Deflection Results

This Guide presents procedures to utilize nondestructive deflection testing results in terms of three factors. They are:

(1) evaluation of the in situ structural capacity of the pavement
(2) rigid pavement joint/load transfer analysis
(3) rigid pavement slab-void detection

Without question, NDT deflection data are primarily associated with the first category, the in situ or "effective" pavement structural capacity. Two approaches for using NDT data in this regard are presented in the next section and discussed in detail in Chapter 5. While methodologies may differ, both approaches use deflection basin measurement (rather than maximum deflection only) to evaluate structural capacity. In addition, both methods rely on dynamic deflections as indicative of performance rather than static-creep deflection response. Thus, data used to evaluate the in situ structural capacity of a pavement

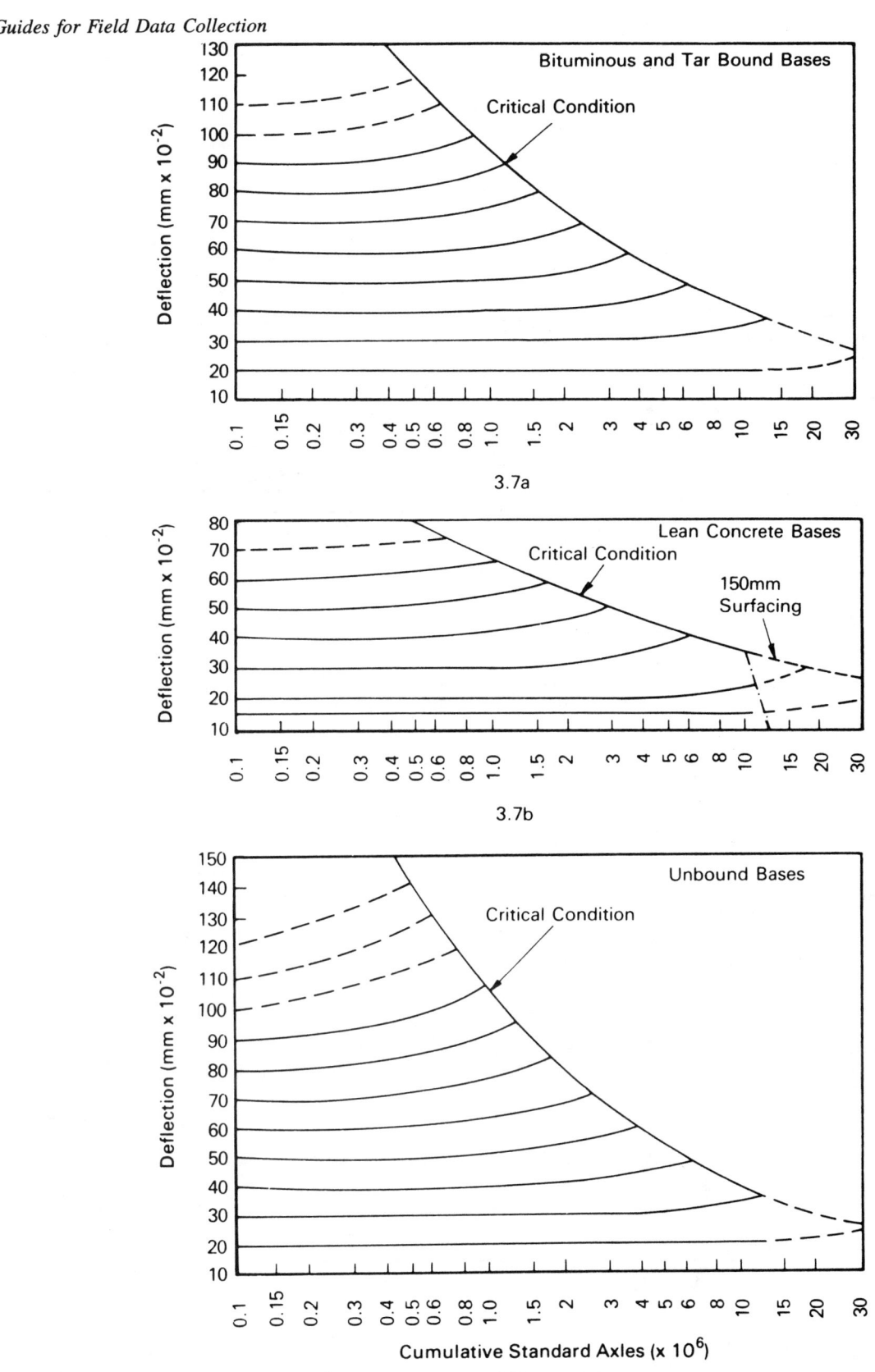

Figure 3.7. Deflection—Life Relationships for Various Pavement Types (Lister and Kennedy—TRRL)

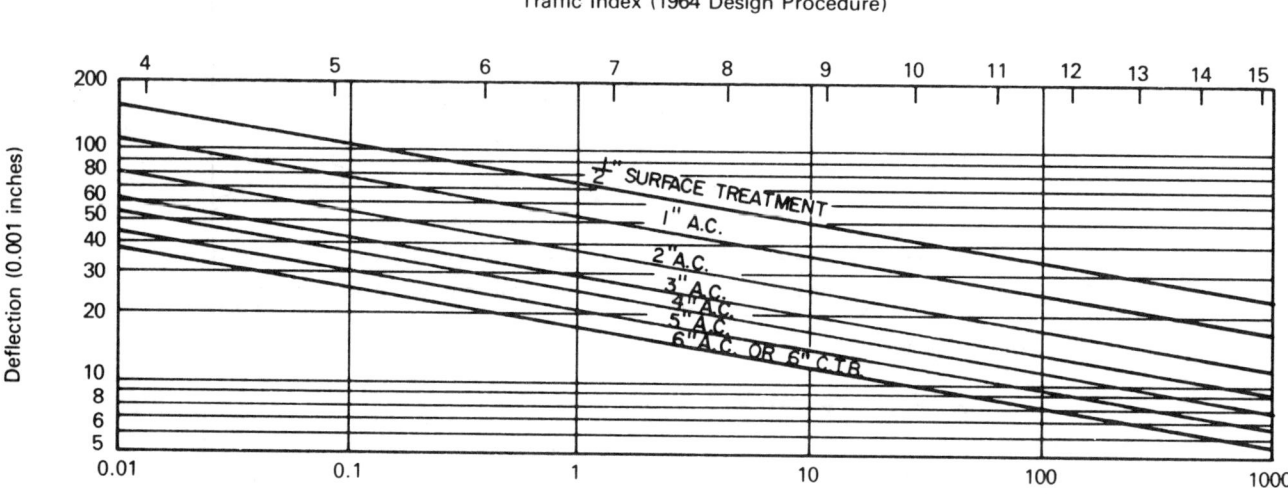

Figure 3.8. Deflection—Life Relationship for Asphaltic Pavements (California Method of Overlay Design)

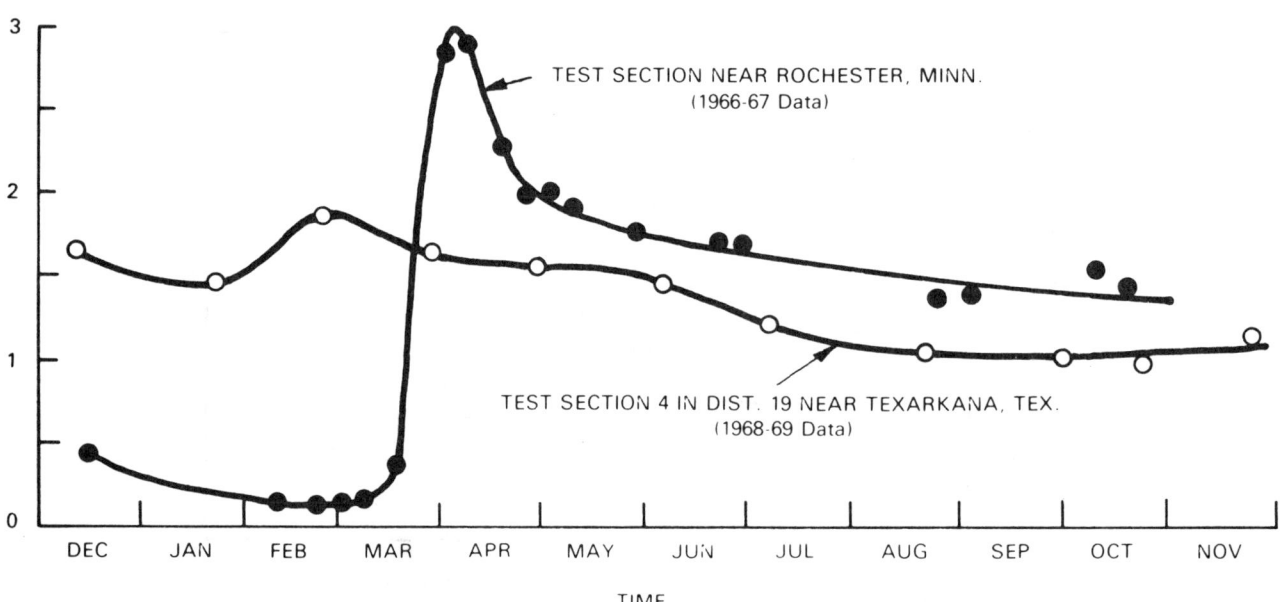

Figure 3.9. Illustration of the Effect of Geographic Location on Seasonal Variations in Deflections

must be gathered from either steady state or impulse devices, according to guidelines presented herein.

Relative to the use of NDT for joint/load transfer studies and slab-void detection, any deflection device may be employed with the stipulation that the deflection equipment meet certain requirements, to be noted herein. However, only dynamic NDT devices (steady state vibratory and impulse) can be used with confidence in evaluating all three factors listed above.

3.5.3 Evaluating the Effective Structural Capacity

General Approaches. Evaluation of the effective structural capacity of a pavement, as set forth in this Guide, requires the use of dynamic (steady state vibratory or impulse) loads, and the subsequent measurement of the deflection basin. Then, using the measured deflection basin, the in situ subgrade modulus is estimated. Within this context, two procedures are available to carry out this evaluation. They are:

(1) pavement layer moduli prediction technique
(2) direct structural capacity prediction technique

While both approaches yield the same value (effective structural capacity, SC_{xeff}), the user should be aware of the advantages and disadvantages of each when making a selection between the two.

Figure 3.10 is a schematic diagram of a typical pavement structure being deflected under a dynamic NDT load. As the load is applied (either steady state or impulse), it spreads through a portion of the pavement system, as represented by the conical zone in the figure. The incline of the sides of this zone, which varies from layer to layer, is related to the relative stiffness or modulus of the material within each layer. As the modulus increases (material becomes stiffer), the stress is spread over a much larger area.

This figure reveals several interesting concepts in NDT pavement analysis. Of significance is the radial distance ($r = a_{3e}$) in which the stress zone intersects the interface of the subbase and subgrade layers. When the deflection basin is measured (via geophones or other measurement devices), any surface deflection obtained at or beyond the a_{3e} value is due only to stresses (deformations) within the subgrade itself. Thus, the outer readings of deflection basin, under dynamic load, primarily reflect the in situ modulus properties of the lower (subgrade) soil. This is the fundamental concept used in either approach to establish the value of the pavement support condition from NDT evaluation. Equally important is the fact that there exists an ideal minimum distance for each pavement type—NDT device combination where the outer geophone should be placed to ensure that the deflection response is not being influenced by upper pavement layers. If the outer geophone is placed beyond this point, predictive estimation errors in the subgrade support or response will occur. Detailed guidance for optimum placement of the outer geophone in NDT evaluation is presented in Chapter 5.

Both procedures for determining effective structural capacity use deflection basin measurements to evaluate subgrade modulus, as described above. It is in the ensuing steps of structural capacity evaluation that the methods differ.

The objective of the Pavement Layer Moduli Prediction Technique is to back calculate, from the measured deflection basin results, all of the in situ-layered elastic moduli. The fundamental premise of this solution is that a unique set of layer moduli exist such that the theoretically predicted deflection basin (using multi-layer theory and the special load characteristics of the NDT device) is equivalent to the measured deflection basin. The general applicability of this approach can be visualized by referring to Figure 3.10. If one views the intersection of the stress zone at the interface of the surface and base/subbase course, the measured surface deflection at this radial offset value must logically be influenced only by the layer moduli of the base/subbase and subgrade layer. Because the subgrade modulus has been determined already, the deflection at this interface/intersection location can be used to determine the modulus of the base/subbase layer. This is the fundamental concept in deflection basin analysis when estimating the in situ layer moduli. In short, the solution initiates at the outer geophone locations (edge of the deflection basin) to determine the moduli of the lowest pavement layer. The sequence progresses by using this "known" material response and deflections at radial offsets approaching the load plate center. In this approach, the values of the thickness, h_i, and Poisson ratio, u_i, must either be known or assumed. This solution is applicable to pavement types of all rigidities (flexible to rigid). Knowledge of the individual properties (i.e., modulus) allows for layer coefficients to be established using the principles found in Part II of the Guide to predict the effective structural capacity (i.e., effective Structural Number or effective PCC Thickness) of the existing pavement.

The second alternative, the Direct Structural Capacity Prediction Technique, employs the fact that the combined stiffness influence of each layer thickness-modulus (thickness-layer coefficient) determines the

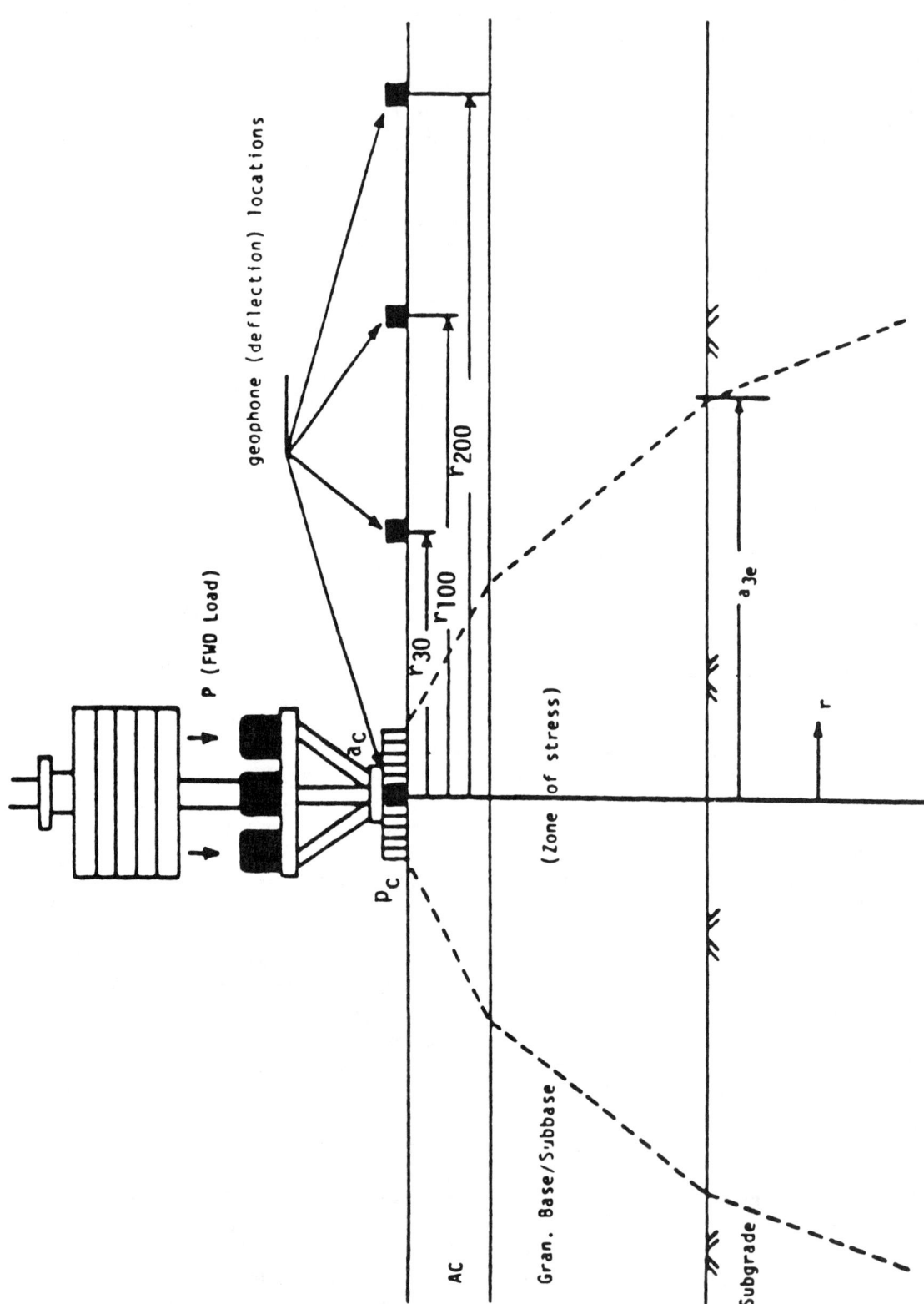

Figure 3.10. Schematic of Stress Zone within Pavement Structure under the FWD Load

overall structural capacity of the pavement. Thus, the maximum NDT deflection (at the load center) may be viewed as the result of two separate pavement parameters: (1) structural capacity, and (2) subgrade modulus. This approach recognizes that structural capacity is a function of the maximum NDT deflection and subgrade modulus. Hence, this technique relies on outer deflection values to estimate the subgrade modulus (support), and the maximum measured NDT deflection to predict the "effective" structural pavement capacity. Detailed procedures for this approach are contained in Chapter 5.

State of the Art. The dynamic NDT-deflection basin methodologies presented and recommended for overlay rehabilitation studies are a technical improvement over approaches that use only a unique deflection-life performance criterion. While present state of the art methods are thought to be technically sound, the engineer must recognize that they are not perfect, nor above future modifications as technology and use advances. Furthermore, the engineer must not blindly use NDT results, but rather assess the reasonableness of any results obtained. If there are excessive differences between NDT-derived estimates and previous agency experience of how local materials behave, an in-depth reevaluation is necessary to clarify the cause of the difference.

The use of dynamic NDT deflection basin techniques provides the user with broad powers in evaluating pavements. One such power is the ability to estimate the in situ support value of the subgrade. This factor is vital to the completion of an accurate overlay analysis.

Another capability of the prediction techniques presented here and in Chapter 5 is the ability to accurately determine the in situ or effective capacity of the existing pavement at the time of measurement to carry load repetitions. Areas of pavement weakness will be recognized by both dynamic NDT processes presented herein. For instance, cracking of pavement layers will manifest itself by an increase in deflection which, in turn, results in lower predicted layer moduli and/or a subsequent reduction in the pavement's load capacity. Also, the impact of moisture increases will be reflected in a change in the deflection basin response of the pavement, as will the influence of the as-constructed/in situ material behavior. In summary, NDT deflection basin analysis is a technique for determining the most accurate estimates of the actual in situ layer properties which collectively define the overall structural capacity of the pavement system.

One final consideration of this approach (particularly the Pavement Layer Moduli Prediction Technique) is the fact that the estimated in situ layer moduli can be used as direct input into the more mechanistic design overlay approaches presented in Part IV and Appendix CC of the Guide. This may be beneficial to agencies that want to conduct more in-depth rehabilitation studies.

NDT Equipment Considerations. As previously noted, the two NDT methodologies to evaluate the effective structural capacity of pavements are considered applicable to any type of dynamic NDT device. Certain fundamentals should be considered by the engineer when selecting an NDT device. They are as follows:

(1) An NDT device that rapidly measures variable load magnitudes at a given location (test point) is desirable. The assessment of deflections under various load levels is useful when nonlinear material response is required and during rigid pavement-void detection studies. (See Section 3.5.5.)

(2) The ability of an NDT device to use dynamic loads approaching actual truck loads is important for several reasons. First, for pavement materials that may exhibit nonlinear behavior (particularly unbound granular and subgrade soils), analysis of pavements with 8-kip to 10-kip loads results in moduli/capacity predictions representative of pavement response under truck traffic. In addition, with the deeper deflection zone caused by larger dynamic loads, additional weaknesses in the pavement structure may be located. Finally, larger dynamic loads examine larger surface/radial locations.

(3) The NDT device should routinely place the deflection sensors at an effective radial distance from the load center. (See Chapter 5 for guidance.)

(4) In general, the NDT device should have a minimum of three or, preferably, four deflection sensors. Please note that six sensors are not twice as effective as three sensors. The actual number of sensors placed depends on the analytical approach used during evaluation.

(5) Whenever practical, the placement of sensors should correspond with the interface intersections of stress zones, as noted in Figure 3.10. This procedure reminds NDT users that the optimal sensor layout for one pavement may

3.5.4 Joint Load Transfer Analysis

Background. In addition to structural capacity evaluation, nondestructive deflection testing can be used to evaluate the in situ load transfer capacity of rigid pavement joints (as well as cracked slabs). With NDT, the engineer can evaluate the actual performance of joints in the field relative to their expected performance in the design phase.

The load size transferred across a rigid pavement joint directly impacts the flexural slab stress at or near the joint. Load transfer capability is measured by the joint efficiency which is commonly expressed in one of two ways: (1) deflection efficiency, and (2) stress efficiency.

Joint load transfer efficiency, d_{je}, based on deflections is represented by:

$$d_{je} = d_u/d_l \times 100$$

In this equation, d_u is the deflection at the joint of the unloaded slab, while d_l is the deflection of the loaded slab. The d_{je} value, determined by deflection ratios of adjacent slabs, directly lends itself to NDT, wherein deflection sensors may be placed on each slab close to the joint. The NDT load plate is also positioned on one of the slabs near the joint. Figure 3.11 illustrates two extreme cases; a joint with excellent load transfer and a joint with no load transfer. Joint deflection efficiency values may range from 0 percent (none) or 100 percent (full).

Joint load transfer efficiency, based upon stress, s_{je}, is represented by:

$$s_{je} = s_u \times s_l \times 100$$

In this equation, s_u and s_l refer to stresses in the unloaded and loaded slabs, respectively. Studies indicate that there is not a one-to-one relationship between deflection efficiency and stress efficiency. Figure 3.12 depicts the relationship between these two joint efficiency parameters.

In the structural evaluation of rigid pavement slab systems, the stress modification factor, J_e, may also be of interest. This factor is related to the stress efficiency (joint) parameter by:

$$J_e = 100/(100 + s_{je})$$

The J_e value is a stress modification factor applied to the theoretically computed free edge slab stress, based upon Westergaard analysis, and yields the actual edge (joint) stress in the slab due to a given level of load transfer efficiency. Thus,

$$s_{act} = J_e(s_{fe})$$

where

s_{act} = the actual slab stress at the edge (joint),
s_{fe} = the theoretically computed free edge stress, and
J_e = the stress modification factor.

Because the J_e value is related to the joint stress efficiency value, which in turn is related to the joint deflection efficiency value, the J_e value (and hence actual-modified free edge stress at a joint) can be determined directly from the d_{je} parameter obtained with NDT. The analysis should not, however, be applied to joints (edges) where the NDT load device is in the immediate vicinity of a slab corner.

Testing should be avoided during midday to minimize the possibility of joint lockup and slab curl. On cool overcast days, deflection may be performed throughout the day.

Procedure. All NDT deflection devices are suitable for evaluating the load transfer efficiency at any joint or crack, provided deflection sensors can be placed close enough to each other across the joint to measure displacement of both the loaded and unloaded slabs. A load approaching 9,000 pounds is preferred because it simulates stress deformations associated with an 18 KSAL.

Once unloaded and loaded slab (joint) deflections are measured, the d_{je} value can be directly determined. Figure 3.12 may then be used to determine the s_{je} (stress) efficiency value, from which J_e (stress modification factor) may be computed. Also, predicted in situ joint slab stress may be estimated from the theoretical free edge stress (only necessary if the stress parameter is to be used in the rehabilitation process).

As an example, assume that the NDT load transfer evaluation of a joint produced these results: joint deflection on the loaded slab, d_e, is 0.030 inches, and joint deflection on the unloaded slab, d_u, is 0.018 inches. For the pavement (slab-foundation) system, a

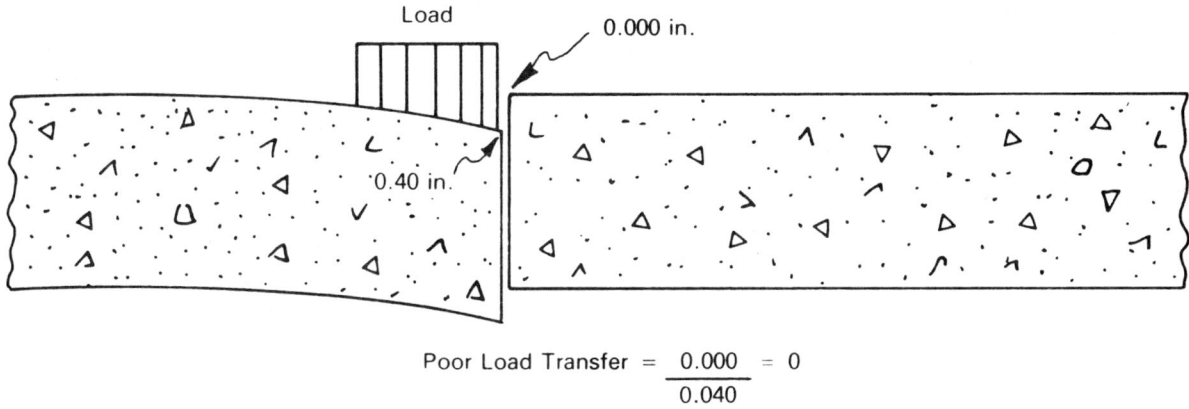

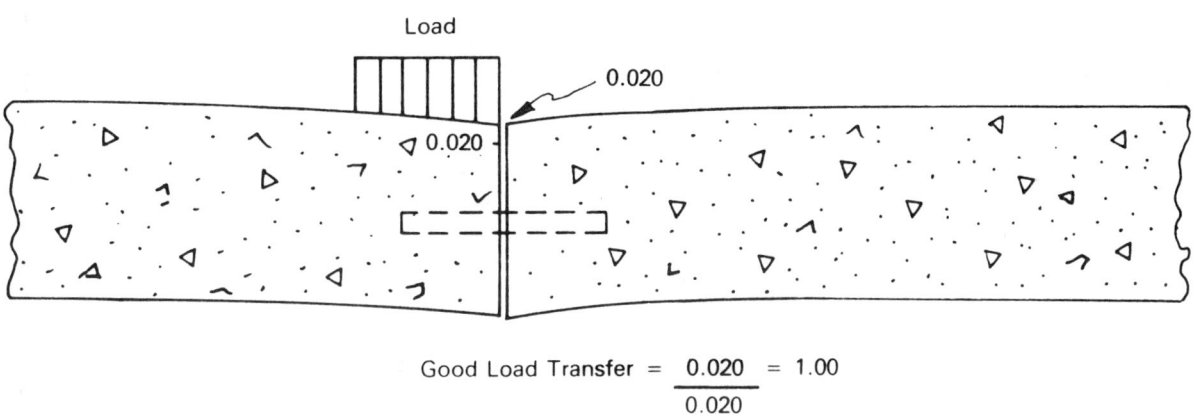

Figure 3.11. Illustration of Poor and Good Load Transfer

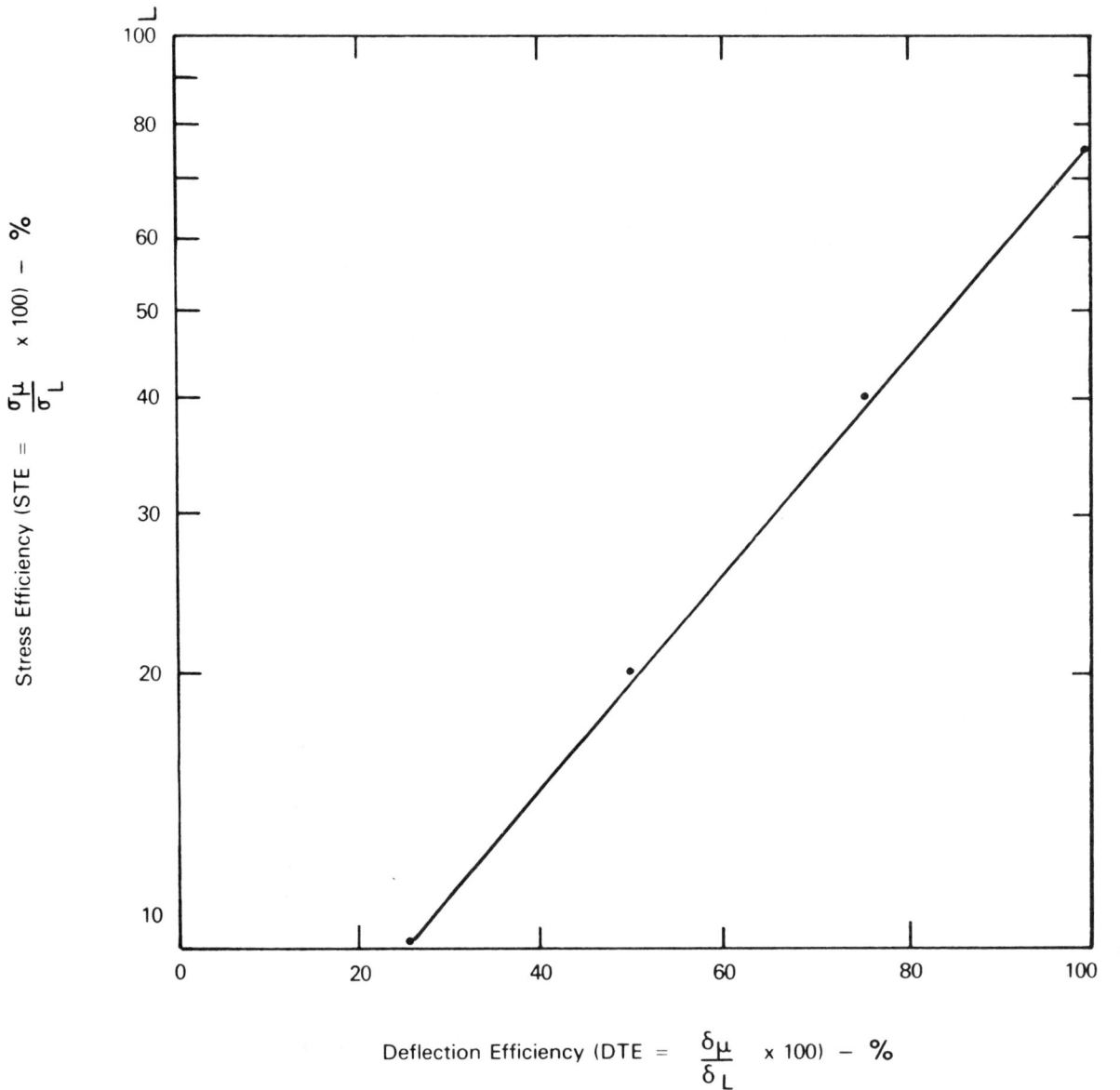

Figure 3.12. Relationship Between Joint Efficiency For Flexural Stress and Deflection Methods of Measurement. (By L. Korbus and E.J. Barenberg: From DOT/FAA/RD-7914, IV), "Longitudinal Joint Systems in Slip-Formed Rigid Pavements—Volume IV."

theoretical free edge stress of $s_{fe} = 525$ psi was analytically computed.

From this data, d_{je} (deflection efficiency) is computed:

$$d_{je} = (0.018/0.030) \times 100$$
$$= 60.0 \text{ percent}$$

Using Figure 3.12, the joint stress efficiency value, s_{je}, is

$$s_{je} = 25.7 \text{ percent } (26\%)$$

Then, the stress modification factor, J_e, is:

$$J_e = 100/(100 + 25.7)$$
$$= .795$$

Finally, the estimated actual (in situ) slab stress at the joint, s_{act}, would be:

$$s_{act} = .795(525)$$
$$= 418 \text{ psi}$$

In theory, the minimum value of J_e equals 0.50 (as d_{je} approaches 100.0 percent) and the maximum value of J_e equals 1.00 (as d_{je} approaches 0 percent or a pure free edge–no load transfer condition).

3.5.5 Use in Slab-Void Detection

Background. The third primary use of nondestructive deflection testing is the detection of voids under joint/crack systems in rigid pavements. In addition, a method exists for estimating the approximate size of voids, which is vital information during slab subsealing rehabilitation work (with or without planned overlay). This nonoverlay method is discussed in Section 4.3.4.

Higher midday temperatures should be avoided during deflection testing to minimize the possibility of joint lockup and slab curl. On cool overcast days, deflection testing may be performed throughout the day. The pattern of testing depends on the method of void detection used. It is recommended that a deflection device capable of simulating heavy truck loads be used. There are no standards in terms of the location and quantity of joints to be tested; this is left to the discretion of the engineer upon conducting a visual field survey.

Presented here are three methods for slab void detection. They are:

(1) Corner Deflection Profile (approximate)
(2) Variable Load Corner Deflection Analysis
(3) Void Size Estimation Procedure

Each successive method of void detection is increasingly detailed.

Corner Deflection Profile (Approximate). This method of void detection requires the measurement of corner deflection under a constant load (preferably 9 kips) along a section of pavement. The approach-and-leave corner deflections are then plotted on a profile and the results inspected for corners with the lowest deflections, as these corners will likely have full support. (See Figure 3.13.) Typically, the approach corner has little or no void. A maximum allowable deflection value, somewhat larger than the apparent full support or no-void value, can then be selected and used as a "field-generated criteria" for corners that may require subsealing (deflection higher than this maximum allowable value). For example, the deflection measurements in Figure 3.13 taken on a doweled JRCP with a Falling Weight Deflectometer show approximately 0.020 inches to be a reasonable maximum deflection. The measurements in Figure 3.14, taken on an undoweled JPCP with a weight truck, show 0.015 inches to be a reasonable maximum deflection. A deflection profile for CRCP is shown in Figure 3.15. Here agin, high deflections identify loss of support or void areas.

If subsealing is undertaken, the deflection at each subsealed location should be measured with the same device and weight used prior to subsealing, and as close to the same temperature as possible. Any corner experiencing deflections in excess of the selected maximum value should be subsealed again. The proportion of slab corners having greater deflections than the maximum allowable deflection can then be computed. This proportion is used in estimating material quantities.

One shortcoming of the corner deflection profile method is that a single value for the maximum allowable deflection may not be appropriate if load transfer varies widely from joint to joint. Because of this and the influence of test temperature upon results, this method, though extremely useful, should be viewed as an approximate approach to void detection. Also, the

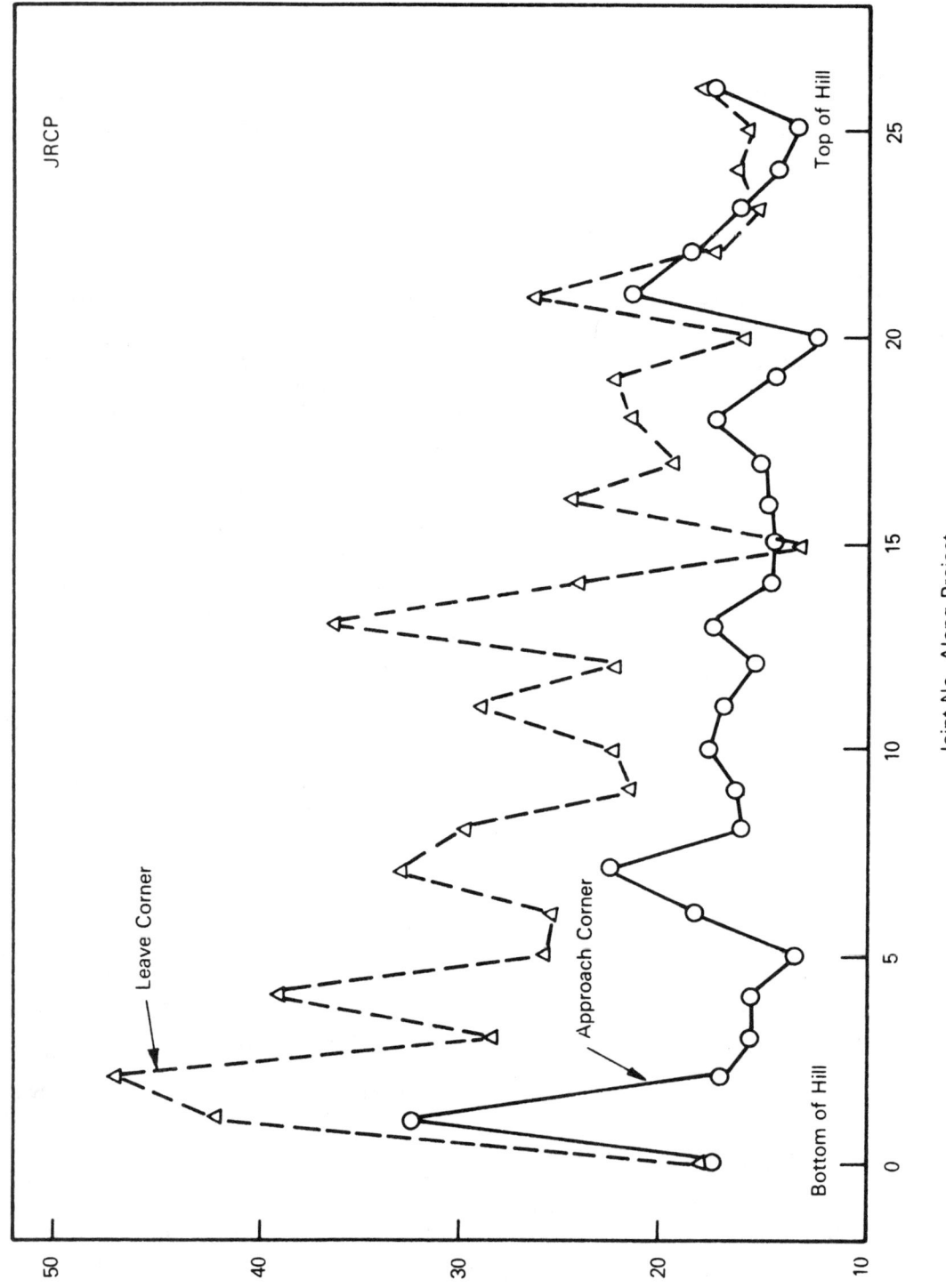

Figure 3.13. Profile of Corner Deflection for JRCP (60 ft. joint space)

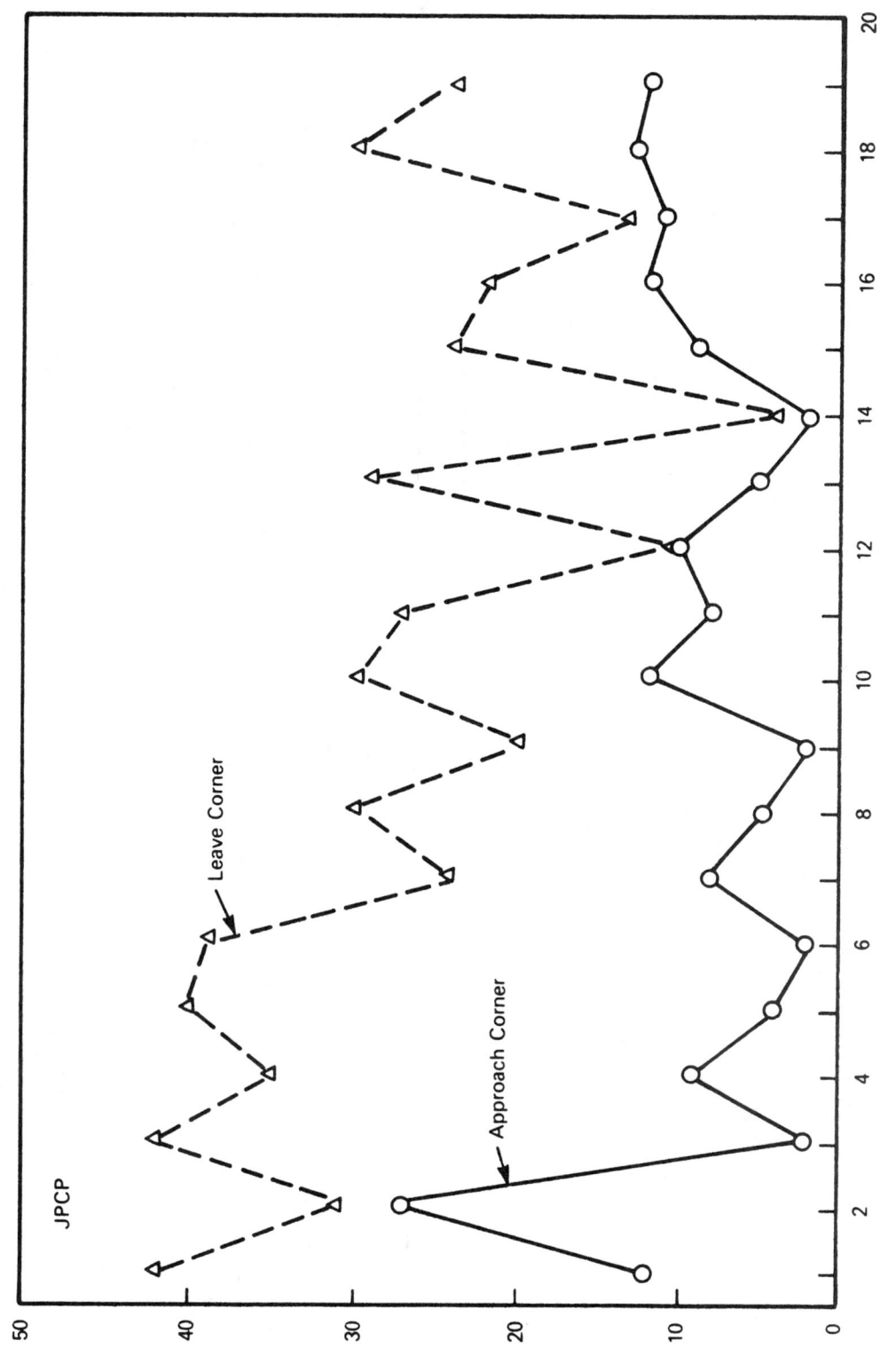

Figure 3.14. 18-Kip Single Axle Corner Deflection Profile of Slab Corners (non-doweled, 20 ft. space)

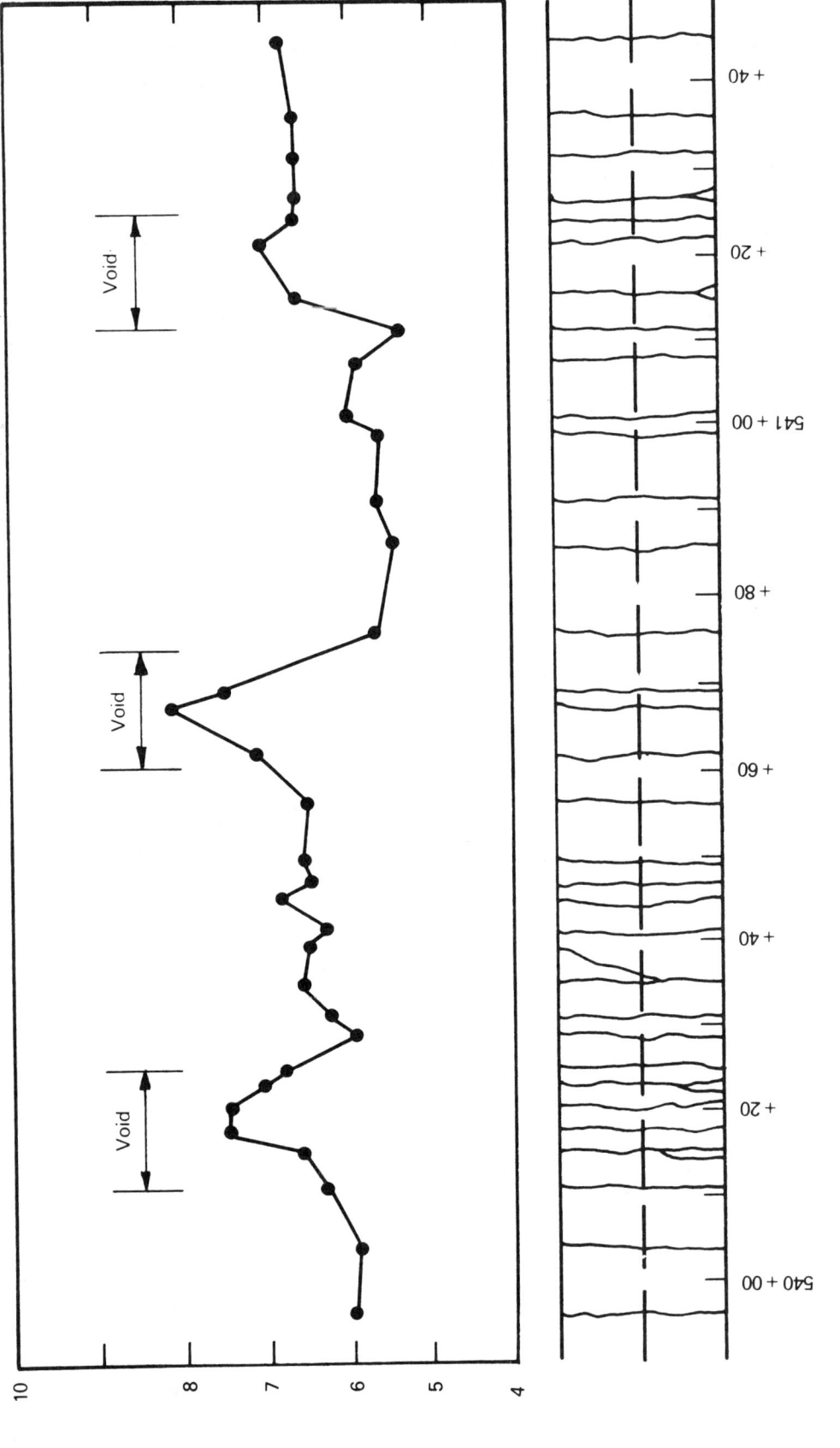

Figure 3.15. Deflection Profile of CRCP at each Crack Showing Void Locations

results give no indication of the probable size of voids that may be presented.

Variable Load Corner Deflection Analysis. This rapid method of void detection may be used while deflection testing is in progress. Corner deflections are measured at three load levels (e.g., 6, 9, 12 kips) to establish the load vs. deflection response for each test location. (See Figures 3.16 and 3.17.) Typically, locations with no voids cross the deflection axis very near the origin (less than or equal to .002 in.). For locations where the load vs. deflection response crosses the axis at points further removed from the origin, voids are indicated. Due to variations in joint load transfer which affects the load vs. deflection response, this method cannot be used to establish the approximate size of the void. However, the percentage of joints having voids may be computed and, thus, indicate the number of joints which will need subsealing. The effects of subsealing at locations where voids are suspected is demonstrated in Figure 3.17.

Void Size Estimation Procedures. A procedure was developed in NCHRP Project 1-21 to estimate the approximate void area under a given slab corner. The procedure requires a 5+ kip plate load (preferably 9 kips), and the ability to measure (and interpret) deflection basins at the slab center, deflection at the slab corner, and transverse joint load transfer. Center slab basin testing results are used to standardize the measured corner deflections (deflection from 9,000-lb. plate load at E = 4,000,000 psi) and the measured load transfer. (Refer to NCHRP 1-21 for details.)

All standardized corner deflections are then entered on a void detection plot according to the adjusted load transfer. (See Figure 3.18.) Deflections plotted in the "zero voids band" indicate joints without voids. Based on the location of this band, deflection levels for all possible load transfer conditions are determined to indicate varying void sizes (4-72 sq. ft. of surface area). Points of deflection falling outside the zero voids band are then used to determine the approximate size and location of voids (in square feet of surface area) at each joint.

Typically, voids can be located on one or both sides of the joint using this method. Subsealing should be performed only at locations where voids exist, with the undersealing hole pattern being adjusted according to the size of the void. The total area of voids can be extrapolated over the project and utilized to estimate material quantities by comparison with other projects.

3.6 FIELD SAMPLING AND TESTING PROGRAMS

3.6.1 Test Types

In general, field testing is categorized into two broad areas: nondestructive testing (NDT) and destructive testing. Destructive tests require the physical removal of pavement layer material in order to obtain a sample (either disturbed or undisturbed) or to conduct an in-place test. Such testing has many disadvantages and limitations, particularly when conducted on moderate to heavily trafficked highway systems. Practical restraints in terms of time and money severely limit the number and variety of destructive tests conducted on routine rehabilitation studies.

Nondestructive testing, on the other hand, does not necessitate physical disturbance of the pavement and, as a result, is preferred for the rehabilitation process. The most widely used form of NDT is associated with the field deflection tests noted in the previous section. However, several additional forms of NDT are now state of the art technologies. The other major type of NDT is associated with layer thickness measurements and void detection under rigid pavement systems. While these tests are not, at present, a part of the routine field testing program, future improvements and advances will undoubtedly occur. Because of the very significant advantages of nondestructive testing over destructive testing, the engineer should continually keep abreast of changes in this technology.

3.6.2 Major Parameters

During the data collection process, the engineer must accumulate enough information on the in-place condition of the pavement system to determine the precise cause of the distress. The parameters of the actual data collected will vary from project to project. To illustrate, if a flexible pavement is experiencing extensive rutting after 15 to 20 years of service, the rehabilitation required is probably routine, and a minimum field sampling and testing program will probably suffice. On the other hand, if a flexible pavement is experiencing extensive rutting after only a few years in service, more extensive field testing and data collection may be necessary to pinpoint the exact cause of the distress and the appropriate rehabilitation measures. Such rutting may be the result of material densification (improper compaction), deformation in the foundation (subgrade), instability in the asphalt layer, etc.

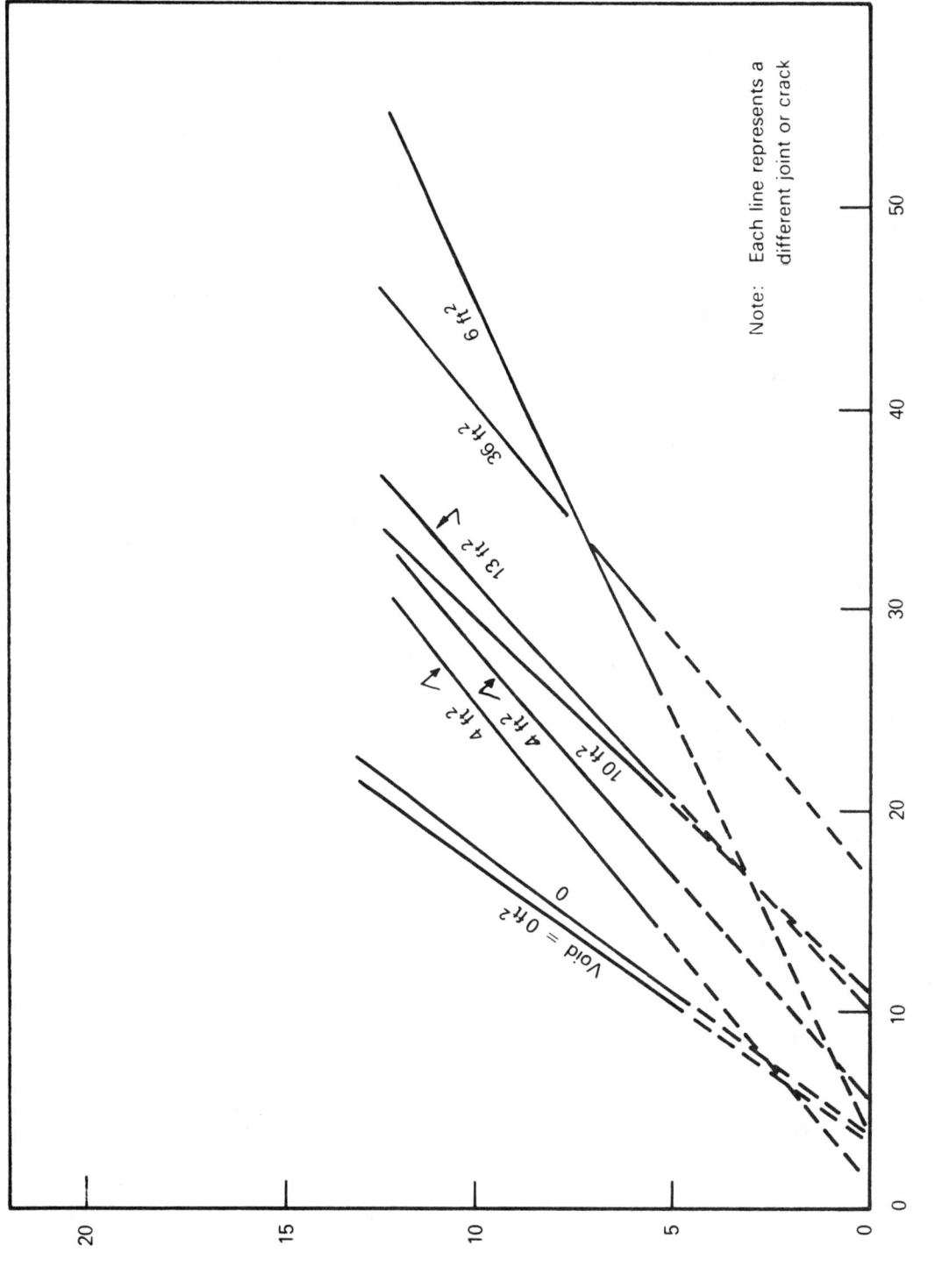

Figure 3.16. Joint Load Deflection (leave side) with Various Sizes of Suspected Voids (I-77 Ohio)

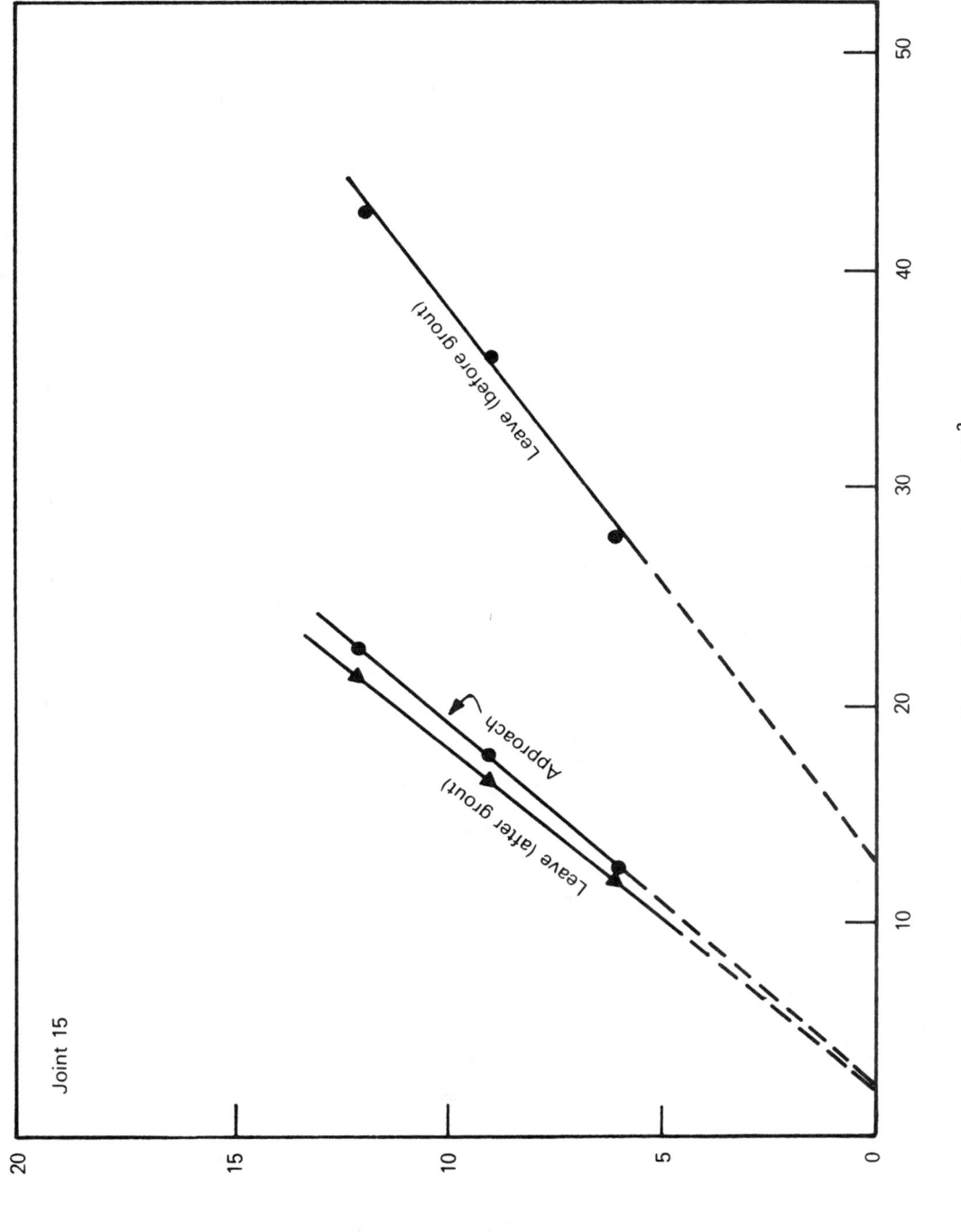

Figure 3.17. Joint Load Deflection where Large Void under Leave Corner was Suspected (Ohio I-77)

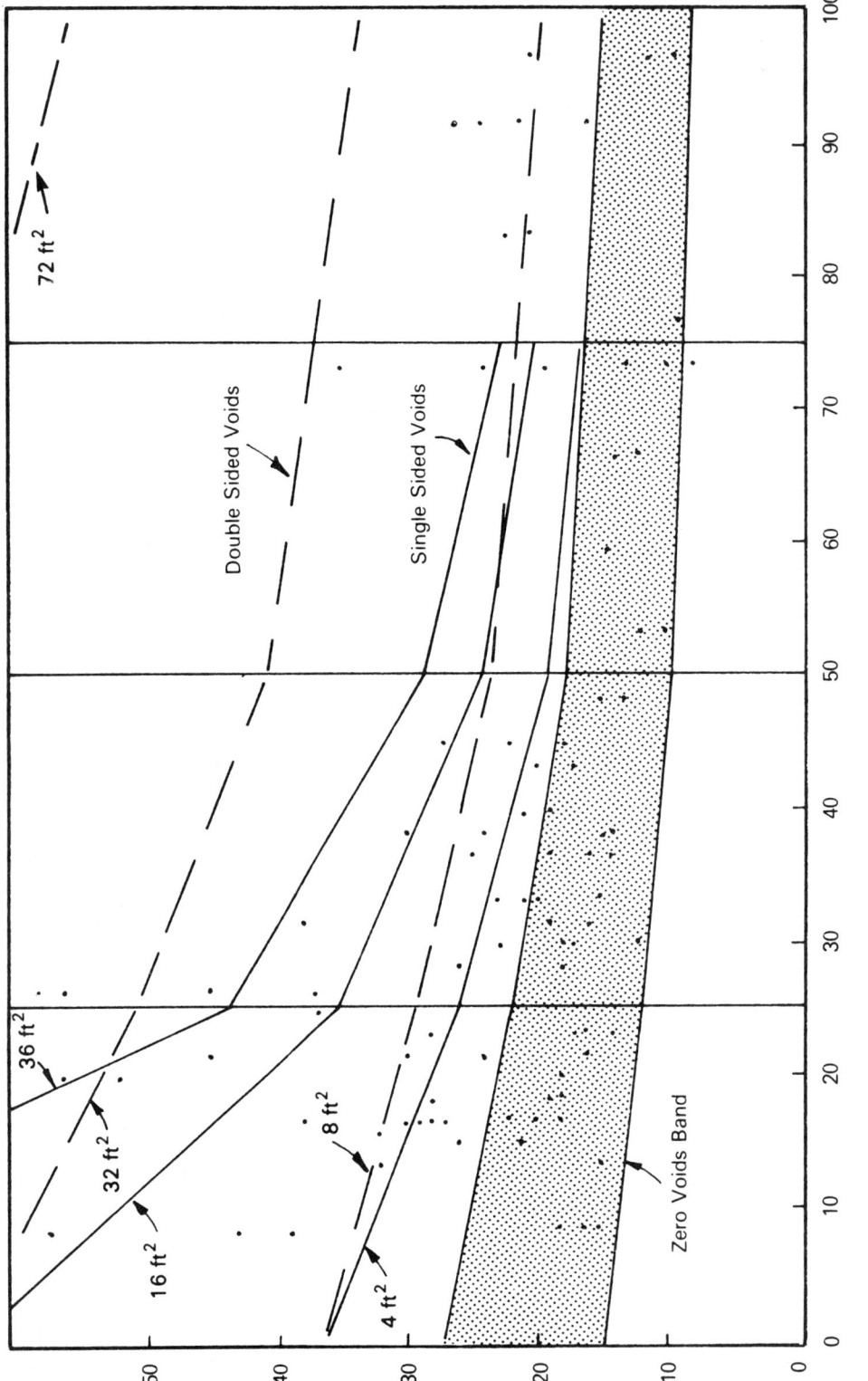

Figure 3.18. Typical Void Detection Plot for Determining Void Size and Location

It is the responsibility of the engineer to determine the scope of the data collection process for a project, and to minimize the cost of the process by avoiding the collection of superfluous information. There are, however, several major parameters that should be viewed as mandatory in any data collection process. They are as follows:

Pavement Deflection Response
In Situ Material Response (Modulus, Strength)
Layer Thicknesses
Layer Material Type

3.6.3 Necessity for Destructive Testing

There are three sources of information available to the engineer during the data collection process: historic data, destructive testing, and nondestructive testing. One or more of these sources may be used to fulfill the data collection parameters listed above. While the emphasis thus far has been on nondestructive testing, destructive testing may play a vital role in field sampling and testing.

In fulfilling the second parameter, in situ material response, NDT is the preferred source of information. However, historic data may be used with the caution that in situ conditions may have altered since the data was gathered. The use of a limited number of destructive tests to verify/modify material properties estimated from either NDT or historic data is sound engineering practice worthy of consideration. Also, these tests may be used to determine drainage conditions and identify problem layers. Test pits may also be of use in this area.

For rigid pavements, one of the more significant material properties influencing performance is the flexural strength (modulus of rupture) of the concrete. General correlations between splitting tensile strength and flexural strength may be used as a source of input since cores can be obtained from the pavement.

Unlike the first two parameters, determination of pavement layer thicknesses and layer material type cannot be made through NDT. While historic information may be available, the extreme importance and sensitivity of this variable calls for the use of destructive testing to verify/modify the available historic information. Layer material type can usually be identified from historic pavement information, unless special circumstances dictate otherwise. A limited amount of coring at randomly selected locations may be used to verify the historic information.

In summary, while NDT is largely preferred to destructive testing, a technically sound engineering field program should include a complementary destructive test program to ensure the accuracy of data obtained. This system of double checking ensures that inaccurate data will not be used in the rehabilitation design.

3.6.4 Selecting the Required Number of Tests

Analysis Unit. While conducting a pavement analysis, the project length should be divided into analysis units. These are pavement segments which exhibit statistically uniform attributes and performance. These units, which are discussed further in Chapter 5 and Appendix J, form the basis for a field sampling and testing program. As shown in Figure 3.1, a certain degree of variability (associated with any parameter) exists within each unit. In addition to its importance from a design reliability viewpoint, this "within unit" variability is helpful in defining a statistically based sampling and testing program.

Limit of Accuracy Curves. Tests conducted on analysis units provide an estimate of the actual mean and standard deviation (or variance) of the property under investigation. As the number of tests increases, the estimated values more closely approximate true values. The principles of statistical confidence levels are very useful in determining how many tests will be necessary to ensure that the estimated mean is within a certain limit of the actual mean. The concept of confidence levels may be explained by the statement that we are $100(1 - \alpha)$ percent confident that the mean (true) value lies within the limits calculated.

Statistical limit of accuracy curves help assess the impact of the number of tests conducted on the precision of the estimate. The limit of accuracy, R, represents the probable range of the true mean from the average obtained by "n" tests, at a given degree of confidence (e.g., 95 percent). Mathematically,

$$R = K_\alpha(\sigma/\sqrt{n})$$

where

K_α = the standardized normal deviate, which is a function of the desired confidence level, $100(1 - \alpha)$, and

σ = true standard deviation of the random variable (parameter) being considered.

For a given variable (deflection thickness, etc.), once a confidence level is selected (e.g., 95 percent), K_α and σ are constants. The R value is inversely proportional to the square root of the number of tests used if randomly selected. Figure 3.19 illustrates the typical schematic plot of R versus n. As illustrated in the figure, there are three zones along the accuracy curve. In Zone I, characterized by a steep slope, the precision of the estimate significantly increases with each additional test or sample. In this zone, the benefit-cost ratios for increasing the number of tests per analysis unit are quite high and worthwhile. On the other hand, Zone III is a region with little slope, where even large increases in the number of tests/samples obtained will not significantly improve the precision of the estimate. In other words, the engineer will certainly not double the accuracy of the estimates within Zone III by doubling the number of tests, and the cost of each additional test outweighs the benefits. Zone II represents the "optimal" range in developing a test program, because it represents the area where accurate estimates will be made using a minimum number of tests.

Application to the Project Example. Figure 3.20 depicts the limit of accuracy curve developed for the example data previously shown in Figures 3.3 and 3.4, and discussed in Appendix J. The standard deviation, developed from the within unit variability, reflects the pooled variance of all 11 analysis units delineated in Figures 3.4. Table 3.5 presents a summary of the number of tests per analysis unit and the resulting limit of accuracy (+R) about the true unit mean value. For the problem conditions noted, if an accuracy of R = ±1.25 is desired, then 6 tests per unit is satisfactory. The entire project would necessitate only 66 tests (6 tests/unit × 11 analysis units). Because an equal interval approach was used to develop the values (Δx = 0.5 mi), 152 tests were actually obtained over the 76-mile project length. Had the statistical test program with random testing been used instead, only 43.4 percent of the tests would have been required. Thus, there are obvious economic advantages to using the statistical approach coupled with the analysis unit concept when developing a field sampling and testing program.

Guidelines for Major Variable Testing/Sampling Program. While the previous example has been based upon confidence estimation of skid resistance, SN(40), the fundamentals can be applied to all pavement variables in the rehabilitation process. Table 3.6 is a summary of typical variability values for a wide variety of parameters.

For all variables, except pavement deflection, variability is expressed by the standard deviation, s, of the unit parameter distribution. Because pavement deflections vary by load magnitude and load plate characteristics, as well as overall pavement structure, the variability is expressed in terms of the Coefficient of Variation (CV) value, defined by:

$$CV = (s/\bar{x})100$$

The summary shown is intended to serve as a general guide to the engineer in assessing the required number of tests (samples) to be obtained in the field program. Whenever possible, design agencies should try to collect their own historic variability data unique to their own materials, environment, and construction practices to supplement the guide data of Table 3.6. Figure 3.21 illustrates typical limit of accuracy curves, at a 95-percent confidence level for the variables and data shown in the table.

Table 3.5. Summary Comparison of Statistically Based Field SN(40) Test Sample

Number of Tests Per Unit	Limit of Accuracy (±R)	Total Number of Tests Required in Project	Actual Tests Conducted in Project	% Tests Needed Relative to Number Used
4	±1.50	44	152	28.9
5	±1.38	55	152	36.2
6	±1.25	66	152	43.4
7	±1.15	77	152	50.7
8	±1.08	88	152	57.9
9	±0.80	176	152	115.8

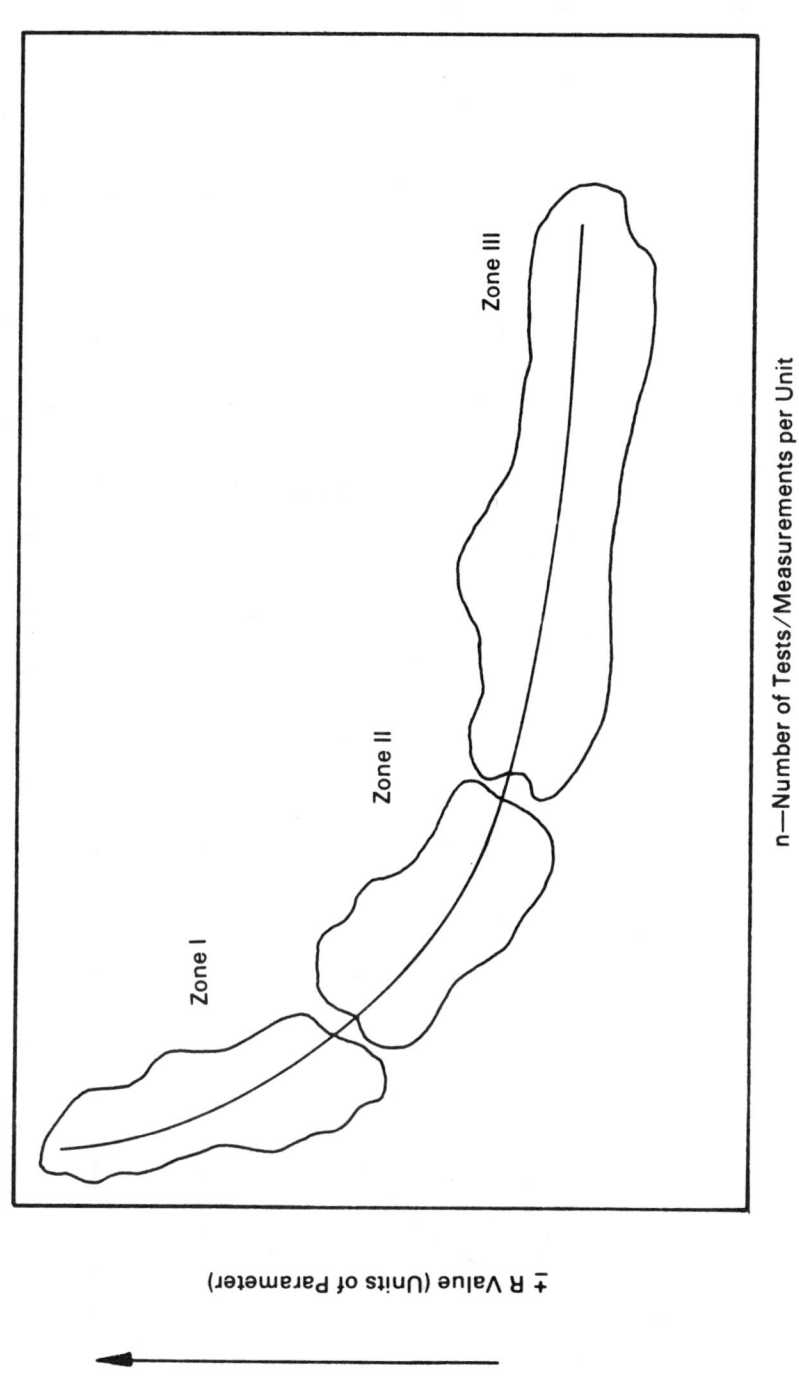

Figure 3.19. Typical Limit of Accuracy Curve for all Pavement Variables Showing General Zones

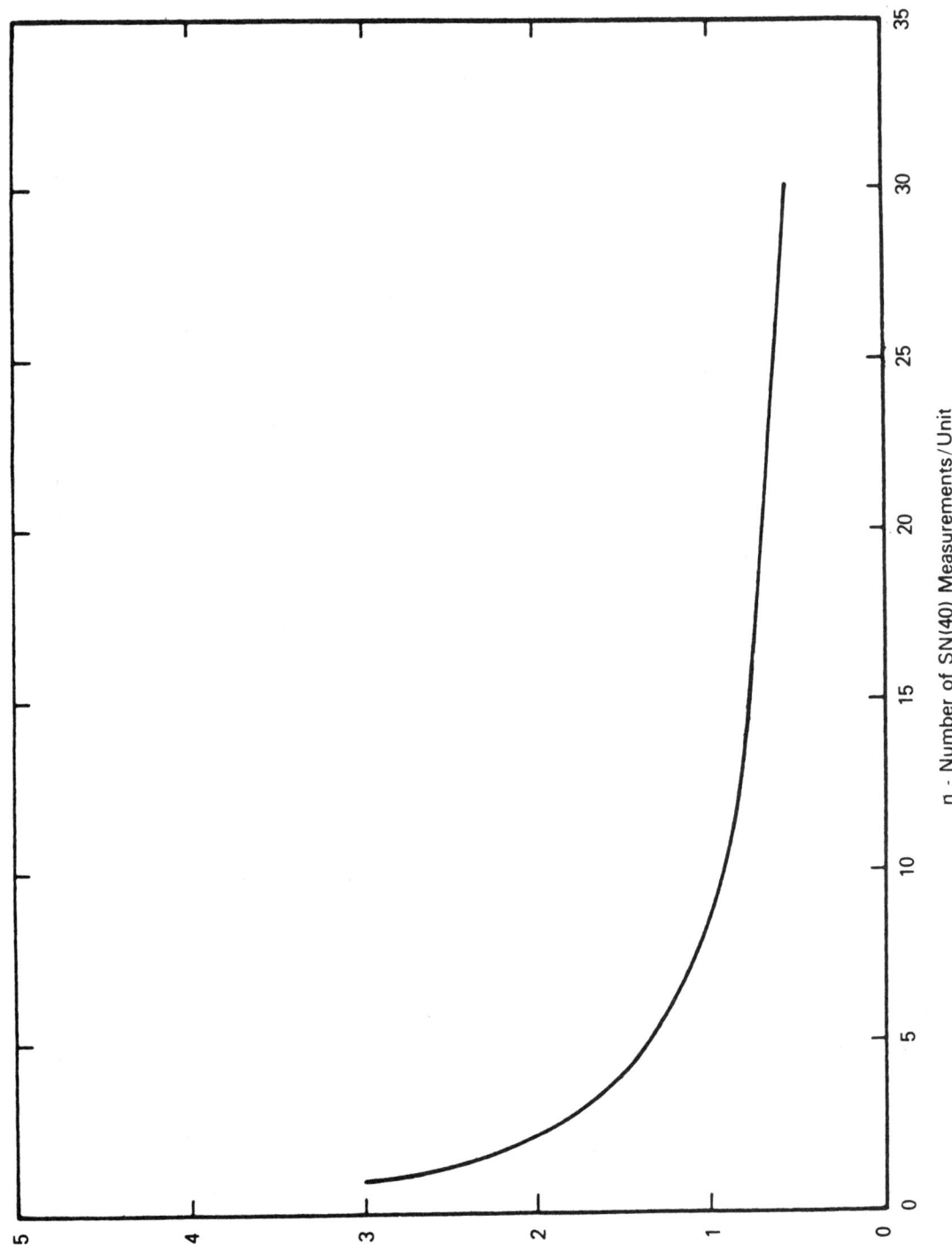

Figure 3.20. Limit of Accuracy Curves for an Example Problem

Table 3.6. Summary of Typical Pavement Parameter Variability

	s—Standard Deviation		
	Low	Average	High
1. Thickness (inches):			
Portland cement concrete	0.1	0.3	0.5
Asphalt concrete	0.3	0.5	0.7
Cement treated base	0.5	0.6	0.7
Granular base	0.6	0.8	1.0
Granular subbase	1.0	1.2	1.5
2. Strength:			
CBR (%)			
Subgrade (4–7)	0.5	1.0	2.0
Subgrade (7–13)	1.0	1.5	2.5
Subgrade (13–20)	2.5	4.0	6.0
Granular subbase (20–50)	5.0	8.0	12.0
Granular base (80+)	10.00	15.0	30.0
PCC flexural strength (psi)	65	100	135
3. Percent compaction (%):			
Embankment/subgrade	2.0	4.5	7.0
Subbase/base	2.0	2.8	3.5
4. Portland cement concrete properties:			
Air content (%)	0.6	1.0	1.5
Slump (inches)	0.6	1.0	1.4
28 Day compressive strength (psi)	400	600	800
5. Asphalt concrete properties:			
Gradation (%)			
3/4 or 1/2	1.5	3.0	4.5
3/8	2.5	4.0	6.0
No. 4	3.2	3.8	4.2
No. 40 or no. 50	1.3	1.5	1.7
No. 200	0.8	0.9	1.0
Asphalt content (%)	0.1	0.25	0.4
Percent compaction (%)	0.75	1.0	1.5
Marshall mix properties			
Stability (lbs)	200	300	400
Flow (in./in.)	1.0	1.3	2.0
Air voids (%)	0.8	1.0	1.4
AC consistency			
Pen (77°F)	2	10	18
Viscosity (149°F)—kilopoise	2	25	100
	CV—Coefficient of Variation		
	Low	Average	High
6. Pavement deflection	15	30	45

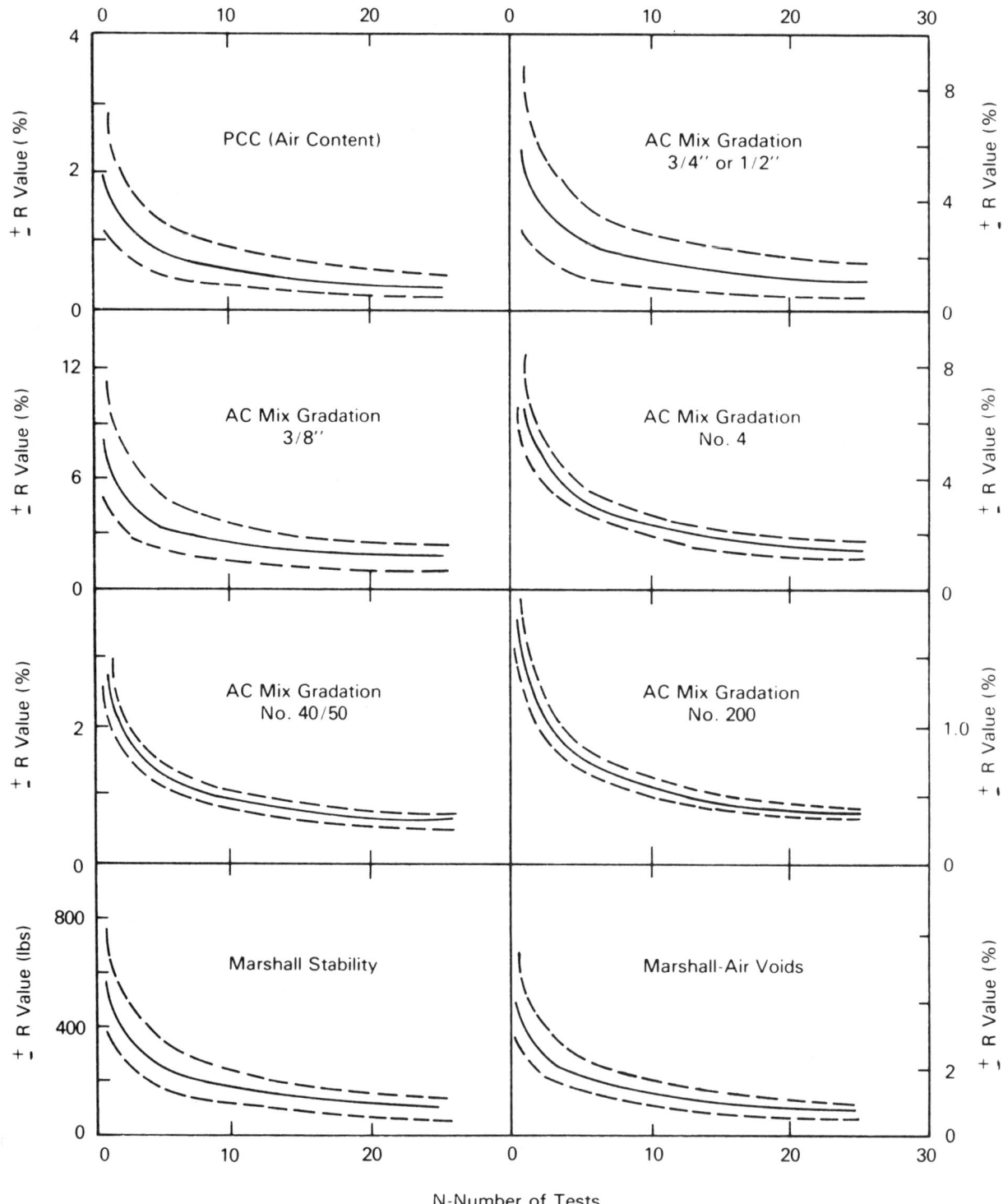

Figure 3.21. Limit of Accuracy Curves: Mean at (95%) Confidence

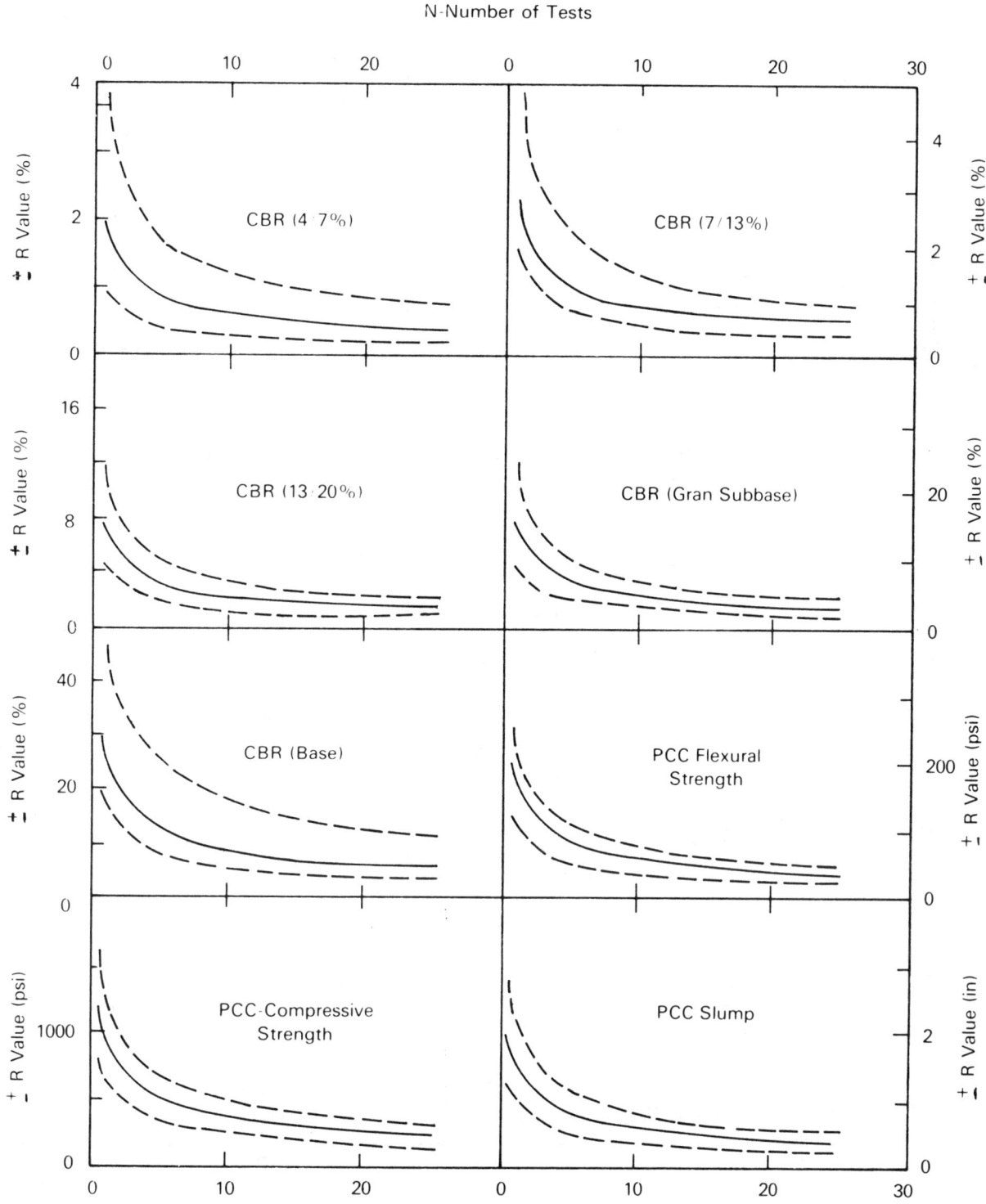

Figure 3.21. Continued—Limit of Accuracy Curves: Mean at (95%) Confidence

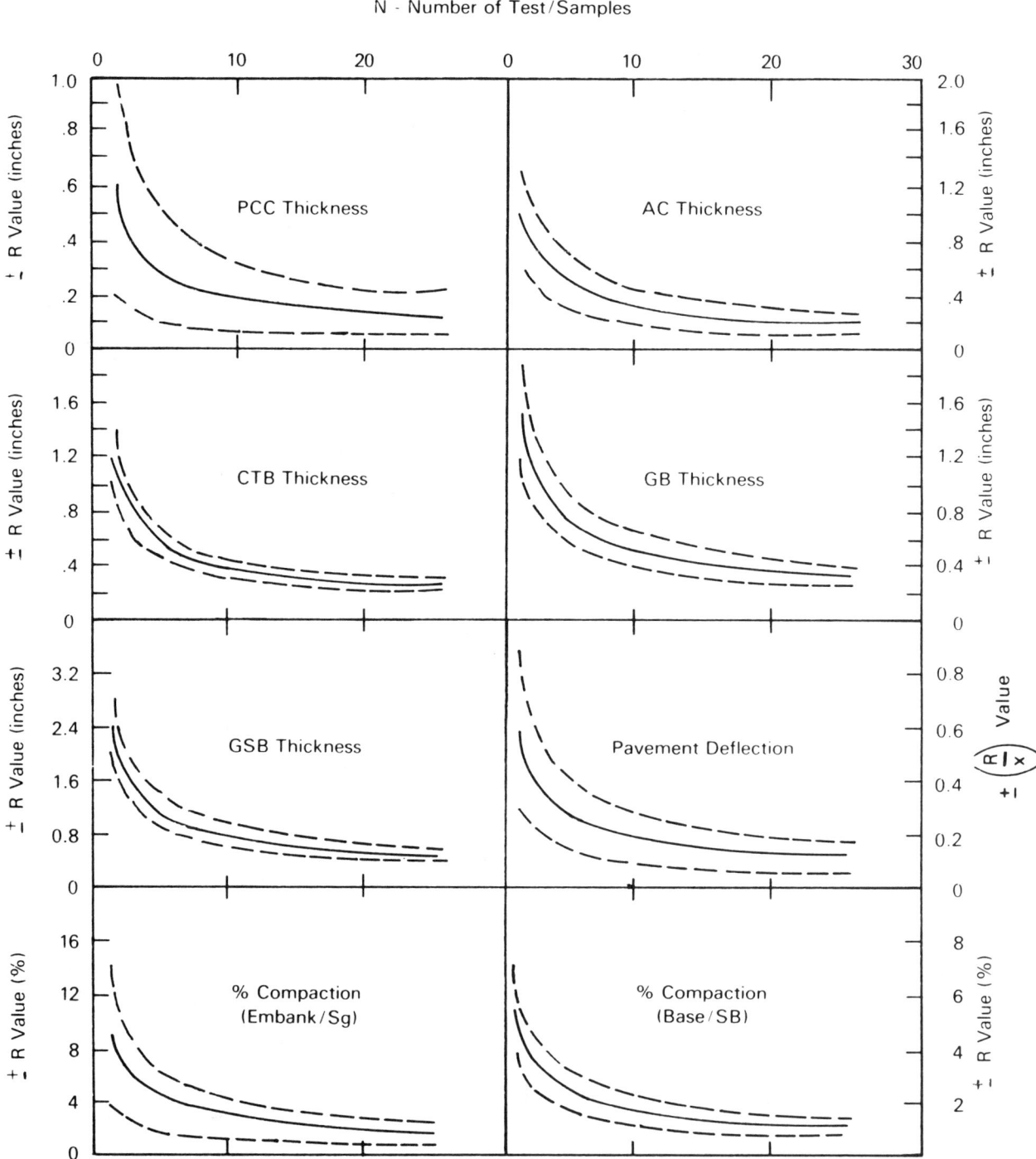

Figure 3.21. Continued—Limit of Accuracy Curves: Mean at (95%) Confidence

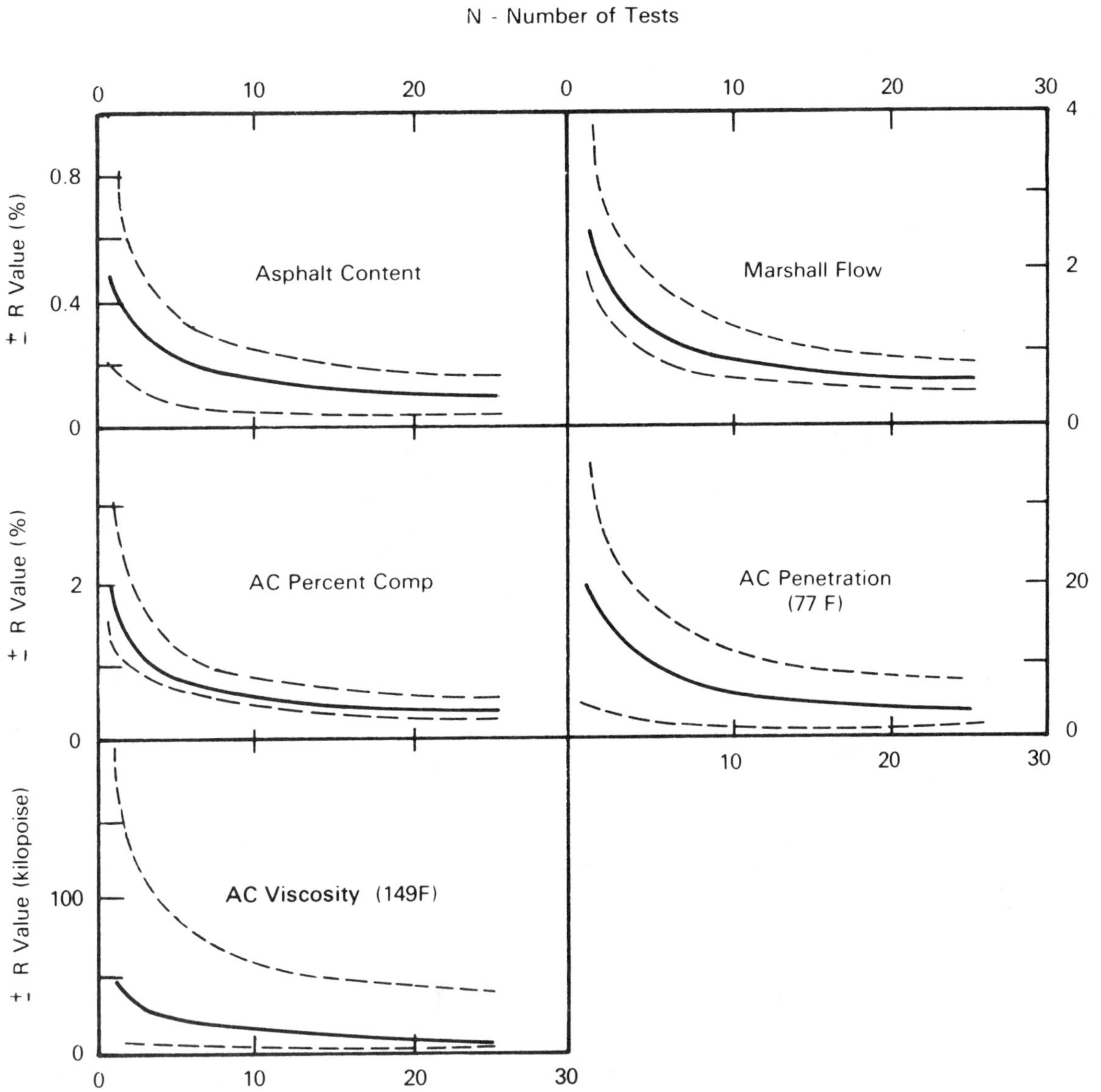

Guides for Field Data Collection

Figure 3.21. Continued—Limit of Accuracy Curves: Mean at (95%) Confidence

CHAPTER 4
REHABILITATION METHODS OTHER THAN OVERLAY

Many different rehabilitation techniques can be applied to pavements to extend their lives without the placement of an overlay. Some of these techniques are applicable prior to an overlay. Use of these techniques is often a cost-effective strategy (in framework of life-cycle cost), and delays the placement of a costly overlay, recycling, or even reconstruction for several years. When evaluating the feasibility and effectiveness of applying rehabilitation methods other than overlays, several factors must be considered, including the surface distress, structural condition, and functional condition of the existing pavement. This chapter describes the background and methodologies associated with these nonoverlay rehabilitation approaches.

4.1 EVALUATION OF PAVEMENT CONDITION

The evaluation of pavement condition (discussed in Chapter 3) includes consideration of specific problems that exist in the pavement. This requires a determination of the types and causes of distress, as well as the extent of pavement deterioration.

4.1.1 Surface Distress

Distress represents a very important and basic measure of current pavement condition. Each type of distress is the result of one or more causes which, when known, provide great insight into the type of rehabilitation work that is required. As enumerated below, distress data are useful in selecting rehabilitation strategies other than overlays.

(1) Distress types that are present at medium or high severity levels and require repair work, can be identified and quantified in the plans and estimates.

(2) An examination of all distress data collected will indicate if pavement condition varies significantly over a given project. Repair can then be varied with pavement condition to minimize costs.

(3) The results of the distress survey can indicate what further testing must be conducted to obtain sufficient data for design.

Distress data are helpful in determining the mechanisms of pavement deterioration. Pavement distresses can be categorized as being caused either by traffic loads or nonload factors, including design, construction, poor-durability materials, and climate factors. This knowledge helps the engineer determine an appropriate rehabilitation technique.

4.1.2 Structural Condition

The most critical area of concern with regard to the feasibility of rehabilitation without overlay is the structural adequacy of the pavement. Only structurally adequate pavements or pavements restored to a structurally adequate state, are candidates for rehabilitation without overlay. The structural evaluation must address whether or not the pavement can support future traffic loadings over the desired design period without structural improvement from an overlay. This analysis is directly addressed in Chapters 3 and 5 using the NDT evaluation.

Existing distress types are an excellent source of information on the impact of past traffic loadings on the pavement. If there is significant load-associated distress, then the structural adequacy of the existing pavement must be questioned. On multi-lane facilities, a difference in distress between the outer and the inner lanes is an indication of the impact of truck traffic on the structural adequacy of the pavement. Historical data on patching and slab replacement are also helpful in ascertaining the rate of deterioration due to structural loadings.

Another method for estimating the structural adequacy of a pavement is to work backward through a design procedure to determine if the pavement structure is adequate to handle past and future traffic loadings. When using this method, the properties of the existing pavement must be determined; it cannot be assumed that they equal the original properties of the pavement at the time of construction. The NDT procedures presented in the Guide provide guidelines for estimating the pavement's remaining structural life.

4.1.3 Functional Condition

The functional condition of a pavement expresses its ability to serve the user, and its major indicators include the following:

> roughness,
> skid resistance/hydroplaning,
> appearance, and
> other safety considerations.

An adequate evaluation of functional condition requires the measurement of roughness and skid resistance along the project in each lane. Areas exhibiting excessive roughness and/or poor skid resistance should then be noted for special consideration in the rehabilitation design.

The overall pavement evaluation should include consideration of the items noted in Figure 2.3 of Chapter 2. Each agency should develop procedures and guidelines for consistently answering the evaluation questions on this list.

4.2 DEVELOPMENT OF FEASIBLE ALTERNATIVES AND STRATEGIES

A feasible alternative is one that addresses the cause of the distress and is effective in both repairing existing deterioration and preventing its recurrence, while satisfying the imposed constraints. Some projects have only one or two feasible nonoverlay alternatives.

Tables 4.1 and 4.2 contain specific recommendations on the selection of candidate methods to repair distress and prevent its recurrence. For each distress type, one or more repair and/or preventive maintenance methods can be applied. If each of the repair and preventive methods meet the pavement's needs and satisfy the imposed constraints (such as available funding and minimum life extension), then they qualify as feasible rehabilitation alternatives.

In order to make the most of limited available funds, the engineer must choose the most cost-effective combination for the project. The following table provides an example by selecting alternative methods for a pavement having both pumping (with loss of support) and faulting:

Existing Distress	Candidate Repair	Candidate Preventive
Pumping (loss of support)	Subseal	Reseal joints Restore load transfer Tied PCC shoulder Subdrainage
Faulting	Grind	All above

Another example is given for a flexible pavement:

Existing Distress	Candidate Repair	Candidate Preventive
Transverse crack	Full-depth patch	Patch joint sealing
Raveling	Chip seal coat	Rejuvenating seal coat Fog seal coat
Rutting	Cold mill ruts Level-up overlay in wheel paths	None known

One repair method and one or more preventive methods must be selected for each distress type. If only repair work is performed, the mechanism causing the distress will immediately begin its destructive work when the pavement is opened to traffic. After each distress type has been treated with an appropriate repair, one or more preventive methods must be applied to provide a cost-effective design. For example, the following alternatives could be developed for the jointed concrete pavement above:

Alternative	Repair Method	Preventive Method
A	Subseal pumping Grind faults	Reseal all joints
B	Subseal pumping Grind faults	Subdrainage
C	Subseal pumping Grind faults	Ties PCC shoulder Reseal all joints
D	Grind faults Restore load transfer	

Table 4.1. Candidate Repair and Preventive Methods for Rigid Pavement Distress

Joint/Crack Distress	Repair Methods	Preventive Methods
Pumping	1. Subseal	1. Reseal joints 2. Restore load* transfer 3. Subdrainage 4. Edge support (PCC shoulder/edge beam)
Faulting	1. Grind 2. Structural overlay	1. Subseal 2. Reseal joints 3. Restore load* transfer 4. Subdrainage 5. Edge support
Slab cracking	1. Full-depth repair 2. Replace/recycle lane	1. Subseal loss of support 2. Restore load* transfer 3. Structural overlay
Joint or crack spalling	1. Full-depth repair 2. Partial-depth repair	1. Reseal joints
Blow-Up	1. Full-depth repair	1. Pressure relief joint 2. Resealing joints/cracks
Punchouts	1. Full-depth repair	1. Polymer or epoxy grouting 2. Subseal loss of support 3. Rigid shoulders

*Drainage analysis required to determine need and benefit.

Table 4.2. Candidate Repair and Preventive Methods for Asphalt Pavement Distress

Distress	Repair Methods	Preventive Methods
Alligator cracking	Full-depth repair	Crack sealing (May slow down alligator cracking)
Bleeding	Apply hot sand	
Block cracking	Seal cracks	
Depression	Level-up overlay	
Polished aggregate	Skid resistant Surface treatment Slurry seal	
Potholes	Full-depth repair	Crack sealing and seal coats
Pumping	Full-depth repair	Crack sealing and seal coats
Raveling and weathering	Seal coats	Rejuvenating seal
Rutting	Level-up overlay and/or cold milling	
Swell	Removal and replacement	Paved shoulder encapsulation

Many projects exhibit several types and severities of distress and, thus, require a combination of several different repair and preventive rehabilitation methods. Very often, several combined repair and preventive maintenance methods are required to return deteriorated pavement to a serviceable condition for a substantial period of time. Each alternative must be evaluated for cost-effectiveness, and a final selection of the most cost-effective is made.

Each agency should develop a comprehensive pavement rehabilitation strategy for every pavement type in their network. The strategy should include procedures for inspection, evaluation, and selection of feasible rehabilitation techniques. The consideration of preventive techniques is most important.

4.3 MAJOR NONOVERLAY METHODS

While numerous nonoverlay rehabilitation methods are utilized, many are experimental in nature. This section provides a description of the following major rehabilitation methods that may be employed as nonoverlay techniques:

(1) Full-Depth Repair
(2) Partial-Depth Patching
(3) Joint-Crack Sealing
(4) Subsealing-Undersealing
(5) Grinding and Milling
(6) Subdrainage
(7) Pressure Relief Joints
(8) Load Transfer Restoration
(9) Surface Treatments

4.3.1 Full-Depth Repair

Full-depth repair has applications to all types of pavement and typically represents a large cost item in a rehabilitation project. Because of the high cost of patching, many agencies tend not to repair distressed areas that should be repaired during pavement rehabilitation. This may result in rapid deterioration and more costly rehabilitation in the future.

The first step in the repair process is determination of locations and boundaries. Specific distress requiring repair must be identified and boundaries selected. Larger areas of extensive distress must be identified for complete removal and replacement of slabs. It is also important that patch boundaries be selected so that most of the significant underlying deterioration is removed. The results of a coring study provide information on additional deterioration that may exist beneath the slab surface. The patch boundary should not be too close to an existing transverse crack or joint, or adjacent slab distress will occur. In general, particularly in freeze-thaw climates, the deterioration near joints and cracks is greater at the bottom of the slab than at the top. Full-depth repairs are discussed in terms of jointed pavements, CRCP, and bituminous patches.

Full-Depth Repair of Jointed Concrete Pavements. The joint design of a full-depth repair is a major determinant of performance. Joint design is still largely an art, although excellent analytical techniques for calculating stresses and deformations are available. A review by the agency of the joint designs and their performance for various levels of traffic will be of great assistance in selecting the required level of load transfer. Poor joint load transfer usually leads to serious spalling, rocking of the patch, faulting, and corner breaks. Following are four techniques used with varying degrees of success to achieve load transfer across transverse patch joints:

(1) Tie bars: deformed rebars are grouted into the existing slab, or slab reinforcement is extended into the patch area where minimum joint movement is desired.
(2) Dowel bars: smooth steel bars are inserted into drilled holes in the existing slab where movement of the joint is desired.
(3) Undercutting: the subbase/roadbed is excavated out from beneath the slab and replaced with concrete. This method should not be used in freeze areas since differential heaving between the patch and existing slab causes severe roughness. Even in nonfrost areas, poor load transfer will be obtained if good concrete consolidation of the lip is not obtained, or if the patch settles.
(4) Aggregate interlock: this can only be used with rough-faced joints and short joint spacings. It is unreliable with heavy truck loads.

There are several types of distress that occur at or near transverse joints that may justify full-depth repair. These are blow-ups, corner breaks, durability "D" cracking, and load transfer-associated distress. Some spalls that extend less than halfway through the slab can be patched with partial-depth patches, as discussed in Section 4.3.2.

Some pavements develop intermediate cracks that deteriorate (e.g., spall and fault) through heavy traffic repeated loading. Locking of the doweled joints accelerates crack deterioration by forcing open the intermediate cracks, which soon lose aggregate interlock under heavy repeated traffic. Cracks that are working should generally be repaired either with a full-depth tied patch or a working joint.

There are also many situations in which existing distress is so extensive that the patching of every deteriorated joint and crack would be either very expensive or impractical. Repair cost can be reduced by simply removing and replacing larger areas of concrete slab. Thus, for a given distressed area, the engineer should estimate the cost for large removal and replacement and for patching of each localized distress using typical costs, and then select the lower cost alternative. If the costs are approximately the same, the large area removal and replacement should be selected, since this will certainly be the most reliable repair (as opposed to numerous patches).

On multiple-lane highways, deterioration may occur only in one lane or across two or more lanes. If distress exists in only one joint, it is not necessary to patch the other lanes. When two or more adjacent lanes contain distress, one lane should generally be patched at a time to maintain traffic flow. This practice also reduces the potential of a blow-up occurring in the other nonpatched lane. For example, on a highway having three lanes in one direction, only one lane should be patched at a time to reduce the potential for blow-ups. If blow-ups occur during the patching of one lane, it may be necessary to cut relief joints at intervals of 600 to 1,200 feet or delay patching until cooler weather occurs.

Full-Depth Repair of Continuously Reinforced Concrete Pavement. This section describes procedures for full-depth repair of CRCP. Only cast-in-place concrete repairs for permanent repair are discussed. Bituminous patches are not recommended for permanent repair of CRCP because they break the continuity of the reinforced concrete system and provide no load transfer across the joint, which results in deterioration of the CRCP. Field experience has shown that adequate load transfer can be obtained when: (a) the reinforcing steel is extended into the repair and tied or welded to additional reinforcement placed in the repair, (b) the subbase is not seriously deteriorated beneath the joint, and (c) the repair face is nearly vertical and rough beneath the reinforcement, and not spalled underneath.

As stated, several types of distress justify full-depth repair such as blow-ups, punchouts, durability "D" cracking, and construction joint problems. Each agency should develop recommendations to closely fit local conditions.

Criteria for repair dimensions should provide adequate lap length and cleanout, and minimize or eliminate patch rocking, pumping, and breakup. The repair boundary should not be too close to an existing transverse crack or joint because adjacent slab distress will develop. Generally, the patch joint should not be closer than 18 inches to the nearest tight crack. However, where cracks are very closely spaced it is sometimes necessary to place the repair as close as 6 inches to an existing tight transverse crack.

There are also some situations where existing distress is so extensive that the repair of every deteriorated area within a short distance would be either very expensive or impractical. Repair costs may be reduced by simply removing and replacing larger areas of the CRCP slab.

When two or more adjacent lanes contain distress, one lane should be repaired at a time so that traffic flow can be maintained. If the distress, such as a wide crack with ruptured steel, occurs across all lanes, special considerations are needed due to the high potential for: (a) blow-ups in the adjacent lane, (b) crushing of the new repair during the first few hours of curing by the expanding CRCP slab, and (c) serious cracking of the repair during the first night as the existing CRCP contracts. The following procedures will minimize problems:

(1) The repair should be placed in the afternoon to avoid being crushed under expansion.
(2) The lane having the lowest truck traffic should be repaired first.

The recommended design of the repair provides for adequate bar/wire laps in the repair area. The new reinforcement can be tied, welded, or secured with mechanical couplers to the existing reinforcement so that the full strength of the bar/wire is developed. The reinforcing steel should be placed so that a minimum of 2.5 inches of cover is provided. The bars should be placed and supported by chairs or by any other means available, such that the steel will not permanently bend down during placement of the concrete.

Patching With Bituminous Mixtures. This section concerns the patching of asphalt and rigid (concrete) surfaced pavements with bituminous patching mixtures. There are two main types of bituminous patching mixtures: (a) those mixed hot and compacted while still hot, and (b) those mixed and then stockpiled for a period before use. These mixtures range widely in quality and costs. The performance of a bituminous patch depends on both the quality of the materials comprising the patching mixture and the quality of construction effort in placing and compacting the mixture. The best bituminous patching mixture will last only a short time if good construction practices are not followed.

Patching is frequently associated with the formation of potholes in flexible pavements which develop through the combined effects of moisture and traffic. Bituminous patching is used as a restoration tool for damaged areas not classified as a pothole area, and is sometimes used as a means of relieving expansive pressure in rigid pavements. The use of bituminous patching is discussed separately in Section 4.3.7.

Localized repair (such as pothole patches) is not the only reason for bituminous patching. Large areas of flexible pavements may develop fatigue cracking, indicating inadequate structural capacity. If these areas are not patched properly, any resurfacing will prove to be a waste of money, since the overlay will deteriorate very rapidly in these areas. The patching of these areas in order to provide a suitable foundation will be discussed later.

Bituminous patches in concrete pavements present unique problems due to the presence of dissimilar materials. Many distress types requiring a concrete patch can be considered as a potential candidate for a bituminous patch. Such a patch should not, however, be considered as a permanent patch. Blow-ups, panel cracks, etc., can all be repaired with bituminous materials if the cause of the distress is also corrected. Special care must be taken if a bituminous patch is to be placed at a joint because the bituminous patch will behave as an expansion joint and allow the concrete pavement to compress the patch. These patches should be continually observed and not be considered as being permanent.

4.3.2 Partial-Depth Pavement Repair

This section describes considerations for the design of partial-depth patches in concrete pavements. The items addressed include: (a) criteria for partial-depth patches by identification of distress types which can be repaired, and (b) description of successful procedures for partial-depth patching.

Criteria. Patching concrete pavement is often necessary to restore the level of serviceability. When applied at appropriate locations, partial-depth patching can be more cost-effective than full-depth patching (e.g., replacing an entire joint to address small spalls). The cost of partial-depth patching is largely dependent on the size, number, and location of the repair areas, as well as the materials used. Lane closure time and traffic volume also affect production rates and costs.

Partial-depth patching can be used to address certain types of distress that do not extend through the full depth of the slab, but instead affect only the top few inches. These distresses include the following:

(1) spalls that have resulted from the use of joint inserts where hardness of aggregates made sawing of the joints difficult or expensive,
(2) spalls caused by the infiltration of incompressible materials into the joints,
(3) spalls caused by misalignment of dowels or other load transfer devices,
(4) localized areas of scaling, and
(5) distress associated with early stages of "D" cracking of alkali reactivity.

Procedures. Many of these distress types occur adjacent to joints. Effective sealing of these joints requires repair of the adjacent distress. Failure to repair these areas prior to placement of an overlay will often result in the appearance of reflective cracks which break down rapidly, causing premature failure of the overlay.

The actual extent of deterioration in the concrete may be greater than the amount showing at the surface. In early stages of spall formation, weakened planes often exist with no visible sign of deterioration. The actual extent of deterioration should be determined since all weak concrete must be removed for effective restoration. After removal, the bottom of the patch is normally checked by "sounding" or other specified methods to ensure complete removal of deteriorated material. The typical depth of concrete removal varies from 1 to 4 inches. Destructive testing (e.g., cores) may be helpful in defining the depth limits. The removal operation should provide a very irregular surface to ensure a high degree of mechanical interlock between the repair material and the existing slab.

If sound concrete cannot be reached, a full-depth patch is required. Small areas of full-depth patching have been combined with partial-depth patches, but generally these have not performed as well as full-depth patches.

Small spall areas along joints generally do not require repair. Areas less than 6 inches and $1^1/_2$ inches wide at the widest point are normally not repaired, but are filled with a sealant (unless a preformed compression seal is to be used in the joint).

Partial-depth patches placed adjacent to transverse, centerline, or shoulder joints require special design and construction considerations. Partial-depth patches placed directly in contact with the adjacent lane frequently develop spalling because of curling stresses. This can be prevented by placing a polyethylene strip (or other thin bond-breaker material) along the centerline joint just prior to placement of the patching material.

Partial-depth patches placed directly against adjacent slabs (across the transverse joint) will be crushed by the compressive forces created when the slabs expand. This may be prevented by placing a strip of styrofoam or asphalt-impregnated fiberboard between the new concrete and the adjoining slab. The patch material must be prevented from infiltrating into the opening since it will result in damaging compressive stresses at lower depths. This step will also guard against damage due to differential vertical movement of the joint when the adjacent lane is trafficked during curing of the patch.

Some patches have been successfully constructed without transverse joint forms by sawing the transverse joint to full depth as soon as the patching material has gained sufficient strength to permit sawing. Any closing of the joint before sawing will fracture the patch. To avoid this problem, joints must be formed in partial-depth patches placed across a joint or crack.

Any patch along the shoulder edge must be formed. If the patch material flows into the shoulder, it may form a "key," restricting longitudinal movement of the slab.

After the surface of the existing concrete has been prepared and just prior to placement of the patch material, the patch area should be coated with a bonding agent to ensure complete bonding of the patch material to the surrounding concrete. Common types of bonding agents include portland cement/sand mixes and epoxy resins. The surface should be surface-dry before the grout is applied, and no free water should be present. Thorough coating of the bottom and all sides of the patch area is essential. The grout should be placed immediately before the patch material is placed so that the grout does not set before it comes in contact with the patching material.

Since partial-depth patches often have large surface areas with respect to their volumes, moisture can be lost quickly. Inadequate attention to curing can result in the development of shrinkage cracks that may cause the patch to fail prematurely. Thus, curing is as important for partial-depth patches as it is for full-depth patches.

4.3.3 Joint and Crack Sealing

Sealing and resealing of joints and cracks in both concrete and asphalt pavements is an important phase of restoration that is often not adequately considered. The effectiveness of sealing joints and cracks in extending the serviceable life of a pavement has long been a matter of controversy among highway agencies in the United States. Inadequate sealing increases distress caused by free water entering the pavement structure, and from the infiltration of incompressibles into the transverse joint. The excess water can accelerate damage in both flexible and rigid pavements, and incompressibles can cause blow-ups and joint deterioration over time in the rigid pavements.

Serviceability and pavement life may be extended through the proper resealing of the joints or cracks which develop in the pavement. These benefits include:

(1) the removal of incompressibles and the prevention of further intrusion, and
(2) the reduction of water infiltration, and the chemicals that may be brought along, into the joint or crack.

In general, resealing can be cost-effective on major highways in all climate regions for one of the two reasons given above. The joint seal need not accomplish both to be effective. It depends on the particular problem existing in the immediate area. If the results of the drainage survey show that moisture in the pavement structure will accelerate, or has accelerated distress, then resealing of the joints or cracks is essential. On low-truck-volume roads, sealing may not be cost-effective, especially in dryer climates. The extent of distress caused or accelerated by free moisture in the pavement structure of the project under consideration should be considered in deciding whether or not to reseal joints or cracks.

Rigid pavements that have experienced blow-ups can be treated with an adequate program of joint or crack cleaning and sealing to keep further incompressibles out of the pavement, slowing the development of further blow-ups. Thermal cracking in asphalt pavements can be kept at low severity levels with adequate sealing to keep out moisture and incompressibles.

To ensure that moisture-accelerated distress will not reduce the life of a pavement, all of the major sources of water infiltration must be sealed. These major sources include: (a) transverse joints in jointed concrete pavements, (b) longitudinal lane/shoulder joints, (c) longitudinal joints between traffic lanes, and (d) cracks in asphalt or concrete pavement surfaces.

The need for sealing the longitudinal lane/shoulder joint is reduced only slightly with properly designed and constructed longitudinal underdrains. The large amount of water entering through the joint may carry fines through the drain. This could result in problems similar to those caused by pumping water upward through the joint.

Joint Sealing. There are a wide variety of sealants on the market today with different properties. The general categories of sealant include:

> field poured sealants—self-leveling
> hot-poured
> cold-poured
> preformed compression seals
> field-poured sealants—nonself-leveling

The factors that influence the performance of a sealant include the movement of the joint or crack, the sealant reservoir shape, the bonding between the sealant and sidewall, and the properties of the sealant. All of these factors must be considered in the design of a joint resealing or sealing project. A procedure for developing the proper dimensions is outlined in Part II, Section 3.3.3, "Joint Sealant Dimensions."

Crack Sealing. Cracks, unlike joints, are irregular in dimension and direction, which makes them more difficult to seal. Fortunately, most cracks will not experience the deformation that joints are subjected to, which potentially allows the sealant to perform better than it would in a joint. Thus, the sealing procedures for cracks are not quite as strict as they are for joints. In some cases, however, joints freeze due to dowel bar corrosion and cause cracks to function as joints. If the distress survey indicates that the distance between cracks is great enough to cause very large movements at the cracks, then the cracks must be considered joints and handled as such.

Thermal cracks in asphalt concrete pavements are a working crack, and can be treated the same as a joint, since they will experience large movements due to temperature variations. The reservoir in the crack will normally not be as clean or as well-formed as that obtained in a joint. The size of the reservoir in the crack should be similar to the size required for a joint undergoing the same movement to minimize the stresses as much as possible.

4.3.4 Subsealing of Concrete Pavements

Pavement subsealing is utilized to fill voids either at the slab-subbase interface or beneath the subbase. These voids are caused by pumping action, generally beneath a concrete pavement slab and/or subbase. In some special cases, flexible (semi-rigid) pavements can also be undersealed. In jointed concrete pavements, voids may develop under transverse joints and cracks. In CRCP, voids can develop anywhere along the slab edge. The loss of support caused by void formation results in large deflections and stresses in the slabs leading to serious problems with JPCP and JRCP, including faulting, corner breaks, diagonal cracks, and finally, complete breakup of the slab. With CRCP, the loss of support is one of the single most serious structural problems leading to a rapid increase in edge punchouts.

Subsealing is performed with a cement grout or asphalt cement. When the subseal material has sufficiently filled the voids, restoration of support to the slabs will be reflected by a reduction in corner deflection in JPCP or JRCP, and edge deflection in CRCP. Subsealing should not be confused with the term "slab jacking" which refers to the lifting of a depressed slab to its original position matching the profile of the road. Subsealing does not correct depressions, increase a pavement's structural capacity, or eliminate faulting. Filling voids restores a pavement's structural integrity, thereby reducing future pumping, faulting, and slab cracking. However, this benefit may diminish over time, in which case, additional subsealing will be required. Where serious pumping has occurred, subsealing should be accompanied by efforts to reduce the amount of water entering the pavement. Subsealing should only be performed at joints/cracks where loss of support exists, where pumping is visibly evident, or where high deflections exist.

Project Analysis. The design of a subsealing project includes: (a) testing the pavement to determine if there are voids, (b) selecting an acceptable grout mixture or asphalt cement, (c) estimating required material quantities, (d) determining an appropriate initial hole pattern, and (e) preparing plans and specifications.

Cement grout mixtures must be capable of penetrating very thin voids, yet have sufficient strength and durability to resist the effects of loading, moisture, and temperature. Two different types of grouts are currently in use: pozzolanic cement grouts and limestone cement grouts. Various additives are available which may be used to alter the behavior of the grout.

Some of the additives are water-reducing agents, fluidifiers, expanding agents (powdered alumina to offset the shrinkage which sometimes occurs with volcanic ashes), and calcium chloride (to accelerate the set of the grout).

Generally, the asphalt used in a grout mixture for undersealing should have a low penetration point and a high softening point. It must also have a viscosity suitable for pumping when heated to temperatures from 400 to 450°F.

Based on the analysis of deflection test results, an initial hole pattern can be recommended that retains the flexibility to meet field conditions. The pattern should consider the general location of voids and their approximate size. Other factors influencing hole patterns include joint/crack condition, joint/crack location, subbase condition, subbase stabilization, etc. Holes should be drilled through the slab and into the nonstabilized subbase a few inches because the depth of the void is uncertain. With stabilized bases, voids are often located in the roadbed below the subbase and, therefore, holes should be drilled through the stabilized subbase a maximum of 3 inches into the roadbed. Close inspection is required by the contracting agency during subsealing to prevent overgrouting and slab lifting, which can create other voids beneath the slab or induce high slab stresses.

The repair effectiveness is determined by remeasuring the deflection of the slab at the same points after subsealing. This testing should also include some joints which were not grouted for use as control joints. Using the methods previously described, if voids are still located after the grouting, the slab should be regrouted.

4.3.5 Diamond Grinding of Concrete Surfaces and Cold Milling of Asphalt Surfaces

This section describes two different techniques that may be used to alter the surface of concrete and asphalt pavements for a variety of purposes. These restoration techniques are commonly used in conjunction with other techniques to restore the pavement to a condition resembling that of a new pavement. However, in certain cases they can be justified as the sole restoration technique performed.

Diamond grinding (texturing) is the use of closely spaced diamond-impregnated blades to cut patterns into hardened concrete. The major purpose of grinding is to remove relatively thin layers of concrete surface material and provide a smooth surface. Cold milling is the use of carbide cutting teeth mounted on a rotary drum to chip off as much as 3 to 4 inches of asphalt concrete surface. The major purpose of cold milling is to remove asphalt material.

Diamond Grinding of Portland Cement Concrete Surfaces. Diamond grinding is an effective technique for: (1) removal of joint and crack faulting, (2) removal of wheel path ruts caused by studded tires, (3) correction of joint unevenness caused by slab warping, and (4) restoration of transverse drainage.

It should be stressed that diamond grinding is a repair technique since it corrects the existing faulting and rutting of concrete pavements, but it does nothing to correct the distress mechanisms. Therefore, grinding is usually performed in combination with other rehabilitation techniques to both repair certain pavement distresses and prevent their recurrence. Diamond grinding is a good example of a rehabilitation technique which significantly improves the rideability of a pavement, but the life extension achieved depends heavily on the effectiveness of the other rehabilitation activities performed concurrently.

Data from condition surveys and roughness measurements should be used to determine when grinding is an appropriate repair technique for the distress existing in a pavement. An important item to measure with regard to diamond grinding is the amount of faulting present. Grinding should be performed only in lanes that have significant faulting, wheel path wear, or other surface roughness or profile problems. Each agency must develop its own criteria for what constitutes significant wear and rutting in order to develop the most timely and cost-effective approach to maintaining and rehabilitating its pavements. By monitoring the rate of faulting increase, an agency can determine when a pavement will need diamond grinding.

It is important to repair the pavement to some minimum level of structural integrity prior to grinding. Placing spall repairs, full-depth patches and new slabs ensures the elimination of construction-related roughness. If the observed roughness is caused by faulting of the joints or cracks, pumping may have occurred beneath the slabs. If nothing is done to reduce pumping, faulting will develop again, and probably more rapidly. Depressions should be leveled up by slab jacking or slab replacement prior to grinding. It is generally not cost-effective to grind out major depressions in the pavement. Any medium or high severity depressions should be removed by slab jacking before grinding begins. Roughness measurements taken along the project and over each lane often provide an excellent indication of depression and swell locations.

In a rehabilitation project involving grinding, the sequence of work is very important. Subsealing, full-depth repair, and spall repair should all be completed before grinding. Joint resealing should follow the grinding operation.

For best results, diamond grinding should be performed continuously along a traffic lane. Continuous grinding is required to provide the riding quality of new pavement. The quality of the grinding job can be determined by remeasuring the roughness using testing equipment and methods commonly used for new pavement construction.

Cold Milling of Asphalt Concrete Surfaces. Cold milling has been successful in removing as much as 3 to 4 inches of asphalt concrete surfacing in a single pass. Cold milling has also been used successfully on concrete pavements to provide a surface for bonding a concrete overlay and for removing deteriorated asphalt overlays. Cold milling is not recommended for concrete pavements that are to be left in service without an overlay because the surface will be extremely rough and the joints will be spalled significantly. Major uses of cold milling include the following:

(1) restoring the curb line of asphalt pavements,
(2) restoring cross slope of asphalt pavements to improve drainage or correcting drainage inlet cover problems,
(3) improving friction resistance of asphalt surfaces,
(4) removing asphalt overlays of concrete pavements,
(5) providing a roughened, clean surface for bonding a concrete overlay,
(6) removing material in conjunction with surface recycling, and
(7) removing material to provide a smoother surface (where the pavement is structurally adequate).

After removal of the surface material through cold milling, most pavements are overlaid. Some projects, however, have been milled and opened to traffic without placement of an overlay but tire noise is high and may generate public complaints. If the pavement is structurally sound, but rough from various nonload-related distresses, this may be a very cost-effective means of delaying overlay placement for a few years. The milled surface is not too rough (except, possibly, for bicyclists) and should provide acceptable service for a few years.

The cold-milled asphalt concrete surface generally provides a friction-resistant surface. After time, the milled pattern will be worn down and the fractured aggregate surface will provide all the friction resistance present. If the aggregate is susceptible to polishing, the friction resistance will eventually be worn away under traffic. This consideration should be investigated when the pavement is left open to traffic.

A uniform texture should be produced throughout the entire length of the project. The longitudinal profile should be held to the same tolerance as new construction.

4.3.6 Subdrainage Design

Subdrainage is an important consideration in the resurfacing, restoration, and rehabilitation of pavement systems. Water is a fundamental variable in most problems associated with pavement performance and is directly or indirectly responsible for many of the distresses found in pavement systems. Appendix AA, "Guidelines for the Design of Highway Internal Drainage Systems," Volume 2, should be referred to for guidance in developing a drainage system for rehabilitation.

A drainage survey may indicate that a subdrainage system is required to control one or more sources of water in the pavement. Pavement construction and maintenance activities often require several types of subsurface drainage. The removal of water will increase the strength or stiffness of the pavement, thereby extending the life. Thus, considerable care is required when designing elaborate and complex drainage systems. The designer must reevaluate the material properties used in design, as outlined in Section 2.3 of Part II, "Material Properties for Structural Design."

Subsurface drainage systems should be designed and constructed with long-term performance and maintenance in mind, including periodic inspections to check performance. Outflow measurements taken at periodic intervals can be compared to those obtained immediately after construction to determine whether or not the drainage system is functioning properly. Substantial decreases may indicate a need for cleaning and/or maintenance activities. The adequacy of the subdrainage installations for an existing pavement can be evaluated by working through a complete "new" drainage analysis of the pavement and assessing its capacity to drain the pavement system.

4.3.7 Pressure Relief Joints

The performance of concrete pavements in many areas of the country may be seriously impaired by expansive pressures caused by net increases in pavement length. These increases in length are the result of one or more factors: infiltration of incompressibles into poorly sealed joints and cracks, pumping of base materials into joints and cracks, or the use of expansive or reactive aggregates in initial construction.

Generally, transverse joints and cracks become filled with incompressible materials when the joints are open and not properly sealed. The joints are widest during the colder seasons in which sand and other de-icing materials are placed on pavement surfaces. Intrusion can also occur from beneath the slab when the vertical deflection at the joint or crack causes pumping of water and base material particles upward into the joint opening. In time, a buildup of incompressibles develops, and the intrusions cause the pavement to "grow." Although the slabs remain the same length, the joints and cracks fill with incompressibles and thereby prevent the pavement from expanding in warmer and wetter periods of the year.

Some agencies have experienced an actual increase in the length of the pavement slab due to a buildup of incompressibles. This generally occurs in areas where reactive or expansive aggregates have been used. The result of the concrete pavement's growth is an increase in compressive stress in the slabs. When this stress exceeds the compressive strength of the slab, spalling or a blow-up occurs. In addition, pavement growth may result in "bridge pushing." As a pavement expands during the warm season, particularly when intrusion is present, it will push against the approach slabs of bridges. In the following sections, the design of pressure relief joints is discussed.

Pressure Relief Joint Design. Pressure relief joints (also known as expansion joints) are full-width and full-depth cuts in the slab used to reduce compressive stresses. Although the exact dimensions vary, pressure relief joints are normally 2 to 4 inches wide when constructed. Due to the potential difficulty of sawing through dowels or other load transfer devices and the danger of encountering unstable subbase conditions near old joints, pressure relief joints are normally placed near mid-slab. Some agencies have placed expansion joints at full-depth patches, but this procedure sometimes produces patch rocking and accelerated failure.

Studies by several agencies have concluded that blow-ups tend to relieve stress for about 500 feet on either side. For this reason, pressure relief joints are typically installed at intervals of 700 to 1,500 feet. When bridge pushing is the only problem, the joints are typically located only near the approach slabs.

The expansion joint is filled with a compressible filler material such as styrofoam or sponge rubber to prevent intrusion of incompressibles. Preformed joint seals have also been used with some success. On highways with very heavy traffic, particularly where a bituminous overlay is to be placed, special heavy-duty expansion joints have been provided.

Some agencies have constructed asphalt cement patches in portland cement concrete pavements to serve as expansion joints. These patches are generally about 4 feet wide and are typically placed in deteriorated areas requiring full-depth patching. They often result in "humping" of the asphalt patch as the concrete pavement expands into the original patch area. This "humping" and the accompanying loss of load transfer, rocking of slabs, and settlement or heaving of concrete patch areas may result in rough pavements and loss of pavement serviceability.

The effect of pressure relief on existing pavement design must be considered prior to installation of pressure relief joints. Short-jointed, undowelled concrete pavements are generally poor candidates for use of pressure relief joints. Pressure relief joints will cause loss of aggregate interlock load transfer in the area of relief, which may result in increased slab cracking and faulting. They will also allow water to enter the pavement structure, resulting in deterioration of the subbase, pumping, rocking of the slab, etc. Similarly, pressure relief joints should not be used in CRCP because they destroy the integrity of the pavement and allow water to enter the subbase more freely, resulting in rapid loss of subgrade support. They should be used on these types of pavements only near bridges when shoving or blow-up has occurred.

The effect of pressure relief on existing joint seals must also be considered. Preformed compression seals may lose contact with joint reservoir walls as the joints open, and the seals will no longer be effective in preventing water and incompressibles from entering the joints. Similarly, the effectiveness of other types of joint seals will be diminished if they are damaged by excessive joint openings. It must be confirmed that transverse joint seals will remain effective after installation of pressure relief joints, or it may become necessary to reseal the joints later.

The optimum time for placement of pressure relief joints has not been determined. It is clear that many jointed concrete pavements perform for over 30 years without ever exhibiting blow-ups or pushing of bridge abutments. Thus, construction of expansion joints is currently recommended only after major blow-ups have occurred, since their placement can result in opening of contraction joints.

Major Considerations and Limitations. Pressure relief joints are almost always installed on pavements having more than one traffic lane, thus, it is generally impossible to install material across the full pavement width on the same day. When relief is provided for one lane only, the other lane(s) can be subjected to substantially higher compressive stresses. A number of major blow-ups have occurred in adjacent lanes when the installation of pressure relief joints across all lanes has been delayed. For this reason, it is necessary to install pressure relief joints in all adjoining lanes as soon as possible. If the joints are constructed during seasons with moderate daily temperature variations, a period of 48 hours between construction of expansion joints in adjacent lanes will normally not be harmful. During the warmest time of the year, or in pavements with expansive aggregates, compressive forces in the pavement may be sufficient to pinch or bind the saw blades during the sawing operation. In addition, the problem of unequal pressure between adjacent lanes is often aggravated during warm weather. For this reason, a temperature range of 40 to 70°F is recommended for installation. Relief joints may be installed during the summer months by sawing at night or early in the morning.

On some pavements, blow-up frequency has increased after overlay with bituminous materials indicating that a need for pressure relief existed prior to overlay. However, when expansion joints are placed prior to overlay, the overlay often deteriorates badly under heavy traffic in the area of the expansion joint. This is due to fatigue that develops as a result of high differential deflections at the joint. It has also been determined that the placement of bituminous overlays aggravates problems inherent in some pavements susceptible to blow-ups by holding moisture in the concrete pavement structure (e.g., the accelerated expansion of moisture-susceptible aggregates resulting in weakened concrete near the joint areas). Installation of pressure relief joints prior to placement of bonded concrete overlays has produced debonding in the area of the joint. This is because it is extremely difficult to saw the joint in the overlay soon enough to prevent the underlying slab from moving independently of the fresh overlay. It is generally recommended that bonded concrete overlays be placed prior to the construction of pressure relief joints.

Because they usually provide no load transfer, pressure relief joints should be used only on pavements that are subject to blow-ups or are pushing bridges. Deflections at the joint will tend to be high, the adjacent slab may deteriorate, and the joint may pump and fault. In wet areas, subdrainage may be necessary to prevent pumping.

Pressure relief joints may completely close over time, making the pavement susceptible to blow-ups and bridge pushing again if the cause of the problem is not remedied. If intrusion of incompressibles into the joints is not stopped or if reactive aggregate problems are present, the construction of pressure relief joints will provide only a temporary solution.

4.3.8 Restoration of Joint Load Transfer in Jointed Concrete Pavements

This section describes different techniques that may be used to restore the load transfer of portland cement concrete (PCC) pavement joints and cracks to reduce pavement stresses and deflections, and thus the rate of deterioration. The ability of a joint or crack to transfer load is a major factor in its structural performance. Load transfer efficiency across a joint or crack is normally defined as the ratio of deflection of the unloaded side of the joint or crack to the deflection of the loaded side. If complete load transfer exists, the ratio will be 1.00 (or 100 percent), and if no load transfer exists, the ratio will be 0.00 (or 0 percent). (See Chapter 3 for further details.) Poor load transfer may cause large increases in slab stresses and deflections, resulting in slab breakup and loss of serviceability.

Dowelled joints normally exhibit very good load transfer (i.e., between 70 and 100 percent). However, repeated heavy loads can cause the dowel sockets to deteriorate, resulting in looseness of the dowels, faulted and spalled joints, and loss of load transfer. Many jointed plain concrete pavements have been constructed without dowels at transverse joints. The load transfer measured at these joints is typically low (except on warm afternoons when joints close tightly). Transverse cracks in both jointed plain and reinforced concrete pavements can also have poor load transfer, particularly when the reinforcing steel has ruptured.

Determining the Need for Load Transfer Restoration. Restoration of load transfer across a transverse joint or crack is performed to retard joint and crack deterioration, pumping, faulting, spalling, and corner breaks. Thus, joints and cracks requiring load transfer restoration must be identified prior to overlay or performance of other rehabilitation work.

Load transfer should be measured during cooler periods, normally in the early morning. Load transfer is often lowest in the outer wheel path and, since most loads will pass over this area, it should be measured at this point. The test load should be applied on one side of the joint or crack. Deflection measurements should be taken on both sides directly adjacent to the joint or crack in accordance with procedures noted in Section 3.5.4. Load transfer restoration should be considered for all transverse joints and cracks that exhibit measured deflection load transfer between 0 and 50 percent. This applies to jointed concrete pavements with or without asphalt concrete overlays.

Design Considerations. Dowels may be installed to restore load transfer of a joint or crack. Dowels placed in slots cut in the pavement are effective in restoring load transfer across joints or cracks. The required number, diameter, and spacing of the dowel bars must be determined. The diameter of the dowels and the number placed in the outer wheel path have a major influence on the prevention of faulting. Dowels should be 18 inches long and at least 1.25 inches in diameter. For a maximum allowable fault of 0.10 inch, the following dowel designs are suggested:

Number of Dowels in Wheelpath	Diameter (inches)
2	1.625
3	1.625
4	1.250
6	2.250

The successful installation of load transfer devices requires sound concrete adjacent to the joint or crack. If the concrete is deteriorated, full-depth repair is more appropriate than load transfer restoration. Joints or cracks having high deflections must be subsealed before load transfer devices are installed. The cause of joint distress should be determined and attempts should be made to correct deficiencies before performing load transfer restoration work.

Additional work to be completed prior to load transfer restoration may include subsealing to fill voids in the pavement foundation, grinding to eliminate faulting, and spall repairs. Work that may be done after load transfer restoration includes joint and crack sealing and subdrain installation.

4.3.9 Surface Treatments

The use of surface treatments or seal coats is a method of pavement rehabilitation for asphalt pavements of all classes, from low-volume roads to Interstate highways. This rehabilitation category is an application of asphalt and/or aggregate to a roadway surface, generally less than 1 inch thick, which improves or protects the surface characteristics of the roadway. In general, there is little or no direct structural improvement of a pavement when a surface treatment is used. However, indirect benefits in increased life and structural capacity can be obtained with this technique.

Seal coats and/or surface treatments have long been used as standard asphalt pavement maintenance and rehabilitation procedures. Historically, they have been used primarily for low-volume streets and roads, extending pavement life at low expense. Because of newer applications, surface treatments and seal coats are discussed to provide the engineer with an understanding of what these applications can do for a pavement, and how to ensure an adequate application of materials for a seal coat or surface treatment. In the following sections, the classification, functions, and design of surface treatments are discussed.

Classification of Seals or Surface Treatments. Seal coats and surface treatments are classified on the basis of their composition, which may be either solely asphalt or, more normally, a combination of asphalt and aggregate. The following are typical categories:

(1) *Open-Graded Friction Courses.* These applications of asphalt and aggregate are designed to drain water off the pavement surface by providing an open, porous structure in the mixture. The rapid removal of water reduces the potential for hydroplaning and, hence, wet weather accidents. These applications are often called plant mix seals or popcorn mixes.

(2) *Asphalt-Aggregate Surface Treatments.* These treatments consist of sequential applications of asphalt and stone chips which can be made either singly or in repetitive layers to build up a structure approaching 1 inch thick (or more), sometimes called armor coating. These applications represent the traditional seal coating done by local agencies. They also serve as the surfacing for low-volume roads.

(3) *Rubberized Asphalt Seal.* This application is a special type of asphalt-aggregate surface treatment. The asphalt material is replaced

with a specialized blend of rubber and asphalt cement. This application has been used as part of a SAMI (Stress-Absorbing Membrane Interlayer) to reduce reflection cracking prior to overlaying. It has been used without overlays recently to take advantage of the added elasticity in the bonding of asphalt to hold the stones more tightly and reduce the tackiness on the surface.

(4) *Slurry Seal.* The slurry seal application consists of a diluted emulsion mixed with a sand-size aggregate in a special mixer. This slurry is then squeegeed onto the pavement surface. The thickness of the slurry seal is generally less than 3/8 inches.

(5) *Fog Seal.* A fog seal is an application of dilute emulsion with no aggregate. It seals the surface and provides a small amount of rejuvenation. It also provides a very distinct delineation between mainline pavement and the shoulder, where they are primarily used, on high-volume roads.

(6) *Sand Seal.* A sand seal consists of a spray application of emulsion with a light covering of sand or screening. This application serves the same function as a fog seal, but it provides a more friction-resistant surface. The appearance of a sand seal surface does not provide the dramatic delineation that a fog seal does.

(7) *Road Oiling.* Road oiling is primarily a dust palliative measure for low-volume, unsurfaced roads. Dilute asphalt materials are applied to hold the dust down on the surface. The oil may be mixed into the surface material with a disc and, after extensive time, provide a weather-resistant surface.

Functions of Seals or Surface Treatments. Surface treatments may provide an extension to the service life of a pavement and reduce required maintenance expenditures until a more cost-effective rehabilitation program can be developed. The major functions of surface treatments are:

(1) *Provide a Wearing Surface.* An asphalt-aggregate surface treatment provides a new aggregate exposed to the traffic, which can furnish better durability and wear characteristics than the original surface. This application generally improves the friction resistance, but improvement in surface durability may not always occur. The aggregate to be used to correct a nondurable surface should be tested thoroughly (L.A. Abrasion and the Sulfate Soundness Test) to ensure that it has satisfactory durability.

(2) *Seal Cracks.* The application of aggregate and/or asphalt in these seal coats provides a large amount of asphalt material that can seal small cracks. The asphalt-aggregate treatments provide the most crack sealing, while the fog seal provides very little crack sealing. The use of rubberized asphalt provides one of the best materials for bridging cracks and maintaining an effective seal. The exclusion of moisture from cracks extends life, and may actually help maintain the structural capacity of the pavement.

(3) *Waterproofing.* The porosity of asphalt pavements varies and may admit water to some extent through the normal void structure. The application of a waterproof surface will restrict moisture infiltration and reduce the rate of deterioration in existing pavements.

(4) *Improve Friction Resistance.* The use of an open-graded friction course reduces the wet weather accident potential of a pavement by reducing the potential for hydroplaning. The aggregate in a standard surface treatment can directly increase the skid resistance of a pavement. The aggregate used in the surface treatment must be controlled to ensure the level of friction resistance remains high following construction.

(5) *Reduce Weathering Effects.* The asphalt application adds softer asphalt material to the oxidized surface of the pavement and retards the hardening of the original asphalt surface. The extra material provided by the asphalt reduces the voids on the surface of the pavement and deters the entry of water and air, which tends to harden the asphalt. Fog and slurry seals are effective in areas where excessive oxidation and hardening of the asphalt in the mixture are a common problem.

(6) *Improve Surface Appearance.* In some instances, the general appearance of the pavement surface may be quite disorderly due to patching and other construction activities. A surface treatment is a simple, effective means of covering these irregularities and providing a uniform appearance to the surface.

(7) *Visual Delineation.* A distinct difference in the visual appearance of the shoulder and the mainline pavement is an aid to motorists. Studies have shown that when this distinction

exists, the drivers avoid driving on the lane/shoulder joint, increasing the life of the pavement. A difference in the appearance and texture of the shoulder is a safety enhancement for the pavement as a whole.

(8) *Structural.* There is no direct structural benefit derived from the application of a surface treatment. Multiple surface treatments of three or more layers appears, however, to provide some structure to the pavement, but no mixture testing is performed. The aggregate used is not uniformly graded to ensure aggregate interlock and, therefore, cannot resist deformation under heavy load repetitions. Thus, a surface treatment cannot properly be considered a structural improvement to a pavement, although some marginal improvements may occur. It can reduce the rate of deterioration of a pavement by sealing cracks and preventing the infiltration of water into the pavement and, thereby, delay the need for structural improvements. Hence, a surface treatment can make an indirect contribution to the structural adequacy of the pavement.

General Design Concepts for Surface Treatments or Seal Coats. There are two components to be considered in the design of a surface treatment or seal coat: the asphalt material and the aggregate. The general design considerations are similar for all types of surface treatments. Actual design procedures for surface treatments are widely available in literature and each agency must evaluate the available procedures to ensure they will work with local materials. In general, the engineer must consider the following to ensure that the surface treatment will perform successfully:

Existing Pavement Structure
Available Materials (Asphalt and Aggregate)
Quantity Selection
Local Condition and Experience

If the existing pavement is not structurally sufficient to carry the projected traffic for 3 to 5 years, a surface treatment should not be considered. The pavement should be scheduled for another form of more extensive rehabilitation because surface treatments are not designed to withstand traffic in excess of the capacity of the underlying pavement. If the underlying pavement has any structural problems resulting from poor drainage or an unstable base, surface treatments should not be considered. These deficiencies should be noted during the survey and evaluation phase of project development.

The physical condition of the surface influences the amount of asphalt material needed in the surface treatment or seal coat. If the surface is flushed or bleeding, the amount of asphalt used should be reduced to compensate for the excess already present. If the surface is oxidized and very porous, the amount used should be increased because the surface will absorb some of the asphalt added during construction. This absorption effectively takes asphalt away from the aggregate in the surface treatment. Recommendations are shown in Table 4.3. These are only recommendations, and will vary depending upon local experience and conditions. If these alterations are ignored, the application rates of the asphalt material will be improper. Factors relative to the type of asphalt and aggregate materials selected, as well as the specific design quantities to use, are a direct function of the material and construction specifications of the user agency. Appropriate user agency information regarding these design, material, and construction specifications should be followed.

4.3.10 Prediction of Life of Rehabilitation Techniques Without Overlay

All pavements deteriorate over time due to traffic loadings, climatic effects, and other causes. It is extremely important to be able to predict the deterioration of pavements (and thus their service life) both in the first performance period after construction or reconstruction, and after one or more rehabilitations.

Need for Performance Prediction Models. The ability to predict the life of a pavement rehabilitation strategy is essential to conduct life-cycle cost analyses

Table 4.3. Recommendations for Changes in Asphalt for Surface Texture

Surface Condition	Increase in Application Rate (gal/yd^2)
Black, flushed, bleeding	−0.01 to −0.06 (−0.3 average)
Smooth, non-porous	0.00
Absorbent:	
Slightly open, oxidized	+0.03
Raveled, open, oxidized	+0.06
Severe weathering, raveling, oxidized	+0.09

and to make any rational decisions as to the best rehabilitation strategy. This section briefly summarizes existing experience and capabilities for predicting the life of rehabilitation techniques without overlay. Prediction of the life of a pavement in its first performance period is described to demonstrate the concepts of life prediction.

Various prediction models have been developed for both new flexible and rigid pavements based on field performance data that relate the key design and climatic factors to several major distress types and serviceability. Each distress and serviceability has a distinct functional relationship.

When a pavement is rehabilitated, the existing pavement condition or amount of deterioration is modified depending on the work performed. Rehabilitation work may repair some or most of the existing deterioration. Therefore, the future rate of deterioration of the rehabilitated pavement depends on all of the previously listed factors for new pavements, plus the amount, type, design, and construction quality of rehabilitation work performed. The performance of a rehabilitated concrete pavement is illustrated in Figure 4.1, which is the same example noted in Table 4.4. It is assumed that concrete pavement restoration work is performed after 10 million 18-kip ESAL's including: subsealing voids, subdrainage, full-depth repairs to joints and cracks, and grinding and resealing of joints. The curves illustrate the potential future performance of the pavement using the same models with some modifications.

Models should be developed to provide this predictive capability to assist the engineer in determining the cost-effectiveness of these rehabilitation strategies. Present state of the art for predicting the performance of pavements rehabilitated without an overlay is essentially limited to engineering judgment, along with a few observations of field performance.

Table 4.4. Inputs and Predicted Outputs for a Specific Pavement Section Using Example Models

Design/Climate Factors	
Type of pavement	Jointed reinforced concrete
Cumulative traffic (ESAL)	1.0 million per year (design lane)
Subgrade type	Fine-grained
Base type	Granular
Modulus of subgrade reaction	200 pci (top of base course)
Slab thickness	9 inches
Durability of PCC aggregates	Non-susceptible to "D" cracking
Design modulus of PCC rupture	650 psi
Reinforcement	0.10 sq. in./ft. width
Joint spacing	40 ft.
Dowel diameter	1.25 inches
Type of joint seal	Hot pour
Shoulders	Asphalt concrete
Subdrainage	None
Average annual precipitation	33 inches (85 cm)
COE Freezing Index	625 degree days below freezing

Predicted—Performance						
Age (years)	ESAL	Pumping	Faulting	Cracking	Joint Deterioration	PSR
0	0	0	0.00	0	0	4.5
5	5	1.7	0.06	301	2	3.3
10	10	2.7	0.11	1,055	11	2.8
15	15	3.0	0.15	1,904	27	2.5
20	20	3.0	0.17	2,628	52	2.2

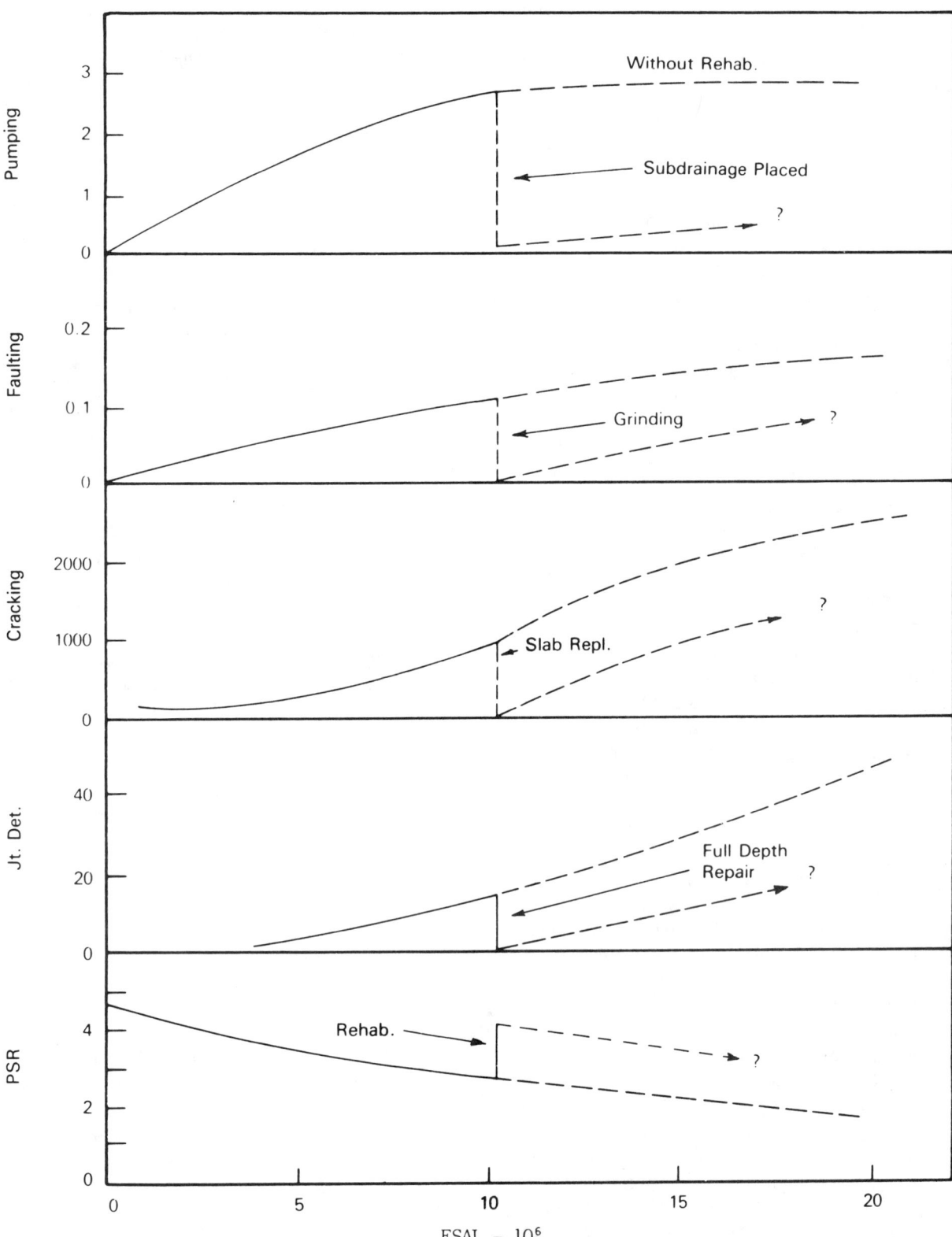

Figure 4.1. Illustration of Pavement Performance after Rehabilitation without Overlay

Life of Selected Rehabilitation Alternatives. Some published information and data exists on the life of different rehabilitation techniques other than overlays. A description of available information is given for a few key techniques:

(1) *Full-Depth Repair.* The life of full-depth repairs depends on the following factors: load transfer capability of the transverse joints, foundation support (pumping of the existing slab near the joints), climate, traffic loading level, quality of construction, length of repair, soundness of existing PCC, and sealing of transverse repair joints. Full-depth repairs, as old as 15 years, have received over 10 million ESAL's. Others have failed in as little as 1 year due to poor design and/or construction. A typical life expectancy is 5 to 15 years under heavy traffic.

(2) *Partial-Depth Repair.* The life of partial-depth patching depends on the following factors: soundness of surrounding PCC, restoration of the transverse joint near the patch, patch materials used, construction procedures (particularly damage to the existing PCC), and bonding agent used. Partial-depth repairs, up to 5 years of age, have been performing satisfactorily. Others have failed in as little as a few months due to poor construction practices. A typical life expectancy is 3 to 8 years.

(3) *Subsealing of Slabs to Fill Voids.* The life of slab subsealing depends on factors such as traffic level, adequacy to which existing voids were filled, quality of subseal material, prevention of free moisture beneath the slab/subbase, annual precipitation, and load transfer in slab which controls corner deflection. Subsealing has been observed to perform satisfactorily for up to 5 years or more. Other jobs have started pumping almost immediately. A typical life expectancy is 4 to 8 years for heavy traffic conditions.

(4) *Joint Resealing.* The life of joint resealing depends primarily on the quality of the sealant (durability, extensibility, etc.), the design of the sealant reservoir, and the construction techniques used. Joint resealing, using the high quality sealants (elastomers, silicones, preformed compression), has performed well from 4 to 10 years when designed and constructed properly. Where typical types of hot-poured sealants or poor construction procedures are used, the life has been less than 1 year. A typical life expectancy for high type sealants and good construction techniques is 4 to 10 years or more.

(5) *Diamond Grinding.* The life of diamond grinding of concrete pavements depends primarily on traffic loading, the hardness of the aggregate, and the lane width of the grinding fins. The oldest diamond grinding projects have lasted more than 15 years in California. The actual life depends more on the deterioration of pavement components other than the surface, such as joints, faulting, and cracking. Thus, the life of the overall pavement depends heavily on concurrent rehabilitation work performed at the time of grinding to prevent or minimize further deterioration. The typical life expectancy for diamond grinding is 8 to 15 years where concurrent rehabilitation work is performed to prevent or minimize future deterioration (e.g., subdrainage, sealing joints, subsealing).

(6) *Surface Treatments.* There are a variety of surface treatments available, and each has its own distinctive performance characteristics and service life. The level of traffic and condition of pavement prior to the placement of the surface treatment are extremely important. Observed life for several surface treatments are provided below for pavements that have typically low-to-medium traffic levels.

Surface Treatment	Observed Life (years)
Single chip seal	3 to 5
Double chip seal	4 to 6
Slurry seal	3 to 5
Rubberized chip seal	3 to 8
Fog or rejuvenation seal	1 to 3
Open-graded friction course	3 to 7

(7) *Pressure Relief Joints.* The life of a pressure relief joint depends on the buildup of pressure in the pavement. This is a function of the amount of incompressibles that have infiltrated into the joints and cracks, and the width of the pressure relief joint. The observed life of pressure relief joints is 1 to 5 years, and then they close tightly and need to be resawed. Typical life is on the order of 4 years.

(8) *Subdrainage.* The life of subdrainage (longitudinal drains) depends greatly on the design of

the filter material and pipe. Drains have been placed that are over 15 years old and still function (assuming that the outlets are kept clean). However, other drains have become clogged within 1 to 2 years due to poor design of the filter material. Actually, assuming proper design of the filter material and piping, drains can be cleaned easily using high-pressure water. Thus, the life expectancy of well-designed subdrains where cleaning is performed is on the order of 10 to 20 years.

Development of Predictive Models for Rehabilitation Pavements. The rate of deterioration of pavements after they have been rehabilitated by methods other than overlay may be greater or less than the rate of deterioration during their first performance cycle. Through the measurements of field performance for different rehabilitation techniques, it should be possible to develop predictive models for distress and serviceability for local regions. As follows, a predictive model was developed for faulting of full-depth repairs using data from repairs located in the Midwest (model has not been validated).

$$\text{FAULT} = (\text{ESAL}/1.3)^{0.478}$$
$$* (-0.3679 + 0.0078\ \text{LENGTH}^{1.537}$$
$$-\ 2.389\ \text{DOWEL} - 2.928\ \text{UCUT}$$
$$+\ 0.285\ \text{PRECIP}^{0.970} + 1.40\text{E-}7\ \text{FREEZE}^{2.256})$$

where

FAULT = Transverse joint faulting, in. × 1,000 (mils),
ESAL = Accumulated equivalent 18-kip single axle loads to lane,
LENGTH = Length of full-depth repair, ft.,
DOWEL = 0, no dowel bars,
 = 1, dowel bars,
UCUT = 0, not an undercut type repair,
 = 1, undercut type repair,
PRECIP = Average annual precipitation, in., and
FREEZE = Freezing Index, degree days below freezing.

Graphs in Figure 4.2 illustrate the functional form and sensitivity of this prediction model. Models could be developed for all types of rehabilitation techniques by selecting and measuring the factors likely to affect their service life, and then developing empirical-mechanistic prediction models. The models must include key factors that affect the service life of rehabilitation techniques. While the state of the art of such predictive models is poor for rehabilitated pavements, each agency is strongly encouraged to initiate data collection schemes to obtain needed information to eventually develop such models. Until such time, attempts to fully utilize life-cycle cost concepts for selecting the most economic rehabilitation alternative must be tempered by a significant amount of subjectiveness.

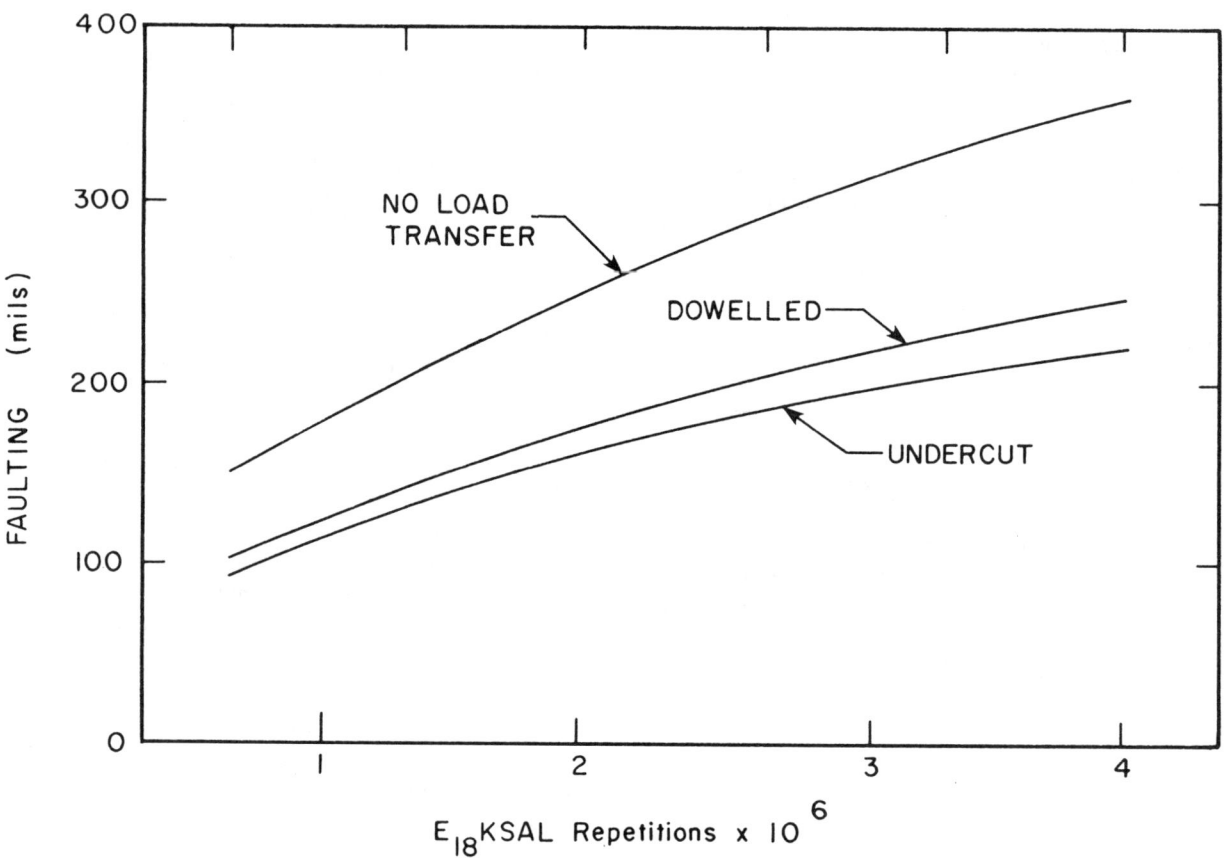

Figure 4.2. Sensitivity and Functional Form of Full-Depth Repair Faulting Model

CHAPTER 5
REHABILITATION METHODS WITH OVERLAYS

Overlays are used to remedy functional or structural deficiencies of existing pavements. It is important that the designer consider the type of deterioration present in determining whether the pavement has a functional or structural deficiency, so that an appropriate overlay type and design can be developed.

Functional deficiency arises from any conditions that adversely affect the highway user. These include poor surface friction and texture, hydroplaning and splash from wheel path rutting, and excess surface distortion (e.g., potholes, corrugation, faulting, blowups, settlements, heaves). The overlay design procedures in this chapter address structural deficiencies. If a pavement has only a functional deficiency, procedures in Part III, Chapter 4 and Section 5.3.2 should be used.

Structural deficiency arises from any conditions that adversely affect the load-carrying capability of the pavement structure. These include inadequate thickness as well as cracking, distortion, and disintegration. It should be noted that several types of distress (e.g., distresses caused by poor construction techniques, low-temperature cracking) are not initially caused by traffic loads but do become more severe under traffic to the point that they also detract from the load-carrying capability of the pavement. Part III, Section 4.1.2 provides descriptions of various structural conditions.

Maintenance overlays and surface treatments are sometimes placed as preventive measures to slow the rate of deterioration of pavements. This type of treatment includes thin AC overlays and various surface treatments which help keep out moisture.

The following abbreviations for pavement and overlay types are used in this chapter:

AC:	Asphalt concrete
PCC:	Portland cement concrete
JPCP:	Jointed plain concrete pavement
JRCP:	Jointed reinforced concrete pavement
CRCP:	Continuously reinforced concrete pavement
AC/PCC:	AC-overlaid Portland cement concrete (JPCP, JRCP, or CRCP)

The procedures described in this chapter address the following types of overlays and existing pavements:

Section	Overlay	Existing Pavement
5.4	AC	AC
5.5	AC	Break/crack and seat and rubblized PCC
5.6	AC	JPCP, JRCP, and CRCP
5.7	AC	AC/JPCP, AC/JRCP, and AC/CRCP
5.8	Bonded PCC	JPCP, JRCP, and CRCP
5.9	Unbonded PCC	JPCP, JRCP, and CRCP
5.10	PCC	AC

5.1 OVERLAY TYPE FEASIBILITY

The feasibility of any type of overlay depends on the following major considerations.

(1) Availability of adequate funds for construction of the overlay. This is basically a constraint, as illustrated in Part III, Figure 2.1.
(2) Construction feasibility of the overlay. This includes several aspects.
    (a) Traffic control
    (b) Materials and equipment availability
    (c) Climatic conditions
    (d) Construction problems such as noise, pollution, subsurface utilities, overhead bridge clearance, shoulder thickness and side slope extensions in the case of limited right-of-way, etc.
    (e) Traffic disruptions and user delay costs
(3) Required future design life of the overlay. Many factors will affect the life of an overlay, such as the following.
    (a) Existing pavement deterioration (specific distress types, severities, and quantities)

(b) Existing pavement design, condition of pavement materials (especially durability problems), and subgrade soil
(c) Future traffic loadings
(d) Local climate
(e) Existing subdrainage situation

All of these factors and others specific to the site need to be considered to determine the suitability of an overlay.

5.2 IMPORTANT CONSIDERATIONS IN OVERLAY DESIGN

Overlay design requires consideration of many different items, including: preoverlay repair, reflection crack control, traffic loadings, subdrainage, milling an existing AC surface, recycling portions of an existing pavement, structural versus functional overlay needs, overlay materials, shoulders, rutting in an existing AC pavement and overlay, durability of PCC slabs, design of joints, reinforcement, and bonding/separation layers for PCC overlays, overlay design reliability level and overall standard deviation, and pavement widening.

These considerations must not be overlooked by the designer. Each of these is briefly described in this section, especially those that are common for all overlay types. They are described in more detail in the sections for each overlay type.

5.2.1 Pre-overlay Repair

Deterioration in the existing pavement includes visible distress as well as damage which is not visible at the surface but which may be detected by other means. How much of this distress should be repaired before an overlay is placed? The amount of pre-overlay repair needed is related to the type of overlay selected. If distress in the existing pavement is likely to affect the performance of the overlay within a few years, it should be repaired prior to placement of the overlay. Much of the deterioration that occurs in overlays results from deterioration that was not repaired in the existing pavements. The designer should also consider the cost tradeoffs of preoverlay repair and overlay type. If the existing pavement is severely deteriorated, selecting an overlay type which is less sensitive to existing pavement condition may be more cost-effective than doing extensive preoverlay repair. Excellent guidelines are available on preoverlay repair techniques (1, 2, 3, 4).

5.2.2 Reflection Crack Control

Reflection cracks are a frequent cause of overlay deterioration. The thickness design procedures in this chapter do not consider reflection cracking. Additional steps must be taken to reduce the occurrence and severity of reflection cracking. Some overlays are less susceptible to reflection cracking than others because of their materials and design. Similarly, some reflection crack control measures are more effective with some pavement and overlay types than with others. Reflection crack control is discussed in more detail in the sections for each overlay type.

5.2.3 Traffic Loadings

The overlay design procedures require the 18-kip equivalent single-axle loads (ESALs) expected over the design life of the overlay in the design lane. The estimated ESALs must be calculated using the appropriate flexible pavement or rigid pavement equivalency factors from Part II of this Guide. The appropriate type of equivalency factors for each overlay type and existing pavement type are given in the following table.

Existing Pavement	Overlay Type	Equivalency Factors to Use
Flexible	AC	Flexible
Rubblized PCC	AC	Flexible
Break/Crack/Seat JRCP, JRCP	AC	Flexible
Jointed PCC	AC or PCC	Rigid
CRCP	AC or PCC	Rigid
Flexible	PCC	Rigid
Composite (AC/PCC)	AC or PCC	Rigid

An approximate correlation exists between ESALs computed using flexible pavement and rigid pavement equivalency factors. Converting from rigid pavement ESALs to flexible pavement ESALs requires multiplying the rigid pavement ESALs by 0.67. For example, 15 million rigid pavement ESALs equal 10 million flexible pavement ESALs. Five million flexible pavement ESALs equal 7.5 million rigid pavement ESALs. Failure to utilize the correct type of ESALs will result in significant errors in the overlay designs. Conversions must be made, for example, when designing an AC overlay of a flexible pavement (flexible ESALs required) and when designing an alternative PCC overlay of the same flexible pavement (rigid ESALs required). Throughout this chapter, ESALs are designated as rigid ESALs or flexible ESALs as appropriate.

The type of ESALs used in the overlay design depends on the pavement performance model (flexible or rigid) being used. In the overlay design procedures presented in this chapter, the flexible pavement model is used in designing AC overlays of AC pavements and fractured slab PCC pavements. The rigid pavement model is used in designing AC and PCC overlays of PCC and ACC/PCC pavements and PCC overlays of AC pavements.

5.2.4 Subdrainage

The subdrainage condition of an existing pavement usually has a great influence on how well the overlay performs. A subdrainage evaluation of the existing pavement should be conducted as described in Part III, Section 3.3. Further guidance is provided in Reference 5. Improving poor subdrainage conditions will have a beneficial effect on the performance of an overlay. Removal of excess water from the pavement cross-section will reduce erosion and increase the strength of the base and subgrade, which in turn will reduce deflections. In addition, stripping in AC pavement and "D" cracking in PCC pavement may be slowed by improved subdrainage.

5.2.5 Rutting in AC Pavements

The cause of rutting in an existing AC pavement must be determined before an AC overlay is designed. An overlay may not be appropriate if severe rutting is occurring due to instability in any of the existing pavement layers. Milling can be used to remove the rutted surface and any underlying rutted asphalt layers.

5.2.6 Milling AC Surface

The removal of a portion of an existing AC surface frequently improves the performance of an AC overlay due to the removal of cracked and hardened AC material. Significant rutting or other major distortion of any layer should be removed by milling before another overlay is placed; otherwise, it may contribute significantly to rutting of the overlay.

5.2.7 Recycling the Existing Pavement

Recycling a portion of an existing AC layer may be considered as an option in the design of an overlay. This has become a very common practice. Complete recycling of the AC layer may also be done (sometimes in conjunction with the removal of a deteriorated base course).

5.2.8 Structural versus Functional Overlays

The overlay design procedures in this chapter provide an overlay thickness to correct a structural deficiency. If no structural deficiency exists, an overlay thickness less than or equal to zero will be obtained. This does not mean, however, that the pavement does not need an overlay to correct a functional deficiency. If the deficiency is primarily functional, then the overlay thickness should be only that which is needed to remedy the functional problem (6). If the pavement has a structural deficiency as well, a structural overlay thickness which is adequate to carry future traffic over the design period is needed.

5.2.9 Overlay Materials

The overlay materials must be selected and designed to function within the specific loading, climatic conditions, and underlying pavement deficiencies present.

5.2.10 Shoulders

Overlaying traffic lanes generally requires that the shoulders be overlaid to match the grade line of the traffic lanes. In selecting an overlay material and thickness for the shoulder, the designer should consider the extent to which the existing shoulder is deteriorated and the amount of traffic that will use the shoulder. For example, if trucks tend to park on the shoulder at certain locations, this should be considered in the shoulder overlay design.

If an existing shoulder is in good condition, any deteriorated areas should be patched. An overlay may

5.2.11 Existing PCC Slab Durability

The durability of an existing PCC slab greatly influences the performance of AC and bonded PCC overlays. If "D" cracking or reactive aggregate exists, the deterioration of the existing slab can be expected to continue after overlay. The overlay must be designed with this progressive deterioration of the underlying slab in mind (7).

5.2.12 PCC Overlay Joints

Bonded or unbonded jointed concrete overlays require special joint design that considers the characteristics (e.g., stiffness) of the underlying pavement. Factors to be considered include joint spacing, depth of saw cut, sealant reservoir shape, and load transfer requirements.

5.2.13 PCC Overlay Reinforcement

Jointed reinforced and continuously reinforced concrete overlays require an adequate amount of reinforcement to hold cracks together. Friction between the overlay slab and the base slab should be considered in the reinforcement design.

5.2.14 PCC Overlay Bonding/Separation Layers

The bonding or separation of concrete overlays must be fully considered. Bonded overlays must be constructed to insure that the overlay remains bonded to the existing slab. Unbonded overlays must be constructed to insure that the separation layer prevents any reflection cracks in the overlay.

5.2.15 Overlay Design Reliability Level and Overall Standard Deviation

An overlay may be designed for different levels of reliability using the procedures described in Part I, Chapter 4 for new pavements. This is accomplished through determination of the structural capacity (SN_f or D_f) required to carry traffic over the design period at the desired level of reliability.

Reliability level has a large effect on overlay thickness. Varying the reliability level used to determine SN_f or D_f between 50 and 99 percent may produce overlay thicknesses varying by 6 inches or more (8). Based on field testing, it appears that a design reliability level of approximately 95 percent gives overlay thicknesses consistent with those recommended for most projects by State highway agencies, when the overall standard deviations recommended in Part I and II are used (8). There are, of course, many situations for which it is desirable to design at a higher or lower level of reliability, depending on the consequences of failure of the overlay. The level of reliability to be used for different types of overlays may vary, and should be evaluated by each agency for different highway functional classifications (or traffic volumes).

The designer should be aware that some sources of uncertainty are different for overlay design than for new pavement design. Therefore, the overall standard deviations recommended for new pavement design may not be appropriate for overlay design. The appropriate value for overall standard deviation may vary by overlay type as well. An additional source of variation is the uncertainty associated with establishing the effective existing structural capacity (SN_{eff} or D_{eff}). However, some sources of variation may be smaller for overlay design than for new pavement design (e.g., estimation of future traffic). Additional research is needed to better establish the standard deviations for overlay design. At the present time it is recommended to use 0.39 for any type of concrete overlay and 0.49 for any type of AC overlay, which is consistent with Part I, Section 4.3.

5.2.16 Pavement Widening

Many AC overlays are placed over PCC pavements in conjunction with pavement widening (either adding lanes or adding width to a narrow lane). If multiple lane widening is to be designed, refer to Part II for guidance. Widening requires coordination between the design of the widened pavement section and the overlay, not only so that the surface will be functionally adequate, but also so that both the existing and widening sections will be structurally adequate. Many lane widening projects have developed serious deterioration along the longitudinal joint due to improper design. The key design recommendations are as follows:

(1) The design "lives" of both the overlay and the new widening construction should be the same to avoid the need for future rehabilitation at significantly different ages.

(2) The widened cross section should generally closely match the existing pavement or cross section in material type, thickness, reinforcement, and joint spacing. However, a shorter joint spacing may be used.

(3) A widened PCC slab section must be tied with deformed bars to the existing PCC slab face. The tie bars should be securely anchored and consistent with ties used in new pavement construction (e.g., No. 5 bars, 30 inches long, grouted and spaced no more than 30 inches apart).

(4) A reflection crack relief fabric may be placed along the longitudinal widening joint.

(5) The overlay should generally be the same thickness over the widening section as over the rest of the traffic lane.

(6) Longitudinal subdrainage should be placed if needed.

5.2.17 Potential Errors and Possible Adjustments to Thickness Design Procedure

The overlay thicknesses obtained using these procedures should be reasonable when the pavement has a structural deficiency. If the overlay thickness appears to be unreasonable, one or more of the following causes may be responsible.

(1) The pavement deterioration may be caused primarily by nonload-associated factors. A computed overlay thickness less than zero or close to zero suggests that the pavement does not need a structural improvement. If a functional deficiency exists, a minimum constructible overlay thickness that addresses the problem could be placed.

(2) Modifications may be needed in the overlay design inputs to customize the procedures to the agency's specific conditions. Each agency should test the overlay design procedures on actual projects to investigate the need for modifications. Reference 8 contains many example overlay designs that illustrate typical inputs and outputs.

   (a) Overlay reliability design level, R. The recommended design reliability levels should be reviewed for overlay designs by each agency, since the recommendations given in Part I are intended for new pavement designs. See Section 5.2.15 for discussion of overlay design reliability.

   (b) Overall standard deviation, S_o. The values recommended for new pavement design may be either too low or too high for overlay design. See Section 5.2.15 for discussion of overall standard deviation.

   (c) Effective slab thickness and structural number adjustment factors. There are many aspects to these that may need agency adjustment.

   (d) Design subgrade resilient modulus and effective k-value. Specifically, a resilient modulus which is consistent with that incorporated into the flexible pavement design equation in Part II, Section 5.4.5 must be used.

   (e) Other design inputs may be in error. Ranges of typical values for inputs are given in the worksheets for overlay design.

5.2.18 Example Designs and Documentation

Reference 8 provides many examples of overlay designs for pavements in different regions of the United States. These may provide the designer with valuable insight into results obtained for actual projects. Reference 9 contains documentation for the concepts involved in the overlay design procedures.

5.3 PAVEMENT EVALUATION FOR OVERLAY DESIGN

It is important that an evaluation of the existing pavement be conducted to identify any functional and structural deficiencies, and to select appropriate pre-overlay repair, reflection crack treatments and overlay designs to correct these deficiencies. This section provides guidance in pavement evaluation for overlay design.

The following sections of Part III of this Guide provide information on pavement evaluation for rehabilitation:

Section 2.3: Selection of Alternative Rehabilitation Methods

Chapter 3: Guides for Field Data Collection

Chapter 4: Rehabilitation Methods Other Than Overlay (portions of this chapter are applicable to preoverlay pavement evaluation and preoverlay repair)

The guidelines and procedures in these chapters are not repeated in this section, but are referenced as needed. This section provides guidelines for pavement evaluation specifically for overlay design purposes. Further details are provided in the sections for design of each overlay type.

5.3.1 Design of Overlay Along Project

Pavement rehabilitation projects involve lengths of pavement that range from a few hundred feet to several miles. There are two approaches to designing an overlay thickness for a project, and both have advantages and disadvantages. The design engineer should select the approach that best fits the specific design situation.

(1) *Uniform Section Approach.* The project is divided into sections of relatively uniform design and condition. Each uniform section is considered independently and overlay design inputs are obtained from each section that represents its average condition (e.g., mean thicknesses, mean number of transverse cracks per mile, mean resilient modulus). Identification of uniform sections is described in Part III, Section 3.2.2. The mean inputs for the section are used to obtain a single overlay thickness for the entire length of the section. The mean inputs must be used in the AASHTO design procedure because design reliability is applied later to give the appropriate safety factor.

(2) *Point-By-Point Approach.* Overlay thicknesses are determined for specific points along the uniform design section (e.g., every 300 feet). All required inputs are determined for each point so that the overlay thickness can be designed. Factors that may change from point to point include deflection, thickness, and condition; other inputs are usually fairly constant along the project. This approach may appear to require much more work; however, in reality it does not require much additional field work, only more runs through the design procedure. This can be done efficiently using a computer.

The point-by-point approach produces a required overlay design thickness for each analysis point along the entire project for a given reliability level. In selecting one thickness for the uniform section, be aware that each overlay thickness has already been increased to account for the design reliability level. Selection of a thickness that is greater than the mean of these values would be designing for a higher level of reliability. The point by point overlay thicknesses can be used to divide the project into different overlay design thickness sections if systematic variation exists along the project, or one design thickness can be selected for the entire project. Areas having unusually high thickness requirements may be targeted for additional field investigation, and may warrant extensive repair or reconstruction.

5.3.2 Functional Evaluation of Existing Pavement

Functional deterioration is defined as any condition that adversely affects the highway user. Some recommended overlay solutions to functional problems are provided (also see table on next page).

(1) Surface Friction and Hydroplaning

All pavement types. Poor wet-weather friction due to polishing of the surface (inadequate macrotexture and/or microtexture). A thin overlay that is adequate for the traffic level may be used to remedy this problem. Guidelines for use of asphalt concrete friction courses are provided in Reference 10.

AC-surfaced pavement. Poor friction due to bleeding of the surface. Milling the AC surface may be required to remove the material that is bleeding to prevent further bleeding through the overlay, and to prevent rutting due to instability. After milling, an open-graded friction course or an overlay thickness adequate for the traffic level may be used to remedy this problem.

AC-surfaced pavement. Hydroplaning and splashing due to wheel path rutting. Determining which layer or layers are rutting and taking appropriate corrective action are important.

(2) Surface Roughness

All pavement types. Long wavelength surface distortion, including heaves and swells. A

Cause of Rutting	Layer(s) Causing Rut	Solution
Total pavement thickness inadequate	Subgrade	Thick overlay
Unstable granular layer due to saturation	Base or subbase	Remove unstable layer or thick overlay
Unstable layer due to low shear strength	Base	Remove unstable layer or thick overlay
Unstable AC mix (including stripping)	Surface	Remove unstable layer
Compaction by traffic	Surface, base, subbase	Surface milling and/or leveling overlay
Studded tire wear	Surface	Surface milling and/or leveling overlay

level-up overlay with varying thickness (adequate thickness on crests) usually corrects these problems.

AC-surfaced pavement. Roughness from deteriorated transverse cracks, longitudinal cracks, and potholes. A conventional overlay will correct the roughness only temporarily, until the cracks reflect through the overlay. Full-depth repair of deteriorated areas and a thicker AC overlay incorporating a reflection crack control treatment may remedy this problem.

AC-surfaced pavement. Roughness from ravelling of surface. A thin AC overlay could be used to remedy this problem. Milling the existing surface may be required to remove deteriorated material to prevent debonding. If the ravelling is due to stripping, the entire layer should be removed because the stripping will continue and may accelerate under an overlay.

PCC-surfaced pavement. Roughness from spalling (including potholes) and faulting of transverse and longitudinal joints and cracks. Spalling can be repaired by full- or partial-depth repairs consisting of rigid materials. Faulting can be alleviated by an overlay of adequate thickness; however, faulting indicates poor load transfer and poor subdrainage. Poor load transfer will lead to spalling of reflected cracks in an AC overlay. Subdrainage improvement may be needed.

Some agencies apply what are called "preventive overlays" that are intended to slow the rate of deterioration. This type of overlay includes thin AC and various surface treatments. These may be applied to pavements which do not present any immediate functional or structural deficiency, but whose condition is expected to deteriorate rapidly in the future.

Overlay designs (including thickness, preoverlay repairs and reflection crack treatments) must address the causes of functional problems and prevent their recurrence. This can only be done through sound engineering, and requires experience in solving the specific problems involved. The overlay design required to correct functional problems should be coordinated with that required to correct any structural deficiencies.

5.3.3 Structural Evaluation of Existing Pavement

Structural deterioration is defined as any condition that reduces the load-carrying capacity of the pavement. The overlay design procedures presented here are based on the concept that time and traffic loadings reduce a pavement's ability to carry loads and an overlay can be designed to increase the pavement's ability to carry loads over a future design period.

Figure 5.1 illustrates the general concepts of structural deficiency and effective structural capacity. The structural capacity of a pavement when new is denoted as SC_o. For flexible pavements, structural capacity is the structural number, SN. For rigid pavements, structural capacity is the slab thickness, D. For existing composite pavements (AC/PCC) the structural capacity is expressed as an equivalent slab thickness.

The structural capacity of the pavement declines with time and traffic, and by the time an evaluation for overlay design is conducted, the structural capacity has decreased to SC_{eff}. The effective structural capacity for each pavement type is expressed as follows:

Flexible pavements: SN_{eff}
Rigid and composite pavements: D_{eff}

If a structural capacity of SC_f is required for the future traffic expected during the overlay design period, an overlay having a structural capacity of SC_{ol} (i.e., $SC_f - SC_{eff}$) must be added to the existing structure. This approach to overlay design is commonly called the structural deficiency approach. Obviously, the required overlay structural capacity can be correct only if the evaluation of existing structural capacity is correct. The primary objective of the structural evalua-

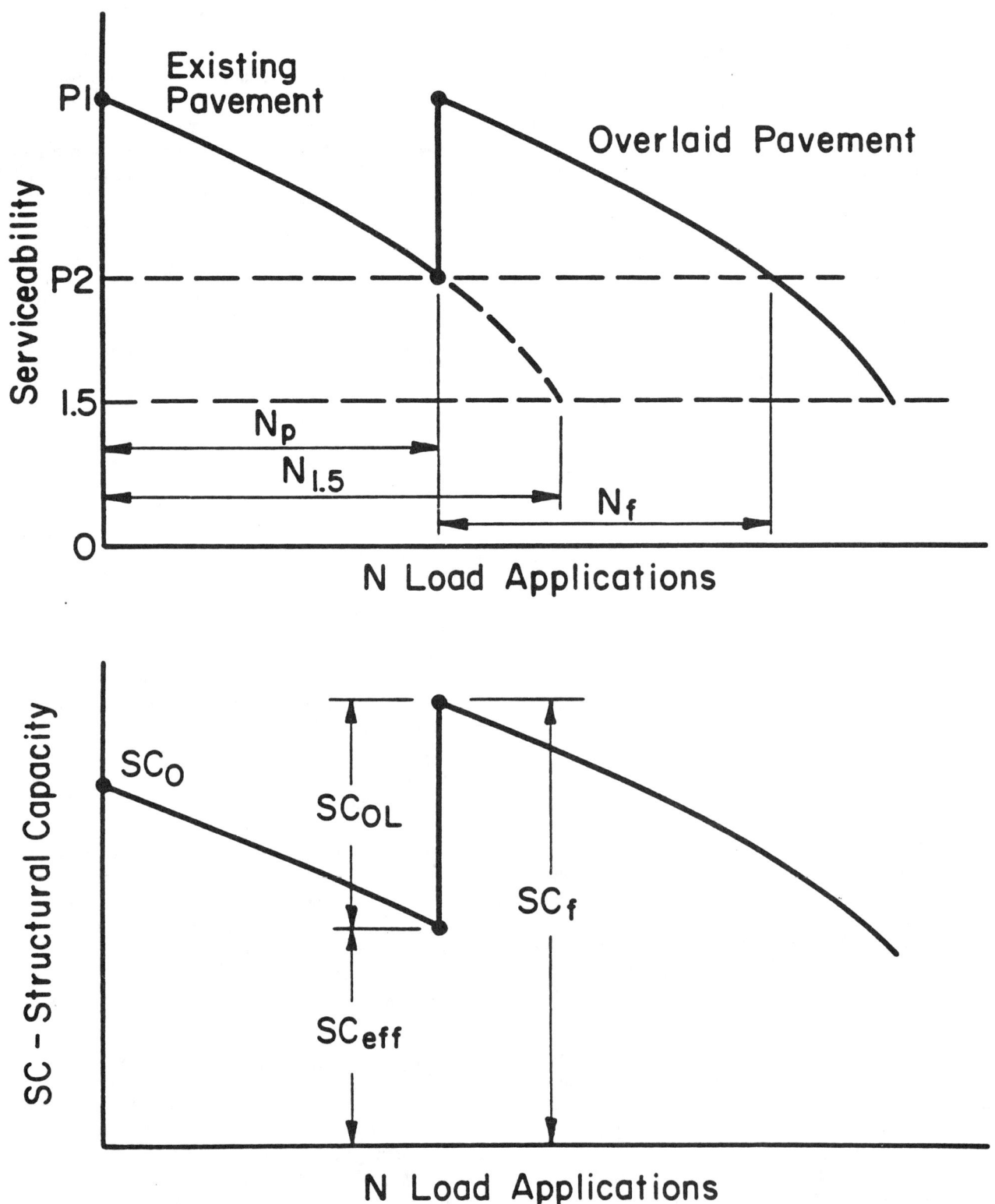

Figure 5.1. Illustration of Structural Capacity Loss Over Time and with Traffic

tion is to determine the effective structural capacity of the existing pavement.

If the declining relationship depicted in Figure 5.1 were well defined, the evaluation of effective structural capacity would be quite easy. This, however, is not the case. No single, specific method exists for evaluating structural capacity. The evaluation of effective structural capacity must consider the current condition of the existing pavement materials, and also consider how those materials will behave in the future. Three alternative evaluation methods are recommended to determine effective structural capacity.

(1) *Structural capacity based on visual survey and materials testing.* This involves the assessment of current conditions based on distress and drainage surveys, and usually some coring and testing of materials.
(2) *Structural capacity based on nondestructive deflection testing (NDT).* This is a direct evaluation of in situ subgrade and pavement stiffness along the project.
(3) *Structural capacity based on fatigue damage from traffic.* Knowledge of past traffic is used to assess the existing fatigue damage in the pavement. The pavement's future remaining fatigue life can then be estimated. The remaining life procedure is most applicable to pavements which have very little visible deterioration.

Because of the uncertainties associated with the determination of effective structural capacity, the three methods cannot be expected to provide equivalent estimates. The designer should use all three methods whenever possible and select the "best" estimate based on his or her judgement. There is no substitute for solid experience and judgment in this selection.

(1) Structural Capacity Based on Visual Survey and Materials Testing

Visual Survey. A key component in the determination of effective structural capacity is the observation of existing pavement conditions. The observation should begin with a review of all information available regarding the design, construction, and maintenance history of the pavement. This should be followed by a detailed survey to identify the type, amount, severity, and location of surface distresses.

Some of the key distress types that are indicators of structural deficiencies are listed below. Some of these are not initially caused by loading, but their severity is increased by loading and thus load-carrying capacity is reduced.

(a) AC-surfaced pavements

Fatigue or alligator cracking in the wheel paths. Patching and a structural overlay are required to prevent this distress from recurring.

Rutting in the wheel paths.

Transverse or longitudinal cracks that develop into potholes.

Localized failing areas where the underlying layers are disintegrating and causing a collapse of the AC surface (e.g., underlying PCC slab with severe "D" cracking, CRCP punchouts, major shear failure of base course/subgrade, stripping of AC base course). This is a very difficult problem to repair and an investigation should be carried out to determine its extent. If it is not extensive, full-depth PCC repair (when a PCC slab exists), and a structural overlay should remedy the problem. If the problem is too extensive for full-depth repair, reconstruction or a structural overlay designed for the weakest area is required.

(b) PCC-surfaced pavements

Deteriorating (spalling or faulting) transverse or longitudinal cracks. These cracks usually must be full-depth repaired, or they will reflect through the overlay. This does not apply to unbonded JPCP or JRCP overlays.

Corner breaks at transverse joints or cracks. Must be full-depth repaired with a full-lane-width repair (this is not required for unbonded JPCP or JRCP overlays).

Localized failing areas where the PCC slab is disintegrating and causing spalls and potholes (e.g., caused by severe "D" cracking, reactive aggregate, or other durability problems). Overlay thickness and preoverlay repair requirements may be prohibitive for some types of overlays.

Localized punchouts, primarily in CRCP. Full-depth repair of existing punchouts and placement of a structural overlay will greatly reduce the likelihood of future punchouts.

Subdrainage Survey. A drainage survey should be coupled with the distress survey. The objective of the drainage survey is to identify moisture-related pavement problems and locations where drainage improvements might be effective in improve the existing structure or reducing the influence of moisture on the performance of the pavement following the overlay.

Coring and Materials Testing Program. In addition to a survey of the surface distress, a coring and testing program is recommended to verify or identify the cause of the observed surface distress. The locations for coring should be selected following the distress survey to assure that all significant pavement conditions are represented. If NDT is used, the data from that testing should also be used to help select the appropriate sites for coring.

The objective of the coring is to determine material thicknesses and conditions. A great deal of information will be gained simply by a visual inspection of the cored material. However, it should be kept in mind that the coring operation causes a disturbance of the material especially along the cut face of AC material.

For example, in some cases coring has been known to disguise the presence of stripping. Consequently, at least some of the asphalt cores should be split apart to check for stripping.

The testing program should be directed toward determining how the existing materials compare with similar materials that would be used in a new pavement, how the materials may have changed since the pavement was constructed, and whether or not the materials are functioning as expected. The types of tests to be performed will depend on the material types and the types of distress observed. A typical testing program might include strength tests for AC and PCC cores, gradation tests to look for evidence of degradation and/or contamination of granular materials, and extraction tests to determine binder contents and gradations of AC mixes. PCC cores exhibiting durability problems may be examined by a petrographer to identify the cause of the problem.

Specific recommendations on estimating the effective structural capacity from the distress survey information are given in the sections for each overlay type.

(2) **Structural Capacity Based on Nondestructive Deflection Testing**

Nondestructive deflection testing is an extremely valuable and rapidly developing technology. When properly applied, NDT can provide a vast amount of information and analysis at a very reasonable expenditure of time, money, and effort. The analyses, however, can be quite sensitive to unknown conditions and must be performed by knowledgeable, experienced personnel.

Within the scope of these overlay design procedures, NDT structural evaluation differs depending on the type of pavement. For rigid pavement evaluation, NDT serves three analysis functions: (1) to examine load transfer efficiency at joints and cracks, (2) to estimate the effective modulus of subgrade reaction (effective k-value), and (3) to estimate the modulus of elasticity of the concrete (which provides an estimate of strength). For flexible pavement evaluation, NDT serves two functions: (1) to estimate the roadbed soil resilient modulus, and (2) to provide a direct estimate of SN_{eff} of the pavement structure. Some agencies use NDT to backcalculate the moduli of the individual layers of a flexible pavement, and then use these moduli to estimate SN_{eff}. This approach is not recommended for use with these overlay design procedures because it implies and requires a level of sophistication that does not exist with the structural number approach to design.

In addition to structural evaluation, NDT can provide other data useful to the design process. Deflection data can be used to quantify variability along the project and to subdivide the project into segments of similar structural strength. The NDT data may also be used in a backcalculation scheme to estimate resilient modulus values for the various pavement layers. Although this procedure does not include the use of these values as a part of the structural condition determination, backcalculation of an unusually low value for any layer should be viewed as a strong indication that a detailed study of the condition of that layer is needed.

The specific methods for estimating effective structural capacity by NDT analysis are discussed within the sections pertaining to the specific overlay types.

(3) **Structural Capacity Based on Remaining Life**

The remaining life approach to structural evaluation relies directly on the concepts illustrated in Figure 5.1. This follows a fatigue damage concept that repeated loads gradually damage the pavement and reduce the number of additional loads the pavement can carry to failure. At any given time, there may be no directly observable indication of damage, but there is a reduction in structural capacity in terms of the future load-carrying capacity (the number of future loads that the pavement can carry).

To determine the remaining life, the designer must determine the actual amount of traffic the pavement has carried to date and the total amount of traffic the pavement could be expected to carry to "failure" (when serviceability equals 1.5, to be consistent with the AASHO Road Test equations). Both traffic amounts must be expressed in 18-kip ESAL. The difference between these values, expressed as a percentage of the total traffic to "failure," is defined as the remaining life:

$$RL = 100 \left[1 - \left(\frac{N_p}{N_{1.5}} \right) \right]$$

where

RL = remaining life, percent
N_p = total traffic to date, 18-kip ESAL
$N_{1.5}$ = total traffic to pavement "failure" (P2 = 1.5), 18-kip ESAL

With RL determined, the designer may obtain a condition factor (CF) from Figure 5.2. CF is defined by the equation:

$$CF = \frac{SC_n}{SC_o}$$

where

SC_n = pavement structural capacity after N_p ESAL
SC_o = original pavement structural capacity

The existing structural capacity may be estimated by multiplying the original structural capacity of the pavement by CF. For example, the original structural number (SN_o) of a flexible pavement may be calculated from material thicknesses and the structural coefficients for those materials in a new pavement. SN_{eff} of the pavement based on a remaining life analysis would be:

$$SN_{eff} = CF * SN_o$$

The structural capacity determined by this relationship does not account for any preoverlay repair. The calculated structural capacity should be viewed as a lower limit value and may require adjustment to reflect the benefits of preoverlay repair.

For the remaining life determination, $N_{1.5}$ can be roughly estimated using the new pavement design equations or nomographs, or other equations based on local agency information. To be consistent with the AASHO Road Test and the development of these equations, a failure PSI equal to 1.5 and a reliability of 50 percent is recommended.

When using this approach, the designer need not be alarmed if the traffic to date (N_p) is found to exceed the expected traffic to failure ($N_{1.5}$) resulting in a calculated negative remaining life. When this happens, the designer could use the minimum value for CF (0.50), or not use the remaining life approach.

The remaining life approach to determine SN_{eff} or D_{eff} has some serious deficiencies associated with it. There are four major sources of error:

(1) The predictive capability of the AASHO Road Test equations,
(2) The large variation in performance typically observed even among pavements of seemingly identical designs,
(3) Estimation of past 18-kip ESALs, and
(4) Inability to account for the amount of preoverlay repair to the pavement. For pavements with considerable deterioration, the SN_{eff} or D_{eff} value obtained from the remaining life method may be much lower than values obtained from other methods that adjust for preoverlay repairs. Thus, the remaining life procedure is most applicable to pavements which have very little visible deterioration.

As a result, this method of determining the remaining life of the pavement can in some cases produce very erroneous results. The following two extreme errors may occur with this approach:

(1) The remaining life estimate may be extremely low even though very little load-associated distress is present. While some fatigue damage

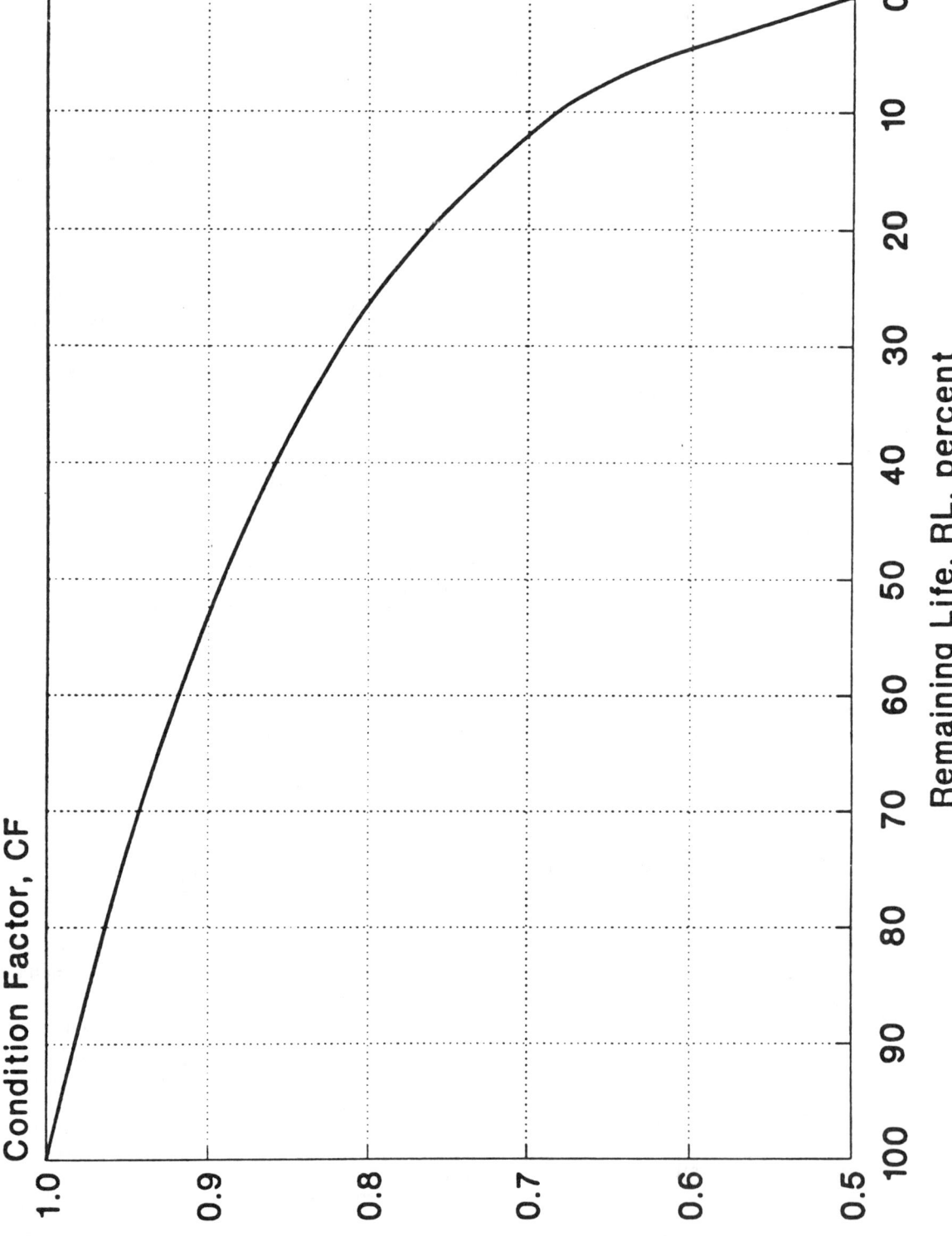

Figure 5.2. Relationship Between Condition Factor and Remaining Life

can exist in a pavement structure before a significant amount of cracking appears, it cannot be a large amount of damage, or it would certainly be evidenced by a significant amount of cracking. If load-related cracking is present in very small amounts and at a low severity level, the pavement has considerable remaining life, regardless of what the traffic-based remaining life calculation suggests.

(2) The remaining life estimate may be extremely high even though a substantial amount of medium- and high-severity load-related cracking is present. In this case, the pavement really has little remaining life.

At any point between these two extremes, the remaining life computed from past traffic may not reflect the amount of fatigue damage in the pavement, but discerning this from observed distress may be more difficult. If the computed remaining life appears to be clearly at odds with the amount and severity of load-associated distress present, do not use the remaining life method to compute the structural capacity of the existing pavement.

The remaining life approach to determining structural capacity is not directly applicable, without modification, to pavements which have already received one or more overlays.

5.3.4 Determination of Design M_R

The design subgrade M_R may be determined by: (1) laboratory testing, (2) NDT backcalculation, (3) estimation from resilient modulus correlation studies, or (4) original design and construction data. Regardless of the method used, the design M_R value must be consistent with the value used in the design performance equation for the AASHO Road Test subgrade. This is especially important when M_R is determined by NDT backcalculation. The backcalculated value is typically too high to be consistent and must be adjusted. If M_R is not adjusted, the SN_f value will be unconservative and poor overlay performance can be expected.

A subgrade M_R may be backcalculated from NDT data using the following equation:

$$M_R = \frac{0.24P}{d_r r}$$

where

M_R = backcalculated subgrade resilient modulus, psi
P = applied load, pounds
d_r = measured deflection at radial distance r, inches
r = radial distance at which the deflection is measured, inches

This equation for backcalculating M_R is based on the fact that, at points sufficiently distant from the center of loading, the measured surface deflection is almost entirely due to deformation in the subgrade, and is also independent of the load radius. For practical purposes, the deflection used should be as close as possible to the loading plate, but must also be sufficiently far from the loading plate to satisfy the assumptions inherent in the above equation. Guidance is provided later in this chapter for selecting the minimum radial distance for determination of M_R.

The recommended method for determination of the design M_R from NDT backcalculation requires an adjustment factor (C) to make the value calculated consistent with the value used to represent the AASHO subgrade. A value for C of no more than 0.33 is recommended for adjustment of backcalculated M_R values to design M_R values. The resulting equation is:

$$\text{Design } M_R = C \left(\frac{0.24P}{d_r r}\right)$$

A subgrade M_R value of 3,000 psi was used for the AASHO Road Test soil in the development of the flexible pavement performance model. This value is consistent with some laboratory tests of soil samples from the AASHO Road Test site, as Figure 5.3 illustrates (*11*). However, these data also show that the resilient modulus of the AASHO Road Test soil is quite stress-dependent, increasing rapidly for deviator stresses less than 6 psi. The subgrade deviator stress at a radial distance appropriate for use in the equation given above for backcalculated M_R will almost always be far less than 6 psi. Thus, the subgrade modulus determined by backcalculation can be expected to be too high to be consistent with the 3,000 psi assumed for the AASHO subgrade.

This was confirmed by two methods. In the first analysis, M_R values backcalculated from deflection data were compared with M_R values obtained from laboratory tests, for the AASHO Road Test and other sites (*12, 13*). The results, which are shown in Figure 5.4, indicate that backcalculated M_R values exceed laboratory M_R values by a factor of three or more. In

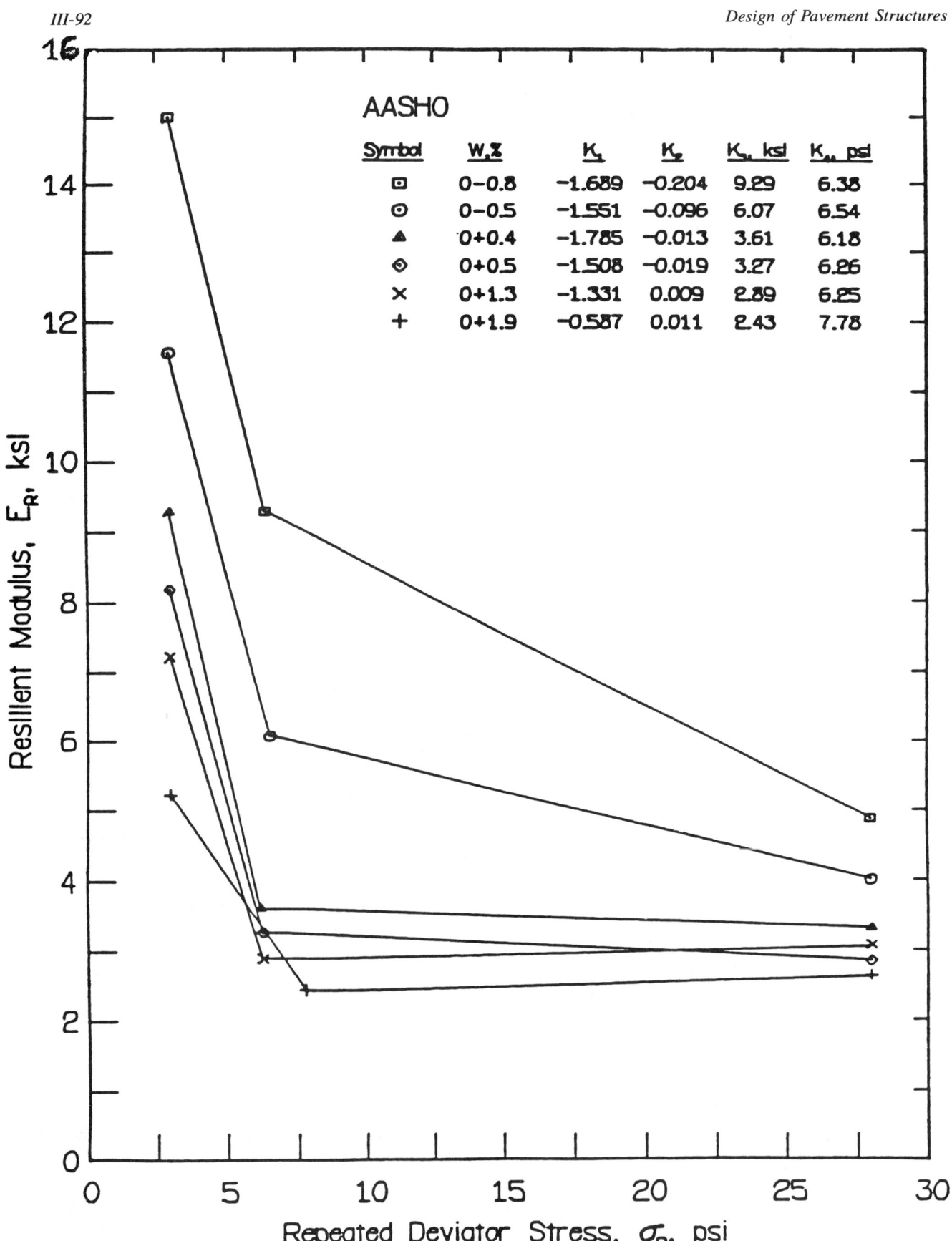

Figure 5.3. AASHO Road Test Subgrade Resilient Modulus Test Results (*11*)

Rehabilitation with Overlays

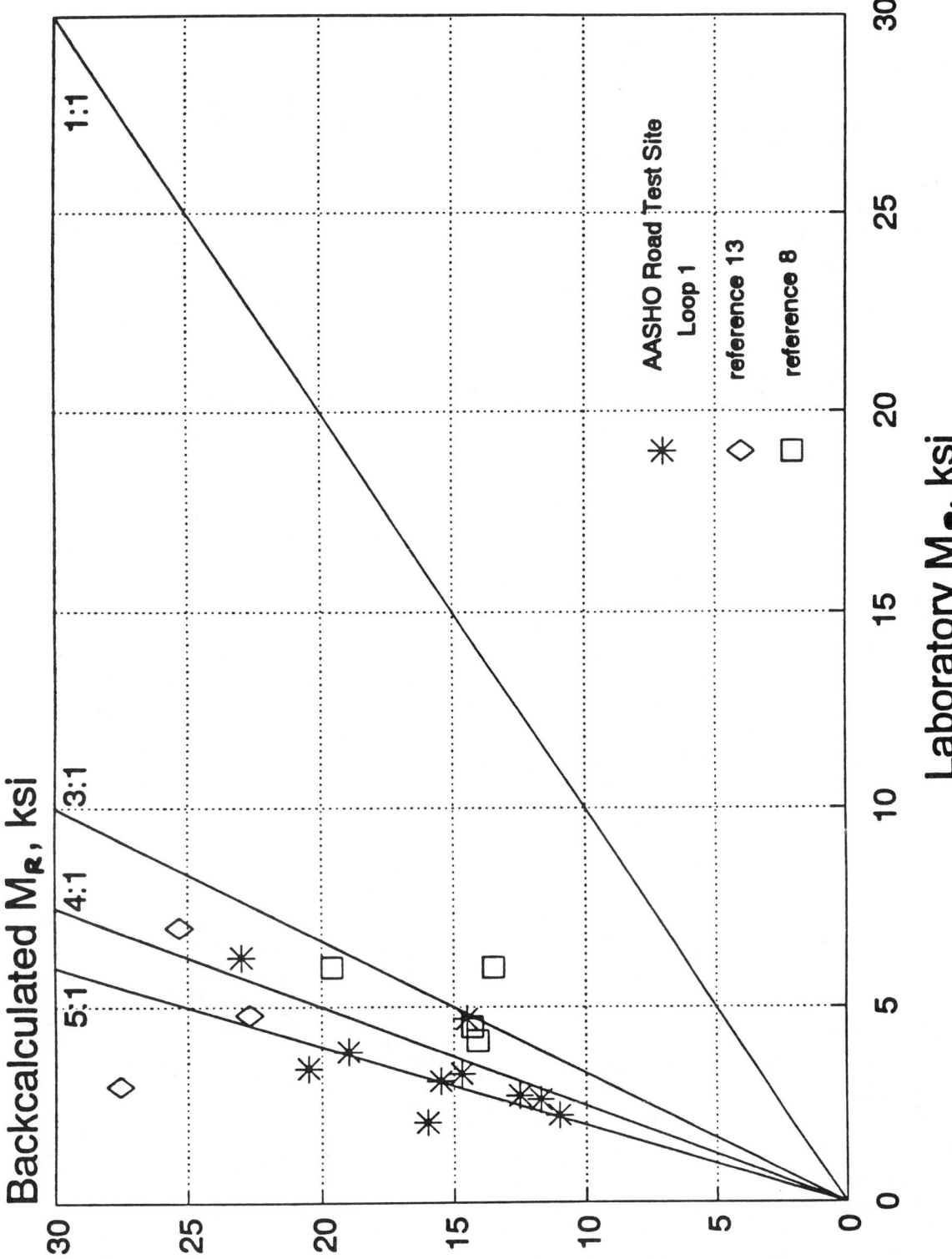

Figure 5.4. Backcalculated Resilient Modulus Versus Laboratory Results on Shelby Tube Samples from the AASHO Road Test Site Plus Data from Two Additional States

the second analysis, the ILLI-PAVE finite element program (*14, 15*) was used to compute M_R values for a variety of pavement structures and subgrade characteristics representative of the AASHO Road Test soil. At radial distances appropriate for backcalculation of M_R, the computed M_R values also exceeded the value of 3,000 psi assumed in development of the AASHO flexible pavement model by a factor of at least three. Similarly, pavement surface deflections computed by ILLI-PAVE produced backcalculated M_R values three or more times greater than 3,000 psi.

All of these analyses suggest that for the soils examined, backcalculated M_R values should be multiplied by an adjustment factor C of no more than 0.33 in order to obtain M_R values appropriate for use in design with the AASHTO flexible pavement model.

The analyses described here pertain to the fine-grained, stress-sensitive soil at the AASHO Road Test site plus fine-grained soil from seven other projects. No attempt has been made in this study to investigate the relationship between backcalculated and laboratory M_R values for granular subgrades. It may be that backcalculated M_R values for granular subgrades would not require a correction factor as large as is required for cohesive subgrades. However, this subject requires further research.

Users are cautioned that the resilient modulus value selected has a very significant effect on the resulting structural number determined. Therefore, users should be very cautious about using high resilient modulus values, or their overlay thickness values will be too thin.

5.4 AC OVERLAY OF AC PAVEMENT

This section covers the design of AC overlays of AC pavements. The following construction tasks are involved in the placement of an AC overlay on an existing AC pavement:

(1) Repairing deteriorated areas and making subdrainage improvements (if needed).
(2) Correcting surface rutting by milling or placing a leveling course.
(3) Constructing widening (if needed).
(4) Applying a tack coat.
(5) Placing the AC overlay (including a reflective crack control treatment if needed).

5.4.1 Feasibility

An AC overlay is a feasible rehabilitation alternative for an AC pavement except when the condition of the existing pavement dictates substantial removal and replacement. Conditions under which an AC overlay would not be feasible include the following.

(1) The amount of high-severity alligator cracking is so great that complete removal and replacement of the existing surface is dictated.
(2) Excessive surface rutting indicates that the existing materials lack sufficient stability to prevent recurrence of severe rutting.
(3) An existing stabilized base shows signs of serious deterioration and would require an inordinate amount of repair to provide uniform support for the overlay.
(4) An existing granular base must be removed and replaced due to infiltration of and contamination by a soft subgrade.
(5) Stripping in the existing AC surface dictates that it should be removed and replaced.

5.4.2 Pre-overlay Repair

The following types of distress should be repaired prior to overlay of AC pavements. If they are not repaired, the service life of the overlay will be greatly reduced.

Distress Type	Required Repair
Alligator Cracking	All areas of high-severity alligator cracking must be repaired. Localized areas of medium-severity alligator cracking should be repaired unless a paving fabric or other means of reflective crack control is used. The repair must include removal of any soft subsurface material.
Linear Cracks	High-severity linear cracks should be patched. Linear cracks that are open greater than 0.25 inch should be filled with a sand-asphalt mixture or other suitable crack filler. Some method of reflective crack control is recommended for transverse cracks that experience significant opening and closing.

Rutting	Remove ruts by milling or placement of a leveling course. If rutting is severe, an investigation into which layer is causing the rutting should be conducted to determine whether or not an overlay is feasible.
Surface Irregularities	Depressions, humps, and corrugations require investigation and treatment of their cause. In most cases, removal and replacement will be required.

5.4.3 Reflection Crack Control

The basic mechanism of reflection cracking is strain concentration in the overlay due to movement in the vicinity of cracks in the existing surface. This movement may be bending or shear induced by loads, or may be horizontal contraction induced by temperature changes. Load-induced movements are influenced by the thickness of the overlay and the thickness and stiffness of the existing pavement. Temperature-induced movements are influenced by daily and seasonal temperature variations, the coefficient of thermal expansion of the existing pavement, and the spacing of cracks.

Pre-overlay repair (patching and crack filling) may help delay the occurrence and deterioration of reflection cracks. Additional reflection crack control measures which have been beneficial in some cases include the following:

(1) Synthetic fabrics and stress-absorbing interlayers (SAMIs) have been effective in controlling reflection of low- and medium-severity alligator cracking. They may also be useful for controlling reflection of temperature cracks, particularly when used in combination with crack filling. They generally do little, however, to retard reflection of cracks subject to significant horizontal or vertical movements.

(2) Crack relief layers greater than 3 inches thick have been effective in controlling reflection of cracks subject to larger movements. These crack relief layers are composed of open-graded coarse aggregate and a small percentage of asphalt cement.

(3) Sawing and sealing joints in the AC overlay at locations coinciding with straight cracks in the underlying AC may be effective in controlling the deterioration of reflection cracks. This technique has been very effective when applied to AC overlays of jointed PCC pavements when the sawcut matches the joint or straight crack within an inch.

(4) Increased AC overlay thickness reduces bending and vertical shear under loads and also reduces temperature variation in the existing pavement. Thus, thicker AC overlays are more effective in delaying the occurrence and deterioration of reflection cracks than are thinner overlays. However, increasing the AC overlay thickness is a costly approach to reflection crack control.

Reflection cracking can have a considerable (often controlling) influence on the life of an AC overlay. Deteriorated reflection cracks detract from a pavement's serviceability and also require frequent maintenance, such as sealing and patching. Reflection cracks also permit water to enter the pavement structure, which may result in loss of bond between the AC overlay and existing AC surface, stripping in either layer, and softening of the granular layers and subgrade. For this reason, reflection cracks should be sealed as soon as they appear and resealed periodically throughout the life of the overlay. Sealing low-severity reflection cracks may also be effective in retarding their progression to medium and high severity levels.

5.4.4 Subdrainage

See Section 5.2.4 for guidelines.

5.4.5 Thickness Design

If the overlay is being placed for the purpose of structural improvement, the required thickness of the overlay is a function of the structural capacity required to meet future traffic demands and the structural capacity of the existing pavement. The required thickness to increase structural capacity to carry future traffic is determined by the following equation.

$$SN_{ol} = a_{ol} * D_{ol} = SN_f - SN_{eff}$$

where

SN_{ol} = Required overlay structural number
a_{ol} = Structural coefficient for the AC overlay
D_{ol} = Required overlay thickness, inches

SN_f = Structural number required to carry future traffic
SN_{eff} = Effective structural number of the existing pavement

The required overlay thickness may be determined through the following design steps. These steps provide a comprehensive design approach that recommends testing the pavement to obtain valid design inputs. If it is not possible to conduct testing (e.g., for a low-volume road), an approximate overlay design may be developed based upon visible distress observation, by skipping Steps 4 and 5 and by estimating other inputs.

Step 1: Existing pavement design and construction.

(1) Thickness and material type of each pavement layer.
(2) Available subgrade soil information (from construction records, soil surveys, county agricultural soils reports, etc.)

Step 2: Traffic analysis.

(1) Past cumulative 18-kip ESALs in the design lane (N_p), for use in the remaining life method of SN_{eff} determination only.
(2) Predicted future 18-kip ESALs in the design lane over the design period (N_f).

Step 3: Condition survey.

Distress types and severities are defined in reference 11. The following distresses are measured during the condition survey and are used in the determination of the structural coefficients. Sampling along the project in the heaviest trafficked lane can be used to estimate these quantities.

(1) Percent of surface area with alligator cracking (class 1, 2, and 3 corresponding to low, medium, and high severities).
(2) Number of transverse cracks per mile (low, medium, and high severities).
(3) Mean rut depth.
(4) Evidence of pumping at cracks and at pavement edges.

Step 4: Deflection testing (strongly recommended).

Measure deflections in the outer wheel path at an interval sufficient to adequately assess conditions. Intervals of 100 to 1,000 feet are typical. Areas that are deteriorated and will be repaired should not be tested. A heavy-load deflection device (e.g., Falling Weight Deflectometer) and a load magnitude of approximately 9,000 pounds are recommended. ASTM D 4694 and D 4695 provide additional guidance on deflection testing. Deflections should be measured at the center of the load and at least one other distance from the load, as described below.

(1) Subgrade resilient modulus (M_R). At sufficiently large distances from the load, deflections measured at the pavement surface are due to subgrade deformation only, and are also independent of the size of the load plate. This permits the backcalculation of the subgrade resilient modulus from a single deflection measurement and the load magnitude, using the following equation:

$$M_R = \frac{0.24P}{d_r r}$$

where

M_R = backcalculated subgrade resilient modulus, psi
P = applied load, pounds
d_r = deflection at a distance r from the center of the load, inches
r = distance from center of load, inches

It should be noted that no temperature adjustment is needed in determining M_R since the deflection used is due only to subgrade deformation.

The deflection used to backcalculate the subgrade modulus must be measured far enough away that it provides a good estimate of the subgrade modulus, independent of the effects of any layers above, but also close enough that it is not too small to measure accurately. The minimum distance may be determined from the following relationship:

$$r \geq 0.7 a_e$$

where

$$a_e = \sqrt{\left[a^2 + \left(D \sqrt[3]{\frac{E_p}{M_R}}\right)^2\right]}$$

a_e = radius of the stress bulb at the subgrade-pavement interface, inches
a = NDT load plate radius, inches
D = total thickness of pavement layers above the subgrade, inches
E_p = effective modulus of all pavement layers above the subgrade, psi (described below)

Before the backcalculated M_R value is used in design, it must be adjusted to make it consistent with the value used in the AASHTO flexible pavement design equation. An adjustment may also be needed to account for seasonal effects. These adjustments are described in Step 6.

(2) Temperature of AC mix. The temperature of the AC mix during deflection testing must be determined. The AC mix temperature may be measured directly, or may be estimated from surface or air temperatures.

(3) Effective modulus of the pavement (E_p). If the subgrade resilient modulus and total thickness of all layers above the subgrade are known or assumed, the effective modulus of the entire pavement structure (all pavement layers above the subgrade) may be determined from the deflection measured at the center of the load plate using the following equation:

$$d_0 = 1.5pa \left\{ \frac{1}{M_R \sqrt{1 + \left(\frac{D}{a}\sqrt[3]{\frac{E_p}{M_R}}\right)^2}} + \frac{\left[1 - \frac{1}{\sqrt{1 + \left(\frac{D}{a}\right)^2}}\right]}{E_p} \right\}$$

where

d_0 = deflection measured at the center of the load plate (and adjusted to a standard temperature of 68°F), inches

p = NDT load plate pressure, psi
a = NDT load plate radius, inches
D = total thickness of pavement layers above the subgrade, inches
M_R = subgrade resilient modulus, psi
E_p = effective modulus of all pavement layers above the subgrade, psi

For a load plate radius of 5.9 inches, Figure 5.5 may be used to determine the ratio E_p/M_R, and E_p may then be determined for a known or assumed value of M_R.

For purposes of comparison of E_p along the length of a project, the d_0 values used to determine E_p should be adjusted to a single reference temperature. Furthermore, if the effective structural number of the existing pavement is to be determined in Step 7 using the values of E_p backcalculated from deflection data, the reference temperature for adjustment of d_0 should be 68°F, to be consistent with the procedure for new AC pavement design described in Part II. Figure 5.6 may be used to adjust d_0 for AC pavements with granular and asphalt-stabilized bases. Figure 5.7 may be used to adjust d_0 for AC pavements with cement- and pozzolanic-stabilized bases.

Step 5: Coring and materials testing (strongly recommended).

(1) *Resilient modulus of subgrade.* If deflection testing is not performed, laboratory testing of samples of the subgrade may be conducted to determine its resilient modulus using AASHTO T 292-91 I with a deviator stress of 6 psi to match the deviator stress used in establishing the 3,000 psi for the AASHO Road Test soil that is incorporated into the flexible design equation. Alternatively, other tests such as R value, CBR or soil classification tests could be conducted and approximate correlations used to estimate resilient modulus. Use of the estimating equation $M_R = 1500 * CBR$ may produce a value that is too large for use in this design procedure. The relationships found in Appendix FF, Figure FF-6 may be more reasonable.

(2) *Samples of AC layers and stabilized base* should be visually examined to assess asphalt stripping, degradation, and erosion.

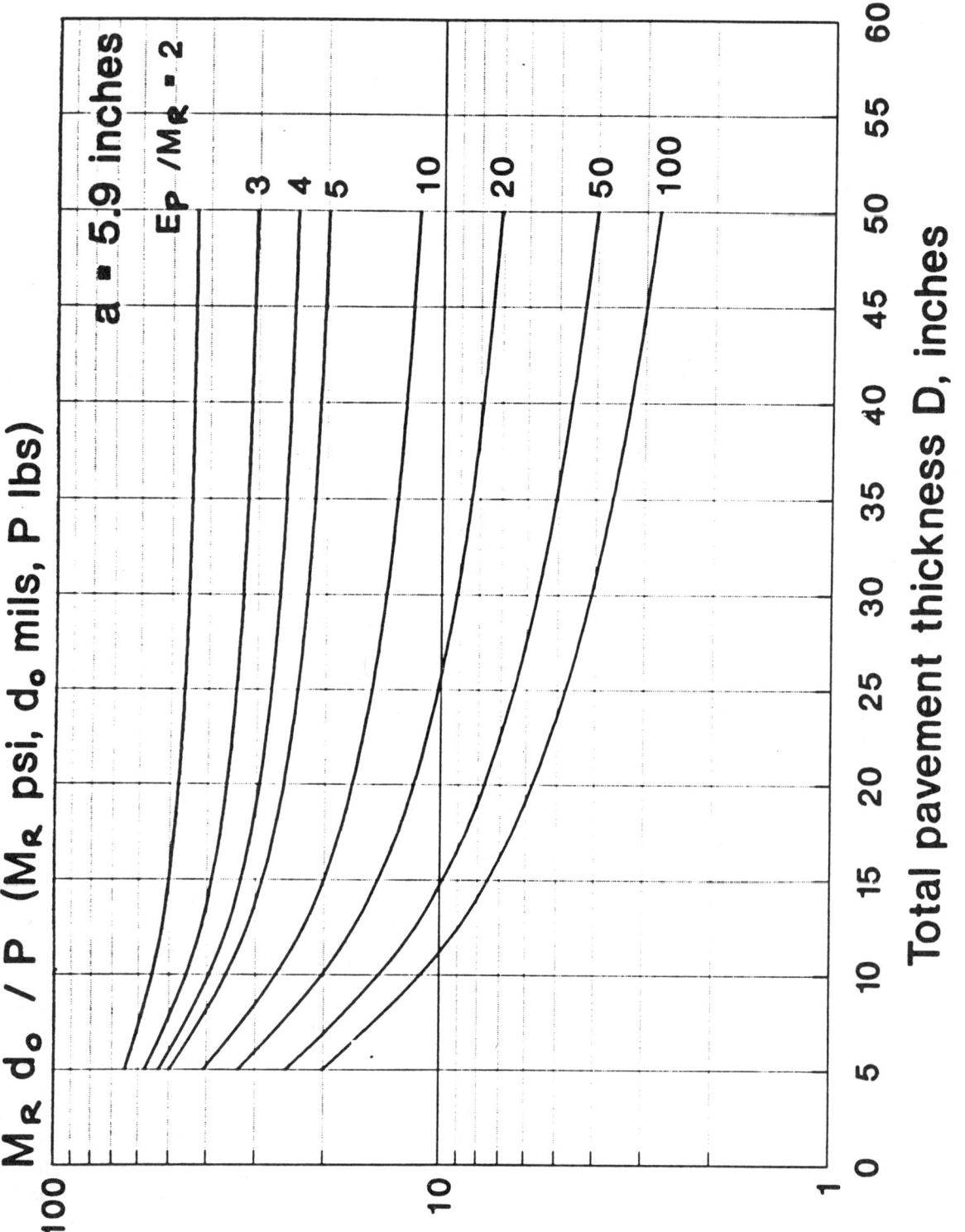

Figure 5.5. Determination of E_P/M_R

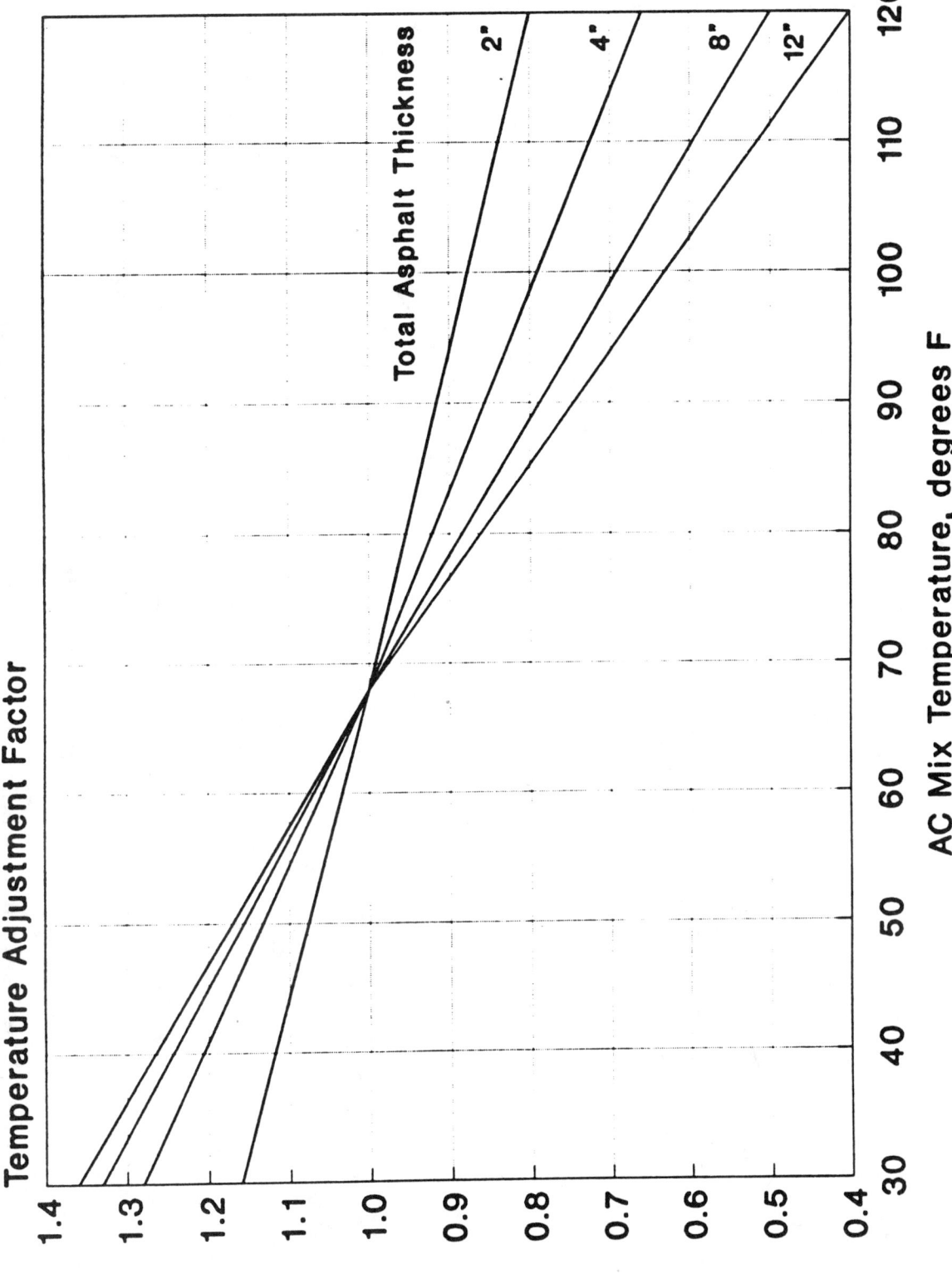

Figure 5.6. Adjustment to d_0 for AC Mix Temperature for Pavement with Granular or Asphalt-Treated Base

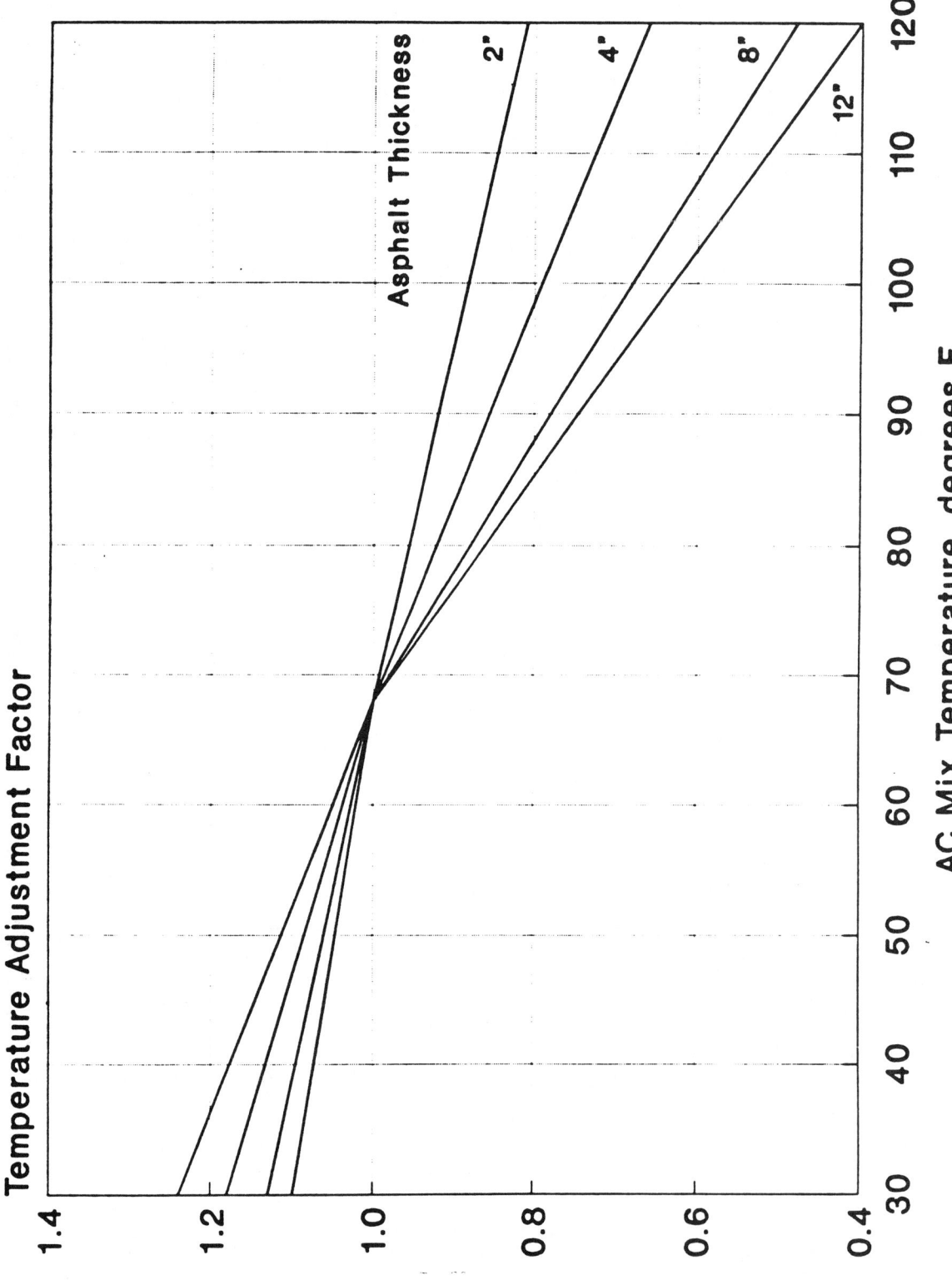

Figure 5.7. Adjustment to d_0 for AC mix Temperature for Pavement with Cement- or Pozzolanic-Treated Base

Rehabilitation with Overlays

(3) *Samples of granular base and subbase* should be visually examined and a gradation run to assess degradation and contamination by fines.

(4) *The thickness of all layers* should be measured.

Step 6: Determination of required structural number for future traffic (SN_f).

(1) *Effective design subgrade resilient modulus.* Determine by one of the following methods:
   (a) Laboratory testing described in Step 5.
   (b) Backcalculation from deflection data. (NOTE: this value must be adjusted to be consistent with the value used in the AASHTO flexible pavement design equation as described below.)
   (c) A very approximate estimate can be made using available soil information and relationships developed from resilient modulus studies. However, if as-constructed soil data are used, the resilient modulus may have changed since construction due to changes in moisture content or other factors.

Regardless of the method used, the effective design subgrade resilient modulus must be (1) representative of the effects of seasonal variation and (2) consistent with the resilient modulus value used to represent the AASHO Road Test soil. A seasonal adjustment, when needed, may be made in accordance with the procedures described in Part II, Section 2.3.1. M_R values backcalculated from deflections must be adjusted to be consistent with the laboratory-measured value used for the AASHO Road Test soil in the development of the flexible pavement design equation. It is recommended that backcalculated M_R values be multiplied by a correction factor C = 0.33 for use in determination of SN_f for design purposes when an FWD load of approximately 9,000 pounds is used (9). This value should be evaluated and adjusted if needed by user agencies for their soil and deflection measurement equipment. Therefore, the following design M_R should be used to determine SN_f:

$$\text{Design } M_R = C \left(\frac{0.24P}{d_r r} \right)$$

where recommended C = 0.33

Note also that the presence of a very stiff layer (e.g., bedrock) within about 15 feet of the top of the subgrade may cause the backcalculated M_R to be high. When such a condition exists, a value less than 0.33 for C may be warranted (9).

The designer is cautioned against using a value of M_R that is too large. The value of M_R selected for design is extremely critical to the overlay thickness. The use of a value greater than 3,000 psi is an indication that the soil is stiffer than the silty-clay A-6 soil at the Road Test site, and consequently will provide increased support and extended pavement life.

(2) *Design PSI loss.* PSI immediately after overlay (P1) minus PSI at time of next rehabilitation (P2).

(3) *Overlay design reliability R (percent).* See Part I, Section 4.2, Part II, Table 2.2, and Part III, Section 5.2.15.

(4) *Overall standard deviation S_o for flexible pavement.* See Part I, Section 4.3.

Compute SN_f for the above design inputs using the flexible pavement design equation or nomograph in Part II, Figure 3.1. When designing an overlay thickness for a uniform pavement section, mean input values must be used. When designing an overlay thickness for specific points along the project, the data for that point must be used. A worksheet for determining SN_f is provided in Table 5.1.

Step 7: Determination of effective structural number (SN_{eff}) of the existing pavement.

Three methods are presented for determining the effective structural number of a conventional AC pavement: an NDT method, a condition survey method, and a remaining fatigue life method. It is suggested that the designer use all three of these to evaluate the pavement, and then select a value for SN_{eff} based on the results, using engineering judgment and the past experience of the agency.

SN_{eff} from NDT for AC Pavements

The NDT method of SN_{eff} determination follows an assumption that the structural capacity of the pavement is a function of its total thickness and overall stiffness. The relationship between SN_{eff}, thickness, and stiffness is:

Table 5.1. Worksheet for Determination of SN_f for AC Pavements

TRAFFIC:

Future 18-kip ESALs in design lane over
design period, N_f = _____

EFFECTIVE ROAD-BED SOIL RESILIENT MODULUS:

Design resilient modulus, M_R = _____ psi

(Adjusted for consistency with flexible pavement model and for seasonal variations. Typical design M_R is 2,000 to 10,000 psi for fine-grained soils, 10,000 to 20,000 for coarse-grained soils. The AASHO Road Test soil value used in the flexible pavement design equation was 3,000 psi.)

SERVICEABILITY LOSS:

Design PSI loss (P1 − P2) (1.2 to 2.5) = _____

DESIGN RELIABILITY:

Overlay design reliability, R (80 to 99 percent) = _____ percent

Overall standard deviation, S_o (typically 0.49) = _____

FUTURE STRUCTURAL CAPACITY:

Required structural number for future traffic is determined from flexible pavement design equation or nomograph in Part II, Figure 3.1.

SN_f = _____

$$SN_{eff} = 0.0045 D \sqrt[3]{E_p}$$

where

D = total thickness of all pavement layers above the subgrade, inches

E_p = effective modulus of pavement layers above the subgrade, psi

E_p may be backcalculated from deflection data as described in Step 4. Figure 5.8 may be used to determine SN_{eff} according to the above equation.

SN_{eff} from Condition Survey for AC Pavements

The condition survey method of SN_{eff} determination involves a component analysis using the structural number equation:

$$SN_{eff} = a_1 D_1 + a_2 D_2 m_2 + a_3 D_3 m_3$$

where

D_1, D_2, D_3 = thicknesses of existing pavement surface, base, and subbase layers

a_1, a_2, a_3 = corresponding structural layer coefficients

m_2, m_3 = drainage coefficients for granular base and subbase

See Part II, Table 2.4, for guidance in determining the drainage coefficients. In selecting values for m_2 and m_3, note that the poor drainage situation for the base and subbase at the AASHO Road Test would be given drainage coefficient values of 1.0.

Depending on the types and amounts of deterioration present, the layer coefficient values assigned to materials in in-service pavement should in most cases be less than the values that would be assigned to the same materials for new construction. An exception to

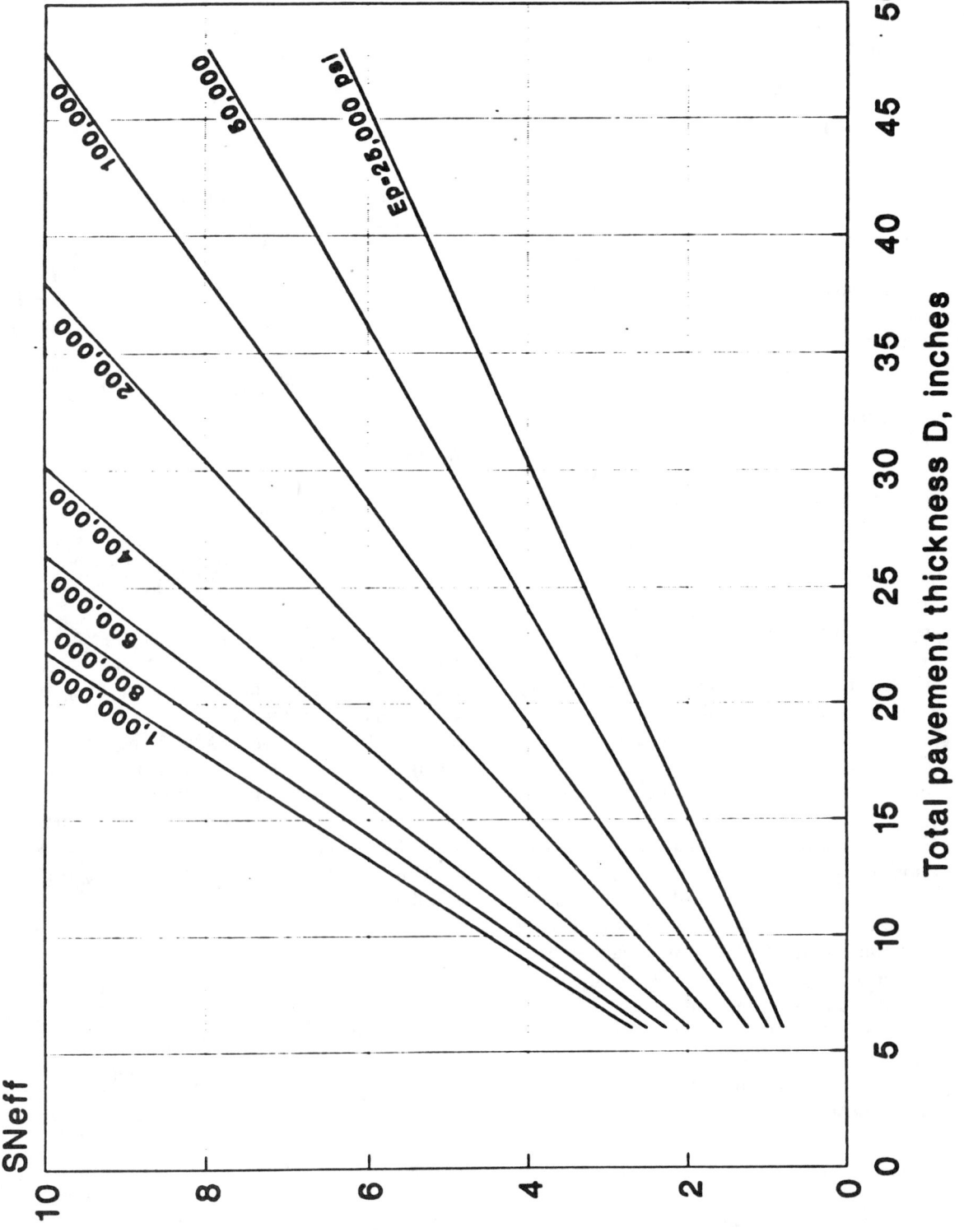

Figure 5.8. SN_{eff} from NDT Method

this general rule might be for unbound granular materials that show no sign of degradation or contamination.

For example, one State uses 0.44 for its new high-quality AC surface, but for overlay design purposes uses a reduced coefficient for the same material in an existing pavement. A value of 0.34 is assigned if the AC layer is in good condition, 0.25 if its condition is fair, and 0.15 if its condition is poor. The condition ratings are made on the basis of the amount of cracking present.

Limited guidance is presently available for the selection of layer coefficients for in-service pavement materials. Each agency must adopt its own set of values. Some suggested layer coefficients for existing materials are provided in Table 5.2.

The following notes apply to Table 5.2:

(1) All of the distress is as observed at the pavement surface.
(2) Patching all high-severity alligator cracking is recommended. The AC surface and stabilized base layer coefficients selected should reflect the amount of high-severity cracking remaining after patching.
(3) In addition to evidence of pumping noted during condition survey, samples of base material should be obtained and examined for evidence of erosion, degradation and contamination by fines, as well as evaluated for drainability, and layer coefficients reduced accordingly.
(4) The percentage of transverse cracking is determined as (linear feet of cracking/square feet of pavement) * 100.
(5) Coring and testing are recommended for evaluation of all materials and are strongly recommended for evaluation of stabilized layers.
(6) There may be other types of distress that, in the opinion of the engineer, would detract from the performance of an overlay. These should be considered through an appropriate decrease of the structural coefficient of the layer exhibiting the distress (e.g., surface raveling of the AC, stripping of an AC layer, freeze-thaw damage to a cement-treated base).

SN_{eff} from Remaining Life for AC Pavements

The remaining life of the pavement is given by the following equation:

$$RL = 100\left[1 - \left(\frac{N_p}{N_{1.5}}\right)\right]$$

where

RL = remaining life, percent
N_p = total traffic to date, ESALs
$N_{1.5}$ = total traffic to pavement "failure," ESALs

$N_{1.5}$ may be estimated using the new pavement design equations or nomographs in Part II. To be consistent with the AASHO Road Test and the development of these equations, a "failure" PSI equal to 1.5 and a reliability of 50 percent is recommended.

SN_{eff} is determined from the following equation:

$$SN_{eff} = CF * SN_o$$

where

CF = condition factor determined from Figure 5.2
SN_o = structural number of the pavement if it were newly constructed

The designer should recognize that SN_{eff} determined by this method does not reflect any benefit for pre-overlay repair. The estimate of SN_{eff} obtained should thus be considered a lower limit value. The SN_{eff} of the pavement will be higher if pre-overlay repair of load-associated distress (alligator cracking) is done. This method for determining SN_{eff} is not applicable, without modification, to AC pavements which have already received one or more AC overlays.

A worksheet for determination of SN_{eff} is provided in Table 5.3.

Step 8: Determination of overlay thickness.

The thickness of AC overlay is computed as follows:

$$D_{ol} = \frac{SN_{ol}}{a_{ol}} = \frac{(SN_f - SN_{eff})}{a_{ol}}$$

where

SN_{ol} = Required overlay structural number
a_{ol} = Structural coefficient for the AC overlay
D_{ol} = Required overlay thickness, inches
SN_f = Structural number determined in Step 6

Table 5.2. Suggested Layer Coefficients for Existing AC Pavement Layer Materials

MATERIAL	SURFACE CONDITION	COEFFICIENT
AC Surface	Little or no alligator cracking and/or only low-severity transverse cracking	0.35 to 0.40
	<10 percent low-severity alligator cracking and/or <5 percent medium- and high-severity transverse cracking	0.25 to 0.35
	>10 percent low-severity alligator cracking and/or <10 percent medium-severity alligator cracking and/or >5–10 percent medium- and high-severity transverse cracking	0.20 to 0.30
	>10 percent medium-severity alligator cracking and/or <10 percent high-severity alligator cracking and/or >10 percent medium- and high-severity transverse cracking	0.14 to 0.20
	>10 percent high-severity alligator cracking and/or >10 percent high-severity transverse cracking	0.08 to 0.15
Stabilized Base	Little or no alligator cracking and/or only low-severity transverse cracking	0.20 to 0.35
	<10 percent low-severity alligator cracking and/or <5 percent medium- and high-severity transverse cracking	0.15 to 0.25
	>10 percent low-severity alligator cracking and/or <10 percent medium-severity alligator cracking and/or >5–10 percent medium- and high-severity transverse cracking	0.15 to 0.20
	>10 percent medium-severity alligator cracking and/or <10 percent high-severity alligator cracking and/or >10 percent medium- and high-severity transverse cracking	0.10 to 0.20
	>10 percent high-severity alligator cracking and/or >10 percent high-severity transverse cracking	0.08 to 0.15
Granular Base or Subbase	No evidence of pumping, degradation, or contamination by fines	0.10 to 0.14
	Some evidence of pumping, degradation, or contamination by fines	0.00 to 0.10

SN_{eff} = Effective structural number of the existing pavement, from Step 7

The thickness of overlay determined from the above relationship should be reasonable when the overlay is required to correct a structural deficiency. See Section 5.2.17 for discussion of factors which may result in unreasonable overlay thicknesses.

5.4.6 Surface Milling

If the AC pavement is to be milled prior to overlay, the depth of milling must be reflected in the SN_{eff} analyses. No adjustment need be made to SN_{eff} values determined by NDT if the depth of milling does not exceed the minimum necessary to remove surface ruts. If a greater depth is milled, the NDT-determined SN_{eff} may be reduced by an amount equal to the depth milled times a structural coefficient for the AC surface based on the condition survey.

5.4.7 Shoulders

See Section 5.2.10 for guidelines.

Table 5.3. Worksheet for Determination of SN_{eff} for AC Pavement

(1) NDT Method For SN_{eff} For AC Pavement:

Total thickness of all pavement layers above subgrade, D = _____ inches

Backcalculated subgrade resilient modulus, M_R = _____ psi

Backcalculated effective pavement modulus, E_p = _____ psi

$$SN_{eff} = 0.0045D \sqrt[3]{E_p} = _____$$

(2) Condition Survey Method For SN_{eff} For AC Pavement:

Thickness of AC surface, D_1 = _____ inches

Structural coefficient of AC surface, a_1, based on condition survey and coring data = _____

Thickness of base, D_2 = _____ inches

Structural coefficient of base, a_2, based on condition survey, material inspection, and testing = _____

Drainage coefficient of base, m_2 = _____

Thickness of subbase, D_3, if present = _____ inches

Structural coefficient of subbase, a_3, based on condition survey, material inspection, and testing = _____

Drainage coefficient of subbase, m_3 = _____

$$SN_{eff} = a_1D_1 + a_2D_2m_2 + a_3D_3m_3 = _____$$

5.4.8 Widening

See Section 5.2.16 for guidelines.

5.5 AC OVERLAY OF FRACTURED PCC SLAB PAVEMENT

This section covers the design of AC overlays placed on PCC pavements after they have been fractured by any of the following techniques: break/seat, crack/seat or rubblize/compact.

> Break/seat consists of breaking a JRCP into pieces larger than about one foot, rupturing the reinforcement or breaking its bond with the concrete, and seating the pieces firmly into the foundation.
>
> Crack/seat consists of cracking a JPCP into pieces typically one to three feet in size and seating the pieces firmly into the foundation.
>
> Seating typically consists of several passes of a 35- to 50-ton rubber-tired roller over a cracked or broken slab.
>
> Rubblize/compact consists of completely fracturing any type of PCC slab (JRCP, JPCP, or CRCP) into pieces smaller than one foot and then compacting the layer, typically with two or more passes of a 10-ton vibratory roller.

The following construction tasks are involved in the placement of an AC overlay on a fractured PCC slab pavement:

(1) Removing and replacing areas that will result in uneven support after fracturing
(2) Making subdrainage improvements if needed

Table 5.3. Worksheet for Determination of SN_{eff} for AC Pavement (continued)

(3) **Remaining Life Method For SN_{eff} for AC Pavement:**

Past 18-kip ESALs in design lane since construction, N_p = _____

18-kip ESALs to failure of existing design, $N_{1.5}$ = _____

$$RL = 100\left[1 - \left(\frac{N_p}{N_{1.5}}\right)\right] = _____$$

Condition factor, CF (Figure 5.2) = _____

Thickness of AC surface, D_1 = _____ inches

Structural coefficient of AC surface, a_1, if newly constructed = _____

Thickness of base, D_2 = _____ inches

Structural coefficient of base, a_2, if newly constructed = _____

Thickness of subbase, D_3, if present = _____ inches

Structural coefficient of subbase, a_3, if newly constructed = _____

$$SN_o = a_1D_1 + a_2D_2m_2 + a_3D_3m_3 = _____$$

$$SN_{eff} = CF * SN_o = _____$$

---

(3) Breaking and seating, crack and seating or rubblizing the PCC slab and rolling to seat or compact
(4) Constructing widening if needed
(5) Applying a tack or prime coat
(6) Placing the AC overlay (including a reflection crack control treatment if needed)

5.5.1 Feasibility

Break/seat, crack/seat and rubblizing techniques are used to reduce the size of PCC pieces to minimize the differential movements at existing cracks and joints, thereby minimizing the occurrence and severity of reflection cracks. The feasibility of each technique is described below.

Rubblizing can be used on all types of PCC pavements in any condition. It is particularly recommended for reinforced pavements. Fracturing the slab into pieces less than 12 inches reduces the slab to a high-strength granular base. Recent field testing of several rubblized projects showed a wide range in backcalculated modulus values among different projects, from less than 100,000 psi to several hundred thousand psi (*16, 17, 18*), and within-project coefficients of variation of as much as 40 percent (*16, 18*).

Crack and seat is used only with JPCP and involves cracking the slab into pieces typically one to three feet in size. Recent field testing of several cracked and seated JPCP projects showed a wide range in backcalculated modulus values among different projects, from a few hundred thousand psi to a few million psi (*16, 19, 20, 21, 22*), and within-project coefficients of variation of 40 percent or more (*16*). Reference 16 recommends that to avoid reflection cracking no more than 5 percent of the fractured slab have a modulus greater than 1 million psi. Effective slab cracking techniques are necessary in order to satisfy this criterion for crack/seat of JPCP.

Break/seat is used only with JRCP and includes the requirement to rupture the reinforcement steel across each crack, or break its bond with the concrete. If the reinforcement is not ruptured and its bond with the concrete is not broken, the differential movements at working joints and cracks will not be reduced and reflection cracks will occur. Recent field testing of several break/seat projects showed a wide range in backcalculated modulus values ranging from a few hundred thousand psi to several million psi (*16, 18, 19, 22*), and within-project coefficients of variation of 40 percent or more (*16, 18*). The wide range in backcalculated moduli reported for break and seat projects suggests a lack of consistency in the technique as performed with past construction equipment. Even though cracks are observed, the JRCP frequently retains a substantial degree of slab action because of failure to either rupture the reinforcing steel or break its bond with the concrete. This may also be responsible for the inconsistency of this technique in reducing reflection cracking. More effective breaking equipment may overcome this problem. This design procedure assumes that the steel will be ruptured or that its bond to the concrete will be broken through an aggressive break/seat process, and that this will be verified in the field through deflection testing before the overlay is placed. The use of rubblization is recommended for JRCP due to its ability to break slab continuity.

These slab fracturing techniques are generally more cost-effective on more deteriorated concrete pavements than on less deteriorated concrete pavements. This is due to the trade-off between the reduction in the amount of pre-overlay repair required for working cracks and deteriorated joints, and the cost of slab fracturing and increased overlay thickness required (*1, 22*).

5.5.2 Pre-overlay Repair

The amount of preoverlay repair needed for break/seat, crack/seat and rubblized projects is not clear. Most projects done prior to 1991 have not included a significant amount of repair. However, the recommended approach is to repair any condition that may provide nonuniform support after the fracturing process so that it will not rapidly reflect through the AC overlay. Also, some AC leveling may be needed for settled areas before the overlay is placed.

5.5.3 Reflection Crack Control

Slab fracturing techniques were developed as methods of reflection crack control. When properly constructed, the crack/seat and rubblizing methods are reasonably effective and should require no additional crack control treatment. However, care must be exercised to assure uniform cracking or rubblizing across the slab width and to firmly seat the cracked slab or compact the rubble. At least one agency that has used crack/seat of JPCP successfully for several years specifies that a fabric be placed in the overlay to aid in controlling reflection cracking. For break/seat of JRCP, reflective cracks will develop if the steel reinforcement is not ruptured and its bond to the concrete is not broken, and if this cannot be guaranteed, it is recommended that JRCP be rubblized.

5.5.4 Subdrainage

See Section 5.2.4 for guidelines. Rubblizing PCC pavement produces fines which may clog the filter materials placed in edge drains. This should be considered in the design of the filter materials. If longitudinal subdrains are to be installed, this should be done prior to fracturing the slab.

5.5.5 Thickness Design

The required thickness of the overlay is a function of the structural capacity required to meet future traffic demands and the structural capacity of the existing slab after fracturing. The required thickness is determined by the following equation:

$$SN_{ol} = a_{ol} * D_{ol} = SN_f - SN_{eff}$$

where

SN_{ol} = Required overlay structural number
a_{ol} = Structural coefficient for the AC overlay
D_{ol} = Required overlay thickness, inches
SN_f = Structural number required to carry future traffic
SN_{eff} = Effective structural number of the existing pavement after fracturing

The required overlay thickness is determined through the following design steps.

Step 1: Existing pavement design and construction.

(1) Thickness and material type of each pavement layer
(2) Available subgrade soil information (from construction records, soil surveys, county agricultural soils reports, etc.)

Step 2: Traffic analysis.

(1) Predicted future 18-kip ESALs in the design lane over the design period (N_f)

Use flexible pavement equivalency factors. If available future traffic estimates are in terms of rigid pavement ESALs, they must be converted to flexible pavement ESALs by dividing by 1.5 (e.g., 15 million rigid pavement ESALs approximately equal 10 million flexible pavement ESALs).

Step 3: Condition survey.

Condition survey data are not used in the determination of overlay thickness. However, condition survey data should be used to determine whether or not fracturing is cost-effective compared to other types of rehabilitation.

Step 4: Deflection testing (recommended).

Deflection measurements are used only for the determination of the design subgrade resilient modulus. Deflections should be measured on the bare PCC slab surface (prior to fracturing) at midslab locations that are not cracked. A heavy-load deflection device (e.g., Falling Weight Deflectometer) and a load magnitude of approximately 9,000 pounds are recommended. ASTM D 4694 and D 4695 provide additional guidance on deflection testing. A deflection measurement at a distance of approximately 4 feet from the center of load is needed.

(1) Subgrade resilient modulus (M_R). At sufficiently large distances from the load, deflections measured at the pavement surface are due to subgrade deformation only, and are also independent of the size of the load plate. This permits the backcalculation of the subgrade resilient modulus from a single deflection measurement and load magnitude, using the following equation.

$$M_R = \frac{0.24P}{d_r r}$$

where

M_R = backcalculated subgrade resilient modulus, psi
P = applied load, pounds
d_r = deflection at a distance r from the center of the load, inches
r = distance from center of load, inches

The deflection used to backcalculate the subgrade modulus must be measured far enough away that it provides a good estimate of the subgrade modulus, independent of the effects of any layers above, but also close enough that it is not too small to measure accurately. The minimum distance may be determined from the following relationship:

$$r \geq 0.7 a_e$$

where

$$a_e = \sqrt{\left[a^2 + \left(D \sqrt[3]{\frac{E_p}{M_R}} \right)^2 \right]}$$

a_e = radius of the stress bulb at the subgrade-pavement interface, inches
a = NDT load plate radius, inches
D = total thickness of pavement layers above the subgrade, inches
E_p = effective modulus of all pavement layers above the subgrade, psi (described below)

Before the backcalculated M_R value is used in design, it must be adjusted to make it consistent with the value used in the AASHTO flexible pavement design equation. An adjustment may also be needed to account for seasonal effects. These adjustments are described in Step 6.

(2) Effective modulus of the pavement (E_p). If the subgrade resilient modulus and total thickness of all layers above the subgrade are known or assumed, the effective modulus of the entire pavement structure (all pavement layers above the subgrade) may be determined from the deflection measured at the center of the load plate using the following equation:

$$d_0 = 1.5 pa \left\{ \frac{1}{M_R \sqrt{1 + \left(\frac{D}{a} \sqrt[3]{\frac{E_p}{M_R}} \right)^2}} + \frac{\left[1 - \frac{1}{\sqrt{1 + \left(\frac{D}{a} \right)^2}} \right]}{E_p} \right\}$$

where

d_0 = deflection measured at the center of the load plate, inches
p = NDT load plate pressure, psi
a = NDT load plate radius, inches
D = total thickness of pavement layers above the subgrade, inches
M_R = subgrade resilient modulus, psi
E_p = effective modulus of all pavement layers above the subgrade, psi

For a load plate radius of 5.9 inches, Figure 5.5 may be used to determine the ratio E_p/M_R, and E_p may then be determined for a known or assumed value of M_R.

Deflection measurements are also useful after the break/seat or crack/seat operations to insure that the slab has been sufficiently fractured (*16*).

Step 5: Coring and material testing.

(1) *Resilient modulus of subgrade.* If deflection testing is not performed, laboratory testing of samples of the subgrade may be conducted to determine its resilient modulus using AASHTO T 292-91 I with a deviator stress of 6 psi to match the deviator stress used in establishing the 3,000 psi for the AASHO Road Test soil that is incorporated into the flexible design equation. Alternatively, other tests such as R value, CBR or soil classification tests could be conducted and approximate correlations used to estimate resilient modulus. Use of the estimating equation $M_R = 1500 * CBR$ may produce a value that is too large for use in this design procedure. The relationships found in Appendix FF, Figure FF-6 may be more reasonable.

(2) *Samples of base layers* should be examined to assess degradation and contamination by fines.

Step 6: Determination of required structural number for future traffic (SN_f).

(1) Effective design subgrade resilient modulus. Determine by one of the following methods:

(a) Laboratory testing as described in Step 5.
(b) Backcalculation from deflection data. (NOTE: this value must be adjusted to be consistent with the value used in the AASHTO flexible pavement design equation as described below.)
(c) A very approximate estimate can be made using available soil information and relationships developed from resilient modulus studies. However, if as-built records are used, it should be noted that the resilient modulus may have changed since construction due to changes in moisture content or other factors.

Regardless of the method used, the effective design subgrade resilient modulus must be (1) representative of the effects of seasonal variation and (2) consistent with the resilient modulus value used to represent the AASHO Road Test soil. A seasonal adjustment, when needed, may be made in accordance with the procedures described in Part II, Section 2.3.1. M_R values backcalculated from deflections must be adjusted to make the values consistent with the laboratory-measured value used for the AASHO Road Test soil in the development of the flexible pavement design equation. For conventional AC pavements, it was recommended that backcalculated M_R values be multiplied by a correction factor $C = 0.33$ for use in determination of SN_f for design purposes when a FWD load of approximately 9,000 pounds is used (*9*). However, because subgrade stresses are much lower under a PCC slab than under a flexible pavement, it is recommended that a smaller correction factor, $C = 0.25$, be used to provide a better estimate of the subgrade M_R. This value should be evaluated and adjusted if needed by user agencies for their soil and deflection measurement equipment. The following design M_R is recommended for use in determining the SN_f for fractured slabs when deflection testing is done on top of the PCC slab:

$$\text{Design } M_R = C \left(\frac{0.24P}{d_r r}\right)$$

where recommended $C = 0.25$.

NOTE also that the presence of a very stiff layer (e.g., bedrock) within about 15 feet of the top of the subgrade may cause the back-calculated M_R to be high. When such a condition exists, a value less than 0.25 for C may be warranted (8, 9).

The designer is cautioned against using a value of M_R that is too large. The value of M_R selected for design is extremely critical to the overlay thickness. The use of a value greater than 3,000 psi is an indication that the soil is stiffer than the silty-clay A-6 soil at the Road Test site, and consequently will provide increased support and extended pavement life.

(2) *Design PSI loss.* PSI immediately after overlay (P1) minus PSI at time of next rehabilitation (P2).

(3) *Overlay design reliability R (percent).* See Part I, Section 4.2, Part II, Table 2.2, and Part III, Section 5.2.15.

(4) *Overall standard deviation S_o for flexible pavement.* See Part I, Section 4.3.

Compute SN_f for the above design inputs using the flexible pavement design equation or nomograph in Part II, Figure 3.1. When designing an overlay thickness for a uniform pavement section, mean input values must be used. When designing an overlay thickness for specific points along the project, the data for that point must be used. A worksheet for determining SN_f is provided in Table 5.4.

Step 7: Determination of effective structural number (SN_{eff}) of the existing fractured slab pavement.

SN_{eff} is determined by component analysis using the structural number equation:

$$SN_{eff} = a_2 D_2 m_2 + a_3 D_3 m_3$$

where

D_2, D_3 = thicknesses of fractured slab and base layers
a_2, a_3 = corresponding structural layer coefficients
m_2, m_3 = drainage coefficients for fractured PCC and granular subbase

See Part II, Table 2.4, for guidance in determining the drainage coefficients. Due to lack of information on drainage characteristics of fractured PCC, a default value of 1.0 for m_2 is recommended. In selecting values for m_3, note that the poor drainage situation for the base and subbase at the AASHO Road Test would be given drainage coefficient values of 1.0.

Suggested layer coefficients for fractured slab pavements are provided in Table 5.5. Each agency should adopt its own set of layer coefficient values for fractured slabs keyed to its construction results on its pavements.

Since the layer coefficient represents the overall performance contribution of that layer, it is likely that it is not related solely to the modulus of that layer, but to other properties as well, such as the load transfer capability of the pieces. The large variability of layer moduli within a project is also of concern. This extra variability should ideally be expressed in an increased overall standard deviation in designing for a given reliability level.

A worksheet for determination of SN_{eff} is provided in Table 5.6.

Step 8: Determination of overlay thickness.

The thickness of AC overlay is computed as follows:

$$D_{ol} = \frac{SN_{ol}}{a_{ol}} = \frac{(SN_f - SN_{eff})}{a_{ol}}$$

where

SN_{ol} = Required overlay structural number
a_{ol} = Structural coefficient for the AC overlay
D_{ol} = Required AC overlay thickness, inches
SN_f = Structural number determined in Step 6
SN_{eff} = Effective structural number of the existing pavement, from Step 7

The thickness of overlay determined from the above relationship should be reasonable when the overlay is required to correct a structural deficiency. See Section 5.2.17 for discussion of factors which may result in unreasonable overlay thicknesses.

5.5.6 Shoulders

See Section 5.2.10 for guidelines.

5.5.7 Widening

See Section 5.2.16 for guidelines.

Table 5.4. Worksheet for Determination of SN_f for Fractured Slab Pavements

TRAFFIC:

Future 18-kip ESALs in design lane over
design period, N_f = _____

EFFECTIVE ROADBED SOIL RESILIENT MODULUS:

Design resilient modulus, M_R = _____ psi

(Adjusted for consistency with flexible pavement model and for seasonal variations. Typical design M_R is 2,000 to 10,000 psi for fine-grained soils, 10,000 to 20,000 for coarse-grained soils. The AASHO Road Test soil value used in the flexible pavement design equation was 3,000 psi.)

SERVICEABILITY LOSS:

Design PSI loss (P1 − P2) (1.2 to 2.5)   = _____

DESIGN RELIABILITY:

Overlay design reliability, R (80 to 99 percent) = _____ percent

Overall standard deviation, S_o (typically 0.49) = _____

FUTURE STRUCTURAL CAPACITY:

Required structural number for future traffic is determined from flexible pavement design equation or nomograph in Part II, Figure 3.1.

SN_f = _____

Table 5.5. Suggested Layer Coefficients for Fractured Slab Pavements

MATERIAL	SLAB CONDITION	COEFFICIENT
Break/Seat JRCP	Pieces greater than one foot with ruptured reinforcement or steel/concrete bond broken	0.20 to 0.35
Crack/Seat JPCP	Pieces one to three feet	0.20 to 0.35
Rubblized PCC (any pavement type)	Completely fractured slab with pieces less than one foot	0.14 to 0.30
Base/subbase granular and stabilized	No evidence of degradation or intrusion of fines	0.10 to 0.14
	Some evidence of degradation or intrusion of fines	0.00 to 0.10

Table 5.6. Worksheet for Determination of SN_{eff} for Break/Seat, Crack/Seat and Rubblized Pavements

Thickness of break/crack or rubblized PCC, D_2	= _____	inches
Structural coefficient of break/crack/seat or rubblized PCC, a_2	= _____	
Drainage coefficient of fractured slab, m_2 (1.0 recommended)	= _____	
Thickness of subbase, D_3, if present	= _____	inches
Structural coefficient of subbase, a_3	= _____	
Drainage coefficient of subbase, m_3	= _____	

$$SN_{eff} = a_2 D_2 m_2 + a_3 D_3 m_3 = _____$$

5.6 AC OVERLAY OF JPCP, JRCP, AND CRCP

This section covers the design of AC overlays of existing JPCP, JRCP, or CRCP. This section may also be used to design an AC overlay if a previous AC overlay is completely removed.

Construction of an AC overlay over JPCP, JRCP, or CRCP consists of the following major activities:

(1) Repairing deteriorated areas and making subdrainage improvements (if needed).
(2) Constructing widening (if needed).
(3) Applying a tack coat.
(4) Placing the AC overlay, including a reflection crack control treatment (if needed).

5.6.1 Feasibility

An AC overlay is a feasible rehabilitation alternative for PCC pavements except when the condition of the existing pavement dictates substantial removal and replacement. Conditions under which an AC overlay would not be feasible include:

(1) The amount of deteriorated slab cracking and joint spalling is so great that complete removal and replacement of the existing surface is dictated.
(2) Significant deterioration of the PCC slab has occurred due to severe durability problems (e.g., "D" cracking or reactive aggregates).
(3) Vertical clearance at bridges is inadequate for required overlay thickness. This may be addressed by reducing the overlay thickness under the bridges (although this may result in early failure at these locations), by raising the bridges, or by reconstructing the pavement under the bridges. Thicker AC overlays may also necessitate raising signs and guardrails, as well as increasing side slopes and extending culverts. Sufficient right-of-way must be available or obtainable to permit these activities.

5.6.2 Preoverlay Repair

The following types of distress in JPCP, JRCP, and CRCP should be repaired prior to placement of an AC overlay.

Distress Type	Repair Type
Working cracks	Full-depth repair or slab replacement
Punchouts	Full-depth PCC repair
Spalled joints	Full-depth or partial-depth repair
Deteriorated repairs	Full-depth repair
Pumping/faulting	Edge drains
Settlements/heaves	AC level-up, slab jacking, or localized reconstruction

Full-depth repairs and slab replacements in JPCP and JRCP should be PCC, dowelled or tied to provide load transfer across repair joints. Some agencies have placed full-depth AC repairs in JPCP and JRCP prior to an AC overlay. However, this has often resulted in rough spots in the overlay, opening of nearby joints and cracks, and rapid deterioration of reflection cracks at AC patch boundaries. (See Part III, Section 4.3.1 and References 1 and 3.)

Full-depth repairs in CRCP should be PCC and should be continuously reinforced with steel which is tied or welded to reinforcing steel in the existing slab to provide load transfer across joints and slab continuity. Full-depth AC repairs should not be used in CRCP prior to placement of an AC overlay, and any existing AC patches in CRCP should be removed and replaced with continuously reinforced PCC. Guidelines on repairs are provided in References 1 and 3.

Installation of edge drains, maintenance of existing edge drains, or other subdrainage improvement should be done prior to placement of the overlay if a subdrainage evaluation indicates a need for such an improvement.

Pressure relief joints should be placed only at fixed structures, and not at regular intervals along the pavement. The only exception to this is where reactive aggregate has caused expansion of the slab. On heavily trafficked routes, pressure relief joints should be of heavy-duty design with dowels (3). If joints contain significant incompressibles, they should be cleaned and resealed prior to placement of the overlay.

5.6.3 Reflection Crack Control

The basic mechanism of reflection cracking is strain concentration in the overlay due to movement in the vicinity of joints and cracks in the existing pavement. This movement may be bending or shear induced by loads, or may be horizontal contraction induced by temperature changes. Load-induced movements are influenced by the thickness of the overlay and the thickness and stiffness of the existing pavement. Temperature-induced movements are influenced by daily and seasonal temperature variations, the coefficient of thermal expansion of the existing pavement, and the spacing of joints and cracks.

In an AC overlay of JPCP or JRCP, reflection cracks typically develop relatively soon after the overlay is placed (often in less than a year). The rate at which they deteriorate depends on the factors listed above as well as the traffic level. Thorough repair of deteriorated joints and working cracks with full-depth dowelled or tied PCC repairs reduces the rate of reflection crack occurrence and deterioration, so long as good load transfer is obtained at the full-depth repair joints. Other preoverlay repair efforts which will discourage reflection crack occurrence and subsequent deterioration include subdrainage improvement, subsealing slabs which have lost support, and restoring load transfer at joints and cracks with dowels grouted in slots.

A variety of reflection crack control measures have been used in attempts to control the rates of reflection crack occurrence and deterioration. Any one of the following treatments may be employed in an effort to control reflection cracking in an AC overlay of JPCP or JRCP:

(1) *Sawing and sealing joints in the AC overlay* at locations coinciding with joints in the underlying JPCP or JRCP. This technique has been very successful when applied to AC overlays of jointed PCC pavements when the sawcut matches the joint or straight crack within an inch.

(2) *Increasing AC overlay thickness.* Reflection cracks will take more time to propagate through a thicker overlay and deteriorate more slowly.

(3) *Placing a bituminous-stabilized granular interlayer (large-sized large stone), prior to or in combination with placement of the AC overlay* has been effective.

(4) *Placing a synthetic fabric or a stress-absorbing interlayer prior to or within the AC overlay.* The effectiveness of this technique is questionable.

(5) *Rubblizing and compacting JPCP, JRCP, or CRCP* prior to placement of the AC overlay. This technique reduces the size of PCC pieces to a maximum of about 12 inches and essentially reduces the slab to a high-strength granular base course. See Section 5.5 for the design procedure for AC overlays of rubblized PCC pavement.

(6) *Cracking and seating JPCP or breaking and seating JRCP* prior to placement of the AC overlay. This technique reduces the size of PCC pieces and seats them in the underlying base, which reduces horizontal (and possibly vertical) movements at cracks. See Section 5.5 for the design procedure for AC overlays of crack/seat JPCP and break/seat JRCP.

Reflection cracking can have a considerable (often controlling) influence on the life of an AC overlay of JPCP or JRCP. Deteriorated reflection cracks detract from a pavement's serviceability and also require frequent maintenance, such as sealing, milling, and patching. Reflection cracks also permit water to enter the pavement structure, which may result in loss of bond between the AC and PCC, stripping in the AC, progression of "D" cracking or reactive aggregate distress in PCC slabs with these durability problems, and softening of the base and subgrade. For this reason, reflection cracks should be sealed as soon as they appear and resealed periodically throughout the life of the overlay. Sealing low-severity reflection cracks may also be effective in retarding their progression to medium and high severity levels.

With an AC overlay of CRCP, permanent repair of punchouts and working cracks with tied or welded reinforced PCC full-depth repairs will delay the occurrence and deterioration of reflection cracks. Improving subdrainage conditions and subsealing in areas where the slab has lost support will also discourage reflection crack occurrence and deterioration. Reflection crack control treatments are not necessary for AC overlays of CRCP, except for longitudinal joints, as long as continuously reinforced PCC repairs are used to repair deteriorated areas and cracks.

5.6.4 Subdrainage

See Section 5.2.4 for guidelines.

5.6.5 Thickness Design

If the overlay is being placed for some functional purpose such as roughness or friction, a minimum thickness overlay that solves the functional problem should be placed. If the overlay is being placed for the purpose of structural improvement, the required thickness of the overlay is a function of the structural capacity required to meet future traffic demands and the structural capacity of the existing pavement. The required overlay thickness to increase structural capacity to carry future traffic is determined by the following equation.

$$D_{ol} = A(D_f - D_{eff})$$

where

D_{ol} = Required thickness of AC overlay, inches
A = Factor to convert PCC thickness deficiency to AC overlay thickness
D_f = Slab thickness to carry future traffic, inches
D_{eff} = Effective thickness of existing slab, inches

The A factor, which is a function of the PCC thickness deficiency, is given by the following equation, and is illustrated in Figure 5.9.

$$A = 2.2233 + 0.0099(D_f - D_{eff})^2 - 0.1534(D_f - D_{eff})$$

AC overlays of conventional JPCP, JRCP, and CRCP have been constructed as thin as 2 inches and as thick as 10 inches. The most typical thicknesses that have been constructed for highways are 3 to 6 inches.

The required overlay thickness may be determined through the following design steps. These design steps provide a comprehensive design approach that recommends testing the pavement to obtain valid design inputs. If it is not possible to conduct this testing (e.g., for a low-volume road), an approximate overlay design may be developed based upon visible distress observations by skipping Steps 4 and 5, and by estimating other inputs.

The overlay design can be done for a uniform section or on a point-by-point basis as described in Section 5.3.1.

Step 1: Existing pavement design.

(1) Existing slab thickness
(2) Type of load transfer (mechanical devices, aggregate interlock, CRCP)
(3) Type of shoulder (tied PCC, other)

Step 2: Traffic analysis.

(1) Past cumulative 18-kip ESALs in the design lane (N_p), for use in the remaining life method of D_{eff} determination only
(2) Predicted future 18-kip ESALs in the design lane over the design period (N_f)
    Use ESALs computed from rigid pavement load equivalency factors

Step 3: Condition survey.

The following distresses are measured during the condition survey for JPCP, JRCP, and CRCP. Sampling along the most heavily trafficked lane of the project may be used to estimate these quantities. Distress types and severities are defined in Reference 23. Deteriorated means medium or higher severity.

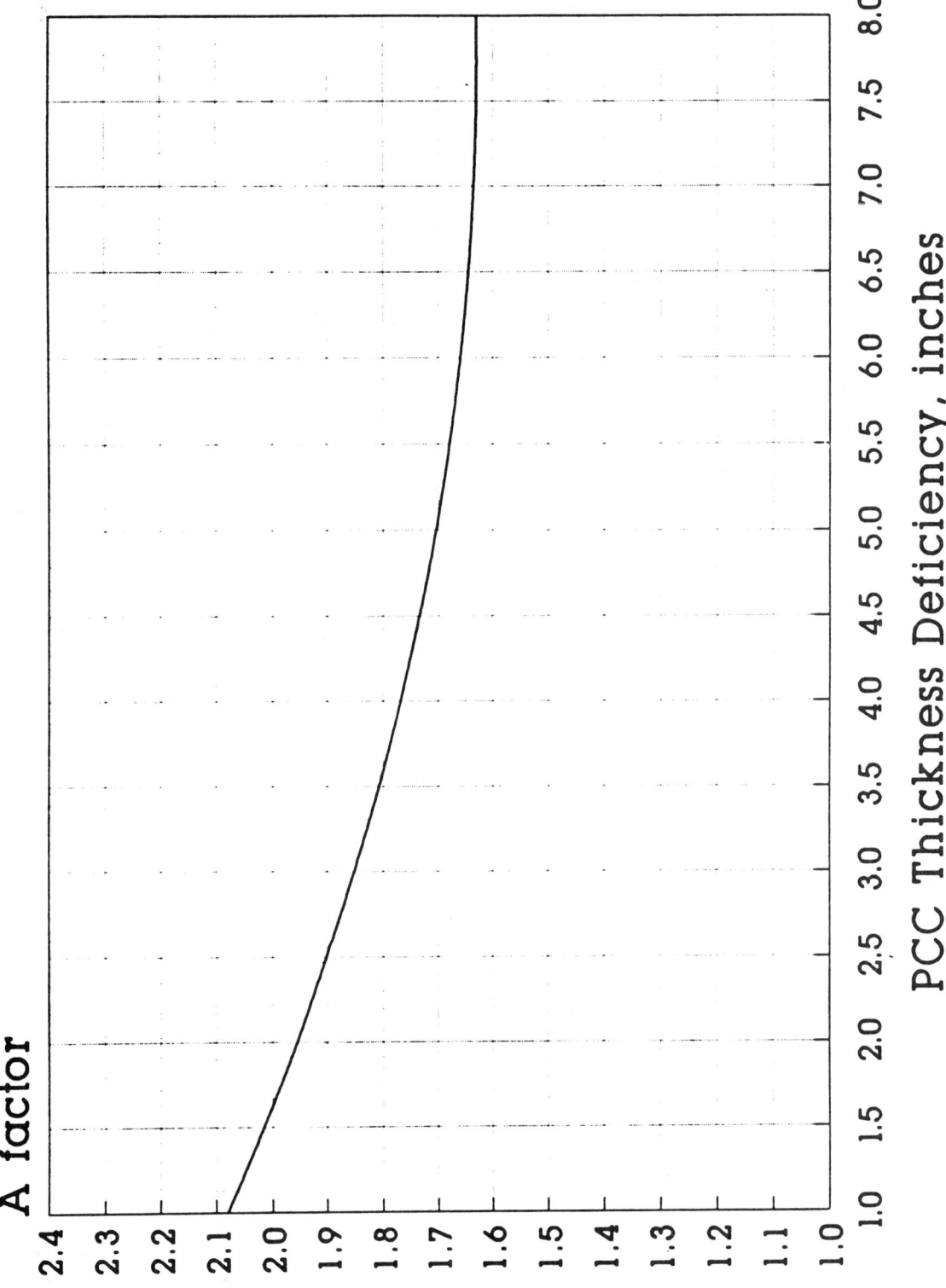

Figure 5.9. A Factor for Conversion of PCC Thickness Deficiency to AC Overlay Thickness

JPCP/JRCP:

(1) Number of deteriorated transverse joints per mile
(2) Number of deteriorated transverse cracks per mile
(3) Number of full-depth AC patches, exceptionally wide joints (greater than 1 inch), and expansion joints per mile (except at bridges)
(4) Presence and overall severity of PCC durability problems
   (a) "D" cracking: low severity (cracks only), medium severity (some spalling), high severity (severe spalling)
   (b) Reactive aggregate cracking: low, medium, high severity
(5) Evidence of faulting, or pumping of fines or water at joints, cracks, and pavement edge

CRCP:

(1) Number of punchouts per mile
(2) Number of deteriorated transverse cracks per mile
(3) Number of full-depth AC patches, exceptionally wide joints (greater than 1 inch) and expansion joints per mile (except at bridges)
(4) Number of existing and new repairs prior to overlay per mile
(5) Presence and general severity of PCC durability problems (NOTE: surface spalling of tight cracks where the underlying CRCP is sound should not be considered a durability problem.)
   (a) "D" cracking: low severity (cracks only), medium severity (some spalling), high severity (severe spalling)
   (b) Reactive aggregate cracking: low, medium, high severity
(6) Evidence of pumping of fines or water

Step 4: Deflection testing (strongly recommended).

Measure slab deflection basins along the project at an interval sufficient to adequately assess conditions. Intervals of 100 to 1,000 feet are typical. Measure deflections with sensors located at 0, 12, 24, and 36 inches from the center of load. Measure deflections in the outer wheel path. A heavy-load deflection device (e.g., Falling Weight Deflectometer) and a load magnitude of 9,000 pounds are recommended. ASTM D 4694 and D 4695 provide additional guidance on deflection testing. For each slab tested, backcalculate the effective k-value and the slab's elastic modulus using Figures 5.10 and 5.11 or a backcalculation program.

The AREA of each deflection basin is computed by the following equation. AREA will typically range from 29 to 32 for sound concrete.

$$\text{AREA} = 6 * \left[1 + 2 \left(\frac{d_{12}}{d_0} \right) + 2 \left(\frac{d_{24}}{d_0} \right) + \left(\frac{d_{36}}{d_0} \right) \right]$$

where

d_0 = deflection in center of loading plate, inches
d_i = deflections at 12, 24, and 36 inches from plate center, inches

(1) *Effective dynamic k-value.* Enter Figure 5.10 with d_0 and AREA to determine the effective dynamic k-value beneath each slab for a circular load radius of 5.9 inches and magnitude of 9,000 pounds. For loads within 2,000 pounds more or less, deflections may be scaled linearly to 9,000-pound deflections.

If a single overlay thickness is being designed for a uniform section, compute the mean effective dynamic k-value of the slabs tested in the uniform section.

(2) *Effective static k-value.*

Effective static k-value

= Effective dynamic k-value/2

The effective static k-value may need to be adjusted for seasonal effects using the approach presented in Part II, Section 3.2.1. However, the k-value can change substantially and have only a small effect on overlay thickness.

(3) *Elastic modulus of PCC slab (E).* Enter Figure 5.11 with AREA, proceed to the effective dynamic k-value curves, and determine a value for ED^3, where D is the slab thickness. Solve for E knowing the slab thickness, D. Typical slab E values range from 3 to 8 million psi. If a slab E value is obtained that is out of this range, an error may exist in the assumed slab thickness, the deflection basin may have been measured over a crack, or the PCC may be significantly deteriorated.

If a single overlay thickness is being designed for a uniform section, compute the mean E value of the slabs tested in the uniform section.

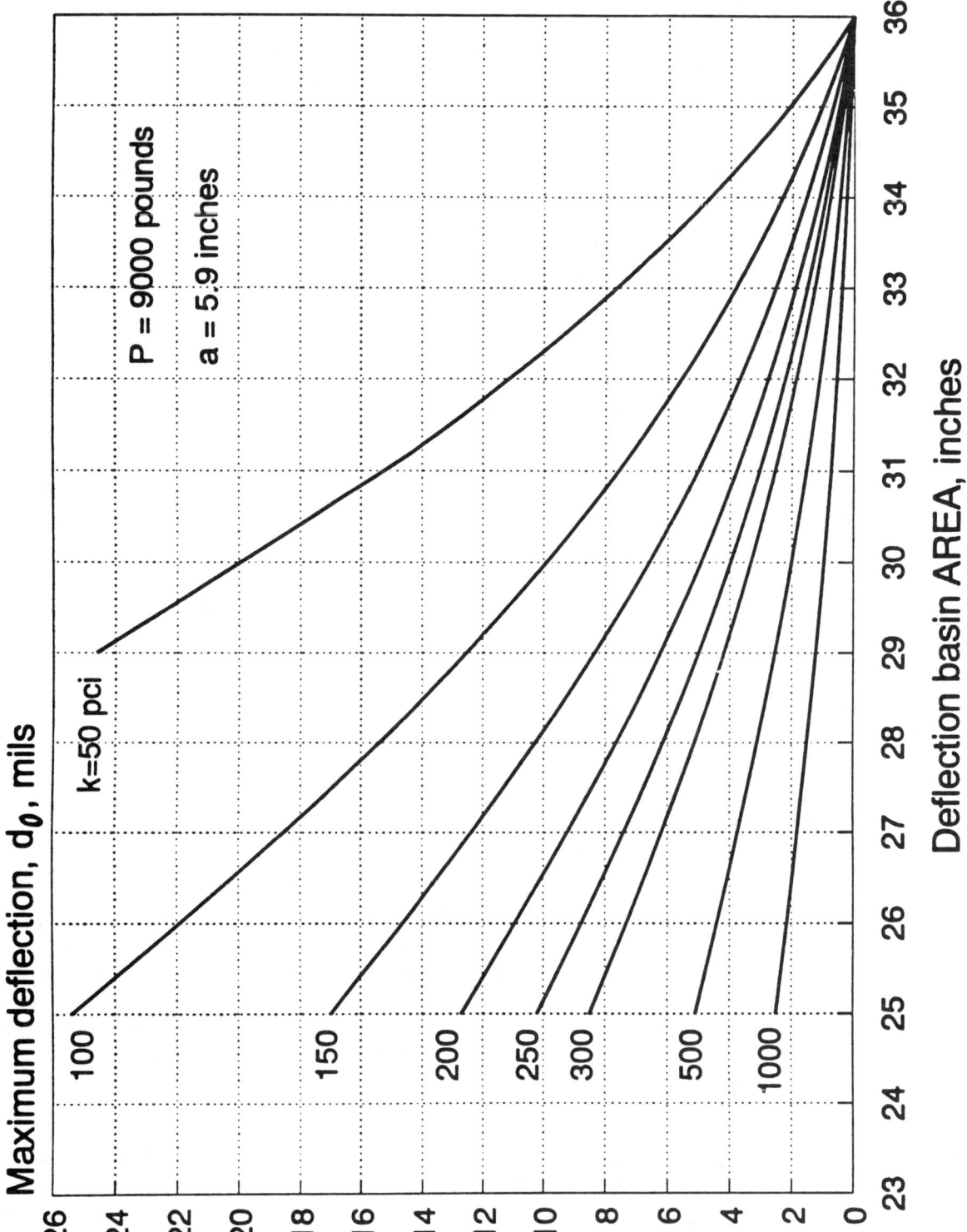

Figure 5.10. Effective Dynamic k-Value Determination from d_0 and AREA

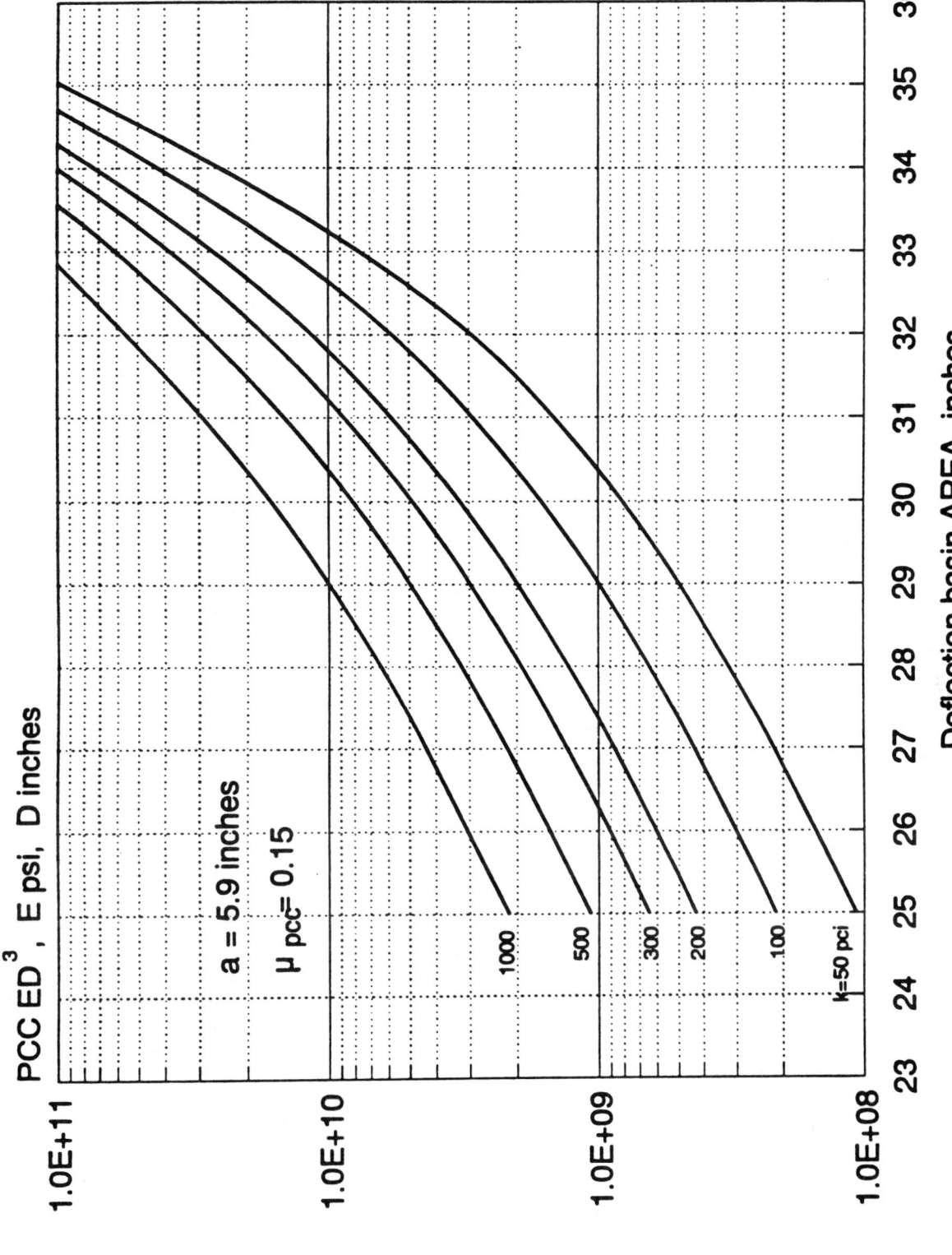

Figure 5.11. PCC Elastic Modulus Determination from k-Value, AREA, and Slab Thickness

Do not use any k-values or E values that appear to be significantly out of line with the rest of the data.

(4) *Joint load transfer.* For JPCP and JRCP, measure joint load transfer in the outer wheelpath at representative transverse joints. Do not measure load transfer when the ambient temperature is greater than 80°F. Place the load plate on one side of the joint with the edge of the plate touching the joint. Measure the deflection at the center of the load plate and at 12 inches from the center. Compute the deflection load transfer from the following equation.

$$\Delta LT = 100 * \left(\frac{\Delta_{ul}}{\Delta_l}\right) * B$$

where

ΔLT = deflection load transfer, percent
Δ_{ul} = unloaded side deflection, inches
Δ_l = loaded side deflection, inches
B = slab bending correction factor

The slab bending correction factor, B, is necessary because the deflections d_0 and d_{12}, measured 12 inches apart, would not be equal even if measured in the interior of a slab. An appropriate value for the correction factor may be determined from the ratio of d_0 to d_{12} for typical center slab deflection basin measurements, as shown in the equation below. Typical values for B are between 1.05 and 1.15.

$$B = \frac{d_{0\ center}}{d_{12\ center}}$$

If a single overlay thickness is being designed for a uniform section, compute the mean deflection load transfer value of the joints tested in the uniform section.

For JPCP and JRCP, determine the J load transfer coefficient using the following guidelines:

Percent Load Transfer	J
>70	3.2
50–70	3.5
<50	4.0

If the rehabilitation will include the addition of a tied concrete shoulder, a lower J factor may be appropriate. See Part II, Table 2.6.

For CRCP, use J = 2.2 to 2.6 for overlay design, assuming that working cracks are repaired with continuously reinforced PCC.

Step 5: *Coring and materials testing (strongly recommended).*

(1) *PCC modulus of rupture (S'_c).* Cut several 6-inch-diameter cores at midslab and test in indirect tension (ASTM C 496). Compute the indirect tensile strength (psi) of the cores. Estimate the modulus of rupture with the following equation.

$$S'_c = 210 + 1.02 IT$$

where

S'_c = modulus of rupture, psi
IT = indirect tensile strength of 6-inch-diameter cores, psi

Step 6: *Determination of required slab thickness for future traffic (D_f).*

The inputs to determine D_f for AC overlays of PCC pavements are representative of the existing slab and foundation properties. This is emphasized because it is the properties of the existing slab (i.e., elastic modulus, modulus of rupture, and load transfer) which will control the performance of the AC overlay.

(1) *Effective static k-value beneath existing PCC slab.* Determine from one of the following methods.
    (a) Backcalculate the effective dynamic k-value from deflection basins. Divide the effective dynamic k-value by 2 to obtain the effective static k-value. The effective static k-value may need to be adjusted for seasonal effects using the approach presented in Part II, Section 3.2.1.
    (b) Conduct plate load tests (ASTM D 1196) after slab removal at a few sites. This alternative is very costly and time-consuming and not often used. The static k-value obtained may need to be adjusted for seasonal effects (see Part II, Section 3.2.1).
    (c) Estimate from soils data and base type and thickness, using Figure 3.3 in Part

II, Section 3.2. This alternative is simple, but the static k-value obtained must be recognized as a rough estimate. The static k-value may need to be adjusted for seasonal effects (see Part II, Section 3.2.1).

(2) *Design PSI loss.* PSI immediately after overlay (P1) minus PSI at time of next rehabilitation (P2).

(3) *J, load transfer factor of existing PCC slab.* See Step 4.

(4) *PCC modulus of rupture of existing slab* determined by one of the following methods:
   (a) Estimated from indirect tensile strength measured from 6-inch-diameter cores as described in Step 5.
   (b) Estimated from the backcalculated E of slab using the following equation.

$$S'_c = 43.5 \left(\frac{E}{10^6}\right) + 488.5$$

where

S'_c = modulus of rupture, psi
E = backcalculated elastic modulus of PCC slab, psi

For CRCP, S'_c may be determined from the backcalculated E values only at points which have no cracks within the deflection basins.

(5) *Elastic modulus of existing PCC slab,* determined by one of the following methods:
   (a) Backcalculated from deflection measurements as described in Step 4.
   (b) Estimated from indirect tensile strength.

(6) *Loss of support of existing slab.* Joint corners that have loss of support may be identified using FWD deflection testing as described in Reference 2. CRCP loss of support may be determined by plotting a slab edge or wheelpath deflection profile and identifying locations with significantly high deflections. Existing loss of support can be corrected with slab stabilization. For overlay thickness design assume a fully supported slab, LS = 0.

(7) *Overlay design reliability, R (percent).* See Part I, Section 4.2, Part II, Table 2.2, and Part III, Section 5.2.15.

(8) *Overall standard deviation (S_o) for rigid pavement.* See Part I, Section 4.3.

(9) *Subdrainage capability of existing slab,* after subdrainage improvements, if any. See Part II, Table 2.5, as well as reference 5, for guidance in determining C_d. Pumping or faulting at joints and cracks determined in Step 3 is evidence that a subdrainage problem exists. In selecting this value, note that the poor subdrainage situation at the AASHO Road Test would be given a C_d of 1.0.

Compute D_f for the above design inputs using the rigid pavement design equation or nomograph in Part II, Figure 3.7. When designing an overlay thickness for a uniform pavement section, mean input values must be used. When designing an overlay thickness for specific points along the project, the data for that point must be used. A worksheet for determining D_f is provided in Table 5.7. Typical values of inputs are provided for guidance. Values outside these ranges should be used with caution.

Step 7: Determination of effective slab thickness (D_{eff}) of existing pavement.

Condition survey and remaining life procedures are presented.

D_{eff} From Condition Survey For PCC Pavements

The effective thickness of the existing slab (D_{eff}) is computed from the following equation:

$$D_{eff} = F_{jc} * F_{dur} * F_{fat} * D$$

where

D = existing PCC slab thickness, inches

(1) *Joints and cracks adjustment factor (F_{jc}).* This factor adjusts for the extra loss in PSI caused by deteriorated reflection cracks in the overlay that will result from any unrepaired deteriorated joints, cracks, and other discontinuities in the existing slab prior to overlay. A deteriorated joint or crack in the existing slab will rapidly reflect through an AC overlay and contribute to loss of serviceability. Therefore, it is recommended that all deteriorated joints and cracks (for non-"D" cracked or reactive aggregate related distressed pavements) and any other major discontinuities in the existing slab be full-depth repaired with dowelled or tied PCC repairs prior to overlay, so that F_{jc} = 1.00.

Table 5.7. Worksheet for Determination of D_f for JPCP, JRCP, and CRCP

SLAB:

Existing PCC slab thickness = _____ inches

Type of load transfer system: mechanical device, aggregate interlock, CRCP

Type of shoulder = tied PCC, other

PCC modulus of rupture (typically 600 to 800 psi) = _____ psi

PCC E modulus (3 to 8 million psi for sound PCC, <3 million for unsound PCC) = _____ psi

J load transfer factor (3.2 to 4.0 for JPCP, JRCP 2.2 to 2.6 for CRCP) = _____

TRAFFIC:

Future 18-kip ESALs in design lane over the design period (N_f) = _____

SUPPORT AND DRAINAGE:

Effective dynamic k-value = _____ psi/inch

Effective static k-value = Effective dynamic k-value/2 (typically 50 to 500 psi/inch) = _____ psi/inch

Subdrainage coefficient, C_d (typically 1.0 for poor subdrainage conditions) = _____

SERVICEABILITY LOSS:

Design PSI loss (P1 − P2) = _____

RELIABILITY:

Design reliability, R (80 to 99 percent) = _____ percent

Overall standard deviation, S_o (typically 0.39) = _____

FUTURE STRUCTURAL CAPACITY:

Required slab thickness for future traffic is determined from rigid pavement design equation or nomograph in Part II, Figure 3.7.

D_f = _____ inches

If it is not possible to repair all deteriorated areas, the following information is needed to determine F_{jc}, to increase the overlay thickness to account for the extra loss in PSI from deteriorated reflection cracks in the design lane:

Pavements with no "D" cracking or reactive aggregate distress:

Number of unrepaired deteriorated joints/mile

Number of unrepaired deteriorated cracks/mile

Number of unrepaired punchouts/mile

Number of expansion joints, exceptionally wide joints (greater than 1 inch), and full-depth, full-lane-width AC patches/mile

Note that tight cracks held together by reinforcement in JRCP or CRCP are not included. However, if a crack in JRCP or CRCP is spalled and faulted the steel has probably ruptured, and the crack should be considered as working. Surface spalling of CRCP cracks is not an indication that the crack is working.

The total number of unrepaired deteriorated joints, cracks, punchouts, and other discontinuities per mile in the design lane is used to determine the F_{jc} from Figure 5.12.

Pavements with "D" cracking or reactive aggregate deterioration:

These types of pavements often have deterioration at the joints and cracks from durability problems. The F_{dur} factor is used to adjust the overlay thickness for this problem. Therefore, when this is the case, the F_{jc} should be determined from Figure 5.12 only using those unrepaired deteriorated joints and cracks that are not caused by durability problems. If all of the deteriorated joints and cracks are spalling due to "D" cracking or reactive aggregate, then $F_{jc} = 1.0$. This will avoid adjusting twice with F_{jc} and F_{dur} factors.

(2) *Durability adjustment factor (F_{dur}).* This factor adjusts for an extra loss in PSI of the overlay when the existing slab has durability problems such as "D" cracking or reactive aggregate distress. Using condition survey data from Step 3, F_{dur} is determined as follows.

- 1.00: No sign of PCC durability problems
- 0.96–0.99: Durability cracking exists, but no spalling
- 0.88–0.95: Substantial cracking and some spalling exists
- 0.80–0.88: Extensive cracking and severe spalling exists

(3) *Fatigue damage adjustment factor (F_{fat}).* This factor adjusts for past fatigue damage that may exist in the slab. It is determined by observing the extent of transverse cracking (JPCP, JRCP) or punchouts (CRCP) that may be caused primarily by repeated loading. Use condition survey data from Step 3 and the following guidelines to estimate F_{fat} in the design lane.

- 0.97–1.00: Few transverse cracks/punchouts exist (none caused by "D" cracking or reactive aggregate distress)
  - JPCP: <5 percent slabs are cracked
  - JRCP: <25 working cracks per mile
  - CRCP: <4 punchouts per mile

- 0.94–0.96: A significant number of transverse cracks/punchouts exist (none caused by "D" cracking or reactive aggregate distress)
  - JPCP: 5–15 percent slabs are cracked
  - JRCP: 25–75 working cracks per mile
  - CRCP: 4–12 punchouts per mile

- 0.90–0.93: A large number of transverse cracks/punchouts exist (none caused by "D" cracking or reactive aggregate distress)
  - JPCP: >15 percent slabs are cracked
  - JRCP: >75 working cracks per mile
  - CRCP: >12 punchouts per mile

D_{eff} From Remaining Life For PCC Pavements

The remaining life of the pavement is given by the following equation:

$$RL = 100 \left[1 - \left(\frac{N_p}{N_{1.5}} \right) \right]$$

where

RL = remaining life, percent
N_p = total traffic to date, ESALs
$N_{1.5}$ = total traffic to pavement "failure," ESALs

$N_{1.5}$ may be estimated using the new pavement design equations or nomographs in Part II. To be consistent with the AASHO Road Test and the development of these equations, a "failure" PSI equal to 1.5 and a reliability of 50 percent are recommended.

D_{eff} is determined from the following equation:

$$D_{eff} = CF * D$$

where

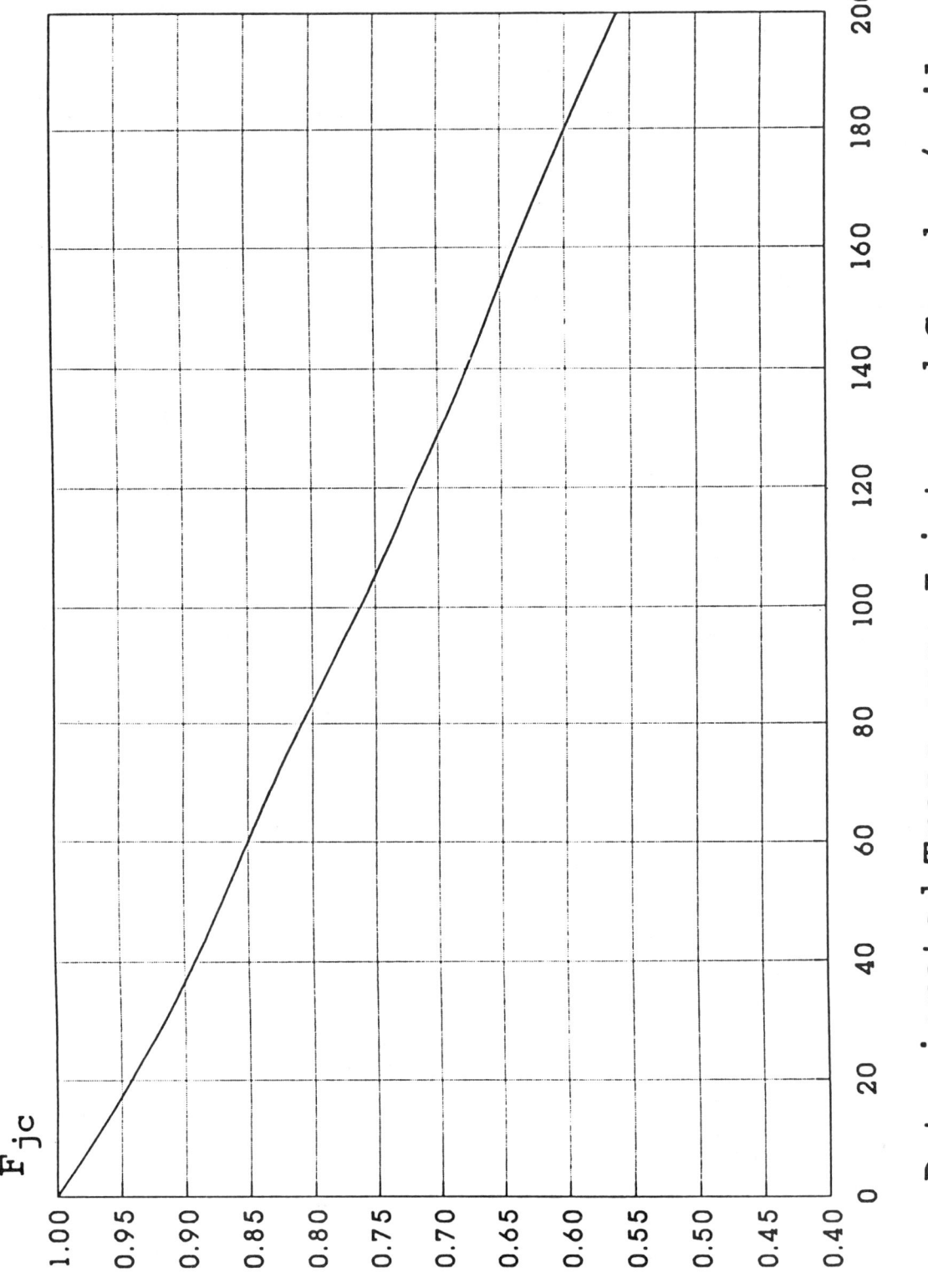

Figure 5.12. F_{jc} Adjustment Factor

Rehabilitation with Overlays

CF = condition factor determined from Figure 5.2
D = thickness of the existing slab

The designer should recognize that D_{eff} determined by this method does not reflect any benefit for pre-overlay repair. The estimate of D_{eff} obtained should thus be considered a lower limit value. The D_{eff} of the pavement will be higher if preoverlay repair of load-associated distress is done. This method for determining D_{eff} is not applicable without modification to pavements which have already received one or more overlays, even if the overlay has been or will be completely milled off.

A worksheet for determination of D_{eff} for JPCP, JRCP, and CRCP is provided in Table 5.8.

Step 8: Determination of Overlay Thickness.

The thickness of AC overlay is computed as follows:

$$D_{ol} = A(D_f - D_{eff})$$

where

D_{ol} = Required thickness of AC overlay, inches
A = Factor to convert PCC thickness deficiency to AC overlay thickness
D_f = Slab thickness determined in Step 6, inches
D_{eff} = Effective thickness of existing slab determined in Step 7, inches

The A factor, which is a function of the PCC thickness deficiency, is given by the following equation and is illustrated in Figure 5.9.

$$A = 2.2233 + 0.0099(D_f - D_{eff})^2 - 0.1534(D_f - D_{eff})$$

The thickness of overlay determined from the above relationship should be reasonable when the overlay is required to correct a structural deficiency. See Section 5.2.17 for discussion of factors which may result in unreasonable overlay thicknesses.

5.6.6 Shoulders

See Section 5.2.10 for guidelines.

5.6.7 Widening

See Section 5.2.16 for guidelines.

5.7 AC OVERLAY OF AC/JPCP, AC/JRCP, AND AC/CRCP

This section covers the design of AC overlays of existing AC/JPCP, AC/JRCP, or AC/CRCP. Although some pavements are newly constructed as AC/PCC, the vast majority of existing AC/PCC pavements are PCC pavements which have been overlaid with AC at least once.

Construction of an AC overlay of AC/JPCP, AC/JRCP, or AC/CRCP consists of the following major activities:

(1) Repairing deteriorated areas and making subdrainage improvements (if needed)
(2) Milling a portion of the existing AC surface
(3) Constructing widening (if needed)
(4) Applying a tack coat
(5) Placing the AC overlay, including a reflection crack control treatment (if needed)

5.7.1 Feasibility

An AC overlay is a feasible rehabilitation alternative for an AC/PCC pavement except when the condition of the existing pavement dictates substantial removal and replacement. Conditions under which another AC overlay would not be feasible include the following.

(1) The amount of deteriorated slab cracking and joint spalling is so great that complete removal and replacement of the existing surface is dictated.
(2) Significant deterioration of the PCC slab has occurred due to severe durability problems (e.g., "D" cracking or reactive aggregates).
(3) Vertical clearance at bridges is inadequate for required overlay thickness. This may be addressed by reducing the overlay thickness under the bridges (although this may result in early failure at these locations), by raising the bridges, or by reconstructing the pavement under the bridges. Thicker AC overlays may also necessitate raising signs and guardrails, as well as increasing side slopes and extending culverts. Sufficient right-of-way must be available or obtainable to permit these activities.

Table 5.8. Calculation of D_{eff} for AC Overlay of JPCP, JRCP, and CRCP in the Design Lane

Condition Survey Method:

F_{jc} Number of unrepaired deteriorated joints/mile = _____

Number of unrepaired deteriorated cracks/mile = _____

Number of unrepaired punchouts/mile = _____

Number of expansion joints, exceptionally wide joints (>1 inch) or AC full-depth patches/mile − _____

Total/mile = _____

F_{jc} = _____ (Figure 5.12)

(Recommended value 1.0, repair all deteriorated areas)

F_{dur} 1.00: No sign of PCC durability problems
0.96–0.99: Some durability cracking exists, but no spalling exists
0.88–0.95: Substantial cracking and some spalling exists
0.80–0.88: Extensive cracking and severe spalling exists

F_{dur} = _____

F_{fat} 0.97–1.00: Very few transverse cracks/punchouts exist
0.94–0.96: A significant number of transverse cracks/punchouts exist
0.90–0.93: A large number of transverse cracks/punchouts exist

F_{fat} = _____

$$D_{eff} = F_{jc} * F_{dur} * F_{fat} * D = _____$$

Remaining Life Method:

N_p = Past design lane ESALs = _____

$N_{1.5}$ = Design lane ESALs to P2 of 1.5 = _____

$$RL = 100 \left[1 - \left(\frac{N_p}{N_{1.5}} \right) \right] = _____$$

CF = _____ (Figure 5.2)

$$D_{eff} = CF * D = _____$$

When another AC overlay of an existing AC/JPCP, AC/JRCP, or AC/CRCP is being considered, the causes of the deterioration in the existing pavement should be carefully investigated. If the PCC slab is sound and in good condition but the existing AC layer is badly rutted or otherwise deteriorated, the AC should be thoroughly repaired or milled off. If, however, distress visible at the AC surface is predominantly a reflection of deterioration in the underlying PCC, the pavement must be repaired through the full depth of the AC and PCC. Otherwise, the distress will reflect rapidly through the new AC overlay. It is strongly recommended that coring and deflection testing be conducted to thoroughly investigate the causes and extent of deterioration in the existing pavement.

5.7.2 Pre-overlay Repair

The following types of distress in AC/JPCP, AC/JRCP, and AC/CRCP should be repaired prior to placement of an AC overlay.

Distress Type	Repair Type
Rutting	Milling
Deteriorated reflection cracks	Full-depth repair or slab replacement
Deteriorated repairs	Full-depth repair
Punchouts	Full-depth repair
Localized distress in AC only	AC patching
Localized distress in PCC	Full-depth repair
Pumping	Edge drains
Settlements/heaves	AC level-up, slab jacking, or localized reconstruction

In AC/JPCP and AC/JRCP, medium- and high-severity reflection cracks in the AC surface are evidence of working cracks, deteriorated joints, or failed repairs in the PCC slab, all of which should be full-depth repaired. Low-severity reflection cracks may exist at regular joints and full-depth repair joints. If these cracks are sealed and do not appear to be deteriorating at a significant rate, they might not warrant pre-overlay repair other than sealing.

In AC/CRCP, reflection cracks of all severities suggest the presence of working cracks, deteriorated construction joints, or failed repairs in the PCC slab, all of which should be repaired. Coring through selected reflection cracks should be conducted to assess the condition of the underlying pavement.

Coring should be conducted at areas of localized distress to determine whether they are caused by a problem in the AC mix or deterioration in the PCC (e.g., "D" cracking). In the latter case, the PCC may be deteriorated to a much greater extent than is evident at the AC surface. Additional coring or removal of portions of the AC may be necessary to select appropriate repair boundaries.

Full-depth repairs to AC/PCC pavements should match the existing cross-section, i.e., the PCC slab should be full-depth repaired with the same thickness of PCC, and then capped with AC to the same thickness as the existing AC. Full-depth repairs and slab replacements in AC/JPCP or AC/JRCP should be AC/PCC, dowelled or tied to provide load transfer across repair joints. Some agencies have placed full-depth AC repairs in AC/JPCP and AC/JRCP prior to an AC overlay. However, this has often resulted in rough spots in the new overlay, opening of nearby joints and cracks, and rapid deterioration of reflection cracks at AC patch boundaries.

AC/CRCP full-depth repairs should be AC/PCC and should be continuously reinforced with steel which is tied or welded to reinforcing steel in the existing slab, to provide load transfer across joints and slab continuity. Full-depth AC repairs should not be used in AC/CRCP prior to placement of an AC overlay, and any existing AC patches in AC/CRCP should be removed and replaced with AC over continuously reinforced PCC. Guidelines on repair are provided in References 1 and 3.

Installation of edge drains, maintenance of existing edge drains, or other subdrainage improvement should be done prior to placement of the overlay if a subdrainage evaluation indicates a need for such an improvement.

Pressure relief joints should be placed only at fixed structures, and not at regular intervals along the pavement. The only exception to this is where reactive aggregate has caused expansion of the slab. On heavily trafficked routes, pressure relief joints should be of heavy-duty design with dowels (*3*).

5.7.3 Reflection Crack Control

Reflection cracking in an AC overlay of AC/JPCP, AC/JRCP, or AC/CRCP occurs over reflection cracks in the first AC overlay, and may also occur over new repairs. The basic mechanism of reflection cracking is strain concentration in the overlay due to movement in the vicinity of joints and cracks in the existing pavement. This movement may be bending or shear induced by loads, or may be horizontal contraction induced by temperature changes. Load-induced movements are influenced by the thickness and stiffness of the AC layers, the thickness of the PCC, the degree of load transfer at the joints and cracks, and the extent of loss of support under the PCC slab. Temperature-induced movements are influenced by daily and seasonal temperature variations, the coefficients of thermal expansion of the existing pavement layers, and the spacing of joints and cracks.

Pre-overlay repair, including full-depth repair, subdrainage improvement, and subsealing, is the most effective means of controlling reflection crack occurrence and deterioration in a second AC overlay of an AC/JPCP or AC/JRCP pavement. Additional reflection crack control treatments may be used as well, including:

(1) *Placing a synthetic fabric, stress-absorbing interlayer, or bituminous-stabilized granular layer prior to or in combination with the AC overlay.*
(2) *Sawing and sealing joints in the AC overlay* at locations coinciding with reflection cracks and repair boundaries in the AC/JPCP or AC/JRCP. This technique has been very successful when applied to AC overlays of jointed PCC pavements when the sawcut matches the joint or straight crack within an inch.
(3) *Increasing the AC overlay thickness.* Reflection cracks will take more time to propagate through a thicker overlay and may deteriorate more slowly.

Reflection cracking can have a considerable (often controlling) influence on the life of an AC overlay of AC/JPCP or AC/JRCP. Deteriorated reflection cracks detract from a pavement's serviceability and also require frequent maintenance, such as sealing, milling, and patching. Reflection cracks also permit water to enter the pavement structure, which may result in loss of bond between the AC and PCC, stripping in the AC layers, progression of "D" cracking or reactive aggregate distress in PCC slabs with these durability problems, and softening of the base and subgrade. For this reason, reflection cracks should be sealed as soon as they appear and resealed periodically throughout the life of the overlay. Sealing low-severity reflection cracks may also be effective in retarding their progression to medium and high severity levels.

Repairing reflection cracks in existing AC/CRCP prior to placement of an AC overlay will delay the occurrence and deterioration of new reflection cracks. Improving subdrainage conditions and subsealing in areas where the slab has lost support will also discourage reflection crack occurrence and deterioration. Reflection crack control treatments are not necessary for AC overlays of AC/CRCP, except for longitudinal joints, as long as continuously reinforced AC/PCC repairs are used to repair deteriorated areas and cracks.

5.7.4 Subdrainage

See Section 5.2.4 for guidelines.

5.7.5 Thickness Design

If the overlay is being placed for some functional purpose such as roughness or friction, a minimum thickness overlay that solves the functional problem should be placed. If the overlay is being placed for the purpose of structural improvement, the required thickness of the overlay is a function of the structural capacity required to meet future traffic demands and the structural capacity of the existing pavement. The required overlay thickness to increase structural capacity to carry future traffic is determined by the following equation.

$$D_{ol} = A(D_f - D_{eff})$$

where

D_{ol} = Required thickness of AC overlay, inches
A = Factor to convert PCC thickness deficiency to AC overlay thickness
D_f = Slab thickness to carry future traffic, inches
D_{eff} = Effective equivalent PCC slab thickness of existing AC/PCC, inches

The A factor, which is a function of the PCC thickness deficiency, is given by the following equation and is illustrated in Figure 5.9.

$$A = 2.2233 + 0.0099(D_f - D_{eff})^2 - 0.1534(D_f - D_{eff})$$

The required overlay thickness may be determined through the following design steps. These design steps

Rehabilitation with Overlays

provide a comprehensive design approach that recommends testing the pavement to obtain valid design inputs. If it is not possible to conduct this testing (e.g., for a low-volume road), an approximate overlay design may be developed based upon visible distress observations by skipping Steps 4 and 5, and by estimating other inputs.

The overlay design can be done for a uniform section or on a point-by-point basis as described in Section 5.3.1.

Step 1: Existing pavement design.

(1) Existing AC surface thickness
(2) Existing PCC slab thickness
(3) Type of load transfer (mechanical devices, aggregate interlock, CRCP)
(4) Type of shoulder (tied PCC, other)

Step 2: Traffic analysis.

(1) Predicted future 18-kip ESALs in the design lane over the design period (N_f)
   Use ESALs computed from rigid pavement load equivalency factors

Step 3: Condition survey.

The following distresses are measured during the condition survey. Sampling along the most heavily trafficked lane of the project may be used to estimate these quantities. Distress types and severities are defined in Reference 23. Deteriorated means medium or higher severity.

AC/JPCP OR AC/JRCP:

(1) Number of deteriorated reflection cracks per mile
(2) Number of full-depth AC patches and expansion joints per mile (except at bridges)
(3) Evidence of pumping of fines or water at cracks and pavement edge
(4) Mean rut depth
(5) Number of localized failures

The following distresses are measured during the condition survey for AC/CRCP. Sampling may be used to estimate these quantities.

AC/CRCP:

(1) Number of unrepaired punchouts per mile
(2) Number of unrepaired reflection cracks per mile
(3) Number of unrepaired existing deteriorated repairs and full-depth AC repairs per mile

(4) Evidence of pumping of fines or water
(5) Mean rut depth

Step 4: Deflection testing (strongly recommended).

Measure slab deflection basins along the project at an interval sufficient to adequately assess conditions. Intervals of 100 to 1,000 feet are typical. Measure deflections with sensors located at 0, 12, 24, and 36 inches from the center of the load. Measure deflections in the outer wheel path, unless rutting of the AC surface interferes with proper seating of the load plate, in which case deflections should be measured between the wheelpaths. A heavy-load deflection device (e.g., Falling Weight Deflectometer) and a load magnitude of 9,000 pounds are recommended. ASTM D 4694 and D 4695 provide additional guidance on deflection testing.

(1) *Temperature of AC mix.* The temperature of the AC mix during deflection testing must be determined. This may be measured directly by drilling a hole into the AC surface, inserting a liquid and a temperature probe, and reading the AC mix temperature when it has stabilized. This should be done at least three times during each day's testing, so that a curve of AC mix temperature versus time may be developed and used to assign a mix temperature to each basin.

If measured AC mix temperatures are not available, they may be approximated from correlations with pavement surface and air temperatures (*24, 25, 26, 27*). Pavement surface temperature may be monitored during deflection testing using a hand-held infrared sensing device which is aimed at the pavement. The mean air temperature for the five days prior to deflection testing, which is an input to some of the referenced methods for estimating mix temperature, may be obtained from a local weather station or other local sources.

(2) *Elastic modulus of AC.* The modulus of the AC layer should be determined for each deflection basin. Two methods are available for determining the AC modulus, E_{ac}.

   (a) *Estimate E_{ac} from AC mix temperature.* The elastic modulus of the AC layer may be estimated from AC mix properties and the AC mix temperature assigned to a deflection basin using the following equation (*26*):

$$\log E_{ac} = 5.553833 + 0.028829 \left(\frac{P_{200}}{F^{0.17033}}\right)$$
$$- 0.03476 V_v + 0.070377 \eta_{70°F,10^6}$$
$$+ 0.000005 t_p^{(1.3+0.49825\log F)} P_{ac}^{0.5}$$
$$- \frac{0.00189}{F^{1.1}} t_p^{(1.3+0.49825\log F)} P_{ac}^{0.5}$$
$$+ 0.931757 \left(\frac{1}{F^{0.02774}}\right)$$

where

E_{ac} = elastic modulus of AC, psi
P_{200} = percent aggregate passing the No. 200 sieve
F = loading frequency, Hz
V_v = air voids, percent
$\eta_{70°F,10^6}$ = absolute viscosity at 70°F, 10^6 poise (e.g., 1 for AC-10, 2 for AC-20)
P_{ac} = asphalt content, percent by weight of mix
t_p = AC mix temperature, °F

This may be reduced to a relationship between AC modulus and AC mix temperature for a particular loading frequency (i.e., approximately 18 Hz for the FWD load duration of 25 to 30 milliseconds) by assuming typical values for the AC mix parameters P_{ac}, V_v, P_{200}, and η. For example, the AC mix design used by one State has the following typical values:

P_{200} = 4 percent
V_v = 5 percent
$\eta_{70°F,10^6}$ = 2 for AC-20
P_{ac} = 5 percent

For these values and an FWD loading frequency of 18 Hz, the following equation for AC elastic modulus versus AC mix temperature is obtained:

$$\log E_{ac} = 6.451235$$
$$- 0.000164671 t_p^{1.92544}$$

Each agency should establish its own relationship for AC modulus versus temperature which is representative of the properties of its AC mixes.

It should be noted that the equation for AC modulus as a function of mix parameters and temperature applies to new mixes. AC which has been in service for some years may have either a higher modulus (due to hardening of the asphalt) or lower modulus (due to deterioration of the AC, from stripping or other causes) at any given temperature.

(b) Diametral resilient modulus testing of AC cores taken from the in-service AC/PCC pavement, as described in Step 5, may be used to establish a relationship between AC modulus and temperature. This relationship may be used to determine the AC modulus of each deflection basin at the time and temperature at which it was measured.

(3) *Effective dynamic k-value beneath PCC slab.* Compute the compression which occurs in the AC overlay beneath the load plate using the following equations.

AC, PCC LAYERS BONDED:

$$d_{0\,compress} = -0.0000328 + 121.5006$$
$$* \left(\frac{D_{ac}}{E_{ac}}\right)^{1.0798}$$

AC, PCC LAYERS UNBONDED:

$$d_{0\,compress} = -0.00002133 + 38.6872$$
$$* \left(\frac{D_{ac}}{E_{ac}}\right)^{0.94551}$$

where

$d_{0\,compress}$ = AC compression at center of load, inches
D_{ac} = AC thickness, inches
E_{ac} = AC elastic modulus, psi

The interface condition is a significant unknown in backcalculation. The AC/PCC interface is fully bonded when the AC layer is first placed, but how well that bond is retained is not known. Examination of cores taken at a later time may show that bond has been reduced or completely lost. This is particularly likely if stripping occurs at the AC/PCC interface. If the current interface bonding condition is not determined by coring, the bonding condition which is considered more representative of the project may be assumed.

Using the above equations, the d_0 of the PCC slab in the AC/PCC pavement may be determined by subtracting the compression which occurs in the AC surface from the d_0 measured at the AC surface.

Compute the AREA of the PCC slab for each deflection basin from the following equation.

$$AREA_{pcc} = 6 * \left[1 + 2 \left(\frac{d_{12}}{d_{0\,pcc}} \right) + 2 \left(\frac{d_{24}}{d_{0\,pcc}} \right) + \left(\frac{d_{36}}{d_{0\,pcc}} \right) \right]$$

where

$d_{0\,pcc}$ = PCC deflection in center of loading plate, inches (surface deflection d_0 minus AC compression $d_{0\,compress}$)
d_i = deflections at 12, 24, and 36 inches from plate center, inches

Enter Figure 5.10 with the $d_{0\,pcc}$ and $AREA_{pcc}$ of the PCC slab to determine the effective dynamic k-value beneath the slab for a circular load radius of 5.9 inches and magnitude of 9,000 pounds. Note that for loads within 2,000 pounds more or less, deflections may be scaled linearly to 9,000-pound deflections.

If a single overlay thickness is being designed for a uniform section, compute the mean effective dynamic k-value of the slabs tested in the uniform section.

(4) *Effective static k-value.*

Effective static k-value

= Effective dynamic k-value/2

The effective static k-value may need to be adjusted for seasonal effects using the approach presented in Part II, Section 3.2.1. However, the k-value can change substantially and have only a small effect on overlay thickness.

(5) *Elastic modulus of PCC slab (E).* Enter Figure 5.11 with the $AREA_{pcc}$ of the top of the PCC slab, proceed to the effective dynamic k-value curves, and determine a value for ED^3, where D is the PCC slab thickness. Solve for E knowing the slab thickness, D. Typical slab E values range from 3 to 8 million psi. If a slab E value is obtained out of this range, an error may exist in the assumed slab thickness, the deflection basin may have been measured over a crack, or the PCC may be significantly deteriorated.

If a single overlay thickness is being designed for a uniform section, compute the mean E value of the slabs tested in the uniform section.

Do not use any k-values or E values that appear to be significantly out of line with the rest of the data.

(6) *Joint load transfer.* For AC/JPCP and AC/JRCP, measure joint load transfer in the outer wheelpath (or between the wheelpaths if the AC is badly rutted) at representative reflection cracks above transverse joints in the PCC slab. Do not measure load transfer when the ambient temperature is greater than 80°F. Place the load plate on one side of the reflection crack with the edge of the plate touching the joint. Measure the deflection at the center of the load plate and at 12 inches from the center. Compute the deflection load transfer from the following equation.

$$\Delta LT = 100 * \left(\frac{\Delta_{ul}}{\Delta_l} \right) * B$$

where

ΔLT = deflection load transfer, percent
Δ_{ul} = unloaded side deflection, inches
Δ_l = loaded side deflection, inches
B = slab bending and AC compression correction factor

The slab bending and AC compression correction factor, B, is necessary because the deflections d_0 and d_{12}, measured 12 inches apart, would not be equal even if measured in the interior of a slab. An appropriate value for the correction factor may be determined from the ratio of d_0 to d_{12} for typical center slab deflection basin measurements, as shown in the equation below.

$$B = \frac{d_{0\,center}}{d_{12\,center}}$$

If a single overlay thickness is being designed for a uniform section, compute the mean deflection load transfer value of the joints tested in the uniform section.

For AC/JPCP and AC/JRCP, determine the J load transfer coefficient using the following guidelines:

Percent Load Transfer	J
>70	3.2
50–70	3.5
<50	4.0

If the rehabilitation will include the addition of a tied concrete shoulder, a lower J factor may be appropriate. See Part II, Table 2.6.

For AC/CRCP, use J = 2.2 to 2.6 for overlay design, assuming that working cracks are repaired with continuously reinforced PCC overlaid with AC.

Step 5: Coring and materials testing (strongly recommended).

(1) *Modulus of AC surface.* Laboratory testing of cores taken from the AC surface in uncracked areas may be used to determine the elastic modulus of the AC surface. This may be done using a repeated-load indirect tension test (ASTM D 4123). The tests should be run at two or more temperatures (e.g., 40, 70, and 90°F) to establish points for a curve of log E_{ac} versus temperature. AC modulus values at any temperature may be interpolated from the laboratory values obtained at any two temperatures. For example, E_{ac} values at 70° and 90°F may be used in the following equation to interpolate E_{ac} at any temperature t°F:

$$\log E_{ac\,t°F} = \left(\frac{\log E_{ac\,70°F} - \log E_{ac\,90°F}}{70 - 90}\right)$$
$$* (t°F - 70°F) + \log E_{ac\,70°F}$$

For purposes of interpreting NDT data, AC modulus values obtained from laboratory testing of cores must be adjusted to account for the difference between the loading frequency of the test apparatus (typically 1 to 2 Hz) and the loading frequency of the deflection testing device (18 Hz for the FWD). This adjustment is made by multiplying the laboratory-determined E_{ac} by a constant value which may be determined for each laboratory testing temperature using the equation given in Step 4 for AC modulus as a function of mix parameters and temperature. Field-frequency E_{ac} values will typically be 2 to 2.5 times higher than lab-frequency values.

Agencies may also wish to establish correlations between resilient modulus and indirect tensile strength for specific AC mixes.

(2) *PCC modulus of rupture (S'_c).* Cut several 6-inch-diameter cores at midslab and test in indirect tension (ASTM C 496). Compute the indirect tensile strength (psi) of the cores. Estimate the modulus of rupture with the following equation.

$$S'_c = 210 + 1.02 IT$$

where

S'_c = modulus of rupture, psi
IT = indirect tensile strength of 6-inch-diameter cores, psi

Step 6: Determination of required slab thickness for future traffic (D_f).

The inputs to determine D_f for AC overlays of AC/PCC pavements are representative of the existing slab and foundation properties. This is emphasized because it is the properties of the existing slab (i.e., elastic modulus, modulus of rupture, and load transfer) which will control the performance of the AC overlay.

(1) *Effective static k-value beneath existing PCC slab.* Determine from one of the following methods.
   (a) Backcalculate effective dynamic k-value from deflection basins as described in Step 4. Divide the effective dynamic k-value by 2 to obtain the effective static k-value. The effective static k-value may

need to be adjusted for seasonal effects using the approach presented in Part II, Section 3.2.1.
  (b) Conduct plate load tests (ASTM D 1196) after slab removal at a few sites. This alternative is very costly and time-consuming and not often used. The static k-value obtained may need to be adjusted for seasonal effects (see Part II, Section 3.2.1).
  (c) Estimate from soils data and base type and thickness, using Figure 3.3 in Part II, Section 3.2. This alternative is simple, but the static k-value obtained must be recognized as a rough estimate. The static k-value obtained may need to be adjusted for seasonal effects (see Part II, Section 3.2.1).
(2) *Design PSI loss.* PSI immediately after overlay (P1) minus PSI at time of next rehabilitation (P2).
(3) *J, load transfer of existing PCC slab.* See Step 4.
(4) *PCC modulus of rupture,* determined by one of the following methods:
  (a) Estimate from indirect tensile strength measured from 6-inch-diameter cores, as described in Step 5.
  (b) For AC/JPCP and AC/JRCP, estimate from the E of the slab, backcalculated as described in Step 4. Use the following equation:

$$S'_c = 43.5 \left(\frac{E}{10^6}\right) + 488.5$$

where

S'_c = modulus of rupture, psi
E = backcalculated elastic modulus of PCC slab, psi

For AC/CRCP, estimating S'_c from backcalculated E values is not recommended since cracks which are not reflected in the existing AC overlay may exist in the CRCP within the deflection basins.

(5) *Elastic modulus of existing PCC slab,* determined by one of the following methods:
  (a) Backcalculated from deflection measurements, as described in Step 4.
  (b) Estimated from indirect tensile strength.
(6) *Loss of support of existing slab* that might exist after rehabilitation. Procedures for use of deflection testing to investigate loss of support beneath AC/PCC pavements have not yet been established. For overlay thickness design assume the slab is fully supported, LS = 0.
(7) *Overlay design reliability, R (percent).* See Part I, Section 4.2, Part II, Table 2.2, and Part III, Section 5.2.15.
(8) *Overall standard deviation, S_o, for PCC pavement.* See Part I, Section 4.3.
(9) *Subdrainage capability of existing slab, after subdrainage improvements, if any.* See Part II, Table 2.5, as well as reference 5, for guidance in determining C_d. Pumping or faulting at reflection cracks is evidence that a subdrainage problem exists. In selecting this value, note that the poor drainage situation at the AASHO Road Test would be given a C_d of 1.0.

Compute D_f for the above design inputs using the rigid pavement design equation or nomograph in Part II, Figure 3.7. When designing an overlay thickness for a uniform pavement section, mean input values must be used. When designing an overlay thickness for specific points along the project, the data for that point must be used. A worksheet for determining D_f is provided in Table 5.9. Typical values of inputs are provided for guidance. Values outside these ranges should be used with caution.

Step 7: Determination of effective slab thickness (D_{eff}) of existing pavement.

A condition survey method for determination of D_{eff} is presented for AC/PCC pavements. The effective thickness of the existing slab (D_{eff}) is computed from the following equation:

$$D_{eff} = (D_{pcc} * F_{jc} * F_{dur}) + \left[\left(\frac{D_{ac}}{2.0}\right) * F_{ac}\right]$$

where

D_{pcc} = thickness of existing PCC slab, inches
D_{ac} = thickness of existing AC surface, inches

(1) *Joints and cracks adjustment factor (F_{jc}).* This factor adjusts for the extra loss in PSI caused by deteriorated reflection cracks that will occur in a second overlay due to unrepaired deteriorated reflection cracks and other dis-

Table 5.9. Worksheet for Determination of D_f for AC/JPCP, AC/JRCP, and AC/CRCP

SLAB:

Existing AC surface thickness = _____ inches

Existing PCC slab thickness = _____ inches

Type of load transfer system: mechanical device, aggregate interlock, CRCP

Type of shoulder = tied PCC, other

PCC modulus of rupture (typically 600 to 800 psi) = _____ psi

PCC E modulus (3 to 8 million psi for sound PCC, <3 million for unsound PCC) = _____ psi

J load transfer factor (3.2 to 4.0 for AC/JPCP, AC/JRCP 2.2 to 2.6 for AC/CRCP) = _____

TRAFFIC:

Future 18-kip ESALs in design lane over the design period (N_f) = _____

SUPPORT AND DRAINAGE:

Effective dynamic k-value = _____ psi/inch

Effective static k-value = Effective dynamic k-value/2 (typically 50 to 500 psi/inch) = _____ psi/inch

Subdrainage coefficient, C_d (typically 1.0 for poor subdrainage conditions) = _____

SERVICEABILITY LOSS:

Design PSI loss (P1 − P2) = _____

RELIABILITY:

Design reliability, R (80 to 99 percent) = _____ percent

Overall standard deviation, S_o (typically 0.39) = _____

FUTURE STRUCTURAL CAPACITY:

Required slab thickness for future traffic is determined from rigid pavement design equation or nomograph in Part II, Figure 3.7.

D_f = _____ inches

---

continuities in the existing AC/PCC pavement prior to overlay. A deteriorated reflection crack in the existing AC/PCC pavement will rapidly reflect through a second overlay and contribute to loss of serviceability. Therefore, it is recommended that all deteriorated reflection cracks and any other major discontinuities in the existing pavement be full-depth repaired with dowelled or tied PCC repairs prior to overlay, so that $F_{jc} = 1.00$.

If it is not possible to repair all deteriorated areas, the following information is needed to determine F_{jc}, to increase the overlay thickness to account for the extra loss in PSI from deteriorated reflection cracks:

Number of unrepaired deteriorated reflection cracks/mile

Number of unrepaired punchouts/mile

Number of expansion joints, exceptionally wide joints (greater than 1 inch), and full-depth, full-lane-width AC patches/mile

The total number of unrepaired deteriorated reflection cracks, punchouts, and other discontinuities per mile is used to determine the F_{jc} from Figure 5.12.

(2) *Durability adjustment factor (F_{dur})*. This factor adjusts for an extra loss in PSI of the overlay when the existing slab has durability problems such as "D" cracking or reactive aggregate distress. Using historical records and condition survey data from Step 3, F_{dur} is determined as follows.

- 1.00: No evidence or history of PCC durability problems
- 0.96–0.99: Pavement is known to have PCC durability problems, but no localized failures or related distresses are visible
- 0.88–0.95: Some durability distress (localized failures, etc.) is visible at pavement surface
- 0.80–0.88: Extensive durability distress (localized failures, etc.) is visible at pavement surface

(3) AC quality adjustment factor (F_{ac}). This factor adjusts the existing AC layer's contribution to D_{eff} based on the quality of the AC material. The value selected should depend only on distresses related to the AC layer (i.e., not reflection cracking) which are not eliminated by surface milling: rutting, stripping, shoving, and also weathering and ravelling if the surface is not milled. Consideration should be given to complete removal of a poor-quality AC layer.

- 1.00: No AC material distress
- 0.96–0.99: Minor AC material distress (weathering, ravelling) not corrected by surface milling
- 0.88–0.95: Significant AC material distress (rutting, stripping, shoving)
- 0.80–0.88: Severe AC material distress (rutting, stripping, shoving)

A worksheet for calculation of D_{eff} is provided in Table 5.10.

Step 8: Determination of Overlay Thickness.

The thickness of AC overlay is computed as follows:

$$D_{ol} = A(D_f - D_{eff})$$

where

D_{ol} = Required thickness of AC overlay, inches
A = Factor to convert PCC thickness deficiency to AC overlay thickness
D_f = Slab thickness determined in Step 6, inches
D_{eff} = Effective thickness of existing slab determined in Step 7, inches

The A factor, which is a function of the PCC thickness deficiency, is given by the following equation and is illustrated in Figure 5.9.

$$A = 2.2233 + 0.0099(D_f - D_{eff})^2 - 0.1534(D_f - D_{eff})$$

The thickness of overlay determined from the above relationship should be reasonable when the overlay is required to correct a structural deficiency. See Section 5.2 for discussion of factors which may result in unreasonable overlay thicknesses.

5.7.6 Surface Milling

If the AC surface is to be milled prior to overlay, the depth of milling should be considered in the determination of D_{eff}. No adjustment need be made to D_{eff} values if the depth of milling does not exceed the minimum necessary to remove surface ruts. If a greater depth is milled, the AC thickness remaining after milling should be used in determining D_{eff}.

5.7.7 Shoulders

See Section 5.2.10 for guidelines.

Table 5.10. Calculation of D_{eff} for AC Overlay of AC/JPCP, AC/JRCP, and AC/CRCP

Condition Survey Method:

F_{jc} Number of unrepaired deteriorated reflection cracks/mile = _____

Number of punchouts/mile = _____

Number of expansion joints, exceptionally wide joints (>1 inch) or full-depth patches/mile = _____

Total/mile = _____

F_{jc} = _____ (Figure 5.12)

(Recommended value 1.0, repair all deteriorated areas)

F_{dur}
1.00: No sign or knowledge of PCC durability problems
0.96–0.99: Pavement is known to have PCC durability problems, but no localized failures or related distresses
0.88–0.95: Some durability distress (localized failures, etc.) is visible at pavement surface
0.80–0.88: Extensive durability distress (localized failures, etc.)

F_{dur} = _____

F_{ac}
1.00: No AC material distress
0.96–0.99: Minor AC material distress (weathering, ravelling) not corrected by surface milling
0.88–0.95: Significant AC material distress (rutting, stripping, shoving)
0.80–0.88: Severe AC material distress (rutting, stripping, shoving)

F_{ac} = _____

$$D_{eff} = (D_{pcc} * F_{jc} * F_{dur}) + \left[\left(\frac{D_{ac}}{2.0}\right) * F_{ac}\right] = _____$$

5.7.8 Widening

See Section 5.2.16 for guidelines.

5.8 BONDED CONCRETE OVERLAY OF JPCP, JRCP, AND CRCP

Bonded concrete overlays have been placed on jointed plain, jointed reinforced and continuously reinforced concrete pavements to improve both structural capacity and functional condition. A bonded concrete overlay consists of the following construction tasks:

(1) Repairing deteriorated areas and making subdrainage improvements (if needed)
(2) Constructing widening (if needed)
(3) Preparing the existing surface to ensure a reliable bond
(4) Placing the concrete overlay
(5) Sawing and sealing the joints

5.8.1 Feasibility

A bonded overlay of JPCP, JRCP, or CRCP is a feasible rehabilitation alternative for PCC pavements except when the conditions of the existing pavement dictate substantial removal and replacement or when

Rehabilitation with Overlays

durability problems exist (*28*). Conditions under which a PCC bonded overlay would not be feasible include:

(1) The amount of deteriorated slab cracking and joint spalling is so great that a substantial amount of removal and replacement of the existing surface is dictated.
(2) Significant deterioration of the PCC slab has occurred due to durability problems (e.g., "D" cracking or reactive aggregates). This will affect performance of the overlay.
(3) Vertical clearance at bridges is inadequate for required overlay thickness. This is not usually a problem because bonded overlays are usually fairly thin.

If construction duration is critical, PCC overlays may utilize high-early-strength PCC mixes. PCC overlays have been opened within 6 to 24 hours after placement using these mixtures.

5.8.2 Pre-overlay Repair

The following types of distress should be repaired prior to placement of the bonded PCC overlay.

Distress Type	Repair Type
Working cracks	Full-depth repair or slab replacement
Punchouts	Full-depth repair
Spalled joints	Full- or partial-depth repair
Deteriorated patches	Full-depth repair
Pumping/faulting	Edge drains
Settlements/heaves	Slab jack or reconstruct area

Full-depth repairs and slab replacements in JPCP and JRCP should be PCC, dowelled or tied to provide load transfer across repair joints. Full-depth repairs in CRCP should be PCC and should be continuously reinforced with steel which is tied or welded to reinforcing steel in the existing slab, to provide load transfer across joints and slab continuity. Full-depth AC repairs should not be used prior to placement of a bonded PCC overlay, and any existing AC patches should be removed and replaced with PCC. Guidelines on repairs are provided in References 1 and 3.

Installation of edge drains, maintenance of existing edge drains, or other subdrainage improvement should be done prior to placement of the overlay if a subdrainage evaluation indicates a need for such an improvement.

Pressure relief joints should be done only at fixed structures, and not at regular intervals along the pavement. The only exception to this is where a reactive aggregate has caused expansion of the slab. On heavily trafficked routes, expansion joints should be of the heavy-duty type with dowels (*3*). If joints contain significant incompressibles, they should be cleaned and resealed prior to overlay placement.

5.8.3 Reflection Crack Control

Any working (spalled) cracks in the existing JPCP, JRCP, or CRCP slab may reflect through the bonded concrete overlay within one year. Reflection cracks can be controlled in bonded overlays by full-depth repair of working cracks in the existing pavement, and for JPCP or JRCP, sawing and sealing joints through the overlay directly over the repair joints. Tight nonworking cracks do not need to be repaired because not all will reflect through the overlay and those that do will usually remain tight. Tight cracks in CRCP will take several years to reflect through, and even then will remain tight.

5.8.4 Subdrainage

See Section 5.2.4 for guidelines.

5.8.5 Thickness Design

If the overlay is being placed for some functional purpose such as roughness or friction, a minimum thickness overlay that solves the functional problem should be placed.

If the overlay is being placed for the purpose of structural improvement, the required thickness of the overlay is a function of the structural capacity required to meet future traffic demands and the structural capacity of the existing pavement. The required overlay thickness to increase structural capacity to carry future traffic is determined by the following equation.

$$D_{ol} = D_f - D_{eff}$$

where

D_{ol} = Required thickness of bonded PCC overlay, inches
D_f = Slab thickness to carry future traffic, inches
D_{eff} = Effective thickness of existing slab, inches

Bonded concrete overlays have been successfully constructed as thin as 2 inches and as thick as 6 inches or more. Three to 4 inches has been typical for most highway pavement overlays (28). If the bonded overlay is being placed only for a functional purpose such as roughness or friction, a thickness of 3 inches should be adequate.

The required overlay thickness may be determined through the following design steps. These design steps provide a comprehensive design approach that recommends testing the pavement to obtain valid design inputs. If it is not possible to conduct this testing, an approximate overlay design may be developed based upon visible distress observations by skipping Steps 4 and 5, and by estimating other inputs.

The overlay design can be done for a uniform section or on a point-by-point basis as described in Section 5.3.1.

Step 1: Existing pavement design.

(1) Existing slab thickness
(2) Type of load transfer (mechanical devices, aggregate interlock, CRCP)
(3) Type of shoulder (tied, PCC, other)

Step 2: Traffic analysis.

(1) Past cumulative 18-kip ESALs in the design lane (N_p), for use in the remaining life method of D_{eff} determination only
(2) Predicted future 18-kip ESALs in the design lane over the design period (N_f)

Step 3: Condition survey.

The following distresses are measured during the condition survey for JPCP, JRCP, and CRCP. Sampling along the project may be used to estimate these quantities in the most heavily trafficked lane. Distress types and severities are defined in Reference 23. Deteriorated means medium or higher severity.

JPCP/JRCP:

(1) Number of deteriorated transverse joints per mile
(2) Number of deteriorated transverse cracks per mile
(3) Number of existing expansion joints, exceptionally wide joints (>1 inch) or AC full-depth patches
(4) Presence and general severity of PCC durability problems
   (a) "D" cracking: low severity (cracks only), medium severity (some spalling), high severity (severe spalling)
   (b) Reactive aggregate cracking: low, medium, high severity
(5) Evidence of faulting, pumping of fines or water at joints, cracks and pavement edge

CRCP:

(1) Number of punchouts per mile
(2) Number of deteriorated transverse cracks per mile
(3) Number of existing expansion joints, exceptionally wide joints (>1 inch) or AC full-depth patches
(4) Number of existing and new repairs prior to overlay per mile
(5) Presence and general severity of PCC durability problems (NOTE: surface spalling of tight cracks where the underlying CRCP is sound should not be considered a durability problem)
   (a) "D" cracking: low severity (cracks only), medium severity (some spalling), high severity (severe spalling)
   (b) Reactive aggregate cracking: low, medium, high severity
(6) Evidence of pumping of fines or water

Step 4: Deflection testing (strongly recommended).

Measure slab deflection basins in the outer wheel path along the project at an interval sufficient to adequately assess conditions. Intervals of 100 to 1,000 feet are typical. Measure deflections with sensors located at 0, 12, 24, and 36 inches from the center of load. A heavy-load deflection device (e.g., Falling Weight Deflectometer) and a load magnitude of 9,000 pounds are recommended. ASTM D 4694 and D 4695 provide additional guidance on deflection testing.

For each slab tested, backcalculate the effective k-value and the slab's elastic modulus using Figures 5.10 and 5.11 or a backcalculation procedure. The AREA of each deflection basin is computed as follows:

$$\text{AREA} = 6 * \left[1 + 2\left(\frac{d_{12}}{d_0}\right) + 2\left(\frac{d_{24}}{d_0}\right) + \left(\frac{d_{36}}{d_0}\right) \right]$$

where

d_0 = deflection in center of loading plate, inches
d_i = deflections at 12, 24, and 36 inches from plate center, inches

AREA will typically range from 29 to 32 for sound concrete.

(1) *Effective dynamic k-value.* Enter Figure 5.10 with d_0 and AREA to determine the effective dynamic k-value beneath each slab for a circular load radius of 5.9 inches and magnitude of 9,000 pounds. Note that for loads within 2,000 pounds more or less, deflections may be scaled linearly to 9,000-pound deflections.

If a single overlay thickness is being designed for a uniform section, compute the mean effective dynamic k-value of the slabs tested in the uniform section.

(2) *Effective static k-value.*

Effective static k-value

= Effective dynamic k-value/2

The effective k-value may need to be adjusted for seasonal effects using the approach presented in Part II, Section 3.2.1. However, the k-value can change substantially and have only a small effect on overlay thickness.

(3) *Elastic modulus of PCC slab (E).* Enter Figure 5.11 with AREA, proceed to the effective dynamic k-value curves, and determine a value for ED^3, where D is the slab thickness. Solve for E knowing the slab thickness, D. Typical slab E values range from 3 to 8 million psi. If a slab E value is obtained that is out of this range, an error may exist in the assumed slab thickness, the deflection basin may have been measured over a crack, or the PCC may be significantly deteriorated.

If a single overlay thickness is being designed for a uniform section, compute the mean E value of the slabs tested in the uniform section.

Do not use any k-values or E values that appear to be significantly out of line with the rest of the data.

(4) *Joint load transfer.* For JPCP and JRCP, measure joint load transfer in the outer wheelpath at representative transverse joints. Do not measure load transfer when the ambient temperature is greater than 80°F. Place the load plate on one side of the joint with the edge of the plate touching the joint. Measure the deflection at the center of the load plate and at 12 inches from the center. Compute the deflection load transfer from the following equation.

$$\Delta LT = 100 * \left(\frac{\Delta_{ul}}{\Delta_l}\right) * B$$

where

ΔLT = deflection load transfer, percent
Δ_{ul} = unloaded side deflection, inches
Δ_l = loaded side deflection, inches
B = slab bending correction factor

The slab bending correction factor, B, is necessary because the deflections d_0 and d_{12}, measured 12 inches apart, would not be equal even if measured in the interior of a slab. An appropriate value for the correction factor may be determined from the ratio of d_0 to d_{12} for typical center slab deflection basin measurements, as shown in the equation below. Typical values for B are between 1.05 and 1.15.

$$B = \frac{d_{0\,center}}{d_{12\,center}}$$

If a single overlay thickness is being designed for a uniform section, compute the mean deflection load transfer value of the joints tested in the uniform section.

For JPCP and JRCP, determine the J load transfer coefficient using the following guidelines:

Percent Load Transfer	J
>70	3.2
50–70	3.5
<50	4.0

If the rehabilitation will include the addition of a tied concrete shoulder, a lower J factor may be appropriate. See Part II, Table 2.6.

For CRCP, use J = 2.2 to 2.6 for overlay design, assuming that working cracks and punchouts are repaired with continuously reinforced PCC.

Step 5: Coring and materials testing (strongly recommended).

(1) *PCC modulus of rupture (S'_c).* Cut several 6-inch-diameter cores at mid-slab and test in indirect tension (ASTM C 496). Compute the indirect tensile strength (psi) of the cores. Estimate the modulus of rupture with the following equation:

$$S'_c = 210 + 1.02 IT$$

where

S'_c = modulus of rupture, psi
IT = indirect tensile strength of 6-inch diameter cores, psi

Step 6: Determination of required slab thickness for future traffic (D_f).

The inputs to determine D_f for bonded PCC overlays of PCC pavements are representative of the existing slab and foundation properties. This is emphasized because it is the properties of the existing slab (i.e., elastic modulus, modulus of rupture, and load transfer) which will control the performance of the bonded overlay.

(1) *Effective static k-value.* Determine from one of the following methods.
  (a) Backcalculate the effective dynamic k-value from deflection basins as described in Step 4. Divide the effective dynamic k-value by 2 to obtain the effective static k-value.
  (b) Conduct plate load tests (ASTM D 1196) after slab removal at a few sites. This alternative is very costly and time-consuming and not often used. The static k-value obtained may need to be adjusted for seasonal effects using the approach presented in Part II, Section 3.2.1.
  (c) Estimate from soils data and base type and thickness, using Figure 3.3 in Part II, Section 3.2. This alternative is simple, but the static k-value obtained must be recognized as a rough estimate. The static k-value obtained may need to be adjusted for seasonal effects using the approach presented in Part II, Section 3.2.1.

(2) *Design PSI loss.* PSI immediately after overlay (P1) minus PSI at time of next rehabilitation (P2).

(3) *J, load transfer factor.* See Step 4.

(4) *PCC modulus of rupture* determined by one of the following methods:
  (a) Estimated from indirect tensile strength measured from 6-inch diameter cores as described in Step 5.
  (b) Estimated from the backcalculated E of slab using the following equation:

$$S'_c = 43.5 \left(\frac{E}{10^6}\right) + 488.5$$

where

S'_c = modulus of rupture, psi
E = backcalculated elastic modulus of PCC slab, psi

For CRCP, S'_c may be determined from the backcalculated E values only at points which have no cracks within the deflection basins.

(5) *Elastic modulus of existing PCC slab,* determined by one of the following methods:
  (a) Backcalculate from deflection measurements as described in Step 4.
  (b) Estimate from indirect tensile strength.

(6) *Loss of support of existing slab.* Joint corners that have loss of support may be identified using FWD deflection testing as described in Reference 2. CRCP loss of support can be determined by plotting a slab edge or wheel path deflection profile and identifying locations with significantly high deflections. Existing loss of support can be improved with slab stabilization. For thickness design, assume a fully supported slab, LS = 0.

(7) *Overlay design reliability, R (percent).* See Part I, Section 4.2, Part II, Table 2.2, and Part III, Section 5.2.15.

(8) *Overall standard deviation (S_o) for rigid pavement.* See Part I, Section 4.3.

(9) *Subdrainage capability of existing slab, after subdrainage improvements, if any.* See Part II,

Rehabilitation with Overlays

Table 2.5, as well as Reference 5, for guidance in determining C_d. Pumping or faulting at joints and cracks determined in Step 3 is evidence that a subdrainage problem exists. In selecting this value, note that the poor subdrainage situation at the AASHO Road Test would be given a C_d of 1.0.

Compute D_f for the above design inputs using the rigid pavement design equation or nomograph in Part II, Figure 3.7. When designing an overlay thickness for a uniform pavement section, mean input values must be used. When designing an overlay thickness for specific points along the project, the data for that point must be used. A worksheet for determining D_f is provided in Table 5.11. Typical values of inputs are provided for guidance. Values outside these ranges should be used with caution.

Step 7: Determination of effective slab thickness (D_{eff}) of existing pavement.

The condition survey and remaining life procedures are presented.

D_{eff} From Condition Survey For PCC Pavements

The effective thickness of the existing slab (D_{eff}) is computed from the following equation:

$$D_{eff} = F_{jc} * F_{dur} * F_{fat} * D$$

where

    D = existing PCC slab thickness, inches

(1) *Joints and cracks adjustment factor (F_{jc}).* This factor adjusts for the extra loss in PSI caused by deteriorated reflection cracks in the overlay that will result from any unrepaired deteriorated joints, cracks, and other discontinuities in the existing slab prior to overlay. A deteriorated joint or crack in the existing slab will rapidly reflect through an AC overlay and contribute to loss of serviceability. Therefore, it is recommended that all deteriorated joints and cracks (for non-"D" cracked or reactive aggregate related distressed pavements) and any other major discontinuities in the existing slab be full-depth repaired with dowelled or tied PCC repairs prior to overlay, so that $F_{jc} = 1.00$.

If it is not possible to repair all deteriorated areas, the following information is needed to determine F_{jc}, to increase the overlay thickness to account for the extra loss in PSI from deteriorated reflection cracks (per design lane):

Pavements with no "D" cracking or reactive aggregate distress:

    Number of unrepaired deteriorated joints/mile

    Number of unrepaired deteriorated cracks/mile

    Number of unrepaired punchouts/mile

    Number of expansion joints, exceptionally wide joints (greater than 1 inch), and full-depth, full-lane-width AC patches/mile

    NOTE that tight cracks held together by reinforcement in JRCP or CRCP are not included. However, if a crack in JRCP or CRCP is spalled and faulted the steel has probably ruptured, and the crack should be considered as working. Surface spalling of CRCP cracks is not an indication that the crack is working.

    The total number of unrepaired deteriorated joints, cracks, punchouts, and other discontinuities per mile is used to determine the F_{jc} from Figure 5.12.

Pavements with "D" cracking or reactive aggregate deterioration:

    These types of pavements often have deterioration at the joints and cracks from durability problems. The F_{dur} factor is used to adjust the overlay thickness for this problem. Therefore, when this is the case, the F_{jc} should be determined from Figure 5.12 only using those unrepaired deteriorated joints and cracks that are not caused by durability problems. If all of the deteriorated joints and cracks are spalling due to "D" cracking or reactive aggregate, then $F_{jc} = 1.0$. This will avoid adjusting twice with the F_{jc} and F_{dur} factors.

(2) *Durability adjustment factor (F_{dur}).* This factor adjusts for an extra loss in PSI of the overlay when the existing slab has durability problems such as "D" cracking or reactive aggregate distress. Using condition survey data from Step 3, F_{dur} is determined as follows.

    1.00: No sign of PCC durability problems

    0.96–0.99: Durability cracking exists, but no spalling

Table 5.11. Worksheet for Determination of D_f for JPCP, JRCP, and CRCP

SLAB:

Existing PCC slab thickness = _____ inches

Type of load transfer system: mechanical device, aggregate interlock, CRCP

Type of shoulder = tied PCC, other

PCC modulus of rupture (typically 600 to 800 psi) = _____ psi

PCC E modulus (3 to 8 million psi for sound PCC, <3 million for unsound PCC) = _____ psi

J load transfer factor (3.2 to 4.0 for JPCP, JRCP 2.2 to 2.6 for CRCP) = _____

TRAFFIC:

Future 18-kip ESALs in design lane over the design period (N_f) = _____

SUPPORT AND DRAINAGE:

Effective dynamic k-value = _____ psi/inch

Effective static k-value = effective dynamic k-value/2 (typically 50 to 500 psi/inch) = _____ psi/inch

Subdrainage coefficient, C_d (typically 1.0 for poor subdrainage conditions) = _____

SERVICEABILITY LOSS:

Design PSI loss (P1 − P2) = _____

RELIABILITY:

Design reliability, R (80 to 99 percent) = _____ percent

Overall standard deviation, S_o (typically 0.39) = _____

FUTURE STRUCTURAL CAPACITY:

Required slab thickness for future traffic is determined from rigid pavement design equation or nomograph in Part II, Figure 3.7.

D_f = _____ inches

0.80–0.95: Cracking and spalling exist (normally a bonded PCC overlay is not recommended under these conditions)

(3) *Fatigue damage adjustment factor (F_{fat}).* This factor adjusts for past fatigue damage that may exist in the slab. It is determined by observing the extent of transverse cracking (JPCP, JRCP) or punchouts (CRCP) that may be caused primarily by repeated loading. Use condition survey data from Step 3 and the following guidelines to estimate F_{fat} for the design lane.

0.97–1.00: Few transverse cracks/punchouts exist (none caused by "D" cracking or reactive aggregate distress)

JPCP: <5 percent slabs are cracked
JRCP: <25 working crack per mile
CRCP: <4 punchouts per mile

0.94-0.96: A significant number of transverse cracks/punchouts exist (none caused by "D" cracking or reactive aggregate distress)
JPCP: 5-15 percent slabs are cracked
JRCP: 25-75 working cracks per mile
CRCP: 4-12 punchouts per mile

0.90-0.93: A large number of transverse cracks/punchouts exist (none caused by "D" cracking or reactive aggregate distress)
JPCP: >15 percent slabs are cracked
JRCP: >75 working cracks per mile
CRCP: >2 punchouts per mile

D_{eff} From Remaining Life For PCC Pavements

The remaining life of the pavement is given by the following equation:

$$RL = 100 \left[1 - \left(\frac{N_p}{N_{1.5}} \right) \right]$$

where

RL = remaining life, percent
N_p = total traffic to date, ESALs
$N_{1.5}$ = total traffic to pavement "failure," ESALs

$N_{1.5}$ may be estimated using the new pavement design equations or nomographs in Part II. To be consistent with the AASHO Road Test and the development of these equations, a "failure" PSI equal to 1.5 and a reliability of 50 percent is recommended.

D_{eff} is determined from the following equation:

$$D_{eff} = CF * D$$

where

CF = condition factor determined from Figure 5.2
D = thickness of the existing slab, inches

The designer should recognize that D_{eff} determined by this method does not reflect any benefit for pre-overlay repair. The estimate of D_{eff} obtained should thus be considered a lower limit value. The D_{eff} of the pavement will be higher if pre-overlay repair of load-associated distress is done.

A worksheet for determination of D_{eff} for JPCP, JRCP, and CRCP is provided in Table 5.12.

Step 8: Determination of Overlay Thickness.

The thickness of bonded PCC overlay is computed as follows:

$$D_{ol} = D_f - D_{eff}$$

where

D_{ol} = Required thickness of bonded PCC overlay, inches
D_f = Slab thickness determined in Step 6, inches
D_{eff} = Effective thickness of existing slab determined in Step 7, inches

The thickness of overlay determined from the above relationship should be reasonable when the overlay is required to correct a structural deficiency. See Section 5.2.17 for discussion of factors which may result in unreasonable overlay thicknesses.

5.8.6 Shoulders

See Section 5.2.10 for guidelines.

5.8.7 Joints

Existing JPCP and JRCP. Transverse and longitudinal joints should be saw cut completely through the overlay thickness (plus 0.5-inch depth) as soon as curing allows after overlay placement. Failure to saw joints soon after placement may result in debonding and cracking at the joints. No dowels or reinforcing steel should be placed in these joints. An appropriate sealant reservoir should be sawed and sealant should be placed as soon as possible.

Existing CRCP. Transverse joints must not be cut in the bonded overlay, as they are not needed. Transverse joints are also not needed for the end joints for full-

Table 5.12. Calculation of D_{eff} for Bonded PCC Overlay of JPCP, JRCP, and CRCP

Condition Survey Method:

F_{jc} Number of unrepaired deteriorated joints/mile = _____

Number of unrepaired deteriorated cracks/mile = _____

Number of unrepaired punchouts/mile = _____

Number of expansion joints, exceptionally wide joints (>1 inch) or AC full-depth patches/mile = _____

Total/mile = _____

F_{jc} = _____ (Figure 5.12)

(Recommended value 1.0, repair all deteriorated areas)

F_{dur}
- 1.00: No sign of PCC durability problems
- 0.96–0.99: Some durability cracking exists, but no spalling exists
- 0.88–0.95: Cracking and spalling exist

F_{dur} = _____

F_{fat}
- 0.97–1.00: Very few transverse cracks/punchouts exist
- 0.94–0.96: A significant number of transverse cracking/punchouts exist
- 0.90–0.93: A large amount of transverse cracking/punchouts exist

F_{fat} = _____

$$D_{eff} = F_{jc} * F_{dur} * F_{fat} * D = _____$$

Remaining Life Method:

N_p = Past design lane ESALs = _____

$N_{1.5}$ = Design lane ESALs to P2 of 1.5 = _____

$$RL = 100\left[1 - \left(\frac{N_p}{N_{1.5}}\right)\right] = _____$$

CF = _____ (Figure 5.2)

$$D_{eff} = CF * D = _____$$

depth reinforced tied concrete patches. Longitudinal joints should be sawed in the same manner as for JPCP and JRCP.

5.8.8 Bonding Procedures and Material

The successful performance of the bonded overlay depends on a reliable bond with the existing surface (*28*). The following guidelines are provided:

(1) The existing surface must be cleaned and roughened, through a mechanical process that removes a thin layer of concrete, but does not damage (crack) the surface. Shot blasting is the most used system. Cold milling has been used, but may cause damage to the surface and thus requires sand blasting afterward to remove any loose particles.
(2) A bonding agent is recommended to help achieve a more reliable bond. Water, cement, and sand mortar; water and cement slurry; and low-viscosity epoxy have been used for this purpose. Bonded overlays constructed without a bonding agent have performed well in some instances.

5.8.9 Widening

See Section 5.2.16 for guidelines.

5.9 UNBONDED JPCP, JRCP, AND CRCP OVERLAY OF JPCP, JRCP, CRCP, AND AC/PCC

An unbonded JPCP, JRCP, or CRCP overlay of an existing JPCP, JRCP, CRCP, or composite (AC/PCC) pavement can be placed to improve both structural capacity and functional condition. An unbonded concrete overlay consists of the following construction tasks:

(1) Repairing only badly deteriorated areas and making subdrainage improvements (if needed)
(2) Constructing widening (if needed)
(3) Placing a separation layer (this layer may also serve as a leveling course)
(4) Placing the concrete overlay
(5) Sawing and sealing the joints

5.9.1 Feasibility

An unbonded overlay is a feasible rehabilitation alternative for PCC pavements for practically all conditions. They are most cost-effective when the existing pavement is badly deteriorated because of reduced need for pre-overlay repair. Conditions under which a PCC unbonded overlay would not be feasible include:

(1) The amount of deteriorated slab cracking and joint spalling is not large and other alternatives would be much more economical.
(2) Vertical clearance at bridges is inadequate for required overlay thickness. This may be addressed by reconstructing the pavement under the overhead bridges or by raising the bridges. Thicker unbonded overlays may also necessitate raising signs and guardrails, as well as increasing side slopes and extending culverts. Sufficient right-of-way must be available or obtainable to permit these activities.
(3) The existing pavement is susceptible to large heaves or settlements.

If construction duration is critical, PCC overlays may utilize high-early-strength PCC mixes. PCC overlays have been opened within 6 to 24 hours after placement using these mixtures.

5.9.2 Pre-overlay Repair

One major advantage of an unbonded overlay is that the amount of repairs to the existing pavement are greatly reduced. However, unbonded overlays are not intended to bridge localized areas of nonuniform support. The following types of distress (on the next page) should be repaired prior to placement of the overlay to prevent reflection cracks that may reduce its service life.

Guidelines on repairs are provided in References 1 and 3. Other forms of pre-overlay treatment for badly deteriorated pavements include slab fracturing (break/seat, crack/seat, or rubblizing) the existing PCC slab prior to placement of the separation layer. Fracturing and seating the existing slab may provide more uniform support for the overlay.

5.9.3 Reflection Crack Control

When an AC separation layer of 1 to 2 inches is used, there should be no problem with reflection of cracks through unbonded overlays. However, this sep-

Distress Type	Overlay Type	Repair
Working crack	JPCP or JRCP	No repair needed
	CRCP	Full-depth dowelled repair if differential deflection is significant
Punchout	JPCP, JRCP, CRCP	Full-depth repair
Spalled joint	JPCP or JRCP	No repair needed
	CRCP	Full-depth repair of severely deteriorated joints
Pumping	JPCP, JRCP, CRCP	Edge drains (if needed)
Settlement	JPCP, JRCP, CRCP	Level-up with AC
Poor joint/crack load transfer	JPCP, JRCP, CRCP	No repair needed; if pavement has many joints or cracks with poor load transfer, consider a thicker AC separation layer

aration layer thickness may not be adequate for an unbonded overlay when the existing pavement has poor load transfer and high differential deflections across transverse cracks or joints.

5.9.4 Subdrainage

See Section 5.2.4 for guidelines.

5.9.5 Thickness Design

The required thickness of the unbonded overlay is a function of the structural capacity required to meet future traffic demands and the structural capacity of the existing pavement. The required overlay thickness to increase structural capacity to carry future traffic is determined by the following equation.

$$D_{ol} = \sqrt{D_f^2 - D_{eff}^2}$$

where

D_{ol} = Required thickness of unbonded PCC overlay, inches
D_f = Slab thickness to carry future traffic, inches
D_{eff} = Effective thickness of existing slab, inches

Unbonded concrete overlays have been successfully constructed as thin as 5 inches and as thick as 12 inches or more. Thicknesses of seven to 10 inches have been typical for most highway pavement unbonded overlays.

The required overlay thickness may be determined through the following design steps. These design steps provide a comprehensive design approach that recommends testing the pavement to obtain valid design inputs. If it is not possible to conduct this testing, an approximate overlay design may be developed based upon visible distress observations by skipping Steps 4 and 5, and by estimating other inputs.

The overlay design can be done for a uniform section or on a point-by-point basis as described in Section 5.3.1.

Step 1: Existing pavement design.

(1) Existing slab thickness
(2) Type of load transfer (mechanical devices, aggregate interlock, CRCP)
(3) Type of shoulder (tied, PCC, other)

Step 2: Traffic analysis.

(1) Past cumulative 18-kip ESALs in the design lane (N_p), for use in the remaining life method of D_{eff} determination only
(2) Predicted future 18-kip ESALs in the design lane over the design period (N_f)

Step 3: Condition survey.

The following distresses are measured during the condition survey for JPCP, JRCP, and CRCP. Sampling along the project may be used to estimate these quantities in the most heavily trafficked lane. Distress types and severities are defined in Reference 23. Deteriorated means medium or higher severity.

JPCP/JRCP:

(1) Number of deteriorated transverse joints per mile

(2) Number of deteriorated transverse cracks per mile
(3) Number of existing expansion joints, exceptionally wide joints (more than 1 inch) or full-depth, full-lane-width AC patches
(4) Presence and general severity of PCC durability problems
   (a) "D" cracking: low severity (cracks only), medium severity (some spalling), high severity (severe spalling)
   (b) Reactive aggregate cracking: low, medium, high severity
(5) Evidence of faulting, pumping of fines or water at joints, cracks and pavement edge

CRCP:

(1) Number of punchouts per mile
(2) Number of deteriorated transverse cracks per mile
(3) Number of existing expansion joints, exceptionally wide joints (>1 inch) or full-depth, full-lane-width AC patches
(4) Number of existing and new repairs prior to overlay per mile
(5) Presence and general severity of PCC durability problems (NOTE: surface spalling of tight cracks where the underlying CRCP is sound should not be considered a durability problem)
   (a) "D" cracking: low severity (cracks only), medium severity (some spalling), high severity (severe spalling)
   (b) Reactive aggregate cracking: low, medium, high severity
(6) Evidence of pumping of fines or water

Step 4: Deflection testing (strongly recommended).

When designing an unbonded overlay for existing JPCP, JRCP, or CRCP, follow the guidelines given below for deflection testing and determination of the effective static k-value. When designing an unbonded overlay for existing AC/PCC, follow the guidelines given in Section 5.7, Step 4, for deflection testing and determination of the effective static k-value.

Measure slab deflection basins in the outer wheel path along the project at an interval sufficient to adequately assess conditions. Intervals of 100 to 1,000 feet are typical. Measure deflections with sensors located at 0, 12, 24, and 36 inches from the center of load. A heavy-load deflection device (e.g., Falling Weight Deflectometer) and a load magnitude of 9,000 pounds are recommended. ASTM D 4694 and D 4695 provide additional guidance on deflection testing.

For each slab tested, backcalculate the effective k-value using Figure 5.10 or a backcalculation procedure. The AREA of each deflection basin is computed from the following equation.

$$\text{AREA} = 6 * \left[1 + 2\left(\frac{d_{12}}{d_0}\right) + 2\left(\frac{d_{24}}{d_0}\right) + \left(\frac{d_{36}}{d_0}\right) \right]$$

where

d_0 = deflection in center of loading plate, inches
d_i = deflections at 12, 24, and 36 inches from plate center, inches

AREA will typically range from 29 to 32 for sound concrete.

(1) *Effective dynamic k-value.* Enter Figure 5.10 with d_0 and AREA to determine the effective dynamic k-value beneath each slab for a circular load radius of 5.9 inches and magnitude of 9,000 pounds. NOTE that for loads within 2,000 pounds more or less, deflections may be scaled linearly to 9,000-pound deflections.

If a single overlay thickness is being designed for a uniform section, compute the mean effective dynamic k-value of the slabs tested in the uniform section.

(2) *Effective static k-value.*

Effective static k-value

= Effective dynamic k-value/2

The effective static k-value may need to be adjusted for seasonal effects using the approach presented in Part II, Section 3.2.1. However, the k-value can change substantially and have only a small effect on overlay thickness.

Step 5: Coring and materials testing.

When designing an unbonded overlay for existing JPCP, JRCP, or CRCP, coring and materials testing of the existing PCC slab are not needed for overlay thickness design. When designing an unbonded overlay for existing AC/PCC, follow the guidelines given in Section 5.7, Step 5, for determination of the AC modulus by coring and materials testing.

Step 6: Determination of required slab thickness for future traffic (D_f).

The elastic modulus, modulus of rupture, and load transfer inputs to determine D_f for unbonded PCC overlays of PCC and AC/PCC pavements are representative of the new PCC overlay to be placed rather than of the existing slab. This is emphasized because it is the properties of the overlay slab (i.e., elastic modulus, modulus of rupture, and load transfer), which will control the performance of the unbonded overlay.

(1) *Effective static k-value beneath the existing pavement.* Determine from one of the following methods.
   (a) Backcalculate the effective dynamic k-value from deflection basins as described in Step 4. Divide the effective dynamic k-value by 2 to obtain the effective static k-value. The static k-value obtained may need to be adjusted for seasonal effects (see Part II, Section 3.2.1).
   (b) Conduct plate load tests (ASTM D 1196) after slab removal at a few sites. This alternative is very costly and time-consuming and not often used. The static k-value obtained may need to be adjusted for seasonal effects (see Part II, Section 3.2.1).
   (c) Estimate from soils data and base type and thickness, using Figure 3.3 in Part II, Section 3.2. This alternative is simple, but the static k-value obtained must be recognized as a rough estimate. The static k-value obtained may need to be adjusted for seasonal effects (see Part II, Section 3.2.1).

(2) *Design PSI loss.* PSI immediately after overlay (P1) minus PSI at time of next rehabilitation (P2).

(3) *J, load transfer factor for joint design of the unbonded PCC overlay.* See Part II, Section 2.4.2, Table 2.6.

(4) *PCC modulus of rupture of unbonded PCC overlay.*

(5) *Elastic modulus of unbonded PCC overlay.*

(6) *Loss of support.* Use LS = 0 for unbonded PCC overlay.

(7) *Overlay design reliability, R (percent).* See Part I, Section 4.2, Part II, Table 2.2, and Part III, Section 5.2.15.

(8) *Overall standard deviation (S_o) for rigid pavement.* See Part I, Section 4.3.

(9) *Subdrainage capability of existing slab,* after subdrainage improvements, if any. See Part II, Table 2.5, as well as Reference 5, for guidance in determining C_d. Pumping or faulting at joints and cracks determined in Step 3 is evidence that a subdrainage problem exists. In selecting this value, note that the poor drainage situation at the AASHO Road Test would be given a C_d of 1.0.

Compute D_f for the above design inputs using the rigid pavement design equation or nomograph in Part II, Figure 3.7. A worksheet for determining D_f is provided in Table 5.13.

Step 7: Determination of effective slab thickness (D_{eff}) of existing pavement.

The condition survey and remaining life procedures are presented.

D_{eff} From Condition Survey

The effective thickness (D_{eff}) of an existing PCC or AC/PCC pavement is computed from the following equation:

$$D_{eff} = F_{jcu} * D$$

where

D = existing PCC slab thickness, inches (NOTE: maximum D for use in unbonded concrete overlay design is 10 inches even if the existing D is greater than 10 inches)

F_{jcu} = joints and cracks adjustment factor for unbonded concrete overlays

NOTE that the existing AC surface is neglected in determining the effective slab thickness of an existing AC/PCC pavement.

Field surveys of unbonded jointed concrete overlays have shown very little evidence of reflection cracking or other problems caused by the existing slab. Therefore, the F_{dur} and F_{fat} are not used for unbonded concrete overlays. The F_{jcu} factor is modified to show a reduced effect of deteriorated cracks and joints in the existing slab, and is given in Figure 5.13.

(1) *Joints and cracks adjustment factor (F_{jcu}).* This factor adjusts for the extra loss in PSI caused by deteriorated reflection cracks or punchouts in the overlay that result from any unrepaired

Rehabilitation with Overlays

Table 5.13. Worksheet for Determination of D_f for Unbonded PCC Overlay

SLAB:

Type of load transfer system: mechanical device, aggregate interlock, CRCP

Type of shoulder = tied PCC, other

PCC modulus of rupture of unbonded overlay
(typically 600 to 800 psi) = _____ psi

PCC E modulus of unbonded overlay (3 to 5 million psi) = _____ psi

J load transfer factor of unbonded overlay
(2.5 to 4.4 for jointed PCC, 2.3 to 3.2 for CRCP) = _____

TRAFFIC:

Future 18-kip ESALs in design lane over
the design period (N_f) = _____

SUPPORT AND DRAINAGE:

Effective dynamic k-value = _____ psi/inch

Effective static k-value = Effective dynamic k-value/2
(typically 50 to 500 psi/inch) = _____ psi/inch

Subdrainage coefficient, C_d
(typically 1.0 for poor subdrainage conditions) = _____

SERVICEABILITY LOSS:

Design PSI loss (P1 − P2) = _____

RELIABILITY:

Design reliability, R (80 to 99 percent) = _____ percent

Overall standard deviation, S_o (typically 0.39) = _____

FUTURE STRUCTURAL CAPACITY:

Required slab thickness for future traffic is determined from rigid pavement design equation or nomograph in Part II, Figure 3.7.

D_f = _____ inches

---

deteriorated joints, cracks and other discontinuities in the existing slab prior to overlay. Very little such loss in PSI has been observed for JPCP or JRCP unbonded overlays.

The following information is needed to determine F_{jcu} to adjust overlay thickness for the extra loss in PSI from deteriorated reflection cracks that are not repaired:

Number of unrepaired deteriorated joints/mile

Number of unrepaired deteriorated cracks/mile

Number of expansion joints, exceptionally wide joints (greater than 1 inch) or full-depth, full-lane-width AC patches/mile

The total number of unrepaired deteriorated joints/cracks and other discontinuities per mile

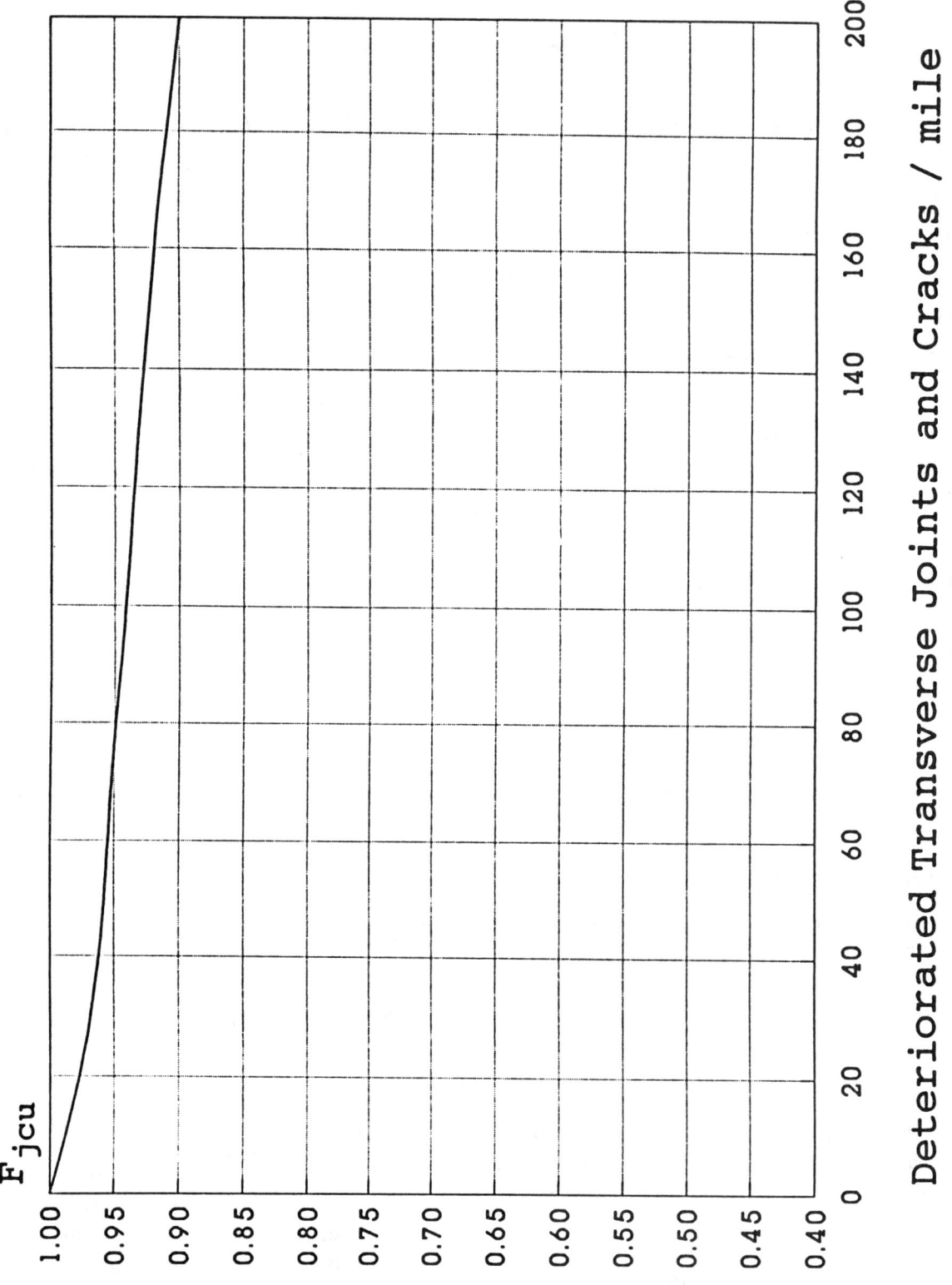

Figure 5.13. F_{jcu} Adjustment Factor for Unbonded JPCP, JRCP, and CRCP Overlays

prior to overlay is used to determine the F_{jcu} from Figure 5.13 for the appropriate type of PCC overlay. As an alternative to extensive full-depth repair for an unbonded overlay to be placed on a badly deteriorated pavement, a thicker AC interlayer should eliminate any reflection cracking problem, so that $F_{jcu} = 1.0$.

D_{eff} From Remaining Life For PCC Pavements

The remaining life of the pavement is given by the following equation:

$$RL = 100 \left[1 - \left(\frac{N_p}{N_{1.5}}\right)\right]$$

where

RL = remaining life, percent
N_p = total traffic to date, ESALs
$N_{1.5}$ = total traffic to pavement "failure," ESALs

$N_{1.5}$ may be estimated using the new pavement design equations or nomographs in Part II. To be consistent with the AASHO Road Test and the development of these equations, a "failure" PSI equal to 1.5 and a reliability of 50 percent are recommended.

D_{eff} is determined from the following equation:

$$D_{eff} = CF * D$$

where

CF = condition factor determined from Figure 5.2
D = thickness of the existing slab, inches (NOTE: maximum D for use in unbonded concrete overlay design is 10 inches even if the existing D is greater than 10 inches)

The designer should recognize that D_{eff} determined by this method does not reflect any benefit for preoverlay repair. The estimate of D_{eff} obtained should thus be considered a lower limit value. The D_{eff} of the pavement will be higher if preoverlay repair of load-associated distress is done. It is also emphasized that this method of determining D_{eff} is not applicable to AC/PCC pavements.

A worksheet for determination of D_{eff} is provided in Table 5.14.

Step 8: Determination of Overlay Thickness.

The thickness of unbonded PCC overlay is computed as follows:

$$D_{ol} = \sqrt{D_f^2 - D_{eff}^2}$$

where

D_{ol} = Required thickness of unbonded PCC overlay, inches
D_f = Slab thickness determined in Step 6, inches
D_{eff} = Effective thickness of existing slab determined in Step 7, inches

The thickness of overlay determined from the above relationship should be reasonable when the overlay is required to correct a structural deficiency. See Section 5.2.17 for discussion of factors which may result in unreasonable overlay thicknesses.

5.9.6 Shoulders

See Section 5.2.10 for guidelines.

5.9.7 Joints

Transverse and longitudinal joints must be provided in the same manner as for new pavement construction, except for the following joint spacing guidelines for JPCP overlays. Due to the unusually stiff support beneath the slab, it is advisable to limit joint spacing to the following to control thermal gradient curling stress:

Maximum joint spacing (feet)

= 1.75 * Slab thickness (inches)

Example: slab thickness = 8 inches

joint spacing = 8 * 1.75 = 14 feet

Table 5.14. Calculation of D_{eff} for Unbonded PCC Overlay of JPCP, JRCP, CRCP, and AC/PCC

Condition Survey Method:

  JPCP, JRCP, or CRCP Overlay:

    F_{jcu} Number of unrepaired deteriorated joints/mile   = _____

         Number of unrepaired deteriorated cracks/mile   = _____

         Number of unrepaired deteriorated punchouts/mile   = _____

         Number of expansion joints, exceptionally wide joints (>1 inch) or full-depth, full-lane-width AC patches/mile   = _____

         Total/mile   = _____

    F_{jcu} = _____ (Figure 5.13)

Effective Slab Thickness:

$$D_{eff} = F_{jcu} * D = _____$$

NOTES: Maximum D allowed is 10 inches for use in calculating D_{eff} for unbonded overlays. Existing AC surface is neglected in calculating D_{eff} for existing AC/PCC pavement when designing an unbonded PCC overlay.

Remaining Life Method:

    N_p = Past design lane ESALs   = _____

    $N_{1.5}$ = Design lane ESALs to P2 of 1.5   = _____

$$RL = 100 \left[1 - \left(\frac{N_p}{N_{1.5}} \right) \right] = _____$$

    CF = _____

$$D_{eff} = CF * D = _____$$

NOTE: Maximum D allowed is 10 inches for use in calculating D_{eff} for unbonded overlays.

5.9.8 Reinforcement

Unbonded JRCP and CRCP overlays must contain reinforcement to hold cracks tightly together. The design of the reinforcement would follow the guidelines given for new pavement construction, except that the friction factor would be high (e.g., 2 to 4) due to bonding between the AC separation layer and the new PCC overlay (see Part II, Section 3.4).

5.9.9 Separation Interlayer

A separation interlayer is needed between the unbonded PCC overlay and the existing slab to isolate the overlay from the cracks and other deterioration in the existing slab. The most common and successfully used separation interlayer material is an AC mixture placed one inch thick. If a level-up is needed the AC interlayer may also be used for that purpose (*29, 30*).

Some thin materials that have been used as bondbreakers have not performed well. Other thin layers have been used successfully, including surface treatments, slurry seals, and asphalt with sand cover for existing pavements without a large amount of faulting or slab breakup. For heavily trafficked highways, the potential problem of erosion of the interlayer must be considered. A thin surface treatment may erode faster than an AC material. There is no reason that a permeable open-graded interlayer cannot be used, provided a drainage system is designed to collect the water from this layer. This type of interlayer would provide excellent reflective crack control as well as preventing pumping and erosion of the interlayer.

5.9.10 Widening

See Section 5.2.16 for guidelines.

5.10 JPCP, JRCP, AND CRCP OVERLAY OF AC PAVEMENT

JPCP, JRCP, and CRCP overlays of AC pavement can be placed to improve both structural capacity and functional conditions. This type of overlay consists of the following major construction tasks:

(1) Repairing deteriorated areas and making subdrainage improvements (if needed)
(2) Constructing widening (if needed)
(3) Milling the existing surface if major distortion or inadequate cross-slope exists
(4) Placing an AC leveling course (if needed)
(5) Placing the concrete overlay
(6) Sawing and sealing the joints

5.10.1 Feasibility

A PCC overlay is a feasible rehabilitation alternative for AC pavements for practically all conditions. They are most cost-effective when the existing pavement is badly deteriorated. Conditions under which a PCC overlay would not be feasible include:

(1) The amount of deterioration is not large and other alternatives would be much more economical.
(2) Vertical clearance at bridges is inadequate for required overlay thickness. This may be addressed by reconstructing the pavement under the overhead bridges or by raising the bridges. Thicker PCC overlays may also necessitate raising signs and guardrails, as well as increasing side slopes and extending culverts. Sufficient right-of-way must be available or obtainable to permit these activities.
(3) The existing pavement is susceptible to large heaves or settlements.

If construction duration is critical, PCC overlays may utilize high-early-strength PCC mixes. PCC overlays have been opened within 6 to 24 hours after placement using these mixtures.

5.10.2 Pre-overlay Repair

One major advantage of a JPCP, JRCP, or CRCP overlay over AC pavement is that the amount of repair required for the existing pavement is greatly reduced. However, the following types of distress (on the next page) should be repaired prior to placement of the overlay to prevent reflection cracks that may reduce its service life. Guidelines on repairs are provided in References 1 and 3.

5.10.3 Reflection Crack Control

Reflection cracking is generally not a problem for JPCP, JRCP, or CRCP overlays of AC pavement. However, if the existing AC pavement has severe transverse thermal cracks, it may be desirable to place some type of separation layer over the transverse cracks to reduce the potential for reflection cracking.

Distress Type	Overlay Type	Repair Type
Alligator cracking	JPCP or JRCP	No repair needed
	CRCP	Patch areas with high deflections
Transverse cracks	JPCP, JRCP, CRCP	No repair needed
Pumping, stripping	JPCP, JRCP, CRCP	Edge drains (if needed)
		Remove stripping layer if severe
Settlement/heave	JPCP, JRCP, CRCP	Level-up with AC

5.10.4 Subdrainage

See Section 5.2.4 for guidelines.

5.10.5 Thickness Design

The required thickness of the PCC overlay is a function of the structural capacity required to meet future traffic demands and the support provided by the underlying AC pavement. The required overlay thickness to increase structural capacity to carry future traffic is determined by the following equation.

$$D_{ol} = D_f$$

where

D_{ol} = Required thickness of PCC overlay, inches
D_f = Slab thickness to carry future traffic, inches

PCC overlays of AC pavement have been successfully constructed as thin as 5 inches and as thick as 12 inches or more. Seven to 10 inches has been typical for most highway pavement overlays.

The required overlay thickness may be determined through the following design steps. These design steps provide a comprehensive design approach that recommends testing the pavement to obtain valid design inputs. If it is not possible to conduct this testing, an approximate overlay design may be developed based upon visible distress observations by skipping Steps 4 and 5, and by estimating other inputs.

The overlay design can be done for a uniform section or on a point-by-point basis as described in Section 5.3.1.

Step 1: Existing pavement design.

(1) Existing material types and layer thicknesses.

Step 2: Traffic analysis.

(1) Predicted future 18-kip ESALs in the design lane over the design period (N_f).

Step 3: Condition survey.

A detailed survey of distress conditions is not required. Only a general survey that identifies any of the following distresses that may affect the performance of a PCC overlay is needed:

(1) Heaves and swells.
(2) Signs of stripping of the AC. This could become even more serious under a PCC overlay.
(3) Large transverse cracks that, without a new separation layer, may reflect through the PCC overlay.

Step 4: Deflection testing (strongly recommended).

Measure deflection basins in the outer wheel path along the project at an interval sufficient to adequately assess conditions. Intervals of 100 to 1,000 feet are typical. A heavy-load deflection device (e.g., Falling Weight Deflectometer) and a load magnitude of 9,000 pounds are recommended. ASTM D 4694 and D 4695 provide additional guidance on deflection testing. Deflections should be measured at the center of the load and at least one other distance from the load, as described in Section 5.4.5, Step 4.

For each point tested, backcalculate the subgrade modulus (M_R) and the effective pavement modulus (E_p) according to the procedures described in Section 5.4 for AC pavements.

(1) *Effective dynamic k-value.* Estimate the effective dynamic k-value from Figure 3.3 in Part II, Section 3.2, using the backcalculated subgrade resilient modulus (M_R), the effective modulus of the pavement layers above the subgrade (E_p), and the total thickness of the pavement layers above the subgrade (D). It is emphasized that the backcalculated subgrade

resilient modulus value used to estimate the effective dynamic k-value should *not* be adjusted by the C factor (e.g., 0.33) which pertains to establishing the design M_R for AC overlays of AC pavements.

If a single overlay thickness is being designed for a uniform section, compute the mean effective dynamic k-value of the uniform section.

Step 5: Coring and materials testing.

Unless some unusual distress condition exists, coring and materials testing are not required.

Step 6: Determination of required slab thickness for future traffic (D_f).

(1) *Effective static k-value (at bottom of PCC overlay over an existing AC pavement).* Determine from one of the following methods.
   (a) Determine the effective dynamic k-value from the backcalculated subgrade modulus M_R, pavement modulus E_p, and pavement thickness D as described in Step 4. Divide the effective dynamic k-value by 2 to obtain the static k-value. The static k-value may need to be adjusted for seasonal effects (see Part II, Section 3.2.1).
   (b) Estimate from soils data and pavement layer types and thicknesses, using Figure 3.3 in Part II, Section 3.2. The static k-value obtained may need to be adjusted for seasonal effects (see Part II, Section 3.2.1).
(2) *Design PSI loss.* PSI immediately after overlay (P1) minus PSI at time of next rehabilitation (P2).
(3) *J, load transfer factor for joint design of the PCC overlay.* See Part II, Section 2.4.2, Table 2.6.
(4) *Modulus of rupture of PCC overlay.* Use mean 28-day, third-point-loading modulus of rupture of the overlay PCC.
(5) *Elastic modulus of PCC overlay.* Use mean 28-day modulus of elasticity of overlay PCC.
(6) *Loss of support.* See Part II.
(7) *Overlay design reliability, R (percent).* See Part I, Section 4.2, Part II, Table 2.2, and Part III, Section 5.2.15.
(8) *Overall standard deviation (S_o) for rigid pavement.* See Part I, Section 4.3.
(9) *Subdrainage capability of existing AC pavement,* after subdrainage improvements, if any.

See Part II, Table 2.5, as well as Reference 5, for guidance in determining C_d. In selecting this value, note that the poor drainage situation at the AASHO Road Test would be given a C_d of 1.0.

Compute D_f for the above design inputs using the rigid pavement design equation or nomograph in Part II, Figure 3.7. When designing an overlay thickness for a uniform pavement section, mean input values must be used. When designing an overlay thickness for specific points along the project, the data for that point must be used. A worksheet for determining D_f is provided in Table 5.15.

Step 7: Determination of Overlay Thickness.

The PCC overlay thickness is computed as follows:

$$D_{ol} = D_f$$

The thickness of overlay determined from the above relationship should be reasonable when the overlay is required to correct a structural deficiency. See Section 5.2.17 for discussion of factors which may result in unreasonable overlay thicknesses.

5.10.6 Shoulders

See Section 5.2.10 for guidelines.

5.10.7 Joints

See Section 5.8.7 for guidelines.

5.10.8 Reinforcement

See Section 5.8.8 for guidelines.

5.10.10 Widening

See Section 5.2.16 for guidelines.

Table 5.15. Worksheet for Determination of D_f for PCC Overlay of AC Pavement

SLAB:

    Type of load transfer system: mechanical device, aggregate interlock, CRCP

    Type of shoulder = tied PCC, other

    PCC modulus of rupture of unbonded overlay
(typically 600 to 800 psi) = _____ psi

    PCC E modulus of unbonded overlay (3 to 5 million psi) = _____ psi

    J load transfer factor of unbonded overlay
(2.5 to 4.4 for jointed PCC, 2.3 to 3.2 for CRCP) = _____

TRAFFIC:

    Future 18-kip ESALs in design lane over
the design period (N_f) = _____

SUPPORT AND DRAINAGE:

    Effective dynamic k-value = _____ psi/inch

    Effective static k-value = Effective dynamic k-value/2
(typically 50 to 500 psi/inch) = _____ psi/inch

    Subdrainage coefficient, C_d
(typically 1.0 for poor subdrainage conditions) = _____

SERVICEABILITY LOSS:

    Design PSI loss (P1 − P2) = _____

RELIABILITY:

    Design reliability, R (80 to 99 percent) = _____ percent

    Overall standard deviation, S_o (typically 0.39) = _____

FUTURE STRUCTURAL CAPACITY:

    Required slab thickness for future traffic is determined from rigid pavement design equation or nomograph in Part II, Figure 3.7.

    D_f = _____ inches

REFERENCES FOR CHAPTER 5

1. Federal Highway Administration, "Pavement Rehabilitation Manual," Pavement Division, Office of Highway Operations, Washington, D.C. (current edition).
2. Darter, M.I., Barenberg, E.J., and Yrjanson, W.A., "Joint Repair Methods For Portland Cement Concrete Pavements," NCHRP Report No. 281, Transportation Research Board, 1985.
3. "Techniques for Pavement Rehabilitation," Training Course Participants Notes, National Highway Institute, Federal Highway Administration, 3d Edition, 1987.
4. Snyder, M.B., Reiter, M.J., Hall, K.T., and Darter, M.I., "Rehabilitation of Concrete Pavements, Volume I—Repair Rehabilitation Techniques," Report No. FHWA-RD-88-071, Federal Highway Administration, 1989.
5. Smith, K.D., Peshkin, D.G., Darter, M.I., Mueller, A.L., and Carpenter, S.H., "Performance of Jointed Concrete Pavements, Phase I, Volume 5, Data Collection and Analysis Procedures," Federal Highway Administration Report No. FHWA/RD/89/140, March 1990.
6. Finn, F.N., and Monismith, C.L., "Asphalt Overlay Design Procedures," NCHRP Synthesis No. 116, Transportation Research Board, 1984.
7. Vespa, J.W., Hall, K.T., Darter, M.I., and Hall, J.P., "Performance of Resurfacing of JRCP and CRCP on the Illinois Interstate Highway System," Illinois Highway Research Report No. 517-5, Federal Highway Administration Report No. FHWA-IL-UI-229, 1990.
8. Darter, M.I., Elliott, R.P., and Hall, K.T., "Revision of AASHTO Pavement Overlay Design Procedures, Appendix: Overlay Design Examples," NCHRP Project 20-7/Task 39, Final Report, April 1992.
9. Darter, M.I., Elliott, R.P., and Hall, K.T., "Revision of AASHTO Pavement Overlay Design Procedures, Appendix: Documentation of Design Procedures," NCHRP Project 20-7/Task 39, April 1992.
10. Halstead, W.J., "Criteria For Use of Asphalt Friction Surfaces," NCHRP Synthesis No. 104, Transportation Research Board, 1983.
11. Thompson, M.R. and Robnett, Q.L., "Resilient Properties of Subgrade Soils," Final Report—Data Summary, Transportation Engineering Series No. 14, Illinois Cooperative Highway Research and Transportation Program Series No. 160, University of Illinois at Urbana-Champaign, 1976.
12. Taylor, M.L., "Characterization of Flexible Pavements by Nondestructive Testing," Ph.D. thesis, University of Illinois at Urbana-Champaign, 1979.
13. Carpenter, S.H., "Layer Coefficients for Flexible Pavements," ERES Consultants, Inc., report for Wisconsin DOT, August 1990.
14. Figueroa, J.L., "Resilient-Based Flexible Pavement Design Procedure for Secondary Roads," Ph.D. thesis, University of Illinois at Urbana-Champaign, 1979.
15. Raad, L. and Figueroa, J.L., "Load Response of Transportation Support Systems," *Transportation Engineering Journal*, American Society of Civil Engineers, Volume 106, No. TE1, 1980.
16. Pavement Consultancy Services/Law Engineering, "Guidelines and Methodologies for the Rehabilitation of Rigid Highway Pavements Using Asphalt Concrete Overlays," for National Asphalt Paving Association, June 1991.
17. Pavement Consultancy Services/Law Engineering, "FWD Analysis of PA I-81 Rubblization Project," for Pennsylvania Department of Transportation, February 1992.
18. Hall, K.T., "Performance, Evaluation, and Rehabilitation of Asphalt Overlaid Concrete Pavements," Ph.D. thesis, University of Illinois at Urbana-Champaign, 1991.
19. Schutzbach, A.M., "Crack and Seat Method of Pavement Rehabilitation," *Transportation Research Record* No. 1215, 1989.
20. Kilareski, W.P. and Bionda, R.A., "Performance/Rehabilitation of Rigid Pavements, Phase II, Volume 2—Crack and Seat and AC Overlay

of Rigid Pavements," Federal Highway Administration Report No. FHWA-RD-89-143, 1989.
21. Ahlrich, R.C., "Performance and Structural Evaluation of Cracked and Seated Concrete," *Transportation Research Record* No. 1215, 1989.
22. Thompson, M.R., "Breaking/Cracking and Seating Concrete Pavements," NCHRP Synthesis No. 144, 1989.
23. Smith, K.D., Darter, M.I., Rauhut, J.B., and Hall, K.T., "Distress Identification Manual for the Long-Term Pavement Performance (LTPP) Studies," Strategic Highway Research Program, 1988.
24. Southgate, H.F., "An Evaluation of Temperature Distribution Within Asphalt Pavements and its Relationship to Pavement Deflection," Kentucky Department of Highways, Research Report KYHPR-64-20, 1968.
25. Shell International Petroleum Company, "Pavement Design Manual," London, England, 1978.
26. Asphalt Institute, "Research and Development of the Asphalt Institute's Thickness Design Manual (MS-1) Ninth Edition," Research Report 82-2, 1982.
27. Hoffman, M.S. and Thompson, M.R., "Mechanistic Interpretation of Nondestructive Pavement Testing Deflections," Transportation Engineering Series No. 32, Illinois Cooperative Highway and Transportation Research Series No. 190, University of Illinois at Urbana-Champaign, 1981.
28. Peshkin, D.G., Mueller, A.L., Smith, K.D., and Darter, M.I., "Structural Overlay Strategies for Jointed Concrete Pavements, Vol. 3: Performance Evaluation and Analysis of Thin Bonded Concrete Overlays," Report No. FHWA-RD-89-144, Federal Highway Administration, 1990.
29. Hutchinson, R.L., "Resurfacing With Portland Cement Concrete," NCHRP Synthesis No. 99, Transportation Research Board, 1982.
30. Voigt, G.F., Carpenter, S.H., and Darter, M.I., "Rehabilitation of Concrete Pavements, Volume II—Overlay Rehabilitation Techniques," Report No. FHWA-RD-88-072, Federal Highway Administration, 1989.

PART IV
MECHANISTIC-EMPIRICAL DESIGN PROCEDURES

PART IV
MECHANISTIC-EMPIRICAL DESIGN PROCEDURES

1.1 INTRODUCTION

Part IV of the Guide is a brief overview of the use of analytical and mechanistic procedures for the design and evaluation of pavement structures.

The use of analytical methods to estimate the stress, strain, or deflection state of pavements is not new. For portland cement concrete pavements, the use of such methods for design dates back to at least 1938 when Bradbury (1) published his paper on design of reinforced concrete pavements. Friberg (2), Newmark (3), Pickett (4), and Ray (5) were among the early contributors in this field through the 1940's and 1950's. Since that time there have been extensive contributions by many investigators from government, industry, and academia. For asphalt concrete pavements, the publications of Burmister (6), McLeod (7), Acum and Fox (8), and Palmer (9), beginning in 1940, have provided some of the basic theories applicable to this type of pavement.

For purposes of this Guide, the use of analytical methods refers to the numerical capability to calculate the stress, strain, or deflection in a multi-layered system, such as a pavement, when subjected to external loads, or the effects of temperature or moisture. Mechanistic methods or procedures will refer to the ability to translate the analytical calculations of pavement response to performance. Performance, for the majority of procedures used, refers to physical distress such as cracking or rutting. For rigid pavements, the procedures have been applied to determination of dowel sizes, reinforcement requirements, and joint spacing. For flexible pavements, the mechanistic procedures have also been applied to roughness predictions.

Mechanistic design procedures are based on the assumption that a pavement can be modeled as a multi-layered elastic or visco-elastic structure on an elastic or visco-elastic foundation. Assuming that pavements can be modeled in this manner, it is possible to calculate the stress, strain, or deflection (due to traffic loadings and/or environments) at any point within or below the pavement structure. However, researchers recognize that pavement performance will likely be influenced by a number of factors which will not be precisely modeled by mechanistic methods. It is, therefore, necessary to calibrate the models with observations of performance, i.e., empirical correlations. Thus, the procedure is referred to in the Guide as a mechanistic-empirical design procedure.

Researchers in this field have hypothesized that modeling the pavement, as described above, should improve the reliability of the design equations which are, in effect, prediction models. For example, in Part II of the Guide, the design nomographs estimate the thickness of the pavement structure required to maintain an acceptable level of service for a specific number of traffic loadings. In a similar way, mechanistic procedures would predict the occurrence of distress or pavement deterioration as a function of traffic and environment or environment alone.

A state of the knowledge summary of mechanistic design procedures has been prepared as a working document and can be found in Volume 3 of the Guide. Volume 3 contains nine chapters and an appendix which describe in detail the current status of the development and use of these procedures for design of new and rehabilitated pavements.

Most current methods of design for flexible pavements make no direct use of mechanistic-design procedures. There are a few exceptions; for example, The Kentucky Department of Transportation (10), The Asphalt Institute (11), and Shell International (12) all have developed such procedures for general application to a variety of design considerations.

Most methods for structural design of rigid pavements do not include mechanistic design concepts. The method of the Portland Cement Association for fatigue cracking of PCCP is a representative example (13).

The design methodology incorporated in the 1972 issue of the AASHTO Design Guide for flexible pavements did not incorporate mechanistic procedures, although the supporting work, included in NCHRP Report 128 (14), did introduce such concepts for possible future use in the Guide. This issue (1986) of the

Guide has indirectly used mechanistic procedures for evaluating seasonal damage and to establish coefficients for drainage and load transfer. Also, the use of the resilient modulus to represent material properties introduces the concept that paving materials can be represented by a quasi-elastic modulus.

In summary, while mechanistic-empirical design procedures are still somewhat limited for use with flexible pavements, there is a significant body of research to draw from if an agency is interested in developing such design procedures. For rigid pavements the use of analytical methods and mechanistic procedures has been the standard of the industry for over 40 years.

1.2 BENEFITS

Researchers in working to develop mechanistic-empirical design procedures hypothesize that these methods, which are based on long-established theory, will model a pavement more correctly than the empirical equations which have been traditionally used for flexible pavements and for some aspects of rigid pavements. The primary benefits which could accrue from the successful application of mechanistic procedures will be: (1) improved reliability for design, (2) ability to predict specific types of distress, and (3) the ability to extrapolate from limited field and laboratory results.

The ability or lack of ability to design a pavement for site-specific conditions influences the amount of conservatism to be included in design. The consequences of increased conservatism will result in less than optimum use of funds. For example, the more conservatism built into each project limits the number of projects that can be constructed in any given time period. Thus, more reliable design methods would result in optimum use of available funds.

A second major benefit of mechanistic procedures is the ability to predict specific types of distress; e.g., cracking, faulting, rutting, etc. Pavement management systems require the ability to predict the occurrence of distress in order to minimize the costs of maintenance and rehabilitation. Mechanistic procedures offer the best opportunity to meet this requirement for PMS.

The third major benefit would be the ability to extrapolate from limited amounts of field or laboratory data before attempting full-scale, long-term demonstration projects. This screening process could save money and time by eliminating those concepts which are judged to have very little merit.

A subset of benefits which could result from the development of mechanistic procedures are summarized as follows:

(1) Estimates of the consequences of new loading conditions can be evaluated. For example, the damaging effects of increased loads, high tire pressures, multiple axles, etc., can be modeled using mechanistic procedures.
(2) Better utilization of available materials can be estimated. For example, the use of stabilized materials in both rigid and flexible pavements can be simulated to predict future performance.
(3) Improved procedures to evaluate premature distress can be developed or conversely to analyze why some pavements exceed their design expectations. In effect, better diagnostic techniques can be developed.
(4) Aging can be included in estimates of performance, e.g., asphalts harden with time which, in turn, affects both fatigue cracking and rutting.
(5) Seasonal effects such as thaw-weakening can be included in estimates of performance.
(6) Consequences of subbase erosion under rigid pavements can be evaluated.
(7) Methods can be developed to better evaluate the long-term benefits of providing improved drainage in the roadway section.

In summary, while the application of mechanistic-empirical design procedures have had only limited application for flexible pavements, there is a consensus among most researchers that such methods offer the best opportunity to improve pavement technology for this type of construction for the next several decades. The application of analytical and mechanistic procedures are now used for the design of rigid pavements and have proven to be effective. The extension of these techniques to new designs and different applications is considered to be a viable objective in the years to come.

1.3 FRAMEWORK FOR DEVELOPMENT AND APPLICATION

Figure 1.1 illustrates the framework for the development of mechanistic-empirical design procedures for new designs and for rehabilitation. Figure 1.2 illustrates the application to an overlay for flexible pavements.

The inputs required for the system include traffic, roadbed soil properties, environment, material char-

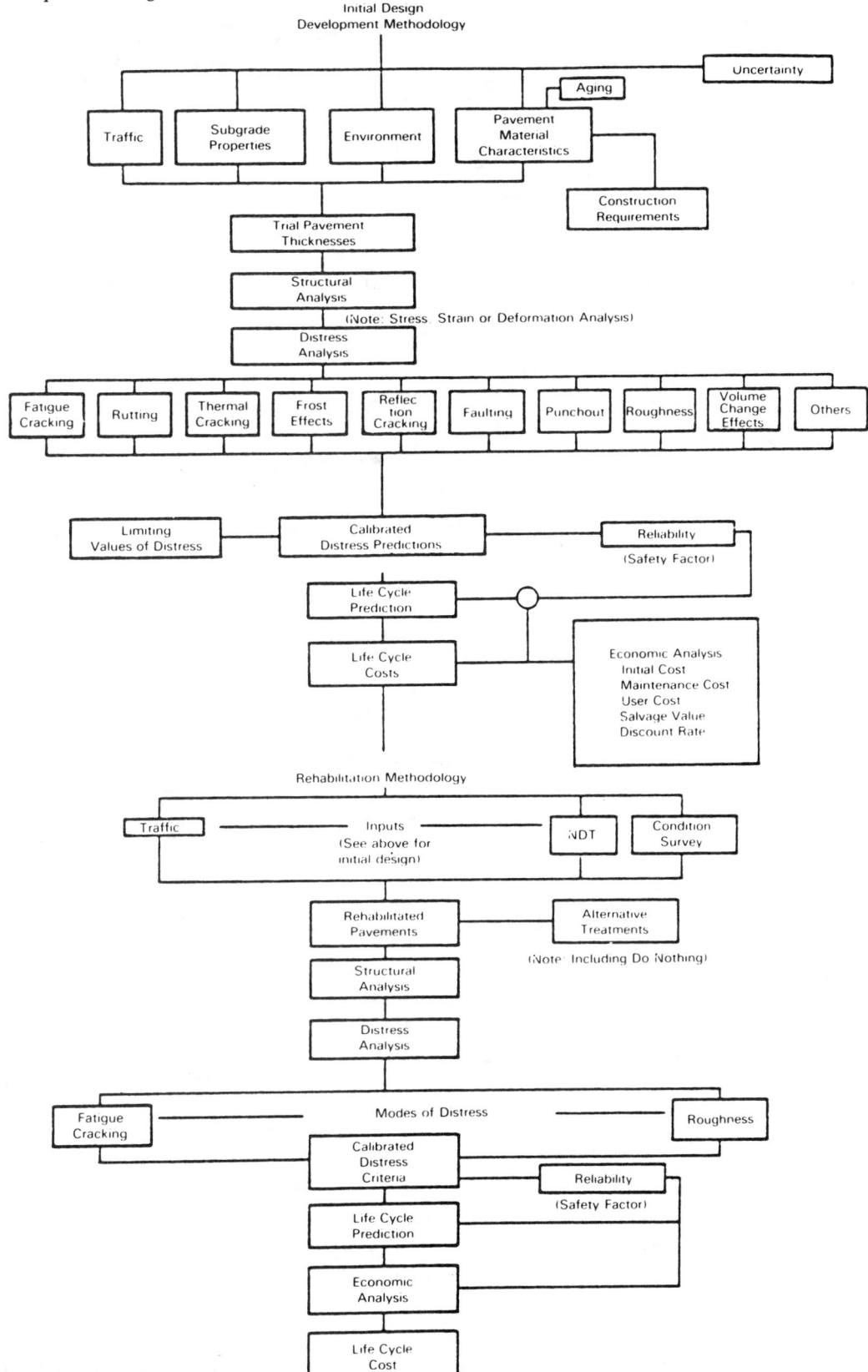

Figure 1.1. Example Approach for the Development of a Mechanistic-Empirical Design Model

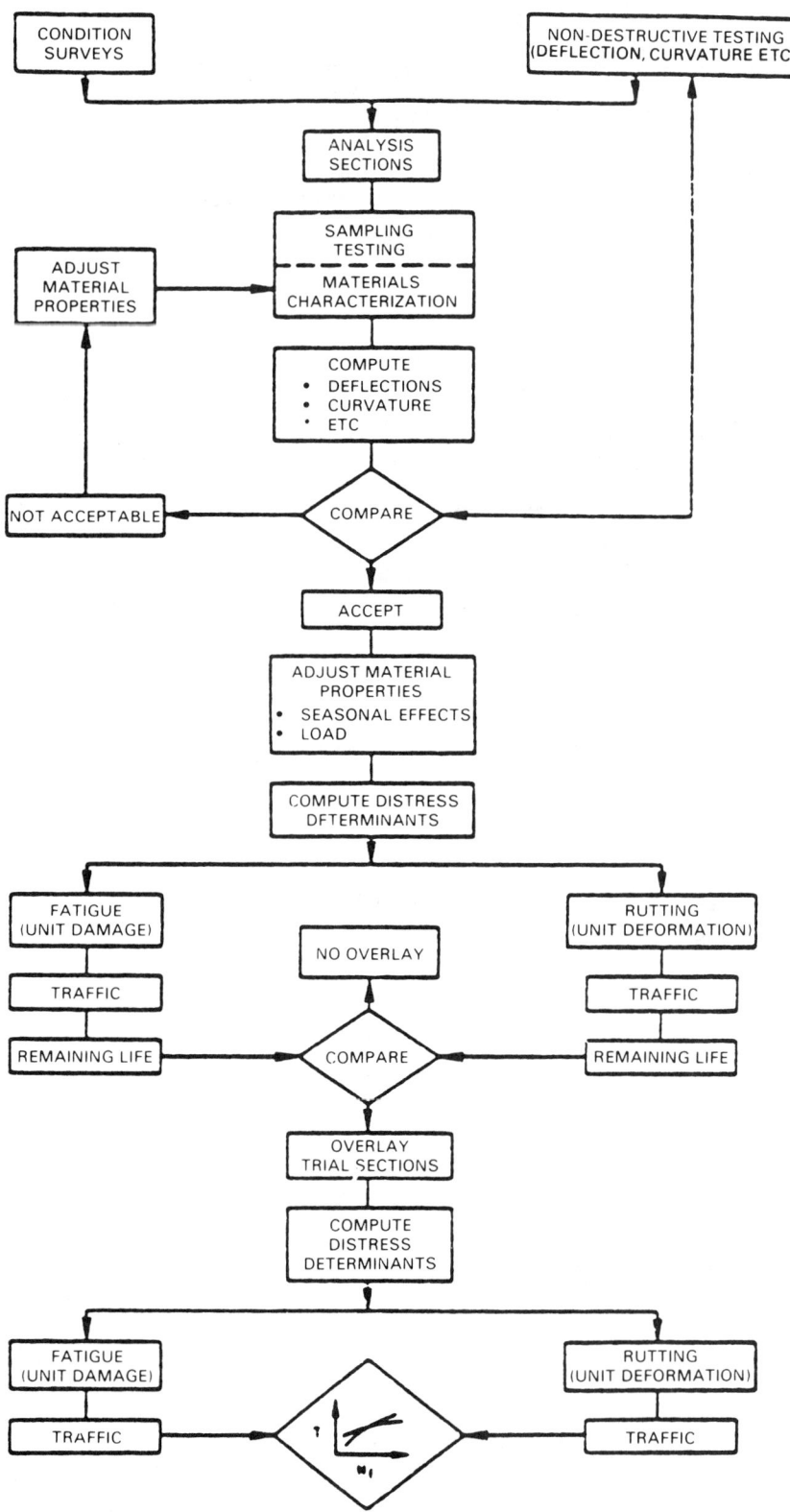

Figure 1.2. Overlay Design Procedure Using Elastic Layered Theory to Represent Pavement Response

Mechanistic-Empirical Design Procedures

acteristics and uncertainty, i.e., variance on each of the inputs. Aging of materials and construction requirements can be included in material characteristics.

The process requires the selection of a series of trial pavement sections, either rigid or flexible, which are considered to include a range of thicknesses and materials appropriate to the design problem. A structural (analytical) analysis is made for each trial section to calculate the stress, strain, and deflection at specific locations depending on distress criteria.

A working hypothesis or distress criteria must be developed for each of the distress types to be predicted by the procedure. For example, for fatigue cracking the distress criteria for rigid pavements is based on the maximum tensile stress in the slab and for flexible pavements is based on the maximum tensile strain in the asphalt concrete. Similar criteria have been developed for each of the distress types shown on Figure 1.1, i.e., fatigue cracking, rutting, faulting, punchouts. The "others" refer to future developments which can or will be developed.

Since Figure 1.1 represents the development framework, it is necessary to calibrate the distress models to field observations. This step will provide information regarding the relationship of the pavement response to various levels and amounts of distress. From these correlations, it has been possible to establish mechanistic-empirical relationships for estimating the type and amount of deterioration as a function of the input variables.

Based on life-cycle predictions, it is possible to calculate life-cycle costs according to procedures described in Part I, Chapter 3, of this Guide. The framework requires that several trial analyses be completed in order to be able to interpolate for specific design conditions and to select a best solution based on performance and cost.

A similar procedure is illustrated for rehabilitation. In this case, the method would require trial designs appropriate for rehabilitation as a starting point. However, as with new construction, the inputs would include traffic, roadbed soil properties, construction requirements, etc.

Figure 1.1 represents the general framework which has been used by most researchers to develop mechanistic design procedures and which could be applied by user agencies (state highway departments) as a guide for in-house development.

In Figure 1.2, the general methodology proposed by some investigators (*15*) for overlays is outlined. The method is summarized briefly as follows:

(1) Condition surveys and nondestructive testing (deflections, curvature, etc.) information is collected in order to identify analysis sections.
(2) Sampling and testing of in-place material is completed on a limited basis in order to adjust material properties to field conditions.
(3) Seasonal variations of material properties and traffic are determined in order to calculate distress. The information used in this step would come from research and development as represented in Figure 1.1.
(4) The two types of distress used for design of the asphalt overlay are rutting and fatigue cracking. For example, if the original pavement had been portland cement concrete, the criteria used might be reflection cracking and rutting.
(5) Based on the distress analysis, a determination is made as to the need for a structural overlay; if none is required, the worn-out pavement will be renewed by recycling, milling, or a thin overlay at minimum cost. If an overlay is required, trial sections will be analyzed which produce plots of traffic versus thickness requirements needed to satisfy specific performance requirements, e.g., rutting not to exceed 0.5 inches and cracking, not more than 25 percent.

In summary, Figure 1.1 provides a set of guidelines, a framework, for developing mechanistic design procedures. Other more detailed procedures may be used; however, the general requirements will be the same. Figure 1.2 illustrates the application of these guidelines to a specific case, i.e., overlay of flexible pavement. It is emphasized that these procedures and applications are not new. The technical literature; American Society of Civil Engineers, Transportation Research Board, American Concrete Institute, American Society for Testing and Materials, the Association of Asphalt Paving Technologists, as well as various international conferences, all have a wealth of background information pertinent to mechanistic-empirical design developments. The technical information is available; however, field trials are limited.

1.4 IMPLEMENTATION

The implementation of a mechanistic pavement design procedure requires that consideration be given to the following items:

(1) determination of types of design considerations, i.e., cracking, roughness, etc.,
(2) development of a plan to obtain input information, i.e., moduli,
(3) equipment acquisition,
(4) computer hardware and software,
(5) training personnel,
(6) development and calibration of prediction models, and
(7) testing.

The implementation of the mechanistic procedures could take several forms.

(1) The procedures could be used to develop design curves similar to those developed by The Asphalt Institute, Shell International, or the Kentucky Department of Transportation. In this form, the analyst will presolve a larger number of problems sufficient to develop design curves. The user is not required to do any analytical work in order to prepare design recommendations. A relatively simple step-by-step procedure can be specified for design which is very similar to procedures in Parts II and III of the Guide.
(2) The procedures could be used in site-specific cases to predict performance when conditions exceed normal design criteria, e.g., excessive loads on standard vehicles or any load on non-standard vehicles.
(3) The procedures could be used to answer "what if" questions, e.g., what would be the effect of increasing the legal axle load on performance or what would be the effect of increased tire pressure or what are the likely consequences of noncompliance with specifications.

Once a user agency has the capability to use mechanistic design procedures, it can be anticipated that many additional applications will be found.

An agency should be aware that the development implementation of mechanistic design procedures will require a commitment of resources and, in general, it will require special training for persons involved in the implementation phase.

Principles of mechanistic design procedures for flexible pavements have been developed through more than 20 years of research. However, for implementation, some special attention will need to be given to the seven items previously enumerated. A brief discussion of these items follows; agency personnel should be aware that this is an overview and that variations are likely for any specific case.

1.4.1 Design Considerations

Mechanistic design procedures can be applied to a wide variety of pavement performance conditions as described in Section 1.3. The dominant types of distress which can be predicted by mechanistic design procedures relate to physical conditions caused by traffic loadings or environment. Only one mechanistic design procedure relates directly to the prediction of ride quality, e.g., present serviceability index (PSI). The VESYS program developed for FHWA has included this capability but requires careful calibration. For ride quality, other agencies have developed empirical methods using calculations of stress, strain, or deformation as independent variables for correlating with field observations.

Thus, design considerations are best suited to physical distress. Only those distress types which control performance or trigger some kind of maintenance or rehabilitation need be considered. For example, for asphalt concrete-surfaced pavements, fatigue cracking, rutting, and possibly low temperature cracking would be likely candidates. If one or more of these is not a problem for the developing agency, it can be eliminated. For rigid pavements, fatigue cracking, faulting, and pumping would be considered.

A well-thought-out experiment design should be prepared in obtaining input information. For example, it will be desirable to develop regression equations to predict modulus relationships for soils and paving materials. A well-planned experiment to include material index values and in-place conditions will reduce the amount of testing required for routine design. Similar experiment designs for collecting pertinent temperature, rainfall, and traffic information will facilitate development of simplified procedures to satisfy input requirements.

In summary, decide what types of distress should control design and establish threshold values for each type, i.e., how much cracking or faulting is considered acceptable before maintenance or rehabilitation costs become excessive.

1.4.2 Input Data

In order to make the necessary calculations, it will be necessary to obtain information pertaining to material properties, environment, and traffic patterns.

When obtaining information concerning material properties, it will be necessary to acquire special testing equipment if such equipment is not already on hand. This equipment for laboratory and field testing is commercially available and some laboratory equipment can be fabricated in-house if preferred.

Equipment requirements will depend on data requirements as follows:

Resilient Modulus of Soil and Granular Materials. Refers to laboratory repeated-load testing equipment for cohesive and noncohesive soil and granular materials. Both confined and unconfined test capabilities will be required.

Resilient Modulus of Stabilized Materials. Refers to laboratory repeated-load testing equipment for asphalt concrete, portland cement concrete, and materials stabilized with asphalt, cement, lime, lime-flyash, or cement-flyash combinations.

Deflection Basin Measurement Equipment. Refers to field testing equipment capable of measuring the deflection basin several feet from the loading plate, preferably at different load magnitudes including the design truck wheel loading (e.g., 9,000 pounds). The deflection basin data can then be used in conjunction with computer programs to "back calculate" the in situ resilient moduli of the pavement layers and the roadbed soil.

1.4.3 Equipment Acquisition

This refers to special equipment designed to facilitate field data collection, e.g., nondestructive testing equipment, road meters, etc.

It will be important in obtaining the necessary equipment to ensure the equipment is designed to furnish the information needed by the mechanistic procedure. In some situations, equipment owned by an agency may not be suitable. There is a tendency to try and establish correlations between pieces of equipment to avoid replacement. Such correlations should be discouraged since they will inevitably introduce error into the procedure. If such correlations are considered to be imperative, they should be made using well-planned experiment designs.

1.4.4 Computer Hardware and Software

Nearly all mechanistic design procedures will require some type of computer hardware and software to perform the detailed computations that are necessary. Most current procedures require a mainframe computer. However, the capability of the latest personal computer models will shortly make it possible to run most mechanistic design procedures and structural analysis programs on a desk-top personal computer. This capability will have a profound impact on making mechanistic design procedures much more user friendly and accessible to the practicing highway pavement design engineer.

The computer software for the new mechanistic pavement design procedures must be installed on the agency mainframe and/or personal computers. This is normally not difficult with a well-documented program. However, it is most important to verify carefully that the program is performing as required. Test programs should be run where the correct solutions are available for verification. In addition, the design engineer should make numerous runs of the program after changing the design inputs over a practical range to evaluate the sensitivity of the design procedure to changes in design inputs. This will not only help to provide confidence in the design procedure, but also show which inputs are most significant; that is, which values should be determined accurately and which ones can be estimated.

1.4.5 Training Personnel

Mechanistic design procedures may utilize concepts, procedures, and equipment with which the practicing pavement design engineer is not familiar. Thus, it is often important that some training accompany the implementation of the mechanistic design procedure. This training could cover some or all of the following:

(1) structural analysis of pavements,
(2) procedures for estimating distress or pavement damage,
(3) effects of climate on pavement performance,
(4) nondestructive testing and interpretation,
(5) use of computers (particularly personal computers),
(6) laboratory repeated-load testing methods and interpretation,
(7) knowledge of basic materials properties related to pavement design, and
(8) pavement evaluation.

Some of the above information is available from training courses provided by the National Highway Institute of the Federal Highway Administration. Also, a few universities offer short courses in pavement design and rehabilitation that may cover several of the above topics. Consultants are also available to assist with "hands-on" training programs if needed.

1.4.6 Field Testing and Calibration

The most important step in the implementation process is field testing and calibration of the predictive models that are utilized in the mechanistic design procedure. Even though a mechanistic design procedure is developed using basic material properties and structural analysis techniques, there are still numerous assumptions and simplifications that must be made in its development. In fact, most mechanistic procedures actually include a combination of mechanistic and empirical predictive models that are used in the design process. For example, there is also the problem of climate, which is so complex that it will never be completely modeled in pavement design. Thus, climate, aging, and other factors must be considered empirically.

It is necessary to ensure that the predictive models used in the mechanistic design procedure (e.g., fatigue cracking, rutting, joint deterioration, faulting) actually give reasonable predictions for the geographic regions under consideration. Thus, climate, materials, thickness combinations, and traffic should be included in the experiment design for calibration. If this verification/calibration testing is not accomplished, there is a risk that the mechanistic design procedures will provide results that are not acceptable or accurate.

An example calibration is summarized as follows:

(1) Obtain data from at least 20 actual field test pavements that have been selected with known design, materials, traffic, and climate data. The sections should range from extensive distress to very little distress or overall deterioration.
(2) The inputs to the structural analysis model should be obtained as specified in the design procedure (e.g., strain, stress, strength, resilient modulus, number of applied traffic loads, climate, etc.) for each of the field pavement sections.
(3) Distress estimates should be computed for each section using the appropriate output from the structural analysis combined with damage criteria.
(4) The estimates are compared with actual field observations of distress to determine calibration factors. A calibration procedure such as this will result in realistic pavement designs and will provide the needed confidence and credibility for the mechanistic approach.

1.4.7 Testing

After the calibration process has been completed, the prediction models developed for each distress type should be tested on a wide range of projects for which performance information is available. Some final adjustments in the distress models may be necessary as part of this final step. The agency should maintain an on-going program of data acquisition to continually improve the system.

At the completion of this final testing, the agency will have a verified, reliable mechanistic-empirical design system with capabilities beyond the usual empirical design methods.

1.5 SUMMARY

The benefits of implementing mechanistic-empirical design procedures for new pavement construction, reconstruction, and/or rehabilitation are many. The benefits occur on both the network pavement management level as well as the project level. The key benefit is in providing the designer with powerful tools to evaluate the performance (specific distress types) of different pavement designs, instead of relying solely on limited empirical correlations or opinions.

Many different pavement design factors can be examined using a mechanistic design approach. Those which improve the performance of individual pavement components can be identified and a life-cycle cost analysis conducted to determine the cost-effectiveness of the design modification. Thus, mechanistic design has the potential to improve pavement design and to provide more reliable design procedures.

The widespread acquisition and use of personal computers that are capable of handling mechanistic design programs will also provide a much more user-friendly and practical design environment for the pavement designer. This is expected to greatly increase the potential for use of mechanistic design procedures.

REFERENCES FOR PART IV

1. Bradbury, R.D., "Reinforced Concrete Pavements," Wire Reinforcement Institute, Washington, D.C., 1938.
2. Friberg, B.F., "Load and Deflection Characteristics of Dowels in Transverse Joints of Concrete Pavements," *Proceedings*, Highway Research Board, 1938.
3. Newmark, N.M., "Influence Charts for Computation of Stresses in Elastic Foundations," University of Illinois Engineering Experiment Station Bulletin 338, 1942.
4. Pickett, G., Raville, M.E., Janes, W.C., and McCormick, F.J., "Deflection, Moments and Reactive Pressures for Concrete Pavements," Kansas State College Experiment Station Bulletin 65, 1951.
5. Pickett, G., and Ray, G.K., "Influence Charts for Rigid Pavements," Transactions, ASCE, 1951.
6. Burmister, D.M., "The Theory of Stresses and Displacements in Layered Systems and Application to the Design of Airport Runways," *Proceedings*, Highway Research Board, 1943.
7. McLeod, N.W., "Some Basic Problems in Flexible Pavement Design," *Proceedings*, Highway Research Board, 1953.
8. Acum, W.E.A., and Fox, L., "Computation of Load Stresses in a Three-Layer Elastic System," Geotechnique, Volume 2, pp. 293–300, 1951.
9. Palmer, L.A., "The Evaluation of Wheel Load Bearing Capacities of Flexible Types of Pavements," *Proceedings*, Highway Research Board, 1946.
10. Southgate, H.F., Deen, R.C., and Havens, J.H., "Development of a Thickness Design System for Bituminous Concrete Pavements," Research Report UKTRP-81-20, Kentucky Transportation Research Program, University of Kentucky, November 1981.
11. "Thickness Design—Asphalt Pavements for Highways and Streets," The Asphalt Institute, Manual Series No. 1, September 1981.
12. "Shell Pavement Design Manual," Shell International Petroleum Company, 1978.
13. "Thickness Design for Concrete Pavement," Portland Cement Association, 1966; with revisions in 1984.
14. Van Til, C.J., McCullough, B.F., Vallerga, B.A., and Hicks, R.G., "Evaluation of Interim Guides for Design of Pavement National Cooperative Highway Program, 128, 1972.
15. Finn, F.N., and Monismith, C.L., "Asphalt Overlay Design Procedures;" National Cooperative Highway Research Synthesis Report, Project 20-5 (in publication, 1985).

APPENDICES

APPENDIX A
GLOSSARY OF TERMS

Analysis Period The period of time for which the economic analysis is to be made; ordinarily will include at least one rehabilitation activity.

Base Course The layer or layers of specified or selected material of designed thickness placed on a subbase or a subgrade to support a surface course.

Composite Pavement A pavement structure composed of an asphalt concrete wearing surface and portland cement concrete slab; an asphalt concrete overlay on a PCC slab is also referred to as a composite pavement.

Construction Joint A joint made necessary by a prolonged interruption in the placing of concrete.

Contraction Joint A joint normally placed at recurrent intervals in a rigid slab to control transverse cracking.

Deformed Bar A reinforcing bar for rigid slabs conforming to "Requirements for Deformations," in AASHTO Designations M 31, M 42, or M 53.

Dowel A load transfer device in a rigid slab, usually consisting of a plain round steel bar.

Drainage Coefficients Factors used to modify layer coefficients in flexible pavements or stresses in rigid pavements as a function of how well the pavement structure can handle the adverse effect of water infiltration.

Equivalent Single Axle Loads (ESAL's) Summation of equivalent 18,000-pound single axle loads used to combine mixed traffic to design traffic for the design period.

Expansion Joint A joint located to provide for expansion of a rigid slab, without damage to itself, adjacent slabs, or structures.

Flexible Pavement A pavement structure which maintains intimate contact with and distributes loads to the subgrade and depends on aggregate interlock, particle friction, and cohesion for stability.

Layer Coefficient (a_1, a_2, a_3) The empirical relationship between structural number (SN) and layer thickness which expresses the relative ability of a material to function as a structural component of the pavement.

Load Transfer Device A mechanical means designed to carry loads across a joint in a rigid slab.

Longitudinal Joint A joint normally placed between traffic lanes in rigid pavements to control longitudinal cracking.

Low-Volume Road A roadway generally subjected to low levels of traffic; in this Guide, structural design is based on a range of 18-kip ESAL's from 50,000 to 1,000,000 for flexible and rigid pavements, and from 10,000 to 100,000 for aggregate-surfaced roads.

Maintenance The preservation of the entire roadway, including surface, shoulders, roadsides, structures, and such traffic control devices as are necessary for its safe and efficient utilization.

Modulus of Subgrade Reaction (k) Westergaard's modulus of subgrade reaction for use in rigid pavement design (the load in pounds per square inch on a loaded area of the roadbed soil or subbase divided by the deflection in inches of the roadbed soil or subbase, psi/in.).

Panel Length The distance between adjacent transverse joints.

Pavement Performance The trend of serviceability with load applications.

Pavement Rehabilitation Work undertaken to extend the service life of an existing facility. This includes placement of additional surfacing material and/or other work necessary to return an existing roadway, including shoulders, to a condition of structural or functional adequacy. This could include the complete removal and replacement of the pavement structure.

Pavement Structure A combination of subbase, base course, and surface course placed on a subgrade to support the traffic load and distribute it to the roadbed.

Performance Period The period of time that an initially constructed or rehabilitated pavement structure will last (perform) before reaching its terminal serviceability; this is also referred to as the *design period*.

Prepared Roadbed In-place roadbed soils compacted or stabilized according to provisions of applicable specifications.

Present Serviceability Index (PSI, p) A number derived by formula for estimating the serviceability rating from measurements of certain physical features of the pavement.

Pumping The ejection of foundation material, either wet or dry, through joints or cracks, or along edges of rigid slabs resulting from vertical movements of the slab under traffic.

Reinforcement Steel embedded in a rigid slab to resist tensile stresses and detrimental opening of cracks.

Resilient Modulus A measure of the modulus of elasticity of roadbed soil or other pavement material.

Rigid Pavement A pavement structure which distributes loads to the subgrade, having as one course a portland cement concrete slab of relatively high-bending resistance.

Roadbed The graded portion of a highway between top and side slopes, prepared as a foundation for the pavement structure and shoulder.

Roadbed Material The material below the subgrade in cuts and embankments and in embankment foundations, extending to such depth as affects the support of the pavement structure.

Selected Material A suitable native material obtained from a specified source such as a particular roadway cut or borrow area, of a suitable material having specified characteristics to be used for a specific purpose.

Serviceability The ability at time of observation of a pavement to serve traffic (autos and trucks) which use the facility.

Single Axle Load The total load transmitted by all wheels of a single axle extending the full width of the vehicle.

Structural Number (SN) An index number derived from an analysis of traffic, roadbed soil conditions, and environment which may be converted to thickness of flexible pavement layers through the use of suitable layer coefficients related to the type of material being used in each layer of the pavement structure.

Subbase The layer or layers of specified or selected material of designed thickness placed on a subgrade to support a base course (or in the case of rigid pavements, the portland cement concrete slab).

Subgrade The top surface of a roadbed upon which the pavement structure and shoulders are constructed.

Surface Course One or more layers of a pavement structure designed to accommodate the traffic load, the top layer of which resists skidding, traffic abrasion, and the disintegrating effects of climate. The top layer of flexible pavements is sometimes called "wearing course."

Tandem Axle Load The total load transmitted to the road by two consecutive axles extending across the full width of the vehicle. (NOTE: The spacing of the tandem axles used at the AASHTO Road Test was 48 inches.)

Tie Bar A deformed steel bar or connector embedded across a joint in a rigid slab to prevent separation of abutting slabs.

Traffic Equivalence Factor (e) A numerical factor that expresses the relationship of a given axle load to another axle load in terms of their effect on the serviceability of a pavement structure. In this Guide, all axle loads are equated in terms of the equivalent number of repetitions of an 18-kip axle.

Triple (Tridem) Axle Load The total load transmitted to the road by three consecutive axles extending across the full width of the vehicle. (NOTE: There were no tridem axles at the AASHTO Road Test; however, the spacing that may be inferred for consecutive triple axles (based on the tandem axle spacings) is 48 inches.)

Welded Wire Fabric (WWF) A Two-way reinforcement system for rigid slabs, fabricated from cold-drawn steel wire, having parallel longitudinal wires welded at regular intervals to parallel transverse wires. The wires may be either smooth or deformed. Deformed wire (used in deformed wire fabric, DWF) is that which has uniformly spaced deformations which inhibit longitudinal movement of the wire and which conform to "Specifications for Welded Deformed Steel Wire Fabric for Concrete Reinforcement," AASHTO Designation—M 221.

APPENDIX B
PAVEMENT TYPE SELECTION GUIDELINES

B.1 GENERAL

The highway engineer or administrator does not have at his disposal an absolute or undisputable method for determining the type of pavement which should be selected for a given set of conditions. However, the selection of pavement type should be an integral part of any pavement management program.

The selection of pavement type is not an exact science but one in which the highway engineer or administrator must make a judgement on many varying factors such as traffic, soils, weather, materials, construction, maintenance, and environment. The pavement type selection may be dictated by an overriding consideration for one or more of these factors.

The selection process may be facilitated by comparison of alternate structural designs for one or more pavement types using theoretical or empirically derived methods. However, such methods are not so precise as to guarantee a certain level of performance from any one alternate or comparable service for all alternates.

Also, comparative cost estimates can be applied to alternate pavement designs to aid in the decision-making process. The cost for the service of the pavement would include not only the initial cost but also subsequent cost to maintain the service level desired. It should be recognized that such procedures are not precise since reliable data for maintenance, subsequent stages for construction, or corrective work and salvage value are not always available, and it is usually necessary to project costs to some future point in time. Also, economic analyses are generally altruistic in that they do not consider the present or future financial capabilities of the contracting agency.

Even if structural design and cost comparative procedures were perfected, they would not by their nature encompass all factors which should be considered in pavement type determination. Such a determination should properly be one of professional engineering judgement based on the consideration and evaluation of all factors applicable to a given highway section.

The factors which may have some influence in the decisionmaking process are discussed below. They are generally applicable to both new and reconstructed pavements. One group includes those factors which may have major influence and may dictate the pavement type in some instances. Some of the major factors are also incorporated in the basic design procedures and influence the structural requirements of the pavement design or subgrade and embankment treatments. In such cases they are assigned an economic value for comparative purposes. The second group includes those factors which have a lesser influence and are usually taken into account when there are no overriding considerations or one type is not clearly superior from an economic standpoint. A flow chart of pavement selection procedure incorporating the major and secondary factors is shown in Figure B.1.

B.2 PRINCIPAL FACTORS

1. **Traffic**

    While the total volume of traffic affects the geometric requirements of the highway, the percentage of commercial traffic and frequency of heavy load applications generally have the major effect on the structural design of the pavement.

    Traffic forecasts for design purposes have generally projected normal growth in the immediate corridor with an appropriate allowance for changes in land use and potential commercial and industrial development. However, experience over the past several decades has shown that the construction of new major highway facilities diverts large amounts of heavy traffic from other routes in a broad traffic corridor. This, coupled with a decline in the quantity of railroad services, has resulted in a considerable underestimation of traffic growth, particularly commercial traffic. Also, the future availability and cost of motor fuels could result in increased legal loads to which pavement structures could be subjected during their design period.

    For these reasons, pavement designs for major facilities should incorporate an appropriate mar-

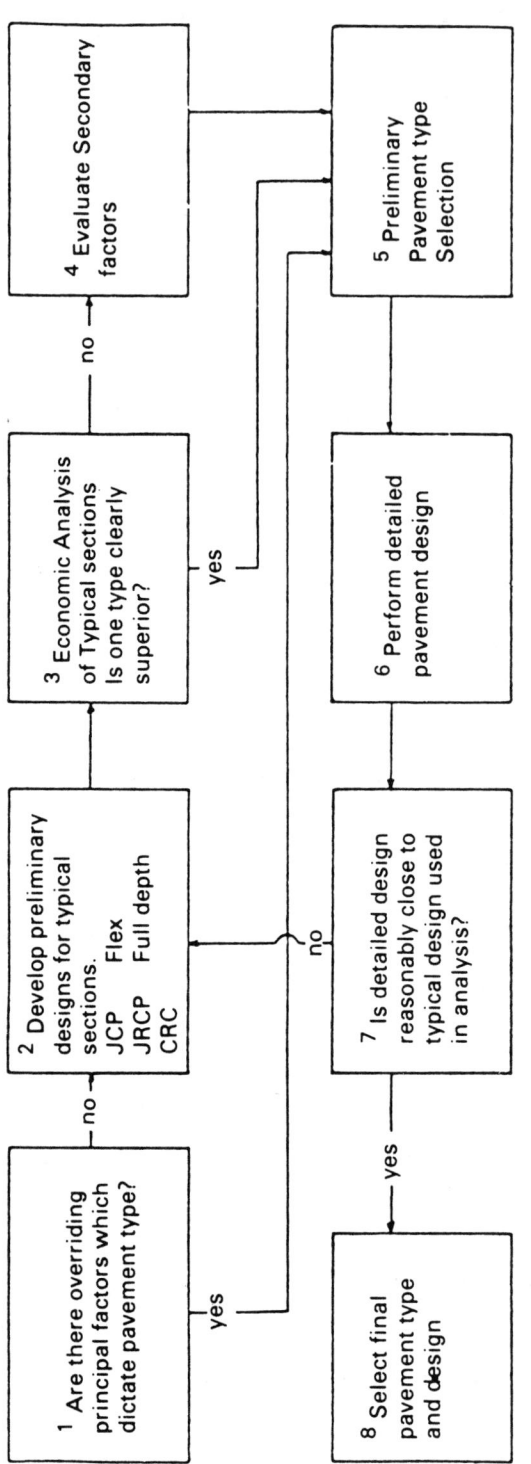

Figure B.1. Pavement Type Selection Process

gin of safety in the traffic factor. Agencies may choose to establish minimum structural requirements for all alternate pavement types to ensure adequate performance and service life for major facilities. Alternate strategies, or combinations of initial design, rehabilitation and maintenance, can be developed to provide equivalent service over a given period of time although the initial designs are not equivalent. For heavily traveled facilities in congested locations, the need to minimize the disruptions and hazards to traffic may dictate the selection of those strategies having long initial service life with little maintenance or rehabilitation regardless of relative economics.

2. **Soils Characteristics**

The load-carrying capability of a native soil, which forms the subgrade for the pavement structure, is of paramount importance in pavement performance. Even in given limited areas the inherent qualities of such native soils are far from uniform, and they are further subjected to variations by the influence of weather.

The characteristics of native soils not only directly affect the pavement structure design but may, in certain cases, dictate the type of pavement economically justified for a given location. As an example, problem soils that change volume with time frequently require stage construction to provide an acceptable riding surface.

3. **Weather**

Weather affects subgrade as well as the pavement wearing course. The amount of rainfall, snow and ice, and frost penetration will seasonally influence the bearing capacity of subgrade materials. Moisture, freezing and thawing, and winter clearing operations will affect pavement wearing surfaces as to performance and maintenance costs. These surfaces, in turn, will have some effect on the ease of winter clearing operations due to differences in thermal absorption or to the ability of the pavement to resist damage from snow and ice control equipment or materials.

In drawing upon the performance record of pavements elsewhere, it is most important to take into consideration the conditions pertaining to the particular climatic belt.

4. **Construction Considerations**

Stage construction of the pavement structure may dictate the type of pavement selected. Other considerations such as speed of construction, accommodating traffic during construction, safety to traffic during construction, ease of replacement, anticipated future widening, seasons of the year when construction must be accomplished, and perhaps others may have a strong influence on paving type selections in specific cases.

5. **Recycling**

The opportunity to recycle material from an existing pavement structure or other sources may dictate the use of one pavement type. Future recycling opportunities may also be considered.

6. **Cost Comparison**

Where there are no overriding factors and several alternate pavement types would serve satisfactorily, cost comparison can be used to assist in determining pavement type.

Unavoidably, there will be instances where financial circumstances are such as to make first costs the dominate factor in selection, even though higher maintenance or repair costs may be involved at a later date. Where circumstances permit, a more realistic measure is cost on the basis of service life or service rendered by a pavement structure. Such costs should include the initial construction cost, the cost of subsequent stages or corrective work, anticipated life, maintenance costs, and salvage value. Costs to road users during periods of reconstruction or maintenance operations are also appropriate for consideration. Although pavement structures are based on an initial design period, few are abandoned at the end of that period and continue to serve as part of the future pavement structure. For this reason, the analysis period should be of sufficient duration to include a representative reconstruction of all pavement types.

B.3 SECONDARY FACTORS

1. Performance of Similar Pavements in the Area

To a large degree, the experience and judgement of the highway engineer is based on the performance of pavements in the immediate area of his jurisdiction. Past performance is a valuable guide, provided there is good correlation between conditions and service requirements between the reference pavements and the designs under study. Caution must be urged against reliance on short-term performance records, and on those long-term records of pavements which may have been subjected to much lighter loadings for a large portion of their present life. The need for periodic reanalysis of performance is apparent.

2. Adjacent Existing Pavements

Provided there is no radical change in conditions, the choice of paving type on a highway may be influenced by adjacent existing sections which have given adequate service. The resultant continuity of pavement type will also simplify maintenance operations.

3. Conservation of Materials and Energy

Pavement selection may be influenced by the pavement type which contains less of a scarce critical material or the type whose material production, transportation, and placement requires less energy consumption.

4. Availability of Local Materials or Contractor Capabilities

The availability and adaptability of local material may influence the selection of pavement type. Also, the availability of commercially produced mixes and the equipment capabilities of area contractors may influence the selection of pavement type, particularly on small projects.

5. Traffic Safety

The particular characteristics of a wearing course surface, the need for delineation through pavement and shoulder contrast, reflectivity under highway lighting, and the maintenance of a nonskid surface as affected by the available materials may each influence the paving type selection in specific locations.

6. Incorporation of Experimental Features

In some instances, the performance of new materials or design concepts must be determined by field testing under actual construction, environmental, or traffic conditions. Where the material or concept is adaptable to only one paving type, the incorporation of such experimental features may dictate pavement type selection.

7. Stimulation of Competition

It is desirable that monopoly situations be avoided, and that improvement in products and methods be encouraged through continued and healthy competition among industries involved in the production of paving materials.

Where alternative pavement designs have comparable initial costs, including the attendant costs of earthwork, drainage facilities, and other appurtenances, and provide comparable service life or life-cycle cost, the highway agency may elect to take alternate bids to stimulate competition and obtain lower prices. If this procedure is used, it is essential that good engineering practices and product improvement are not abandoned for the purpose of making cost more competitive.

Where several materials will adequately serve as a component within the pavement structure, such as a subbase or a base course, contractors should be permitted the option of using any of the approved materials.

8. Municipal Preference, Participating Local Government Preference and Recognition of Local Industry

While these considerations seem outside the realm of the highway engineer, they cannot always be ignored by the highway administrator, especially if all other factors involved are indecisive as to the pavement type to select.

B.4 CONCLUSION

In the foregoing, there have been listed and discussed those factors and considerations which influence, to various degrees, the determination of paving types. This has brought to the fore the need, in certain areas, for the development of basic information that is not available at present. It has also served to point out, in general, that conditions are so variable and influences sufficiently different from locality to locality to necessitate a study of individual projects in most instances.

APPENDIX C
ALTERNATE METHODS OF DESIGN
FOR PAVEMENT STRUCTURES

Each of the 50 states have developed methods of design for new construction and for overlays. Many of the states have used information and procedures contained in the 1972 version of the AASHTO Interim Guide for the Design of Pavement Structures. It can be expected that these states will adapt information from this revision (1986) to the Guide as the basis for updating design procedures as seems appropriate. The latest state procedures are available by contacting the appropriate authorities in each of the states.

In addition to methods adopted by the states, pavement design procedures have been developed by private industry, often through related associations. This Appendix includes a representative listing of such methods.

The U.S. Army Corps of Engineers and U.S. Forest Service (Department of Agriculture) have developed design methods for use on military installations and Forest Service roads and are referenced herein.

The Federal Highway Administration sponsors research and development related to pavement design and has recently issued reports indicative of activities in this area. These methods are considered to be in the development stage and have not been officially adopted by FHWA.

This Appendix is divided into five (5) sections; C.1 and C.2 for the design of new flexible- and rigid-type pavements, respectively, C.3 for design of overlays, C.4 for structural analysis, and C.5 for low-volume roads.

The reports on structural analysis describe analytical procedures to calculate stress, strain, or deformation in pavement structures.

The design procedures included in this Appendix are not intended as a complete list; however, they do provide representative information from a variety of sources.

C.1 FLEXIBLE PAVEMENT DESIGN

1. The Asphalt Institute
   (a) Thickness Design—Asphalt Pavements for Highways and Streets, 1981 (MS-1)
   (b) Research and Development of The Asphalt Institute's Thickness Design Manual (MS-1) Ninth Edition, Research Report No. 82-2-RR-82-2 August 1982
   (c) Computer Program DAMA, User's Manual CP-1, October 1983
   (d) The Asphalt Institute
       Asphalt Institute Building
       College Park, Maryland 20740
2. The National Crushed Stone Association
   (a) Flexible Pavement Design Guide of Highways, 1972
   (b) National Crushed Stone Association
       1415 Elliot Place, N.W.
       Washington, D.C. 20007
3. U.S. Army Corps of Engineers
   Roads, Streets, Walks and Open Storage Areas, Flexible Pavement Design, U.S. Army Technical Manual, TM-5-822-5, May 1980
4. Shell Method
   (a) Shell Pavement Design Method, 1978
       Shell International Petroleum Company Limited, London
   (b) Koole, R.C. and Visser, W., "The New Shell Method for Asphalt Pavements and Overlay Design," *Proceedings*, Canadian Technical Asphalt Association, Volume XXIII, November 1978, p. 2.42

C.2 RIGID PAVEMENT DESIGN

1. Portland Cement Association
   (a) Thickness Design for Concrete Highway and Street Pavements, 1984
   (b) Portland Cement Association
       5420 Old Orchard Road
       Skokie, Illinois 60077-4321
2. Concrete Reinforcing Steel Institute
   "Design of Continuously Reinforced Concrete for Highways," Concrete

Reinforcing Steel Institute, 933 North Plum Grove Road; Schaumburg, Illinois 60195, 1981

C.3 OVERLAY DESIGN

1. The Asphalt Institute
   (a) "Asphalt Overlays for Highways and Street Rehabilitation," The Asphalt Institute, Manual Series No. 17, 1983
       (1) Deflection Procedure
       (2) Effective Thickness Procedure
   (b) The Asphalt Institute
       Asphalt Institute Building
       College Park, Maryland 20740
2. Portland Cement Association
   (a) "Design of Concrete Overlays for Pavements," ACI 324.1R-67 Journal ACI, August 1967
   (b) "Guide to Concrete Resurfacing Design and Selection Criteria," 1981 PCA, 1981
   (c) Portland Cement Association
       5240 Old Orchard Road
       Skokie, Illinois 60077-4321
3. Shell Method
   (a) Claessen, A.I.M. and Ditmarsch, R., "Pavement Evaluation and Overlay Design, The Shell Method," *Proceedings*, Fourth International Conference on the Structural Design of Asphalt Pavements, University of Michigan, August 1977
   (b) Koole, R.C., "Overlay Design Based on Falling Weight Deflectometer Measurements," *Transportation Research Record No. 700*, Transportation Research Board, Washington, D.C., 1979, pp. 59–72.
4. Federal Highway Administration
   (a) Majidzadeh, K. and Ilves, G., "Flexible Pavement Overlay Design Procedures," Volume 1, Evaluation and Modification of the Design Methods, FHWA/RD-81/032, Resource International, Worthington, Ohio 1981.
   (b) Austin Research Engineers, Inc., "Asphalt Concrete Overlays of Flexible Pavements," Volume I, Development of New Design Criteria, FHWA Report No. FHWA-RD-75-75, August 1975
   (c) Majidzadeh, K. and Ilves, G., "Evaluation of Rigid Pavement Overlay Design Procedure, Development of the OAR Procedure," FHWA/RD-83/090, Resource International, Worthington, Ohio, July 1983

C.4 STRUCTURAL ANALYSIS

1. VESYS II for flexible pavements
   Rauhut, J.B., O'Quin, J.C., and Hudson, W.R., "Sensitivity Analysis of FHWA Structural Model VESYS II," Volume 2, Sensitivity Analysis, Report No. FHWA-RD-76-24, 1976
2. ILLIPAVE for flexible pavements
   Road, L. and Figueroa, J.L., "Load Response of Transportation Support Systems," Journal, ASCE TE-1, 1980
3. National Cooperative Highway Research Program, Project 1-10B
   "The Use of Distress Prediction Subsystems for the Design of Pavement Structures," *Proceedings*, Fourth International Conference on Structural Design of Asphalt Pavements, University of Michigan, Ann Arbor, Michigan, 1977
4. ILLISLAB for concrete pavements
   Tabatabaie, A.M., Barenberg, E.J., and Smith, R.E., "Longitudinal Joint Systems in Slip-Formed Rigid Pavements," Volume II, Analysis of Load Transfer Pavements, Report No. FAA-RD-79-4-11, 1979
5. SLAB-49 for concrete pavements
   Hudson, W.R. and Matlock, H., "Discrete-Element Analysis for Discontinuous Plates," ASCE, Volume ST 10, 1968

C.5 LOW-VOLUME ROAD SURFACE DESIGN

1. U.S. Department of Agriculture
   (a) Surfacing Handbook
       FSH 7709.56a
       Forest Service
   (b) USDA-Forest Service
       P.O. Box 2417
       Washington, D.C. 20013

APPENDIX D
CONVERSION OF MIXED TRAFFIC TO EQUIVALENT SINGLE AXLE LOADS FOR PAVEMENT DESIGN

D.1 GENERAL CONSIDERATIONS

Part I of this Guide outlines the fact that estimates of the amount of traffic and its characteristics play a primary role in the pavement design and analysis process. Parts II and III require traffic information for design of pavement structures. This Appendix provides guidelines for estimating the number of equivalent single axle loads which can be expected to be applied to a pavement during a specified design period or to estimate equivalent axle load applications that have been applied to existing pavements. Although typical and historical traffic parameters are furnished in this Appendix for illustrative purposes, pavement designers and analysts are cautioned to use the best locally available data to represent specific site conditions. Such traffic data should be available from the designing agency as part of its regular traffic monitoring effort. As the science of pavement design and management matures, it is vital that a close working relationship exists among these groups.

There are currently major initiatives underway to improve the quality of traffic data. Statistically based programs for traffic monitoring are being adopted in many states. Microcomputer technology is rapidly improving the ability of planners to assemble better traffic data using automatic vehicle classifiers and weigh-in-motion (WIM) installations.

History has clearly shown that while it may be possible to accurately measure today's traffic, the characteristics of this traffic change over time. With the exception of interruptions during petroleum shortages in recent years, a rather constant increase in traffic is evident. This type of information, plus forecasts of population, land use, economic factors, etc., are used by transportation planners to forecast future travel. At the local level, such forecasts are generally developed on a system basis and on most high level highways for specific corridors. These should be used in the pavement design process.

From 1970 to 1983, the percent of the total volume made up of passenger cars and buses (on rural Interstate highways) decreased from 77 to 63, while the percent of the traffic stream made up of 5-axle or more combinations increased from 9 to 17. Between 1970 and 1983, the total equivalent single axle loads increased by 105 percent. The significant point is that if pavements had been designed in 1970, assuming a constant traffic growth for all types of vehicles, a serious underdesign of pavements would have resulted.

Users of this Guide are cautioned that what are discussed are nationwide summary data. Trends within a given state, or corridor within a state, may vary significantly. This can happen for a number of reasons, including economic conditions, industry locational patterns, truck weight laws, enforcement intensity, equipment changes by the trucking industry, etc. Pavement designers should be particularly sensitive to the changes which will likely take place on the nationwide basis as a result of the Surface Transportation Assistance Act (STAA) of 1982. As a result of this legislation, there may well be (1) significant changes in both truck weights within particular vehicle categories and shifts to different equipment (twin trailers), (2) changes in position of load application due to wider trucks, and (3) increased intensity of use on certain routes designated for these new vehicle configurations. Additionally, deregulation of the trucking industry will likely change the portion of trucks traveling empty in many corridors.

These discussions highlight the need for each state to be conducting a comprehensive program of traffic counting, vehicle classification, and truck weighing. These changing traffic trends can be expected to have significant influences on the lives of existing pavements and on the design of new pavements.

To use the pavement design procedures presented in this Guide, mixed traffic must be converted to an equivalent number of 18-kip single axle loads. The procedure for accomplishing this conversion includes:

(1) derivation of load equivalence factors,
(2) conversion of mixed traffic to 18-kip equivalent single axle load (ESAL) applications, and
(3) lane distribution considerations.

To express varying axle loads in terms of a single design parameter, it is necessary to develop axle load equivalence factors. These factors, when multiplied by the number of axle loads within a given weight category, give the number of 18-kip single axle load applications which will have an equivalent effect on the performance of the pavement structures.

Load equivalency factors represent the ratio of the number of repetitions of any axle load and axle configuration (single, tandem, tridem) necessary to cause the same reduction in PSI as one application of an 18-kip single axle load. Load equivalency factors are presented later in Tables D.1 through D.18 for a range of pavement structural combinations, axle configurations, and terminal serviceability values of 2.0, 2.5, and 3.0. Appendix MM of Volume 2 presents the AASHO Road Test-based equations that were used to generate these tables. It also provides some support for extending the tables to tridem axle loadings.

The prediction of traffic (ESAL's) for design purposes must rely on information from past traffic, modified by factors for growth or other expected changes. Most states, in cooperation with FHWA, accumulate past traffic information in the form of truck weight study data W-4 tables. Typical information includes: (1) axle weight distributions in 2,000-lb. intervals, (2) ESAL's for all trucks weighed, (3) ESAL's per 1,000 trucks weighed by truck class, (4) ESAL's for all trucks counted, and (5) percent distribution of ESAL's by truck class.

To arrive at the design ESAL's, it is necessary to assume a structural number (SN) for flexible pavements or slab thickness (D) for rigid pavements, and then select the equivalence factors listed in Tables D.1 through D.18. The use of an SN of 5.0 or a D of 9 inches for the determination of 18-kip single axle equivalence factors will normally give results that are sufficiently accurate for design purposes, even though the final design may be somewhat different. If in error, this assumption will usually result in an overestimation of 18-kip equivalent single axles. When more accurate results are desired and the computed design is appreciably different (1 inch of PCC for rigid or 1 inch of asphalt concrete for flexible) from the assumed value, a new value should be assumed, the design 18-kip ESAL traffic (w_{18}) recomputed, and the structural design determined for the new w_{18}. The procedure should be continued until the assumed and computed values are as close as desired.

If the number of equivalent axle loads represents the total for all lanes and both directions of travel, this number must be distributed by direction and by lanes for design purposes. Directional distribution is usually made by assigning 50 percent of the traffic to each direction, unless special considerations (such as more loaded trucks moving in one direction and more empty trucks in the other) warrant some other distribution. In regard to lane distribution, most states assign 100 percent of the traffic in each direction (i.e., 50 percent of the total) to the design lane. Some states have developed lane distribution factors for multilane facilities. The range of factors used is presented below.

Number of Lanes in Both Directions	Percent of 18-kip ESAL Traffic in Design Lane
1	100
2	80–100
3	60–80
4 or more	50–75

If lane or directional distribution factors are utilized and pavements are designed on the basis of distributed traffic, consideration should be given to the use of variable cross sections. Heavier structural sections in the outside lanes should be considered if warranted by the lane distribution analysis.

In view of the increased emphasis on improved traffic monitoring made possible by weigh-in-motion (WIM) and automatic vehicle classification and counting, it is recommended that each state develop appropriate factors for multilane facilities.

D.2 CALCULATING ESAL APPLICATIONS

When calculating ESAL's for the design of a particular project, it is convenient to convert the estimated traffic distribution into truck load factors. Two methods of calculating truck load factors from W-4 information are summarized in the following paragraphs of this section.

Where axle load information is available from a weigh station that can be assumed to be representative of traffic for the pavement to be designed, the truck load factor can be calculated directly. For example, assume that the data in Figure D.1 illustrates the weighing of 5-axle, tractor semi-trailer trucks at a specific weigh station. Traffic (load) equivalency factors are obtained from Table D.4, the number of axles represents the grouping or distribution of weights

Table D.1. Axle Load Equivalency Factors for Flexible Pavements, Single Axles and p_t of 2.0

Axle Load (kips)	Pavement Structural Number (SN)					
	1	2	3	4	5	6
2	.0002	.0002	.0002	.0002	.0002	.0002
4	.002	.003	.002	.002	.002	.002
6	.009	.012	.011	.010	.009	.009
8	.030	.035	.036	.033	.031	.029
10	.075	.085	.090	.085	.079	.076
12	.165	.177	.189	.183	.174	.168
14	.325	.338	.354	.350	.338	.331
16	.589	.598	.613	.612	.603	.596
18	1.00	1.00	1.00	1.00	1.00	1.00
20	1.61	1.59	1.56	1.55	1.57	1.59
22	2.49	2.44	2.35	2.31	2.35	2.41
24	3.71	3.62	3.43	3.33	3.40	3.51
26	5.36	5.21	4.88	4.68	4.77	4.96
28	7.54	7.31	6.78	6.42	6.52	6.83
30	10.4	10.0	9.2	8.6	8.7	9.2
32	14.0	13.5	12.4	11.5	11.5	12.1
34	18.5	17.9	16.3	15.0	14.9	15.6
36	24.2	23.3	21.2	19.3	19.0	19.9
38	31.1	29.9	27.1	24.6	24.0	25.1
40	39.6	38.0	34.3	30.9	30.0	31.2
42	49.7	47.7	43.0	38.6	37.2	38.5
44	61.8	59.3	53.4	47.6	45.7	47.1
46	76.1	73.0	65.6	58.3	55.7	57.0
48	92.9	89.1	80.0	70.9	67.3	68.6
50	113.	108.	97.	86.	81.	82.

Table D.2. Axle Load Equivalency Factors For Flexible Pavements, Tandem Axles and p_t of 2.0

Axle Load (kips)	Pavement Structural Number (SN)					
	1	2	3	4	5	6
2	.0000	.0000	.0000	.0000	.0000	.0000
4	.0003	.0003	.0003	.0002	.0002	.0002
6	.001	.001	.001	.001	.001	.001
8	.003	.003	.003	.003	.003	.002
10	.007	.008	.008	.007	.006	.006
12	.013	.016	.016	.014	.013	.012
14	.024	.029	.029	.026	.024	.023
16	.041	.048	.050	.046	.042	.040
18	.066	.077	.081	.075	.069	.066
20	.103	.117	.124	.117	.109	.105
22	.156	.171	.183	.174	.164	.158
24	.227	.244	.260	.252	.239	.231
26	.322	.340	.360	.353	.338	.329
28	.447	.465	.487	.481	.466	.455
30	.607	.623	.646	.643	.627	.617
32	.810	.823	.843	.842	.829	.819
34	1.06	1.07	1.08	1.08	1.08	1.07
36	1.38	1.38	1.38	1.38	1.38	1.38
38	1.76	1.75	1.73	1.72	1.73	1.74
40	2.22	2.19	2.15	2.13	2.16	2.18
42	2.77	2.73	2.64	2.62	2.66	2.70
44	3.42	3.36	3.23	3.18	3.24	3.31
46	4.20	4.11	3.92	3.83	3.91	4.02
48	5.10	4.98	4.72	4.58	4.68	4.83
50	6.15	5.99	5.64	5.44	5.56	5.77
52	7.37	7.16	6.71	6.43	6.56	6.83
54	8.77	8.51	7.93	7.55	7.69	8.03
56	10.4	10.1	9.3	8.8	9.0	9.4
58	12.2	11.8	10.9	10.3	10.4	10.9
60	14.3	13.8	12.7	11.9	12.0	12.6
62	16.6	16.0	14.7	13.7	13.8	14.5
64	19.3	18.6	17.0	15.8	15.8	16.6
66	22.2	21.4	19.6	18.0	18.0	18.9
68	25.5	24.6	22.4	20.6	20.5	21.5
70	29.2	28.1	25.6	23.4	23.2	24.3
72	33.3	32.0	29.1	26.5	26.2	27.4
74	37.8	36.4	33.0	30.0	29.4	30.8
76	42.8	41.2	37.3	33.8	33.1	34.5
78	48.4	46.5	42.0	38.0	37.0	38.6
80	54.4	52.3	47.2	42.5	41.3	43.0
82	61.1	58.7	52.9	47.6	46.0	47.8
84	68.4	65.7	59.2	53.0	51.2	53.0
86	76.3	73.3	66.0	59.0	56.8	58.6
88	85.0	81.6	73.4	65.5	62.8	64.7
90	94.4	90.6	81.5	72.6	69.4	71.3

Table D.3. Axle Load Equivalency Factors for Flexible Pavements, Triple Axles and p_t of 2.0

Axle Load (kips)	Pavement Structural Number (SN)					
	1	2	3	4	5	6
2	.0000	.0000	.0000	.0000	.0000	.0000
4	.0001	.0001	.0001	.0001	.0001	.0001
6	.0004	.0004	.0003	.0003	.0003	.0003
8	.0009	.0010	.0009	.0008	.0007	.0007
10	.002	.002	.002	.002	.002	.001
12	.004	.004	.004	.003	.003	.003
14	.006	.007	.007	.006	.006	.005
16	.010	.012	.012	.010	.009	.009
18	.016	.019	.019	.017	.015	.015
20	.024	.029	.029	.026	.024	.023
22	.034	.042	.042	.038	.035	.034
24	.049	.058	.060	.055	.051	.048
26	.068	.080	.083	.077	.071	.068
28	.093	.107	.113	.105	.098	.094
30	.125	.140	.149	.140	.131	.126
32	.164	.182	.194	.184	.173	.167
34	.213	.233	.248	.238	.225	.217
36	.273	.294	.313	.303	.288	.279
38	.346	.368	.390	.381	.364	.353
40	.434	.456	.481	.473	.454	.443
42	.538	.560	.587	.580	.561	.548
44	.662	.682	.710	.705	.686	.673
46	.807	.825	.852	.849	.831	.818
48	.976	.992	1.015	1.014	.999	.987
50	1.17	1.18	1.20	1.20	1.19	1.18
52	1.40	1.40	1.42	1.42	1.41	1.40
54	1.66	1.66	1.66	1.66	1.66	1.66
56	1.95	1.95	1.93	1.93	1.94	1.94
58	2.29	2.27	2.24	2.23	2.25	2.27
60	2.67	2.64	2.59	2.57	2.60	2.63
62	3.10	3.06	2.98	2.95	2.99	3.04
64	3.59	3.53	3.41	3.37	3.42	3.49
66	4.13	4.05	3.89	3.83	3.90	3.99
68	4.73	4.63	4.43	4.34	4.42	4.54
70	5.40	5.28	5.03	4.90	5.00	5.15
72	6.15	6.00	5.68	5.52	5.63	5.82
74	6.97	6.79	6.41	6.20	6.33	6.56
76	7.88	7.67	7.21	6.94	7.08	7.36
78	8.88	8.63	8.09	7.75	7.90	8.23
80	9.98	9.69	9.05	8.63	8.79	9.18
82	11.2	10.8	10.1	9.6	9.8	10.2
84	12.5	12.1	11.2	10.6	10.8	11.3
86	13.9	13.5	12.5	11.8	11.9	12.5
88	15.5	15.0	13.8	13.0	13.2	13.8
90	17.2	16.6	15.3	14.3	14.5	15.2

Table D.4. Axle Load Equivalency Factors for Flexible Pavements, Single Axles and p_t of 2.5

Axle Load (kips)	Pavement Structural Number (SN)					
	1	2	3	4	5	6
2	.0004	.0004	.0003	.0002	.0002	.0002
4	.003	.004	.004	.003	.002	.002
6	.011	.017	.017	.013	.010	.009
8	.032	.047	.051	.041	.034	.031
10	.078	.102	.118	.102	.088	.080
12	.168	.198	.229	.213	.189	.176
14	.328	.358	.399	.388	.360	.342
16	.591	.613	.646	.645	.623	.606
18	1.00	1.00	1.00	1.00	1.00	1.00
20	1.61	1.57	1.49	1.47	1.51	1.55
22	2.48	2.38	2.17	2.09	2.18	2.30
24	3.69	3.49	3.09	2.89	3.03	3.27
26	5.33	4.99	4.31	3.91	4.09	4.48
28	7.49	6.98	5.90	5.21	5.39	5.98
30	10.3	9.5	7.9	6.8	7.0	7.8
32	13.9	12.8	10.5	8.8	8.9	10.0
34	18.4	16.9	13.7	11.3	11.2	12.5
36	24.0	22.0	17.7	14.4	13.9	15.5
38	30.9	28.3	22.6	18.1	17.2	19.0
40	39.3	35.9	28.5	22.5	21.1	23.0
42	49.3	45.0	35.6	27.8	25.6	27.7
44	61.3	55.9	44.0	34.0	31.0	33.1
46	75.5	68.8	54.0	41.4	37.2	39.3
48	92.2	83.9	65.7	50.1	44.5	46.5
50	112.	102.	79.	60.	53.	55.

Table D.5. Axle Load Equivalency Factors for Flexible Pavements, Tandem Axles and p_t of 2.5

Axle Load (kips)	Pavement Structural Number (SN)					
	1	2	3	4	5	6
2	.0001	.0001	.0001	.0000	.0000	.0000
4	.0005	.0005	.0004	.0003	.0003	.0002
6	.002	.002	.002	.001	.001	.001
8	.004	.006	.005	.004	.003	.003
10	.008	.013	.011	.009	.007	.006
12	.015	.024	.023	.018	.014	.013
14	.026	.041	.042	.033	.027	.024
16	.044	.065	.070	.057	.047	.043
18	.070	.097	.109	.092	.077	.070
20	.107	.141	.162	.141	.121	.110
22	.160	.198	.229	.207	.180	.166
24	.231	.273	.315	.292	.260	.242
26	.327	.370	.420	.401	.364	.342
28	.451	.493	.548	.534	.495	.470
30	.611	.648	.703	.695	.658	.633
32	.813	.843	.889	.887	.857	.834
34	1.06	1.08	1.11	1.11	1.09	1.08
36	1.38	1.38	1.38	1.38	1.38	1.38
38	1.75	1.73	1.69	1.68	1.70	1.73
40	2.21	2.16	2.06	2.03	2.08	2.14
42	2.76	2.67	2.49	2.43	2.51	2.61
44	3.41	3.27	2.99	2.88	3.00	3.16
46	4.18	3.98	3.58	3.40	3.55	3.79
48	5.08	4.80	4.25	3.98	4.17	4.49
50	6.12	5.76	5.03	4.64	4.86	5.28
52	7.33	6.87	5.93	5.38	5.63	6.17
54	8.72	8.14	6.95	6.22	6.47	7.15
56	10.3	9.6	8.1	7.2	7.4	8.2
58	12.1	11.3	9.4	8.2	8.4	9.4
60	14.2	13.1	10.9	9.4	9.6	10.7
62	16.5	15.3	12.6	10.7	10.8	12.1
64	19.1	17.6	14.5	12.2	12.2	13.7
66	22.1	20.3	16.6	13.8	13.7	15.4
68	25.3	23.3	18.9	15.6	15.4	17.2
70	29.0	26.6	21.5	17.6	17.2	19.2
72	33.0	30.3	24.4	19.8	19.2	21.3
74	37.5	34.4	27.6	22.2	21.3	23.6
76	42.5	38.9	31.1	24.8	23.7	26.1
78	48.0	43.9	35.0	27.8	26.2	28.8
80	54.0	49.4	39.2	30.9	29.0	31.7
82	60.6	55.4	43.9	34.4	32.0	34.8
84	67.8	61.9	49.0	38.2	35.3	38.1
86	75.7	69.1	54.5	42.3	38.8	41.7
88	84.3	76.9	60.6	46.8	42.6	45.6
90	93.7	85.4	67.1	51.7	46.8	49.7

Table D.6. Axle Load Equivalency Factors for Flexible Pavements, Triple Axles and p_t of 2.5

Axle Load (kips)	Pavement Structural Number (SN)					
	1	2	3	4	5	6
2	.0000	.0000	.0000	.0000	.0000	.0000
4	.0002	.0002	.0002	.0001	.0001	.0001
6	.0006	.0007	.0005	.0004	.0003	.0003
8	.001	.002	.001	.001	.001	.001
10	.003	.004	.003	.002	.002	.002
12	.005	.007	.006	.004	.003	.003
14	.008	.012	.010	.008	.006	.006
16	.012	.019	.018	.013	.011	.010
18	.018	.029	.028	.021	.017	.016
20	.027	.042	.042	.032	.027	.024
22	.038	.058	.060	.048	.040	.036
24	.053	.078	.084	.068	.057	.051
26	.072	.103	.114	.095	.080	.072
28	.098	.133	.151	.128	.109	.099
30	.129	.169	.195	.170	.145	.133
32	.169	.213	.247	.220	.191	.175
34	.219	.266	.308	.281	.246	.228
36	.279	.329	.379	.352	.313	.292
38	.352	.403	.461	.436	.393	.368
40	.439	.491	.554	.533	.487	.459
42	.543	.594	.661	.644	.597	.567
44	.666	.714	.781	.769	.723	.692
46	.811	.854	.918	.911	.868	.838
48	.979	1.015	1.072	1.069	1.033	1.005
50	1.17	1.20	1.24	1.25	1.22	1.20
52	1.40	1.41	1.44	1.44	1.43	1.41
54	1.66	1.66	1.66	1.66	1.66	1.66
56	1.95	1.93	1.90	1.90	1.91	1.93
58	2.29	2.25	2.17	2.16	2.20	2.24
60	2.67	2.60	2.48	2.44	2.51	2.58
62	3.09	3.00	2.82	2.76	2.85	2.95
64	3.57	3.44	3.19	3.10	3.22	3.36
66	4.11	3.94	3.61	3.47	3.62	3.81
68	4.71	4.49	4.06	3.88	4.05	4.30
70	5.38	5.11	4.57	4.32	4.52	4.84
72	6.12	5.79	5.13	4.80	5.03	5.41
74	6.93	6.54	5.74	5.32	5.57	6.04
76	7.84	7.37	6.41	5.88	6.15	6.71
78	8.83	8.28	7.14	6.49	6.78	7.43
80	9.92	9.28	7.95	7.15	7.45	8.21
82	11.1	10.4	8.8	7.9	8.2	9.0
84	12.4	11.6	9.8	8.6	8.9	9.9
86	13.8	12.9	10.8	9.5	9.8	10.9
88	15.4	14.3	11.9	10.4	10.6	11.9
90	17.1	15.8	13.2	11.3	11.6	12.9

Table D.7. Axle Load Equivalency Factors for Flexible Pavements, Single Axles and p_t of 3.0

Axle Load (kips)	Pavement Structural Number (SN)					
	1	2	3	4	5	6
2	.0008	.0009	.0006	.0003	.0002	.0002
4	.004	.008	.006	.004	.002	.002
6	.014	.030	.028	.018	.012	.010
8	.035	.070	.080	.055	.040	.034
10	.082	.132	.168	.132	.101	.086
12	.173	.231	.296	.260	.212	.187
14	.332	.388	.468	.447	.391	.358
16	.594	.633	.695	.693	.651	.622
18	1.00	1.00	1.00	1.00	1.00	1.00
20	1.60	1.53	1.41	1.38	1.44	1.51
22	2.47	2.29	1.96	1.83	1.97	2.16
24	3.67	3.33	2.69	2.39	2.60	2.96
26	5.29	4.72	3.65	3.08	3.33	3.91
28	7.43	6.56	4.88	3.93	4.17	5.00
30	10.2	8.9	6.5	5.0	5.1	6.3
32	13.8	12.0	8.4	6.2	6.3	7.7
34	18.2	15.7	10.9	7.8	7.6	9.3
36	23.8	20.4	14.0	9.7	9.1	11.0
38	30.6	26.2	17.7	11.9	11.0	13.0
40	38.8	33.2	22.2	14.6	13.1	15.3
42	48.8	41.6	27.6	17.8	15.5	17.8
44	60.6	51.6	34.0	21.6	18.4	20.6
46	74.7	63.4	41.5	26.1	21.6	23.8
48	91.2	77.3	50.3	31.3	25.4	27.4
50	110.	94.	61.	37.	30.	32.

Table D.8. Axle Load Equivalency Factors for Flexible Pavements, Tandem Axles and p_t of 3.0

Axle Load (kips)	Pavement Structural Number (SN)					
	1	2	3	4	5	6
2	.0002	.0002	.0001	.0001	.0000	.0000
4	.001	.001	.001	.000	.000	.000
6	.003	.004	.003	.002	.001	.001
8	.006	.011	.009	.005	.003	.003
10	.011	.024	.020	.012	.008	.007
12	.019	.042	.039	.024	.017	.014
14	.031	.066	.068	.045	.032	.026
16	.049	.096	.109	.076	.055	.046
18	.075	.134	.164	.121	.090	.076
20	.113	.181	.232	.182	.139	.119
22	.166	.241	.313	.260	.205	.178
24	.238	.317	.407	.358	.292	.257
26	.333	.413	.517	.476	.402	.360
28	.457	.534	.643	.614	.538	.492
30	.616	.684	.788	.773	.702	.656
32	.817	.870	.956	.953	.896	.855
34	1.07	1.10	1.15	1.15	1.12	1.09
36	1.38	1.38	1.38	1.38	1.38	1.38
38	1.75	1.71	1.64	1.62	1.66	1.70
40	2.21	2.11	1.94	1.89	1.98	2.08
42	2.75	2.59	2.29	2.19	2.33	2.50
44	3.39	3.15	2.70	2.52	2.71	2.97
46	4.15	3.81	3.16	2.89	3.13	3.50
48	5.04	4.58	3.70	3.29	3.57	4.07
50	6.08	5.47	4.31	3.74	4.05	4.70
52	7.27	6.49	5.01	4.24	4.57	5.37
54	8.65	7.67	5.81	4.79	5.13	6.10
56	10.2	9.0	6.7	5.4	5.7	6.9
58	12.0	10.6	7.7	6.1	6.4	7.7
60	14.1	12.3	8.9	6.8	7.1	8.6
62	16.3	14.2	10.2	7.7	7.8	9.5
64	18.9	16.4	11.6	8.6	8.6	10.5
66	21.8	18.9	13.2	9.6	9.5	11.6
68	25.1	21.7	15.0	10.7	10.5	12.7
70	28.7	24.7	17.0	12.0	11.5	13.9
72	32.7	28.1	19.2	13.3	12.6	15.2
74	37.2	31.9	21.6	14.8	13.8	16.5
76	42.1	36.0	24.3	16.4	15.1	17.9
78	47.5	40.6	27.3	18.2	16.5	19.4
80	53.4	45.7	30.5	20.1	18.0	21.0
82	60.0	51.2	34.0	22.2	19.6	22.7
84	67.1	57.2	37.9	24.6	21.3	24.5
86	74.9	63.8	42.1	27.1	23.2	26.4
88	83.4	71.0	46.7	29.8	25.2	28.4
90	92.7	78.8	51.7	32.7	27.4	30.5

Table D.9. Axle Load Equivalency Factors for Flexible Pavements, Triple Axles and p_t of 3.0

Axle Load (kips)	Pavement Structural Number (SN)					
	1	2	3	4	5	6
2	.0001	.0001	.0001	.0000	.0000	.0000
4	.0005	.0004	.0003	.0002	.0001	.0001
6	.001	.001	.001	.001	.000	.000
8	.003	.004	.002	.001	.001	.001
10	.005	.008	.005	.003	.002	.002
12	.007	.014	.010	.006	.004	.003
14	.011	.023	.018	.011	.007	.006
16	.016	.035	.030	.018	.013	.010
18	.022	.050	.047	.029	.020	.017
20	.031	.069	.069	.044	.031	.026
22	.043	.090	.097	.065	.046	.039
24	.059	.116	.132	.092	.066	.056
26	.079	.145	.174	.126	.092	.078
28	.104	.179	.223	.168	.126	.107
30	.136	.218	.279	.219	.167	.143
32	.176	.265	.342	.279	.218	.188
34	.226	.319	.413	.350	.279	.243
36	.286	.382	.491	.432	.352	.310
38	.359	.456	.577	.524	.437	.389
40	.447	.543	.671	.626	.536	.483
42	.550	.643	.775	.740	.649	.593
44	.673	.760	.889	.865	.777	.720
46	.817	.894	1.014	1.001	.920	.865
48	.984	1.048	1.152	1.148	1.080	1.030
50	1.18	1.23	1.30	1.31	1.26	1.22
52	1.40	1.43	1.47	1.48	1.45	1.43
54	1.66	1.66	1.66	1.66	1.66	1.66
56	1.95	1.92	1.86	1.85	1.88	1.91
58	2.28	2.21	2.09	2.06	2.13	2.20
60	2.66	2.54	2.34	2.28	2.39	2.50
62	3.08	2.92	2.61	2.52	2.66	2.84
64	3.56	3.33	2.92	2.77	2.96	3.19
66	4.09	3.79	3.25	3.04	3.27	3.58
68	4.68	4.31	3.62	3.33	3.60	4.00
70	5.34	4.88	4.02	3.64	3.94	4.44
72	6.08	5.51	4.46	3.97	4.31	4.91
74	6.89	6.21	4.94	4.32	4.69	5.40
76	7.78	6.98	5.47	4.70	5.09	5.93
78	8.76	7.83	6.04	5.11	5.51	6.48
80	9.84	8.75	6.67	5.54	5.96	7.06
82	11.0	9.8	7.4	6.0	6.4	7.7
84	12.3	10.9	8.1	6.5	6.9	8.3
86	13.7	12.1	8.9	7.0	7.4	9.0
88	15.3	13.4	9.8	7.6	8.0	9.6
90	16.9	14.8	10.7	8.2	8.5	10.4

Table D.10. Axle Load Equivalency Factors for Rigid Pavements, Single Axles and p_t of 2.0

Axle Load (kips)	Slab Thickness, D (inches)								
	6	7	8	9	10	11	12	13	14
2	.0002	.0002	.0002	.0002	.0002	.0002	.0002	.0002	.0002
4	.002	.002	.002	.002	.002	.002	.002	.002	.002
6	.011	.010	.010	.010	.010	.010	.010	.010	.010
8	.035	.033	.032	.032	.032	.032	.032	.032	.032
10	.087	.084	.082	.081	.080	.080	.080	.080	.080
12	.186	.180	.176	.175	.174	.174	.173	.173	.173
14	.353	.346	.341	.338	.337	.336	.336	.336	.336
16	.614	.609	.604	.601	.599	.599	.598	.598	.598
18	1.00	1.00	1.00	1.00	1.00	1.00	1.00	1.00	1.00
20	1.55	1.56	1.57	1.58	1.58	1.59	1.59	1.59	1.59
22	2.32	2.32	2.35	2.38	2.40	2.41	2.41	2.41	2.42
24	3.37	3.34	3.40	3.47	3.51	3.53	3.54	3.55	3.55
26	4.76	4.69	4.77	4.88	4.97	5.02	5.04	5.06	5.06
28	6.58	6.44	6.52	6.70	6.85	6.94	7.00	7.02	7.04
30	8.92	8.68	8.74	8.98	9.23	9.39	9.48	9.54	9.56
32	11.9	11.5	11.5	11.8	12.2	12.4	12.6	12.7	12.7
34	15.5	15.0	14.9	15.3	15.8	16.2	16.4	16.6	16.7
36	20.1	19.3	19.2	19.5	20.1	20.7	21.1	21.4	21.5
38	25.6	24.5	24.3	24.6	25.4	26.1	26.7	27.1	27.4
40	32.2	30.8	30.4	30.7	31.6	32.6	33.4	34.0	34.4
42	40.1	38.4	37.7	38.0	38.9	40.1	41.3	42.1	42.7
44	49.4	47.3	46.4	46.6	47.6	49.0	50.4	51.6	52.4
46	60.4	57.7	56.6	56.7	57.7	59.3	61.1	62.6	63.7
48	73.2	69.9	68.4	68.4	69.4	71.2	73.3	75.3	76.8
50	88.0	84.1	82.2	82.0	83.0	84.9	87.4	89.8	91.7

Table D.11. Axle Load Equivalency Factors for Rigid Pavements, Tandem Axles and p_t of 2.0

Axle Load (kips)	Slab Thickness, D (inches)								
	6	7	8	9	10	11	12	13	14
2	.0001	.0001	.0001	.0001	.0001	.0001	.0001	.0001	.0001
4	.0006	.0005	.0005	.0005	.0005	.0005	.0005	.0005	.0005
6	.002	.002	.002	.002	.002	.002	.002	.002	.002
8	.006	.006	.005	.005	.005	.005	.005	.005	.005
10	.014	.013	.013	.012	.012	.012	.012	.012	.012
12	.028	.026	.026	.025	.025	.025	.025	.025	.025
14	.051	.049	.048	.047	.047	.047	.047	.047	.047
16	.087	.084	.082	.081	.081	.080	.080	.080	.080
18	.141	.136	.133	.132	.131	.131	.131	.131	.131
20	.216	.210	.206	.204	.203	.203	.203	.203	.203
22	.319	.313	.307	.305	.304	.303	.303	.303	.303
24	.454	.449	.444	.441	.440	.439	.439	.439	.439
26	.629	.626	.622	.620	.618	.618	.618	.618	.618
28	.852	.851	.850	.850	.850	.849	.849	.849	.849
30	1.13	1.13	1.14	1.14	1.14	1.14	1.14	1.14	1.14
32	1.48	1.48	1.49	1.50	1.51	1.51	1.51	1.51	1.51
34	1.90	1.90	1.93	1.95	1.96	1.97	1.97	1.97	1.97
36	2.42	2.41	2.45	2.49	2.51	2.52	2.53	2.53	2.53
38	3.04	3.02	3.07	3.13	3.17	3.19	3.20	3.20	3.21
40	3.79	3.74	3.80	3.89	3.95	3.98	4.00	4.01	4.01
42	4.67	4.59	4.66	4.78	4.87	4.93	4.95	4.97	4.97
44	5.72	5.59	5.67	5.82	5.95	6.03	6.07	6.09	6.10
46	6.94	6.76	6.83	7.02	7.20	7.31	7.37	7.41	7.43
48	8.36	8.12	8.17	8.40	8.63	8.79	8.88	8.93	8.96
50	10.00	9.69	9.72	9.98	10.27	10.49	10.62	10.69	10.73
52	11.9	11.5	11.5	11.8	12.1	12.4	12.6	12.7	12.8
54	14.0	13.5	13.5	13.8	14.2	14.6	14.9	15.0	15.1
56	16.5	15.9	15.8	16.1	16.6	17.1	17.4	17.6	17.7
58	19.3	18.5	18.4	18.7	19.3	19.8	20.3	20.5	20.7
60	22.4	21.5	21.3	21.6	22.3	22.9	23.5	23.8	24.0
62	25.9	24.9	24.6	24.9	25.6	26.4	27.0	27.5	27.7
64	29.9	28.6	28.2	28.5	29.3	30.2	31.0	31.6	31.9
66	34.3	32.8	32.3	32.6	33.4	34.4	35.4	36.1	36.5
68	39.2	37.5	36.8	37.1	37.9	39.1	40.2	41.1	41.6
70	44.6	42.7	41.9	42.1	42.9	44.2	45.5	46.6	47.3
72	50.6	48.4	47.5	47.6	48.5	49.9	51.4	52.6	53.5
74	57.3	54.7	53.6	53.6	54.6	56.1	57.7	59.2	60.3
76	64.6	61.7	60.4	60.3	61.2	62.8	64.7	66.4	67.7
78	72.5	69.3	67.8	67.7	68.6	70.2	72.3	74.3	75.8
80	81.3	77.6	75.9	75.7	76.6	78.3	80.6	82.8	84.7
82	90.9	86.7	84.7	84.4	85.3	87.1	89.6	92.1	94.2
84	101.	97.	94.	94.	95.	97.	99.	102.	105.
86	113.	107.	105.	104.	105.	107.	110.	113.	116.
88	125.	119.	116.	116.	116.	118.	121.	125.	128.
90	138.	132.	129.	128.	129.	131.	134.	137.	141.

Table D.12. Axle Load Equivalency Factors for Rigid Pavements, Triple Axles and p_t of 2.0

Axle Load (kips)	Slab Thickness, D (inches)								
	6	7	8	9	10	11	12	13	14
2	.0001	.0001	.0001	.0001	.0001	.0001	.0001	.0001	.0001
4	.0003	.0003	.0003	.0003	.0003	.0003	.0003	.0003	.0003
6	.0010	.0009	.0009	.0009	.0009	.0009	.0009	.0009	.0009
8	.002	.002	.002	.002	.002	.002	.002	.002	.002
10	.005	.005	.005	.005	.005	.005	.005	.005	.005
12	.010	.010	.009	.009	.009	.009	.009	.009	.009
14	.018	.017	.017	.016	.016	.016	.016	.016	.016
16	.030	.029	.028	.027	.027	.027	.027	.027	.027
18	.047	.045	.044	.044	.043	.043	.043	.043	.043
20	.072	.069	.067	.066	.066	.066	.066	.066	.066
22	.105	.101	.099	.098	.097	.097	.097	.097	.097
24	.149	.144	.141	.139	.139	.138	.138	.138	.138
26	.205	.199	.195	.194	.193	.192	.192	.192	.192
28	.276	.270	.265	.263	.262	.262	.262	.262	.261
30	.364	.359	.354	.351	.350	.349	.349	.349	.349
32	.472	.468	.463	.460	.459	.458	.458	.458	.458
34	.603	.600	.596	.594	.593	.592	.592	.592	.592
36	.759	.758	.757	.756	.755	.755	.755	.755	.755
38	.946	.947	.949	.950	.951	.951	.951	.951	.951
40	1.17	1.17	1.18	1.18	1.18	1.18	1.18	1.18	1.19
42	1.42	1.43	1.44	1.45	1.46	1.46	1.46	1.46	1.46
44	1.73	1.73	1.75	1.77	1.78	1.78	1.79	1.79	1.79
46	2.08	2.07	2.10	2.13	2.15	2.16	2.16	2.16	2.17
48	2.48	2.47	2.51	2.55	2.58	2.59	2.60	2.60	2.61
50	2.95	2.92	2.97	3.03	3.07	3.09	3.10	3.11	3.11
52	3.48	3.44	3.50	3.58	3.63	3.66	3.68	3.69	3.69
54	4.09	4.03	4.09	4.20	4.27	4.31	4.33	4.35	4.35
56	4.78	4.69	4.76	4.89	4.99	5.05	5.08	5.09	5.10
58	5.57	5.44	5.51	5.66	5.79	5.87	5.91	5.94	5.95
60	6.45	6.29	6.35	6.53	6.69	6.79	6.85	6.88	6.90
62	7.43	7.23	7.28	7.49	7.69	7.82	7.90	7.94	7.97
64	8.54	8.28	8.32	8.55	8.80	8.97	9.07	9.13	9.16
66	9.76	9.46	9.48	9.73	10.02	10.24	10.37	10.44	10.48
68	11.1	10.8	10.8	11.0	11.4	11.6	11.8	11.9	12.0
70	12.6	12.2	12.2	12.5	12.8	13.2	13.4	13.5	13.6
72	14.3	13.8	13.7	14.0	14.5	14.9	15.1	15.3	15.4
74	16.1	15.5	15.4	15.7	16.2	16.7	17.0	17.2	17.3
76	18.2	17.5	17.3	17.6	18.2	18.7	19.1	19.3	19.5
78	20.4	19.6	19.4	19.7	20.3	20.9	21.4	21.7	21.8
80	22.8	21.9	21.6	21.9	22.6	23.3	23.8	24.2	24.4
82	25.4	24.4	24.1	24.4	25.0	25.8	26.5	26.9	27.2
84	28.3	27.1	26.7	27.0	27.7	28.6	29.4	29.9	30.2
86	31.4	30.1	29.6	29.9	30.7	31.6	32.5	33.1	33.5
88	34.8	33.3	32.8	33.0	33.8	34.8	35.8	36.6	37.1
90	38.5	36.8	36.2	36.4	37.2	38.3	39.4	40.3	40.9

Table D.13. Axle Load Equivalency Factors for Rigid Pavements, Single Axles and p_t of 2.5

Axle Load (kips)	Slab Thickness, D (inches)								
	6	7	8	9	10	11	12	13	14
2	.0002	.0002	.0002	.0002	.0002	.0002	.0002	.0002	.0002
4	.003	.002	.002	.002	.002	.002	.002	.002	.002
6	.012	.011	.010	.010	.010	.010	.010	.010	.010
8	.039	.035	.033	.032	.032	.032	.032	.032	.032
10	.097	.089	.084	.082	.081	.080	.080	.080	.080
12	.203	.189	.181	.176	.175	.174	.174	.173	.173
14	.376	.360	.347	.341	.338	.337	.336	.336	.336
16	.634	.623	.610	.604	.601	.599	.599	.599	.598
18	1.00	1.00	1.00	1.00	1.00	1.00	1.00	1.00	1.00
20	1.51	1.52	1.55	1.57	1.58	1.58	1.59	1.59	1.59
22	2.21	2.20	2.28	2.34	2.38	2.40	2.41	2.41	2.41
24	3.16	3.10	3.22	3.36	3.45	3.50	3.53	3.54	3.55
26	4.41	4.26	4.42	4.67	4.85	4.95	5.01	5.04	5.05
28	6.05	5.76	5.92	6.29	6.61	6.81	6.92	6.98	7.01
30	8.16	7.67	7.79	8.28	8.79	9.14	9.35	9.46	9.52
32	10.8	10.1	10.1	10.7	11.4	12.0	12.3	12.6	12.7
34	14.1	13.0	12.9	13.6	14.6	15.4	16.0	16.4	16.5
36	18.2	16.7	16.4	17.1	18.3	19.5	20.4	21.0	21.3
38	23.1	21.1	20.6	21.3	22.7	24.3	25.6	26.4	27.0
40	29.1	26.5	25.7	26.3	27.9	29.9	31.6	32.9	33.7
42	36.2	32.9	31.7	32.2	34.0	36.3	38.7	40.4	41.6
44	44.6	40.4	38.8	39.2	41.0	43.8	46.7	49.1	50.8
46	54.5	49.3	47.1	47.3	49.2	52.3	55.9	59.0	61.4
48	66.1	59.7	56.9	56.8	58.7	62.1	66.3	70.3	73.4
50	79.4	71.7	68.2	67.8	69.6	73.3	78.1	83.0	87.1

Table D.14. Axle Load Equivalency Factors for Rigid Pavements, Tandem Axles and p_t of 2.5

Axle Load (kips)	Slab Thickness, D (inches)								
	6	7	8	9	10	11	12	13	14
2	.0001	.0001	.0001	.0001	.0001	.0001	.0001	.0001	.0001
4	.0006	.0006	.0005	.0005	.0005	.0005	.0005	.0005	.0005
6	.002	.002	.002	.002	.002	.002	.002	.002	.002
8	.007	.006	.006	.005	.005	.005	.005	.005	.005
10	.015	.014	.013	.013	.012	.012	.012	.012	.012
12	.031	.028	.026	.026	.025	.025	.025	.025	.025
14	.057	.052	.049	.048	.047	.047	.047	.047	.047
16	.097	.089	.084	.082	.081	.081	.080	.080	.080
18	.155	.143	.136	.133	.132	.131	.131	.131	.131
20	.234	.220	.211	.206	.204	.203	.203	.203	.203
22	.340	.325	.313	.308	.305	.304	.303	.303	.303
24	.475	.462	.450	.444	.441	.440	.439	.439	.439
26	.644	.637	.627	.622	.620	.619	.618	.618	.618
28	.855	.854	.852	.850	.850	.850	.849	.849	.849
30	1.11	1.12	1.13	1.14	1.14	1.14	1.14	1.14	1.14
32	1.43	1.44	1.47	1.49	1.50	1.51	1.51	1.51	1.51
34	1.82	1.82	1.87	1.92	1.95	1.96	1.97	1.97	1.97
36	2.29	2.27	2.35	2.43	2.48	2.51	2.52	2.52	2.53
38	2.85	2.80	2.91	3.03	3.12	3.16	3.18	3.20	3.20
40	3.52	3.42	3.55	3.74	3.87	3.94	3.98	4.00	4.01
42	4.32	4.16	4.30	4.55	4.74	4.86	4.91	4.95	4.96
44	5.26	5.01	5.16	5.48	5.75	5.92	6.01	6.06	6.09
46	6.36	6.01	6.14	6.53	6.90	7.14	7.28	7.36	7.40
48	7.64	7.16	7.27	7.73	8.21	8.55	8.75	8.86	8.92
50	9.11	8.50	8.55	9.07	9.68	10.14	10.42	10.58	10.66
52	10.8	10.0	10.0	10.6	11.3	11.9	12.3	12.5	12.7
54	12.8	11.8	11.7	12.3	13.2	13.9	14.5	14.8	14.9
56	15.0	13.8	13.6	14.2	15.2	16.2	16.8	17.3	17.5
58	17.5	16.0	15.7	16.3	17.5	18.6	19.5	20.1	20.4
60	20.3	18.5	18.1	18.7	20.0	21.4	22.5	23.2	23.6
62	23.5	21.4	20.8	21.4	22.8	24.4	25.7	26.7	27.3
64	27.0	24.6	23.8	24.4	25.8	27.7	29.3	30.5	31.3
66	31.0	28.1	27.1	27.6	29.2	31.3	33.2	34.7	35.7
68	35.4	32.1	30.9	31.3	32.9	35.2	37.5	39.3	40.5
70	40.3	36.5	35.0	35.3	37.0	39.5	42.1	44.3	45.9
72	45.7	41.4	39.6	39.8	41.5	44.2	47.2	49.8	51.7
74	51.7	46.7	44.6	44.7	46.4	49.3	52.7	55.7	58.0
76	58.3	52.6	50.2	50.1	51.8	54.9	58.6	62.1	64.8
78	65.5	59.1	56.3	56.1	57.7	60.9	65.0	69.0	72.3
80	73.4	66.2	62.9	62.5	64.2	67.5	71.9	76.4	80.2
82	82.0	73.9	70.2	69.6	71.2	74.7	79.4	84.4	88.8
84	91.4	82.4	78.1	77.3	78.9	82.4	87.4	93.0	98.1
86	102.	92.	87.	86.	87.	91.	96.	102.	108.
88	113.	102.	96.	95.	96.	100.	105.	112.	119.
90	125.	112.	106.	105.	106.	110.	115.	123.	130.

Table D.15. Axle Load Equivalency Factors for Rigid Pavements, Triple Axles and p_t of 2.5

Axle Load (kips)	Slab Thickness, D (inches)								
	6	7	8	9	10	11	12	13	14
2	.0001	.0001	.0001	.0001	.0001	.0001	.0001	.0001	.0001
4	.0003	.0003	.0003	.0003	.0003	.0003	.0003	.0003	.0003
6	.001	.001	.001	.001	.001	.001	.001	.001	.001
8	.003	.002	.002	.002	.002	.002	.002	.002	.002
10	.006	.005	.005	.005	.005	.005	.005	.005	.005
12	.011	.010	.010	.009	.009	.009	.009	.009	.009
14	.020	.018	.017	.017	.016	.016	.016	.016	.016
16	.033	.030	.029	.028	.027	.027	.027	.027	.027
18	.053	.048	.045	.044	.044	.043	.043	.043	.043
20	.080	.073	.069	.067	.066	.066	.066	.066	.066
22	.116	.107	.101	.099	.098	.097	.097	.097	.097
24	.163	.151	.144	.141	.139	.139	.138	.138	.138
26	.222	.209	.200	.195	.194	.193	.192	.192	.192
28	.295	.281	.271	.265	.263	.262	.262	.262	.262
30	.384	.371	.359	.354	.351	.350	.349	.349	.349
32	.490	.480	.468	.463	.460	.459	.458	.458	.458
34	.616	.609	.601	.596	.594	.593	.592	.592	.592
36	.765	.762	.759	.757	.756	.755	.755	.755	.755
38	.939	.941	.946	.948	.950	.951	.951	.951	.951
40	1.14	1.15	1.16	1.17	1.18	1.18	1.18	1.18	1.18
42	1.38	1.38	1.41	1.44	1.45	1.46	1.46	1.46	1.46
44	1.65	1.65	1.70	1.74	1.77	1.78	1.78	1.78	1.79
46	1.97	1.96	2.03	2.09	2.13	2.15	2.16	2.16	2.16
48	2.34	2.31	2.40	2.49	2.55	2.58	2.59	2.60	2.60
50	2.76	2.71	2.81	2.94	3.02	3.07	3.09	3.10	3.11
52	3.24	3.15	3.27	3.44	3.56	3.62	3.66	3.68	3.68
54	3.79	3.66	3.79	4.00	4.16	4.26	4.30	4.33	4.34
56	4.41	4.23	4.37	4.63	4.84	4.97	5.03	5.07	5.09
58	5.12	4.87	5.00	5.32	5.59	5.76	5.85	5.90	5.93
60	5.91	5.59	5.71	6.08	6.42	6.64	6.77	6.84	6.87
62	6.80	6.39	6.50	6.91	7.33	7.62	7.79	7.88	7.93
64	7.79	7.29	7.37	7.82	8.33	8.70	8.92	9.04	9.11
66	8.90	8.28	8.33	8.83	9.42	9.88	10.17	10.33	10.42
68	10.1	9.4	9.4	9.9	10.6	11.2	11.5	11.7	11.9
70	11.5	10.6	10.6	11.1	11.9	12.6	13.0	13.3	13.5
72	13.0	12.0	11.8	12.4	13.3	14.1	14.7	15.0	15.2
74	14.6	13.5	13.2	13.8	14.8	15.8	16.5	16.9	17.1
76	16.5	15.1	14.8	15.4	16.5	17.6	18.4	18.9	19.2
78	18.5	16.9	16.5	17.1	18.2	19.5	20.5	21.1	21.5
80	20.6	18.8	18.3	18.9	20.2	21.6	22.7	23.5	24.0
82	23.0	21.0	20.3	20.9	22.2	23.8	25.2	26.1	26.7
84	25.6	23.3	22.5	23.1	24.5	26.2	27.8	28.9	29.6
86	28.4	25.8	24.9	25.4	26.9	28.8	30.5	31.9	32.8
88	31.5	28.6	27.5	27.9	29.4	31.5	33.5	35.1	36.1
90	34.8	31.5	30.3	30.7	32.2	34.4	36.7	38.5	39.8

Table D.16. Axle Load Equivalency Factors for Rigid Pavements, Single Axles and p_t of 3.0

Axle Load (kips)	Slab Thickness, D (inches)								
	6	7	8	9	10	11	12	13	14
2	.0003	.0002	.0002	.0002	.0002	.0002	.0002	.0002	.0002
4	.003	.003	.002	.002	.002	.002	.002	.002	.002
6	.014	.012	.011	.010	.010	.010	.010	.010	.010
8	.045	.038	.034	.033	.032	.032	.032	.032	.032
10	.111	.095	.087	.083	.081	.081	.080	.080	.080
12	.228	.202	.186	.179	.176	.174	.174	.174	.173
14	.408	.378	.355	.344	.340	.337	.337	.336	.336
16	.660	.640	.619	.608	.603	.600	.599	.599	.599
18	1.00	1.00	1.00	1.00	1.00	1.00	1.00	1.00	1.00
20	1.46	1.47	1.52	1.55	1.57	1.58	1.58	1.59	1.59
22	2.07	2.06	2.18	2.29	2.35	2.38	2.40	2.41	2.41
24	2.90	2.81	3.00	3.23	3.38	3.47	3.51	3.53	3.54
26	4.00	3.77	4.01	4.40	4.70	4.87	4.96	5.01	5.04
28	5.43	4.99	5.23	5.80	6.31	6.65	6.83	6.93	6.98
30	7.27	6.53	6.72	7.46	8.25	8.83	9.17	9.36	9.46
32	9.59	8.47	8.53	9.42	10.54	11.44	12.03	12.37	12.56
34	12.5	10.9	10.7	11.7	13.2	14.5	15.5	16.0	16.4
36	16.0	13.8	13.4	14.4	16.2	18.1	19.5	20.4	21.0
38	20.4	17.4	16.7	17.7	19.8	22.2	24.2	25.6	26.4
40	25.6	21.8	20.6	21.5	23.8	26.8	29.5	31.5	32.9
42	31.8	26.9	25.3	26.0	28.5	32.0	35.5	38.4	40.3
44	39.2	33.1	30.8	31.3	33.9	37.9	42.3	46.1	48.8
46	47.8	40.3	37.2	37.5	40.1	44.5	49.8	54.7	58.5
48	57.9	48.6	44.8	44.7	47.3	52.1	58.2	64.3	69.4
50	69.6	58.4	53.6	53.1	55.6	60.6	67.6	75.0	81.4

Table D.17. Axle Load Equivalency Factors for Rigid Pavements, Tandem Axles and p_t of 3.0

Axle Load (kips)	Slab Thickness, D (inches)								
	6	7	8	9	10	11	12	13	14
2	.0001	.0001	.0001	.0001	.0001	.0001	.0001	.0001	.0001
4	.0007	.0006	.0005	.0005	.0005	.0005	.0005	.0005	.0005
6	.003	.002	.002	.002	.002	.002	.002	.002	.002
8	.008	.006	.006	.006	.005	.005	.005	.005	.005
10	.018	.015	.013	.013	.013	.012	.012	.012	.012
12	.036	.030	.027	.026	.026	.025	.025	.025	.025
14	.066	.056	.050	.048	.047	.047	.047	.047	.047
16	.111	.095	.087	.083	.081	.081	.081	.080	.080
18	.174	.153	.140	.135	.132	.131	.131	.131	.131
20	.260	.234	.217	.209	.205	.204	.203	.203	.203
22	.368	.341	.321	.311	.307	.305	.304	.303	.303
24	.502	.479	.458	.447	.443	.440	.440	.439	.439
26	.664	.651	.634	.625	.621	.619	.618	.618	.618
28	.859	.857	.853	.851	.850	.850	.850	.849	.849
30	1.09	1.10	1.12	1.13	1.14	1.14	1.14	1.14	1.14
32	1.38	1.38	1.44	1.47	1.49	1.50	1.51	1.51	1.51
34	1.72	1.71	1.80	1.88	1.93	1.95	1.96	1.97	1.97
36	2.13	2.10	2.23	2.36	2.45	2.49	2.51	2.52	2.52
38	2.62	2.54	2.71	2.92	3.06	3.13	3.17	3.19	3.20
40	3.21	3.05	3.26	3.55	3.76	3.89	3.95	3.98	4.00
42	3.90	3.65	3.87	4.26	4.58	4.77	4.87	4.92	4.95
44	4.72	4.35	4.57	5.06	5.50	5.78	5.94	6.02	6.06
46	5.68	5.16	5.36	5.95	6.54	6.94	7.17	7.29	7.36
48	6.80	6.10	6.25	6.93	7.69	8.24	8.57	8.76	8.86
50	8.09	7.17	7.26	8.03	8.96	9.70	10.17	10.43	10.58
52	9.57	8.41	8.40	9.24	10.36	11.32	11.96	12.33	12.54
54	11.3	9.8	9.7	10.6	11.9	13.1	14.0	14.5	14.8
56	13.2	11.4	11.2	12.1	13.6	15.1	16.2	16.9	17.3
58	15.4	13.2	12.8	13.7	15.4	17.2	18.6	19.5	20.1
60	17.9	15.3	14.7	15.6	17.4	19.5	21.3	22.5	23.2
62	20.6	17.6	16.8	17.6	19.6	22.0	24.1	25.7	26.6
64	23.7	20.2	19.1	19.9	22.0	24.7	27.3	29.2	30.4
66	27.2	23.1	21.7	22.4	24.6	27.6	30.6	33.0	34.6
68	31.1	26.3	24.6	25.2	27.4	30.8	34.3	37.1	39.2
70	35.4	29.8	27.8	28.2	30.6	34.2	38.2	41.6	44.1
72	40.1	33.8	31.3	31.6	34.0	37.9	42.3	46.4	49.4
74	45.3	38.1	35.2	35.4	37.7	41.8	46.8	51.5	55.2
76	51.1	42.9	39.5	39.5	41.8	46.1	51.5	56.9	61.3
78	57.4	48.2	44.3	44.0	46.3	50.7	56.6	62.7	67.9
80	64.3	53.9	49.4	48.9	51.1	55.8	62.1	68.9	74.9
82	71.8	60.2	55.1	54.3	56.5	61.2	67.9	75.5	82.4
84	80.0	67.0	61.2	60.2	62.2	67.0	74.2	82.4	90.3
86	89.0	74.5	67.9	66.5	68.5	73.4	80.8	89.8	98.7
88	98.7	82.5	75.2	73.5	75.3	80.2	88.0	97.7	107.5
90	109.	91.	83.	81.	83.	88.	96.	106.	117.

Table D.18. Axle Load Equivalency Factors for Rigid Pavements, Triple Axles and p_t of 3.0

Axle Load (kips)	Slab Thickness, D (inches)								
	6	7	8	9	10	11	12	13	14
2	.0001	.0001	.0001	.0001	.0001	.0001	.0001	.0001	.0001
4	.0004	.0003	.0003	.0003	.0003	.0003	.0003	.0003	.0003
6	.001	.001	.001	.001	.001	.001	.001	.001	.001
8	.003	.003	.002	.002	.002	.002	.002	.002	.002
10	.007	.006	.005	.005	.005	.005	.005	.005	.005
12	.013	.011	.010	.009	.009	.009	.009	.009	.009
14	.023	.020	.018	.017	.017	.016	.016	.016	.016
16	.039	.033	.030	.028	.028	.027	.027	.027	.027
18	.061	.052	.047	.045	.044	.044	.043	.043	.043
20	.091	.078	.071	.068	.067	.066	.066	.066	.066
22	.132	.114	.104	.100	.098	.097	.097	.097	.097
24	.183	.161	.148	.143	.140	.139	.139	.138	.138
26	.246	.221	.205	.198	.195	.193	.193	.192	.192
28	.322	.296	.277	.268	.265	.263	.262	.262	.262
30	.411	.387	.367	.357	.353	.351	.350	.349	.349
32	.515	.495	.476	.466	.462	.460	.459	.458	.458
34	.634	.622	.607	.599	.595	.594	.593	.592	.592
36	.772	.768	.762	.758	.756	.756	.755	.755	.755
38	.930	.934	.942	.947	.949	.950	.951	.951	.951
40	1.11	1.12	1.15	1.17	1.18	1.18	1.18	1.18	1.18
42	1.32	1.33	1.38	1.42	1.44	1.45	1.46	1.46	1.46
44	1.56	1.56	1.64	1.71	1.75	1.77	1.78	1.78	1.78
46	1.84	1.83	1.94	2.04	2.10	2.14	2.15	2.16	2.16
48	2.16	2.12	2.26	2.41	2.51	2.56	2.58	2.59	2.60
50	2.53	2.45	2.61	2.82	2.96	3.03	3.07	3.09	3.10
52	2.95	2.82	3.01	3.27	3.47	3.58	3.63	3.66	3.68
54	3.43	3.23	3.43	3.77	4.03	4.18	4.27	4.31	4.33
56	3.98	3.70	3.90	4.31	4.65	4.86	4.98	5.04	5.07
58	4.59	4.22	4.42	4.90	5.34	5.62	5.78	5.86	5.90
60	5.28	4.80	4.99	5.54	6.08	6.45	6.66	6.78	6.84
62	6.06	5.45	5.61	6.23	6.89	7.36	7.64	7.80	7.88
64	6.92	6.18	6.29	6.98	7.76	8.36	8.72	8.93	9.04
66	7.89	6.98	7.05	7.78	8.70	9.44	9.91	10.18	10.33
68	8.96	7.88	7.87	8.66	9.71	10.61	11.20	11.55	11.75
70	10.2	8.9	8.8	9.6	10.8	11.9	12.6	13.1	13.3
72	11.5	10.0	9.8	10.6	12.0	13.2	14.1	14.7	15.0
74	12.9	11.2	10.9	11.7	13.2	14.7	15.8	16.5	16.9
76	14.5	12.5	12.1	12.9	14.5	16.2	17.5	18.4	18.9
78	16.2	13.9	13.4	14.2	15.9	17.8	19.4	20.5	21.1
80	18.2	15.5	14.8	15.6	17.4	19.6	21.4	22.7	23.5
82	20.2	17.2	16.4	17.2	19.1	21.4	23.5	25.1	26.1
84	22.5	19.1	18.1	18.8	20.8	23.4	25.8	27.6	28.8
86	25.0	21.2	19.9	20.6	22.6	25.5	28.2	30.4	31.8
88	27.6	23.4	21.9	22.5	24.6	27.7	30.7	33.2	35.0
90	30.5	25.8	24.1	24.6	26.8	30.0	33.4	36.3	38.3

Axle Load	Traffic Equivalency Factor		Number of Axles		A18 Kip EAL's
Single Axles	P = 2.5, SN = 5				
Under 3,000	0.0002	X	0	=	0.000
3,000 - 6,999	0.0050	X	1	=	0.005
7,000 - 7,999	0.0320	X	6	=	0.192
8,000 - 11,999	0.0870	X	144	=	12.528
12,000 - 15,999	0.3600	X	16	=	5.760
26,000 - 29,999	5.3890	X	1	=	5.3890
Tandem Axle Groups					
Under 6,000	0.0100	X	0	=	0.000
6,000 - 11,993	0.0100	X	14	=	0.140
12,000 - 17,999	0.0440	X	21	=	0.924
18,000 - 23,999	0.1480	X	44	=	6.512
24,000 - 29,999	0.4260	X	42	=	17.892
30,000 - 32,000	0.7530	X	44	=	33.132
32,001 - 32,500	0.8850	X	21	=	18.585
32,501 - 33,999	1.0020	X	101	=	101.202
34,000 - 35,999	1.2300	X	43	=	52.890
18 Kip EAL's for all trucks wieghed				=	255.151

$$\text{Truck Load Factor} = \frac{\text{18 Kip EAL's for all trucks weighed}}{\text{Number of trucks weighed 165}} = \frac{255.151}{165} = 1.5464$$

Figure D.1. Computation of the Truck Load Factor for 5 Axle or Greater Trucks on Flexible Pavements with an SN = 5 and a Terminal Serviceability of 2.5

within the axle load intervals indicated. The ESAL's by axle load intervals is summed to give a total ESAL's for 165 trucks of this type which were weighed. The truck load factor is found to be 1.5464. A similar set of calculations can be made for each truck classification included in the W-4 tables.

It should be noted that this truck load factor was based on an assumed terminal serviceability of 2.5 and a structural number (SN) of 5.0. In most cases, such an assumption will provide information sufficiently accurate for design purposes. When more accuracy is required, it will be necessary to recalculate the truck load factor with the new equivalency factors as previously discussed.

When information is not available directly from weigh station loadings, it is necessary to use representative values for each of the various truck classifications. No adjustments can be made for serviceability or thicknesses using this alternate. This method is likely to be the one used most often.

The work sheet in Table D.19 may be used to calculate ESAL's using truck load factors obtained directly or based on representative values furnished by the design agency.

The first column (A) represents the base year daily volume counts of each vehicle type taken from data collected at classification count stations representative of the design location.

The second column (B) indicates the growth factor assigned to each of the various vehicle types. The calculations should take into account the fact that growth factors normally vary from one vehicle type to another. Table D.20 provides appropriate multipliers for a given growth rate and design period. Any growth factor selected should reflect consideration of the variables mentioned in Section D.1.

The third column (C) is basically a product of the first two columns multiplied again by 365. The result is the accumulated applications of specific vehicle types during the analysis period.

The fourth column (D) indicates the individual ESAL values for each of the vehicle types.

The fifth column (E) is an extension of columns (C) and (D) indicating the total ESAL's (by vehicle type) that might be applied to the sample section during the analysis period. The summation of these values then is the total 18-kip ESAL traffic that should be used for pavement structural design.

The number of equivalent axle loads derived using the procedure represents the total for all lanes and both directions of travel. This number must then be distributed by direction and by lanes, as discussed in Section D.1.

D.3 EXAMPLE ESAL CALCULATIONS

In order to illustrate more specifically how this procedure works, a number of sample calculations follow. Table D.21 shows the calculations of 18-kip ESAL's for a facility having traffic typical of a rural arterial. Data for this example comes from the W-2 and W-4 tables and are assumed to be representative of the design facility. In developing the Example 1 calculation, the following assumptions were made:

(1) Traffic volumes (for all vehicle types) will increase at a rate of 2 percent per year, compounded annually (as previously noted, a poor assumption).
(2) The axle weights of the various vehicle types will remain constant over the analysis period.
(3) Terminal serviceability (p_t) is 2.5.
(4) Analysis period is 20 years; since stage construction is not considered, the performance period is also 20 years.
(5) Slab thickness (D) is equal to 9 inches.

From the W-2 table, the number of passenger cars (5,925) is entered in Column A, followed by the number for buses (35). From the W-4 table, the total number of vehicles counted is used for the balance of the Column A entries, using only the numbers for the current year's data. For this example, 1,135 panel and pickup trucks, 3 other two-axle/four-tire trucks, 372 two-axle/six-tire trucks, etc., have been entered to complete Column A.

Table D.20 provides the criteria for selection of values for Column B. For the 20-year analysis period and the fixed growth factor of 2 percent per year for all vehicle types, a value of 24.30 is obtained. Multiplying Column A by Column B and then multiplying this number by 365 to annualize it, Column C can be completed.

Returning to the W-4 tables, summary information is provided for the average ESAL's per 1,000 trucks weighed. For this example, under panel and pickup trucks, the value is 12.2 ESAL per 1,000 trucks, or 0.0122 per vehicle. The W-4 table will provide similar information for each truck classification shown in Table D.21.

Finally, by multiplying the numbers in Column C by the values in Column D, Column E can be completed. The summation of the numbers in Column E, then, is the total design 18-kip ESAL value. In the first example, it has been predicted that this sample section will experience 43.8 million 18-kip ESAL applications over the next 20 years assuming only a 2-percent annual growth in traffic with no change in axle

Appendix D D-23

Table D.19. Worksheet for Calculating 18-kip Equivalent Single Axle Load (ESAL) Applications

Location _____

Analysis Period = _____ Years

Assumed SN or D = _____

Vehicle Types	Current Traffic (A)	Growth Factors (B)	Design Traffic (C)	E.S.A.L. Factor (D)	Design E.S.A.L. (E)
Passenger Cars Buses					
Panel and Pickup Trucks Other 2-Axle/4-Tire Trucks 2-Axle/6-Tire Trucks 3 or More Axle Trucks All Single Unit Trucks					
3 Axle Tractor Semi-Trailers 4 Axle Tractor Semi-Trailers 5+ Axle Tractor Semi-Trailers All Tractor Semi-Trailers					
5 Axle Double Trailers 6+ Axle Double Trailers All Double Trailer Combos					
3 Axle Truck-Trailers 4 Axle Truck-Trailers 5+ Axle Truck-Trailers All Truck-Trailer Combos					
All Vehicles				Design E.S.A.L.	

Table D.20. Traffic Growth Factors*

Analysis Period Years (n)	Annual Growth Rate, Percent (g)							
	No Growth	2	4	5	6	7	8	10
1	1.0	1.0	1.0	1.0	1.0	1.0	1.0	1.0
2	2.0	2.02	2.04	2.05	2.06	2.07	2.08	2.10
3	3.0	3.06	3.12	3.15	3.18	3.21	3.25	3.31
4	4.0	4.12	4.25	4.31	4.37	4.44	4.51	4.64
5	5.0	5.20	5.42	5.53	5.64	5.75	5.87	6.11
6	6.0	6.31	6.63	6.80	6.98	7.15	7.34	7.72
7	7.0	7.43	7.90	8.14	8.39	8.65	8.92	9.49
8	8.0	8.58	9.21	9.55	9.90	10.26	10.64	11.44
9	9.0	9.75	10.58	11.03	11.49	11.98	12.49	13.58
10	10.0	10.95	12.01	12.58	13.18	13.82	14.49	15.94
11	11.0	12.17	13.49	14.21	14.97	15.78	16.65	18.53
12	12.0	13.41	15.03	15.92	16.87	17.89	18.98	21.38
13	13.0	14.68	16.63	17.71	18.88	20.14	21.50	24.52
14	14.0	15.97	18.29	19.16	21.01	22.55	24.21	27.97
15	15.0	17.29	20.02	21.58	23.28	25.13	27.15	31.77
16	16.0	18.64	21.82	23.66	25.67	27.89	30.32	35.95
17	17.0	20.01	23.70	25.84	28.21	30.84	33.75	40.55
18	18.0	21.41	25.65	28.13	30.91	34.00	37.45	45.60
19	19.0	22.84	27.67	30.54	33.76	37.38	41.45	51.16
20	20.0	24.30	29.78	33.06	36.79	41.00	45.76	57.28
25	25.0	32.03	41.65	47.73	54.86	63.25	73.11	98.35
30	30.0	40.57	56.08	66.44	79.06	94.46	113.28	164.49
35	35.0	49.99	73.65	90.32	111.43	138.24	172.32	271.02

*Factor $= \dfrac{(1 + g)^n - 1}{g}$, where $g = \dfrac{\text{rate}}{100}$ and is not zero. If annual growth rate is zero, the growth factor is equal to the analysis period.

NOTE: The above growth factors multiplied by the first year traffic estimate will give the total volume of traffic expected during the analysis period.

Appendix D

Table D.21. Worksheet for Calculating 18-kip Equivalent Single Axle Load (ESAL) Applications

Location __Example 1__   Analysis Period = __20__ Years

Assumed SN or D = __9″__

Vehicle Types	Current Traffic (A)	Growth Factors (B)	Design Traffic (C)	E.S.A.L. Factor (D)	Design E.S.A.L. (E)
Passenger Cars	5,925	2% 24.30	52,551,787	.0008	42,041
Buses	35	24.30	310,433	.6806	211,280
Panel and Pickup Trucks	1,135	24.30	10,066,882	.0122	122,816
Other 2-Axle/4-Tire Trucks	3	24.30	26,609	.0052	138
2-Axle/6-Tire Trucks	372	24.30	3,299,454	.1890	623,597
3 or More Axle Trucks	34	24.30	301,563	.1303	39,294
All Single Unit Trucks					
3 Axle Tractor Semi-Trailers	19	24.30	168,521	.8646	145,703
4 Axle Tractor Semi-Trailers	49	24.30	434,606	.6560	285,101
5+ Axle Tractor Semi-Trailers	1,880	24.30	16,674,660	2.3719	39,550,626
All Tractor Semi-Trailers					
5 Axle Double Trailers	103	24.30	913,559	2.3187	2,118,268
6+ Axle Double Trailers	0	24.30			
All Double Trailer Combos					
3 Axle Truck-Trailers	208	24.30	1,844,856	.0152	28,042
4 Axle Truck-Trailers	305	24.30	2,705,198	.0152	41,119
5+ Axle Truck-Trailers	125	24.30	1,108,688	.5317	589,489
All Truck-Trailer Combos					
All Vehicles	10,193		90,406,816	Design E.S.A.L.	43,772,314

weights for each vehicle type and no change in vehicle type distribution over the analysis period.

In a second example, Table D.22 assumes the same base year traffic. However, a 2-percent estimate is assumed for passenger cars and buses, as well as single-unit trucks, a 4-percent growth in tractor semi-trailer and truck full trailer combinations, and a 5-percent growth in double trailer combinations. Past experience has shown that these estimates are not uncommon. By using the appropriate growth factors in Column B and going through the exercise exactly as before, it is estimated that the total design 18-kip ESAL value is now 53.7 million, about 23 percent more than in the first example.

The third example shown in Table D.23 also uses the same base year traffic. The assumed growth rate is increased to 4 percent for passenger vehicles and single-unit trucks, to 6 percent for tractor semi-trailers and truck full trailer combinations, and to 7 percent for double trailer combinations. This example results in a total design ESAL value of about 66.4 million, or an increase of about 50 percent over the first example.

If, in these examples, it is assumed that the facility is a four-lane rural Interstate highway and that the directional and lane distribution factors are 0.5 and 0.9, respectively, the design lane traffic estimates are calculated as follows:

Example 1: .5 × .9 × 43,772,314 = 19,697,541 18-kip ESAL

Example 2: .5 × .9 × 53,726,060 = 24,176,727 18-kip ESAL

Example 3: .5 × .9 × 66,376,294 = 29,869,332 18-kip ESAL

D.4 HIGHLIGHTS FROM THE EXAMPLE WORKSHEETS

Sample location is a rural Interstate site in the midwest with an average daily traffic of 10,193. It is comprised of 58 percent passenger cars, 30 percent commercial trucks, and 12 percent light trucks and buses. The 18-kip ESAL values for passenger cars and buses were developed from recent weight studies of actual vehicles traveling on the highway using traffic equivalency factors for each axle load group derived at the AASHO Road Test.

When using a conservative growth factor of 2 percent per year over the entire traffic stream, about 44 million 18-kip ESAL will be applied to the highway. Increasing the growth rate from 2 percent to 4 or 5 percent for other than single-unit vehicles (the heavier trucks) will increase the ESAL totals to about 54 million, an increase of about 23 percent. By increasing the growth rate to a more aggressive value of 4 percent for light vehicles and 7 percent for heavier vehicle types, the total 18-kip ESAL value becomes about 66 million, an increase of some 52 percent over the conservative growth rate of 2 percent.

It is interesting to note that in the moderate growth rate of 2 percent to 5 percent, with large tractor semi-trailers estimated at 4-percent growth rate, just one vehicle can make a difference of almost 26,000 18-kip ESAL when extended from the daily traffic stream. Also, the five-axle or more tractor semi-trailer makes up about 18 percent of the traffic stream, but is estimated to apply about 90 percent of the ESAL's. Even with an aggressive growth rate of 6 percent, this vehicle type will comprise about 21 percent of the total vehicles expected to use the facility during the analysis period, and also apply 90 percent of the ESAL's. A 6-percent growth rate may seem high, but on some Interstate routes, growth rates in excess of 9 percent for trucks with five or more axles have been reported.

Appendix D

Table D.22. Worksheet for Calculating 18-kip Equivalent Single Axle Load (ESAL) Applications

Location: Example 2
Analysis Period = 20 Years
Assumed SN or D = 9″

Vehicle Types	Current Traffic (A) Daily	Growth Factors (B)	Design Traffic (C) Annual	E.S.A.L. Factor (D)	Design E.S.A.L. (E)
Passenger Cars	5,925	2% 24.30	52,551,787	.0008	42,041
Buses	35	24.30	310,433	.6806	211,280
Panel and Pickup Trucks	1,135	2% 24.30	10,066,882	.0122	122,816
Other 2-Axle/4-Tire Trucks	3	24.30	26,609	.0052	138
2-Axle/6-Tire Trucks	372	24.30	3,299,454	.1890	623,597
3 or More Axle Trucks	34	24.30	301,563	.1303	39,294
All Single Unit Trucks					
3 Axle Tractor Semi-Trailers	19	4% 29.78	206,524	.8646	178,561
4 Axle Tractor Semi-Trailers	49	29.78	532,615	.6560	349,396
5+ Axle Tractor Semi-Trailers	1,880	29.78	20,435,036	2.3719	48,469,861
All Tractor Semi-Trailers					
5 Axle Double Trailers	103	5% 33.06	1,242,891	2.3187	2,881,891
6+ Axle Double Trailers	0				
All Double Trailer Combos					
3 Axle Truck-Trailers	208	4% 29.78	2,260,898	.0152	34,366
4 Axle Truck-Trailers	305	29.78	3,315,259	.0152	50,392
5+ Axle Truck-Trailers	125	29.78	1,358,713	.5317	722,427
All Truck-Trailer Combos					
All Vehicles	10,193		95,908,664	Design E.S.A.L.	53,726,060

Table D.23. **Worksheet for Calculating 18-kip Equivalent Single Axle Load (ESAL) Applications**

Location: Example 3
Analysis Period = 20 Years
Assumed SN or D = 9"

Vehicle Types	Current Traffic (A)	Growth Factors (B)	Design Traffic (C)	E.S.A.L. Factor (D)	Design E.S.A.L. (E)
Passenger Cars	5,925	4% 29.78	64,402,972	.0008	51,522
Buses	35	29.78	380,440	.6806	258,927
Panel and Pickup Trucks	1,135	4% 29.78	12,337,109	.0122	150,513
Other 2-Axle/4-Tire Trucks	3	29.78	32,609	.0052	170
2-Axle/6-Tire Trucks	372	29.78	4,043,528	.1890	764,227
3 or More Axle Trucks	34	29.78	369,570	.1303	48,155
All Single Unit Trucks					
3 Axle Tractor Semi-Trailers	19	6% 36.79	255,139	.8646	220,593
4 Axle Tractor Semi-Trailers	49	36.79	657,989	.6560	431,641
5+ Axle Tractor Semi-Trailers	1,880	36.79	25,245,298	2.3719	59,879,322
All Tractor Semi-Trailers					
5 Axle Double Trailers	103	7% 41.00	1,541,395	2.3187	3,574,033
6+ Axle Double Trailers	0	41.00			
All Double Trailer Combos					
3 Axle Truck-Trailers	208	6% 36.79	2,793,097	.0152	42,455
4 Axle Truck-Trailers	305	36.79	4,095,647	.0152	62,254
5+ Axle Truck-Trailers	125	36.79	1,678,544	.5317	892,482
All Truck-Trailer Combos					
All Vehicles	10,193		117,833,337	Design E.S.A.L.	66,376,294

APPENDIX E
POSITION PAPER ON SHOULDER DESIGN

Prepared by the AASHTO Joint Task Force on Pavements, June 1983.

During the early years of highway construction, the need for first class shoulders was perhaps a secondary item. But now with the tremendous increase in both number and speed of vehicles, the need for adequate shoulders has greatly increased.

As defined by AASHTO, a highway shoulder is the "portion of roadway contiguous with the traveled way for accommodation of stopped vehicles for emergency use, and for lateral support of base and surface courses." The definition is now almost universally accepted by all concerned with highway design, construction, maintenance, and operations.

A Michigan study revealed that there is a wide variance of practices related to shoulder design. This study disclosed that California has developed a formal design procedure for shoulders, 14 other states have documented policies, 28 states have no policy, and 5 states pave their shoulders integrally with the mainline pavement. It is apparent that a definite need exists to develop criteria to improve methodology throughout the United States in Shoulder Design.

In the past, the design for the structural adequacy of the shoulder was not considered to be critical because the number of applications of heavy loads was limited. However, recent studies by Emery, Hicks, Barksdale, and others have shown truck encroachment to be one of the major causes of shoulder distress; therefore, the relationship between traffic loading and shoulder distress is much greater than was first realized. This is only one factor or design variable that must be considered before attempting to recommend development of a national shoulder design policy. Other factors to be considered in the design of shoulders and the significant effects these factors have on the overall design of a pavement are:

(1) Thickness
(2) Width
(3) Shoulder materials
(4) Seal at pavement and shoulder interface
(5) Maintenance
(6) Permeability of shoulder
(7) Environmental factors
(8) Location (cut, fill, grade sections, etc.)
(9) Construction techniques
(10) Subgrade condition
(11) Design features of adjacent mainline pavement

Some of the most pressing problems which can be observed with present-day shoulders and which may need to be considered in the design are:

(1) Design thickness as related to load-carrying requirements.
(2) Cost and service criteria for stabilization of shoulder aggregate base courses.
(3) Problems associated with the pavement edge and shoulder edge interface.
(4) Abrasive effects of traffic.
(5) Permeability or degree of imperviousness required for shoulder aggregate base courses.
(6) Relationship between shoulder performance and subgrade support.
(7) Relationship of shoulder drainage subsystem to overall subsurface drainage system.
(8) Construction and maintenance methods and operations which result in adverse shoulder performance.
(9) Effects of environment on shoulder performance.
(10) Type and texture of shoulder surface for waterproofing and delineating purposes.
(11) Effects of shoulder geometrics on performance.
(12) Warrants for paved shoulders.

It is noted that many authors, both researchers and highway engineers knowledgeable in the field of pavement design, suggest the need for developing the necessary criteria for a unified and widely accepted structural design guide for shoulders.

Regarding the specific design of shoulders, the following recommendations are listed for review:

(1) Predicate shoulder thickness design upon criteria which will reflect the magnitude and frequency of loads to which the shoulder will be subjected.

(2) Integrate shoulder drainage with the overall pavement subdrainage design.
(3) Avoid the use of aggregate base courses having a significant percentage of minus 200 mesh sieve materials to prevent frost heaving, pumping, clogging of the shoulder drainage system, and base instability.
(4) Take advantage of the desirable performance features of plant-mixed bituminous and various stabilized shoulder materials as opposed to bituminous surface-treated, unbound shoulder aggregate bases.
(5) Have a definite program of shoulder maintenance.
(6) Take advantage of the desirable performance of rigid shoulders adjacent to rigid main line pavements.
(7) Incorporate criteria for paving shoulders.

It is recognized that the listing of problems associated with shoulders and the recommendations given are by no means comprehensive and should be supplemented with information from other sources. The procedures and assumptions used to develop the design equations for both rigid and flexible could be utilized in the design of shoulders. AASHTO has provided the necessary groundwork; therefore, it appears that definite design criteria must be agreed upon to complete the methodology for the Structural Design of Shoulders.

This position paper was submitted to the states for comment in 1981 prior to adoption in 1983 by the Joint Task Force on Pavements. The comments tend to reflect the preferences of individual agencies. There was no general consensus expressed; however, in order to provide some indication of the views of the respondents, a listing of pertinent comments, as related to this Guide, are summarized as follows:

(1) Design mainline and shoulders as a single unit; allows for future additions of new traveled lanes.
(2) Paved shoulders should be of the same material as the mainline and concrete shoulders should be tied to mainline.
(3) Provide for means to seal joint between shoulder and traveled way.
(4) Consider low-volume roads in developing shoulder design criteria and investigate advantages and economics of 28- to 30-foot mainline section with an aggregate shoulder.
(5) Give proper consideration to full-depth shoulder alternatives.
(6) For shoulder design, consider use of shoulders for detouring traffic and/or as an extra lane during peak hours.

BIBLIOGRAPHY

1. American Association of State Highway Officials, "AASHO Highway Definitions," June 1968.
2. American Concrete Paving Association, "Concrete Shoulders: Performance Construction Design Details," Technical Bulletin No. 12, 1972.
3. Barksdale, R.D., and Hicks, R.G., "Improved Pavement Shoulder Joint Design," NCHRP Report 202, June 1979.
4. "Current Practices in Shoulder Design, Construction, Maintenance and Operations," Highway Research Circular, HRB No. 142, April 1972.
5. Emery, D.K., Jr., "Transverse Lane Placement for Design Trucks on Rural Freeways," Preliminary Report, Georgia Department of Transportation, 1974.
6. Federal Highway Administration, "Portland Cement Concrete Shoulders," FHWA Technical Advisory T 5040.11, September 20, 1979.
7. Goulden, W., "Pavement Faulting Study, Extent and Severity of Pavement Faulting in Georgia," Georgia Department of Transportation, Research Project No. 7104, August 1972.
8. Haven, J.H., and Rahal, A., "Experimental Portland Cement Concrete Shoulder Design and Construction," Kentucky Department of Transportation, Research Report No. 403.
9. Hicks, R.G., Barksdale, R.D., and Emery, D.K., "Design Practices for Paved Shoulders," Transportation Research Record 594, 1976.
10. Highway Design Manual, State of California, Department of Public Works, 1972.
11. Highway Research Circular Number 142, "Current Practices in Shoulder Design, Construction, Maintenance and Operations," April 1973.
12. Hughes, R.D., "Concrete Shoulders," Kentucky Department of Transportation, Bureau of Highways, Research Division.
13. Illinois Division of Highways, "Portland Cement Concrete Shoulders," Research and Development Report No. 27, July 1970.

14. Illinois Division of Highways, "Paved Shoulder Problems on Stevenson Expressway," Research and Development Report No. 19, July 1967.
15. Lokken, E.C., "What We Have Learned to Date from Experimental Concrete Shoulder Projects," presented to Highway Research Board Committee A2A07 on Shoulder Design, January 1973.
16. McKenzie, L.J., "Experimental Paved Shoulders on Frost Susceptible Soils," Illinois Department of Transportation, Research and Development Reports No. 24, December 1969, and No. 39, March 1972.
17. National Cooperative Highway Research Program, Syntheses of Highway Practice 63, "Design and Use of Highway Shoulders," August 1979.
18. Novak, E.C., Jr., "Study of Frost Action in Class AA Shoulders Near Pontiac, Michigan," Michigan Department of State Highways and Transportation, Research Report No. 671, April 1968.
19. O'Toole, M.L., "Highway Shoulders: Their Construction and Maintenance Problems," *Proceedings*, Fifty-eighth Michigan Highway Conference, 1973.
20. Portigo, J.M., "State of Art Review of Paved Shoulders," Transportation Research Record 594, 1976.
21. Smith, H.A., "Pavement Design and the Decision-making Progress," Proceedings from a Conference on Utilization of Graded Aggregate Base Materials in Flexible Pavements, March 1974.
22. Spellman, D.C., Stoker, J.R., and Neal, B.F., "Faulting of Portland Cement Concrete Pavements," State of California, Department of Public Works, Division of Highways, Materials and Research Department, Research Report No. 635167-2, January 1972.
23. Treybig, H.J., Hudson, W.R., and Abou-Ayyash, A., "Application of Slab Analysis Method to Rigid Pavements," Texas University, Center for Highway Research, Texas Highway Department, FHWA Contract No. RESST 3-5-63-56.
24. Arnold, D.J., "Experimental Concrete and Bituminous Shoulders," Michigan Department of Highways and Transportation.
25. FHWA Report, "Structural Analysis and Design of PCC Shoulders," April 1982.
26. FHWA Technical Advisory T5040.18, "Paved Shoulders," July 29, 1982.

APPENDIX F
LIST OF TEST PROCEDURES

1. CBR, California Bearing Ratio (ASTM D 1883, AASHTO T 193, MilStd 621A): To determine the load-bearing capacity. The results are used to approximate the resilient modulus.
2. Dynamic Modulus of Asphalt Mixtures (ASTM D 3497): To determine the dynamic modulus of bituminous material under standard compressive loading conditions.
3. Elastic Modulus of Portland Cement Concrete (ASTM C 469): To determine the chord modulus of elasticity in compression.
4. Hveem Stability (ASTM D 1560, AASHTO T 246*): To determine resistance to deformation/cohesion, of compacted bituminous mixtures.
5. Marshall Stability (ASTM D 1559, AASHTO T 245): To determine the plastic flow rate of bituminous mixtures.
6. Modulus of Rupture:
    Center Point Loading
        (ASTM C 293, AASTHO T 177)
    Third Point Loading
        (ASTM C 78, AASHTO T 97)
    These methods cover the determination of concrete strength under flexural loading conditions.
7. Plasticity Index (ASTM D 424, AASHTO T 90): To find the range of water contents over which the soil is in a plastic state.
8. R-value (ASTM D 2844, AASHTO T 190): To determine the load-bearing capacity of a material.
9. Resilient Modulus of Asphalt Concrete from Diametral Strain (ASTM D 4123): To estimate the modulus of asphalt concrete and other relatively low-strength materials under simulated field-loading conditions.
11. Splitting-Tensile Strength, Concrete (ASTM C 496, AASHTO T 198): To determine the splitting tensile strength of cylindrical concrete specimens such as molded cylinders and drilled cores.
12. Unconfined Compressive Strength:
    For cohesive soils
        (ASTM D 2166, AASHTO T 208)
    For cement-treated materials
        (ASTM D 1633)
    To find the unconfined compressive strength of soils using molded cylinders as test specimens.
13. Joint Sealants for Concrete:
    Cold applied specifications
        (ASTM D 1850)
    Hot poured specifications
        (ASTM D 3405, D 1190, D 3406)
    Preformed compression
        (ASTM D 2628)
    Cork filler for expansion joints
        (AASHTO M 153)

*AASHTO M____ and T____ specifications are contained in the *Standard Specifications for Transportation Materials and Methods of Sampling and Testing*.

APPENDIX G
TREATMENT OF ROADBED SWELLING AND/OR FROST HEAVE IN DESIGN

This appendix provides the procedures and graphs to predict the direct effect of roadbed swelling and frost heave on serviceability loss. It should be understood that both the design models presented herein treat swelling and frost heave in terms of their *differential* effects on the longitudinal profile of the road surface. Consequently, if experience indicates that either swelling or frost heave will occur (relatively) uniformly along the length of the roadway (thus having little effect on road roughness and loss of serviceability), then these models should not be applied. These design models should also not be applied if it is anticipated that an improved drainage system and/or the use of frost control procedures (e.g., placement of non-frost-susceptible material) will eliminate the potential for differential swelling or frost heaving.

G.1 ROADBED SWELLING

To generate the swelling curve, it is first necessary to estimate three variables which affect the rate and potential magnitude of serviceability loss due to swelling (1) swell rate constant, (2) potential vertical rise, and (3) swell probability. Generally, swelling need only be considered for fine-grained soils such as clays and silts. It should also be recognized that all clays or silts are not swelling materials. If a rehabilitation project is being considered, then the designer should also recognize that much of the swelling may have already occurred since much of the swell occurs in the first few years after initial construction. Thus, a low overall swell may be anticipated after overlay even though the area has active swelling. Recognition of this should lead the designer to take advantage of stage construction alternatives. Figure G.1 is provided to identify regions in the United States that are susceptible to swelling clay conditions.

The *swell rate constant* is a factor used to estimate the rate at which swelling will take place. This constant can vary anywhere between 0.04 and 0.20. A higher value should be used when the soil is exposed to a large moisture supply from either high rainfall, poor drainage, or other sources of moisture. Lower values should be used when the roadbed soil has less access to moisture. Figure G.2 provides a chart for subjectively estimating the rate of roadbed soil swelling, considering the available moisture supply and the fabric of the roadbed soil. A less subjective approach to establishing values for this factor will be derived once practitioners become more familiar with its application and are able to calibrate it to actual field observations.

The *potential vertical rise* (V_R) represents the amount of vertical expansion that can occur in the roadbed soil under extreme swell conditions (i.e., high plasticity and extended moisture availability). The designer may obtain V_R from laboratory test results, an empirical procedure, or by experience. Figure G.3 provides a chart that can be used to estimate the potential vertical rise at a particular location given the swelling layer's plasticity index (ASTM Test No. D 424), moisture condition, and overall thickness of the layer. The moisture condition is a subjective decision based on an estimate of how close the soil moisture conditions during construction are to the in situ moisture conditions at a later date.

Swell probability represents the proportion (expressed as a percent) of the project length that is subject to swell. The probability of swelling at a given location is considered to be 100 percent if the roadbed soil plasticity index (AASHTO T 90) is greater than 30 and the layer thickness is greater than 2 feet (or if the V_R is greater than 0.20 inches). Thus, the overall swell probability can be estimated from the roadbed soil boring and laboratory test program. If the project length is separated into swelling and nonswelling materials and they are treated separately, then a probability of 100 percent is used for the swelling sections.

Table G.1 presents a form that can be used to tabulate this data when developing the serviceability loss versus time chart. Each bore hole is representative of conditions over a specific section length. Roadbed thickness represents the thickness of the layer subject to swell (for thicknesses greater than 30 feet, use 30 feet). The plasticity index (PI) is determined from

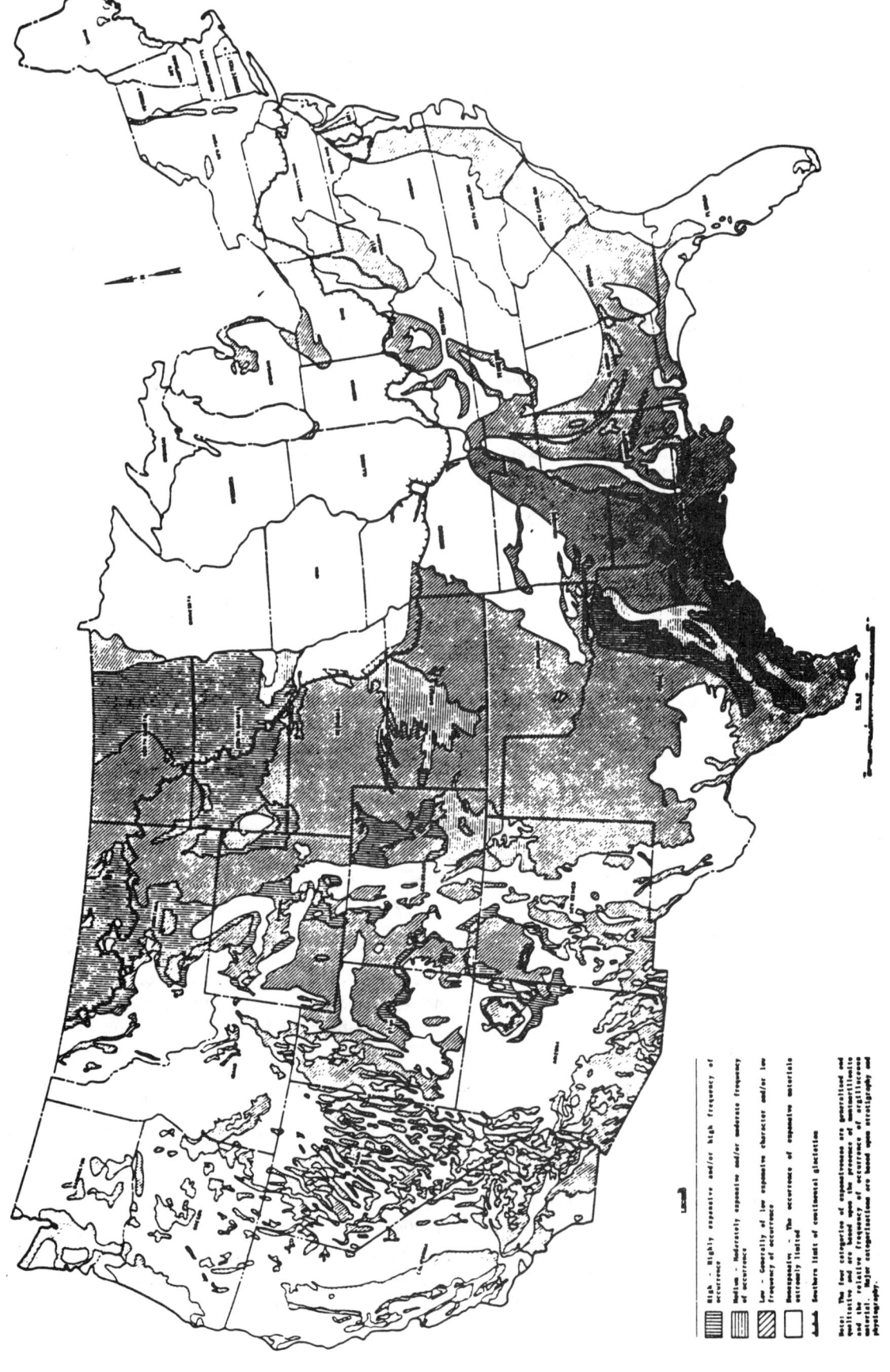

Figure G.1. Occurrence and Distribution of Potentially Expansive Soils in the United States, Part I (8)

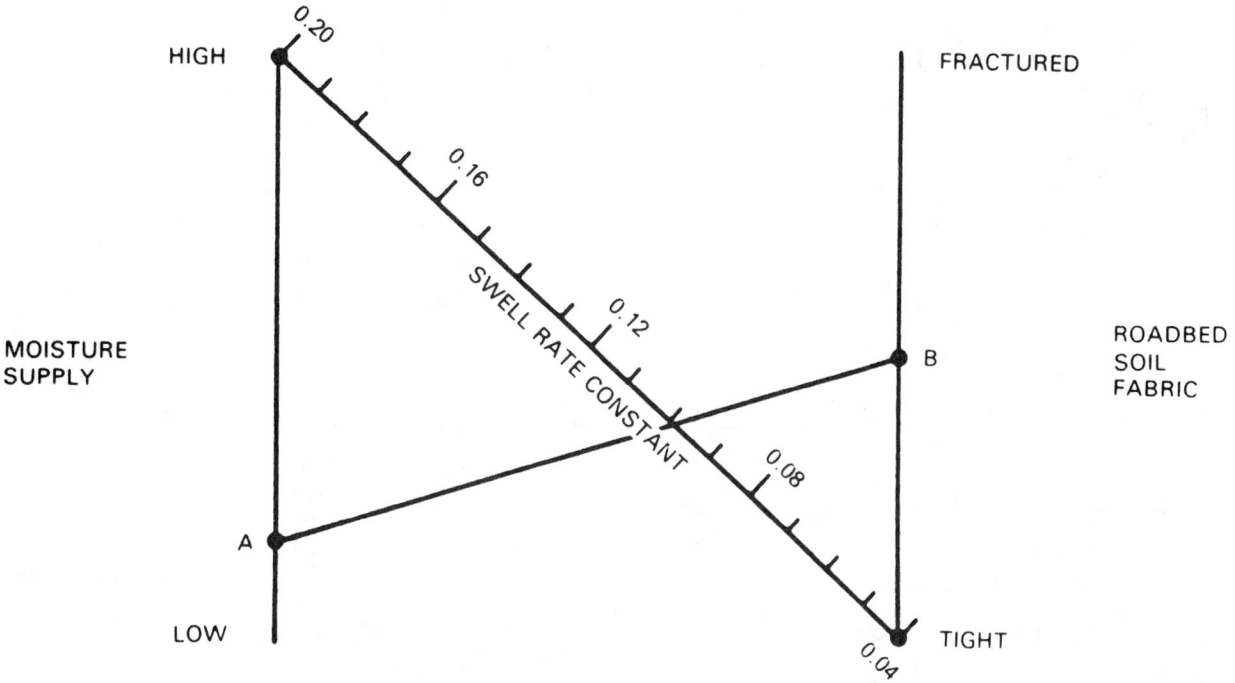

NOTES:
a) LOW MOISTURE SUPPLY:

   Low rainfall
   Good drainage

b) HIGH MOISTURE SUPPLY

   High rainfall
   Poor drainage
   Vicinity of culverts, bridge abutments, inlet leads

c) SOIL FABRIC CONDITIONS (self explanatory)

d) USE OF THE NONOGRAPH

1) Select the appropriate moisture supply condition which may be somewhere between low and high (such as A).

2) Select the appropriate soil fabric (such as B). This scale must be developed by each individual agency.

3) Draw a straight line between the selected points (A to B).

4) Read swell rate constant from the diagonal axis (read 0.10).

Figure G.2. Nomograph for Estimating Swell Rate Constant, Part II (1)

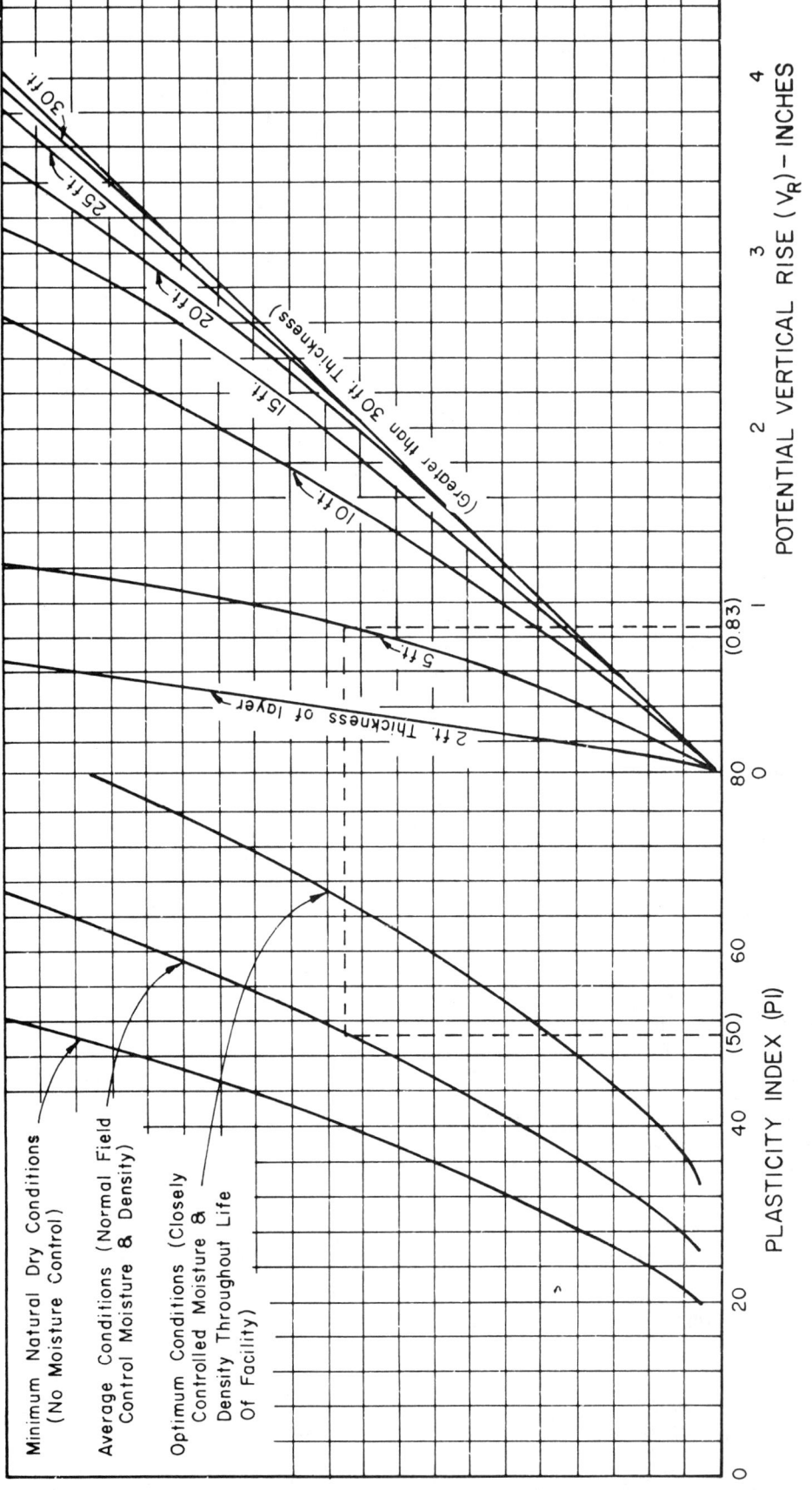

Figure G.3. Chart for Estimating the Approximate Potential Vertical Rise of Natural Soils, Part II (2)

Appendix G

Table G.1. Table for Estimating Swell Parameters to Use for Design

Bore Hole Number	Section Length (ft)	Roadbed Thickness (ft)	Soil Plasticity Index (PI)	Moisture Condition	Potential Vertical Rise (in.)	Soil Fabric	Swell Rate Constant
1							
2							
3							
4							
5							
6							
7							
8							
9							
10							
11							
12							
13							
14							
15							
16							

Atterberg Limits Tests (AASHTO T 258) run on the fine-grained portion of the roadbed soil samples obtained from the bore holes. Moisture condition (or supply) refers to the availability of moisture for roadbed soil absorption and V_R represents the potential vertical rise at a given site as determined from Figure G.3 or AASHTO T 258 Test Method. Soil fabric represents the capability of moisture to infiltrate the soil; most fine-grained soils beneath a pavement are neither fractured nor extremely tight. The swell rate constant at a given site is estimated from Figure G.2.

The design values for potential vertical rise and swell rate constant are determined by calculating a weighted average (based on section length). Likewise, for swelling probability, a weighted average percent should be calculated based on the lengths of all the sections having a V_R greater than 0.20 inches.

With the three major swelling factors defined, it is now possible to develop the swelling curve illustrated in Figure 2.2 in Part II. This is accomplished by solving the swelling serviceability loss equation (see Figure G.3) for several time periods. Although the actual equation may be better suited for generating this curve, the nomograph in Figure G.4 is helpful in identifying the overall effects of the individual swell parameters. The time period used with the graph should be equal to the analysis period, except where stage construction or rehabilitation design strategies are considered. For these latter conditions the performance period should be used.

G.2 FROST HEAVE

This section provides preliminary guidelines for identifying the serviceability loss due to differential frost heave. As an agency develops its own criteria, it may wish to replace the material herein. If an agency uses a procedure to reduce frost heave such as replacement or placing susceptible materials below the frost line, then the frost heave rate will be low or approach zero. The frost heave phenomenon is very similar to roadbed swelling in that it can result in a significant loss of road serviceability due to differential expansion (the differential values of expansion are of interest not the total value). Frost heaving occurs when free water in the roadbed soil collects and freezes to form ice lenses. The accumulation of thickness from these ice lenses causes localized heaving of the pavement surface during extended frozen periods. Obviously, frost heaving will not be a problem in areas which are arid or have a minimum frost penetration into the roadbed. Frost heaving can also be minimized by providing drainage to reduce the availability of free water. Figure G.5 shows the distribution of seasonal frost and permafrost in North America.

The model for frost heave is almost identical to that for roadbed swelling. It was derived from the performance of 18 experimental sections in the State of Michigan. There are three factors, each of which corresponds to a similar factor for swelling: (1) frost heave rate, (2) maximum potential serviceability loss due to frost heave, and (3) frost heave probability.

Frost heave rate defines the rate of increase of frost heave roughness (in millimeters per day). The rate of heave depends on the type of roadbed material and its percentage of fine-grained material. Figure G.6 presents a chart that may be used to estimate the rate of heave based on the roadbed soil's Unified Soil classification and percent (by weight) of material finer than 0.02 mm (AASHTO T 88 or agency's correlation with the minus 200 material).

The *maximum potential serviceability loss* due to frost heave is dependent on the quality of drainage and the depth of frost penetration. Figure G.5 provides a graph that may be used to estimate maximum serviceability loss based on these two factors. Note that the distinction between levels of drainage quality is the same as that defined in the treatment of drainage effects on material properties (Section 2.4.1; Part II):

Drainage Quality	**Water Removed Within**
Excellent	1/2 day
Good	1 day
Fair	1 week
Poor	1 month
Very Poor	(never removed)

Because of the relationship between drainage, depth of frost penetration, and maximum serviceability loss, Figure G.7 may also be used to identify the quality of drainage (or thickness of nonfrost-susceptible material) needed to control the maximum serviceability loss. Caution is advised, however, since this graph represents more of a qualitative than quantitative relationship between the three factors.

Frost heave probability should basically be the designer's estimate of the percent area of the project that will experience frost heave. Obviously, this is affected by several factors including the extent of frost-susceptible roadbed material, moisture availability, drainage quality, number of freeze-thaw cycles during the year, and depth of frost penetration. The designer should also rely heavily on past experience since, unlike

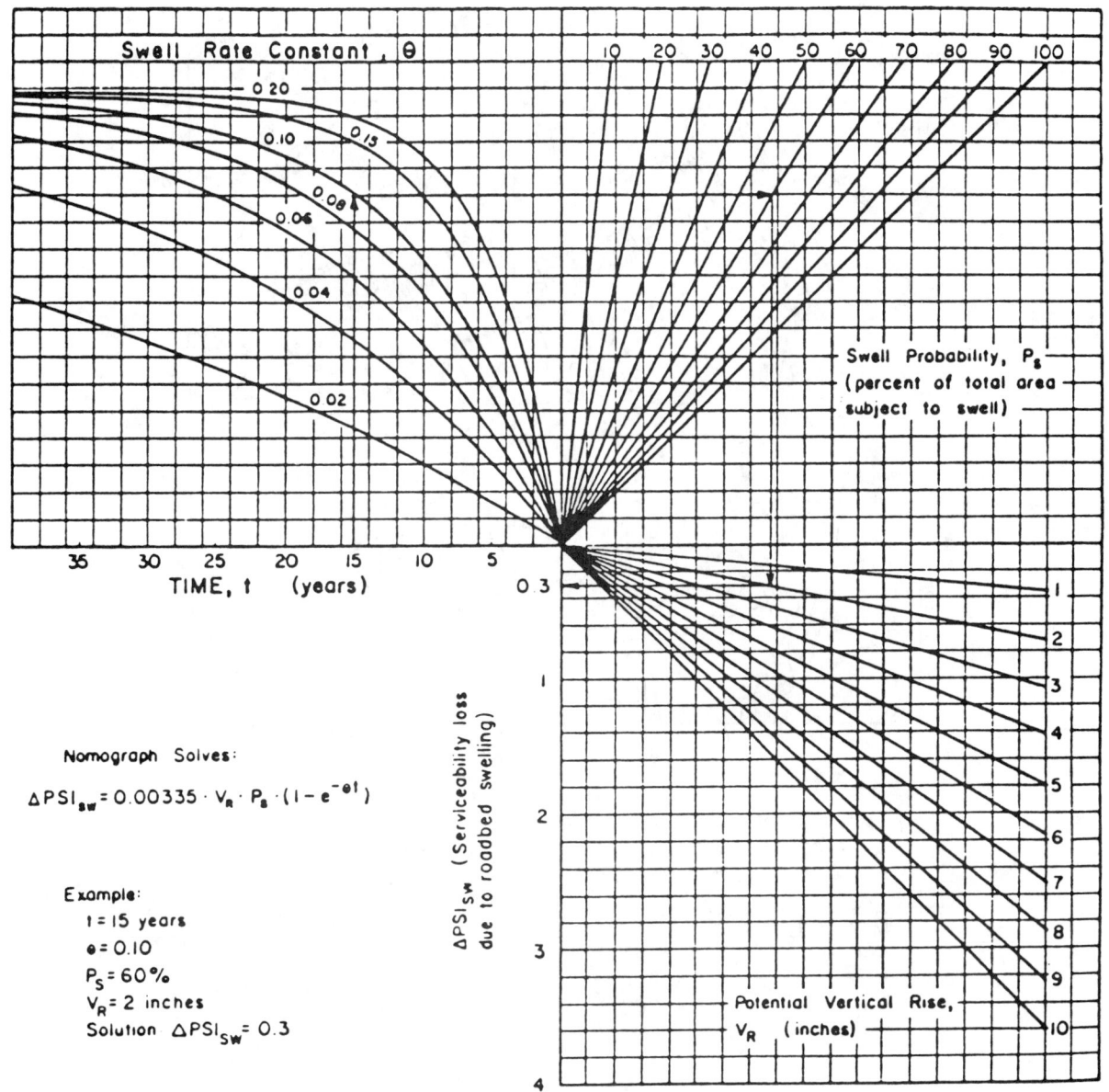

Figure G.4. Chart for Estimating Serviceability Loss Due to Roadbed Swelling

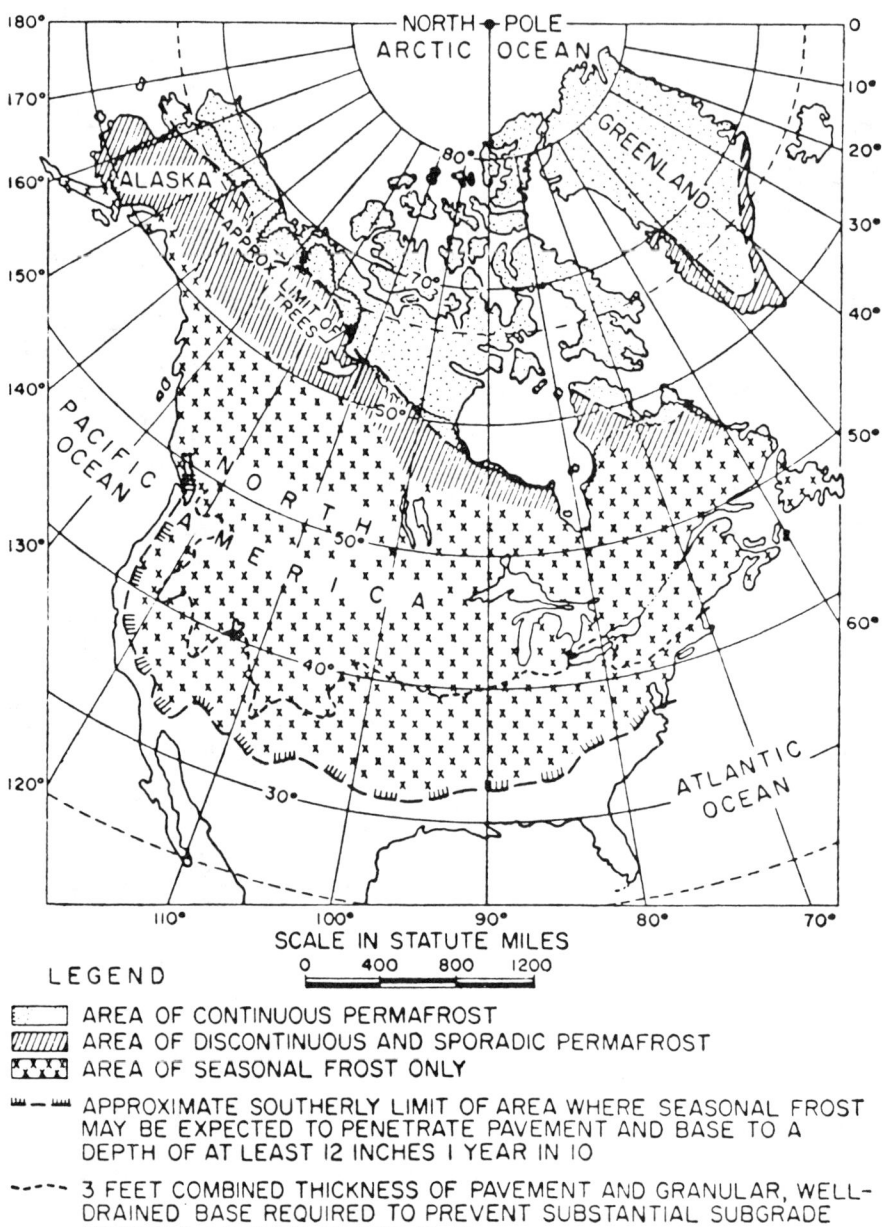

NOTE: Patches and islands of permafrost may be found in areas south of crosshatched zone, particularly in elevated mountain locations.

Figure G.5. Seasonal Frost and Permafrost in North America, Part II (43)

Appendix G

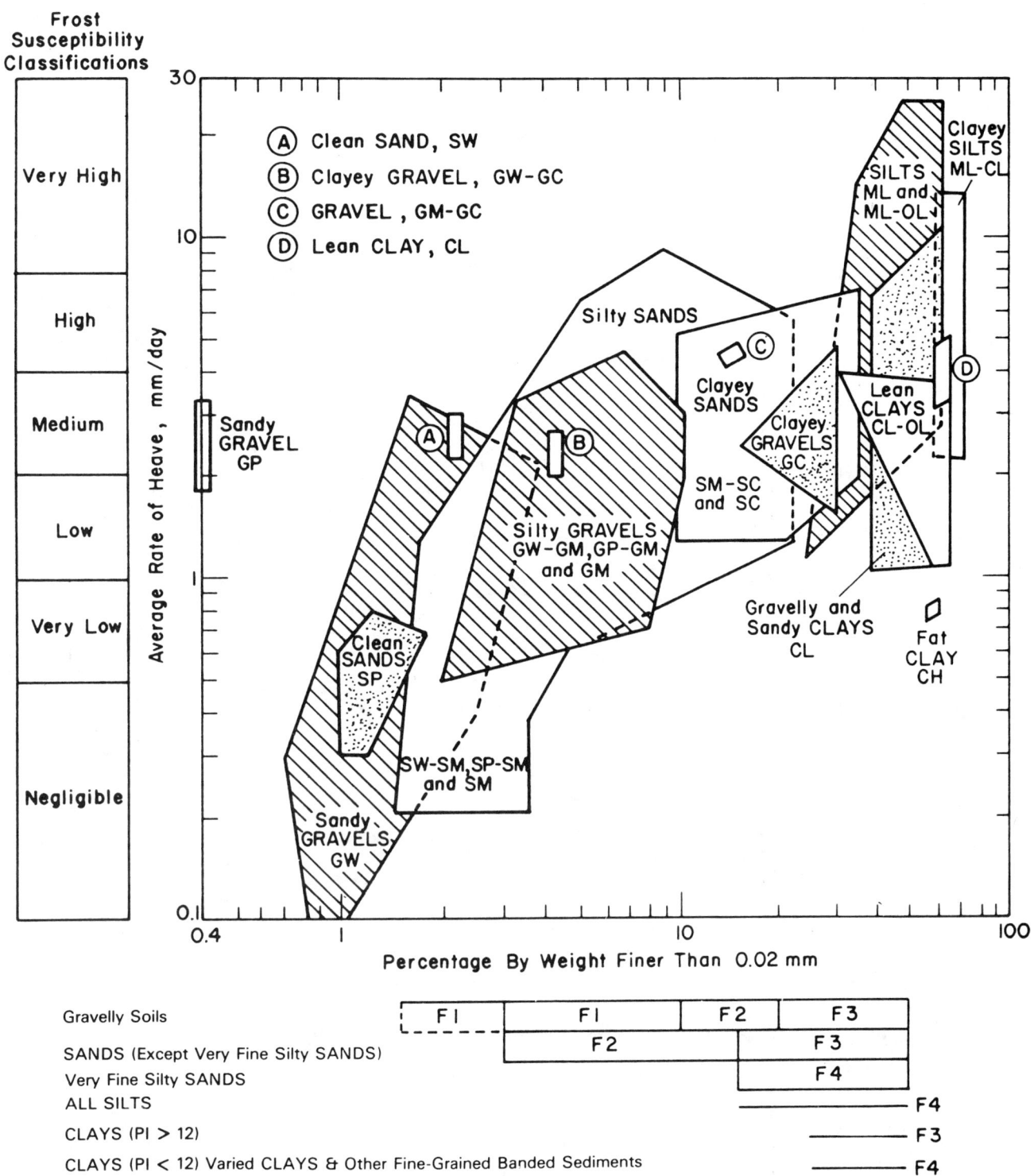

Figure G.6. Chart for Estimating Frost Heave Rate for a Roadbed Soil, Part II (*11*)

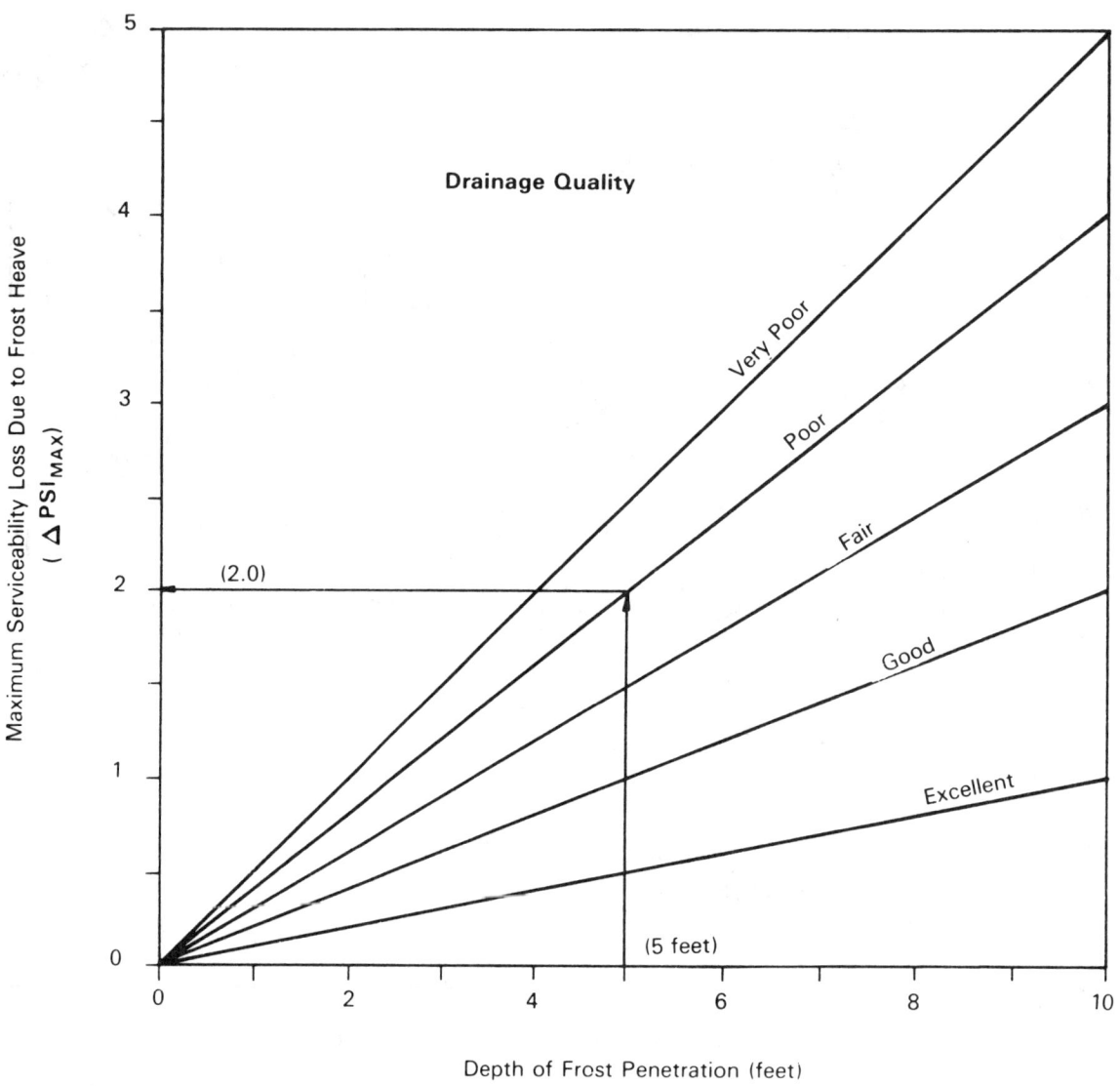

Figure G.7. Graph for Estimating Maximum Serviceability Loss Due to Frost Heave

Appendix G

swelling probability, there is no clear-cut method for approximating frost heave probability.

Once values for the three frost heave factors are defined, the equation for serviceability loss (presented in Figure G.8) should be used to generate a frost heave serviceability loss curve similar to that presented in Figure 2.2 (Part II). The time, t, used with Figure G.8 should be equal to the analysis period. For stage construction and rehabilitation strategies, the performance period is used. The frost heave serviceability loss curve should then be combined with the swelling serviceability loss curve (if applicable) to produce a total serviceability loss versus time curve. This curve will then be used as a component of the design procedure discussed in Chapter 3, Part II.

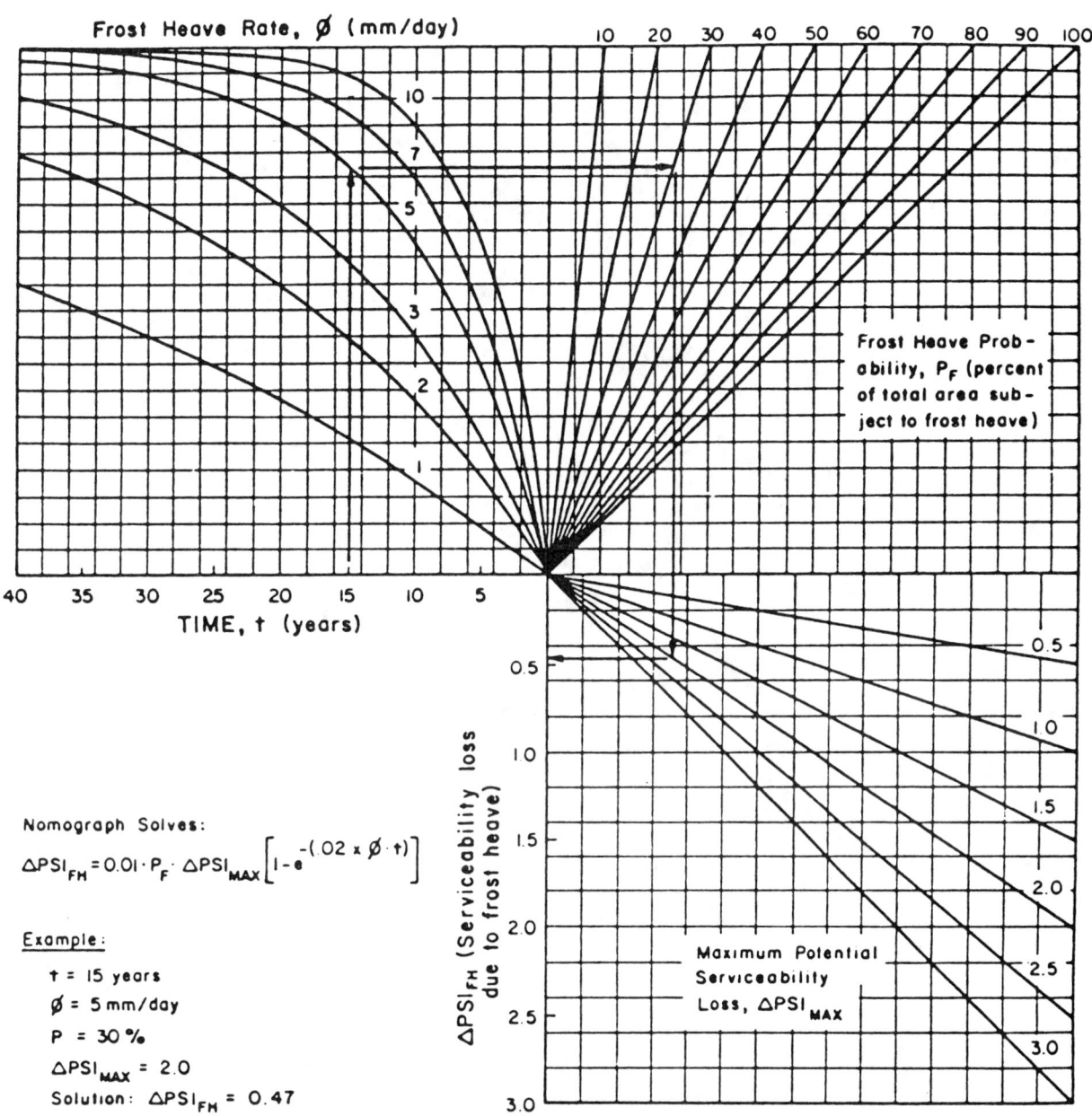

Figure G.8. Chart for Estimating Serviceability Loss Due to Frost Heave

APPENDIX H
FLEXIBLE PAVEMENT DESIGN EXAMPLE

The following example is provided to illustrate the flexible pavement design procedure presented in Section 3.1 of Part II. The design requirements for this example are described here in the same order as they are in Part II, Chapter 2.

H.1 DESIGN REQUIREMENTS

Time Constraints

The analysis period selected for this design example is 20 years. The maximum performance period (or service life) selected for the initial flexible pavement structure in this example is 15 years. Thus, it will be necessary to consider stage construction (i.e., planned rehabilitation) alternatives to develop design strategies which will last the analysis period.

Traffic

Based on average daily traffic and axle weight data from the planning group, the estimated two-way 18-kip equivalent single axle load (ESAL) applications during the first year of the pavement's life is 2.5×10^6 and the projected (compound) growth rate is 3 percent per year. The directional distribution factor (D_D) is assumed to be 50 percent and the lane distribution factor (D_L) for the facility (assume three lanes in one direction) is 80 percent. Thus, the traffic, during the first year (in the design lane) is $2.5 \times 10^6 \times 0.80 \times 0.50$ or 1.0×10^6 18-kip ESAL applications. Figure H.1 provides a plot of the cumulative 18-kip ESAL traffic over the 20-year analysis period. The curve and equation for future traffic (w_{18}) are reflective of the assumed exponential growth rate (g) of 3 percent.

Reliability

Although the facility will be a heavily trafficked state highway, it is in a rural situation where daily traffic volumes should never exceed half of its capacity. Thus, a 90-percent overall reliability level was selected for design. This means that for a two-stage strategy (initial pavement plus one overlay), the design reliability for each stage must be $0.90^{1/2}$ or 95 percent. Similarly, for a three-stage strategy (initial pavement plus two overlays), the design reliability for all three stages must be $0.90^{1/3}$ or 96.5 percent.

Another criteria required for the consideration of reliability is the overall standard deviation (S_o). Although it is possible to estimate this parameter through an analysis of variance of all the design factors (see Volume 2, Appendix EE), an approximate value of 0.35 will be used for the purposes of this example problem.

Environmental Impacts

Eighty bore holes were obtained along the 16-mile length of the project (approximately one every thousand feet). Based on an examination of the borehole samples and subsequent soil classifications, it was determined that the soil at the first twelve bore hole sites (approximately 12,000 feet) was basically of the same composition and texture. Significantly different results were obtained from examinations at the other bore hole sites. Based on this type of unit delineation, this 12,000-foot section of the project will be designed separately from all the rest.

The site of this highway construction project is in a location that can be environmentally classified as U.S. Climatic Region II, i.e., wet with freeze-thaw cycling. The soil is considered to be a highly active swelling clay. Because of this and the availability of moisture from high levels of precipitation, a drainage system will be constructed which is capable of removing excess moisture in less than 1 day. The duration of below-freezing temperatures in this environment, however, is not sufficient to result in any problems with frost heaving.

Table H.1 summarizes the data used to consider the effects of roadbed swelling on future loss of serviceability. Columns 1 and 2 indicate the bore hole number and length of the corresponding section (or

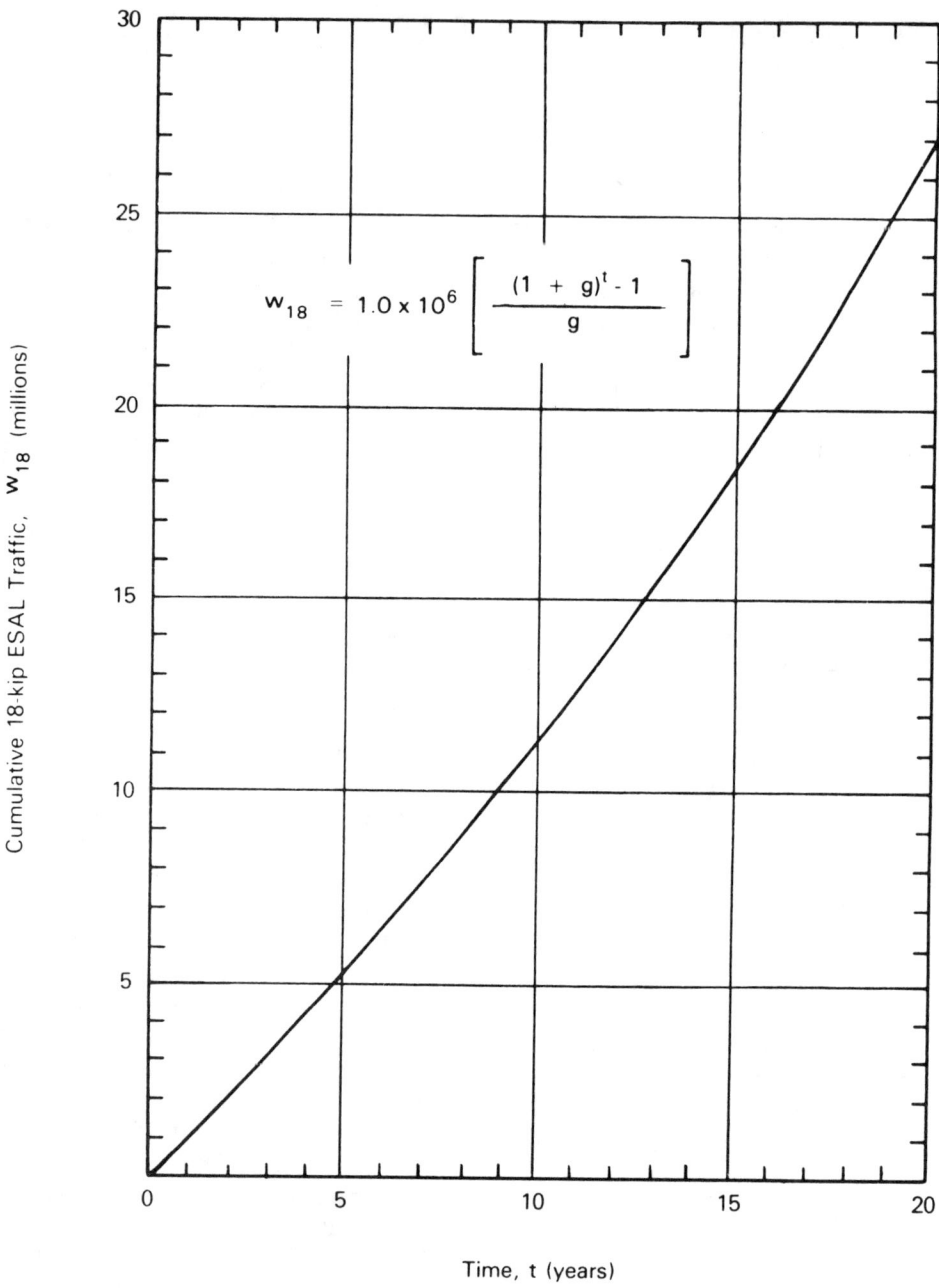

Figure H.1. Plot of Cumulative 18-kip ESAL Traffic Versus Time for Assumed Conditions

Appendix H

Table H.1. Table for Estimating Swell Parameters for Flexible Pavement Design Example

(1) Bore Hole Number	(2) Section Length (ft)	(3) Roadbed Thickness (ft)	(4) Soil Plasticity Index (PI)	(5) Moisture Condition	(6) Potential Vertical Rise (in.)	(7) Soil Fabric	(8) Swell Rate Constant
1	900	>30	48	Optimum	0.82	rel. tight	0.07
2	1,200	>30	56	"	1.34	"	"
3	800	>30	67	"	2.20	"	"
4	1,000	>30	15	"	0.00	"	0.10
5	1,000	>30	46	"	0.70	"	0.07
6	1,100	>30	62	"	1.86	"	"
7	1,000	>30	65	"	2.00	"	"
8	900	>30	71	"	2.60	"	"
9	1,200	>30	38	"	0.28	"	"
10	800	>30	60	"	1.80	"	"
11	900	>30	19	"	0.00	"	0.10
12	1,200	>30	51	"	1.04	"	0.07
13							
14							
15							
16							

Total 12,000

segment) of the project. The depth to any rigid foundation at the site is, for all practical purposes, semi-infinite. (Roadbed soil thicknesses greater than 30 feet are considered to be semi-infinite.)

Column 4 shows the average plasticity index (PI) of the soil at each bore hole location. PI values above 40 are indicative of potential high volume change of the material.

Column 5 represents the estimated moisture condition of the roadbed material after pavement construction. Because of the plan to construct a "good" drainage system, the future moisture conditions are considered to be "optimum" throughout the project length.

Column 6 presents the results of applying the chart in Figure G.3 of Appendix G to estimate the potential vertical rise (V_R) at each bore hole location.

Column 7 represents a qualitative estimate of the fabric of the soil or the rate at which it can take on moisture. The natural impermeability of clay materials means that the soil at this site tends towards "tight." This, combined with the relatively low moisture supply (due to the installation of a drainage system), means that the swell rate constant at each location having a "tight" fabric (i.e., plasticity index (PI) greater than about 20) can be estimated at 0.07 (see Figure G.2 in Appendix G). For the occasions where PI was less than 20, a value of 0.10 was used because of the likelihood of greater permeability.

Based on the data in Table H.1, the overall swell rate constant and potential vertical rise are determined by calculating a weighted average; thus,

Swell Rate Constant = 0.075

Potential Vertical Rise (V_R) = 1.2 inches

The swelling probability is simply the percent of the length of the project which has a potential vertical rise greater than 0.2 inches. 10,100 feet out of the total 12,000 have a V_R greater than 0.2 inches, thus the swelling probability is 84 percent.

These factors were then used to generate the serviceability loss versus time curve presented in Figure H.2. The curve shown was generated using the equation presented in Figure G.4 of Appendix G. This represents a graph of the estimated total environmental serviceability loss versus time, since frost heave is not a consideration.

Serviceability

Based on the traffic volume and functional classification of the facility (6-lane state highway), a terminal serviceability (p_t) of 2.5 was selected. Past experience indicates (for the purposes of this hypothetical example) that the initial serviceability (p_o) normally achieved for flexible pavements in the state is significantly higher than that at the AASHO Road Test (4.6 compared to 4.2). Thus, the overall design serviceability loss for this problem is:

$$\Delta PSI = p_o - p_t = 4.6 - 2.5 = 2.1$$

Effective Roadbed Soil Resilient Modulus

Figure H.3 summarizes the data used to characterize the effective resilient modulus of the roadbed soil. Individual moduli are specified for 24 half-month intervals to define the seasonal effects. These values are also reflective of the roadbed support that would be expected under the improved moisture conditions provided by the "good" drainage system:

Roadbed Moisture Condition	Roadbed Soil Resilient Modulus (psi)
Wet	5,000
Dry	6,500
Spring-Thaw	4,000
Frozen	20,000

The frozen season (from mid-January to mid-February) is 1 month long, the spring-thaw season (mid-February to March) is 0.5 months long, the wet periods (March through May and mid-September through mid-November) total 5 months, and the dry periods (June through mid-September and mid-November through mid-January) total 5.5 months. Application of the effective roadbed soil M_R estimation procedure results in a value of 5,700 psi.

Pavement Layer Materials Characterization

Three types of pavement materials will constitute the individual layers of the structure. The moduli for

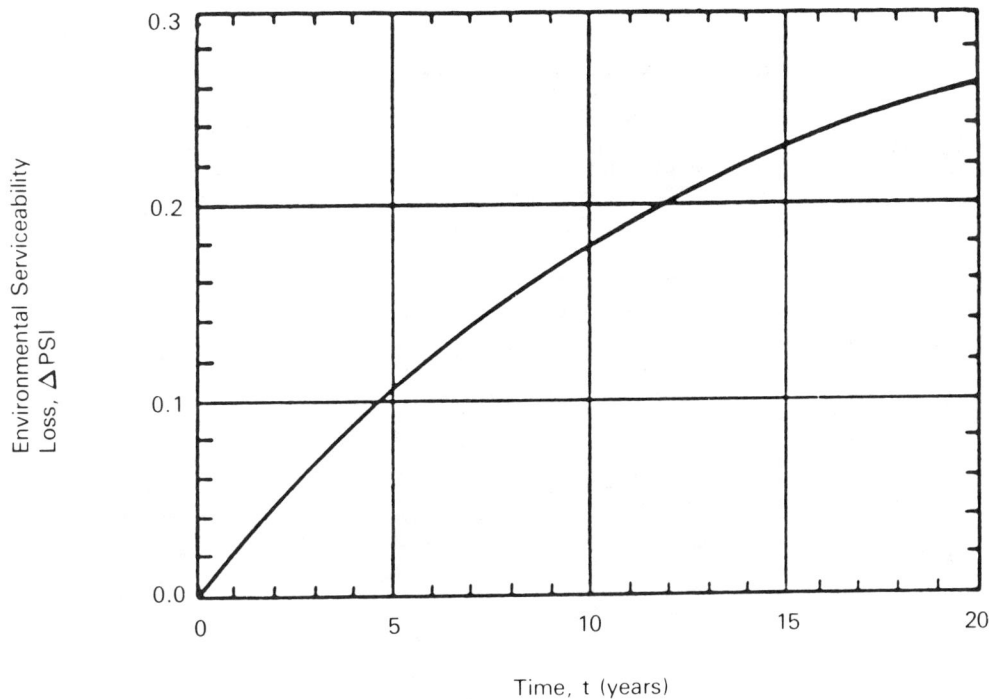

Figure H.2. Graph of Environmental Serviceability Loss Versus Time for Swelling Conditions Considered

each, determined using the recommended laboratory test procedures, are as follows:

Asphalt Concrete: $E_{AC} = 400,000$ psi

Granular Base: $E_{BS} = 30,000$ psi

Granular Subbase: $E_{SB} = 11,000$ psi

These values correspond to the average year-round moisture conditions that would be expected without any type of pavement drainage system. (The effects of positive drainage on material requirements are considered in a later section.)

Layer Coefficients

The structural layer coefficients (a_i-values) corresponding to the moduli defined in the previous section are as follows:

Asphalt Concrete: $a_1 = 0.42$ (Figure 2.5, Part II)

Granular Base: $a_2 = 0.14$ (Figure 2.6, Part II)

Granular Subbase: $a_3 = 0.08$ (Figure 2.7, Part II)

Drainage Coefficient

The only item that is considered under the heading "Pavement Structural Characteristics" (Part II, Section 2.4) in the design of a flexible pavement is the method of drainage. The drainage coefficient (m-value) corresponding to the granular base and subbase materials for a "good" drainage system (i.e., water removed within 1 day) and the balanced wet-dry climate of U.S. Climatic Region II is 1.20. (The range in Part II, Table 2.4, for 1 to 5 percent moisture exposure time is 1.15 to 1.25.)

H.2 DEVELOPMENT OF INITIAL STAGE OF A DESIGN ALTERNATIVE

Since the estimated maximum performance period (15 years) is less than the design analysis period (20 years), any initial structure selected will require an overlay to last the analysis period. The thickest recommended initial structure (evaluated here) is that corresponding to the maximum 15-year performance period. Thinner initial structures, selected for the purpose of life-cycle cost analyses, will require thicker

Month	Roadbed Soil Modulus, M (psi)	Relative Damage u
Jan.	6,500	0.167
	20,000	0.012
Feb.	20,000	0.012
	4,000	0.515
Mar.	5,000	0.307
	5,000	0.307
Apr.	5,000	0.307
	5,000	0.307
May	5,000	0.307
	5,000	0.307
June	6,500	0.167
	6,500	0.167
July	6,500	0.167
	6,500	0.167
Aug.	6,500	0.167
	6,500	0.167
Sept.	6,500	0.167
	5,000	0.307
Oct.	5,000	0.307
	5,000	0.307
Nov.	5,000	0.307
	6,500	0.167
Dec.	6,500	0.167
	6,500	0.167
Summation: u =		5.446

Average: $= \dfrac{\Sigma}{n} = \dfrac{5.46}{24} = 0.227$

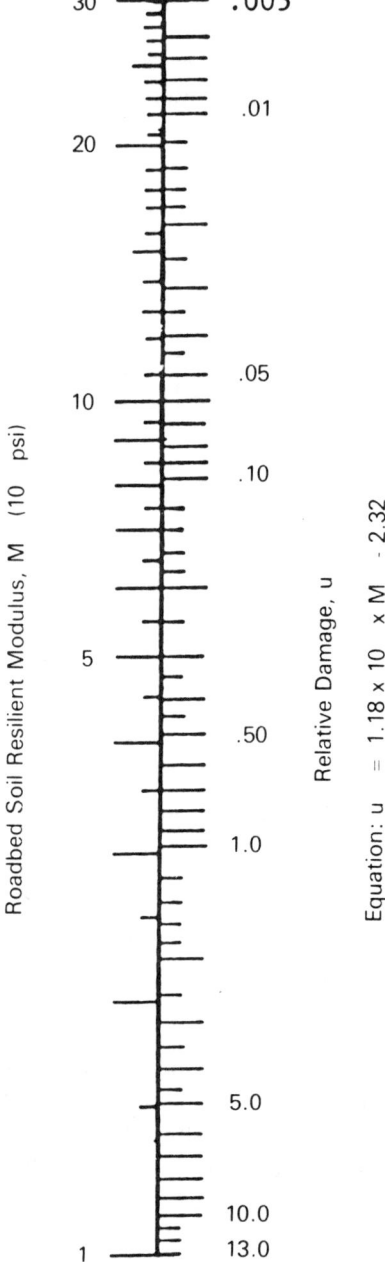

Equation: $u = 1.18 \times 10^8 \times M^{-2.32}$

Effective Roadbed Soil Resilient Modulus, M_R (psi) 5,700
(corresponds to u_f)

Figure H.3. Estimation of Effective Roadbed Soil Resilient Modulus

overlays (at an earlier date) to last the same analysis period.

The strategy with the maximum recommended initial structure number is determined using the effective roadbed soil resilient modulus of 5,700 psi, a reliability of 95 percent, an overall standard deviation of 0.35, a design serviceability loss of 2.1 and the cumulative traffic at the maximum performance period, 18.6×10^6 18-kip ESAL (from Figure H.1 for a time of 15 years). Applying Figure 3.1 from Part II, the result is a maximum initial structure number (SN) of 5.6. Because of serviceability loss due to swelling however, an overlay will be required before the end of the 15-year design performance period. Using the step-by-step procedure described in Part II, Section 3.1.3, the service life that can actually be expected is about 13 years (see Table H.2). Thus, the overlay that must be designed will need to carry the remaining 18-kip ESAL traffic over the last 7 years of the analysis period.

H.3 DETERMINATION OF STRUCTURAL LAYER THICKNESSES FOR INITIAL STRUCTURE

The thicknesses of each layer above the roadbed soil or subgrade are determined using the procedure described in Part II, Section 3.1.4. (See Figure 3.2.) For the design SN of 5.6 developed in this example, the determination of the layer thicknesses is demonstrated below:

Solve for the SN required above the base material by applying Figure 3.1 (in Part II) using the resilient modulus of the base material (rather than the effective roadbed soil resilient modulus). Values of E_{BS} equal to 30,000 psi, first stage reliability (R) equal to 95 percent, w_{18} equal to 16.0×10^6 and ΔPSI_{TR} equal to 1.89 (the latter two are from Table H.2) result in an SN_1 of 3.2. Thus, the asphalt concrete surface thickness required is:

$$D_1^* = SN_1/a_1 = 3.2/0.42 = 7.6 \text{ (or 8 inches)}$$

$$SN_1^* = a_1 D_1^x = 0.42 \times 8 = 3.36$$

Similarly, using the subbase modulus of 11,000 psi as the effective roadbed soil resilient modulus, SN_2 is equal to 4.5 and the thickness of base material required is:

$$D_2^* = (SN_2 - SN_1^*)/(a_2 m_2)$$

$$= (4.5 - 3.36)/(0.14 \times 1.20)$$

$$= 6.8 \text{ (or 7 inches)}$$

$$SN_2^* = 7 \times 0.14 \times 1.20 = 1.18$$

Finally, the thickness of subbase required is:

$$D_3^* = (SN_3 - (SN_1^* + SN_2^*))/(a_3 m_3)$$

$$= (5.6 - (3.36 + 1.18))/(0.08 \times 1.20)$$

$$= 11 \text{ inches}$$

Table H.2. Reduction in Performance Period (Service Life) of Initial Pavement Arising From Swelling Considerations

Initial SN _____5.6_____

Maximum Possible Performance Period (years) _____15_____

Design Serviceability Loss, $\Delta PSI = p_o - p_t =$ _____4.6 − 2.5 = 2.1_____

(1) Iteration No.	(2) Trial Performance Period (years)	(3) Serviceability Loss Due to Swelling ΔPSI_{SW}	(4) Corresponding Serviceability Loss Due to Traffic ΔPSI_{TR}	(5) Allowable Cumulative Traffic (18-kip ESAL)	(6) Corresponding Performance Period (years)
1*	13	0.21	1.89	16.0×10^6	13.2

*Convergence achieved after only one iteration.

APPENDIX I
RIGID PAVEMENT DESIGN EXAMPLE

The following example is provided to illustrate the rigid pavement design procedure presented in Part II, Section 3.2. The design requirements for this example are described here in the same order as they appear in Part II, Chapter 2.

I.1 DESIGN REQUIREMENTS

Time Constraints

The analysis period selected for this design example is 35 years. The maximum initial performance period (or service life) selected for the initial rigid pavement structure in this example is 25 years. Thus, it will be necessary to consider stage construction (i.e., planned rehabilitation) to develop design strategies which will last the analysis period.

Traffic

Based on average daily traffic and axle weight data, the estimated two-way 18-kip equivalent single axle load (ESAL) applications during the first year of the pavement's life is 357,000 and the projected (compound) growth rate is 3 percent per year. The directional distribution factor (D_D), is assumed to be 50 percent and the lane distribution factor (D_L), for the facility (assume three lanes in one direction) is 80 percent. Thus, the traffic during the first year of the analysis period (in the design lane) is 357,000 × 0.80 × 0.50 or 142,800 18-kip ESAL applications. Figure I.1 provides a plot of the cumulative 18-kip ESAL traffic over the 35-year analysis period. The curve and equation for future traffic (w_{18}) are reflective of the assumed exponential growth rate (g) of 3 percent.

Reliability

The highway facility being designed is a heavily trafficked state highway in a rural region where the expected daily traffic volumes should never exceed half of the capacity. Based on this information, an overall reliability level of 90 percent was chosen for design. This means that for a two-stage strategy (initial pavement plus one overlay), the design reliability for design. This means that for a two-stage strategy (initial pavement plus one overlay), the design reliability for each stage must be $0.90^{1/2}$ or 95 percent. Similarly, for a three-stage strategy (initial pavement plus two overlays), the design reliability for all three stages must be $0.90^{1/3}$ or 96.5 percent.

Another criteria required for the consideration of reliability is the overall standard deviation (S_o). Although it is possible to estimate this parameter through an analysis of variance of all the design factors (see Volume 2, Appendix EE), an approximate value of 0.29 will be used for the purposes of this example.

Environmental Impacts

A soil survey consisting of bore holes taken at approximate 1,000-foot intervals shows that soil conditions do not vary substantially along the length of the project. Thus, one pavement cross-section design will serve for the entire project length.

The site where this project is to be constructed is located in U.S. Climatic Region I, i.e., wet with freeze-thaw cycling. The roadbed soil is a clay material which is susceptible to swelling. Because of this and the high level of available moisture that is common to this region, a drainage system will be constructed capable of removing excessive moisture in 1 day or less. The duration of below-freezing temperatures in this region, however, is not sufficient to warrant frost-heave consideration in the design process.

Table I.1 summarizes the data used to characterize the parameters of roadbed swelling that influence future serviceability loss. Columns 1 and 2 indicate the bore hole number and length of the corresponding section (or segment) of the project. The depth to any rigid foundation at the site is, for all practical purposes, semi-infinite (i.e., greater than 30 feet). Column 4 shows the average plasticity index (PI) of the soil at each bore hole site. Values of PI above 40

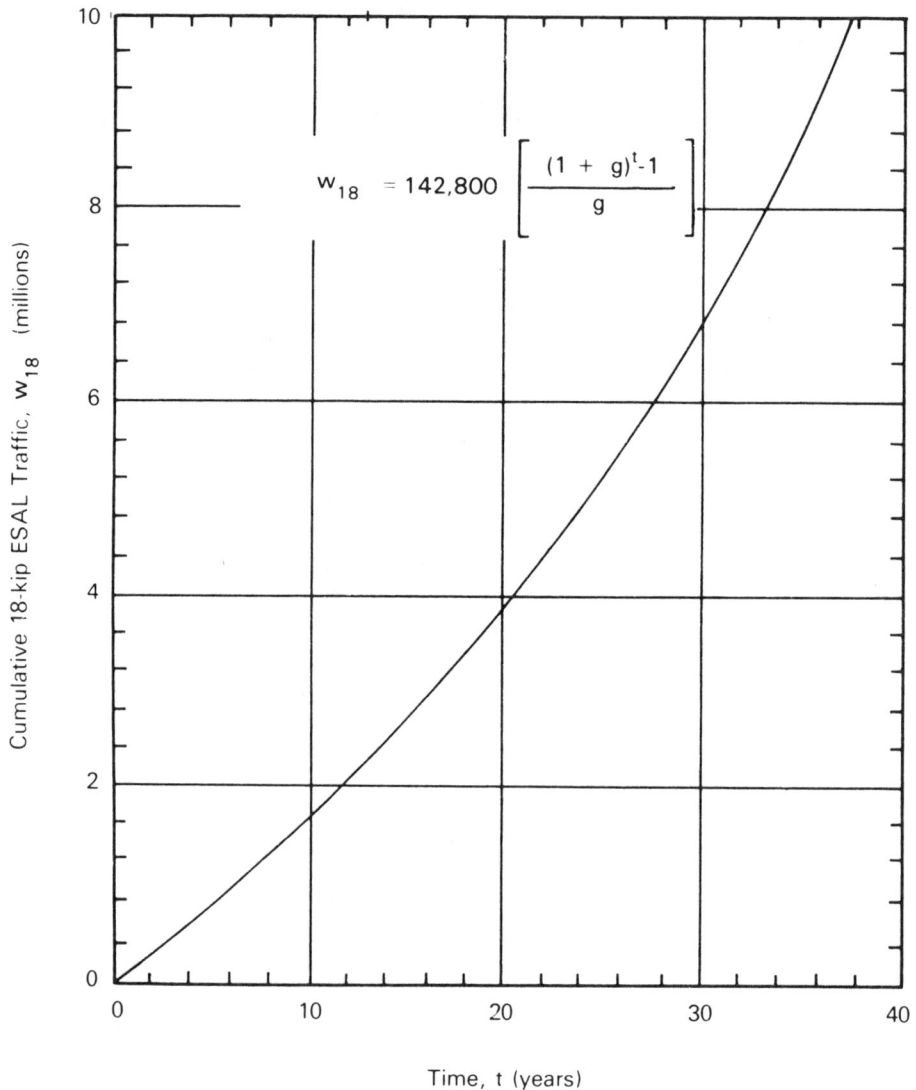

Figure I.1. Plot of Cumulative 18-kip ESAL Traffic Versus Time for Assumed Conditions

are indicative of soils with a high potential for volume change.

Column 5 represents the estimated moisture condition of the roadbed material after pavement construction. Because of the plan to construct a good drainage system, the future moisture conditions are considered to be "optimum" throughout the project length.

Column 6 presents the results of applying the chart in Figure G.3 (Appendix G) to estimate the potential vertical rise (V_R) at each bore hole location.

Column 7 represents a qualitative estimate of the soil fabric or the rate at which the soil can absorb moisture. The natural impermeability of clay means that the soil fabric at this site is relatively tight (i.e., low permeability). This, combined with the relatively low moisture supply (due to the installation of a drainage system), means that the swell rate constant at each location having a "tight" fabric (i.e., plasticity index (PI) greater than about 20) can be estimated at 0.07 (see Figure G.2 in Appendix G). For the occasions where PI was less than 20, a value of 0.10 was used because of the likelihood of greater permeability.

The overall swell rate constant and potential vertical rise are determined by calculating a weighted average based on the data from Table I.1, thus:

Swell Rate Constant = 0.072

Potential Vertical Rise (V_R) = 1.23 inches

Table I.1. Table for Estimating Swell Parameters for Rigid Pavement Design Example

(1) Bore Hole Number	(2) Section Length (ft)	(3) Roadbed Thickness (ft)	(4) Soil Plasticity Index (PI)	(5) Moisture Condition	(6) Potential Vertical Rise (in.)	(7) Soil Fabric	(8) Swell Rate Constant
1	950	>30	51	Optimum	1.04	rel. tight	0.07
2	1,000	>30	56	"	1.34	"	0.07
3	1,000	>30	61	"	1.81	"	0.07
4	950	>30	27	"	0.00	"	0.07
5	1,200	>30	49	"	0.87	"	0.07
6	1,100	>30	70	"	2.59	"	0.07
7	1,050	>30	60	"	1.80	"	0.07
8	950	>30	55	"	1.30	"	0.07
9	1,000	>30	47	"	0.71	"	0.07
10	1,000	>30	65	"	2.00	"	0.07
11	1,100	>30	27	"	0.00	"	0.07
12	900	>30	51	"	1.04	"	0.07
13	900	>30	70	"	2.59	"	0.07
14	1,100	>30	65	"	2.00	"	0.07
15	1,200	>30	14	"	0.00	"	0.10
16	900	>30	47	"	0.71	"	0.07

Total 16,300

The swelling probability is simply the percent of the project's length that has a potential vertical rise greater than 0.2 inches. 13,050 feet out of the total 16,300-foot project length have a V_R greater than 0.2 inches, thus the swelling probability is 80 percent.

These factors were then used to generate the serviceability loss versus time curve presented in Figure I.2. The curve was generated using the equation shown in Figure G.4 (Appendix G). Since frost heave is not considered, this curve represents the total estimated environmental serviceability loss with time.

Serviceability

Based on the traffic volume and functional classification of the facility, a terminal serviceability (p_t) of 2.5 was selected for design. Since the initial serviceability (p_o) is expected to be 4.5, the overall design serviceability loss for this problem is:

$$\Delta PSI = p_o - p_t = 4.5 - 2.5 = 2.0$$

Effective Modulus of Subgrade Reaction

Because of the effects of subbase characteristics on the effective modulus of subgrade reaction (k), its calculation is included as a step in the iterative design procedure described in a later section.

Pavement Layer Materials Characterization

Portland cement concrete (PCC) pavements are generally laid on either a modified roadbed soil or subbase material. This design example will consider a PCC pavement placed on a granular subbase material. The modulus for both the PCC slab and subbase layer, determined using the recommended laboratory test procedures (see Section 2.3.3 in Part II), are as follows:

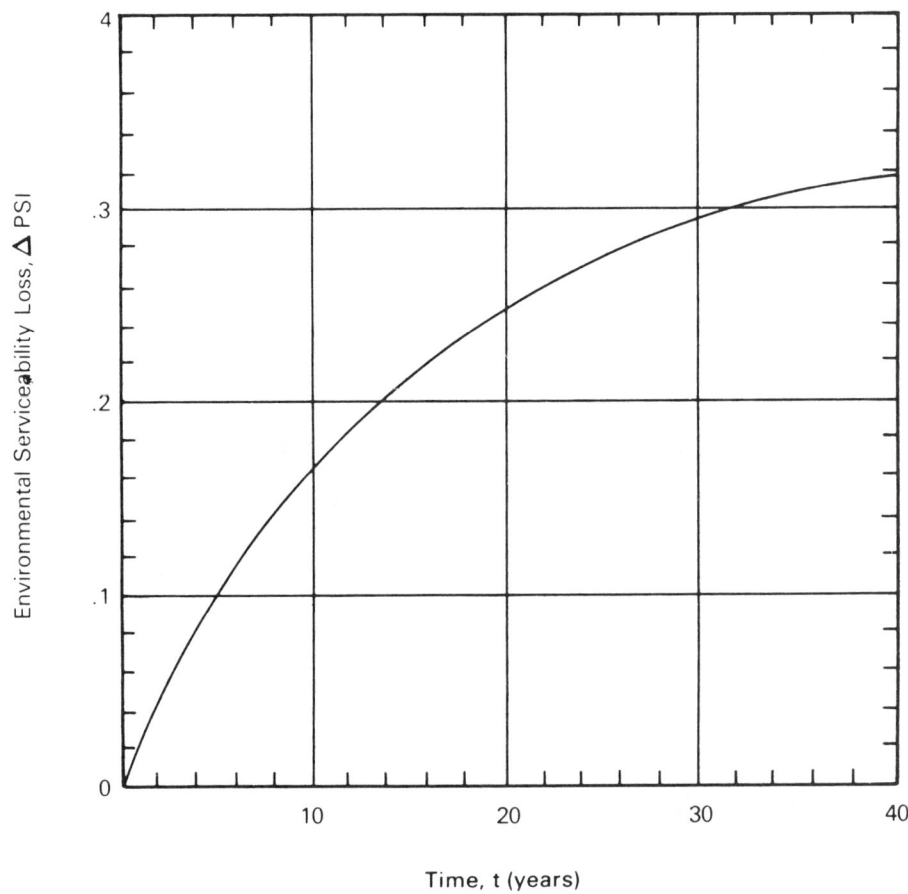

Figure I.2. Plot of Environmental Serviceability Loss Versus Time for Swelling Conditions Considered

Portland Cement Concrete:

$$E_c = 4{,}200{,}000 \text{ psi}$$

Granular Subbase:

$$E_{SB} = 15{,}000 \text{ psi (wet conditions) and}$$
$$25{,}000 \text{ psi (dry conditions)}$$

PCC Modulus of Rupture

Using the recommended flexural beam test procedures, the mean value for the modulus of rupture (flexural strength) of the portland cement concrete was determined to be 578 psi. This represents the average 28-day strength from numerous specimens testing using third point loading.

Drainage

The drainage coefficient for rigid pavements (C_d), is a function of the prevailing drainage condition and the average yearly rainfall. For this design example, the quality of drainage is assumed to be "good" (i.e., water removed within 1 day) and the percent of time the pavement is exposed to moisture levels approaching saturation is from 5 to 25 percent of the year. Based on this criteria, a value of 1.05 for C_d was selected from Table 2.5 in Part II.

Load Transfer

The pavement being designed is a jointed reinforced concrete pavement with tied PCC shoulders (10 feet wide). The joint spacing is 30 feet with dowel bars providing load transfer at each joint. The load transfer coefficient (J) selected for this condition (from Part II, Section 2.4.4) is 2.8. This compares to the value of 3.2 that would normally be used for a jointed pavement without tied shoulders.

Loss of Support

Table 2.6 in Part II provides recommended loss of support factors (LS) indicative of the potential for voids to form beneath the slab. This table indicates a range of LS between 1.0 and 3.0 for the conditions specified in this example (i.e., unbound granular material). Because of the planned construction of a "good" drainage system, there will be less potential for pumping to occur; therefore, the minimum value of LS equal 1.0 was selected for this design example.

Reinforcement for Jointed Concrete Pavement

The reinforcement required is a function of the steel working stress, the joint spacing, and the slab-base friction factor. The steel that will be used in this design example is Grade 60 billet steel; therefore, the working stress is equal to 60,000 psi × 0.75 or 45,000 psi. The joint spacing is 30 feet and the friction factor for a slab on a granular material is approximately 1.5.

I.2 DEVELOPMENT OF INITIAL STAGE OF A DESIGN ALTERNATIVE

Since the estimated maximum performance period (25 years) is less than the analysis period (35 years), any initial structure selected will require an overlay to last the analysis period. The thickest recommended initial structure is that corresponding to the maximum performance period (25 years). Thinner initial structures, selected for the purpose of life-cycle cost analyses, will require thicker overlays (at an earlier date) to last the analysis period.

Develop Effective Modulus of Subgrade Reaction

Slab support is estimated using a step-by-step procedure to define alternate levels of the effective modulus of subgrade reaction (k) based on the subbase characteristics and the seasonal variation of roadbed soil resilient modulus. This is accomplished using Figures 3.3, 3.4, 3.5, and 3.6 from Part II, Section 3.2. The results for one subbase design (6 inches of granular material) are illustrated in Table I.2.

The first step is to estimate the combinations (or levels) that are to be considered and enter them in the heading of the table. For each subbase combination evaluated, a separate table is required. This example considers only the combination shown in Table I.2.

The second step of the process is to estimate the seasonal roadbed soil resilient modulus values. For this example problem, the year is divided into 12 consecutive 1-month time intervals with an appropriate seasonal modulus value defined for each. These values (shown below) reflect the enhanced support that would be expected under the improved moisture conditions provided by a "good" drainage system.

Table I.2. Estimation of Effective Modulus of Subgrade Reaction for Design Example

Trial Subbase: Type __Granular__   Depth to Rigid Foundation (feet) __30__

Thickness (inches) __6"__   Projected Slab Thickness (inches) __10__

Loss of Support, LS __1.0__

(1)	(2)	(3)	(4)	(5)	(6)
Month	Roadbed Modulus, M_R (psi)	Subbase Modulus, E_{SB} (psi)	Composite k-Value (pci) (Fig. II-3.3)	k-Value (pci) on Rigid Foundation (Fig. II-3.4)	Relative Damage, u_r (Fig. II-3.5)
Jan.	5,000	15,000	260	—	1.2
Feb.	5,000	15,000	260	—	1.2
Mar.	5,000	15,000	260	—	1.2
Apr.	5,000	15,000	260	—	1.2
May	5,000	15,000	260	—	1.2
June	6,500	25,000	390	—	1.0
July	6,500	25,000	390	—	1.0
Aug.	6,500	25,000	390	—	1.0
Sept.	6,500	25,000	390	—	1.0
Oct.	6,500	25,000	390	—	1.0
Nov.	5,000	15,000	260	—	1.2
Dec.	5,000	15,000	260	—	1.2

Average: $\bar{u}_r = \dfrac{\Sigma u_r}{n} = \dfrac{13.4}{12} = 1.12$ Summation: $\Sigma u_r =$ __13.4__

Effective Modulus of Subgrade Reaction, k (pci) = __320__

Corrected for Loss of Support: k (pci) = __105__

Roadbed Soil Condition	Roadbed Soil Resilient Modulus (psi)
Wet	5,000
Dry	6,500
Spring-thaw	4,000
Frozen	20,000

(The frozen and spring-thaw seasons are practically nonexistent for the environment considered in this design example. The wet season is 7 months and the dry season is 5 months.)

The third step in estimating the effective k-value is to record the subbase elastic (resilient) modulus (E_{SB}) values for each season. These values, 15,000 psi for the dry season and 25,000 psi for the wet season, were entered in Column 3 of Table I.2. The seasons for subbase modulus variation are the same as those used for the roadbed soil resilient modulus.

The fourth step is to estimate the composite modulus of subgrade reaction for each 1-month interval using Part II, Figure 3.3, and the values from the first two steps. The results for the design example are entered in Column 4 of Table I.2.

The fifth step is to develop a k-value which reflects the presence of a rigid foundation at some close depth below the surface. For this example, the rigid foundation is more than 30 feet below the surface of the subgrade and, according to criteria in Part II, has no influence on the composite k-value.

The sixth step is to project the slab thickness and use Figure 3.5 in Part II to estimate the relative damage (u_r) corresponding to one 18-kip equivalent single axle load in each season. For this example, the projected slab thickness is 10 inches. The u_r values are entered in Column 6.

The seventh step is to add all the relative damage values in Column 6 and determine the average (1.12 in the example). The effective k-value corresponding to this average relative damage (from Figure 3.5 in Part II) is 320 pci.

The last step of this process is to correct the effective k-value for loss of support. Given that the loss of support factor (LS) is 1.0, the corrected effective k-value (from Figure 3.6 in Part II) is 105 pci.

Slab Thickness

Because swelling will lead to serviceability loss in this design example, an iterative procedure is required. The objective of this iterative process is to identify (for a trial slab thickness) when the combined serviceability loss due to traffic and environment reach the design level. Application of this process for the design example is shown in Table I.3 and described below:

Step 1. The application of planned stage construction in the design process means that there is a range of slab thicknesses that can be initially constructed. The minimum thickness is that which would provide the minimum acceptable service life. The maximum practical slab thickness is that which would provide the maximum performance period (25 years) without any consideration of roadbed soil swelling. Using the latter case in this design example results in a slab thickness of 9 inches* (from Figure 3.7 in Part II). The following steps are to estimate the expected service life of the 9-inch slab when swelling is considered.

[*NOTE: Although the 9-inch slab apparently differs significantly from the 10-inch projected value used in Table I.2, the difference in terms of their effects on the effective k-value is not significant. The effective k-value for a 10-inch slab (prior to correction for loss of support) was 320 pci; the effective k-value assuming a 9-inch slab is 300 pci.]

Step 2. The initial trial performance period must be less than the maximum possible performance period (25 years). For this step, a trial period of 20 years was selected (Column 2).

Step 3. Using the graph of cumulative environmental serviceability loss versus time (Figure I.2), an estimate of serviceability loss due to roadbed swelling corresponding to the trial period was determined. For this example, ΔPSI_{sw} is equal to 0.25 at 20 years (Column 3).

Step 4. Subtract the serviceability loss due to swelling from the design total serviceability loss ($\Delta PSI = 12.0$) to estimate the corresponding serviceability loss due to traffic. The result is $\Delta PSI_{TR} = 2.0 - 0.25 = 1.75$ (Column 4).

Step 5. Using Figure 3.7 from Part II, the estimated allowable cumulative 18-kip ESAL traffic is determined. For this example, the result is 4.6×10^6 18-kip ESAL (Column 5).

Table I.3. Reduction in Performance Period (Service Life) Arising from Swelling Consideration

Initial Pavement Thickness __10.0__
Maximum Possible Performance Period (years) __25__
Design Serviceability Loss, $\Delta PSI = p_o - p_t =$ __4.5 − 2.5 = 2.0__

(1) Iteration No.	(2) Trial Performance Period (years)	(3) Serviceability Loss Due to Roadbed Swelling ΔPSI_{SW}	(4) Corresponding Serviceability Loss Due to Traffic ΔPSI_{TR}	(5) Allowable Cumulative Traffic (18-kip ESAL)	(6) Corresponding Performance Period (years)
1	20.0	0.25	1.75	4.6×10^6	22.9
2	21.5	0.26	1.74	4.5×10^6	22.8

Step 6. Using Figure I.1, the time corresponding to 4.6×10^6 18-kip ESAL applications is approximately 22.9 years (Column 6).

Step 7. Since the pavement life calculated in Step 6 is not within 1 year of the trial performance period, the iterative process must continue. The trial performance period is now 21.5 years and the process returns to Step 3. The results of the second iteration indicate that regardless of the trial estimate for the performance period, the outcome in Column 6 will always be about 23 years. Thus, no more iterations are required.

For this particular example design, the pavement cross section consists of a 9-inch jointed reinforced concrete slab with 6 inches of granular subbase and a drainage system that removes water in less than 1 day. This structure will reach its terminal serviceability in approximately 23 years. Thus, to complete the design strategy, an overlay must be designed to carry the remaining 18-kip ESAL traffic over the last 12 years of the analysis period.

I.3 REINFORCEMENT DESIGN

The nomograph for estimating the percent of steel reinforcement required in a jointed reinforced concrete pavement is presented in Figure 3.8 in Part II. The inputs to this nomograph for this design example are as follows:

(1) slab length, L = 30 feet
(2) steel working stress, f_s = 45,000 psi
(3) friction factor, F = 1.5

Application of the nomograph for these conditions results in a required longitudinal steel reinforcing percentage of 0.05 percent. Since there are three 12-foot lanes and a 10-foot-wide PCC shoulder (all tied at the longitudinal joints), the transverse steel percentage required is somewhat higher (0.075 percent).

Tie Bar Design

Since the pavement will consist of three 12-foot-wide PCC lanes with a 10-foot-wide (tied) PCC shoulder on the outside lane, the distances to the nearest free edge (as illustrated in Figure I.3) are 12, 22, and 10 feet for longitudinal joints 1, 2, and 3, respectively. Thus, for the 9-inch slab, the maximum recommended tie bar spacing for each joint (as determined from Part II, Figures 3.13 and 3.14) are as follows:

Long. Joint No.	Distance to the Closest Free Edge, x (feet)	Maximum Spacing (inches) ½-inch Bars	⅝-inch Bars
1	12	36	48
2	22	20	30
3	10	42	48

If ½-inch tie bars are used, the minimum overall length should be 25 inches. If ⅝-inch tie bars are used, then the minimum overall length should be 30 inches.

Appendix I

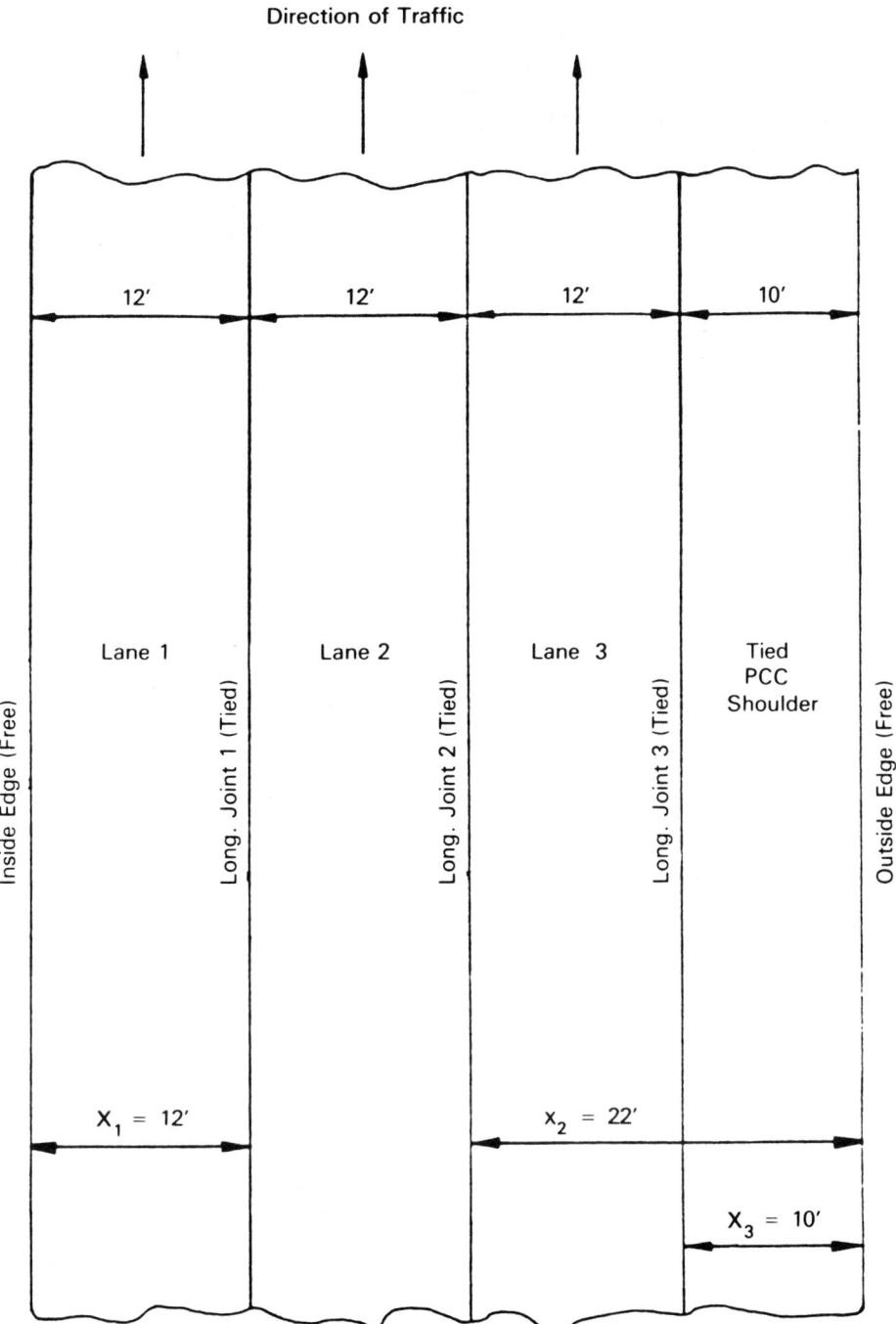

Figure I.3. Plan View of Three-Lane Facility Showing Longitudinal Joint Positions and Corresponding Distances to Nearest Free Edge

Dowel Bar Design

Dowel bar design is described in Section 2.4.4 of Part II. For this design example, the dowel spacing is 12 inches and the dowel length is 18 inches. The dowel diameter is equal to slab thickness (9 inches) multiplied by $1/8$, or 1 and $1/8$ inches.

APPENDIX J
ANALYSIS UNIT DELINEATION BY CUMULATIVE DIFFERENCES

J.1 APPROACH FUNDAMENTALS

A relatively straightforward and powerful analytical method for delineating statistically homogenous units from pavement response measurements along a highway system is the cumulative difference approach. While the methodology presented is fundamentally easy to visualize, the manual implementation for large data bases becomes very time-consuming and cumbersome. However, the approach is presented because it is readily adaptable to a computerized (microcomputer) solution and graphic analysis. This approach can be used for a wide variety of measured pavement response variables such as deflection, serviceability, skid resistance, pavement distress-severity indices, etc.

Figure J.1 illustrates the overall approach concept using the initial assumptions of a continuous and constant response value (r_i) within various intervals (0 to x_1; x_1 to x_2; x_2 to x_3) along a project length. From this figure, it is obvious that *three* unique units having different response magnitudes (r_1, r_2, and r_3) exist along the project. Figure J.1(a) illustrates such a response-distance result. If one were to determine the trend of the cumulative area under the response-distance plot, Figure J.1(b) would result. The solid line indicates the results of the actual response curves. Because the functions are continuous and constant within a unit, the cumulative area, at any x, is simply the integral or

$$A = \int_0^{x_1} r_1 \, dx + \int_{x_1}^{x} r_2 \, dx \qquad (J.1)$$

with each integral being continuous within the respective intervals:

$$(0 \leq x \leq x_1) \text{ and } (x_1 \leq x \leq x_2)$$

In Figure J.1(b), the dashed line represents the cumulative area caused by the overall *average* project response. It should be recognized that the slopes (derivatives) of the cumulative area curves are simply the response value for each unit (r_1, r_2, and r_3) while the slope of the dashed line is the overall average response value of the entire project length considered. At the distance, x, the cumulative area of the average project response is:

$$A_x = \int_0^x r \, dx \qquad (J.2)$$

with

$$\bar{r} = \frac{\int_0^{x_1} r_1 \, dx + \int_{x_1}^{x_2} r_2 \, dx + \int_{x_2}^{x_3} r_3 \, dx}{L_p} = \frac{A_T}{L_p}$$

and therefore

$$\bar{A}_x = L_p \times A_T$$

Knowing both A_x and $\bar{A}_x$ allows for the determination of the cumulative difference variable Z_x from:

$$Z_x = A_x - \bar{A}_x$$

As noted in Figure J.1(b), Z_x is simply the difference in cumulative area values, at a given x, between the actual and project average lines. If the Z_x value is, in turn, plotted against distance, x, Figure J.1(c) results. An examination of this plot illustrates that the location of unit boundaries always coincides with the location (along x) where the slope of the Z_x function changes algebraic signs (i.e., from negative to positive or vice versa). This fundamental concept is the ultimate basis used to analytically determine the boundary location for the analysis units.

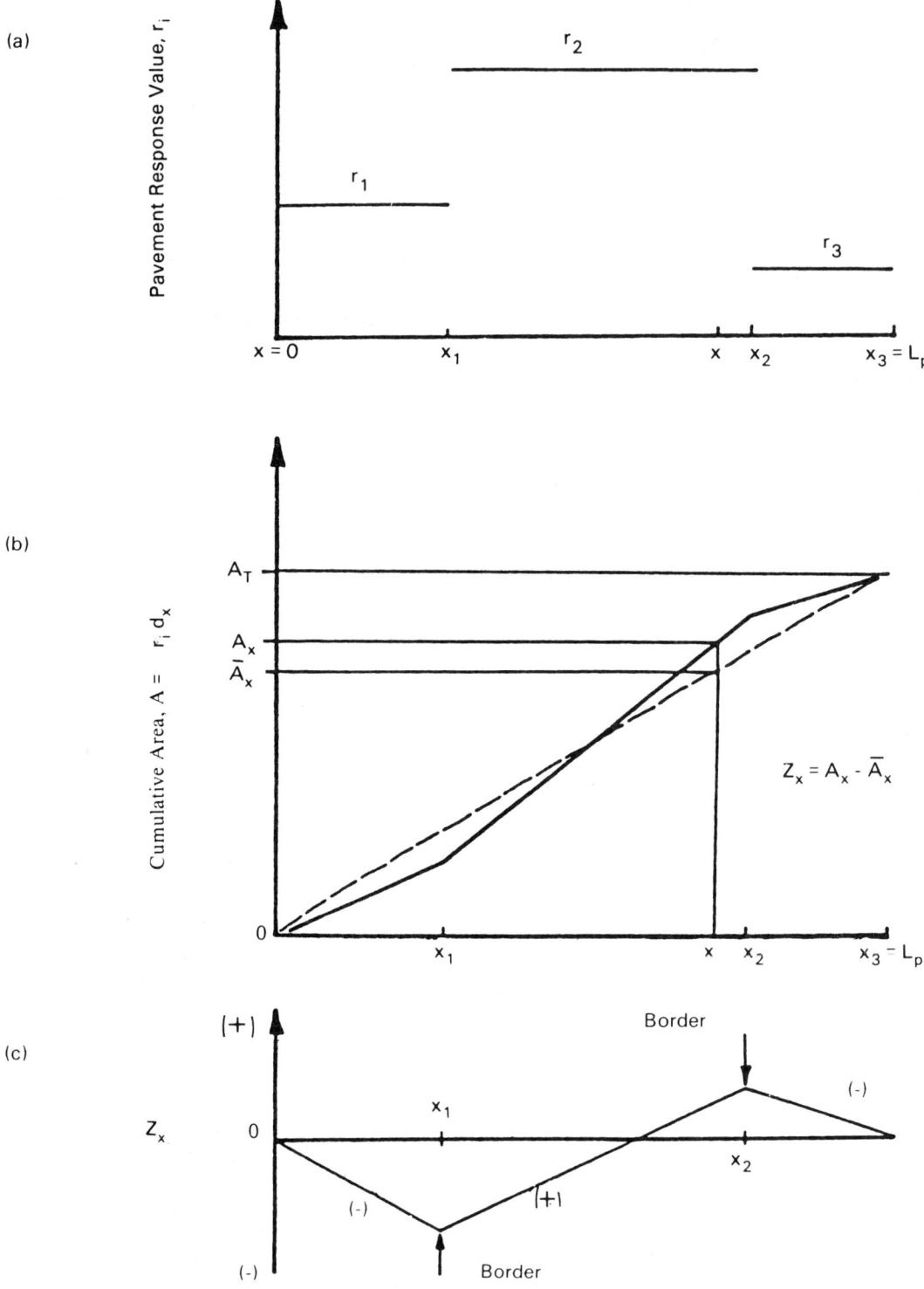

Figure J.1. Concepts of Cumulative Difference Approach to Analysis Unit Delineation

J.2 APPLICATION TO DISCONTINUOUS VARIABLES

The schematic figures shown in Figure J.1 are obviously highly idealized. In practice, measurements are normally discontinuous (point measurements), frequently obtained at unequal intervals and never constant, even within a unit. In order to apply the foregoing principles into a solution methodology capable of dealing with these conditions, a numerical difference approach must be used. The form of the Z_x function is:

$$Z_X = \sum_{i=1}^{n} a_i - \frac{\sum_{i=1}^{n} a_i}{L_p} \sum_{i=1}^{n} x_i$$

with

$$a_i = \frac{(r_{i-1} + r_i) \times x_i}{2} = \bar{r}_i \times x_i \quad (J.6)$$

(NOTE: let $r_o = r_1$ for first interval)

where

- n = the nth pavement response measurement,
- n_t = total number of pavement response measurements taken in project,
- r_i = pavement response value of the ith measurement,
- $\bar{r}_i$ = average of the pavement response values between the $(i - 1)$ and i^{th} tests, and
- L_p = total project length.

If equal pavement testing intervals are used:

$$Z_x = \sum_{i=1}^{n} a_i - \frac{n}{n_t} \sum_{i=1}^{n_t} a_i$$

J.3 TABULAR SOLUTION SEQUENCE

Table J.1 is a table illustrating how the solution sequence progresses and the necessary computational steps required for an unequal interval analysis. The table and entries should be self-explanatory.

J.4 EXAMPLE ANALYSIS

In Part III, Chapter 3, actual results were shown for an analysis unit delineation based upon a field Skid Number test survey: SN(40). Table J.2 is a partial summary of the analysis, indicating only the initial and final portions of the analysis for brevity. This tabular data and solution forms the basis of the information shown in Part III, Figures 3.3 and 3.4.

Table J.1. Tabular Solution Sequence—Cumulative Difference Approach

Col. (1) Station (Distance)	Col. (2) Pavement Response Value (r_i)	Col. (3) Interval Number (n)	Col. (4) Interval Distance (Δx_i)	Col. (5) Cumulative Interval Distance $(\Sigma \Delta x_i)$	Col. (6) Average Interval Response $(\bar{r}_i)$	Col. (7) Actual Interval Area (a_i)	Col. (8) Cumulative Area Σa_i	Col. (9) Z_x Value $Z_x =$ Col. (8) $-$ F* Col. (5)
1	r_1							
		1	Δx_1	Δx_1	$\bar{r}_1 = r_1$	$a_1 = \bar{r}_1 \Delta x_1$	a_1	$Z_{x_1} = a_1 - F^* \Delta x_1$
2	r_2							
		2	Δx_2	$(\Delta x_1 + \Delta x_2)$	$\bar{r}_2 = \dfrac{(r_1 + r_2)}{2}$	$a_2 = \bar{r}_2 \Delta x_2$	a	$Z_{x_2} = (a_1 + a_2)$ $- F^*(\Delta x_1 + \Delta x_2)$
3	r_3							
		3	Δx_3	$(\Delta x_1 + \Delta x_2 + \Delta x_3)$	$\bar{r}_3 = \dfrac{(r_2 + r_3)}{2}$	$a_3 = \bar{r}_3 \Delta x_3$	$a_1 + a_2 + a_3$	
L_p	r_n							
		N_t	Δx_{nt}	$(\Delta x_1 + \cdots + \Delta x_{nt})$	$\bar{r}_{nt} = \dfrac{(r_{n-1} + r_n)}{2}$	$a_{nt} = \bar{r}_{nt} \Delta x_{nt}$	$a_1 + \cdots + a_{2t}$	$Z_{x_{nt}} = (a_1 + \cdots + a_{nt})$ $- F^*(\Delta x_1 + \cdots + \Delta x_{nt})$

$$A_t = \sum_{i=1}^{n_t} a_i$$

$$F^* = \frac{A_t}{L_p}$$

Table J.2. Cumulative Difference Example Problem (SN − 40)

Col. (1) Station (Distance)	Col. (2) SN (40) Value	Col. (3) Interval Number	Col. (4) Interval Distance	Col. (5) Cumulative Distance	Col. (6) Avg. Interval SN (40)	Col. (7) Actual Interval Area	Col. (8) Cumulative Area	Col. (9) Z_x Value
0.5 mi	23	1	0.5	0.5	23	11.50	11.50	$11.50 - 31.49(0.5) = -4.25$
1.0	26	2	0.5	1.0	24.5	12.25	23.75	$23.75 - 31.49(1.0) = -7.74$
1.5	23	3	0.5	1.5	24.5	12.25	36.00	$36.00 - 31.49(1.5) = -11.24$
2.0	24	4	0.5	2.0	23.5	11.75	47.75	$47.75 - 31.49(2.0) = -15.23$
2.5	26	5	0.5	2.5	25.0	12.50	60.25	$60.25 - 31.49(2.5) = -18.48$
⋮	⋮	⋮	⋮	⋮	⋮	⋮	⋮	⋮
74.0	27	148	0.5	74.0	29.5	14.75	2340.00	$2340 - 31.49(74) = +9.74$
74.5	25	149	0.5	74.5	26.0	13.00	2353.00	$2353.00 - 31.49(74.5) = +7.00$
75.0	28	150	0.5	75.0	26.5	13.25	2366.25	$2366.25 - 31.49(75.0) = +4.50$
75.5	26	151	0.5	75.5	27.0	13.50	2379.75	$2379.75 - 31.49(75.5) = +2.26$
76.0	28	152	0.5	76.0	27.0	13.50	2393.25	$2393.25 - 31.49(76.0) = +0.00$

$A_t = 2393.25$
$L_p = 76.0$
$F^* = 31.49$

APPENDIX K
TYPICAL PAVEMENT DISTRESS
TYPE-SEVERITY DESCRIPTIONS

TYPICAL PAVEMENT DISTRESS TYPE-SEVERITY DESCRIPTIONS

This appendix contains general descriptions of the major types of distress that may be encountered in both flexible (asphalt concrete) and rigid pavements. Also noted is a typical description of three distress severity levels associated with each distress. This information has been obtained from FHWA/RD-81/080 study "A Pavement Moisture Accelerated Distress Identification System." These descriptions are provided as a guide to user agencies only and should not be viewed as a standard method for distress type-severity identification. This information, along with an estimate of the amount of each distress-severity combination, represents an example of the minimum information needs required for a thorough condition (distress) survey.

NOTE: In presenting the distress types and severity descriptions, the following letters refer to different levels of severity:

L—Low   M—Medium   H—High

K.1 DISTRESS TYPES (ASPHALT SURFACED PAVEMENTS)

Name of Distress: Alligator or Fatigue Cracking

Description:

Alligator or fatigue cracking is a series of interconnecting cracks caused by fatigue failure of the asphalt concrete surface (or stabilized base) under repeated traffic loading. The cracking initiates at the bottom of the asphalt surface (or stabilized base) where tensile stress and strain is highest under a wheel load. The cracks propagate to the surface initially as one or more longitudinal parallel cracks. After repeated traffic loading, the cracks connect, forming many-sided, sharp-angled pieces that develop a pattern resembling chicken wire or the skin of an alligator. The pieces are usually less than 1 foot on the longest side. Alligator cracking occurs only in areas that are subjected to repeated traffic loadings. Therefore, it would not occur over an entire area unless the entire area was subjected to traffic loading. Alligator cracking does not occur in asphalt overlays over concrete slabs. Pattern-type cracking which occurs over an entire area that is *not* subjected to loading is rated as block cracking which is not a load-associated distress. Alligator cracking is considered a major structural distress.

Severity Levels:

L—Longitudinal disconnected hairline cracks running parallel to each other. The cracks are not spalled. Initially there may only be a single crack in the wheel path (defined as Class 1 cracking at AASHO Road Test).

M—Further development of low-severity alligator cracking into a pattern of pieces formed by cracks that may be lightly surface-spalled. Cracks may be sealed (defined as Class 2 cracking at AASHO Road Test).

H—Medium alligator cracking has progressed so that pieces are more severely spalled at the edges and loosened until the cells rock under traffic. Pumping may exist (defined as Class 3 cracking at AASHO Road Test).

How to Measure:

Alligator cracking is measured in square feet or square meters of surface area. The major difficulty in measuring this type of distress is that many times, two or three levels of severity exist within one distressed area. If these portions can be easily distinguished from each other, they should be measured and recorded separately. However, if the different levels

Table K.1. Identification of Distress Types

Asphalt Surfaced Pavements	Jointed Reinforced Concrete Pavements
1. Alligator or Fatigue Cracking	1. Blow-Up
2. Bleeding	2. Corner Break
3. Block Cracking	3. Depression
4. Corrugation	4. Durability ("D") Cracking
5. Depression	5. Faulting-Transverse Joints/Cracks
6. Joint Reflection Cracking from PCC Slab	6. Joint Load Transfer System Deterioration
7. Lane/Shoulder Dropoff or Heave	7. Seal Damage-Transverse Joints
8. Lane/Shoulder Joint Separation	8. Lane/Shoulder Dropoff or Heave
9. Longitudinal and Transverse Cracking (Non-PCC Slab Joint Reflective)	9. Lane/Shoulder Joint Separation
10. Patch Deterioration	10. Longitudinal Cracks
11. Polished Aggregate	11. Longitudinal Joint Faulting
12. Potholes	12. Patch Deterioration
13. Pumping and Water Bleeding	13. Patch Adjacent Slab Deterioration
14. Raveling and Weathering	14. Popouts
15. Rutting	15. Pumping and Water Bleeding
16. Slippage Cracking	16. Reactive Aggregate Distress
17. Swell	17. Scaling and Map Cracking
	18. Spalling (Transverse and Longitudinal Joint/Crack)
	19. Spalling (Corner)
	20. Swell
	21. Transverse and Diagonal Cracks

of severity cannot be easily divided, the entire area should be rated at the highest severity level present.

Name of Distress: Bleeding

Description:

Bleeding is a film of bituminous material on the pavement surface which creates a shiny, glass-like, reflecting surface that usually becomes quite sticky. Bleeding is caused by excessive amounts of asphalt cement in the mix and/or low air void contents. It occurs when asphalt fills the voids of the mix during hot weather and then expands out onto the surface of the pavement. Since the bleeding process is not reversible during cold weather, asphalt will accumulate on the surface.

Severity Levels:

No degrees of severity are defined. Bleeding should be noted when it is extensive enough to cause a reduction in skid resistance.

How to Measure:

Bleeding is measured in square feet or square meters of surface area.

Name of Distress: Block Cracking

Description:

Block cracks divide the asphalt surface into approximately *rectangular* pieces. The blocks range in size from approximately 1 ft^2 to 100 ft^2. Cracking into larger blocks are generally rated as longitudinal and transverse cracking. Block cracking is caused mainly by shrinkage of the asphalt concrete and daily temperature cycling (which results in daily stress/strain cycling). *It is not load-associated*, although load can increase the severity of individual cracks from low to medium to high. The occurrence of block cracking usually indicates that the asphalt has hardened significantly. Block cracking normally occurs over a large proportion of pavement area, but sometimes will occur only in nontraffic areas. This type of distress differs from alligator cracking in that alligator cracks form smaller, many-sided pieces with sharp angles. Also unlike block cracks, alligator cracks are caused by repeated traffic loadings and are, therefore, located only in trafficked areas (i.e., wheel paths).

Severity Levels:

L—Blocks are defined by (1) nonsealed cracks that are nonspalled (sides of the crack are vertical) or only minor spalling with a ¼-inch (6 mm) or less mean width; or (2) sealed cracks have a sealant in satisfactory condition to prevent moisture infiltration.

M—Blocks are defined by either (1) sealed or nonsealed cracks that are moderately spalled; (2) nonsealed cracks that are not spalled or have only minor spalling, but have a mean width greater than approximately ¼-inch (6 mm); or (3) sealed cracks that are not spalled or have only minor spalling, but have sealant in unsatisfactory condition.

H—Blocks are well defined by cracks that are severely spalled.

How to Measure:

Block cracking is measured in square feet or square meters of surface area. It usually occurs at one severity level in a given pavement section; however, any areas of the pavement section having distinctly different levels of severity should be measured and recorded separately.

Name of Distress: Corrugation

Description:

Corrugation is a form of plastic movement typified by ripples across the asphalt pavement surface. It occurs usually at points where traffic starts and stops. Corrugation usually occurs in asphalt layers that lack stability in warm weather, but may also be attributed to excessive moisture in a subgrade, contamination of the mix, or lack of aeration of liquid asphalt mixes.

Severity Levels:

L—Corrugations cause some vibration of the vehicle which creates no discomfort.
M—Corrugations cause significant vibration of the vehicle which creates some discomfort.
H—Corrugations cause excessive vibration of the vehicle which creates substantial discomfort, and/or a safety hazard, and/or vehicle damage, requiring a reduction in speed for safety.

How to Measure:

Corrugation is measured in square feet or square meters of surface area. Severity levels are determined by riding in a mid- to full-sized sedan weighing approximately 3,000 to 3,800 lb. (13.3–16.9 kN) over the pavement inspection unit at the posted speed limit.

Name of Distress: Depression

Description:

Depressions are localized pavement surface areas having elevations slightly lower than those of the surrounding pavement. In many instances, light depressions are not noticeable until after a rain, when ponding water creates "birdbath" areas; but the depressions can also be located without rain because of strains created by oil droppings from vehicles. Depressions can be caused by settlement of the foundation soil or can be "built in" during construction. Depressions cause roughness and when filled with water of sufficient depth could cause hydroplaning of vehicles.

Severity Levels:

L—Depressions cause some bounce of the vehicle which creates no discomfort.
M—Depressions cause significant bounce of the vehicle which creates some discomfort.
H—Depressions cause excessive bounce of the vehicle which creates substantial discomfort, and/or safety hazard, and/or vehicle damage, requiring a reduction in speed for safety.

How to Measure:

Depressions are measured in square feet or meters in each inspection unit. Each depression is rated according to its level of severity. Severity level is determined by riding in a mid- to full-sized sedan weighing approximately 3,000 to 3,800 lb. (13.3–16.9 kN) over the pavement inspection unit at the posted speed limit.

Name of Distress: Joint Reflection Cracking from PCC Slab

Description:

This distress occurs only on pavements having an asphalt concrete surface over a jointed portland cement concrete (PCC) slab and they occur at transverse and longitudinal joints (i.e., widening joints). This distress does not include reflection cracking away from a joint or from any other type of base (i.e., cement stabilized, lime stabilized) as these cracks are identified as "Longitudinal and Transverse Cracking." Joint reflection cracking is caused mainly by movement of the PCC slab beneath the asphalt concrete (AC) surface because of thermal and moisture changes; it is generally not load-initiated. However, traffic loading may cause a breakdown of the AC near the initial crack, resulting in spalling. A knowledge of slab dimensions beneath the AC surface will help to identify these cracks.

Severity Levels:

L—Cracks have either minor spalling or no spalling and can be sealed or nonsealed. If nonsealed, the cracks have a mean width of ¼-inch (6 mm) or less; sealed cracks are of any width, but their sealant material is in satisfactory condition to substantially prevent water infiltration. No significant bump occurs when a vehicle crosses the crack.

M—One of the following conditions exists: (1) cracks are moderately spalled and can be either sealed or nonsealed of any width; (2) sealed cracks are not spalled or have only minor spalling, but the sealant is in a condition so that water can freely infiltrate; (3) nonsealed cracks are not spalled or are only lightly spalled, but the mean crack width is greater than ¼-inch (6 mm); (4) low-severity random cracking exists near the crack or at the corners of intersecting cracks; or (5) the crack causes a significant bump to a vehicle.

H—(1) Cracks are severely spalled and/or there exists medium or high random cracking near the crack or at the corners of intersecting cracks, or (2) the crack causes a severe bump to a vehicle.

How to Measure:

Joint reflection cracking is measured in linear feet or meters. The length and severity level of each crack should be identified and recorded. If the crack does not have the same severity level along its entire length, each general portion should be recorded separately. The vehicle used to determine bump severity is a mid- to full-sized sedan weighing approximately 3,000 to 3,800 lb. (13.3–16.9 kN) over the pavement inspection unit at the posted speed limit.

Name of Distress: Lane/Shoulder Drop-off or Heave

Description:

Lane/shoulder drop-off or heave occurs wherever there is a difference in elevation between the traffic lane and the shoulder. Typically, the outside shoulder settles due to consolidation or a settlement of the underlying granular or subgrade material or pumping of the underlying material. Heave of the shoulder may occur due to frost action or swelling soils. Drop-off of granular or soil shoulder is generally caused from blowing away of shoulder material from passing trucks.

Severity Levels:

Severity level is determined by computing the mean difference in elevation between the traffic lane and shoulder:

L	¼–½ in.	(6–13 mm)
M	½–1 in.	(3–25 mm)
H	>1 in.	(>25 mm)

How to Measure:

Lane/shoulder drop-off or heave is measured every 100 feet (30 m) in inches (or mm) along the joint. The

Appendix K

mean difference in elevation is computed from the data and used to determine severity level.

Name of Distress: Lane/Shoulder Joint Separation

Description:

Lane/shoulder joint separation is the widening of the joint between the traffic lane and the shoulder, generally due to movement in the shoulder. If the joint is tightly closed or well sealed so water cannot enter (or if there is no joint due to full-width paving), then lane/shoulder joint separation is not considered a distress. If the shoulder is not paved (i.e., gravel or grass), then the severity should be rated as high. If a curbing exists, then it should be rated according to the width of the joint between the asphalt surface and curb.

Severity Levels:

Severity level is determined by the mean joint opening. No severity level is counted if the joint is well sealed to prevent moisture intrusion.

L	0.04–.12 in.	(1–3 mm)
M	>.12–.40 in.	(>3–10 mm)
H	>.40 in.	(>10 mm) (also a nonpaved shoulder)

How to Measure:

Lane/shoulder joint separation is measured in inches (or millimeters) at about 50 feet (15.2 m) intervals along the sample unit. The mean separation is used to determine severity level.

Name of Distress: Longitudinal and Transverse Cracking (Non-PCC Slab Joint Reflective)

Description:

Longitudinal cracks are parallel to the pavement's centerline or laydown direction. They may be caused by (1) a poorly constructed paving lane joint, (2) shrinkage of the AC surface due to low temperatures or hardening of the asphalt, or (3) a reflective crack caused by cracks beneath the surface course, including cracks in PCC slabs (but not at PCC slab joints). *Transverse* cracks extend across the pavement centerline or direction of laydown. They may be caused by items (2) or (3) above. These types of cracks are not usually load-associated.

Severity Levels:

- L—Cracks have either minor spalling or no spalling, and cracks can be sealed or nonsealed. If sealed, cracks have a mean width of ¼ inch (6 mm) or less; sealed cracks are of any width, but their sealant material is in satisfactory condition to substantially prevent water infiltration. No significant bump occurs when a vehicle crosses the crack.
- M—One of the following conditions exists: (1) cracks are moderately spalled and can either be sealed or nonsealed of any width; (2) sealed cracks are not spalled or have only minor spalling, but the sealant is in a condition so that water can freely infiltrate; (3) nonsealed cracks are not spalled or have only minor spalling, but mean crack width is greater than ¼ inch (6 mm); (4) low severity random cracking exists near the crack or at the corners of intersecting cracks; or (5) the crack causes a significant bump to a vehicle.
- H—(1) Cracks are severely spalled; and/or medium- or high-random cracking exists near the crack or at the corners of intersecting cracks, or (2) the crack causes a severe bump to a vehicle.

How to Measure:

Longitudinal and transverse cracks are measured in linear feet or linear meters. The length and severity of each crack should be identified and recorded. If the crack does not have the same severity level along its entire length, each general portion of the crack having a different severity level should be recorded separately. The vehicle used to determine bump severity is a mid- to full-sized sedan weighing approximately 3,000 to 3,800 lb. (13.3–16.9 kN) over the pavement inspection unit at the posted speed limit.

Name of Distress: Patch Deterioration

Description:

A patch is an area where the original pavement has been removed and replaced with either similar or different material.

Severity Levels:

L—Patch is in very good condition and is performing satisfactorily.
M—Patch is somewhat deteriorated, having low to medium levels of any types of distress.
H—Patch is badly deteriorated and soon needs replacement.

How to Measure:

Each patch is measured in square feet or square meters of surface area. Even if a patch is in excellent condition, it is still rated low severity.

Name of Distress: Polished Aggregate

Description:

Aggregate polishing is caused by repeated traffic applications. Polished aggregate is present when close examination of a pavement reveals that the portion of aggregate extending above the asphalt is either very small or there are no rough or angular aggregate particles to provide good skid resistance.

Severity Levels:

No degrees of severity are defined. However, the degree of polishing should be significant in reducing skid resistance before it is included as a distress.

How to Measure:

Polished aggregate is measured in square feet or square meters of surface area. The existence of polishing can be detected by both visually observing and running the fingers over the surface.

Name of Distress: Potholes

Description:

A bowl shaped hole of various sizes in the pavement surface. The surface has broken into small pieces by alligator cracking or by localized disintegration of the mixture and the material is removed by traffic. Traffic loads force the underlying materials out of the hole, increasing the depth.

Severity Levels:

Area			
(ft^2)	<1	1–3	>3
(m^2)	<1/3	1/3–1	>1
Depth—in (mm)			
<1 (<25)	L	L	M
1–2 (25–50)	M	M	H
>2 (>51)	M	H	H

How to Measure:

Portholes are counted in number of holes of each severity level in the inspection unit.

Name of Distress: Pumping and Water Bleeding

Description:

Pumping is the ejection of water and fine materials under pressure through cracks under moving loads. As the water is ejected, it carries fine material resulting in progressive material deterioration and loss of support. Several cases of pumping of stabilized base materials have been observed for example. Surface staining or accumulation of material on the surface close to cracks is evidence of pumping. Water bleeding occurs where water seeps slowly out of cracks in the pavement surface.

Severity Levels:

L—Water bleeding exists or water pumping can be observed when heavy loads pass over the pavement; however, no fines (or only a very small amount) can be seen on the surface of the pavement.

Appendix K

M—Some pumped material can be observed near cracks in the pavement surface.
H—A significant amount of pumped material exists on the pavement surface near the cracks.

How to Count:

If pumping or water bleeding exists anywhere in the sample unit, it is counted as occurring.

Name of Distress: Raveling and Weathering

Description:

Raveling and weathering are the wearing away of the pavement surface caused by the dislodging of aggregate particles (raveling) and loss of asphalt binder (weathering). They generally indicate that the asphalt binder has hardened significantly.

Severity Levels:

L—Aggregate or binder has started to wear away but has not progressed significantly.
M—Aggregate and/or binder has worn away and the surface texture is moderately rough and pitted. Loose particles generally exist.
H—Aggregate and/or binder has worn away and the surface texture is severely rough and pitted.

How to Measure:

Raveling and weathering are measured in square feet or square meters of surface area.

Name of Distress: Rutting

Description:

A rut is a surface depression in the wheel paths. Pavement uplift may occur along the sides of the rut; however, in many instances, ruts are noticeable only after a rainfall, when the wheel paths are filled with water. Rutting stems from a permanent deformation in any of the pavement layers or subgrade, usually caused by consolidation or lateral movement of the materials due to traffic loads. Rutting may be caused by plastic movement in the mix in hot weather or inadequate compaction during construction. Significant rutting can lead to major structural failure of the pavement and hydroplaning potential. Wear of the surface in the wheel paths from studded tires can also cause a type of "rutting."

Severity Levels:

Severity	Mean Rut Depth Criteria
L	$1/4–1/2$ in. (6–13 mm)
M	$>1/2–1$ in. (13–25)
H	>1 in. (>25 mm)

How to Measure:

Rutting is measured in square feet or square meters of surface area, and its severity is determined by the mean depth of the rut. To determine the mean rut depth, a 4-foot (1.2 m) straightedge should be laid across the rut and the maximum depth measured. The mean depth should be computed from measurements taken every 20 feet (6 m) along the length of the rut.

Name of Distress: Slippage Cracking

Description:

Slippage cracks are crescent- or half-moon-shaped cracks generally having two ends pointed into the direction of traffic. They are produced when braking or turning wheels cause the pavement surface to slide and deform. This usually occurs when there is a low-strength surface mix or poor bond between the surface and next layer of pavement structure.

Severity Levels:

No degrees of severity are defined. It is sufficient to indicate that a slippage crack exists.

How to Measure:

Slippage cracking is measured in square meters or in square feet of surface area within the inspection unit.

Name of Distress: Swell

Description:

Swell is characterized by an upward bulge in the pavement's surface. A swell may occur sharply over a small area or as a longer, gradual wave. Either type of swell can be accompanied by surface cracking. A swell is usually caused by frost action in the subgrade or by swelling soil, but a swell can also occur on the surface of an asphalt overlay (over PCC) as a result of a blow-up in the PCC slab. They can often be identified by oil droppings on the surface.

Severity Levels:

L—Swell causes some bounce of the vehicle which creates no discomfort.
M—Swell causes significant bounce of the vehicle which creates some discomfort.
H—Swell causes excessive bounce of the vehicle which creates substantial discomfort, and/or a safety hazard, and/or vehicle damage, requiring reduction in speed for safety.

How to Measure:

Swells within the inspection unit are measured in square feet or meters. Severity level is determined by riding in a mid- to full-sized sedan weighing approximately 3,000 to 3,800 lb. (13.3–16.9 kN) over the pavement inspection unit at the posted speed limit.

K.2 DISTRESS TYPES (JOINTED REINFORCED CONCRETE PAVEMENTS)

Name of Distress: Blow-up

Description:

Most blow-ups occur during the spring and hot summer at a transverse joint or wide crack. Infiltration of incompressible materials into the joint or crack during cold periods results in high compressive stresses in hot periods. When this compressive pressure becomes too great, a localized upward movement of the slab or shattering occurs at the joint or crack. Blow-ups are accelerated due to a spalling away of the slab at the bottom creating reduced joint contact area. The presence of "D" cracking or freeze-thaw damage also weakens the concrete near the joint resulting in increased spalling and blow-up potential.

Severity Levels:

L—Blow-up has occurred, but only causes some bounce of the vehicle which creates no discomfort.
M—Blow-up causes a significant bounce of the vehicle which creates some discomfort. Temporary patching may have been placed because of the blow-up.
H—Blow-up causes excessive bounce of the vehicle which creates substantial discomfort, and/or a safety hazard, and/or vehicle damage, requiring a reduction in speed for safety.

How to Measure:

Blow-ups are measured by counting the number existing in each uniform section. Severity level is determined by riding in a mid- to full-sized sedan weighing approximately 3,000 to 3,800 lb. (13.3–16.9 kN) over the uniform section at the posted speed limit. The number is not as important as the fact that initial blow-ups signal a problem with "lengthening" or gradual downhill movement—and others should be expected to occur until the maximum distance is down to 1,000 feet between blow-ups, the distance required to develop full restraint of an interior section.

Name of Distress: Corner Break

Description:

A corner break is a crack that intersects the joints at a distance less than 6 feet (1.8 m) on each side measured from the corner of the slab. A corner break extends vertically through the entire slab thickness. It should not be confused with a corner spall, which intersects the joint at an angle through the slab and is typically within 1 foot (0.3 m) from the slab corner. Heavy repeated loads combined with pumping, poor load transfer across the joint, and thermal curling and moisture warping stresses result in corner breaks.

Severity Levels:

L—Crack is tight (hairline). Well-sealed cracks are considered tight. No faulting or break-up of broken corner exists. Crack is not spalled.

M—Crack is working and spalled at medium severity, but break-up of broken corner has not occurred. Faulting of crack or joint is less than 1/2 inch (13 mm). Temporary patching may have been placed because of corner break.

H—Crack is spalled at high severity, the corner piece has broken into two or more pieces, or faulting of crack or joint is more than 1/2 inch (13 mm).

How to Measure:

Corner breaks are measured by counting the number that exists in the uniform section. Different levels of severity should be counted and recorded separately. *Corner breaks adjacent to a patch will be counted as "patch adjacent slab deterioration."*

Name of Distress: Depression

Description:

Depressions in concrete pavements are localized settled areas. There is generally significant slab cracking in these areas due to uneven settlement. The depressions can be located by stains caused by oil droppings from vehicles and by riding over the pavement. Depressions can be caused by settlement or consolidation of the foundation soil or can be "built-in" during construction. They are frequently found near culverts. This is usually caused by poor compaction of soil around the culvert during construction. Depressions cause slab cracking, roughness, and hydroplaning when filled with water of sufficient depth.

Severity Levels:

L—Depression causes a distinct bounce of vehicle which creates no discomfort.

M—Depression causes significant bounce of the vehicle which creates some discomfort.

H—Depression causes excessive bounce of the vehicle which creates substantial discomfort, and/or a safety hazard, and/or vehicle damage, requiring a reduction in speed for safety.

How to Measure:

Depressions are measured by counting the number that exists in each uniform section. Each depression is rated according to its level of severity. Severity level is determined by riding in a mid- to full-sized sedan weighing approximately 3,000 to 3,800 lb. (13.3–16.9 kN) over the uniform section at the posted speed limit.

Name of Distress: Durability ("D") Cracking

Description:

"D" cracking is a series of closely spaced crescent-shaped hairline cracks that appear at a PCC pavement slab surface adjacent and roughly parallel to transverse and longitudinal joints, transverse and longitudinal cracks, and the free edges of pavement slab. The fine surface cracks often curve around the intersection of longitudinal joints/cracks and transverse joints/cracks. These surface cracks often contain calcium hydroxide residue which causes a dark coloring of the crack and immediate surrounding area. This may eventually lead to disintegration of the concrete within 1 to 2 feet (0.30–0.6 m) or more of the joint or crack, particularly in the wheelpaths. "D" cracking is caused by freeze-thaw expansive pressures of certain types of coarse aggregates and typically begins at the bottom of the slab which disintegrates first. Concrete durability problems caused by reactive aggregates are rated under "Reactive Aggregate Distress."

Severity Levels:

L—The characteristic pattern of closely spaced fine cracks has developed near joints, cracks, and/or free edges; however, the width of the affected area is generally <12 inch (30 cm) wide at the center of the lane in transverse cracks and joints. The crack pattern may fan out at the intersection of transverse cracks/joints with longitudinal cracks/joints. No joint/crack spalling has occurred, and no patches have been placed for "D" cracking.

M—The characteristic pattern of closely spaced cracks has developed near the crack, joint, or free edge and: (1) is generally wider than 12 inch (30 cm) at the center of the lane in transverse cracks and/or joints; or (2) low- or medium-severity joint/crack or corner spalling has developed in the affected area; or (3) temporary patches have been placed due to "D" cracking-induced spalling.

H—The pattern of fine cracks has developed near joints or cracks and (1) a high severity level of spalling at joints/cracks exists and considerable material is loose in the affected area; or (2) the crack pattern has developed generally over the entire slab area between cracks and/or joints.

How to Measure:

"D" cracking is measured by counting the number of joints or cracks (including longitudinal) affected. Different severity levels are counted and recorded separately. "D" cracking adjacent to a patch is rated as patch-adjacent slab deterioration. "D" cracking should not be counted if the fine crack pattern has not developed near cracks, joints, and free edges. Pop-outs and discoloration of joints, cracks, and free edges may occur without "D" cracking.

Name of Distress: Faulting of Transverse Joints and Cracks

Description:

Faulting is the difference of elevation across a joint or crack. Faulting is caused in part by a buildup of loose materials under the approach slab near the joint or crack as well as depression of the leave slab. The buildup of eroded or infiltrated materials is caused by pumping from under the leave slab and shoulder (free moisture under pressure) due to heavy loadings. The warp and/or curl upward of the slab near the joint or crack due to moisture and/or temperature gradient contributes to the pumping condition. Lack of load transfer contributes greatly to faulting.

Severity Levels:

Severity is determined by the average faulting over the joints within the sample unit.

How to Measure:

Faulting is determined by measuring the difference in elevation of slabs at transverse joints for the slabs in the sample unit. Faulting of cracks are measured as a guide to determine the distress level of the crack. Faulting is measured 1 foot in from the outside (right) slab edge on all lanes except the innermost passing lane. Faulting is measured 1 foot in from the inside (left) slab edge on the inner passing lane. If temporary patching prevents measurement, proceed on to the next joint. Sign convention: + when approach slab is higher than departure slab, − when the opposite occurs. Faulting never occurs in the opposite direction.

Name of Distress: Joint Load Transfer System Associated Deterioration (Second Stage Cracking)

Description:

This distress develops as a transverse crack a short distance (e.g., 9 inches (23 cm)) from a transverse joint at the end of joint dowels. This usually occurs when the dowel system fails to function properly due to extensive corrosion or misalignment. It may also be caused by a combination of smaller diameter dowels and heavy traffic loadings.

Severity Levels:

L—Hairline (tight) crack with no spalling or faulting or well-sealed crack with no visible faulting or spalling.

M—Any of the following conditions exist; the crack has opened to a width less than 1 inch (25 mm); the crack has faulted less than $1/2$ inch (13 mm); the crack may have spalled to a low- or medium-severity level; the area between the crack and joint has started to break up, but pieces have not been dislodged to the point that a tire damage or safety hazard is present; or temporary patches have been placed due to this joint deterioration.

H—Any of the following conditions exist: a crack with width of opening greater than 1 inch (25 mm); a crack with a high-severity level of spalling; a crack faulted $1/2$ inch (13 mm) or more; or the area between the crack and joint has broken up and pieces have been dislodged to the point that a tire damage or safety hazard is present.

How to Measure:

The number of joints with each severity level are counted in the uniform section.

Name of Distress: Joint Seal Damage of Transverse Joints

Description:

Joint seal damage exists when incompressible materials and/or water can infiltrate into the joints. This infiltration can result in pumping, spalling, and blow-ups. A joint sealant bonded to the edges of the slabs protects the joints from accumulation of incompressible materials and also reduces the amount of water seeping into the pavement structure. Typical types of joint seal damage are: (1) stripping of joint sealant, (2) extrusion of joint sealant, (3) weed growth, (4) hardening of the filler (oxidation), (5) loss of bond to the slab edges, and (6) lack or absence of sealant in the joint.

Severity Levels:

L—Joint sealant is in good condition throughout the section with only a minor amount of any of the above types of damage present. Little water and no incompressibles can infiltrate through the joint.

M—Joint sealant is in fair condition over the entire surveyed section, with one or more of the above types of damage occurring to a moderate degree. Water can infiltrate the joint fairly easily; some incompressibles can infiltrate the joint. Sealant needs replacement within 1 to 3 years.

H—Joint sealant is in poor condition over most of the sample unit, with one or more of the above types of damage occurring to a severe degree. Water and incompressibles can freely infiltrate the joint. Sealant needs immediate replacement.

How to Measure:

Joint sealant damage of transverse joints is rated based on the overall condition of the sealant over the entire sample unit.

Name of Distress: Lane/Shoulder Drop-off or Heave

Description:

Lane/shoulder drop-off or heave occurs when there is a difference in elevation between the traffic lane and shoulder. Typically, the outside shoulder settles due to consolidation or a settlement of the underlying granular or subgrade material or pumping of the underlying material. Heave of the shoulder may occur due to frost action or swelling soils. Drop-off of granular or soil shoulder is generally caused from blowing away of shoulder material from passing trucks.

Severity Levels:

Severity level is determined by computing the mean difference in elevation between the traffic lane and shoulder.

How to Measure:

Lane/shoulder drop-off or heave is measured in the sample unit at all joints when joint spacing is >50 feet (15 m), at every third joint when spacing is <50 feet (15 m). It is also measured at mid-slab in each slab measured at the joint. The mean difference in elevation is computed from the data and used to determine severity level. Measurements at joints are made 1 foot (0.3 m) from the transverse joint on the departure slab only on the outer lane/shoulder.

Name of Distress: Lane/Shoulder Joint Separation

Description:

Lane/shoulder joint separation is the widening of the joint between the traffic lane and the shoulder, generally due to movement in the shoulder. If the joint is tightly closed or well sealed so that water cannot easily infiltrate, then lane/shoulder joint separation is not considered a distress.

Severity Levels:

No severity-level is recorded if the joint is tightly sealed.

L—Some opening, but less than or equal to 0.12 inch (3 mm).

M—More than 0.12 inch (3 mm) but equal to or less than 0.4 inch (10 mm) opening.

H—More than 0.4 (10 mm) opening. Gravel or sod shoulders are rated as high.

How to Measure:

Lane/shoulder joint separation is measured and recorded in inches (or mm) near transverse joints and at mid-slab. The mean separation is used to determine the severity level.

Name of Distress: Longitudinal Cracks

Description:

Longitudinal cracks occur generally parallel to the centerline of the pavement. They are often caused by improper construction of longitudinal joints or by a combination of heavy load repetition, loss of foundation support, and thermal and moisture gradient stresses.

Severity Levels:

L—Hairline (tight) crack with no spalling or faulting, or a well-sealed crack with no visible faulting or spalling.
M—Working crack with a moderate or less severity spalling and/or faulting less than $1/2$ inch (12 mm).
H—A crack with width greater than 1 inch (25 mm); a crack with a high-severity level of spalling; or a crack faulted $1/2$ inch (13 mm) or more.

How to Measure:

Cracks are measured in linear feet (or meters) for each level of distress. The length and average severity of each crack should be identified and recorded.

Name of Distress: Longitudinal Joint Faulting

Description:

Longitudinal joint faulting is a difference in elevation of two traffic lanes measured at the longitudinal joint. It is caused primarily by heavy truck traffic and settlement of the foundation.

Severity Levels:

Severity level is determined by measuring the maximum fault.

How to Measure:

Where the longitudinal joint has faulted, the length of the affected area and the maximum joint faulting is recorded.

Name of Distress: Patch Deterioration (including replaced slabs)

Description:

A patch is an area where a portion or all of the original slab has been removed and replaced with a permanent type of material (e.g., concrete or hot-mixed asphalt). *Only permanent patches should be considered.*

Severity Levels:

L—Patch has little or no deterioration. Some low severity spalling of the patch edges may exist. Faulting across the slab-patch joints must be less than $1/4$ inch (6 mm). Patch is rated low severity even if it is in excellent condition.
M—Patch has cracked (low-severity level) and/or some spalling of medium-severity level exists around the edges. Minor rutting may be present. Faulting of $1/4$ to $3/4$ inch (6–19 mm) exists. Temporary patches may have been placed because of permanent patch deterioration.
H—Patch has deteriorated by spalling, rutting, or cracking within the patch to a condition which requires replacement.

How to Measure:

The number of patches within each uniform section is recorded. Patches at different severity levels are counted and recorded separately. Additionally, the approximate square footage (or meters) of each patch and type (i.e., PCC or asphalt) is recorded. All patches are rated either L, M, or H.

Appendix K

Name of Distress: Patch Adjacent Slab Deterioration

Description:

Deterioration of the original concrete slab adjacent to a permanent patch is given the above name. This may be in the form of spalling of the slab at the slab/patch joint, "D" cracking of the slab adjacent to the patch, a corner break in the adjacent slab, or a second permanent patch placed adjacent to the original patch.

Severity Levels:

Severity levels are the same as that described for the particular distress found. A second permanent patch placed adjacent to a previously placed permanent patch will be rated here as medium severity. Temporary patches placed because of this deterioration will also be rated here as medium severity.

How to Measure:

The number of permanent patches with distress in the original slab adjacent to the patch at each severity level will be counted and recorded separately. Additionally, the type of patch (AC or PCC) and distress will be recorded separately.

Name of Distress: Popouts

Description:

A popout is a small piece of concrete that breaks loose from the surface due to freeze-thaw action, expansive aggregates, and/or nondurable materials popouts may be indicative of unsound aggregates and "D" cracking. Popouts typically range from approximately 1 inch (25 mm) to 4 inches (10 cm) in diameter and from $1/2$-inch to 2-inches (13–51 mm) deep.

Severity Levels:

No degrees of severity are defined for popouts. The average popout density must exceed approximately one popout per square yard (square meter) over the entire slab area before they are counted as a distress.

How to Measure:

The density of popouts can be determined by counting the number of popouts per square yard (square meter) of surface in areas having typical amounts.

Name of Distress: Pumping and Water Bleeding

Description:

Pumping is the movement of material by water pressure beneath the slab when it is deflected under a heavy moving wheel load. Sometimes the pumped material moves around beneath the slab, but often it is ejected through joints and/or cracks (particularly along the longitudinal lane/shoulder joint with an asphalt shoulder). Beneath the slab there is typically particle movement counter to the direction of traffic across a joint or crack that results in a buildup of loose materials under the approach slab near the joint or crack. Many times some fine materials (silt, clay, sand) are pumped out, leaving a thin layer of relatively loose clean sand and gravel beneath the slab, along with voids causing loss of support. Pumping occurs even in pavement sections containing stabilized subbases.

Water bleeding occurs when water seeps out of joints and/or cracks. Many times it drains out over the shoulder in low areas.

Severity Levels:

L—Water is forced out of a joint or crack when trucks pass over the joints or cracks; water is forced out of the lane/shoulder longitudinal joint when trucks pass along the joint; or water bleeding exists. No fines can be seen on the surface of the traffic lanes or shoulder.

M—A small amount of pumped material can be observed near some of the joints or cracks on the surface of the traffic lane or shoulder. Blow holes may exist.

H—A significant amount of pumped materials exist on the pavement surface of the traffic lane or shoulder along the joints or cracks.

How to Measure:

If pumping or water bleeding exists anywhere in the sample unit, it is counted as occurring at highest severity level as defined above.

Name of Distress: Reactive Aggregate Distresses

Description:

Reactive aggregates either expand in alkaline environments or develop prominent siliceous reaction rims in concrete. It may be an alkali-silica reaction or an alkali-carbonate reaction. As expansion occurs, the cement matrix is disrupted and cracks. It appears as a map-cracked area; however, the cracks may go deeper into the concrete than in normal map cracking. It may affect most of the slab or it may first appear at joints and cracks.

Severity Levels:

Only one level of severity is defined. If alkali-aggregate cracking occurs anywhere in the slab, it is counted. If the reaction has caused spalling or map cracking, these are also counted.

How to Measure:

Reactive-aggregate distress is measured in square feet or square meters.

Name of Distress: Scaling and Map Cracking or Crazing

Description:

Scaling is the deterioration of the upper $1/8$ to $1/2$ inch (3–13 mm) of the concrete slab surface. Map cracking or crazing is a series of fine cracks that extend only into the upper surface of the slab surface. Map cracking or crazing is usually caused by over-finishing of the slab and may lead to scaling of the surface. Scaling can also be caused by reinforcing steel being too close to the surface.

Severity Levels:

L—Crazing or map cracking exists; the surface is in good condition with no scaling.
M/H—Scaling exists.

How to Measure:

Scaling and map cracking or crazing are measured by area of slab in square feet or meters.

Name of Distress: Spalling (Transverse and Longitudinal Joint/Crack)

Description:

Spalling of cracks and joints is the cracking, breaking, or chipping (or fraying) of the slab edges within 2 feet (0.6 m) of the joint/crack. A spall usually does not extend vertically through the whole slab thickness but extends to intersect the joint at an angle. Spalling usually results from (1) excessive stresses at the joint or crack caused by infiltration of incompressible materials and subsequent expansion, (2) disintegration of the concrete from freeze-thaw action of "D" cracking, (3) weak concrete at the joint (caused by honeycombing), (4) poorly designed or constructed load transfer device (misalignment, corrosion), and/or (5) heavy repeated traffic loads.

Severity Levels:

L—The spall or fray does not extend more than 3 inches (8 cm) on either side of the joint or crack. No temporary patching has been placed to repair the spall.
M—The spall or fray extends more than 3 inches (8 cm) on either side of the joint or crack. Some pieces may be loose and/or missing, but the spalled area does not present a tire damage or safety hazard. Temporary patching may have been placed because of spalling.
H—The joint is severely spalled or frayed to the extent that a tire damage or safety hazard exists.

How to Measure:

Spalling is measured by counting and recording separately the number of joints with each severity level. If more than one level of severity exists along a joint, it will be recorded as containing the highest severity level present. Although the definition and severity levels are the same, spalling of cracks should not be recorded. *The spalling of cracks is included in rating severity levels of cracks.* Spalling of transverse and longitudinal joints will be recorded separately. Spalling of the slab edge adjacent to a permanent patch will be recorded as patch adjacent slab deterioration. If spalling is caused by "D" cracking, it is counted as both spalling and "D" cracking at appropriate severity levels.

Name of Distress: Spalling (Corner)

Description:

Corner spalling is the raveling or breakdown of the slab within approximately 1 foot (0.3 m) of the corner. However, corner spalls with both edges less than 3 inches (8 cm) long will not be recorded. A corner spall differs from a corner break in that the spall usually angles downward at about 45° to intersect the joint, while a break extends vertically through the slab. Corner spalling can be caused by freeze-thaw deterioration, "D" cracking, and other factors.

Severity Levels:

L—Spall is not broken into pieces and is in place and not loose.
M—One of the following conditions exists: Spall is broken into pieces; cracks are spalled; some or all pieces are loose or absent but do not present tire damage or safety hazard; or spall is patched.
H—Pieces of the spall are missing to the extent that the hole presents a tire damage or safety hazard.

How to Measure:

Corner spalling is measured by counting and recording separately the number of corners spalled at each severity level within the sample unit.

Name of Distress: Swell

Description:

A swell is an upward movement or heave of the slab surface resulting in a sometimes sharp wave. The swell is usually accompanied by slab cracking. It is usually caused by frost heave in the subgrade or by an expansive soil. Swells can often be identified by oil droppings on the surface as well as riding over the pavement in a vehicle.

Severity Levels:

L—Swell causes a distinct bounce of the vehicle which creates no discomfort.
M—Swell causes significant bounce of the vehicle which creates some discomfort.
H—Swell causes excessive bounce of the vehicle which creates substantial discomfort, and/or a safety hazard, and/or vehicle damage, requiring a reduction in speed for safety.

How to Measure:

The number of swells within the uniform section are counted and recorded by severity level. Severity levels are determined by riding in a mid- to full-sized sedan weighing approximately 3,000 to 3,800 lb. (13.3–16.9 kN) over the uniform section at the posted speed limit.

Name of Distress: Transverse and Diagonal Cracks

Description:

Linear cracks are caused by one or a combination of the following: heavy load repetition, thermal and moisture gradient stresses, and drying shrinkage stresses. Medium- or high-severity cracks are working cracks and are considered major structural distresses. They may sometimes be due to deep-seated differential settlement problems. (NOTE: Hairline cracks that are less than 6 feet (1.8 m) long are not rated.)

Severity Levels:

L—Hairline (tight) crack with no spalling or faulting, a well-sealed crack with no visible faulting or spalling.

M—Working crack with low- to medium-severity level of spalling, and/or faulting less than 1/2 inch (13 mm). Temporary patching may be present.

H—A crack with width of greater than 1 inch (25 mm); a crack with a high-severity level of spalling; or a crack faulted 1/2 inch (13 mm) or more.

How to Measure:

The number and severity level of each crack should be identified and recorded. If the crack does not have the same severity level along the entire length, the crack is rated at the highest severity level present. Cracks in patches are recorded under patch deterioration.

APPENDIX L
DOCUMENTATION OF DESIGN PROCEDURES

L1.0 INTRODUCTION

This appendix provides documentation for the development of the revised AASHTO overlay design procedures. This work was conducted under NCHRP Project 20-7/Task 39.

L1.1 PROBLEM STATEMENT

A need has been identified to modify Chapter 5 of Part III of the 1986 AASHTO *Guide for Design of Pavement Structures*. (*1*) This chapter of the Guide addresses the subject of overlay design for pavement rehabilitation.

Pavement overlay design procedures presented in the current Chapter 5 of the Guide are not being used by most State highway agencies. Because the development of the procedures was not fully documented, they are perceived to be complex and confusing. When applied to specific pavements that are candidates for overlay, the procedures yield inconsistent and questionable results. Among the concerns expressed about the current procedures are:

(1) The "remaining life" factor, which has a significant influence on overlay thickness, is extremely complex and has a questionable basis.
(2) No guidelines are given for determination of future required structural capacity (SN_f or D_f) for a specific project design.
(3) Limited or no provisions or guidelines are given for reflective crack control, pre-overlay repair, overlay type feasibility, subdrainage, widening and lane additions, reinforcement and joint design for concrete overlays, separation layers for unbonded overlays, and overlay design for composite pavements.
(4) No guidelines are given for relating backcalculated subgrade resilient modulus values to the resilient modulus value included in the AASHTO flexible pavement design equation.

The following improvements to Chapter 5 are necessary if the Guide's overlay design procedures are to be accepted by State and local agencies:

(1) Simplification of the procedures for practical use by practicing engineers,
(2) Clearer descriptions of the procedures for easier understanding and implementation,
(3) Improved adaptability of the overlay thickness design procedures to local conditions to produce more reasonable results,
(4) Addition of guidelines on such items as reflection crack control, joint design, and pre-overlay repair needs, overlay design for composite pavements, and
(5) Complete documentation of the procedures.

L1.2 OBJECTIVE

The objective of this research work is to modify Chapter 5 of Part III of the AASHTO *Guide for Design of Pavement Structures* so that the Guide's pavement overlay design procedures will yield valid and acceptable designs.

L1.3 APPROACH

The overlay design procedures presented in Chapter 5 utilize the structural deficiency approach, in which the effective structural capacity of the existing pavement is determined and then subtracted from the future required structural capacity as determined from the AASHTO flexible and rigid pavement design equations. This concept was retained to maintain compatibility between Parts II and III of the Guide and to keep the procedure relatively simple. Development of a more sophisticated mechanistic approach to overlay design was not within the scope of this limited contract. Nondestructive deflection testing for characterization of the existing pavement is recommended, to the extent appropriate within the framework of these empirical design procedures.

The primary focus of the revision effort was to modify the overlay design procedures to make them simpler and more complete. It is essential, however, that agencies desiring to use these overlay design procedures calibrate and adjust them as necessary to produce designs which are appropriate for their local conditions. As stated in Part I, Chapter 1, "The Guide by its very nature cannot possibly include all the site-specific conditions that occur in each region of the United States. It is therefore necessary for the user to adapt local experience to the use of the Guide." This statement applies even more to overlay design than to new pavement design.

Several States submitted overlay design projects to assist in verifying and improving the procedures. A total of seventy-four examples were developed to demonstrate and validate the procedures. These results were very useful in improving many aspects of the overlay design procedures. The overlay design examples are documented in Reference 37.

L2.0 OVERVIEW OF THE AASHTO DESIGN MODELS AND ASSUMPTIONS

The current AASHTO overlay design procedure has its roots in the original prediction models developed at the AASHO Road Test for new asphalt concrete (AC) and jointed Portland cement concrete (PCC) pavements. A knowledge of the original models and the subsequent modifications made to them is very important in an overlay design procedure which is based upon them.

The AASHTO Guide's approach to overlay design is a "structural deficiency" approach, in which the structural capacity of the overlay must satisfy a deficiency between the structural capacity required to support future traffic over a specified design period and the existing pavement's effective structural capacity. For AC pavement, for example, the structural overlay thickness is determined from the following equation:

$$SN_{ol} = a_{ol} * D_{ol} = SN_f - SN_{eff} \quad (2.1)$$

The overlay thickness required to satisfy the structural deficiency of the existing pavement is highly dependent upon SN_f for flexible pavement or D_f for rigid pavement, which is determined according to the procedures for new design given in **Part II of the Guide**, as if a new pavement were being constructed (of the existing and overlay material) on the existing subgrade. Any errors made in determining SN_f will produce errors in the determination of the required overlay thickness.

Tables L2.1 and L2.2 provide a summary of the original and extended performance models for AC pavement and for jointed reinforced (JRCP) and jointed plain (JPCP) concrete pavement (referred to collectively as JCP). Continuously reinforced concrete pavement (CRCP) was not tested at the AASHO Road Test. It is included by assuming an appropriate value for the J load transfer factor.

L2.1 ORIGINAL AASHO ROAD TEST PERFORMANCE MODELS (1960)

At the end of the AASHO Road Test, the performance data were used to develop empirical regression models that predicted the number of axle loads (single and tandem) of a given weight that a pavement could carry from construction (Initial Serviceability Index) to the end of its service life (Terminal Serviceability Index). Tables L2.1 and L2.2 show the AASHO Road Test variables in the original AC and PCC pavement models.

L2.1.1 Quality Control

Construction quality control was very good at the AASHO Road Test.

L2.1.2 Section Length

Pavement sections were very short (120 to 240 feet). This length would not include the normal variations in subgrade along a typical highway project.

L2.1.3 Materials

A single source of AC, PCC, crushed limestone base and gravel-sand subbase was utilized. The properties of these materials had a major effect on the performance of the pavements. For example, the subbase contained a high percentage of plastic fines, which resulted in substantial pumping for the concrete pavements, and a reduction in subbase modulus for the asphalt pavements. Some special short sections of asphalt and cement-treated base were constructed and tested.

Table L2.1. Summary of Concrete Pavement Design Factors Included in Original and Extended Performance Prediction Models in the AASHTO Guide

Model	Design Factors
1960 Original AASHO Road Test	1. Slab thickness 2. Number and magnitude single- or tandem-axle loads 3. Initial serviceability index 4. Terminal serviceability index
1961 Extension	5. Modulus of subgrade reaction 6. PCC modulus of elasticity 7. PCC Poisson's ratio 8. PCC modulus of rupture 9. Axle load equivalency factors
1972 Extension	10. J factor recommended for CRCP and unprotected corner design 11. Joint design recommendations 12. Reinforcement design procedures
1981 Extension	13. Safety factor to reduce design M_R
1986 Extension	14. Drainage adjustment factor 15. Loss of support adjustment factor 16. J factor for different load transfer systems 17. Design reliability level 18. Resilient modulus for subgrade 19. Environmental serviceability loss

Table L2.2. Summary of Asphalt Concrete Pavement Design Factors Included in Original and Extended Performance Prediction Models in the AASHTO Guide

Model	Design Factors
1960 Original Road Test	1. Thickness of AC surface, crushed stone base, and sand-gravel subbase 2. Number and magnitude of single- or tandem-axle loads 3. Initial serviceability 4. Terminal serviceability index 5. Structural coefficients for AC surface layer 6. Structural coefficients for granular, asphalt-treated and cement-treated base
1961 Extension	7. Soil support scale 8. Axle load equivalency factors 9. Regional climatic factor
1986 Extension	10. Granular layer drainage adjustment factor 11. Correlations between moduli and structural coefficients for surface, base, and subbase 12. Design reliability level 13. Resilient modulus for subgrade 14. Environmental serviceability loss

L2.1.4 Combinations of Layers

For both AC and PCC pavements, specific combinations of materials and orders of layers were used. The performance models based upon these data relate specifically to those material combinations. Other combinations or orders, such as very thick AC over subgrade or a single base layer, or PCC over a stabilized base, were not included at the AASHO Road Test.

L2.1.5 Subgrade

The subgrade was an A-6 soil with CBR ranging from 2 to 4 and k-value of about 45 psi in the spring. The resilient modulus of this soil as determined in the laboratory using unconfined repeated loading at a deviator stress of 6 psi was approximately 3,000 psi when tested at about one percent wet of optimum moisture content. (7).

L2.1.6 Climate

The climate of northern Illinois, the site of the AASHO Road Test, is characterized by about 30 inches of rain annually and depth of frost penetration of about 30 inches. The number of freeze-thaw cycles at the subbase level is about 12 per year.

L2.1.7 Dowels and Reinforcement

All concrete pavement joints had uncoated dowels and were spaced at 15 feet for JPCP and 40 feet for JRCP. Reinforcement in the JRCP was welded wire fabric. There were no undowelled joints and no skewed joints.

L2.1.8 Time and Traffic

The length of test was only two years and the number of actual load applications was 1.1 million or less in each lane. The pavements experienced mainly load-associated damage (cracking and rutting). No long-term observation of the effects of age and climate on pavement condition (e.g., cracking and ravelling on AC, "D" cracking and joint spalling on JCP) was possible given the short duration of the test.

L2.1.9 Axle Loads and Tire Pressure

Only one type and weight of axle was applied to each traffic lane. Maximum tire pressure for the heaviest loads was about 85 psi.

L2.1.10 Subdrainage

No subdrainage system existed at the AASHO Road Test for either concrete or asphalt pavements. Base and subbase layers were daylighted. Many of the AC pavements at the Road Test failed during the spring thaw period, and many of the PCC pavements failed due to extensive pumping and loss of support.

L2.2 LIMITATIONS OF ORIGINAL PERFORMANCE MODELS

The following describes some limitations of the original performance models developed from the Road Test data.

L2.2.1 Serviceability-Traffic Relationship

A fundamental assumption behind the AASHO Road Test model is that loss in PSI is only caused by traffic. While this may be reasonable for the two-year duration of the Road Test, pavement performance for actual in-service pavements is not always dependent solely upon traffic, particularly when environmental, material and construction factors are contributing significantly to loss of serviceability.

L2.2.2 Initial Serviceability

Pavement performance started from an Initial PSI averaging 4.2 for AC and 4.5 for JCP at the AASHO Road Test. Pavements constructed today can be built smoother.

L2.2.3 Terminal Serviceability

Pavement performance was assumed to end when a pavement reached a PSI of 1.5.

L2.2.4 Present Serviceability Index (PSI) Equations

The PSI of all pavements was calculated from models for AC and JCP that included the following factors:

(1) Mean of the slope variance, SV, in the two wheelpaths
(2) JCP cracking, C, as total linear feet of class 3 and class 4 cracks per 1,000 square feet of pavement area (class 3 is a crack opened or spalled 0.25 inch or more over at least half of the crack length, and class 4 is a crack which has been sealed)
(3) AC cracking, C, as linear feet per 1,000 square feet of pavement area
(4) AC rutting, RD, in inches, in the two wheelpaths
(5) Patching in square feet per 1,000 square feet of pavement area

L2.2.5 Site-Specific Conditions

The models represent the combinations of loads, pavement layer thicknesses, materials properties, and joint and reinforcement designs that were studied at the AASHO Road Test.

L2.2.6 PSI Loss for Thicker Pavements

Serviceability trends for the lowest and next-to-lowest levels of slab thicknesses were generally well defined. In certain cases, at least the beginning of a loss trend could be detected for sections whose PCC slab thickness or AC thickness was at the next-to-highest level. If the serviceability history of a section did not show any definite loss, then there was no way to determine the magnitude of the rate of loss. Such was the case for many PCC and AC pavements whose thicknesses were at the highest level in any Road Test traffic lane.

L2.2.7 Short-Term Versus Long-Term Performance

Reinforcement in concrete slabs was not found to significantly affect JRCP performance over the two-year period of the Road Test. However, the presence of reinforcement and longer joint spacings proved to be important factors after these pavements had been in service for 14 years on I-80 during the extended AASHO Road Test. Many of the transverse cracks opened up (from joint lock-up due to dowel corrosion) and deteriorated, even on the very thick slabs. This was not observed on the shorter-jointed JPCP.

L2.2.8 Scatter of Data

Because of random variations in the observed data, there were differences between the average predictions from the models and the actual performance of individual sections. Analysis of the residuals showed that for PCC pavements the scatter corresponds approximately to ± 12 percent of the slab thickness given by the performance curves. The mean absolute residual for log W (18-kip ESALs) was equal to 0.17. That is, the 90-percent confidence limits on log W are log W $-$ 0.34 and log W $+$ 0.34. A similar statistic was obtained for AC pavements.

L2.3 EXTENDED PERFORMANCE MODELS (1961)

Since the original AASHO models were limited to only the original AASHO Road Test conditions, they were not directly applicable to pavement design. They had to be "extended" to make it possible for them to handle additional design and traffic factors. The additional factors added are listed in Tables 1 and 2. The following extensions were incorporated into the 1961 Interim Guide. (2, 3)

L2.3.1 Regional Factor

A regional factor was added to account for differences in climate from that in which the AASHO Road Test was conducted. This regional factor was only developed for AC pavements, and was not retained in the 1986 version.

L2.3.2. Soil Support Value

A soil support scale was added to the AC performance model to make it possible to design for different soils than the AASHO Road Test soil. This scale was replaced by resilient modulus in the 1986 version.

L2.3.3 Structural Number

The structural number (SN) concept was formalized (formerly called the thickness index in the original models) along with the structural layer coefficients:

$$SN = a_1 D_1 + a_2 D_2 + a_3 D_3$$

where

a_1, a_2, a_3 = structural coefficients of AC surface, base and subbase
D_1, D_2, D_3 = thickness of surface, base and subbase, inches

The SN is actually a weighted thickness of flexible pavement. The structural coefficients are regression coefficients for the particular data analyzed. Although the mean values were used for design, they varied across the traffic loops. For example, the a_1 coefficient was 0.30 for Loop 6 and 0.80 for Loop 2. They have also been shown to vary with layer thickness and position within the pavement structure. (4) They are not constant for any given material, and since they relate to overall loss of serviceability, they probably represent many different material properties and other environmental and construction conditions for a given pavement layer (e.g., for AC layers, permanent deformation and fatigue properties).

Mean values for the crushed stone base and gravel-sand subbase were recommended; however, these varied widely over the different loops. Tentative mean values for asphalt-treated and cement-treated bases were obtained from the few special wedge thickness sections constructed with these materials.

Selection of the appropriate values for a deteriorated pavement layers is one of the challenges of overlay design.

L2.3.4 PCC Material Properties

Using the Spangler corner stress equation, the Road Test model was modified and extended to include material properties for PCC pavements. An assumption was made that any change in the tensile stress/strength ratio resulting from changes in physical constants E, k, D, and S'_c would have the same effect on W as varying slab thickness would have on W according to the original model.

L2.3.5 PCC Load Transfer Factor

The joint load transfer term (J factor) was assumed to be 3.2 for the AASHO Road Test dowelled pavements. Selection of different values of J and application of this equation to unprotected (undowelled) corners, and cracks with deformed reinforcement (CRCP) has been recommended in the 1986 Guide.

L2.3.6 Load Equivalency Factors

The concept of equivalency factors was developed and incorporated into the 1962 Interim Guide. It is assumed that mixed traffic load applications, including various tandem-axle configurations on the road today, can be combined, by equivalency factors, to give equivalent 18-kip single-axle load applications which can be used in the performance prediction models.

L2.4 REVISED PERFORMANCE MODELS (1971, 1981)

A few additions were made during these years, as shown in Tables 1 and 2. In 1971, the Guide was revised and some additional recommendations on obtaining inputs were provided. Further guidance on joints and reinforcement were given. (5) In 1981, a safety factor for reducing the concrete modulus of rupture was added to increase the safety factor in designs. (6)

L2.5 REVISED AND EXTENDED PERFORMANCE MODELS (1986)

Some major revisions were made in 1986 to both the JCP and AC predictive models. (1) Several major factors were added to the performance models, as shown in Tables 1 and 2. The 1986 models are those used in the current Chapter 5 AASHTO overlay design procedure.

L2.5.1 Resilient Modulus and k-Value

One major change was the incorporation of the resilient modulus as the subgrade soil property for design. This replaced the soil support scale for AC

Appendix L

pavements. The method used to develop the resilient modulus scale is not rigorous and is subject to question.

Examination of the revised AASHTO model indicates that the resilient modulus of the Road Test roadbed soil was assumed to be 3,000 psi. Research by Elliott and Thompson (7) shows that the resilient modulus varied widely above and below 3,000 psi throughout the year at the Road Test. Values of resilient modulus backcalculated from deflection data must be adjusted to fit within the context of the 3,000 psi used in the flexible pavement prediction model.

L2.5.2 Reliability Level

Another major addition was the incorporation of the reliability concept into the design of both AC and JCP. The design reliability and overall standard deviation were included in an attempt to account for all errors and uncertainty associated with various assumptions, random variation of performance prediction, uncertainty in input values (such as traffic loadings and material properties), and the desire to be able to design for a higher confidence level. The reliability concept is only applied to the number of ESALs a pavement can carry. For example, a pavement designed with a 50-percent reliability level would have a 50-percent chance of not failing before the design traffic was applied. Designing for two times the design traffic increases the reliability level to roughly 90 percent, which results in a thicker pavement cross-section. This does not, however, result in a twofold improvement in all other aspects of the pavement's performance. Climatic factors, for example, may cause a substantial loss in PSI.

L2.5.3 Drainage Factor

Drainage coefficients were added to the JCP and AC design models. A drainage coefficient value of 1.0 is meant to reflect Road Test drainage conditions. As mentioned, the subdrainage conditions at the Road Test were generally poor. Values greater than 1.0 mean better drainage of the pavement, while values less than 1.0 mean worse drainage than existed at the AASHO Road Test.

L2.5.4 Load Transfer Factor

The load transfer J factor is recommended to vary for different joint designs. This must be carefully assessed when designing overlays for jointed concrete pavements.

L2.5.5 Loss of Support Considerations

Loss of support procedures are provided that can result in very low k-values. For overlay design, the use of a large loss of support value may result in very thick overlays.

L2.5.6 Guidelines for Layer Coefficients

Further guidelines are given for obtaining structural coefficients of AC layers. These are closely tied to the resilient moduli of those materials.

L3.0 DETERMINATION OF SN_f FOR FLEXIBLE PAVEMENTS

The total flexible pavement structural number required to carry future traffic (SN_f) is determined using the procedure for new pavements presented in Part II, Chapter 3 of the 1986 AASHTO Guide. (1) The inputs required for SN_f are:

(1) Estimated future traffic ESALs, W_{18}
(2) Design reliability, R
(3) Overall standard deviation, S_o
(4) Effective roadbed resilient modulus, M_R
(5) Design serviceability loss, ΔPSI

For overlay design, special consideration is required for selection of R, S_o, and M_R.

L3.1 DESIGN RELIABILITY AND OVERALL STANDARD DEVIATION

An overlay may be designed for different levels of reliability using the procedures described in Part I, Chapter 4 for new pavements. This is accomplished through determination of the structural capacity (SN_f or D_f) required to carry traffic over the design period at the desired level of reliability.

Reliability level has a large effect on overlay thickness as is shown in the many examples of overlay design. (37) Based on field testing projects, it appears that a design reliability level of approximately 95 percent gives overlay thicknesses consistent with those recommended for most projects by State highway agencies, when the overall standard deviations recommended in Parts I and II are used. (37) There were many projects where 95 percent did not give the recommended overlay thickness, however. There are many situations for which it is desirable to design at a higher or lower level of reliability, depending on the consequences of failure of the overlay. The level of reliability to be used for different types of overlays may vary, and should be evaluated by each agency for different highway functional classifications (or traffic volumes).

The designer should be aware that some sources of uncertainty are different for overlay design than for new pavement design. Therefore, the overall standard deviations recommended for new pavement design may not be appropriate for overlay design. In fact, the appropriate value for overall standard deviation may vary by overlay type as well. An additional source of variation is the uncertainty associated with establishing the effective structural capacity (SN_{eff} or D_{eff}) of the existing pavement structure. However, some sources of variation may be less significant for overlay design than for new pavement design (e.g., estimation of future traffic).

The sources of variation differ depending on the specific methodology used for overlay design. There are two general approaches to overlay design: a uniform section approach and a point-by-point approach. The uniform section approach is used when the entire project (or each uniform section of the project) is treated as having a single SN_{eff}, a single M_R, and a single required SN_f, and a single overlay thickness is determined. Generally, the overall approach is used with the visual inspection/testing and the remaining life methods of determining SN_{eff}. However, the uniform section approach can also be used with NDT data when SN_{eff} and M_R are determined using a "representative" deflection basin, or when a mean SN_{eff} and mean M_R are determined for the deflection basins measured. The point-by-point approach is used when overlay thicknesses are determined for each location at which deflections are measured.

S_o will differ for these two approaches to design. The prediction errors associated with traffic estimation and pavement performance will be the same for both approaches; and both approaches will include some variation due to the uncertainty associated with the determination of SN_{eff}. However, for the point-by-point approach, the error associated with variation of M_R and SN_{eff} will be significantly less. As a result, S_o

should be less in the point-by-point method. No attempt has been made in the development of this procedure to identify appropriate values of S_o for these two approaches.

L3.2 DETERMINATION OF DESIGN M_R

The design subgrade M_R may be determined by: (1) laboratory testing, (2) NDT backcalculation, (3) estimation from resilient modulus correlation studies, or (4) original design and construction data. Regardless of the method used, the design M_R value must be consistent with the value used in the design performance equation for the AASHO Road Test subgrade. This is especially important when M_R is determined by NDT backcalculation. The backcalculated value is typically too high to be consistent and must be adjusted. If M_R is not adjusted, the SN_f value will not be conservative and poor overlay performance can be expected.

A subgrade M_R may be backcalculated from NDT data using the following equation:

$$M_R = \frac{0.24P}{d_r r} \quad (3.1)$$

where

M_R = backcalculated subgrade resilient modulus, psi
P = applied load, pounds
d_r = measured deflection at radial distance r, inches
r = radial distance at which deflection is measured, inches

The derivation of this equation is presented in Chapter 5. This equation for backcalculating M_R is based on the fact that, at points sufficiently distant from the center of loading, the measured surface deflection is almost entirely due to deformation in the subgrade, and is also independent of the load radius. For practical purposes, the deflection used should be as close as possible to the loading plate. As the distance increases, the magnitude of the deflection decreases and the effects of measurement error are magnified. However, the deflection used must also be sufficiently far from the loading plate to satisfy the assumptions inherent in the equation. Analyses presented in Chapter 5 show that r should be at least 0.7 times a_e, the effective radius of the stress bulb at the pavement/subgrade interface. The equation for a_e is presented in Chapter 5.

The recommended method for determination of the design M_R from NDT backcalculation requires an adjustment factor (C) added to make the value consistent with the value used to represent the AASHO subgrade. A value for C of no more than 0.33 is recommended for adjustment of backcalculated M_R values to design M_R values.

The resulting equation is:

$$\text{Design } M_R = C \left(\frac{0.24P}{d_r r} \right) \quad (3.2)$$

A subgrade M_R value of 3,000 psi was used for the AASHO Road Test soil in the development of the flexible pavement performance model. The AASHTO Guide and appendices (1, 8) do not indicate how or why this value was selected, but it is consistent with some of the test data on the AASHO soil reported by Thompson and Robnett (9). Their data are shown in Figure L3.1. From this it may be concluded that 3,000 psi is appropriate for the AASHO soil when it is about 1 percent wet of optimum and subjected to a deviator stress of about 6 psi or more. However, the resilient modulus of the soil is shown to be quite stress-dependent, increasing rapidly for deviator stresses less than 6 psi. The subgrade deviator stress at the radial distance used with Equation 3.1 will almost always be far less than 6 psi. Thus, the subgrade modulus determined by backcalculation can be expected to be too high to be consistent with the 3,000 psi used for the AASHO subgrade.

This was confirmed by two methods. The first analysis involved deflection data and resilient modulus tests on Shelby tube subgrade samples from the AASHO Road Test site. The second analysis used the ILLI-PAVE finite element program. (10, 11)

The deflection data and tests on the Shelby tube samples were reported by Traylor. (12) The deflections were measured several years after the Road Test, on the Loop 1 pavements which were not trafficked and are still in place. The NDT device was the FHWA Thumper used in the impulse load mode. The magnitude of loading was about 4,000 pounds. The Shelby tube samples were taken shortly after the deflection measurements. The resilient moduli of the samples were measured in the laboratory using a deviator stress of 6 psi (essentially consistent with the AASHO subgrade value).

For the first analysis, subgrade resilient modulus values were calculated for each deflection site using the deflection measured closest to the loading plate at which r was equal to or greater than $0.7a_e$. Figure L3.2 is a plot of the calculated M_R values versus the laboratory values from the Shelby tube samples. The calculated values are greater than the laboratory values by a factor of 4.8 on the average.

For the second analysis, ILLI-PAVE was used to examine the effects of the AASHO subgrade stress-dependency illustrated in Figure L3.1. ILLI-PAVE models the stress-dependency of cohesive soils as two intersecting lines, as shown in Figure L3.3. Through an iterative process, ILLI-PAVE selects modulus values for each subgrade element that match the predicted stress conditions for the element. The model inputs are the slopes of the two lines (K_1 and K_2), the point of intersection (E_{Ri} and σ_{Di}) and a lower limit deviator stress that sets a maximum limit on the resilient modulus. Using the data from Figure L3.1, the following values were selected to model the AASHO subgrade: $E_{Ri} = 3,000$ psi, $\sigma_{Di} = 6$ psi, $K_1 = 1.4$ ksi/psi, $K_2 = 0.01$ ksi/psi, and lower limit deviator stress = 2 psi (maximum possible $M_R = 8,600$ psi).

The pavements analyzed by ILLI-PAVE had 3-inch and 5-inch AC surfaces ($E_{ac} = 500$ ksi) and aggregate bases ranging in thickness from 8 inches to 22 inches (base $M_R = 9,000 \, \theta^{0.33}$). The loading was 9,000 pounds on a 5.9-inch circular area (equivalent to a typical FWD deflection test). Examination of the ILLI-PAVE output showed that, at radial distances that would be used for backcalculation, the final resilient modulus of most subgrade elements was 8,600 psi, the maximum possible. The lowest modulus was 7,130 psi at a radial distance of 16 inches under the 3-inch AC/8-inch base pavement.

Surface deflections computed by ILLI-PAVE for the pavements described above were also used to predict subgrade modulus values using Equation 3.1. Deflections at radial distances ranging from 12 inches to 57 inches were used. The calculated subgrade modulus values ranged from 9,280 to 11,800 psi. There was no pattern of the modulus increasing or decreasing with the radial distance. Using only the modulus for the lowest r/a_e ratio greater than 0.7, the mean calculated modulus was 9,806 psi.

All of these results suggest that if appropriate deflection data were available from the AASHO Road Test the backcalculated subgrade resilient modulus would be greater than 3,000 psi by a factor of at least 3. Therefore, the value used for C in backcalculating M_R for design should be no greater than 0.33 for cohesive soils.

Further comparative data were obtained during the field testing phase of this study. (37) Data were ob-

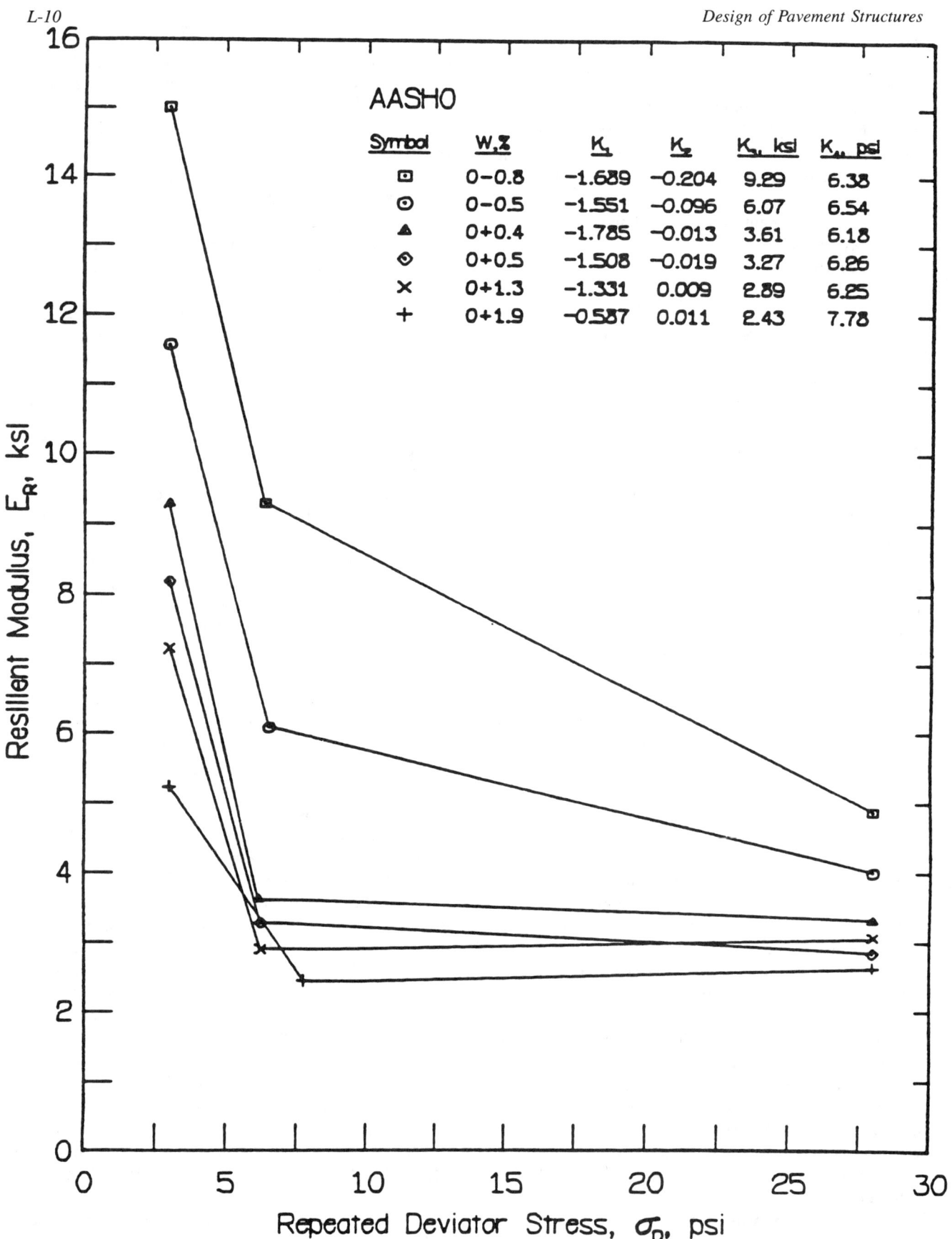

Figure L3.1. AASHO Road Test Subgrade Resilient Modulus Reported by Thompson and Robnett (9)

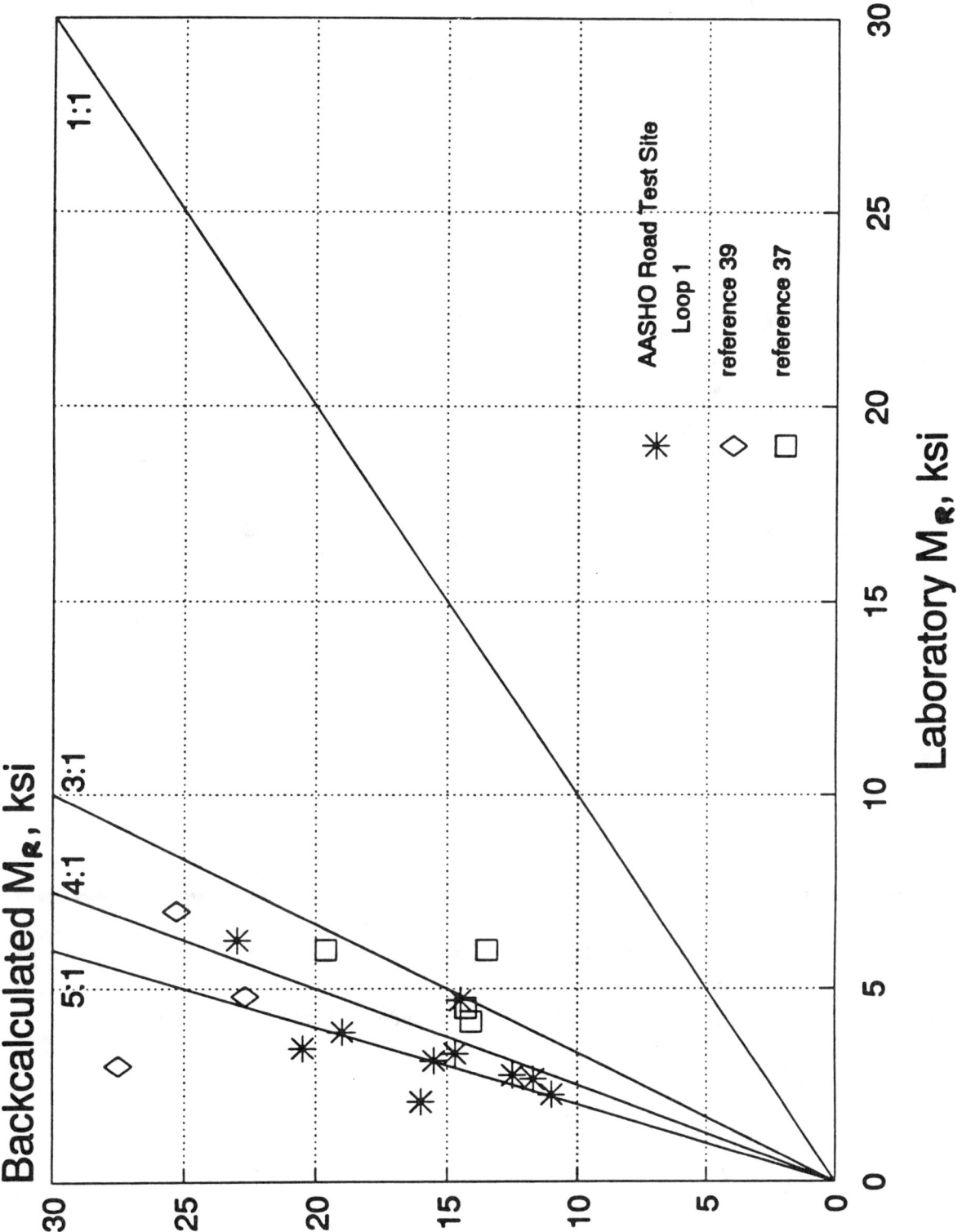

Figure L3.2. Backcalculated Resilient Modulus Versus Laboratory Results on Shelby Tube Samples from the AASHO Road Test Site Plus Data from Two Additional States

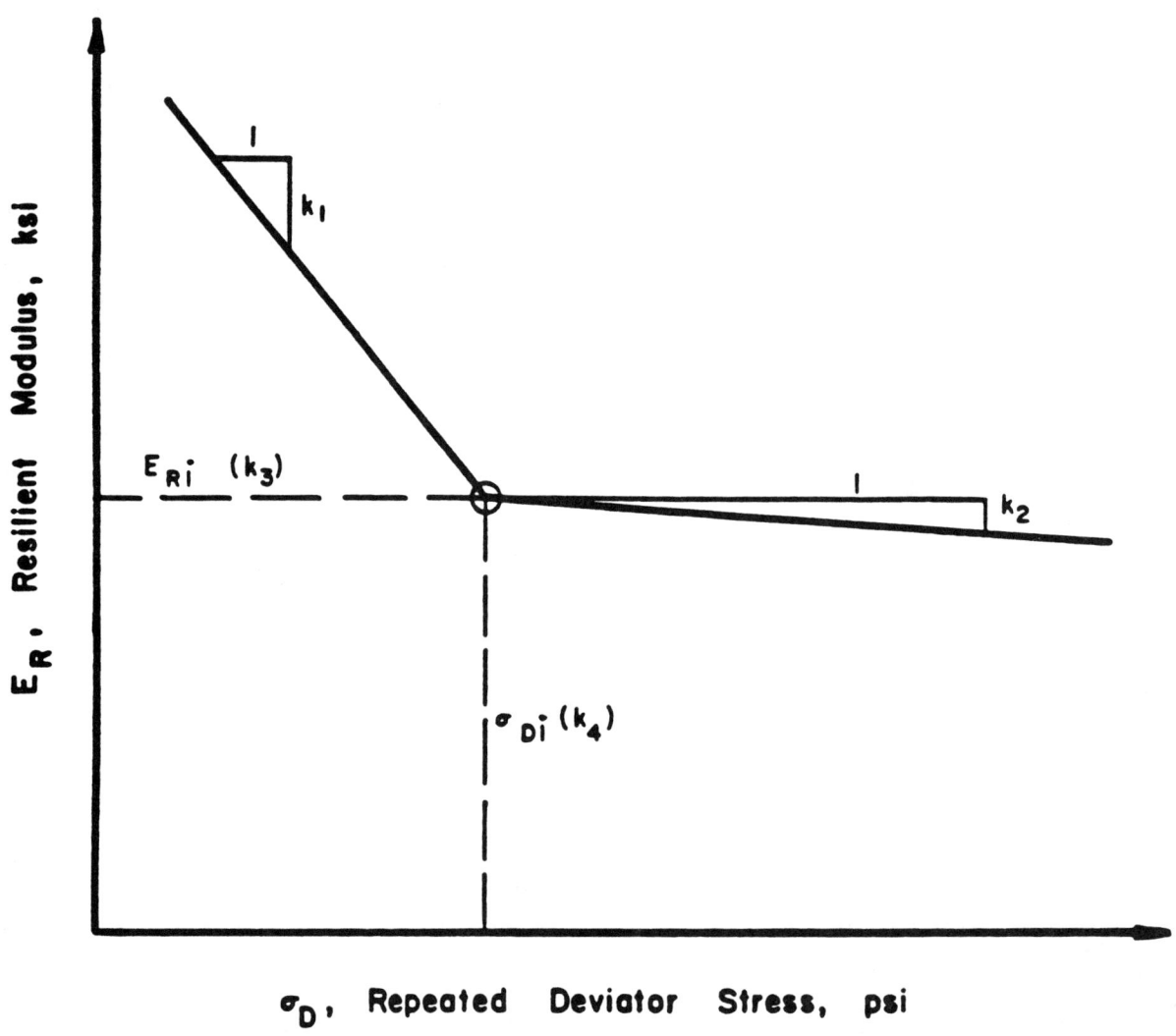

Figure L3.3. Subgrade Resilient Modulus Stress-Dependency Model Used in ILLI-PAVE

tained from several projects located in two States where the FWD was used at a load of 9,000 pounds and the subgrade resilient modulus was then backcalculated using equation 3.1. Subgrade samples were then taken from the pavement and tested in the laboratory for resilient modulus at a deviator stress of about 6 psi. The additional data are given below.

Soil Type	Lab M_R (psi)	Backcalculated M_R, Eqn. 3.1 (psi)	Ratio (Backcalculated/Lab)
A-2, A-4, A-6	7,000	25,000	3.6
A-2, A-6	4,800	22,700	4.7
A-4	3,000	27,500	9.2
A-4, A-2-5	6,000	13,500	2.3
A-7-6	6,000	19,600	3.3
A-2-4	4,150	14,100	3.4
A-4	4,500	14,300	3.2

The mean ratio of field to lab moduli from these tests is 4.2, or C = 0.24. These results are also plotted on Figure L3.2.

The analyses described here pertain to the fine-grained, stress-sensitive soil at the AASHO Road Test site plus fine-grained soil from seven other projects. No attempt has been made in this study to investigate the relationship between backcalculated and laboratory M_R values for granular subgrades. It may be that backcalculated M_R values for granular subgrades would not require a correction factor as large as is required for cohesive subgrades. However, this subject requires further research.

Users are cautioned that the resilient modulus value selected has a very significant effect on the resulting structural number determined. Therefore, users should be very cautious about using high resilient modulus values, or their overlay thickness values will be too thin.

L4.0 DETERMINATION OF D_f FOR RIGID AND COMPOSITE PAVEMENTS

The slab thickness required to carry future traffic (D_f) was determined using the procedure for new pavement design presented in Part II, Chapter 3 of the Guide. The inputs required for D_f are:

(1) Estimated future traffic ESALs, W_{18}
(2) Effective static k-value
(3) Design PSI loss, ΔPSI
(4) Load transfer factor, J
(5) PCC modulus of rupture, S'_c
(6) PCC elastic modulus, E_{pcc}
(7) Loss of support factor
(8) Design reliability, R
(9) Overall standard deviation, S_o
(10) Drainage factor, C_d

For overlay design, special consideration is required for the selection of most of these inputs.

L4.1 DESIGN RELIABILITY AND OVERALL STANDARD DEVIATION

The discussion given in Section 3.1 on design reliability and overall standard deviation for flexible pavement overlay design also applies to overlay design for rigid and composite pavements.

L4.2 DYNAMIC k-VALUE AND PCC ELASTIC MODULUS FOR BARE PCC PAVEMENTS

The new rigid pavement design equation given in Part II of the Guide requires characterization of the subgrade and base layers supporting the PCC slab by an effective k-value. The overlay design procedures presented in the 1986 Guide proposed that this effective k-value be determined using backcalculated values for the elastic moduli of the base and subgrade. This approach is not recommended in the revised overlay design procedures. Rather, direct backcalculation of the effective k-value from NDT data is recommended, and a simple procedure for doing so is provided.

A simple two-parameter approach to backcalculation of surface and foundation moduli for a two-layer pavement system was proposed by Hoffman and Thompson in 1981 for flexible pavements. (13) They proposed the AREA, given by the following equation, to characterize the deflection basin:

$$AREA = 6 * \left[1 + 2\left(\frac{d_{12}}{d_0}\right) + 2\left(\frac{d_{24}}{d_0}\right) + \left(\frac{d_{36}}{d_0}\right)\right]$$

(4.1)

where

d_0 = maximum deflection at the center of the load plate, inches

d_i = deflections at 12, 24, and 36 inches from plate center, inches

AREA has units of length, rather than area, since each of the deflections is normalized with respect to d_0 in order to remove the effect of different load levels and to restrict the range of values obtained. AREA and d_0 are thus independent parameters, from which the surface and foundation elastic moduli may be determined. Hoffman and Thompson developed a nomograph for backcalculation of flexible pavement surface and subgrade moduli from d_0 and AREA.

The AREA concept was subsequently applied to backcalculation of PCC slab elastic modulus values and subgrade k-values. (14, 15) Further investigation of this concept by Barenberg and Petros (45) and by Ioannides (16) has produced a forward solution procedure to replace the iterative and graphical procedures used previously. This solution is based on the fact that, for a given load radius and sensor arrangement, a unique relationship exists between AREA and the dense liquid radius of relative stiffness (ℓ_k) of the pavement system, in which the subgrade is characterized by a k-value (18):

$$\ell_k = \sqrt[4]{\frac{E_{pcc} D_{pcc}^3}{12(1 - \mu_{pcc}^2)k}} \quad (4.2)$$

where

ℓ_k = dense liquid radius of relative stiffness, inches

E_{pcc} = PCC elastic modulus, psi

D_{pcc} = PCC thickness, inches

μ_{pcc} = PCC Poisson's ratio

k = effective k-value, psi/inch

Figure L4.1 illustrates the relationship between AREA and ℓ_k for a = 5.9 in, the radius of an FWD load plate. The following equation for ℓ_k as a function of AREA was developed by Hall (20):

$$\ell_k = \left[\frac{\ln\left(\frac{36 - \text{AREA}}{1812.279133}\right)}{-2.559340} \right]^{4.387009} \quad (4.3)$$

With AREA calculated from measured deflections using Equation 4.1, ℓ_k may be obtained from Equation 4.3 or Figure L4.1. The effective k-value may then be obtained from Westergaard's deflection equation (18):

$$k = \left(\frac{P}{8d_0 \ell_k^2}\right) \left\{ 1 + \left(\frac{1}{2\pi}\right) \right. \\ \left. * \left[\ln\left(\frac{a}{2\ell_k}\right) + \gamma - 1.25 \right] \left(\frac{a}{\ell_k}\right)^2 \right\} \quad (4.4)$$

where

d_0 = maximum deflection, inches

P = load, pounds

γ = Euler's constant, 0.57721566490

Figure L4.2 was developed from Equation 4.4, for a load P = 9,000 pounds and a load radius a = 5.9 inches. For loads within about 2,000 lbs of this value, the deflections d_0, d_{12}, d_{24}, and d_{36} may be scaled linearly to 9,000-lb deflections.

With the effective k-value known, the slab ED^3 may then be computed from the definition of ℓ_k (Equation 4.2), and for a known or assumed slab thickness D, the concrete elastic modulus E may be determined. Figure L4.3 was developed for determination of the slab E, assuming a Poisson's ratio μ = 0.15 for the PCC and load radius a = 5.9 inches.

L4.3 STATIC k-VALUE

The k-value backcalculated from NDT data is a dynamic k-value, whereas the required input to the new pavement design equation in Part II of the Guide is a static k-value. In an analysis of AASHO Road Test data, dynamic repeated-load k-values were found to exceed static values by a factor of 1.77 on the average. (19) Foxworthy's analysis of data collected on seven Air Force base pavements indicated that dynamic k-values exceeded static values by a factor of 2.3 on the average. (15) Reducing backcalculated k-values by 2 has been found to produce very reasonable values for static k-values. It is recommended in the overlay design procedures that backcalculated k-values be divided by 2 to obtain static k-values for use in determining D_f with Part II's new pavement design equation.

L4.4 EFFECTIVE k-VALUE AND PCC ELASTIC MODULUS FOR AC/PCC PAVEMENTS

In order to apply the backcalculation procedure described in the preceding section to an existing AC/

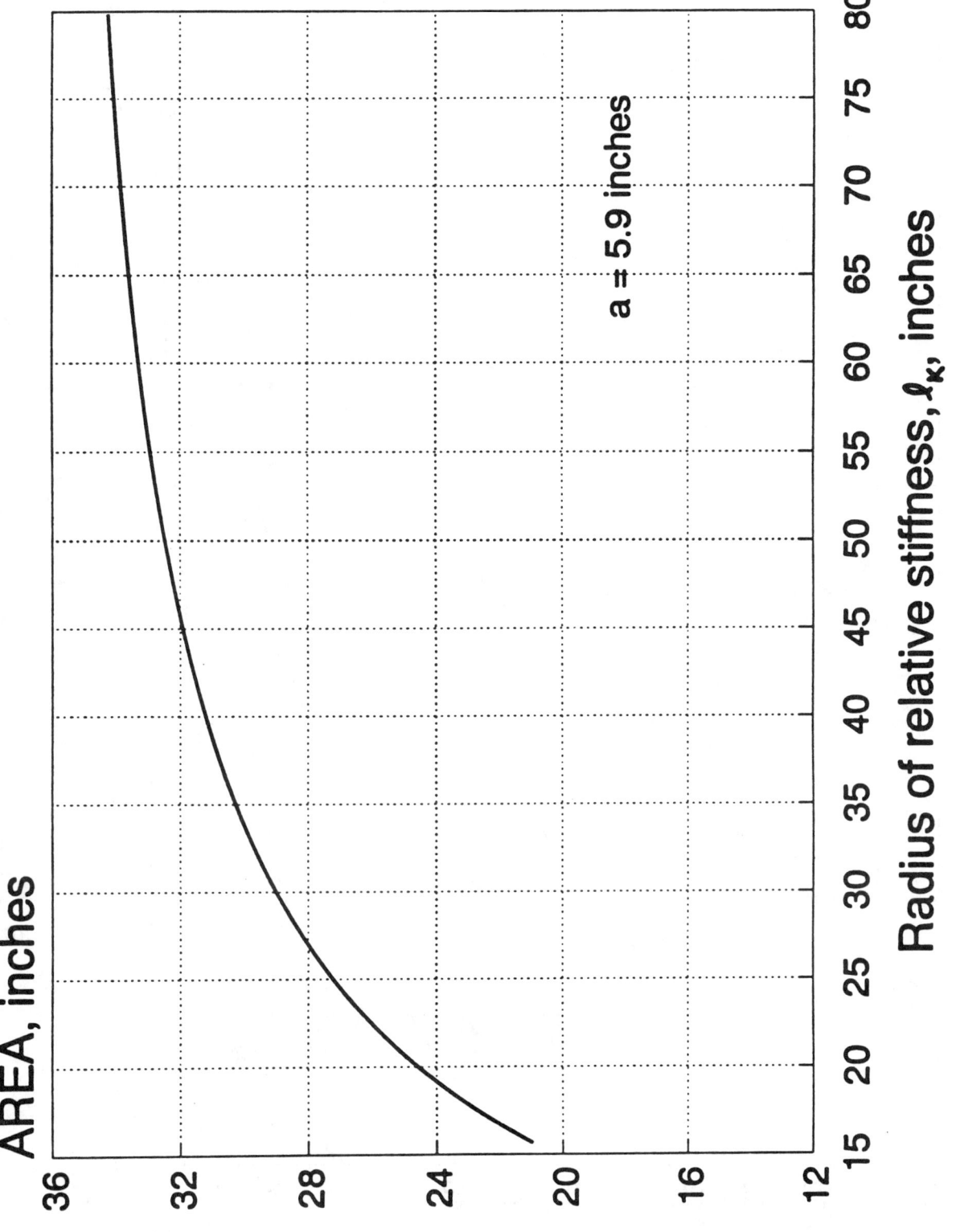

Figure L4.1. Relationship of AREA to ℓ_k (20)

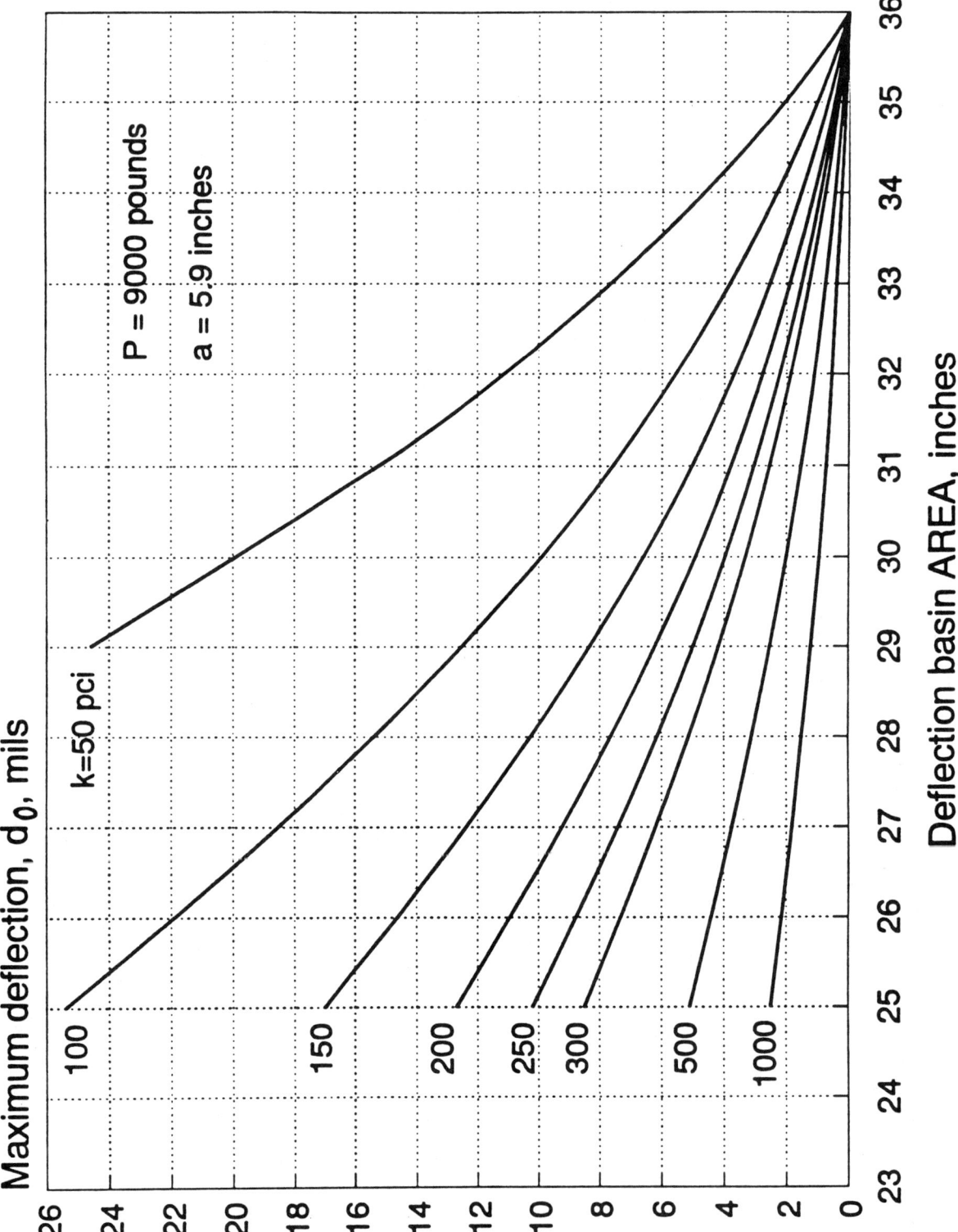

Figure L4.2. Effective Dynamic k-Value Determination from d_0 and AREA

Appendix L

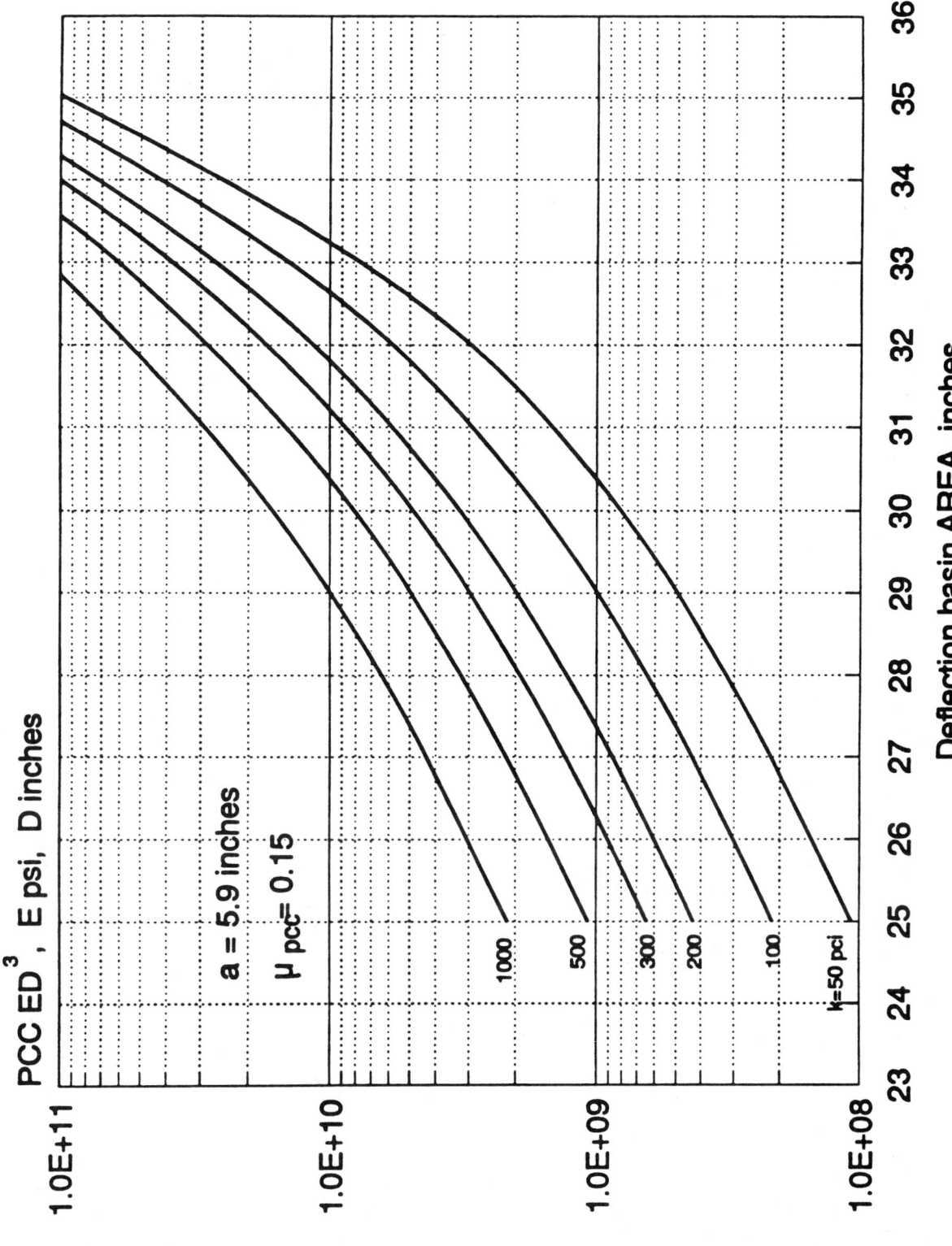

Figure L4.3. PCC Elastic Modulus Determination from k-Value, AREA, and Slab Thickness

PCC, deflections measured on the existing AC surface must be adjusted to account for the influence of the AC layer. The procedure for doing so, which is taken from Reference 20, is described in this section.

L4.4.1 AC Elastic Modulus

An existing AC/PCC pavement cannot properly be modelled as a slab on grade, since the AC overlay exhibits not only bending but also compression. To determine the amount of compression that occurs in the AC overlay, the elastic modulus of the AC layer must be determined. The recommended method for determining E_{ac} is to monitor the temperature of the AC mix during deflection testing, and to use a relationship between E_{ac} and temperature to assign a modulus value to each deflection basin.

The AC mix temperature may be measured directly during deflection testing by drilling a hole to the mid-depth of the overlay, inserting a liquid and a temperature probe into the hole, and reading the AC mix temperature when it has stabilized. This should be done at least three times during each day's testing, so that a curve of AC mix temperature versus time may be developed and used to assign a mix temperature to each basin.

If measured AC mix temperatures are not available, they may be approximately estimated from pavement surface and air temperatures using procedures developed by Southgate (21), Shell (22), the Asphalt Institute (23), or Hoffman and Thompson (13). Pavement surface temperature may be monitored during deflection testing using a hand-held infrared sensing device which is aimed at the pavement. The mean air temperature for the five days prior to deflection testing, which is an input to some of the referenced methods for estimating mix temperature, may be obtained from a local weather station or other local sources.

Two methods for determining the AC elastic modulus as a function of mix temperature are provided in the overlay design procedure. The first method uses the Asphalt Institute's equation for AC modulus as a function of mix parameters, mix temperature, and loading frequency, given by Equation 4.6. This equation, developed by Witczak for use in the Asphalt Institute's Design Manual (MS-1) (23), is a refinement of work originally done for the Asphalt Institute by Kallas and Shook. (43) It is considered highly reliable for dense-graded AC mixes with gravel or crushed stone aggregates. (44)

$$\log E_{ac} = 5.553833 + 0.028829 \left(\frac{P_{200}}{F^{0.17033}}\right)$$
$$- 0.03476 V_v + 0.070377 \eta_{70°F,10^6}$$
$$+ 0.000005 t_p^{(1.3 + 0.49825 \log F)} P_{ac}^{0.5} \quad (4.5)$$
$$- \frac{0.00189}{F^{1.1}} t_p^{(1.3 + 0.49825 \log F)} P_{ac}^{0.5}$$
$$+ 0.931757 \left(\frac{1}{F^{0.02774}}\right)$$

where

E_{ac} = elastic modulus of AC, psi
P_{200} = percent aggregate passing the No. 200 sieve
F = loading frequency, Hz
V_v = air voids, percent
$\eta_{70°F,10^6}$ = absolute viscosity at 70°F, 10^6 poise (e.g., 1 for AC-10, 2 for AC-20)
P_{ac} = asphalt content, percent by weight of mix
t_p = AC mix temperature, °F

This can be reduced to a relationship between AC modulus and AC mix temperature for a particular loading frequency (i.e., approximately 18 Hz for the FWD load duration of 25 to 30 milliseconds) by assuming typical values for the AC mix parameters P_{ac}, V_v, P_{200}, and η. For example, the AC mix design used by one state has the following typical values:

P_{200} = 4 percent
V_v = 5 percent
$\eta_{70°F,10^6}$ = 2 for AC-20
P_{ac} = 5 percent

For these values and an FWD loading frequency of 18 Hz, the following equation for AC elastic modulus versus AC mix temperature is obtained:

$$\log E_{ac} = 6.451235 - 0.000164671 t_p^{1.92544} \quad (4.6)$$

Each agency should establish its own relationship for AC modulus versus temperature which is representative of the properties of its AC mixes.

It should be noted that the Asphalt Institute's equation for AC modulus applies to new mixes. AC which has been in service for some years may have either a higher modulus (due to hardening of the asphalt) or lower modulus (due to deterioration of the AC, from stripping or other causes) at any given temperature.

The second method for establishing a relationship between E_{ac} and mix temperature involves repeated-load indirect tension testing (ASTM D 4123) of AC cores taken from the in-service AC/PCC pavement. Testing at two or more temperatures (e.g., 40, 70, and 90°F) is recommended to establish points for a curve of log E_{ac} versus temperature. AC modulus values at any temperature may be interpolated from the laboratory values obtained at any two temperatures. For example, E_{ac} values at 70° and 90°F may be used in the following equation to interpolate E_{ac} at any temperature t°F:

$$\log E_{ac\,t°F} = \left(\frac{\log E_{ac\,70°F} - \log E_{ac\,90°F}}{70 - 90}\right) * (t°F - 70°F) + \log E_{ac\,70°F} \quad (4.7)$$

For purposes of interpreting NDT data, AC modulus values obtained from laboratory testing of cores must be adjusted to account for the difference between the loading frequency of the test apparatus (typically 1 to 2 Hz) and the loading frequency of the deflection testing device (18 Hz for the FWD). This adjustment is made by multiplying the laboratory-determined E_{ac} by a constant value which may be determined for each laboratory testing temperature using the Asphalt Institute's equation. Field-frequency E_{ac} values will typically be 2 to 2.5 times higher than lab-frequency values.

L4.4.2 Correction to d_0

An elastic layer program (BISAR) was used to model AC/PCC pavement structures over a broad range of parameters:

AC thickness:	3, 5, and 7 inches
AC modulus:	250, 500, 750, 1,000, and 1,250 ksi
PCC thickness:	6, 9, and 12 in.
PCC modulus:	3, 5, and 7 million psi
Subgrade modulus:	6, 24, and 42 ksi
AC/PCC interface:	bonded and unbonded

A load magnitude of 9,000 pounds and a load radius of 5.9 inches were used. Poisson's ratio values used for the AC, PCC, and subgrade were 0.35, 0.15, and 0.50 respectively. The PCC/subgrade interface was modelled as unbonded.

Deflections were computed at the surface of the AC and the surface of the PCC at radial offsets of 0, 12, 24, and 36 inches. Compression in the AC layer, as indicated by the change in d_0 between the AC and PCC surfaces, often accounted for a significant portion of the total deflection, depending primarily on the thickness and modulus of the AC, and to a lesser extent on the AC/PCC interface condition. For example, in systems with a thick AC layer (7 inches) and a low AC modulus (250 ksi), more than 50 percent of the total deflection in the pavement occurred in the AC layer.

The change in d_0 is significantly greater when the AC is not bonded to the PCC than when it is bonded. For each interface bonding condition, it was found that the change in d_0 could be predicted very reliably as a function of the ratio of the AC thickness to AC modulus (D_{ac}/E_{ac}). These relationships were found to be very insensitive to the ranges of other parameters investigated. The following equations were obtained for these relationships:

AC/PCC BONDED:

$$d_{0\,compress} = -0.0000328 + 121.5006 \left(\frac{D_{ac}}{E_{ac}}\right)^{1.0798}$$

(4.8)

AC/PCC UNBONDED:

$$d_{0\,compress} = -0.00002132 + 38.6872 \left(\frac{D_{ac}}{E_{ac}}\right)^{0.94551}$$

where

d_0 compress	= AC compression at center of load, inches
D_{ac}	= AC thickness, inches
E_{ac}	= AC elastic modulus, psi

Using Equation 4.8, the d_0 of the PCC slab in the AC/PCC pavement may be determined by subtracting the compression which occurs in the AC surface from the d_0 measured at the AC surface.

The interface condition is a significant unknown in backcalculation. The AC/PCC interface is fully bonded when the AC layer is first placed, but how well that bond is retained is not known. Examination of cores taken at a later time may show that bond has been reduced or completely lost. This is particularly likely if stripping occurs at the AC/PCC interface. If the current interface bonding condition is not determined by coring, the bonding condition which is considered more representative of the project may be assumed.

L4.4.3 Computed AREA of PCC

In the elastic layer analyses conducted, only d_0 was found to change significantly between the AC and PCC layers. Differences in d_{12}, d_{24}, and d_{36} were very close to zero over the entire range of parameters. Therefore, the AREA of the PCC slab may be computed from Equation 4.1 using the d_0 of the PCC slab determined as described above, and d_{12}, d_{24}, and d_{36} measured at the AC surface.

$$\text{AREA}_{pcc} = 6 * \left[1 + 2\left(\frac{d_{12}}{d_{0\,pcc}}\right) + 2\left(\frac{d_{24}}{d_{0\,pcc}}\right) + \left(\frac{d_{36}}{d_{0\,pcc}}\right) \right] \quad (4.9)$$

The PCC d_0 and AREA, determined as described above, may then be used to determine the PCC elastic modulus and effective dynamic k-value. The effective k-value obtained is a dynamic k-value, and should thus be divided by 2 to obtain an appropriate static k-value for use in determining D_f from the rigid pavement design equation in Part II of the Guide.

L4.5 LOAD TRANSFER FACTOR

This factor relates to the ability of a joint to transfer shear load. Table L2.6 in Section II of the Guide provides recommended values for J for new pavement design that depend on the use of dowels, AC or tied PCC shoulders and pavement type (jointed versus CRCP). For a concrete pavement being overlaid, the J factor selected must reflect the ability of the existing joints (or cracks for CRCP) to transfer load. This ability may be measured through use of NDT as described in Part III, Section 3.5.4 of the Guide.

If NDT is used to measure deflection load transfer across representative joints, the results may be used to select a J factor from some recommended values given in Section 5.5.5, Step 4 of the overlay design procedure. These values were selected based on the knowledge that a new doweled pavement shows a measured deflection load transfer across transverse joints of greater than 70 percent, and thus are assigned a value of 3.2 as for a protected corner. Measured load transfer values of less than 50 percent are typical of joints having only aggregate interlock with no mechanical load transfer, and thus are assigned a value of 4.0, or similar to an unprotected corner. These values represent the extremes, and others may be chosen between these values.

For CRCP, cracks are held reasonably tight with reinforcement and should provide excellent load transfer, corresponding to a J factor of 2.2 to 2.6. If, however, the CRCP has been patched with either AC repairs or with unreinforced or poorly constructed PCC repairs, the patch joints will have poor load transfer. Depending on the amount of patching of this type present, assignment of a much higher J factor (as for an unprotected corner) is warranted.

L4.6 k-VALUE FOR PCC/AC OVERLAY DESIGN

The k-value that a PCC overlay slab will actually experience in support when the slab is placed on an AC pavement will typically be much lower than one that might be calculated directly from deflections measured on the AC surface (dividing plate pressure by total deflection). Ideally, backcalculation of k-values from NDT deflections measured on PCC overlays of AC pavements would give the best indication of the k-values that such overlays experience. However, since this type of field data is not available, other methods must be employed to select a k-value for use in design of PCC overlays of AC pavements.

The method provided in the overlay design procedure for estimation of a design static k-value involves backcalculation of the subgrade M_R and effective pavement modulus E_p according to the procedure described for AC pavements in Chapter 5. The effective dynamic k-value is estimated from Figure 3.3 in Part II, Section 3.2, using the backcalculated subgrade resilient modulus (M_R), the effective modulus of the pavement layers above the subgrade (E_p), and the total thickness of the pavement layers above the subgrade (D). It is emphasized that the backcalculated subgrade resilient modulus value used to estimate the effective dynamic k-value should *not* be adjusted by the C factor (e.g., 0.33) which pertains to establishing the design M_R for AC overlays of AC pavements. The effective dynamic k-value must be divided by 2 to obtain the static k-value for design.

The engineer should be aware that there are some significant limitations to this approach to determining the design static k-value for PCC/AC design. Figure 3.3 in Part II, Section 3.2 was developed using an elastic layer computer program, without verification

Appendix L

with field deflection data. While the approach described here for determining k-value may yield reasonable values in some instances, it may yield unreasonably high values in other instances. Further research of the subject of support for PCC overlays, including deflection testing on in-service PCC/AC pavements and backcalculating effective k-values, is strongly encouraged.

L5.0 DETERMINATION OF SN_{eff} FOR FLEXIBLE PAVEMENTS

The design of AC overlays for AC pavements by the procedures presented in Chapter 5 requires the determination of the effective structural number (SN_{eff}) of the existing pavement. Three methods of determination have been adopted: (1) a visual survey/material testing method, (2) a remaining life method, and (3) an NDT method. Because of the uncertainties associated with the determination of SN_{eff}, the designer should not expect the three methods to give identical estimates. It is recommended that the designer use all three methods whenever possible and select the "best" estimate based on engineering judgement. There is no substitute for solid experience and sound judgement in overlay design.

L5.1 SN_{eff} BASED ON VISUAL SURVEY/MATERIAL TESTING

This method of SN_{eff} determination involves a component analysis using the structural number equation:

$$SN_{eff} = a_1 D_1 + a_2 D_2 m_2 + a_3 D_3 m_3 \quad (5.1)$$

where

D_i = thicknesses of existing surface, base, and subbase layers, inches
a_i = corresponding structural layer coefficients
m_i = drainage coefficients for granular base and subbase

Guidance in determining drainage coefficients is given in Part II, Table 2.4 of the AASHTO Guide. In selecting values for m_2 and m_3, it should be noted that the poor drainage conditions for the base and subbase at the AASHO Road Test would be given drainage coefficient values of 1.0.

Little guidance is presently available for the selection of layer coefficients for in-service pavement materials. It is generally accepted that the coefficients should be less than the coefficients used for the same materials in a new pavement and should reflect the amount of distress present. An exception to this might be for unbound granular bases that show no signs of degradation or contamination of fines.

Despite the lack of guidance in this area, Table L5.1 presents some suggested ranges for layer coefficients of typical materials. These values were selected based on limited information on values used by some agencies and organizations. Each highway agency should review these values in light of its own conditions and experience and adopt its own set of values.

The following notes apply to Table L5.1:

(1) All of the distress is as observed at the pavement surface.
(2) Patching all high-severity alligator cracking is recommended. The AC surface and stabilized base layer coefficients selected should reflect the amount of high-severity cracking remaining after patching.
(3) In addition to evidence of pumping noted during condition survey, samples of base material should be obtained and examined for evidence of erosion, degradation, and contamination by fines, as well as evaluated for drainability, and layer coefficients reduced accordingly.
(4) The percentage of transverse cracking is determined as (linear feet of cracking/square feet of pavement) * 100.
(5) Coring and testing are recommended for evaluation of all materials and are strongly recommended for evaluation of stabilized layers.
(6) There may be other types of distress that, in the opinion of the engineer, would detract from the performance of an overlay. These should be considered through an appropriate decrease of the structural coefficient of the layer exhibiting the distress (e.g., surface raveling of the AC, stripping of an AC layer, freeze-thaw damage to a cement-treated base).

L5.2 SN_{eff} BASED ON REMAINING LIFE

The remaining life approach to the determination of pavement structural capacity follows the fatigue concept that repeated loads gradually damage the pavement and reduce the number of additional loads that can be carried to failure. At any given time, there may

Table L5.1. Suggested Layer Coefficients for Existing AC Pavement Layer Materials

Material	Surface Condition	Coefficient
AC Surface	Little or no alligator cracking and/or only low-severity transverse cracking	0.35 to 0.40
	<10 percent low-severity alligator cracking and/or <5 percent medium- and high-severity transverse cracking	0.25 to 0.35
	>10 percent low-severity alligator cracking and/or <10 percent medium-severity alligator cracking and/or >5–10 percent medium- and high-severity transverse cracking	0.20 to 0.30
	>10 percent medium-severity alligator cracking and/or <10 percent high-severity alligator cracking and/or >10 percent medium- and high-severity transverse cracking	0.14 to 0.20
	>10 percent high-severity alligator cracking and/or >10 percent high-severity transverse cracking	0.08 to 0.15
Stabilized Base	Little or no alligator cracking and/or only low-severity transverse cracking	0.20 to 0.35
	<10 percent low-severity alligator cracking and/or <5 percent medium- and high-severity transverse cracking	0.15 to 0.25
	>10 percent low-severity alligator cracking and/or <10 percent medium-severity alligator cracking and/or >5–10 percent medium- and high-severity transverse cracking	0.15 to 0.20
	>10 percent medium-severity alligator cracking and/or <10 percent high-severity alligator cracking and/or >10 percent medium- and high-severity transverse cracking	0.10 to 0.20
	>10 percent high-severity alligator cracking and/or >10 percent high-severity transverse cracking	0.08 to 0.15
Granular Base or Subbase	No evidence of pumping, degradation, or contamination by fines	0.10 to 0.14
	Some evidence of pumping, degradation, or contamination by fines	0.00 to 0.10

be no directly observable indication of damage, but there is a reduction in structural capacity in terms of future load-carrying capacity. This reduced load-carrying capacity must be considered in overlay design.

A remaining life consideration was included in the 1986 AASHTO Guide, but the concept and application differed significantly from the approach used with the current procedures. In the 1986 Guide, the remaining life was not used to determine the existing structural capacity. Instead, a remaining life factor (F_{RL}) was applied in the overlay thickness determination equation independent of and in addition to SN_{eff}. The flexible pavement overlay equation was:

$$SN_{ol} = SN_f - F_{RL} * SN_{eff} \quad (5.2)$$

where

SN_{ol} = required structural number of overlay
SN_f = required structural number to carry future traffic
F_{RL} = remaining life factor
SN_{eff} = effective structural number of existing pavement

Elliott (26) examined the remaining life factor as used in the 1986 Guide and demonstrated that the application was flawed, as a result of a compounding

Appendix L

of assumptions. The result was that use of the F_{RL} term resulted in design inconsistencies. Elliott suggested an alternative solution which followed most of the original development of the remaining life concept but eliminated the need for one assumption. Using this alternate approach, the F_{RL} value is always 1.0. He subsequently recommended that the F_{RL} term be removed from the overlay design equation. Based on these findings, the F_{RL} term was not included in the revised overlay design procedures. Elliott's paper is reproduced in Appendix M.

Nevertheless, the general concept of decreasing structural capacity and remaining life is valid. Therefore, a remaining life method for determining effective structural capacity was adopted.

The remaining life approach adopted for these procedures utilizes the work done in the initial development of the remaining life concept for the 1986 Guide. That work introduced the idea of a condition factor defined by the following equation:

$$CF = \frac{SC_n}{SC_o} \quad (5.3)$$

where

CF = condition factor
SC_n = pavement's structural capacity after n ESAL applications
SC_o = pavement's structural capacity when it was new

For flexible pavements, the SC terms are replaced by the structural number (SN). If at any point in time CF is known, the effective structural number (SN_{eff}) may be calculated:

$$SN_{eff} = CF * SN_o \quad (5.4)$$

To make use of this, a relationship between CF and remaining life (RL) is needed. Such a relationship was developed for the 1986 Guide using the AASHTO pavement design equations. Elliott's investigation confirmed this relationship. The CF-RL relationship was:

$$CF = RL^{0.165} \quad (5.5)$$

Although developed for the 1986 Guide, this specific relationship was not used in the 1986 Guide. The discussion in Appendix CC of the 1986 Guide (8) indicates that CF from the equation is reasonable for all values of RL greater than 0.005 (CF = 0.42). However, when RL is zero, CF is also zero, which was not considered to be realistic. As a result, a different relationship was assumed for the 1986 Guide.

Nevertheless, this relationship, with one slight modification, is used in the current overlay design procedures for the determination of the effective structural capacity based on remaining life. The modification consists of setting a minimum CF value of 0.5. Equation 5.5 may be used to calculate CF for all values of RL greater than 0.05. For RL less than 0.05, CF may be calculated using a straight line interpolation between CF at RL = 0.05 and CF = 0.5 at RL = 0.00. Figure L5.1 is a plot of both Equation 5.5 and the proposed CF-RL curve for determination of effective structural capacity.

L5.3 SN_{eff} BASED ON NONDESTRUCTIVE TESTING

Implicit in the determination of structural number from NDT data is an assumption of a relationship between pavement stiffness and layer coefficients. Such an assumption must be recognized as being a substantial simplification of a complex problem. Thus, the structural number determined from the analysis should be viewed as only one approximation of the "true" structural capacity of the existing pavement.

The procedure recommended for NDT determination of SN_{eff} is based on the "equal stiffness" approach described in Appendix NN of the 1986 Guide. (8) However, instead of requiring the backcalculation of the modulus value of each pavement layer, the recommended procedure uses the "effective" modulus of the total pavement structure above the subgrade. The advantages to this approach are that it is simpler to apply and it does not suggest a level of sophistication that does not actually exist within the context of the structural number concept.

Based on Appendix NN of the 1986 Guide, the equation for the effective structural number is:

$$SN_{eff} = 0.0045 D \sqrt[3]{E_p} \quad (5.6)$$

where

D = total thickness of surface, base and subbase, inches
E_p = effective modulus of the pavement, psi

Figure L5.2 was developed from Equation 5.6.

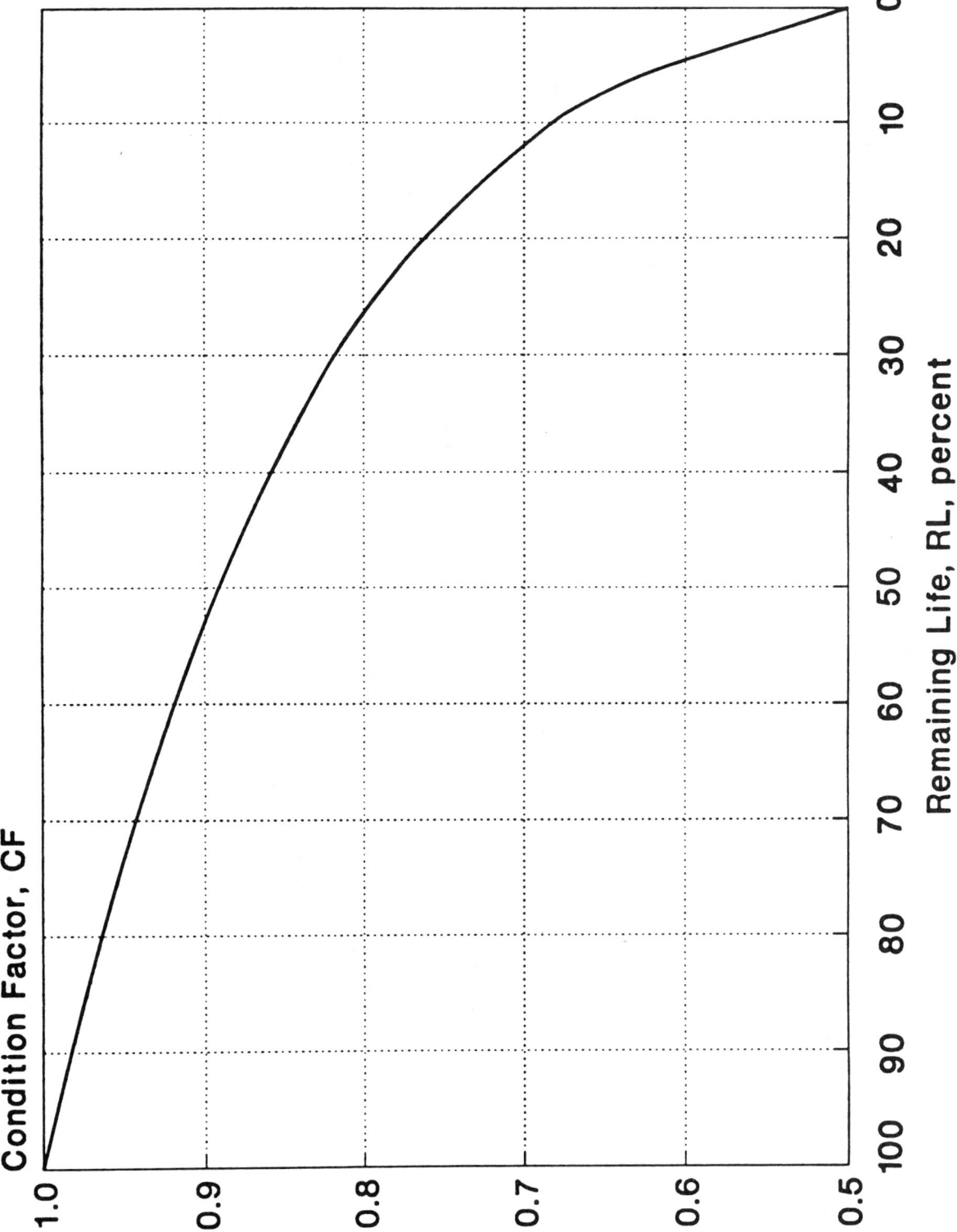

Figure L5.1. Relationship of Condition Factor to Remaining Life

Appendix L

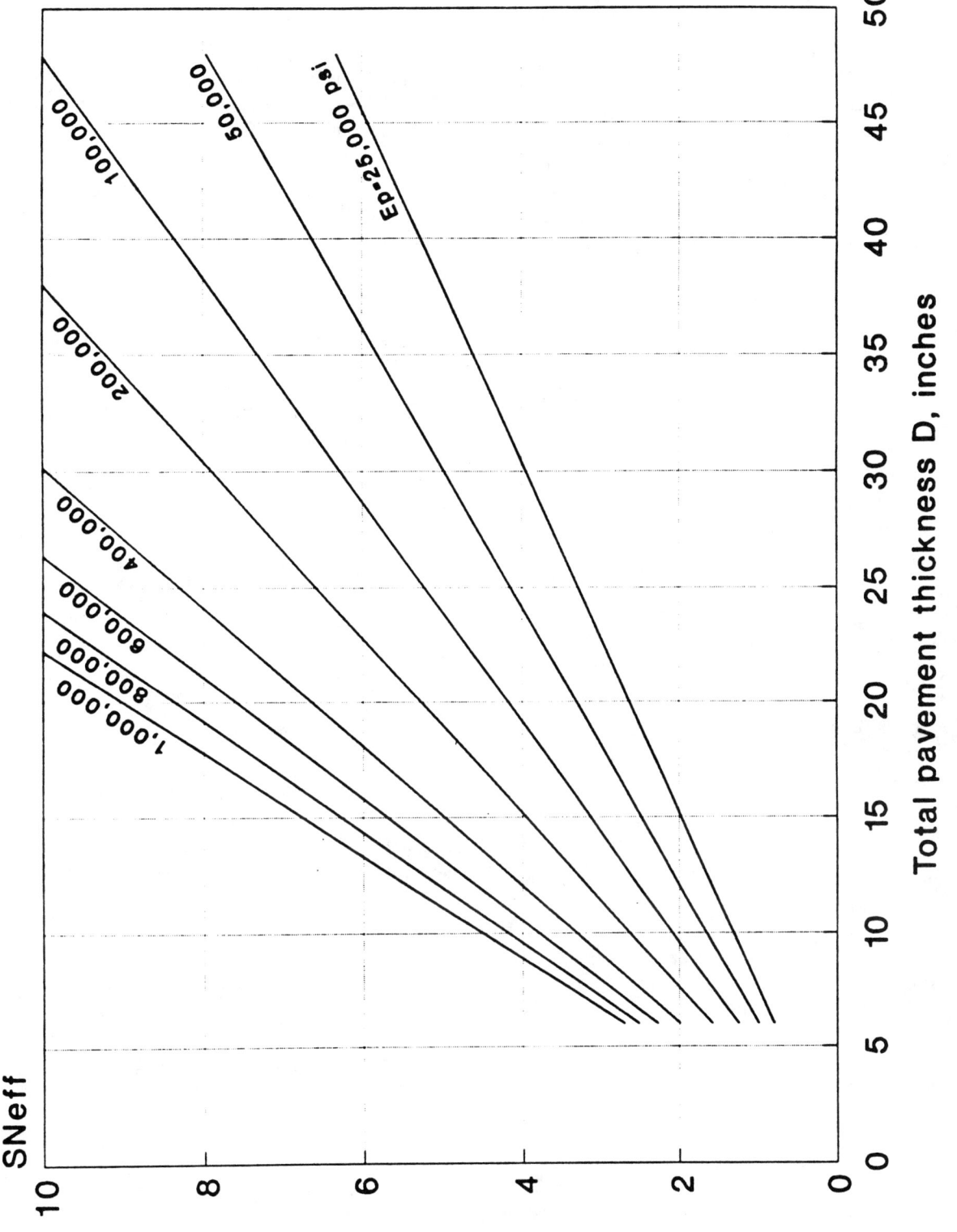

Figure L5.2. SN_{eff} from NDT Method

The method for determining the value of E_p is based on the Boussinesq equation (27) for deflection in an elastic half-space, as well as an assumption that the pavement system can be represented as two layers—a subgrade of infinite depth and having a modulus M_R, and a pavement having a total thickness D and an effective modulus (E_p). To simplify the equations with no significant loss of accuracy, it is further assumed that Poisson's ratio (μ) for both the subgrade and pavement materials is 0.5. The Boussinesq equation for deflection at any depth (z) in an elastic half-space assuming $\mu = 0.5$, as presented by Yoder (28), is:

$$d = \frac{1.5pa}{E} F_b(z) \quad (5.7)$$

where

p = contact pressure, psi
a = circular load radius, inches
E = elastic modulus, psi

$$F_b(z) = \frac{1}{\sqrt{\left[1 + \left(\frac{z}{a}\right)^2\right]}} \quad (5.8)$$

Boussinesq's equation applies to deflection in a half-space, that is, a one-layer system. In 1949, Odemark (29) presented an approximate method for determining deflection in a two-layer system, using Boussinesq's one-layer equation and the concept of "equivalent thickness" described in 1940 by Barber. (30) The deflection (d) measured at the surface at the center of loading is assumed to be the sum of the subgrade and pavement deflections.

The deflection in the top layer is computed by transforming the two-layer system into a one-layer system of pavement material, that is, a homogeneous half-space with modulus E_p. The deflection at the surface is given by Boussinesq's equation with $z = 0$. From Equation 5.7:

$$d_0 = \frac{1.5pa}{E_p} \quad (5.9)$$

If the total thickness of the pavement is denoted by D, the deflection at depth D is given by Boussinesq's equation with z = D:

$$d_D = \frac{1.5pa}{E_p} * \frac{1}{\sqrt{\left[1 + \left(\frac{D}{a}\right)^2\right]}} \quad (5.10)$$

The deflection in the pavement between $z = 0$ and $z = D$ may then be determined by subtracting Equation 5.10 from Equation 5.9:

$$d_p = d_0 - d_D = \frac{1.5pa}{E_p} * \left\{ 1 - \frac{1}{\sqrt{\left[1 + \left(\frac{z}{a}\right)^2\right]}} \right\}$$

(5.11)

The deflection in the subgrade is computed by transforming the two-layer system into an equivalent one-layer system of subgrade material with modulus M_R. To do so, the pavement of thickness D and modulus E_p is represented by an equivalent thickness D_e of subgrade material. The deflection at the top of the subgrade is given by Boussinesq's equation with $z = D_e$:

$$d_s = \frac{1.5pa}{M_R} * \frac{1}{\sqrt{\left[1 + \left(\frac{D_e}{a}\right)^2\right]}} \quad (5.12)$$

where

$$D_e = D * \sqrt[3]{\frac{E_p}{M_R}} \quad (5.13)$$

The total deflection (measured at the pavement surface) is then obtained by adding the pavement and subgrade deflections (Equations 5.11 and 5.12) and substituting in the definition of D_e (Equation 5.13):

$$d = d_s + d_p \quad (5.14)$$

$$d_0 = 1.5pa \left\{ \frac{1}{M_R \sqrt{1 + \left(\frac{D}{a}\sqrt[3]{\frac{E_p}{M_R}}\right)^2}} + \frac{\left[1 - \frac{1}{\sqrt{1 + \left(\frac{D}{a}\right)^2}}\right]}{E_p} \right\}$$

(5.15)

If the pavement thickness D and subgrade resilient modulus M_R are known or assumed, the only un-

Appendix L

known quantity in this equation is E_p. E_p can quickly be solved by iteration using a computer or spreadsheet. Figure L5.3 was developed from Equation 5.15 for load radius a = 5.9 inches. For other load radius values, Equation 5.15 should be used to determine E_p.

L5.4 Temperature Adjustment

Because the stiffness of AC materials change significantly with temperature, d_0 will vary depending upon the temperature of the AC layers at the time of testing. For purposes of comparison of E_p along the length of a project, the measured d_0 values should be adjusted to a single reference temperature. Furthermore, if SN_{eff} is to be determined by the NDT method, the reference temperature for adjustment of d_0 should be 68°F, to be consistent with the procedure for new AC pavement design described in Part II of the Guide. The adjustment to d_0 is based on the ratio of predicted deflections:

$$T(t) = \frac{d_0(68)}{d_0(t)} \qquad (5.16)$$

where

$T(t)$ = temperature adjustment factor
$d_0(68)$ = d_0 at 68°F
$d_0(t)$ = d_0 at testing temperature t°F

Elastic layer analyses were used to develop adjustment factors for the following pavement parameters:

AC thickness: 2, 4, 6, 8, and 12 inches
AC modulus (temperature):
  2,000 ksi   (30°F)
  1,300 ksi   (50°F)
    670 ksi   (68°F)
    360 ksi   (85°F)
    132 ksi   (105°F)
     55 ksi   (120°F)
Granular base: 0 and 12 inches, E = 30 ksi
Cement-stabilized base: 0 and 10 inches, E = 850 ksi
Subgrade modulus: 5, 10, and 20 ksi

The moduli of the AC were estimated using the equation developed by the Asphalt Institute (*23*) and assuming typical mix properties. The resulting adjustment factors were plotted versus the AC temperature. These plots were subsequently used to develop two figures for temperature adjustment, one for granular and asphalt-stabilized base pavements (Figure L5.4) and one for cement- and pozzolanic-base pavements (Figure L5.5).

L5.5 DETERMINATION OF SUBGRADE MODULUS FOR SN_{eff} DETERMINATION

The subgrade modulus that is to be used in equation 5.15 should also be determined from the NDT data. However, this value of M_R is not necessarily the same as the value that is used to determine the total required structural number of the pavement (SN_f). The design M_R must be consistent with the value used in the design performance equation for the AASHO Road Test subgrade. The reader is referred to Chapter 3 for a discussion of the determination of the design M_R from NDT data.

A simple method for estimating the subgrade modulus from deflections measured at the surface of a layered pavement structure has been proposed by Ullidtz. (*31, 32*) The method is based on the following two observations:

(1) As distance away from the load increases, compression of the layers above the subgrade becomes less significant to the measured deflection at the pavement surface.
(2) As distance away from the load increases, the approximation of a distributed load by a point load improves.

The first observation may be restated to say that at some sufficiently large radial distance from the applied load, the deflection measured at the pavement surface is equal to the deflection at the top of the subgrade, and thus depends entirely on the elastic properties of the subgrade, regardless of the number, thickness, and elastic properties of the overlying layers.

The second observation resulted from a comparison of deflections predicted by the elastic layer program BISAR in a layered pavement structure at various radial distances from the center of a distributed circular load with deflections predicted at the

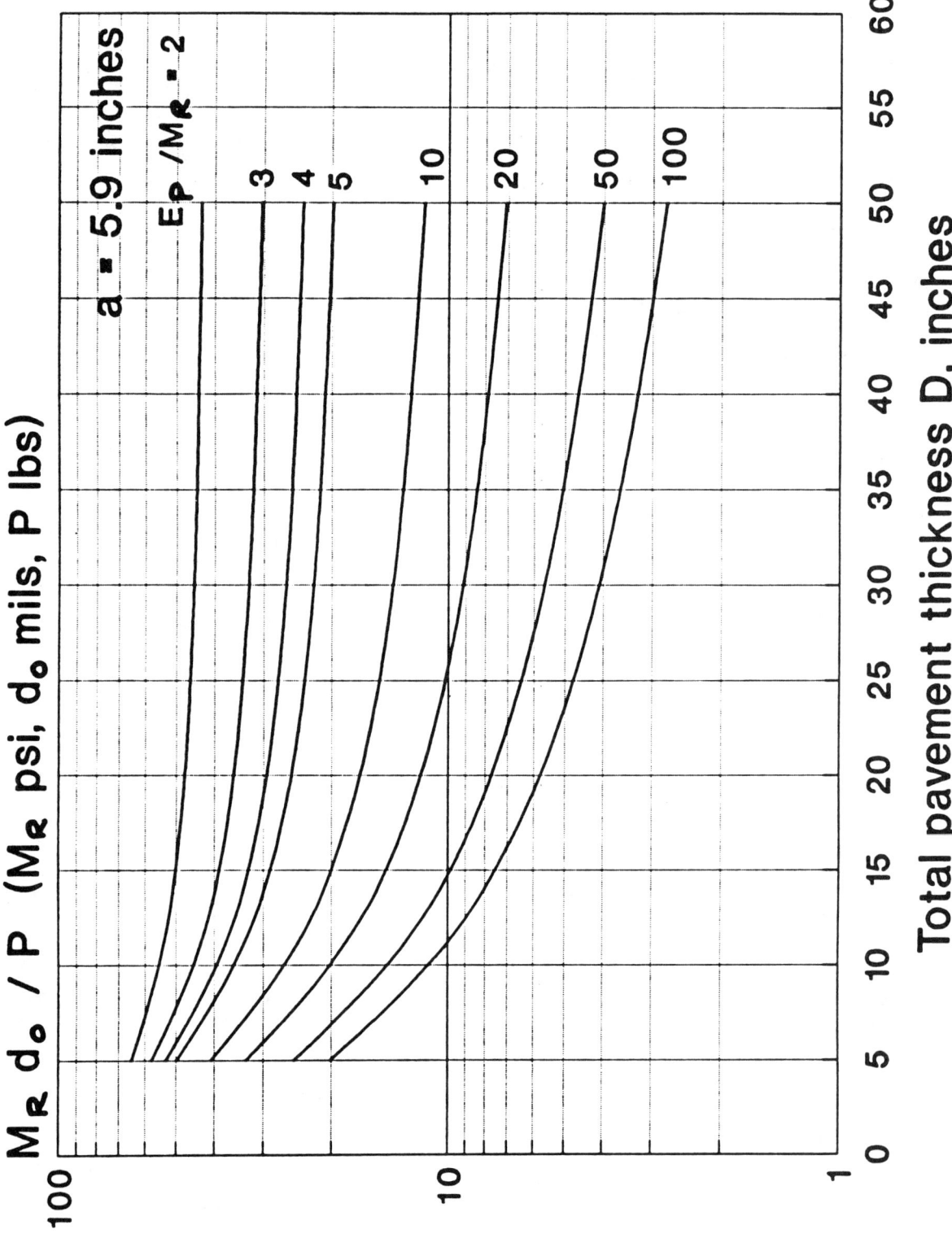

Figure L5.3. Determination of E_p/M_R from Maximum Deflection d_0

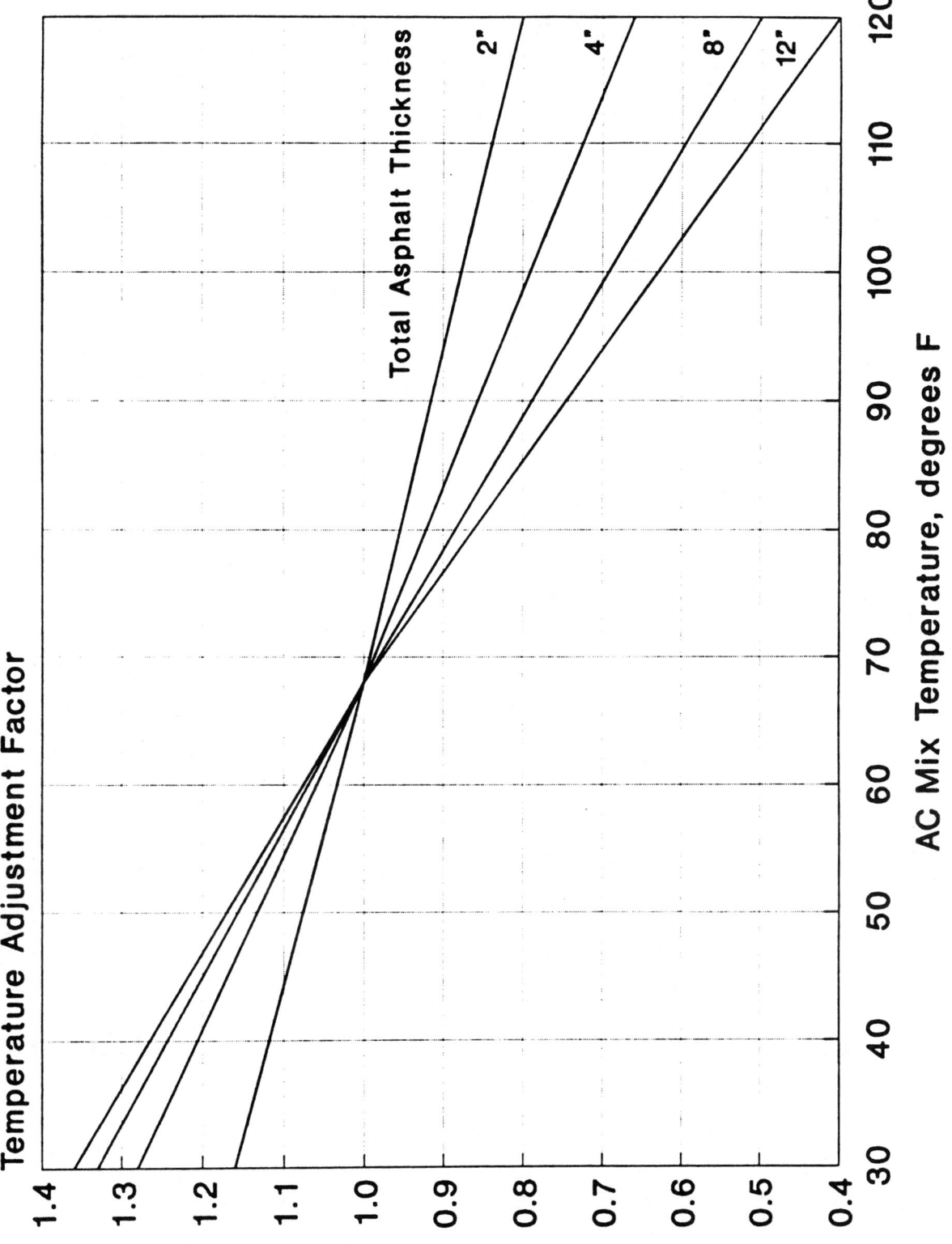

Figure L5.4. d_0 Adjustment for AC Mix Temperature for Granular and Asphalt-Treated Base Pavements

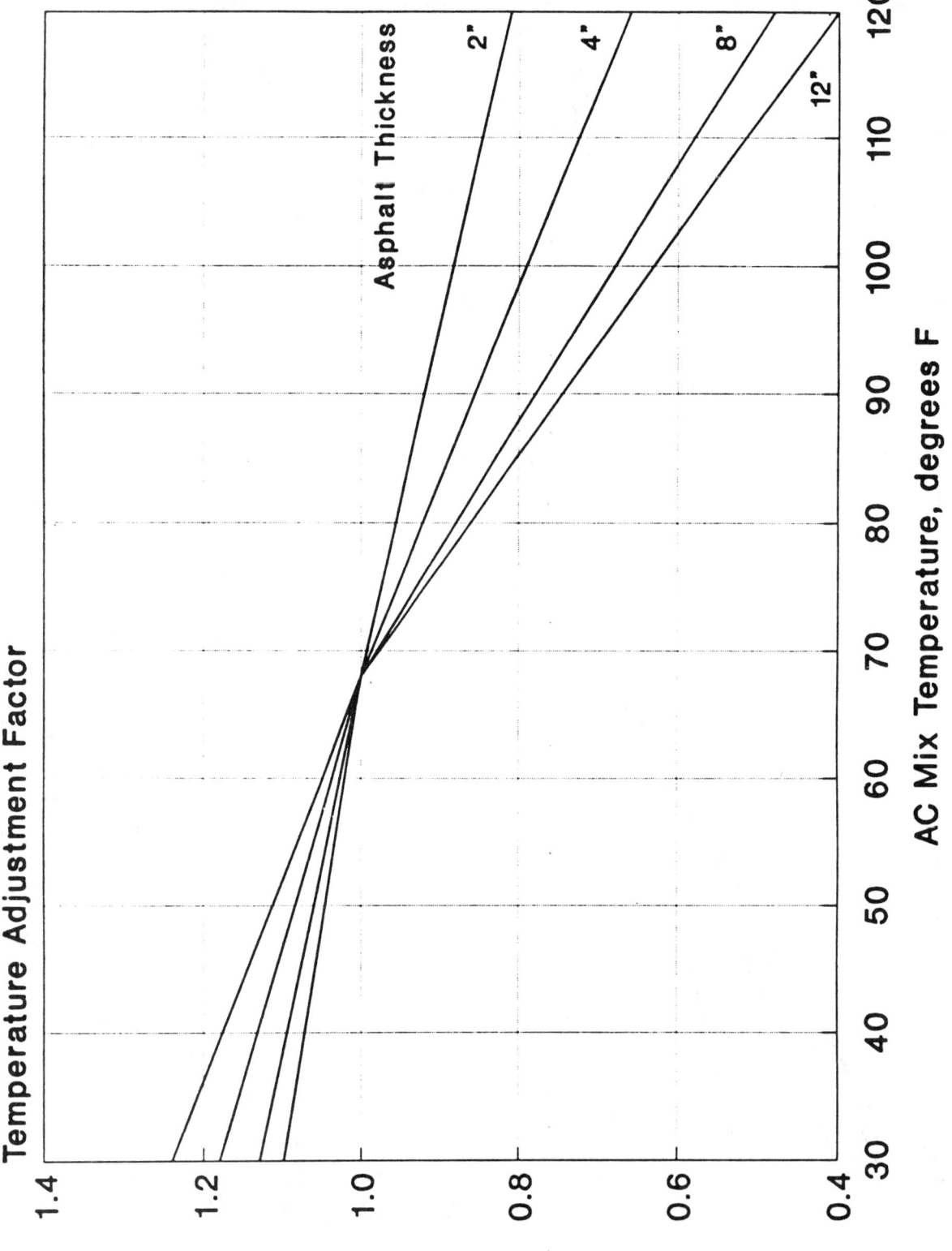

Figure L5.5. d_0 Adjustment for AC Mix Temperature for Cement- and Pozzolanic-Treated Base Pavements

Appendix L

same distance by the Boussinesq equation for deflection in a one-layer system at points away from the center of the load:

$$d_r = \frac{P(1 + \mu)}{2\pi R M_R} * [2(1 - \mu) + \cos^2\theta] \quad (5.17)$$

where

- d_r = deflection at distance r from the applied load, inches
- μ = Poisson's ratio of subgrade
- P = load, pounds
- M_R = elastic modulus of subgrade, psi
- r = radial distance from load, inches
- z = depth, inches

$$R^2 = z^2 + r^2 \quad (5.18)$$

$$\cos\theta = \frac{z}{R} \quad (5.19)$$

Substituting in the values shown above for R and cos θ, Equation 5.17 may be written as:

$$d_r = \frac{1}{M_R} * \frac{P(1 + \mu)}{2\pi\sqrt{z^2 + r^2}} * \left[2(1 - \mu) + \frac{z^2}{z^2 + r^2}\right] \quad (5.20)$$

Since the pavement surface deflection d_r at this radial distance r is the same as the deflection of the subgrade, the load P may be considered to be applied at the surface of the subgrade. With z = 0, Equation 5.20 reduces to:

$$d_r = \frac{P(1 - \mu^2)}{\pi r M_R} \quad (5.21)$$

which may then be rearranged to solve for the subgrade modulus M_R:

$$M_R = \frac{P(1 - \mu^2)}{\pi r d_r} \quad (5.22)$$

Assuming the subgrade Poisson's ratio $\mu = 0.5$, this equation reduces to:

$$M_R = \frac{0.24P}{d_r r} \quad (5.23)$$

Ullidtz stated that this equation should be used with deflections measured at distances greater than the "effective" thickness (D_e) of the pavement. The 1986 AASHTO Guide contained the same equation and recommended that the distance be greater than the effective radius (a_e) of the stress bulb at the subgrade/pavement interface. With the subgrade Poisson's ratio equal to 0.5, the equation for the effective radius is:

$$a_e = \sqrt{a^2 + \left(D \sqrt[3]{\frac{E_p}{M_R}}\right)^2} \quad (5.24)$$

For practical purposes, the deflection used should be as close as possible to the loading plate. As the distance increases, the magnitude of the deflection decreases and the effect of measurement error is magnified. Analyses were performed to determine how close the deflection could be without introducing a serious error in the subgrade modulus determination. Deflection basins were generated using an elastic layer program. The total pavement thicknesses ranged from 10 to 36 in. and the AC thicknesses ranged from 2 to 16 in. Deflections at various radial distances were used to calculate a subgrade modulus using Equation 5.23. The calculated modulus values were compared with the values used in the elastic layer analyses.

The ratios of calculated/actual modulus values are plotted in Figure L5.6 versus the ratio of radial distance to the effective radius of the stress bulb (r/a_e). Based on the analysis it is concluded that the deflection used for subgrade modulus determination should be from a distance greater than or equal to 0.7 times a_e. It should further be noted that no temperature adjustment is needed in determining M_R since the deflection used is due only to subgrade deformation.

L5.6 AC OVERLAY OF FRACTURED PCC SLABS

All three different fractured slab techniques were placed in a separate section to be designed using flexible pavement design procedures. The selection of overlay thickness for break/seat, crack/seat and rubblize/compact techniques presents a relatively new challenge for the pavement designer. The use of the AASHTO pavement performance equations for the se-

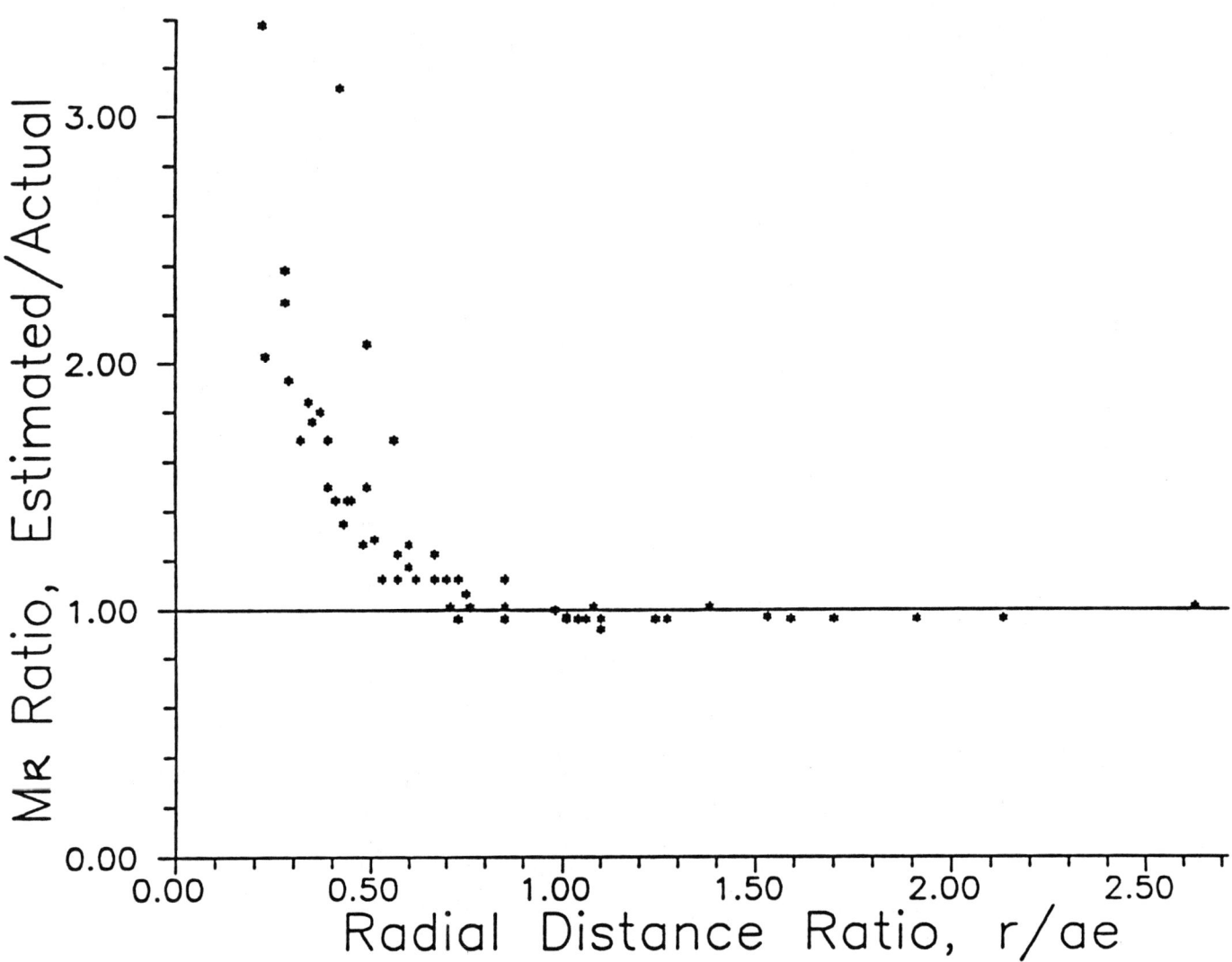

Figure L5.6. Influence of Radial Distance on the Accuracy of Backcalculated M_R Values

lection of the overlay thickness for these projects results in another level of extrapolation beyond the AASHO Road Test database.

Rubblizing can be used on all types of PCC pavements in any condition. It is particularly recommended for reinforced pavements. Fracturing the slab into pieces less than 12 inches reduces the slab to a high-strength granular base. Recent field testing of several rubblized projects showed a wide range in backcalculated modulus values among different projects, from less than 100,000 psi to several hundred thousand psi (20, 38, 46), and within-project coefficients of variation of as much as 40 percent. (20, 38)

Crack and seat is used only with JPCP and involves cracking the slab into pieces typically one to three feet in size. Recent field testing of several cracked and seated JPCP projects showed a wide range in backcalculated modulus values among different projects, from a few hundred thousand psi to a few million psi (34, 36, 38, 47, 48), and within-project coefficients of variation of 40 percent or more. (38) Reference 38 recommends that to avoid reflection cracking no more than 5 percent of the fractured slab have a modulus greater than 1 million psi. Effective slab cracking techniques are necessary in order to satisfy this criterion for crack/seat of JPCP.

Break/seat is used only with JRCP and includes the requirement to rupture the reinforcement steel across each crack, or break its bond with the concrete. If the reinforcement is not ruptured and its bond with the concrete is not broken, the differential movements at working joints and cracks will not be reduced and reflection cracks will occur. Recent field testing of several break/seat projects showed a wide range in backcalculated modulus values ranging from a few hundred thousand psi to several million psi (20, 34, 38, 47), and within-project coefficients of variation of 40 percent or more. (20, 38) The wide range in backcalculated moduli reported for break and seat projects suggests a lack of consistency in the technique as performed with past construction equipment. Even though cracks are observed, the JRCP frequently retains a substantial degree of slab action because of failure to either rupture the reinforcing steel or break its bond with the concrete. This may also be responsible for the inconsistency of this technique in reducing reflection cracking. More effective breaking equipment may overcome this problem. This design procedure assumes that the steel will be ruptured or that its bond to the concrete will be broken through an aggressive break/seat process, and that this will be verified in the field through deflection testing before the overlay is placed. The use of rubblization is recommended for JRCP due to its ability to break slab continuity.

L6.0 DETERMINATION OF D_{eff} FOR RIGID AND COMPOSITE PAVEMENTS

L6.1 D_{eff} BASED ON VISUAL SURVEY/MATERIAL TESTING

The effective slab thickness, D_{eff}, represents a slab thickness that has been adjusted to consider several important factors that will affect the life of the overlay. The overlay design procedure attempts to "protect" the existing pavement and subgrade from future traffic load damage by providing increased structural capacity roughly equal to a new pavement required to carry the anticipated traffic. The AASHTO rigid performance equation was developed at the AASHO Road Test from performance of jointed concrete pavements, not for overlays. To utilize this equation for overlays, there are some additional factors that must be considered.

The most significant of these additional factors that are under the control of the designer include the deterioration of transverse reflection cracks caused by underlying deteriorated joint and cracks, PCC disintegration caused from poor PCC durability and PCC slab past fatigue damage. Rutting of an AC overlay and aging of an AC overlay causing shrinkage cracks and raveling/weathering are additional factors that affect overlay life but are materials related problems.

These additional factors have a much more significant effect on the performance of AC and bonded PCC overlays than unbonded jointed concrete overlays. The following sections describe the background for the adjustment factors for AC and bonded PCC overlays. A following section describes their use for unbonded overlays.

Values for the three main factors are described. It is emphasized that user agencies should carefully examine the values for these condition factors and modify them as needed on the basis of local conditions and experience.

L6.1.1 Condition Factor F_{jc}

The AASHTO rigid pavement design equation does not consider the loss of serviceability caused by deteriorated transverse reflection cracks through the overlay. In fact, the AASHTO design for future structural capacity has absolutely nothing to do with reflection cracking. The procedure does consider loss of serviceability due to regular transverse joints that create some roughness due to faulting and minor spalling, which is similar to transverse reflection cracks that do not deteriorate (spall, depress) significantly. A direct way to consider the loss of serviceability of the overlay caused by deteriorated transverse reflection cracks must be added to the procedure.

Transverse joints and cracks (and punchouts in CRCP) that are wide and/or deteriorated and showing poor load transfer will typically result in rapid reflection of cracks in AC or PCC bonded overlays. The deterioration of those reflection cracks usually causes a serious loss of serviceability.

This potential failure mechanism must be considered directly either through treatments (e.g., fabrics, saw and seal, rubblizing, break/seat, crack/seat or through increasing overlay thickness through the joints and cracks adjustment factor (F_{jc}). Overlay thickness increases will have two effects. The first is to somewhat delay the occurrence of the initial crack, but the most important effect is to reduce the severity of the reflected crack and retard its rate of deterioration which is very important in maintaining serviceability.

It may not be cost-effective to increase slab thickness to reduce reflection crack deterioration. This Guide strongly recommends that all existing deteriorated joints and cracks are repaired with full-depth doweled PCC repairs, so that $F_{jc} = 1.00$. However, if this is not possible, then the F_{jc} can be adjusted to account for this extra loss in PSI.

It is strongly recommended that all "deteriorated" joints and cracks (punchouts in CRCP should be full-depth repaired with tied reinforced repairs) be full-depth repaired with doweled or tied PCC repairs to avoid serious deterioration of reflection cracks. If they are, then $F_{jc} = 1.0$. Due to funding and other constraints, this may not be possible and the F_{jc} is then determined from Figure L6.1 which is based upon the number of unrepaired transverse joints and cracks (or punchouts or other similar failed areas) that are considered to be "deteriorated" and would cause the reflection crack to spall or settle. Any of the following conditions should be counted as deteriorated transverse joints/cracks:

(1) medium- or high-severity spalled transverse joints
(2) medium- or high-severity spalled transverse cracks (including any joint where the spalls are patched with AC mixture)
(3) existing expansion joints, exceptionally wide joints (greater than 1 inch) or full-depth, full-lane-width AC patches

Pavements with "D" cracking or reactive aggregate deterioration often have deterioration at the joints and cracks from durability problems. The F_{dur} factor is used to adjust the overlay thickness for this problem. Therefore, when this is the case, the F_{jc} should be determined from Figure L6.1 only using those non-repaired deteriorated joints and cracks that are not caused by durability problems. If all of the deteriorated joints and cracks are spalling due to "D" cracking or reactive aggregate, then $F_{jc} = 1.0$. This will avoid adjusting twice with F_{jc} and F_{dur} factors.

If these are repaired using a full-depth PCC repair that contains dowel or tie bars for load transfer, then these are no longer serious problems that cause rapid deterioration of the reflection cracks. The cracks will still occur but their deterioration will be far less severe, and can be controlled through proper crack sealing.

The extent of joint load transfer will also affect the deterioration of reflection cracks in the overlay. This is considered to some extent by the J factor in determining D_f.

The F_{jc} curve was developed by considering the effect that deteriorated reflection cracks has upon reducing the serviceability of the overlaid pavement, and thus reducing its service life. A major portion of the loss of serviceability is due to deteriorated reflection cracks (75 percent was assumed). Therefore, an adjustment factor, F_{jc}, must be applied so that a thicker overlay will result to adjust for this loss in serviceability from reflection cracks. The following steps were taken in the development of the F_{jc} factor.

The serviceability of an AC-overlaid JCP pavement was related to the number of deteriorated reflection cracks/mile. Deteriorated here means any crack that had spalled, had adjacent broken pieces or had settled significantly to cause roughness. A panel of raters riding in a standard-sized automobile made the ratings given in Figure L6.2. Present serviceability rating loss was calculated as 4.5 (typical new pavement) minus the mean panel PSI rating. The line labelled "100" is the best-fit curve through the data. It was assumed that 75 percent of the total PSI loss was caused by the deteriorated transverse reflection cracks, with the rest caused by other less severe cracks, rutting and foundation movements. The curve labeled "75" represents 75 percent of the best-fit curve loss in PSI.

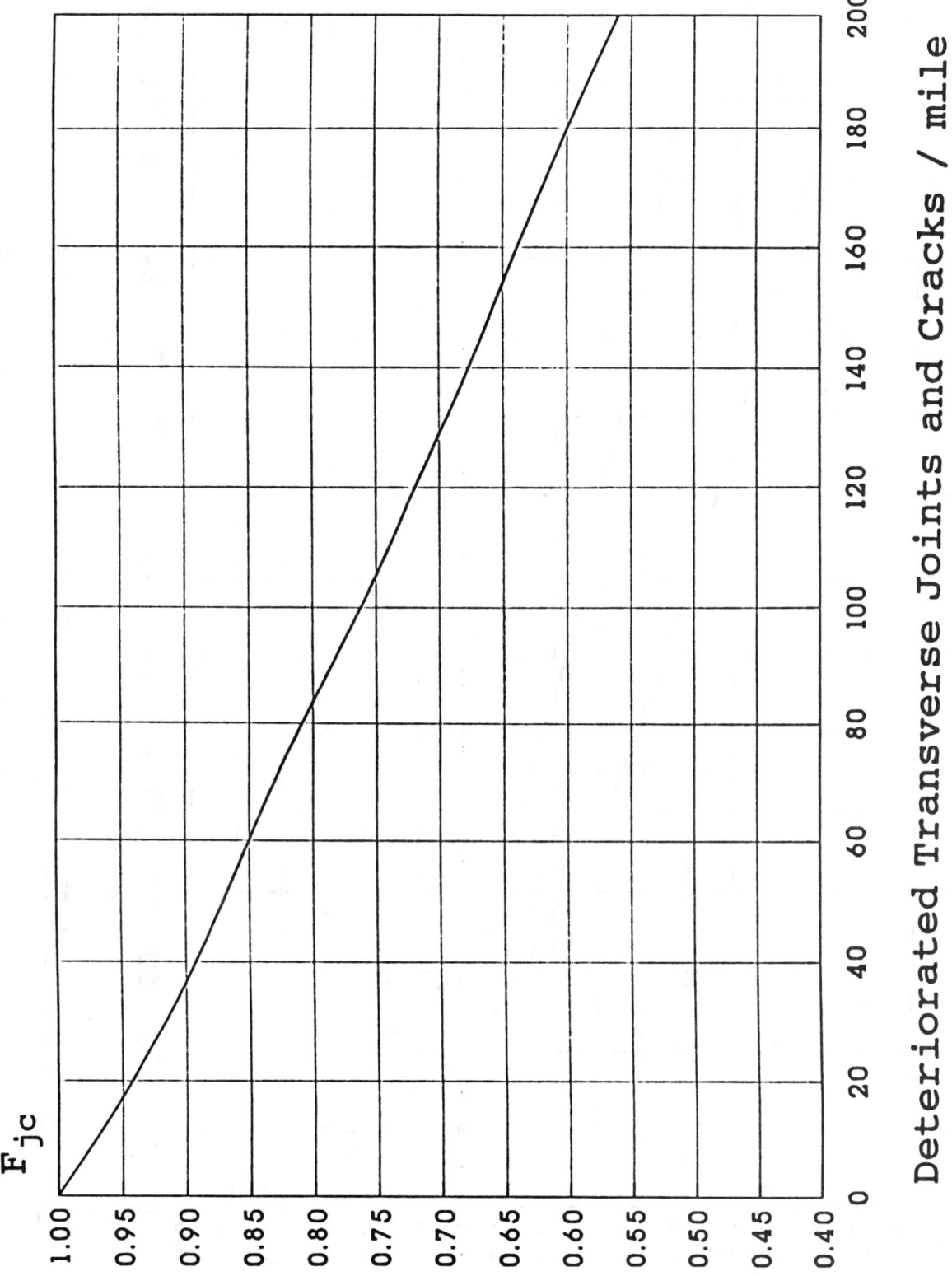

Figure L6.1. F_{jc} Adjustment Factor

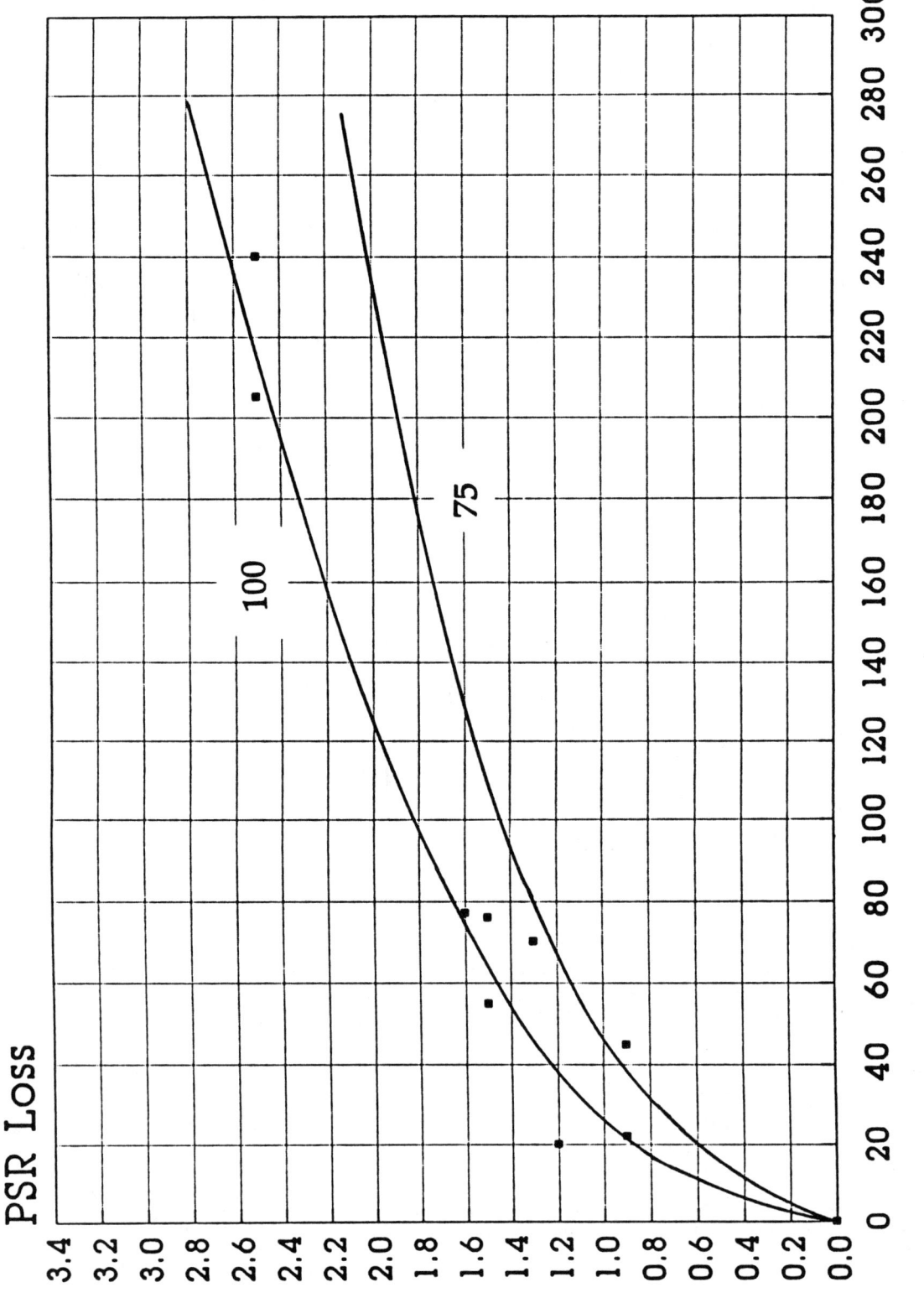

Figure L6.2. PSR Loss Versus Deteriorated Transverse Joints and Cracks per Mile

Figure L6.3 was developed next using the AASHTO rigid pavement design equation and several pavement sections submitted by the states for overlay design. Each point represents a single section of jointed concrete pavement. The amount of loss of PSI was varied from 0 to nearly 100 percent of the total loss of PSI by changing the terminal PSI in the rigid pavement equation and solving for the required overlay thickness. For example, the following results were obtained for one of the sections:

Total PSI Loss	Assumed PSI Loss from Refl. Cracking	PSI Loss from Other Causes	Required AC OL (in)	Computed F_{jc}**
2.0*	0.0	2.0	3.2	1.00
2.0	0.5	1.5	4.1	0.95
2.0	1.0	1.0	5.2	0.88
2.0	1.5	0.5	7.4	0.74
2.0	1.8	0.2	10.2	0.53

*Initial PSI = 4.5, terminal PSI = 2.5, 4.5 − 2.5 = 2.0.
**The computed F_{jc} is that required to give the correct overlay thickness so that the overlaid pavement will be able to carry the design ESALs without reducing the PSI below 2.5.

The final step was to use Figures L6.2 and L6.3 to develop the Figure L6.1 curve relating the number of deteriorated joints/cracks to F_{jc}. This was done point by point by taking 20, 40, 60, etc. cracks/mi from Figure L6.2 and determining the corresponding PSI loss (using the 75-percent curve) and taking this value to Figure L6.3 to determine the F_{jc}. The corresponding F_{jc} and number of deteriorated transverse cracks point was plotted on Figure L6.1. A smooth curve was finally plotted for Figure L6.1.

For existing composite pavements (AC/PCC) that are being considered for a new AC overlay, it can be assumed that any reflection crack that is spalled and deteriorated is probably overlying a joint or crack in similar condition in the base slab. It is recommended that these areas be full-depth repaired. If they are not, then they should also be counted as a deteriorated joint/crack in determining the appropriate F_{jc}.

It is noted at this point that the three adjustment factors (F_{jc}, F_{dur}, and F_{fat} are somewhat interrelated and certain situations could arise that would produce an overlay design that is too conservative. For example, a jointed pavement with all of the joints spalled due to "D" cracking would have a low F_{jc} and F_{dur} resulting in a very thick overlay. To avoid this situation, the following modification was developed.

A point of particular concern is the strong correlation between the F_{jc} and F_{dur}, since any pavement with significant durability problems would very likely have a lot of deteriorated joints and cracks. Therefore, when durability problems exist, the counting of any deteriorated joint or crack to determine F_{jc} is prohibited where the deterioration is caused by "D" cracking or reactive aggregate problems. For example, if all of the joints were spalled from "D" cracking, the $F_{jc} = 1.00$ and the F_{dur} is determined according to the standard guidelines.

The other correlations between F_{jc}, F_{dur}, and F_{fat} are not believed to be particularly significant, and there are numerous things that can make the correlation very poor, especially when any pre-overlay repair is done, or when most of the deterioration is in the joints and not cracks. Therefore, no additional considerations were given to other correlations between the three factors.

L6.1.2 Condition Factor F_{dur}

The disintegration of the existing PCC slab beneath an overlay has resulted in increased deterioration of AC and bonded PCC overlays, reducing their serviceability and service life. The PCC slab disintegrates to the point that it reverts back to granular material in localized areas. These areas of the underlying PCC slab experience something like shear failure under heavy loads resulting in a seriously localized failed area in the wheel paths that cause rapid loss of serviceability. This can be adjusted to some extent by increasing overlay thickness to reduce vertical stresses under wheel loads so that the design life can be achieved.

The types of PCC durability that are most significant are "D" cracking and reactive aggregates damage. The occurrence of either of these is likely to have a significant effect on pavement performance.

For example, in Illinois, survival curves based on hundreds of AC overlays of JRCP and CRCP show that pavements without "D" cracking will last substantially longer and carry substantially many more ESALs to failure than pavements with "D" cracking. This result is used by backing through the overlay design procedure to develop reasonable values for F_{dur}.

A typical JRCP pavement with an AC overlay in Illinois is designed for 15 million rigid

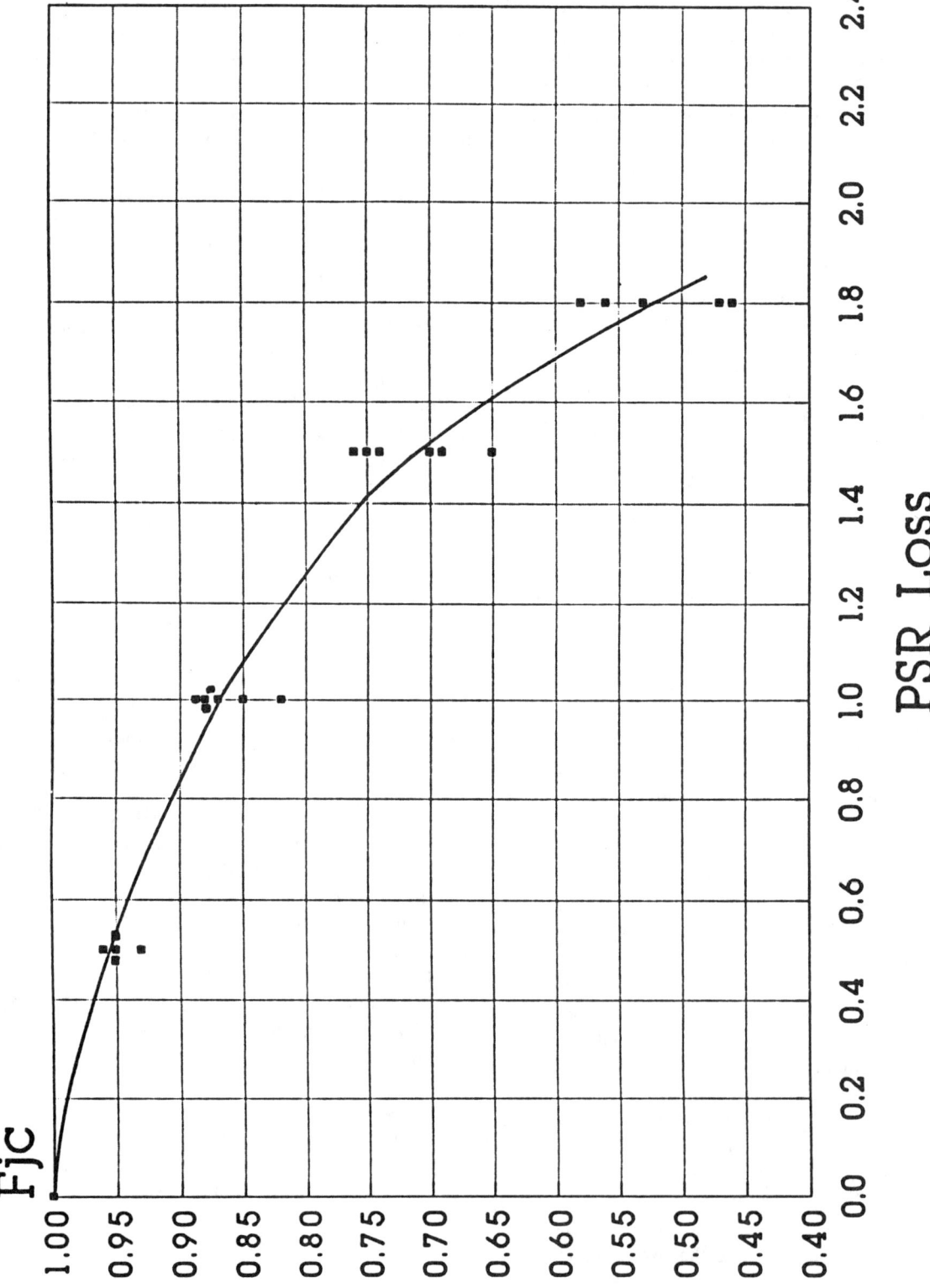

Figure L6.3. F_{jc} Factor Versus PSR Loss

ESALs (over a 15-year period). Setting $F_{jc} = 1.0$, $F_{dur} = 1.0$ and $F_{fat} = 0.95$ (typical), an AC overlay thickness of 3.4 inches is obtained from the design procedure at an R = 95 percent. This assumes sound concrete for the slab.

If the pavement has significant "D" cracking, this 3.4-inch overlay would only carry about 15/2 = 7.5 million ESALs to terminal serviceability. This would correspond to a $F_{dur} = 0.88$ for the same AC overlay thickness. If the pavement is then designed with an $F_{dur} = 0.88$ and ESALs = 15 million, a required AC overlay of 5.3 inches is obtained. This is a reasonable value based upon survival curve data for "D" cracked pavements in Illinois.

Based upon this information, some guidelines for F_{dur} were developed for less severe and more severe concrete durability conditions.

F_{dur} = 1.0, if the PCC slab exhibits no signs of "D" cracking or reactive aggregate damage.

0.96–0.99, if the PCC slab exhibits signs of "D" cracking or reactive aggregate damage (cracks) at joints and cracks, but only a minor amount of spalling has occurred.

0.88–0.95, if the PCC slab exhibits "D" cracking or reactive aggregate cracking and spalling at joints and cracks to the extent that significant amount of spalling has occurred.

0.80–0.88, if the PCC slab exhibits "D" cracking or reactive aggregate cracking and spalling at joints and cracks to an extensive degree.

For bonded PCC overlays, the last two categories are combined: $F_{dur} = 0.80$ to 0.95 for pavements with "D" cracking and spalling. Bonded PCC overlays are normally not recommended under these conditions.

For composite pavements (AC/PCC), the durability of the underlying slab is difficult to determine from visual observations, unless the deterioration is severe and shear failures have actually occurred beneath the AC overlay. This results in localized failure areas that can be quite severe. It is very difficult to locate these areas, let alone to repair them. Selected coring, milling representative areas of the existing AC overlay, or removing a section of pavement across a traffic lane and observing the AC/PCC slab face are some ways to determine the extent of concrete durability problems in the base slab.

L6.1.3 Condition Factor F_{fat}

Most existing concrete slabs have carried traffic for many years and have developed some fatigue damage which affects future cracking and, therefore, should be considered in the overlay design. Consideration of this past fatigue damage will make the overlay a little thicker to retard future fatigue caused cracks developing in the PCC slab that would eventually reflect through the overlay, deteriorate and cause additional loss of serviceability. Therefore, the F_{fat} factor is included to adjust for the existing fatigue damage in the slab. It should be noted that the existing slab can have a very high flexural strength, but it will still have fatigue damage that will reduce its future fatigue life.

This past fatigue damage is usually considered through the calculation of Miner's fatigue damage, and the future allowable damage is 1.0 minus the past damage. For example, if past damage was computed to be 0.4 (or 40-percent fatigue consumed), future damage available until 50 percent of the slabs were cracked would be $1.0 - 0.4 = 0.6$ (or 60 percent), instead of 1.0 (or 100 percent). This results in a somewhat thicker overlay. However, the impact of this factor is not dramatic. (42) For example, calculations were performed that showed if the past Miner's damage varied from 0.1 to 0.75, the AC overlay thickness varied only about 2 inches. A variation of the F_{fat} factor from 1.0 to 0.9 also results in a change in AC overlay requirements of about 2 inches. Based upon this limited information, the following guidelines were developed.

F_{fat} = 0.97–1.00, few transverse cracks/punchouts exist (<5-percent slabs are cracked), very little fatigue damage has occurred.

0.94–0.96, a significant amount of transverse cracks/punchouts exists (5–15 percent slabs cracked), a significant amount of fatigue damage has occurred.

0.90–0.93, a large amount of transverse cracks/punchouts exists (> 15-percent slabs), a large amount of fatigue damage has occurred.

For AC/PCC pavements, the degree of fatigue damage in the PCC slab is very difficult to discern from the number of reflection cracks in the AC surface, since it may be difficult to distinguish reflection cracks caused by working cracks in the PCC from reflection cracks caused by joints or repairs. Field performance has not shown a significant number of "fatigue" type failures on composite pavements, which may be due to the beneficial effect that an overlay has on reducing both load and thermal curling stresses. Because there is no practical way to obtain a value for F_{fat}, it has been eliminated for second AC overlay design for composite pavements.

L6.1.4 Condition Factors for Unbonded Concrete Overlays

Unbonded concrete overlays of existing concrete pavements or of composite (AC/PCC) pavements require different considerations. Field surveys of unbonded jointed concrete overlays, separated by the conventional one-half to one inch of bituminous material layer, have not shown any evidence of reflection cracking or other distress from any of the following conditions (35):

- deteriorated joints or cracks in underlying slabs,
- additional fatigue cracking of the base slab, or
- durability problems in the base slab.

On the other hand, unbonded CRCP overlays have been shown to be more dependent on the supporting slab conditions, particularly unrepaired distress in the base slab. (49, 50, 51) Although the thickness design procedure provided is the same for both jointed and CRC overlays, it is emphasized that unbonded overlays are not intended to bridge areas of poor support, and in particular CRC overlays may require more pre-overlay repair in some situations.

Only one condition factor is used for thickness design of unbonded overlays, F_{jcu}, which is illustrated in Figure L6.4. This condition factor makes a smaller correction to the existing slab thickness D than the F_{jc} factor (Figure L6.1) which is used for design of bonded PCC overlays and AC overlays. This smaller correction reflects the fact that unbonded overlays are less likely to experience reflection cracking due to deteriorated joints or cracks in the existing slab. The F_{dur} and F_{fat} factors are not used at all in the unbonded overlay design procedure, since it has not been shown that unbonded overlay performance is sensitive to either durability problems or fatigue damage in the base slab.

When designing an unbonded overlay of an existing AC/PCC pavement, the existing AC pavement is not considered to make any contribution to D_{eff}, that is, the D used is the thickness of the PCC slab only.

L6.1.5 Condition Factor F_{ac}

Material problems in the existing AC surface layer of an AC/PCC pavement is likely to reflect those problems through a new AC overlay and cause increased loss of serviceability and service life. This can be adjusted to some extent by increasing the new overlay thickness to reduce vertical stresses and strains under wheel loads so that the design life can be achieved. This factor adjusts the existing AC layer's contribution to D_{eff} based on the quality of the AC material. The value selected should depend only on distresses related to the AC material (i.e., rutting, stripping, shoving, weathering and ravelling, but not reflection cracking) which are not eliminated by surface milling. Consideration should be given to complete removal of a poor-quality AC layer. Values of F_{ac} were developed based on engineering judgement for reasonable loss of PSI.

1.00:	No AC material distress
0.96–0.99:	Minor AC material distress (weathering, ravelling) not corrected by surface milling
0.88–0.95:	Significant AC material distress (rutting, stripping, shoving)
0.80–0.88:	Severe AC material distress (rutting, stripping, shoving)

L6.1.6 AC-to-PCC Conversion Factor A

A conversion factor, A is used in the AC overlay design procedures for PCC and AC/PCC pavements to convert PCC thickness deficiency to required AC overlay thickness, as shown below:

$$D_{ol} = A(D_f - D_{eff}) \qquad (6.1)$$

A value for A of about 2.5 is commonly considered reasonable for thin overlays. For example, a 2-inch bonded PCC overlay is roughly equivalent, in terms of stress in the base slab, to a 5-inch AC overlay. However, it is a concern that for greater PCC thickness deficiencies, using a value of 2.5 for A produces AC

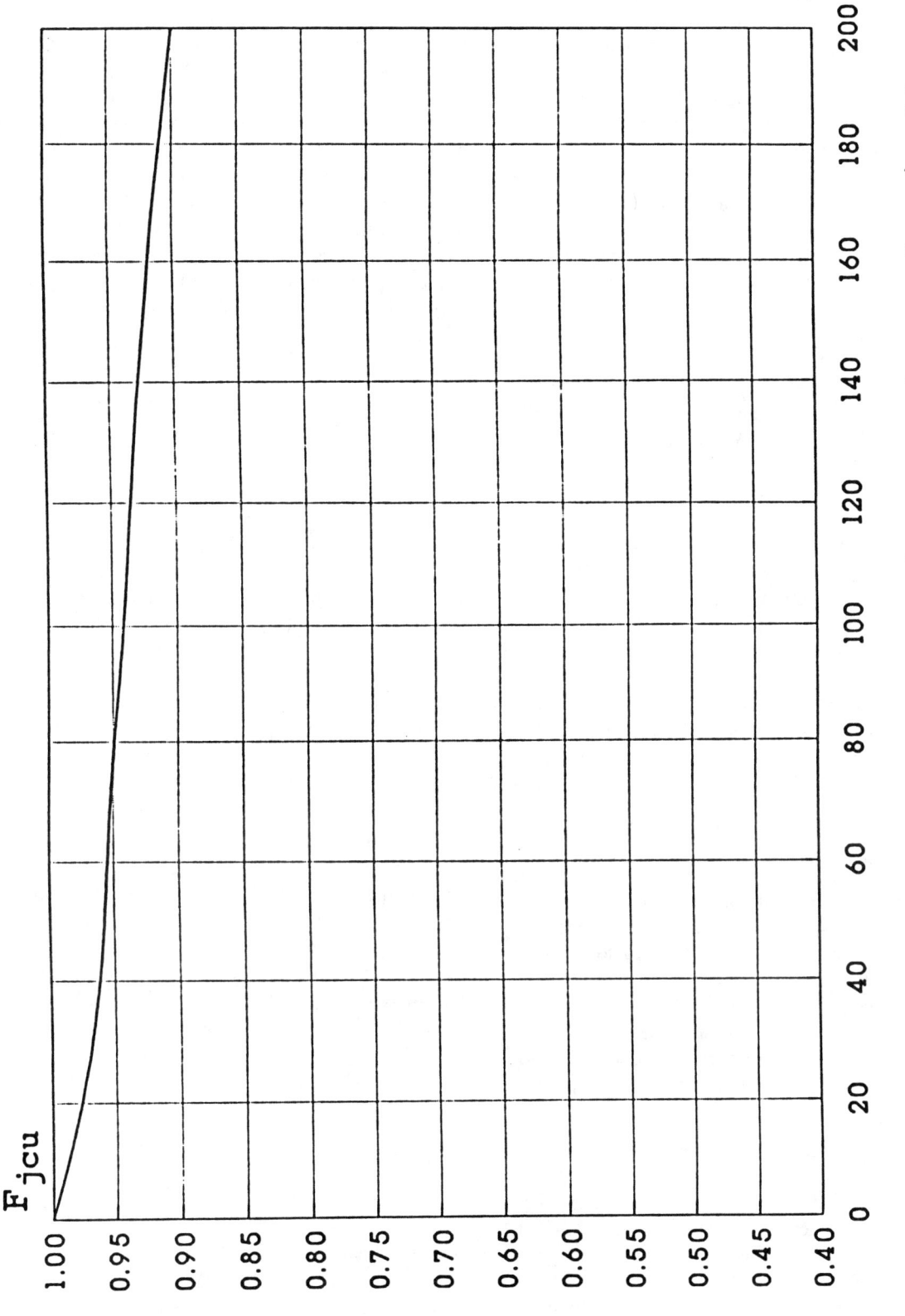

Figure L6.4. F_{jcu} Adjustment Factor for Unbonded JPCP, JRCP, and CRCP Overlays

overlay thicknesses which are not realistic. For example, a 6-inch bonded PCC overlay would correspond to a 15-inch AC overlay if A = 2.5. This concern warrants an investigation and reconsideration of the A factor for design of AC overlays of PCC pavements.

The Corps of Engineers has used a value of A = 2.5 for roadway and airfield AC overlay design for many years. The historical development of the Corps of Engineers' overlay design procedures was described by Chou in 1984. (40) The value of 2.5 for A was determined experimentally in the early 1950's based on accelerated traffic tests on six test tracks. The concrete slab thicknesses ranged from 6 to 12 inches and the AC overlay thicknesses ranged from 3 to 42 inches. The six test tracks also encompassed a range of subgrade strengths. Only 26 of the 53 test sections had AC overlays which were full-depth, dense-graded AC. The remainder of the "flexible" overlays were plant-mix black base with an AC surface, water-bound macadam with an AC surface, stabilized crushed rock with an AC surface, and sand-asphalt base with an AC surface.

The analysis of the results from these field tests was reported by Mellinger and Sale in 1956. (41) Reference 41 contains a plot of the test results for the AC overlays, in which PCC deficiency (full slab design thickness for test loading minus existing thickness) is plotted against AC overlay thickness. A straight line with a 1:2.5 slope was drawn through the plotted points. This line is not a best-fit line, but rather a recommended design line, as described by Mellinger and Sale:

> "A 'design' line, which is on the conservative boundary of the points plotted, has been placed on Figure 7. In fact, the design line, which shows a 2.5 to 1 ratio between flexible overlay thickness and concrete deficiency, encompasses all items that were carried to failure; and only those items which were not failed under the imposed traffic fall below this recommended slope. For these unfailed items, the indication is that less overlay thickness would have carried the same number of coverages."
> (41)

A summary of the results from Reference 41 is given in Table L6.1. It is evident that most of the A factors are closer to 2.0 than 2.5. Of the thicker overlays with higher A factors, nearly all except those at the Sharonville No. 1 test track did not fail during the course of the field tests.

To investigate further what A factor should be used in design of AC overlays of PCC pavements, the elastic layer program BISAR was used to compute stresses in PCC slabs with a range of PCC modulus (1–6 million psi), AC modulus (250–750 ksi), bonded PCC overlay deficiency (1–8 inches) and corresponding AC overlay thicknesses (1–16 inches). The A factor required (for an AC overlay thickness which would produce the same stress in the base slab as a given thickness of bonded PCC overlay) was found to decrease as the PCC overlay thickness increased. The value of the A factor depends on the elastic modulus of the AC, which of course varies daily and seasonally with temperature. For an AC modulus corresponding to the standard temperature of 68°F (450 ksi) used in Part II of the Guide, the following equation was obtained.

$$A = 2.2233 + 0.0099(D_f - D_{eff})^2 - 0.1534(D_f - D_{eff}) \qquad (6.2)$$

where

A = Factor to convert PCC thickness deficiency to AC overlay thickness
D_f = Slab thickness to carry future traffic, inches
D_{eff} = Effective thickness of existing slab, inches
R^2 = 99 percent
standard error = 0.018

In addition to determining AC overlay thickness requirements, an A factor must also be used to convert an existing AC overlay to an equivalent thickness of PCC for determination of the D_{eff} of an AC/PCC pavement. Again, the true value changes with the elastic modulus of the in-service AC overlay, but for design purposes, a single value is required. Based on the results of the field and analytical studies described here, a value of 2.0 was selected, as shown in the following equation.

$$D_{eff} = (D_{pcc} * F_{jc} * F_{dur}) + \left[\left(\frac{D_{ac}}{2.0}\right) * F_{ac}\right] \qquad (6.3)$$

where

D_{pcc} = thickness of existing PCC slab, inches
D_{ac} = thickness of existing AC surface, inches

Table L6.1. Results of Corps of Engineers Study of PCC-vs-AC Overlay A Factor (41)

Test Site	Overlay Type	PCC Deficiency (inches)	AC Overlay Thickness (inches)	A Factor
Sharonville No. 1	All AC binder and surface	2.2	5.6	2.5
		1.6	6.5	4.1
		1.1	8.0	7.3
		1.3	8.1	6.2
		3.6	9.5**	2.6**
		2.3	9.5	4.1
		2.1	9.6	4.6
Sharonville No. 1	Black base and 4-inch AC surface	4.9	9.8	2.0
		6.8	13.2	1.9
		7.2	15.8	2.2
		3.5	8.2	2.3
		5.6	12.2	2.2
		5.3	10.3	1.9
Sharonville No. 1	Waterbound macadam and 4-inch AC surface	6.2	12.1	2.0
		7.9	18.0	2.3
		9.6	24.5**	2.6**
Sharonville No. 2	All AC binder and surface	5.0	9.5	1.9
		7.2	14.5	2.0
		8.6	20.5	2.4
		3.4	6.0	1.8
		4.5	9.5	2.1
		5.8	11.5	2.0
		2.2	4.25	1.9
		3.5	7.0	2.0
		4.6	9.5	2.1
		2.2	4.0	1.8
		1.5	3.0	2.0
Sharonville No. 2	Stabilized crushed rock and 4-inch AC surface	9.3	42.0**	4.5**
		9.3	35.0**	3.8**
		7.3	23.0**	3.2**
		7.3	17.0**	2.3**
		5.4	12.5	2.3
		4.0	9.0	2.2
Maxwell	All AC binder and surface	1.7	6.0**	3.5**
		2.1	3.0**	1.4
		1.8	4.0**	2.2
Lockbourne No. 3	Black base and 3-inch AC surface	2.7	6.0**	2.2**
		2.7	9.0**	3.3**
Lockbourne No. 3	Sand-asphalt base and 3-inch AC surface	2.7	9.0**	3.3**

**AC overlay carried more than the 3,000 coverages applied in the test without failure.

L6.2 D_{eff} BASED ON REMAINING LIFE

The remaining life approach to determination of D_{eff} is the same for rigid pavements as for flexible pavements, except that structural capacity is characterized by slab thickness terms (D and D_{eff}) rather than structural number terms (SN and SN_{eff}). The reader is referred to Section 5.2 for a thorough discussion of this topic. For both rigid and flexible pavements, it is emphasized that the remaining life approach to determining effective structural capacity does not reflect any benefit for pre-overlay repair.

The remaining life approach is applicable only to bare AC and PCC pavements, for which the allowable traffic to failure ($N_{1.5}$) may be determined from the new pavement design equations or nomographs given in Part II of the Guide. It is not applicable to pavements which are constructed as new composites (AC/PCC), since no design equation is given in Part II for this type of pavement. The remaining life approach is also not applicable to AC or PCC pavements which have already been overlaid one or more times. For these reasons, no remaining life approach to D_{eff} determination is given in the design procedure for AC overlays of AC/JCP and AC/CRCP pavements.

L6.3 D_{eff} BASED ON NONDESTRUCTIVE TESTING

Long-term deflection testing of rigid pavements has shown that no change occurs in deflections taken at midslab until after slab cracking occurs at that location. There may be a relationship between pavement deflections and D_{eff} for rigid pavements; however, it is complicated by the fact that the measurements would have to be taken across joints and cracks. This presents difficulties in field testing (which cracks of what severities to test, where to place the load plate and sensors) and also in the analysis (which crack data to use, how to include results from cracks and joints as well as midslab locations). Further research is needed before this approach can be developed. For this reason, no NDT approach to determination of D_{eff} is given in the overlay design procedures for rigid pavements.

L7.0 CONCLUSIONS

The overlay design procedures presented in the revisions to Chapter 5 utilize the concepts of structural deficiency, structural number for flexible pavements, and future required structural capacity determined from the AASHTO flexible and rigid pavement design equations. These concepts were retained to maintain compatibility between Parts II and III of the Guide.

Development of a more sophisticated mechanistic approach to overlay design was not within the scope of this project. Nondestructive deflection testing for characterization of the existing pavement is recommended, to the extent appropriate within the framework of these empirical design procedures.

The 1986 AASHTO overlay design procedures were extensively revised to make them easier to use, more adaptable to calibration by local agencies, and more comprehensive. Key revisions to the overlay design procedures include the following.

(1) Guidelines for overlay type feasibility
(2) Guidelines for several important considerations:
    Pre-overlay repair
    Subdrainage
    Shoulders
    AC rutting
    PCC durability
    Pavement widening
    Reflection crack control
    AC surface milling
    AC surface recycling
    Overlay design reliability level
    PCC overlay bonding/separation layers
    PCC overlay joints and reinforcement
(3) Description of complete step-by-step overlay design procedure for each overlay type:
    AC overlay of ACP
    AC overlay of fractured slab PCC pavements
    AC overlay of JPCP, JRCP, and CRCP
    AC overlay of AC/JPCP, AC/JRCP, and AC/CRCP
    Bonded PCC overlay of JPCP, JRCP, and CRCP
    Unbonded JPCP/JRCP/CRCP overlay of JPCP, JRCP, CRCP, or AC/PCC
    JPCP, JRCP, and CRCP overlay of ACP
(4) Guidelines for nondestructive and visual/coring and testing for overlay design
(5) Guidelines for selecting inputs for determination of required future structural capacity (SN_f, D_f)
(6) Guidelines for characterization of effective structural capacity of existing pavement (SN_{eff}, D_{eff}) using three approaches:

Visual condition survey and materials testing
NDT testing (where appropriate)
Remaining life (where appropriate)
Adjustments to effective structural capacity may be made based upon pre-overlay repair.

(7) Improved adaptability of the overlay thickness design procedures to local conditions to produce more reasonable answers.

Many example overlay designs are provided in Reference 37 that illustrate the application of the procedures under different conditions. The results achieved with the procedures appear to generally provide adequate overlay thickness designs. There exists several inputs that can be adjusted to tailor the procedures to any given highway agency, such as design reliability level, selection of resilient modulus of subgrade, layer coefficients, and joint load transfer J factor.

REFERENCES

1. American Association of State Highway and Transportation Officials, *Guide for Design of Pavement Structures*, Washington, D.C., 1986.
2. American Association of State Highway and Transportation Officials, *Interim Guide for Design of Flexible Pavement Structures*, Committee on Design, 1961.
3. American Association of State Highway and Transportation Officials, *Interim Guide for Design of Rigid Pavement Structures*, Committee on Design, 1962.
4. Wang, M.C. and Kilareski, W.P., "Field Performance of Aggregate-Lime-Pozzolan Base Material," Transportation Research *Record* No. 725, 1979.
5. American Association of State Highway and Transportation Officials, *Interim Guide for Design of Pavement Structures*, 1972.
6. American Association of State Highway and Transportation Officials, *Interim Guide for Design of Pavement Structures*, Chapter III revised, 1981.
7. Elliott, R.P. and Thompson, M.R., "Mechanistic Design Concepts for Conventional Flexible Pavements," Transportation Engineering Series No. 42, Illinois Cooperative Highway Research and Transportation Program Series No. 208, University of Illinois at Urbana-Champaign, 1976.
8. American Association of State Highway and Transportation Officials, *Guide for Design of Pavement Structures*, Volume 2, Appendices, Washington, D.C., 1986.
9. Thompson, M.R. and Robnett, Q.L., "Resilient Properties of Subgrade Soils," Final Report—Data Summary, Transportation Engineering Series No. 14, Illinois Cooperative Highway Research and Transportation Program Series No. 160, University of Illinois at Urbana-Champaign, 1976.
10. Figueroa, J.L., "Resilient-Based Flexible Pavement Design Procedure for Secondary Roads," Ph.D. thesis, University of Illinois at Urbana-Champaign, 1979.
11. Raad, L. and Figueroa, J.L., "Load Response of Transportation Support Systems," *Transportation Engineering Journal*, American Society of Civil Engineers, Volume 106, No. TE1, 1980.
12. Traylor, M.L., "Characterization of Flexible Pavements by Nondestructive Testing," Ph.D. thesis, University of Illinois at Urbana-Champaign, 1979.
13. Hoffman, M.S. and Thompson, M.R., "Mechanistic Interpretation of Nondestructive Pavement Testing Deflections," Transportation Engineering Series No. 32, Illinois Cooperative Highway and Transportation Research Series No. 190, University of Illinois at Urbana-Champaign, 1981.
14. ERES Consultants, Inc., "Nondestructive Structural Evaluation of Airfield Pavements," prepared for U.S. Army Corps of Engineers Waterways Experiment Station, Vicksburg, MS, 1982.
15. Foxworthy, P.T., "Concepts for the Development of a Nondestructive Testing and Evaluation System for Rigid Airfield Pavements," Ph.D. thesis, University of Illinois at Urbana-Champaign, 1985.
16. Ioannides, A.M., "Dimensional Analysis in NDT Rigid Pavement Evaluation," *Transportation Engineering Journal*, American Society of Civil Engineers, Volume 116, No. TE1, 1990.
17. Ioannides, A.M., Barenberg, E.J., and Lary, J.A., "Interpretation of Falling Weight Deflectometer Results Using Principles of Dimensional Analysis," *Proceedings*, Fourth International Conference on Concrete Pavement

Design and Rehabilitation, Purdue University, 1989.
18. Westergaard, H.M., "Stresses in Concrete Runways of Airports," *Proceedings*, Highway Research Board, Volume 19, 1939.
19. Highway Research Board, "The AASHO Road Test, Report 5, Pavement Research," Special Report 61E, 1962.
20. Hall, K.T., "Performance, Evaluation, and Rehabilitation of Asphalt Overlaid Concrete Pavements," Ph.D. thesis, University of Illinois at Urbana-Champaign, 1991.
21. Southgate, H.F., "An Evaluation of Temperature Distribution Within Asphalt Pavements and its Relationship to Pavement Deflection," Kentucky Department of Highways, Research Report KYHPR-64-20, 1968.
22. Shell International Petroleum Company, "Pavement Design Manual," London, England, 1978.
23. Asphalt Institute, "Research and Development of the Asphalt Institute's Thickness Design Manual (MS-1) Ninth Edition," Research Report 82-2, 1982.
24. Thompson, M.R. and Cation, K.A., "A Proposed Full-Depth Asphalt Concrete Thickness Design Procedure," Transportation Engineering Series No. 45, Illinois Cooperative Highway and Transportation Program Series No. 213, University of Illinois at Urbana-Champaign, 1986.
25. Carpenter, S.H. and VanDam, T., "Laboratory Performance Comparisons of Polymer-Modified and Unmodified Asphalt Concrete Mixtures," Transportation Research *Record* No. 1115, 1987.
26. Elliott, R.P., "An Examination of the AASHTO Remaining Life Factor," Transportation Research *Record* No. 1215, 1989.
27. Boussinesq, J., "Application des Potentials à l'Étude de L'Equilibre et du Mouvement des Solides Elastiques," Gauthier-Villars, Paris, France, 1885.
28. Yoder, E.J., *Principles of Pavement Design*, John Wiley and Sons, 1957.
29. Odemark, N., "Investigations as to the Elastic Properties of Soils and Design of Pavements According to the Theory of Elasticity," Meddelande 77, Statens Väginstitut, Stockholm, Sweden, 1949. English translation provided by A.M. Ioannides, 1990.
30. Barber, E.S., author's closure, comments on C.A. Hogentogler, Jr.'s discussion of "Soil Displacement Under a Circular Loaded Area," by L.A. Palmer and E.S. Barber, *Proceedings*, Highway Research Board, Volume 20, 1940.
31. Ullidtz, P., *Pavement Analysis*, Elsevier Science Publishers B.V., 1987.
32. Ullidtz, P. "Overlay and Stage by Stage Design," *Proceedings*, Fourth International Conference on Structural Design of Asphalt Pavements, Ann Arbor, Michigan, 1977.
33. Vespa, J.W., Hall, K.T., Darter, M.I., and Hall, J.P., "Performance of Resurfacing of JRCP and CRCP on the Illinois Interstate Highway System," Transportation Engineering Series No. 61, Illinois Cooperative Highway and Transportation Research Program Series No. 229, 1990.
34. Thompson, M.R., "Breaking/Cracking and Seating Concrete Pavements," NCHRP *Synthesis* No. 144, 1989.
35. Voigt, G.F., Carpenter, S.H., and Darter, M.I., "Rehabilitation of Concrete Pavements, Volume 2—Overlay Rehabilitation Techniques," Federal Highway Administration Report No. FHWA-RD-88-072, 1989.
36. Kilareski, W.P. and Bionda, R.A., "Performance/Rehabilitation of Rigid Pavements, Phase II, Volume 2—Crack and Seat and AC Overlay of Rigid Pavements," Federal Highway Administration Report No. FHWA-RD-89-143, 1989.
37. Darter, M.I., Elliott, R.P., and Hall, K.T., "Revision of AASHTO Pavement Overlay Design Procedures, Appendix: Overlay Design Examples," NCHRP Project 20-7/Task 39, Final Report, April 1992.
38. Pavement Consultancy Services/Law Engineering, "Guidelines and Methodologies for the Rehabilitation of Rigid Highway Pavements Using Asphalt Concrete Overlays," for National Asphalt Paving Association, June 1991.
39. Carpenter, S.H., "Layer Coefficients for Flexible Pavements," ERES Consultants, Inc., for Wisconsin DOT, August 1990.
40. Chou, Y.T., "Asphalt Overlay Design for Airfield Pavements," *Proceedings*, Association of Asphalt Paving Technologists, Volume 53, April 1984, pp. 266–284.
41. Mellinger, F.M. and Sale, J.P., "The Design of Non-Rigid Overlays for Concrete Airfield Pavements," *Air Transport Journal*, American Society of Civil Engineers, Volume 82, Number AT 2, May 1956.
42. Seiler, W.J., "A Knowledge-Base for Rehabilitation of Airfield Concrete Pavements," Ph.D.

thesis, University of Illinois at Urbana-Champaign, 1991.

43. Kallas, B.F. and Shook, J.F., "Factors Influencing Dynamic Modulus of Asphalt Concrete," *Proceedings*, Association of Asphalt Paving Technologists, Volume 38, 1949.

44. Miller, J.S., Uzan, J., and Witczak, M.W., "Modification of the Asphalt Institute Bituminous Mix Modulus Predictive Equation," *Transportation Research Record* No. 911, 1983.

45. Barenberg, E.J. and Petros, K.A., "Evaluation of Concrete Pavements Using NDT Results," Project IHR-512, University of Illinois and Illinois Department of Transportation, Report No. UILU-ENG-91-2006, 1991.

46. Pavement Consultancy Services/Law Engineering, "FWD Analysis of PA I-81 Rubblization Project," for Pennsylvania Department of Transportation, February 1992.

47. Schutzbach, A.M., "Crack and Seat Method of Pavement Rehabilitation," *Transportation Research Record* No. 1215, 1989.

48. Ahlrich, R.C., "Performance and Structural Evaluation of Cracked and Seated Concrete," *Transportation Research Record* No. 1215, 1989.

49. Tyner, H.L., Gulden, W., and Brown, D., "Resurfacing of Plain Jointed Concrete Pavements," *Transportation Research Record* No. 814, 1981.

50. Crawley, A.B. and Sheffield, J.P., "Continuously Reinforced Concrete Overlay of Existing Continuously Reinforced Concrete Pavement," *Transportation Research Record* No. 924, 1983.

51. Turgeon, R. and Ishman, K.D., "Evaluation of Continuously Reinforced Concrete Overlay and Repairs on Interstate Route 90, Erie County, Pennsylvania," Pennsylvania Department of Transportation, Research Report No. 79-01, 1985.

APPENDIX M
AN EXAMINATION OF THE AASHTO REMAINING LIFE FACTOR

Robert P. Elliott

The 1986 AASHTO Pavement Design Guide introduced a remaining life factor that is applied in the design of pavement overlays. An examination of the remaining life concept was made to determine its practicality. The examination revealed inconsistencies in overlay designs determined using the AASHTO remaining life factor. Further investigation revealed that the remaining life factor should have a value of 1.0 for all overlay situations. As a result, it is recommended that the AASHTO overlay design approach be revised to exclude remaining life considerations.

The 1986 AASHTO Pavement Design Guide (*1*) introduced a remaining life concept that is applied in the design of overlays. The concept is based on the rationale that the structural capacity of a pavement decreases with load applications. For a pavement that has been overlaid, the structural capacity of the original pavement is a function of the loads applied before overlay as well as those applied after overlay. As presented by AASHTO, the remaining life concept requires that overlay thicknesses be selected considering both the "remaining" life of the pavement at the time of overlay and the expected "remaining" life when the next overlay will be applied.

For flexible pavement overlay design, the remaining life concept is applied using the equation:

$$SN_{ol} = SN_n - F_{rl} * SN_{eff} \qquad (1)$$

where

SN_{ol} = required structural number for the overlay
SN_n = total structural number required, based on traffic soils, etc.

Department of Civil Engineering, University of Arkansas, 4190 Bell Engineering Center, Fayetteville 72701.

F_{rl} = remaining life factor, a function of pavement condition prior to overlay and the condition predicted at the end of the design traffic
SN_{eff} = the effective structural number of the existing pavement at the time of overlay

The remaining life factor (F_{rl}) is determined using the graph shown as Figure M1. In using the graph, R_{Lx} is the remaining life factor of the existing pavement at the time of overlay, and R_{Ly} is the anticipated future remaining life of the overlaid pavement when it will be overlaid. Concern has been expressed regarding the F_{rl} concept. Of particular concern is the fact that at low values of R_{Lx} and R_{Ly}, the general slope of the F_{rl} curve reverses. This investigation was initiated to study the concept and to establish a rationale for this slope reversal.

The investigation demonstrated inconsistencies in overlay designs using the AASHTO remaining life concept and suggests that for consistent designs F_{rl} should be 1.0 for all values of remaining life.

CONCEPT OF REMAINING LIFE

The AASHTO remaining life concept is discussed in detail elsewhere (*2*). The following abbreviated discussion is presented for those not familiar with that document.

The remaining life concept was developed to be used in a structural deficiency approach to overlay design. In the structural deficiency approach, the structural requirement for the overlay (SN_{ol}) is determined as the difference between the structure needed to support future (design) traffic (SN_n) and the structural capacity of the existing pavement (SN_{eff}). F_{rl} was added to the basic structural deficiency equation to account for future structural damage to the existing pavement.

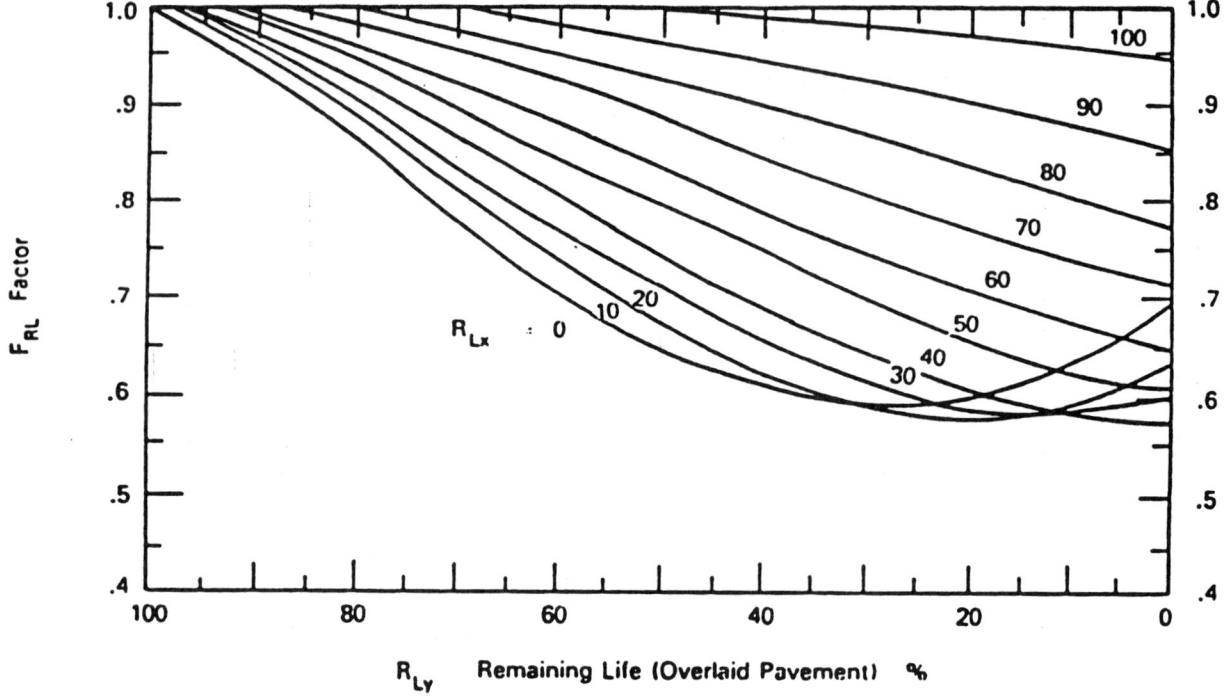

Figure M1. AASHTO Remaining Life Factor Curves (*1*)

The fundamentals of remaining life are illustrated in Figure M2 using the flexible pavement structural number as the measure of structural capacity. The serviceability of a pavement decreases with time and traffic from an initial value, Po. Without rehabilitation, the serviceability would eventually reach a "failure" level, P_f. The total number of traffic applications to "failure" is shown as N_f.

At some point prior to failure, however, an overlay is placed. The traffic applications to that point are x. The remaining life (R_{Lx}) is defined as the additional applications that could have been applied to "failure" expressed as a fraction of the total possible applications. That is:

$$R_{Lx} = (N_f - x)/N_f \qquad (2)$$

The structural capacity of the pavement decreases similarly from SN_o to SN_f. At the time of overlay, the pavement structural capacity is SN_x. A pavement condition factor (C_x) can be defined as:

$$C_x = SN_x/SN_o \qquad (3)$$

Since SN_x is also the effective structural capacity (SN_{eff}) of the pavement at the time of overlay, SN_{eff} can be expressed as a function of C_x and SN_o.

$$SN_{eff} = C_x * SN_o \qquad (4)$$

For the AASHTO Guide, a relationship between C_x and R_{Lx} was developed using the AASHTO flexible pavement design equation. C_x and R_{Lx} values were computed for various designs based on present serviceable indices at "failure" (P_f) of 1.5 to 2.5. These produced a "best-fit" relationship:

$$C_x = R_{Lx}^{0.165} \qquad (5)$$

A first step in this investigation was to attempt to reproduce this relationship. C_x and R_{Lx} values were computed for structural numbers ranging from 6.0 to 2.5, with P_f equal to 1.5 and 1.0. As shown in Figure M3, these values fit the AASHTO relationship reasonably well.

The AASHTO remaining life concept, however, does not use the "best-fit" relationship. Although the C_x values produced by the relationship were viewed as being realistic to R_{Lx} values as low as 0.005, the relationship was abandoned because C_x goes to zero at

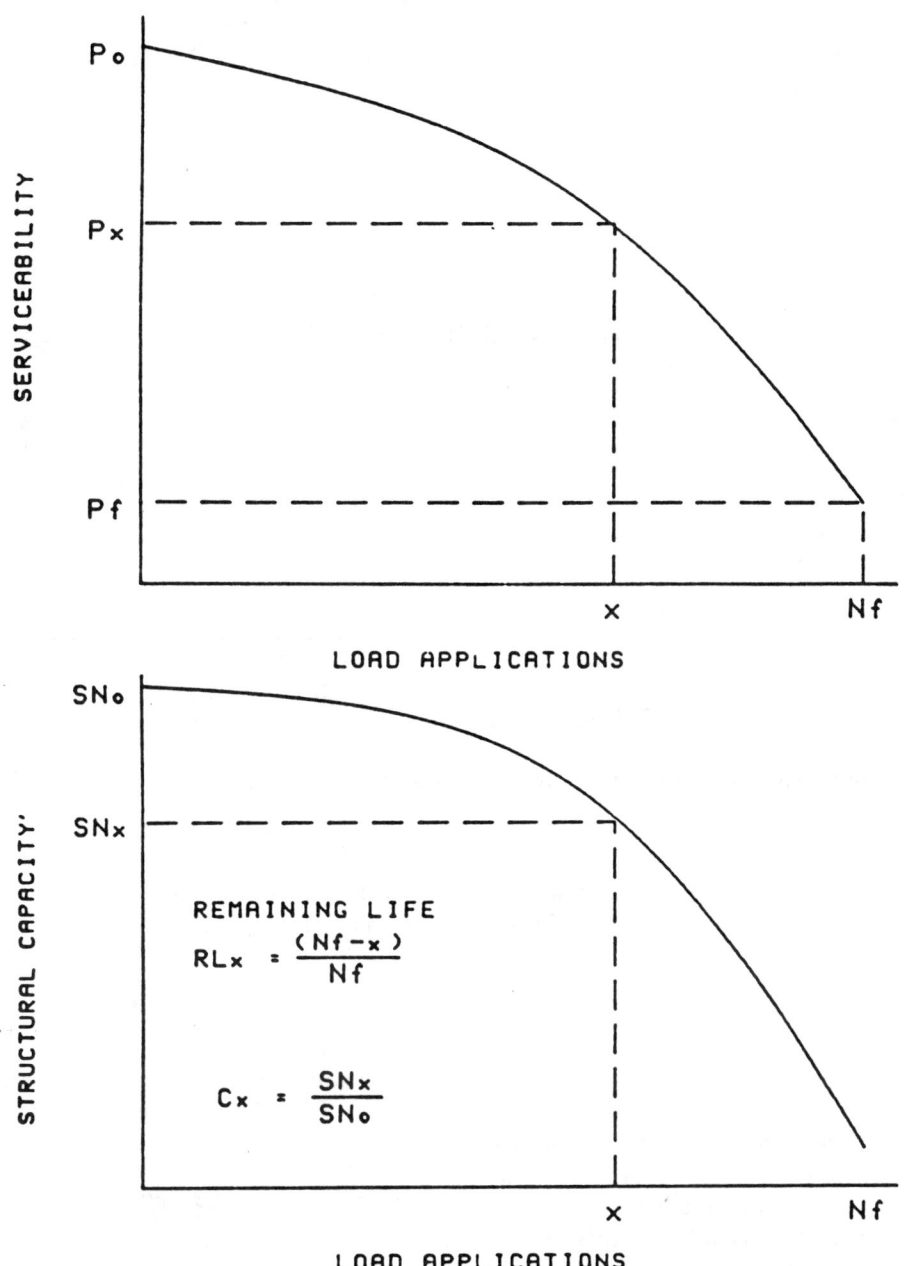

Figure M2. Illustration of the Remaining Life Concept

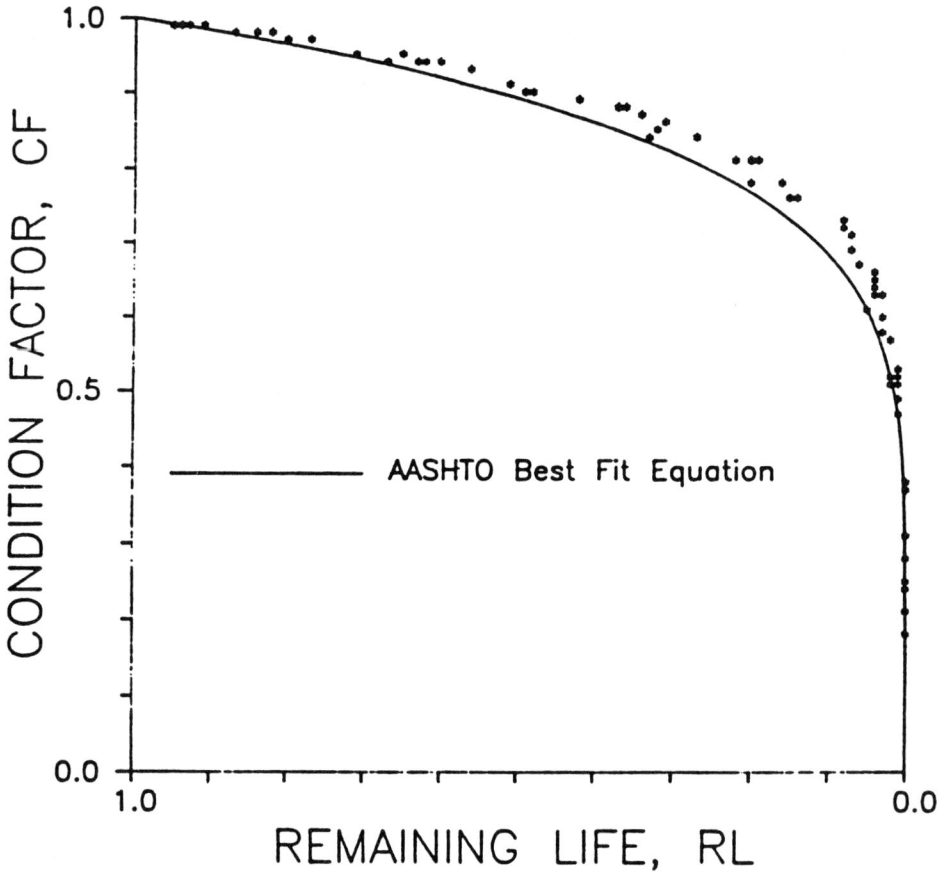

Figure M3. Comparison of Values from this Investigation with the AASHTO "Best-Fit" Equation

"failure" (R_{Lx} = zero). A modified relationship was used by AASHTO. The modified relationship (2) is:

$$C_x = 1 - 0.7 * e^{-(R_{Lx}+0.85)^2} \quad (6)$$

The best-fit and modified relationships are compared in Figure M4. In addition to C_x not going to zero at "failure," the modified relationship provides a C_x value for a negative remaining life. Although the meaning of a negative remaining life is not clear, this feature of the modified relationship is a necessary (although perhaps erroneous) part of the AASHTO application of remaining life.

APPLICATION OF REMAINING LIFE TO OVERLAYS

The reduction in structural capacity of the overlaid pavement is similar to that shown in Figure M2. Thus, if SN_n and y were used in place of the SN_o and x used previously, the structural capacity of the overlaid pavement after y load applications would be:

$$SN_y = C_y * SN_n \quad (7)$$

Without the remaining life factor (F_{rl}), SN_n is $SN_{ol} + SN_{eff}$. Thus, Equation 7 can be written:

$$SN_y = C_y * SN_{ol} + C_y * SN_{eff} \quad (8)$$

AASHTO (2) argued that this equation is incorrect since the existing pavement (SN_{eff}) would lose structural capacity at a greater rate than would the overlay (SN_{ol}). To "correct" the equation, AASHTO stated that $C_y * SN_{eff}$ should be replaced by a similar function that includes the original (new) structural number of the existing pavement (SN_o) and a condition factor (C_{yx}) that is a function of the traffic applications (or remaining life) both before and after the overlay. That is:

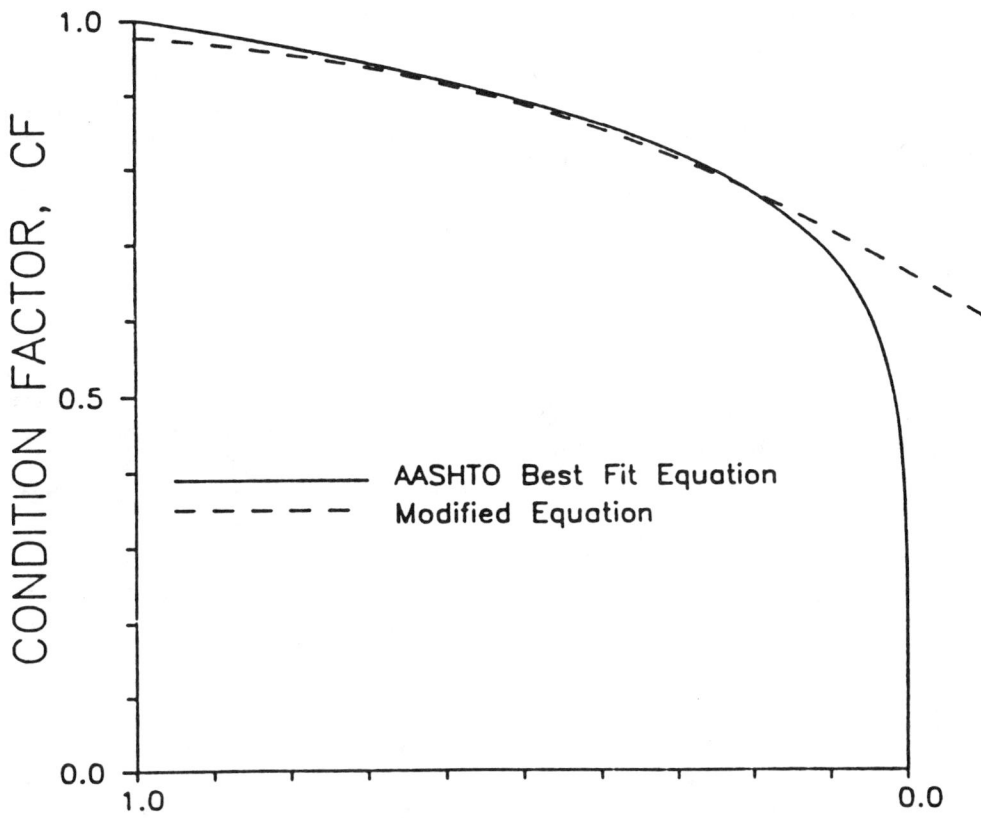

Figure M4. Comparison of the AASHTO "Best-Fit" and Modified Equations

$$C_{yx} = f(R_{Lx}, R_{Ly}) \quad (9)$$

and

$$SN_y = C_y * SN_{ol} + C_{yx} * SN_o \quad (10)$$

From these, AASHTO developed a relationship for F_{rl} in terms of C_{yx}, C_x, and C_y:

$$F_{rl} = C_{yx}/(C_x * C_y) \quad (11)$$

At this point, it should be noted that Equation 8 already included SN_o and a function of the traffic before and after overlay ($C_x * C_y$). Using Equation 4, SN_{eff} in Equation 8 may be replaced by $C_x * SN_o$, resulting in:

$$SN_y = C_y * SN_{ol} + C_x * C_y * SN_o \quad (12)$$

Nevertheless, the introduction of C_{yx} might be viewed as an advance since $C_x * C_y$ specifies the structural loss relationship for the existing pavement, while C_{yx} does not. Yet, in order to apply F_{rl}, it was necessary to assume an arbitrary relationship (Equation 13, below).

REMAINING LIFE FACTOR CURVES

The second step in the current investigation was to verify the remaining life factor curves (Figure M1). These curves were developed using Equations 6 and 11. However, because C_{xy} is a function of R_{Lx} and R_{Ly}, AASHTO has to assume a relationship between the two in order to apply Equation 6. It was assumed that the combined remaining life (R_{Lxy}) would be equal to the remaining life at the time of overlay (R_{Lx}) minus the damage done (d_y) during the period of overlay. That is:

$$R_{Lxy} = R_{Lx} - d_y \quad (13)$$

Since d_y is $1 - R_{Ly}$, this equation may be written:

$$R_{Lxy} = R_{Lx} + R_{Ly} - 1 \quad (14)$$

Initially, this assumption seems reasonable. However, it produces an uneasiness that grows with further reflection. By subtracting the full damage done after overlay, there seems to be no accounting for the reduction in the rate of damage that results from the lower load stresses due to the overlay. Also, because both R_{Lx} and R_{Ly} generally will be less than 0.5, the combined remaining life will be negative. A negative remaining life has no meaning. Finally, because the condition factor relationship itself (Equation 6) is assumed, this assumption (Equation 13) results in a compounding of assumptions.

Nevertheless, application of this assumption together with Equations 6 and 11 verified the mathematical accuracy of Figure M1, including the slope reversals at the lower values of R_{Lx} and R_{Ly}.

INCONSISTENCIES IN APPLICATION

The third step in the current investigation involved application of the F_{rl} factors to a hypothetical design situation to see if reasonable values and trends were produced. The design situation selected involved a design traffic ESAL of 5 million and an effective structural number for the existing pavement (SN_{eff}) of 4.5. The required overlay structural numbers (SN_{ol}s) were determined for terminal Present Serviceability Indices (PSIs) ranging from 3.5 to 1.55. The remaining life of the existing pavement (R_{Lx}) was also varied, using the values 0.0, 0.2, and 0.4.

The total structural number required (SN_n) and remaining life of the overlay (R_{Ly}) were computed using the AASHTO design equation (1) with a "failure" PSI of 1.5. A reliability to 50 percent and subgrade resilient modulus of 3,000 psi were used to reduce the equation to the original AASHO Road Test equation and eliminate any potential effects resulting from assumptions involved in adding reliability and subgrade modulus to the equation. To assure accuracy in application, the F_{rl} values were calculated in lieu of being taken from Figure M1.

The results of the analyses are listed in Table M1 and displayed graphically in Figure M5. The slope reversals seen in Figure M5 clearly illustrate an inconsistency. The major inconsistency, however, is the general negative slope of the curves between terminal PSIs of 2.0 to 3.0. For a given design situation, design to a lower terminal PSI should result in a lower required structural number. This is correctly illustrated by the trend of the SN_n values in Table M1. However, after F_{rl} is applied to establish the overlay requirement, the general trend for SN_{ol} is reversed.

Quite obviously, something is wrong with the AASHTO remaining life approach.

MODIFICATION OF THE REMAINING LIFE APPROACH

The final step in the investigation was to identify the problem with the concept and to develop a recommended correction. The apparent source of the problem is in the compounding of assumptions: first, with the modification of the C_x-R_{Lx} relationship (Equations 5 and 6) and, second, with the combined remaining life relationship (Equation 14).

As an alternative to Equation 14, the following development is suggested. The curve in Figure M6 represents some as yet undefined relationship between C and R_L. At some point (x), the pavement is overlaid and the existing pavement values are C_x and R_{Lx}. After the overlay, C of the existing pavement will continue to decline from C_x, but R_L will now be 100. This is represented on Figure M6 by the revised R_L scale.

At the time of the second resurfacing (y), the respective values are C_{yx} and R_{Ly}. A simple scale transformation of R_{Ly} from the revised scale to the original scale shows that:

$$R_{Lxy} = R_{Lx} * R_{Ly} \quad (15)$$

This equation for R_{Lxy} eliminates the need for a negative remaining life. The philosophy behind it is similar to the concept of the man who each day walks halfway to his destination. He never arrives. As long as the pavement is overlaid prior to "failure," "failure" is not reached in any component. The existing damage condition remains in the existing materials and progresses. However, the overlay is designed to slow the rate of additional damage, so that the "failure" condition is reached for the entire pavement.

Equations 15 and 11 were used to determine F_{rl} values with both the original C-R_L relationship (Equation 5) and the modified version (Equation 6). With the original relationship, F_{rl} is always 1.0:

$$\begin{aligned} F_{rl} &= (R_{Lxy})^{.165}/(R_{Lx}^{.165} * R_{Ly}^{.165}) \\ &= (R_{Lx} * R_{Ly})^{.165}/(R_{LX} * R_{LY})^{.165} \\ &= 1.0 \end{aligned} \quad (16)$$

Table M1. Overlay Computations Using Remaining Life Factors

Design ESAL = 5,000,000							SN$_{eff}$ = 4.5	
Terminal Required			R$_{Lx}$ = 0.0		R$_{Lx}$ = 0.2		R$_{Lx}$ = 0.4	
PSI	SN$_n$	R$_{Ly}$	F$_{rl}$	SN$_{ol}$	F$_{rl}$	SN$_{ol}$	F$_{rl}$	SN$_{ol}$
3.5	6.65	.904	.988	2.20	.999	2.15	1.00	2.15
3.25	6.02	.904	.945	1.77	.967	1.67	.987	1.58
3.00	5.59	.827	.881	1.63	.919	1.45	.955	1.29
2.50	5.03	.603	.711	1.83	.773	1.55	.848	1.21
2.25	4.84	.465	.633	1.99	.689	1.74	.776	1.35
2.00	4.69	.317	.589	2.04	.616	1.92	.703	1.53
1.75	4.57	.167	.605	1.85	.576	1.98	.642	1.68
1.60	4.50	.062	.665	1.51	.578	1.90	.615	1.73
1.55	4.48	.029	.694	1.36	.586	1.84	.610	1.74

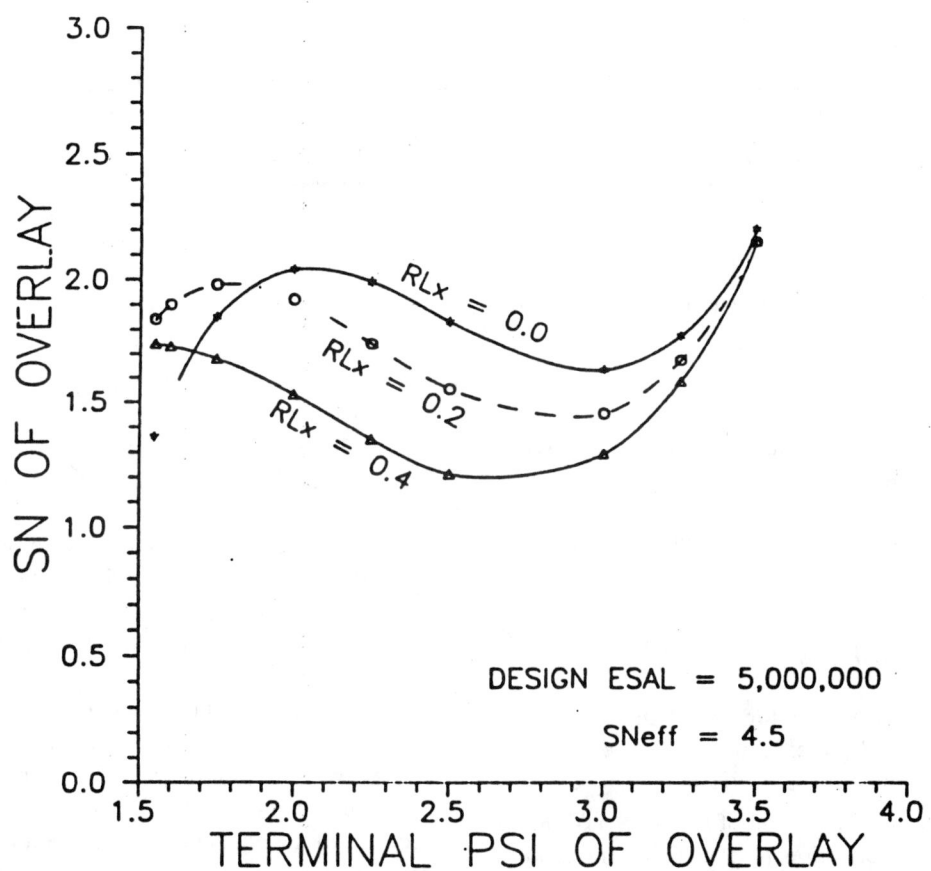

Figure M5. Results of Overlay Analyses Using the AASHTO Remaining Life Factor

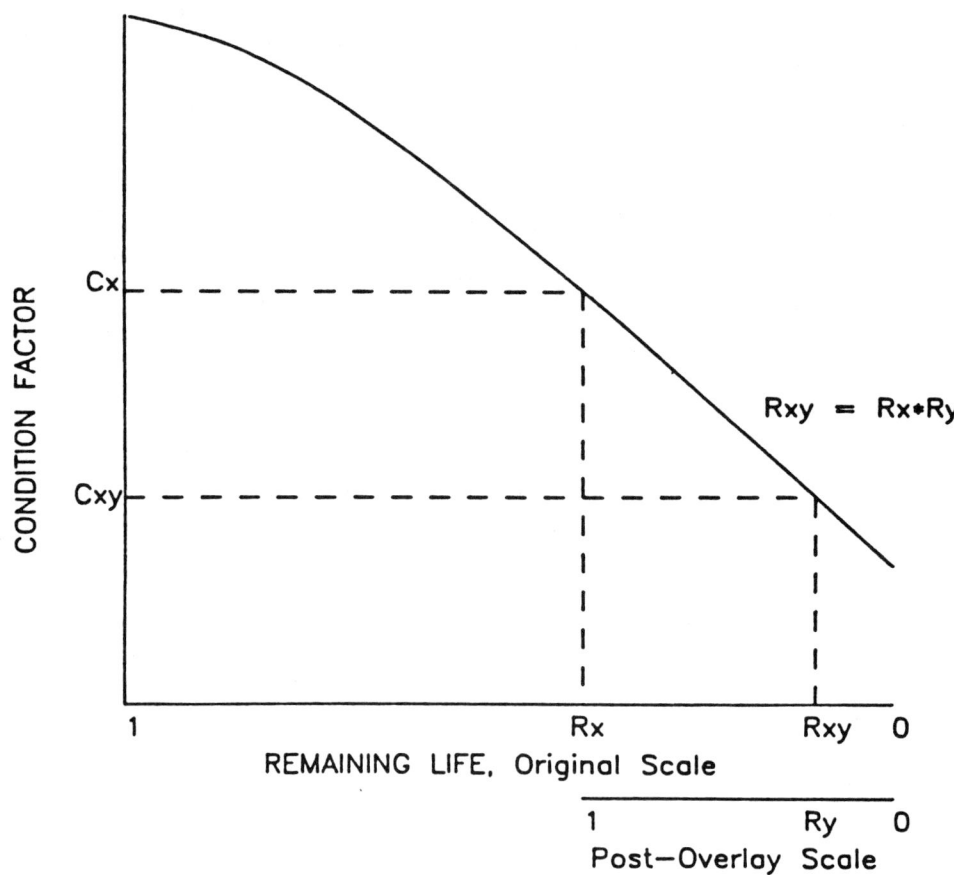

Figure M6. Modified Approach for Determining C_{xy}

With the modified AASHTO relationship (Equation 6), the equation is more complicated. However, except for very low values of both R_{Lx} and R_{Ly}, F_{rl} is generally about 1.0. At very low R_L values, F_{rl} becomes greater than 1.0. (At R_{Lx} and R_{Ly} equal to 0.0, F_{rl} is 1.5.)

OTHER DIFFICULTIES

Inconsistency in application is not the only difficulty with the AASHTO remaining life concept. Other difficulties need to be recognized and researched. The first of these is the application of the AASHTO Road Test performance equation to establish a remaining life-condition relationship.

The Road Test equation is an empirical relationship selected to provide a means of predicting the performance of the research pavements at the Road Test. It is not a theoretical or fundamental performance relationship and may, in fact, not even be the "best-fit" prediction relationship. It is simply the best relationship found by the researchers involved in the Road Test using the analytical tools that were available at that time. To apply the equation in the fashion used relative to remaining life represents a very significant extrapolation beyond the data and original intent of the equation.

Second, as it is being applied, the remaining life concept assumes that all materials will experience damage and structural loss at the same rate. It is conceivable that at "failure" a stabilized layer will be reduced to the equivalency of a granular layer while a granular layer may experience little loss.

The third difficulty is with the reliance on structural number. Many pavement engineers and researchers have expressed concern with the structural number approach to pavement design since it was first introduced. The structural number approach assume that each incremental thickness of a material provides an equal contribution to the structural capacity of the pavement regardless of the total thickness or total pavement configuration. Several studies have shown that this assumption is erroneous (3-6).

These difficulties are mentioned not to suggest abandonment of the AASHTO overlay approach but to remind the pavement design community of their existence, so that the procedures do not become "etched in stone." Additional thought and research in these areas are needed.

CONCLUSION AND RECOMMENDATION

This investigation has demonstrated that the AASHTO remaining life concept produced inconsistent overlay design thicknesses. The cause of the inconsistencies appears to be due to a compounding of assumptions used to produce the remaining life factor (F_{rl}) curves (Figure M1). An alternative approach developed as a part of this investigation found that the appropriate value for F_{rl} is 1.0. As a result, it is recommended that the AASHTO overlay design approach be revised to exclude remaining life considerations.

ACKNOWLEDGMENTS

This paper is based on a project entitled "Development of a Flexible Pavement Overlay Design Procedure Utilizing Nondestructive Testing Data," which is being conducted by the Arkansas Highway and Transportation Research Center, University of Arkansas. The project is sponsored by the Arkansas State Highway and Transportation Department and the U.S. Department of Transportation, Federal Highway Administration.

REFERENCES

1. *AASHTO Guide for Design of Pavement Structures*. American Association of State Highway and Transportation Officials, Washington, D.C., 1986.
2. *Remaining Life Considerations in Overlay Design*. Appendix CC, Vol. 2, AASHTO Guide for Design of Pavement Structures. American Association of State Highway and Transportation Officials, Washington, D.C., 1986.
3. Gomez, M. and Thompson, M., *Structural Coefficients and Thickness Equivalency Ratios*. Transportation Engineering Series 38. University of Illinois, Urbana-Champaign, 1983.
4. Dunn, H.D., Jr., *A Study of Four Stabilized Base Courses*. Ph.D. dissertation. Department of Civil Engineering, Pennsylvania State University, University Park, 1974.
5. Elliott, R.P., *Rehabilitated AASH(T)O Road Test—Analysis of Performance Data Reported in Illinois Physical Research Report 76*. Q1P-101. National Asphalt Pavement Association, Riverdale, Md., 1981.
6. Wang, M.C. and Larson, T.D., Performance Evaluation of Bituminous Concrete Pavements at the Pennsylvania State Test Track. In *Transportation Research Record 632*, TRB, National Research Council, Washington, D.C., 1977, pp. 21-27.

The contents of this paper reflect the view of the author, who is responsible for the facts and accuracy of the data presented herein. The contents do not necessarily reflect the official views of the Arkansas Highway and Transportation Department or the Federal Highway Administration. This paper does not constitute a standard, specification, or regulation.

Publication of this paper sponsored by Committee on Pavement Management Systems.

APPENDIX N
OVERLAY DESIGN EXAMPLES

N1.0 SUMMARY OF RESULTS FROM FIELD TESTING

This Appendix to the revised AASHTO overlay design procedure contains several example overlay designs for each of the pavement and overlay types addressed by the procedure. A total of seventy-four examples were developed to demonstrate and validate the procedures. These results were extremely useful in verifying and improving the overlay design procedures. The example design projects can also be used by future researchers to help verify improved overlay design procedures.

These examples were developed for actual in-service pavements located throughout the United States. Design, traffic, condition, and deflection data were provided for these projects by 10 State highway agencies. State personnel were actively involved in developing these examples during the development of the revised overlay design procedures. The overlay design procedures were evaluated by the highway agency personnel for clarity and ease of use and many comments were incorporated into the procedures.

In addition, the overlay thicknesses indicated by the procedures were evaluated with respect to State highway agencies' recommendations, based on their design procedures and experience with overlay performance.

Each of the example projects in this Appendix is identified by the region of the United States in which it is located and by number within the region. The following regional identifiers are used:

NE	Northeast
SE	Southeast
MW	Midwest
NW	Northwest
SW	Southwest

Each of the regions is represented in the overlay design examples for each pavement and overlay type to the extent possible. Seven separate groupings of overlays designs are included:

Overlay Type	Existing Pavement
AC	AC pavement
AC	Fractured PCC slab
AC and Bonded PCC	JPCP and JRCP
AC and Bonded PCC	CRCP
AC	AC/PCC (composite)
Unbonded PCC	JPCP, JRCP, CRCP
JPCP and JRCP	AC pavement

A summary of results obtained is presented for each of these groups. In addition, a single page spreadsheet showing all of the inputs and outputs for each project is given. Lotus 123 spreadsheets were prepared for each of the above overlay design procedures to aid in the calculations.

Deflection data were used whenever available from the State agency. Note that the spreadsheets only show one to five representative deflection basins so that the number of calculations required would be within reason. The basins chosen are believed to provide an overlay thickness close to the mean for the project. However, this does not imply that any project should be represented by this few a number of basins. On the contrary, the procedures can be programmed to handle any number of deflection basins and corresponding overlay designs very efficiently. To illustrate this approach and some results, four examples were developed using deflection data from several deflection basins along the projects.

The following major points are made relative to field testing of the procedures. Please see the individual summaries for each overlay group for more details.

1. Reliability level has a large effect on overlay thickness. The design reliability level that most often matched the overlay thickness constructed by the agency was approximately 95 percent as illustrated by plots of 95 percent thickness versus agency overlay thickness. However, there exists many design situations for which it is desirable to design at a higher or lower level of reliability.
2. Some overlay projects were designed for huge traffic loadings (more than 25 million ESALs. These projects should be very carefully considered since this is well beyond the limits of this overlay design procedure.
3. Results obtained from designing overlays with NDT deflections vs designing from condition survey techniques produced generally similar results. However, it is believed that the deflection procedure is by far the most accurate overall and is highly recommended. The condition survey method, coupled with materials testing, can be developed to give adequate results.
4. It is apparent from these results that different climatic/geographic zones require different overlay thicknesses, even if all other design inputs are exactly the same. The AASHTO Design Guide does not provide a way to deal with this problem. Therefore, each agency will need to test the procedures on their pavements and determine their reasonableness and required adjustments. There are many ways to adjust the procedure to produce desired overlay thicknesses (e.g., reliability, resilient modulus, J factor, etc.).

N2.0 VARIABILITY OF OVERLAY DESIGN THICKNESS ALONG A PROJECT

The individual overlay design examples given in this Appendix utilize from one to five deflection basins from a project to backcalculate the layer moduli and to then design the overlay. This limited number of deflection basins was used only in the interests of reducing the number of calculations and paperwork involved in reporting the results. For actual projects where deflection data are available, there will often be well over 100 deflection basins taken along the project. The overlay design procedures can handle any number of deflection basins through the development of efficient software. In fact, it is very informative and useful to calculate overlay thicknesses point by point along the project to directly see the variation involved.

The following four examples are provided to illustrate the NDT design approach using all the deflection basins measured on a project. Note that only the first 20 basins from two of the projects were analyzed. These examples show the variability that can be encountered and its influence on the design thickness.

Appendix N

Project 1

Existing pavement:	5 inches AC 5 inches granular base
Design traffic:	642,000 ESALs (8 years)

Overlay Results:	See Figure N1 profile of overlay thickness
	Mean overlay thickness
	    R = 50 percent, 2.3 inches (standard deviation 1.3 inches)
	    R = 90 percent, 3.7 inches (standard deviation 1.5 inches)

Project 2

Existing pavement:	1 inches AC 7 inches granular base
Design traffic:	47,500 ESALs (8 years)

Overlay Results:	See Figure N2 profile of overlay thickness
	Mean overlay thickness
	    R = 50 percent, 2.3 inches (standard deviation 1.1 inches)
	    R = 90 percent, 3.4 inches (standard deviation 1.3 inches)

Project 3

Existing pavement:	3 inches AC 12 inches granular base
Design traffic:	800,000 ESALs (10 years)

Overlay Results:	See Figure N3 profile of overlay thickness
	Mean overlay thickness
	    R = 50 percent, 2.3 inches (standard deviation 1.3 inches)
	    R = 90 percent, 4.2 inches (standard deviation 1.4 inches)

Project 4

Existing pavement:	6 inches AC 8.5 inches granular base
Design traffic:	1,000,000 ESALs (10 years)

Overlay Results:	See Figure N4 profile of overlay thickness
	Mean overlay thickness
	    R = 50 percent, 4.4 inches (standard deviation 1.8 inches)
	    R = 90 percent, 6.6 inches (standard deviation 2.0 inches)

The amount of variation in required AC overlay thickness along a highway pavement from point to point is quite high, having a coefficient of variation of about 50 percent. Therefore, it is important to measure a number of deflection basins along any given project to determine the mean and range of conditions that exist so that a reasonable overlay thickness can be selected for the design section. These profiles can be used to divide the section into two or more overlay design sections, if practical.

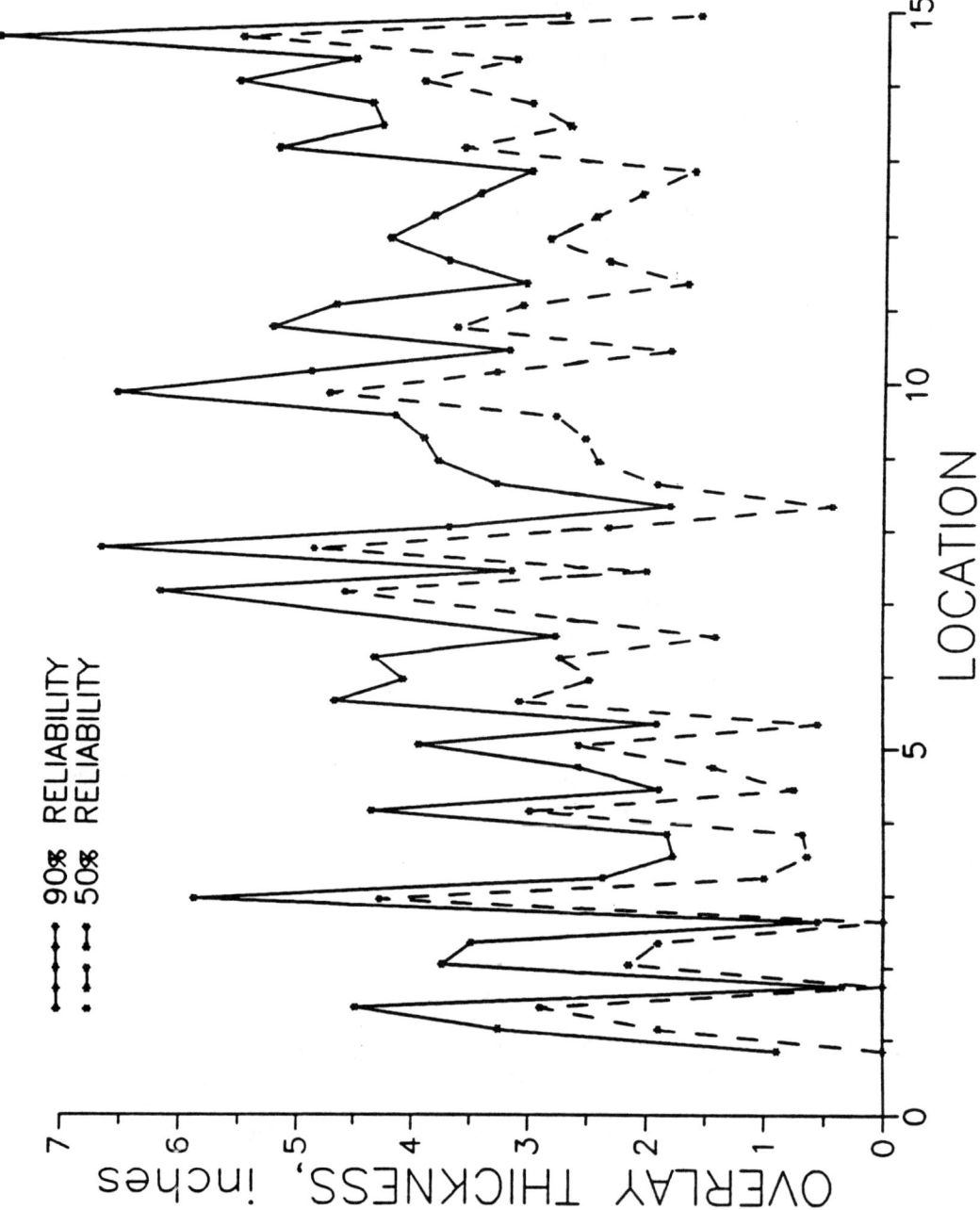

Figure N1. Profile of Design Overlay Thickness for AC Overlay of AC Pavement for Project 1

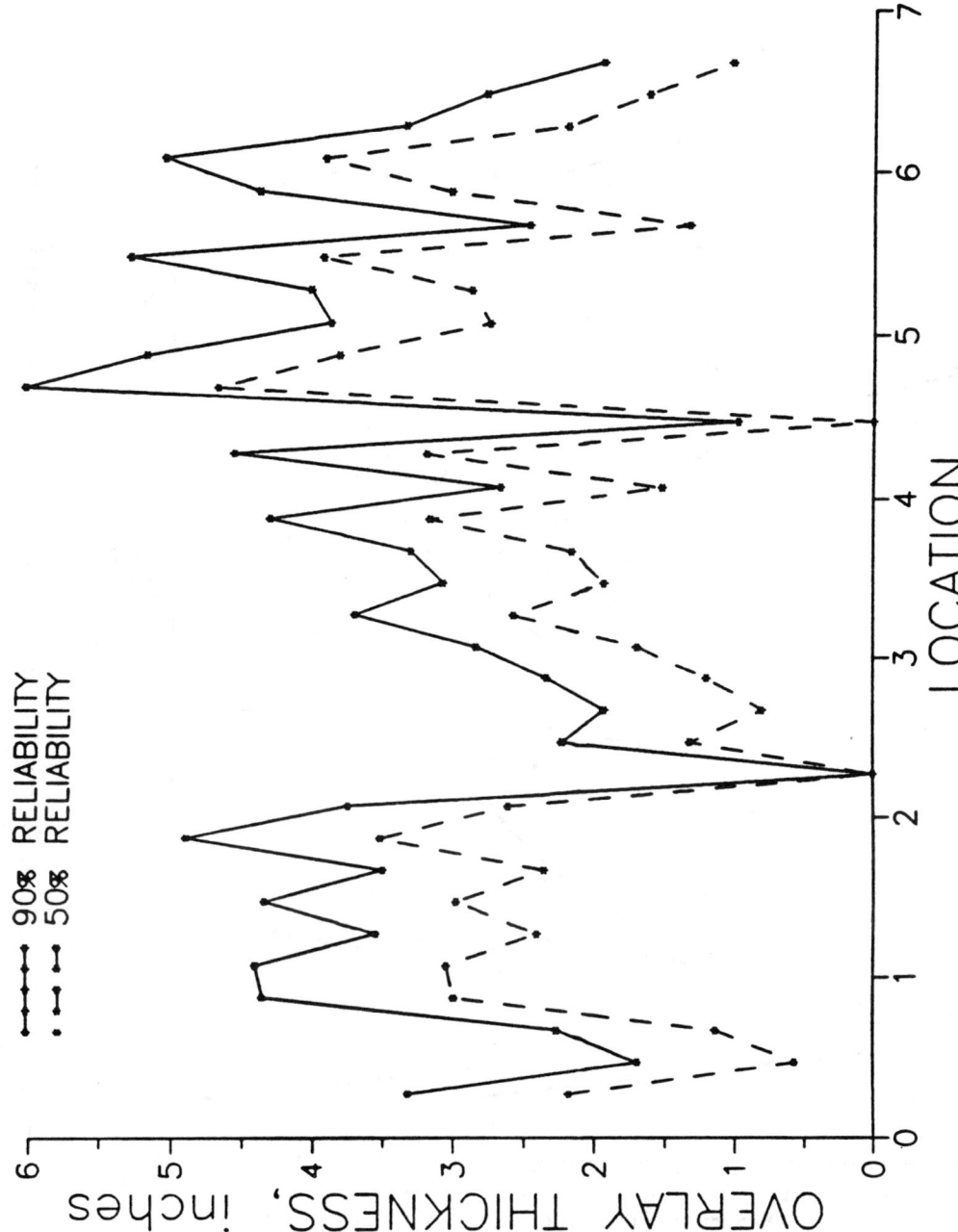

Figure N2. Profile of Design Overlay Thickness for AC Overlay of AC Pavement for Project 2

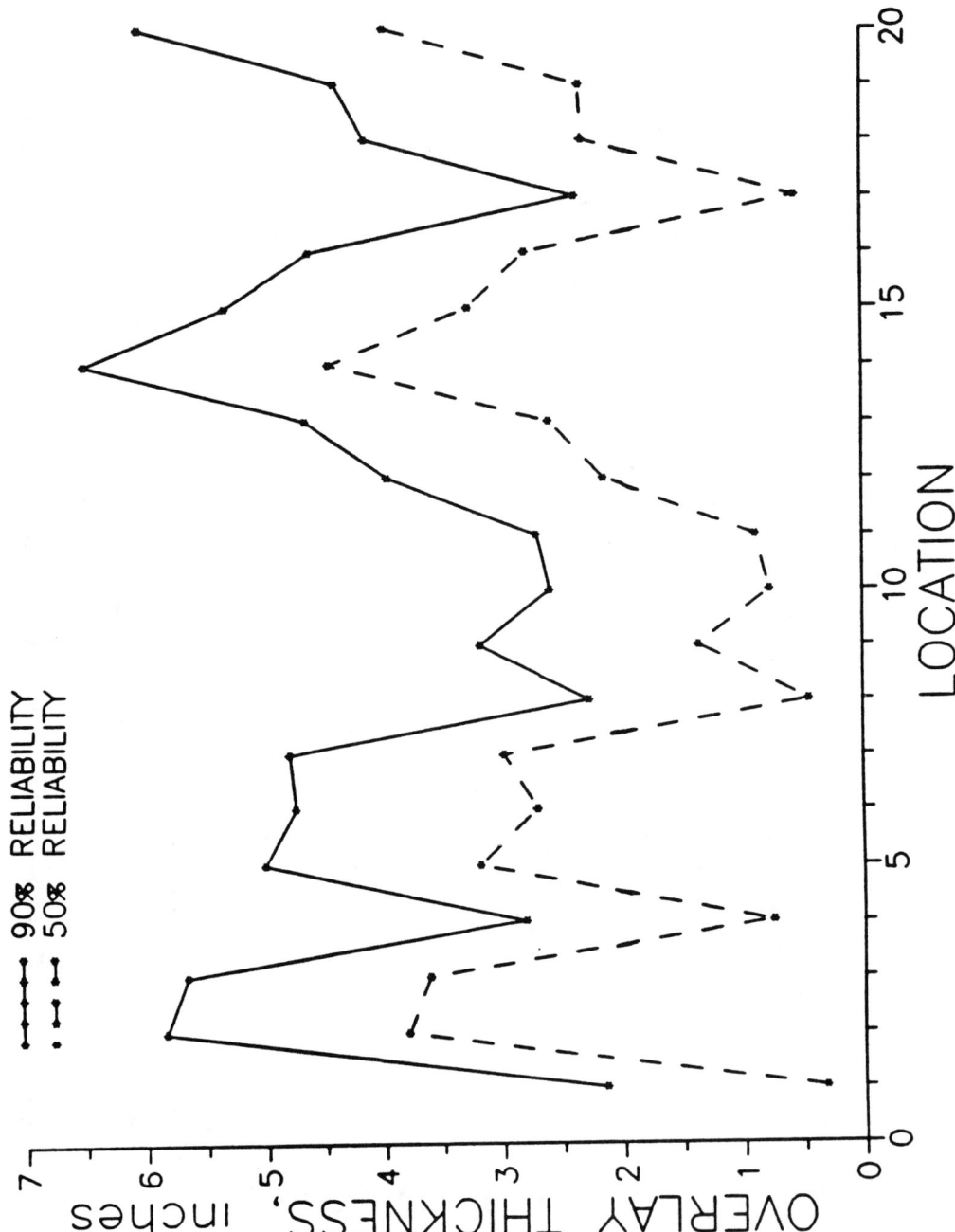

Figure N3. Profile of Design Overlay Thickness for AC Overlay of AC Pavement for Project 3

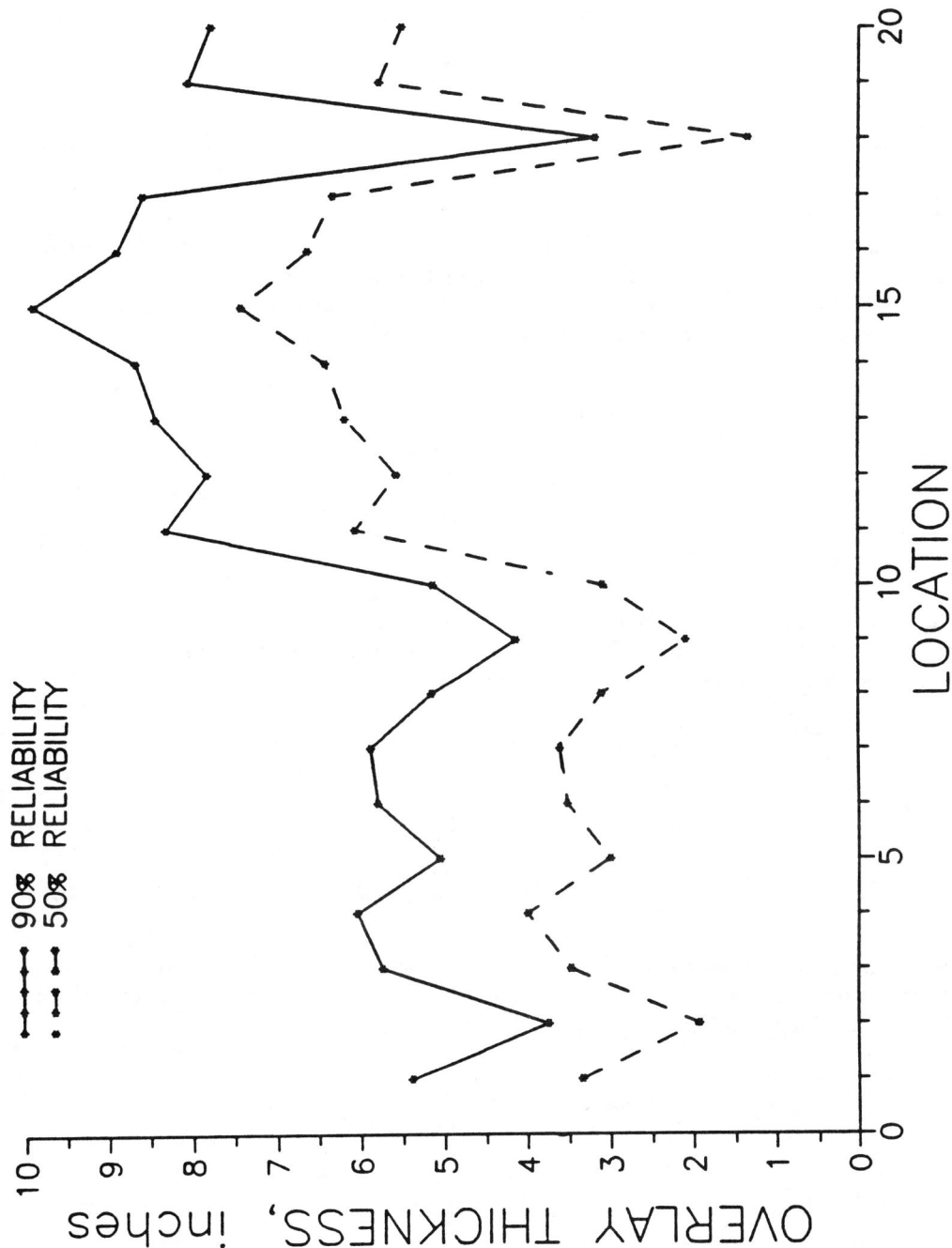

Figure N4. Profile of Design Overlay Thickness for AC Overlay of AC Pavement for Project 4

N3.0 AC OVERLAY OF AC PAVEMENT

Region-Project	Overlay Type	Existing Pavement	Design ESALs	Design Reliability	NDT Method Overlay Thickness (in)	Condition Method Overlay Thickness (in)
SW-1	AC	AC	11,000,000	50	0	0.5
				80	0	1.9
				90*	0.2	2.6
				95	0.9	3.4
				99	2.2	4.6

State design procedure indicates no AC overlay is needed which corresponds to a 90-percent reliability level. The overlay thicknesses shown above reflect one deflection basin which was identified as the highest deflection basin on the project. Therefore, other deflection basins would indicate a thinner overlay requirement. Overlay thicknesses obtained using the condition survey are rough estimates since a condition survey was not performed.

Region-Project	Overlay Type	Existing Pavement	Design ESALs	Design Reliability	NDT Method Overlay Thickness (in)	Condition Method Overlay Thickness (in)
SW-2	AC	AC	11,000,000	50	0	0
				80	0.5	1.3
				90	1.3	2.1
				95	1.9	2.7
				99*	3.3	4.0

State design procedure indicates a 4.2-inch overlay is needed, which corresponds to a 99-percent reliability level. Overlay thicknesses obtained using the condition survey are rough estimates since a condition survey was not performed.

Region-Project	Overlay Type	Existing Pavement	Design ESALs	Design Reliability	NDT Method Overlay Thickness (in)	Condition Method Overlay Thickness (in)
SW-3	AC	AC	11,000,000	50	1.6	1.0
				80	3.1	2.5
				90	4.0	3.3
				95*	4.7	4.0
				99	6.1	5.4

State design procedure indicates a 5.4-inch overlay is needed, which corresponds to a 95- to 99-percent reliability level. Overlay thicknesses obtained using the condition survey are rough estimates since a condition survey was not performed.

Region-Project	Overlay Type	Existing Pavement	Design ESALs	Design Reliability	NDT Method Overlay Thickness (in)	Condition Method Overlay Thickness (in)
SW-4	AC	AC	11,000,000	50	1.4	0
				80*	2.8	1.3
				90	3.6	2.1
				95	4.2	2.8
				99	5.5	4.1

State design procedure indicates a 3-inch overlay is needed, which corresponds to an 80- to 90-percent reliability level. Overlay thicknesses obtained using the condition survey are rough estimates since a condition survey was not performed.

Region-Project	Overlay Type	Existing Pavement	Design ESALs	Design Reliability	NDT Method Overlay Thickness (in)	Condition Method Overlay Thickness (in)
MW-1	AC	AC	100,000	50	0	1.4
				80	0.9	2.3
				90	1.4	2.8
				95*	1.9	3.3
				99	2.8	4.2

A 2-inch overlay is considered reasonable based upon other overlays placed on similar projects in this area. This corresponds to a 95-percent reliability level.

MW-2	AC	AC	150,000	50	1.7	1.8
				80*	2.8	2.9
				90	3.4	3.6
				95	4.0	4.1
				99	5.1	5.2

A 2.5-inch overlay is considered reasonable based upon other overlays placed on similar projects in this area. This corresponds to a 80-percent reliability level.

NW-1	AC	AC	2,400,000	50	1.6	2.2
				80	2.8	3.5
				90	3.5	4.2
				95	4.1	4.7
				99	5.2	5.9

No agency overlay design available.

NW-2	AC	AC	2,808,000	50	2.9	2.6
				80	4.2	3.9
				90	4.9	4.7
				95	5.5	5.3
				99	6.7	6.5

State design procedure gives overlay thicknesses of 2 to 7 inches for different sections of this project. Pavement thickness varies from 14 to 23 inches. The deflection basin used is an average for the project.

Region-Project	Overlay Type	Existing Pavement	Design ESALs	Design Reliability	NDT Method Overlay Thickness (in)	Condition Method Overlay Thickness (in)
NW-3	AC	AC	5,550,000	50	0	0
				80	0.7	0.9
				90	1.4	1.6
				95*	2.0	2.2
				99	3.2	3.4

State design procedure gives an overlay thickness of 2.5 inches, which corresponds to a 95- to 99-percent reliability level. Pavement thickness varies from 22 to 24 inches. The deflection basin used is an average for the project.

NW-4	AC	AC	880,000	50	0.5	0.2
				80	1.7	1.4
				90	2.3	2.0
				95	2.9	2.6
				99*	4.0	3.6

State design procedure gives an overlay thickness of 3.5 inches, which corresponds to a 95- to 99-percent reliability level. Pavement thickness varies from 15 to 26 inches. The deflection basin used is an average for the project.

NW-5	AC	AC	1,360,000	50	3.3	3.2
				80*	4.5	4.4
				90	5.1	5.1
				95	5.7	5.7
				99	6.8	6.8

State design procedure gives an overlay thickness of 4 inches, which corresponds to about a 70-percent reliability level. Pavement thickness varies from 6 to 10 inches. The deflection basin used is an average for the project.

NW-6	AC	AC	1,576,000	50	1.8	2.1
				80*	3.0	3.3
				90	3.7	4.0
				95	4.3	4.6
				99	5.4	5.7

State design procedure gives an overlay thickness of 2.5 inches, which corresponds to about a 70-percent reliability level. Pavement thickness varies from 10 to 26 inches. The deflection basin used is an average for the project.

Region-Project	Overlay Type	Existing Pavement	Design ESALs	Design Reliability	NDT Method Overlay Thickness (in)	Condition Method Overlay Thickness (in)
NE-1	AC	AC	931,327	50		0
				80		0
				90		0.7
				95*		1.3
				99		2.4

State constructed a 1.5-inch AC overlay which corresponds to a 95-percent reliability level using condition survey procedures. No deflection data are available. Subgrade resilient modulus was estimated from CBR using AASHTO Guide Appendix FF.

Region-Project	Overlay Type	Existing Pavement	Design ESALs	Design Reliability	NDT Method Overlay Thickness (in)	Condition Method Overlay Thickness (in)
NE-2	AC	AC	574,900	50		0
				80		0.9
				90		1.4
				95		2.0
				99*		2.9

State constructed a 3-inch AC overlay which corresponds to a 99-percent reliability. No deflection data are available. Subgrade resilient modulus was estimated from CBR using AASHTO Guide Appendix FF.

Region-Project	Overlay Type	Existing Pavement	Design ESALs	Design Reliability	NDT Method Overlay Thickness (in)	Condition Method Overlay Thickness (in)
NE-3	AC	AC	147,816 (10 years)	50		0
				80		0
				90		0
				95*		0.1
				99		0.8

State design procedure indicates 0.25-inch overlay thickness required which corresponds to a 95-percent reliability. State actually constructed a minimum 1-inch AC overlay. No deflection data are available. Subgrade resilient modulus was estimated from CBR using AASHTO Guide Appendix FF.

Region-Project	Overlay Type	Existing Pavement	Design ESALs	Design Reliability	NDT Method Overlay Thickness (in)	Condition Method Overlay Thickness (in)
NE-4	AC	AC	7,040,000 (20 years)	50		0
				80		1.1
				90*		1.7
				95*		2.3
				99		3.4

State recommends a 2-inch AC overlay plus leveling where necessary which corresponds to a 90- to 95-percent reliability. No deflection data are available. Subgrade resilient modulus was estimated from CBR using AASHTO Guide Appendix FF.

Summary of Results for AC Overlay of AC Pavement

1. In general, the revised AASHTO overlay thicknesses agree with State recommendations as shown in Figure N5. Some of the differences are due to the lack of consistent data from some of the projects. For example, some projects had thicknesses that varied widely, and the correlation between pavement thickness and deflection basins was unknown.
2. The revised AASHTO overlay thickness designs based upon NDT are generally consistent with those based on the condition survey results. Figure N6 shows the correlation between overlay thickness at the 95-percent level determined by NDT and condition survey procedures.
3. The subgrade resilient modulus has a large effect on the resulting overlay thicknesses. Therefore, it is of utmost importance to obtain an appropriate modulus value to enter into the AASHTO flexible pavement design equation. The reduction in backcalculated modulus by a factor of three appears reasonable. Use of too high a value will result in inadequate AC overlay thickness.

    Some data available from one State permits a direct comparison between laboratory and backcalculated modulus values:

Project	Lab M_R (psi)	Backcalculated M_R (psi)	Ratio
NW-2	6,000	13,483	2.25
NW-3	6,000	19,608	3.27
NW-4	4,150	14,085	3.39
NW-5	4,500	14,286	3.17
Averages:	5,163	15,365	3.02

    Even though the average ratio is 3.0, there is a wide variation. Each agency will need to evaluate this ratio, as well as other factors, to tailor the design procedure to its own conditions.
4. The design reliability level is very significant. The example projects ranged from collector highways to heavily traveled Interstate-type highways. A design reliability level of approximately 95 percent usually produced reasonable overlay thicknesses.

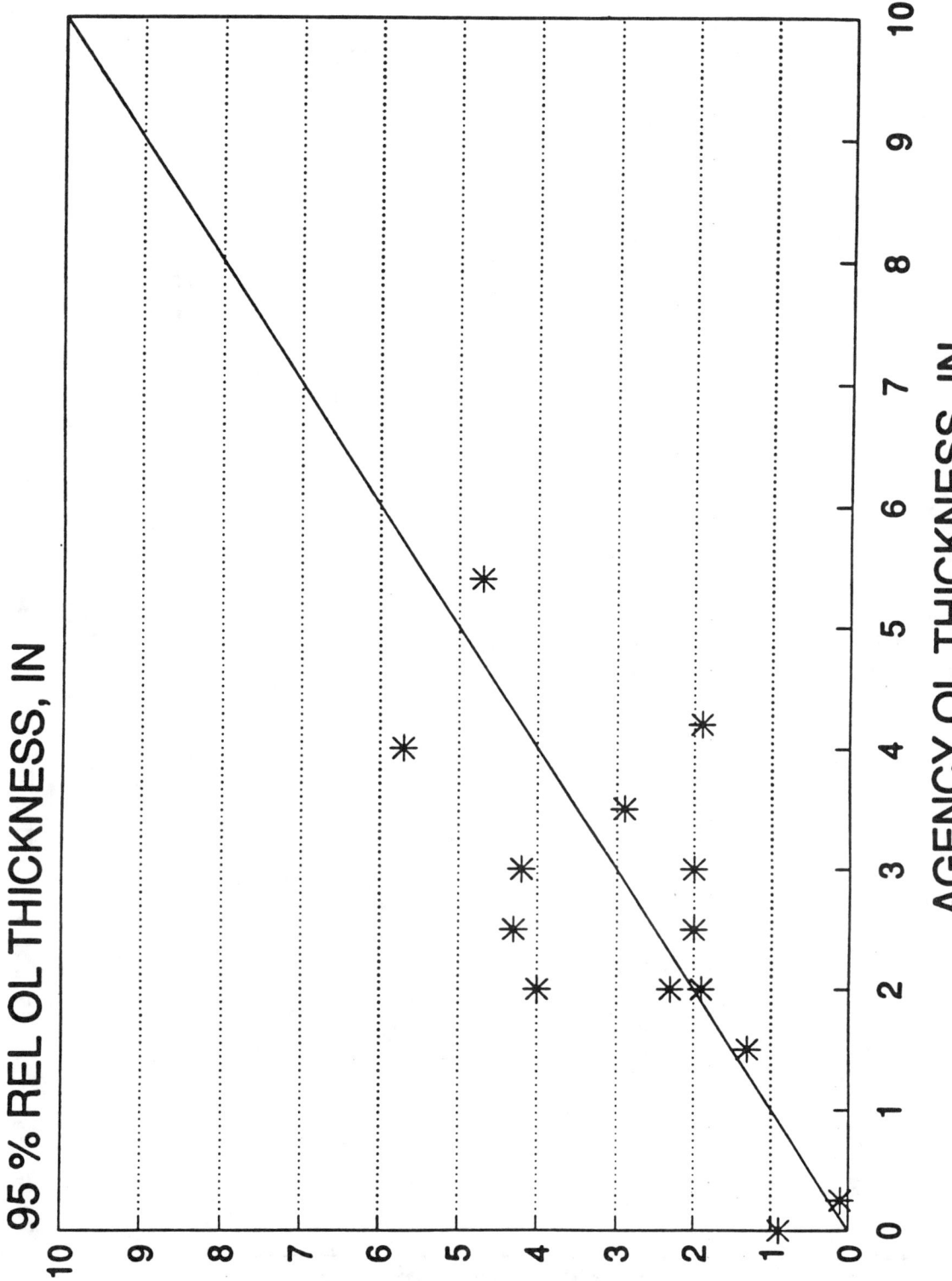

Figure N5. Comparison of Design AC Overlay Thicknesses and Agency AC Overlay Thicknesses for AC/AC Pavements (95-percent reliability level)

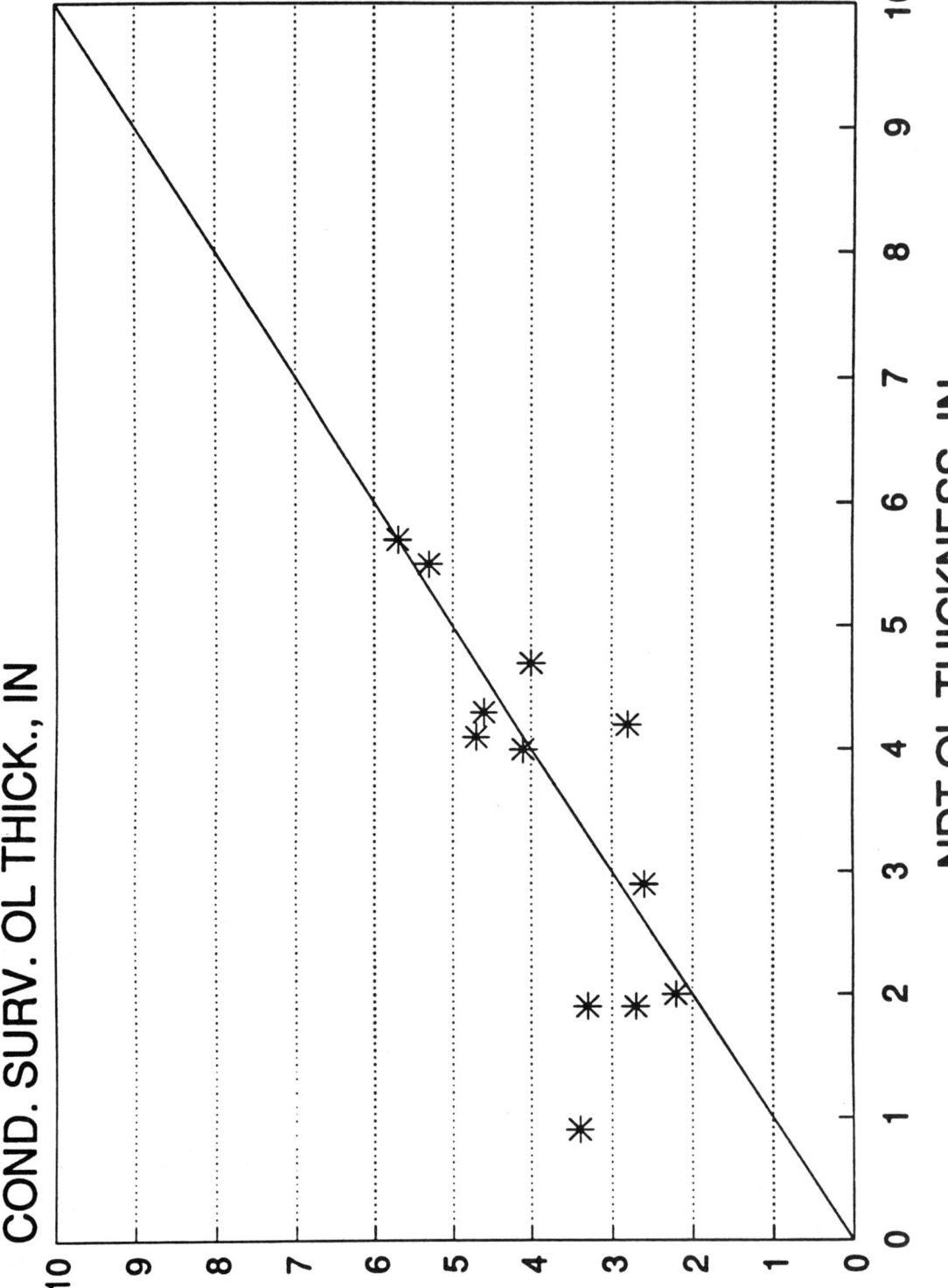

Figure N6. Comparison of Overlay Thicknesses Determined by NDT and Condition Survey Procedures for AC/AC Pavements (95-percent reliability level)

Appendix N

REVISED CHAPTER 5 AASHTO DESIGN GUIDE OVERLAY DESIGN

SW-1   AC OVERLAY OF CONVENTIONAL AC PAVEMENT (PROJ. 6044)

EXISTING PAVEMENT DESIGN

AC SURFACE	8.00 inches	SUBGRADE ?
GRAN BASE	3.00	
GRAN SUBBASE	10.40	
TOTAL THICKNESS	21.40	

Future design lane ESALs = 11,000,000 (FLEXIBLE ESALs)

DETERMINE SN_f

Vary trial SN_f until computed ESALs equal future design ESALs.

SN_f	M_R, psi	R	Z	S_o	P_1	P_2	ESAL
4.57	5,622	50	0	0.45	4.2	2.5	11,060,851
5.20	5,622	80	0.841	0.45	4.2	2.5	11,306,234
5.53	5,622	90	1.282	0.45	4.2	2.5	11,165,565
5.85	5,622	95	1.645	0.45	4.2	2.5	11,624,755
6.40	5,622	99	2.327	0.45	4.2	2.5	11,358,808
TRIAL			INPUT	INPUT	INPUT	INPUT	

DETERMINE SN_{eff} BY NDT METHOD

Vary trial E_p/M_R until computed D_0 equals actual value.

	ACTUAL			SUBGRADE		TRIAL COMPUTED			
STATION	LOAD, lbs	D_0, mils	D_r, mils	M_R, psi	C FACTOR	E_p/M_R	D_0, mils	E_p, psi	SN_{eff}
100	8,222	7.65	3.25	16,866	3	10.57	7.65	178,270	5.42

r = 36 inches
Check r > 0.7 ae = 33.13 inches

DETERMINE SN_{eff} BY CONDITION SURVEY METHOD

LAYER	STR COEF	DRAIN m	SN_{eff}
AC SURFACE	0.35	1.00	2.80
BASE	0.14	1.00	0.42
SUBBASE	0.11	1.00	1.14

SN_{eff} = 4.36

DETERMINE SN_{eff} BY REMAINING LIFE METHOD

Past design lane ESALs = ??? (FLEXIBLE ESALs)

LAYER	THICK, in	NEW ST CF	SN_o
AC SURFACE	8.00	0.44	3.52
BASE	3.00	0.14	0.42
SUBBASE	10.40	0.00	0
TOTAL	21.40		3.94

SN_o	M_R, psi	Z	S_o	P_1	P_2	$N_{1.5}$	R_L, %	CF	SN_{eff}
3.94	5,622	0	0	4.2	1.5	8,375,477			
		INPUT	INPUT	INPUT	INPUT				

DETERMINE OVERLAY THICKNESS

AC OL structural coefficient = 0.44

DESIGN RELIABILITY	NDT METHOD, in	CONDITION METHOD, in	REM LIFE METHOD, in
50	0.00	0.47	
80	0.00	1.90	
90	0.25	2.65	
95	0.98	3.38	
99	2.23	4.63	

REVISED CHAPTER 5 AASHTO DESIGN GUIDE OVERLAY DESIGN

SW-2   AC OVERLAY OF CONVENTIONAL AC PAVEMENT (PROJ. 0512)

EXISTING PAVEMENT DESIGN

AC SURFACE	4.50 inches	SUBGRADE ?
GRAN BASE	7.50	
GRAN SUBBASE	20.00	
TOTAL THICKNESS	32.00	

Future design lane ESALs = 11,000,000 (FLEXIBLE ESALs)

DETERMINE SN_f

Vary trial SN_f until computed ESALs equal future design ESALs.

SN_f	M_R, psi	R	Z	S_o	P_1	P_2	ESAL
4.75	5,007	50	0	0.45	4.2	2.5	10,984,277
5.38	5,007	80	0.841	0.45	4.2	2.5	11,035,686
5.73	5,007	90	1.282	0.45	4.2	2.5	11,091,727
6.02	5,007	95	1.645	0.45	4.2	2.5	11,024,719
6.60	5,007	99	2.327	0.45	4.2	2.5	11,025,739
TRIAL			INPUT	INPUT	INPUT	INPUT	

DETERMINE SN_{eff} BY NDT METHOD

Vary trial E_p/M_R until computed D_0 equals actual value.

	ACTUAL			SUBGRADE		TRIAL COMPUTED			
STATION	LOAD, lbs	D_0, mils	D_r, mils	M_R, psi	C FACTOR	E_p/M_R	D_0, mils	E_p, psi	SN_{eff}
600	9,171	19.27	4.07	15,022	3	3.09	19.28	46,418	5.18

r = 36 inches
Check r > 0.7 ae = 32.89 inches

DETERMINE SN_{eff} BY CONDITION SURVEY METHOD

LAYER	STR COEF	DRAIN m	SN_{eff}
AC SURFACE	0.35	1.00	1.58
BASE	0.14	1.00	1.05
SUBBASE	0.11	1.00	2.20

SN_{eff} = 4.83

DETERMINE SN_{eff} BY REMAINING LIFE METHOD

Past design lane ESALs = ??? (FLEXIBLE ESALs)

LAYER	THICK, in	NEW ST CF	SN_o
AC SURFACE	4.50	0.44	1.98
BASE	7.50	0.14	1.05
SUBBASE	20.00	0.00	0
TOTAL	32.00		3.03

SN_o	M_R, psi	Z	S_o	P_1	P_2	$N_{1.5}$	R_L, %	CF	SN_{eff}
3.03	5,007	0	0	4.2	1.5	952,248			
		INPUT	INPUT	INPUT	INPUT				

DETERMINE OVERLAY THICKNESS

AC OL structural coefficient = 0.44

DESIGN RELIABILITY	NDT METHOD, in	CONDITION METHOD, in	REM LIFE METHOD, in
50	0.00	0.00	
80	0.47	1.26	
90	1.26	2.06	
95	1.92	2.72	
99	3.24	4.03	

Appendix N

REVISED CHAPTER 5 AASHTO DESIGN GUIDE OVERLAY DESIGN

SW-3 AC OVERLAY OF CONVENTIONAL AC PAVEMENT (PROJ. 0515)

EXISTING PAVEMENT DESIGN

AC SURFACE	5.00 inches	SUBGRADE ?
GRAN BASE	6.00	
GRAN SUBBASE	20.00	
TOTAL THICKNESS	31.00	

Future design lane ESALs = 11,000,000 (FLEXIBLE ESALs)

DETERMINE SN_f

Vary trial SN_f until computed ESALs equal future design ESALs.

SN_f	M_R, psi	R	Z	S_o	P_1	P_2	ESAL
5.21	3,806	50	0	0.45	4.2	2.5	11,081,990
5.87	3,806	80	0.841	0.45	4.2	2.5	11,096,518
6.24	3,806	90	1.282	0.45	4.2	2.5	11,166,407
6.54	3,806	95	1.645	0.45	4.2	2.5	11,014,379
7.16	3,806	99	2.327	0.45	4.2	2.5	11,088,213
TRIAL			INPUT	INPUT	INPUT	INPUT	

DETERMINE SN_{eff} BY NDT METHOD

Vary trial E_p/M_R until computed D_0 equals actual value.

	ACTUAL			SUBGRADE		TRIAL COMPUTED			
STATION	LOAD, lbs	D_0, mils	D_r, mils	M_R, psi	C FACTOR	E_p/M_R	D_0, mils	E_p, psi	SN_{eff}
800	8,837	25.70	5.16	11,417	3	2.92	25.69	33,339	4.49

r = 36 inches
Check r > 0.7 ae = 31.29 inches

DETERMINE SN_{eff} BY CONDITION SURVEY METHOD

LAYER	STR COEF	DRAIN m	SN_{eff}
AC SURFACE	0.35	1.00	1.75
BASE	0.14	1.00	0.84
SUBBASE	0.11	1.00	2.20

SN_{eff} = 4.79

DETERMINE SN_{eff} BY REMAINING LIFE METHOD

Past design lane ESALs = ??? (FLEXIBLE ESALs)

LAYER	THICK, in	NEW ST CF	SN_o
AC SURFACE	5.00	0.44	2.2
BASE	6.00	0.14	0.84
SUBBASE	20.00	0.00	0
TOTAL	31.00		3.04

SN_o	M_R, psi	Z	S_o	P_1	P_2	$N_{1.5}$	R_L, %	CF	SN_{eff}
3.04	3,806	0	0	4.2	1.5	515,652			
		INPUT	INPUT	INPUT	INPUT				

DETERMINE OVERLAY THICKNESS

AC OL structural coefficient = 0.44

DESIGN RELIABILITY	NDT METHOD, in	CONDITION METHOD, in	REM LIFE METHOD, in
50	1.64	0.95	
80	3.14	2.45	
90	3.98	3.30	
95	4.66	3.98	
99	6.07	5.39	

REVISED CHAPTER 5 AASHTO DESIGN GUIDE OVERLAY DESIGN

SW-4 AC OVERLAY OF CONVENTIONAL AC PAVEMENT (PROJ. 0517)

EXISTING PAVEMENT DESIGN

AC SURFACE	5.00 inches	SUBGRADE ?
GRAN BASE	6.00	
GRAN SUBBASE	20.00	
TOTAL THICKNESS	31.00	

Future design lane ESALs = 11,000,000 (FLEXIBLE ESALs)

DETERMINE SN_f

Vary trial SN_f until computed ESALs equal future design ESALs.

SN_f	M_R, psi	R	Z	S_o	P_1	P_2	ESAL
4.74	5,065	50	0	0.45	4.2	2.5	11,121,070
5.37	5,065	80	0.841	0.45	4.2	2.5	11,183,641
5.71	5,065	90	1.282	0.45	4.2	2.5	11,100,905
6.00	5,065	95	1.645	0.45	4.2	2.5	11,042,512
6.59	5,065	99	2.327	0.45	4.2	2.5	11,191,828
TRIAL			INPUT	INPUT	INPUT	INPUT	

DETERMINE SN_{eff} BY NDT METHOD

Vary trial E_p/M_R until computed D_0 equals actual value.

	ACTUAL			SUBGRADE		TRIAL COMPUTED			
STATION	LOAD, lbs	D_0, mils	D_r, mils	M_R, psi	C FACTOR	E_p/M_R	D_0, mils	E_p, psi	SN_{eff}
400	9,437	31.73	4.14	15,196	3	1.71	31.77	25,986	4.13

r = 36 inches
Check r > 0.7 ae = 26.28 inches

DETERMINE SN_{eff} BY CONDITION SURVEY METHOD

LAYER	STR COEF	DRAIN m	SN_{eff}
AC SURFACE	0.35	1.00	1.75
BASE	0.14	1.00	0.84
SUBBASE	0.11	1.00	2.20

SN_{eff} = 4.79

DETERMINE SN_{eff} BY REMAINING LIFE METHOD

Past design lane ESALs = ??? (FLEXIBLE ESALs)

LAYER	THICK, in	NEW ST CF	SN_o
AC SURFACE	5.00	0.44	2.2
BASE	6.00	0.14	0.84
SUBBASE	20.00	0.00	0
TOTAL	31.00		3.04

SN_o	M_R, psi	Z	S_o	P_1	P_2	$N_{1.5}$	R_L, %	CF	SN_{eff}
3.04	5,065	0	0	4.2	1.5	1,001,038			
		INPUT	INPUT	INPUT	INPUT				

DETERMINE OVERLAY THICKNESS

AC OL structural coefficient = 0.44

DESIGN RELIABILITY	NDT METHOD, in	CONDITION METHOD, in	REM LIFE METHOD, in
50	1.38	0.00	
80	2.81	1.32	
90	3.59	2.09	
95	4.25	2.75	
99	5.59	4.09	

Appendix N   N-19

REVISED CHAPTER 5 AASHTO DESIGN GUIDE OVERLAY DESIGN

MW-1 AC OVERLAY OF CONVENTIONAL AC PAVEMENT (NEWMARK DR)

EXISTING PAVEMENT DESIGN

AC SURFACE	1.50 inches	SUBGRADE A-6
GRAN BASE	6.00	
GRAN SUBBASE	0.00	
TOTAL THICKNESS	7.50	

Future design lane ESALs = 100,000 (flexible ESALs)

DETERMINE SN_f

Vary trial SN_f until computed ESALs equal future design ESALs

SN_f	M_R, psi	R	Z	S_o	P_1	P_2	ESAL
2.62	3,289	50	0	0.45	4.2	2.5	101,452
3.03	3,289	80	0.841	0.45	4.2	2.5	101,875
3.26	3,289	90	1.282	0.45	4.2	2.5	100,366
3.47	3,289	95	1.645	0.45	4.2	2.5	100,710
3.89	3,289	99	2.327	0.45	4.2	2.5	100,712
TRIAL			INPUT	INPUT	INPUT	INPUT	

DETERMINE SN_{eff} BY NDT METHOD

Vary trial E_p/M_R until computed D_0 equals actual value.

	ACTUAL			SUBGRADE		TRIAL COMPUTED			
STATION	LOAD, lbs	D_0, mils	D_r, mils	M_R, psi	C FACTOR	E_p/M_R	D_0, mils	E_p, psi	SN_{eff}
100	9,000	16.10	6.08	9,868	3	48.20	16.18	475,658	2.63

r = 36 inches
Check r > 0.7 ae = 19.55 inches

DETERMINE SN_{eff} BY CONDITION SURVEY METHOD

LAYER	STR COEF	DRAIN m	SN_{eff}
AC SURFACE	0.35	1.00	0.53
BASE	0.25	1.00	1.50
SUBBASE	0.00	1.00	0.00

SN_{eff} = 2.03

DETERMINE SN_{eff} BY REMAINING LIFE METHOD

Past design lane ESALs = 95,000 (flexible ESALs)

LAYER	THICK, in	NEW ST CF	SN_o
AC SURFACE	1.50	0.44	0.66
BASE	6.00	0.33	1.98
SUBBASE	0.00	0.00	0
TOTAL	7.50		2.64

SN_o	M_R, psi	Z	S_o	P_1	P_2	$N_{1.5}$	R_L, %	CF	SN_{eff}
2.64	3,289	0	0.45	4.2	1.5	138,561	31	0.83	2.18
		INPUT	INPUT	INPUT	INPUT				

DETERMINE OVERLAY THICKNESS

AC OL structural coefficient = 0.44

DESIGN RELIABILITY	NDT METHOD, in	CONDITION METHOD, in	REM LIFE METHOD, in
50	0.00	1.35	1.00
80	0.90	2.28	1.93
90	1.42	2.81	2.45
95	1.90	3.28	2.93
99	2.85	4.24	3.88

REVISED CHAPTER 5 AASHTO DESIGN GUIDE OVERLAY DESIGN

MW-2 AC OVERLAY OF CONVENTIONAL AC PAVEMENT (FIRST STREET)

EXISTING PAVEMENT DESIGN

AC SURFACE	4.00 inches	SUBGRADE A-6
GRAN BASE	8.00	
GRAN SUBBASE	0.00	
TOTAL THICKNESS	12.00	

Future design lane ESALs = 150,000 (flexible ESALs)

DETERMINE SN_f

Vary trial SN_f until computed ESALs equal future design ESALs.

SN_f	M_R, psi	R	Z	S_o	P_1	P_2	ESAL
3.24	2,256	50	0	0.45	4.2	2.5	152,158
3.73	2,256	80	0.841	0.45	4.2	2.5	150,629
4.01	2,256	90	1.282	0.45	4.2	2.5	150,195
4.25	2,256	95	1.645	0.45	4.2	2.5	149,762
4.73	2,256	99	2.327	0.45	4.2	2.5	150,572
TRIAL			INPUT	INPUT	INPUT	INPUT	

DETERMINE SN_{eff} BY NDT METHOD

Vary trial E_p/M_R until computed D_0 equals actual value.

	ACTUAL			SUBGRADE		TRIAL COMPUTED			
STATION	LOAD, lbs	D_0, mils	D_r, mils	M_R, psi	C FACTOR	E_p/M_R	D_0, mils	E_p, psi	SN_{eff}
10+00	9,096	25.61	8.96	6,768	3	14.58	25.62	98,675	2.50

r =   36 inches
Check r > 0.7 ae = 20.93 inches

DETERMINE SN_{eff} BY CONDITION SURVEY METHOD

LAYER	STR COEF	DRAIN m	SN_{eff}
AC SURFACE	0.33	1.00	1.32
BASE	0.14	1.00	1.12
SUBBASE	0.00	1.00	0.00

SN_{eff} = 2.44

DETERMINE SN_{eff} BY REMAINING LIFE METHOD

Past design lane ESALs = 90,000 (flexible ESALs)

LAYER	THICK, in	NEW ST CF	SN_o
AC SURFACE	4.00	0.44	1.76
BASE	8.00	0.14	1.12
SUBBASE	0.00	0.00	0
TOTAL	12.00		2.88

SN_o	M_R, psi	Z	S_o	P_1	P_2	$N_{1.5}$	R_L, %	CF	SN_{eff}
2.88	2,256	0	0.45	4.2	1.5	105,000	14	0.73	2.09
		INPUT	INPUT	INPUT	INPUT				

DETERMINE OVERLAY THICKNESS

AC OL structural coefficient = 0.44

DESIGN RELIABILITY	NDT METHOD, in	CONDITION METHOD, in	REM LIFE METHOD, in
50	1.69	1.82	2.62
80	2.81	2.93	3.73
90	3.44	3.57	4.37
95	3.99	4.11	4.91
99	5.08	5.20	6.00

Appendix N

REVISED CHAPTER 5 AASHTO DESIGN GUIDE OVERLAY DESIGN

NW-1 AC OVERLAY OF CONVENTIONAL AC PAVEMENT

EXISTING PAVEMENT DESIGN

AC SURFACE	4.25 inches	SUBGRADE SANDY SILT, SANDY GRAVEL
GRAN BASE	8.00	
GRAN SUBBASE	0.00	
TOTAL THICKNESS	12.25	

Future design lane ESALs = 2,400,000 (FLEXIBLE ESALs)

DETERMINE SN_f

Vary trial SN_f until computed ESALs equal future design ESALs.

SN_f	M_R, psi	R	Z	S_o	P_1	P_2	ESAL
3.60	5,634	50	0	0.45	4.2	2.5	2,417,312
4.14	5,634	80	0.841	0.45	4.2	2.5	2,430,778
4.44	5,634	90	1.282	0.45	4.2	2.5	2,429,228
4.69	5,634	95	1.645	0.45	4.2	2.5	2,408,097
5.19	5,634	99	2.327	0.45	4.2	2.5	2,403,245
TRIAL			INPUT	INPUT	INPUT	INPUT	

DETERMINE SN_{eff} BY NDT METHOD

Vary trial E_p/M_R until computed D_0 equals actual value.

	ACTUAL			SUBGRADE		TRIAL COMPUTED			
STATION	LOAD, lbs	D_0, mils	D_r, mils	M_R, psi	C FACTOR	E_p/M_R	D_0, mils	E_p, psi	SN_{eff}
100	9,000	12.80	3.55	16,901	3	8.45	12.80	142,817	2.88

r = 36 inches
Check r > 0.7 ae = 17.95 inches

DETERMINE SN_{eff} BY CONDITION SURVEY METHOD

LAYER	STR COEF	DRAIN m	SN_{eff}
AC SURFACE	0.35	1.00	1.49
BASE	0.14	1.00	1.12
SUBBASE	0.00	1.00	0.00

SN_{eff} = 2.61

DETERMINE SN_{eff} BY REMAINING LIFE METHOD

Past design lane ESALs = 400,000 (FLEXIBLE ESALs)

LAYER	THICK, in	NEW ST CF	SN_o
AC SURFACE	4.25	0.44	1.87
BASE	8.00	0.14	1.12
SUBBASE	0.00	0.00	0
TOTAL	12.25		2.99

SN_o	M_R, psi	Z	S_o	P_1	P_2	$N_{1.5}$	R_L, %	CF	SN_{eff}
2.99	5,634	0	0.45	4.2	1.5	1140161	65	0.93	2.78
		INPUT	INPUT	INPUT	INPUT				

DETERMINE OVERLAY THICKNESS

AC OL structural coefficient = 0.44

DESIGN RELIABILITY	NDT METHOD, in	CONDITION METHOD, in	REM LIFE METHOD, in
50	1.63	2.26	1.85
80	2.86	3.48	3.08
90	3.54	4.16	3.76
95	4.11	4.73	4.33
99	5.25	5.87	5.47

REVISED CHAPTER 5 AASHTO DESIGN GUIDE OVERLAY DESIGN

NW-2 AC OVERLAY OF CONVENTIONAL AC PAVEMENT (PELTON DAM ROAD)

EXISTING PAVEMENT DESIGN

AC SURFACE	5.50 inches	SUBGRADE SANDY SILT, SANDY GRAVEL
GRAN BASE	12.00	
GRAN SUBBASE	0.00	
TOTAL THICKNESS	17.50	

Future design lane ESALs = 2,808,000 (FLEXIBLE ESALs)

DETERMINE SN_f

Vary trial SN_f until computed ESALs equal future design ESALs.

SN_f	M_R, psi	R	Z	S_o	P_1	P_2	ESAL
4.01	4,494	50	0	0.45	4.2	2.5	2,805,583
4.58	4,494	80	0.841	0.45	4.2	2.5	2,793,711
4.90	4,494	90	1.282	0.45	4.2	2.5	2,804,877
5.17	4,494	95	1.645	0.45	4.2	2.5	2,806,109
5.70	4,494	99	2.327	0.45	4.2	2.5	2,811,508
TRIAL			INPUT	INPUT	INPUT	INPUT	

DETERMINE SN_{eff} BY NDT METHOD

Vary trial E_p/M_R until computed D_0 equals actual value.

	ACTUAL			SUBGRADE		TRIAL COMPUTED			
STATION	LOAD, lbs	D_0, mils	D_r, mils	M_R, psi	C FACTOR	E_p/M_R	D_0, mils	E_p, psi	SN_{eff}
1	9,000	24.10	4.45	13,483	3	3.08	24.11	41,528	2.73

r = 36 inches
Check r > 0.7 ae = 18.30 inches

DETERMINE SN_{eff} BY CONDITION SURVEY METHOD

LAYER	STR COEF	DRAIN m	SN_{eff}
AC SURFACE	0.30	1.00	1.65
BASE	0.10	1.00	1.20
SUBBASE	0.00	1.00	0.00

$SN_{eff} = 2.85$

DETERMINE SN_{eff} BY REMAINING LIFE METHOD

Past design lane ESALs = ??? (flexible ESALs)

LAYER	THICK, in	NEW ST CF	SN_o
AC SURFACE	5.50	0.44	2.42
BASE	12.00	0.14	1.68
SUBBASE	0.00	0.00	0
TOTAL	17.50		4.10

SN_o	M_R, psi	Z	S_o	P_1	P_2	$N_{1.5}$	R_L, %	CF	SN_{eff}
4.1	4,494	0	0	4.2	1.5	6,715,080			
		INPUT	INPUT	INPUT	INPUT				

DETERMINE OVERLAY THICKNESS

AC OL structural coefficient = 0.44

DESIGN RELIABILITY	NDT METHOD, in	CONDITION METHOD, in	REM LIFE METHOD, in
50	2.92	2.64	
80	4.21	3.93	
90	4.94	4.66	
95	5.55	5.27	
99	6.76	6.48	

Appendix N N-23

REVISED CHAPTER 5 AASHTO DESIGN GUIDE OVERLAY DESIGN

NW-3 AC OVERLAY OF CONVENTIONAL AC PAVEMENT (JOSEPH ST INTERCHANGE)

EXISTING PAVEMENT DESIGN

AC SURFACE	7.00 inches	SUBGRADE A-7-6
GRAN BASE	16.00	
GRAN SUBBASE	0.00	
TOTAL THICKNESS	23.00	

Future design lane ESALs = 5,550,000 (FLEXIBLE ESALs)

DETERMINE SN_f

Vary trial SN_f until computed ESALs equal future design ESALs.

SN_f	M_R, psi	R	Z	S_o	P_1	P_2	ESAL
3.89	6,536	50	0	0.45	4.2	2.5	5,520,953
4.46	6,536	80	0.841	0.45	4.2	2.5	5,578,464
4.77	6,536	90	1.282	0.45	4.2	2.5	5,556,152
5.04	6,536	95	1.645	0.45	4.2	2.5	5,588,225
5.56	6,536	99	2.327	0.45	4.2	2.5	5,580,037
TRIAL			INPUT	INPUT	INPUT	INPUT	

DETERMINE SN_{eff} BY NDT METHOD

Vary trial E_p/M_R until computed D_0 equals actual value.

	ACTUAL			SUBGRADE		TRIAL COMPUTED			
STATION	LOAD, lbs	D_0, mils	D_r, mils	M_R, psi	C FACTOR	E_p/M_R	D_0, mils	E_p, psi	SN_{eff}
1	9,000	14.74	3.06	19,608	3	3.31	14.72	64,902	4.16

r = 36 inches
Check r > 0.7 ae = 24.35 inches

DETERMINE SN_{eff} BY CONDITION SURVEY METHOD

LAYER	STR COEF	DRAIN m	SN_{eff}
AC SURFACE	0.35	1.00	2.45
BASE	0.10	1.00	1.60
SUBBASE	0.00	1.00	0.00

SN_{eff} = 4.05

DETERMINE SN_{eff} BY REMAINING LIFE METHOD

Past design lane ESALs = ??? (flexible ESALs)

LAYER	THICK, in	NEW ST CF	SN_o
AC SURFACE	7.00	0.44	3.08
BASE	16.00	0.14	2.24
SUBBASE	0.00	0.00	0
TOTAL	23.00		5.32

SN_o	M_R, psi	Z	S_o	P_1	P_2	$N_{1.5}$	R_L, %	CF	SN_{eff}
5.32	6,536	0	0	4.2	1.5	*********			
		INPUT	INPUT	INPUT	INPUT				

DETERMINE OVERLAY THICKNESS

AC OL structural coefficient = 0.44

DESIGN RELIABILITY	NDT METHOD, in	CONDITION METHOD, in	REM LIFE METHOD, in
50	0.00	0.00	
80	0.68	0.93	
90	1.39	1.64	
95	2.00	2.25	
99	3.18	3.43	

REVISED CHAPTER 5 AASHTO DESIGN GUIDE OVERLAY DESIGN

NW-4 AC OVERLAY OF CONVENTIONAL AC PAVEMENT (KIWA SPRINGS)

EXISTING PAVEMENT DESIGN

AC SURFACE	4.50 inches	SUBGRADE SILTY SAND
GRAN BASE	16.00	
GRAN SUBBASE	0.00	
TOTAL THICKNESS	20.50	

Future design lane ESALs = 880,000 (FLEXIBLE ESALs)

DETERMINE SN_f

Vary trial SN_f until computed ESALs equal future design ESALs.

SN_f	M_R, psi	R	Z	S_o	P_1	P_2	ESAL
3.27	4,695	50	0	0.45	4.2	2.5	880,995
3.77	4,695	80	0.841	0.45	4.2	2.5	881,307
4.06	4,695	90	1.282	0.45	4.2	2.5	889,890
4.30	4,695	95	1.645	0.45	4.2	2.5	884,951
4.78	4,695	99	2.327	0.45	4.2	2.5	885,927
TRIAL			INPUT	INPUT	INPUT	INPUT	

DETERMINE SN_{eff} BY NDT METHOD

Vary trial E_p/M_R until computed D_0 equals actual value.

	ACTUAL			SUBGRADE		TRIAL COMPUTED			
STATION	LOAD, lbs	D_0, mils	D_r, mils	M_R, psi	C FACTOR	E_p/M_R	D_0, mils	E_p, psi	SN_{eff}
AVE DEF	9,000	25.52	4.26	14,085	3	2.52	25.53	35,493	3.03

r = 36 inches
Check r > 0.7 ae = 19.96 inches

DETERMINE SN_{eff} BY CONDITION SURVEY METHOD

LAYER	STR COEF	DRAIN m	SN_{eff}
AC SURFACE	0.35	1.00	1.58
BASE	0.10	1.00	1.60
SUBBASE	0.00	1.00	0.00

SN_{eff} = 3.18

DETERMINE SN_{eff} BY REMAINING LIFE METHOD

Past design lane ESALs = ??? (flexible ESALs)

LAYER	THICK, in	NEW ST CF	SN_o
AC SURFACE	4.50	0.44	1.98
BASE	16.00	0.14	2.24
SUBBASE	0.00	0.00	0
TOTAL	20.50		4.22

SN_o	M_R, psi	Z	S_o	P_1	P_2	$N_{1.5}$	R_L, %	CF	SN_{eff}
4.22	4,695	0	0	4.2	1.5	9,237,517			
		INPUT	INPUT	INPUT	INPUT				

DETERMINE OVERLAY THICKNESS

AC OL structural coefficient = 0.44

DESIGN RELIABILITY	NDT METHOD, in	CONDITION METHOD, in	REM LIFE METHOD, in
50	0.54	0.22	
80	1.68	1.35	
90	2.34	2.01	
95	2.88	2.56	
99	3.97	3.65	

REVISED CHAPTER 5 AASHTO DESIGN GUIDE OVERLAY DESIGN

NW-5 AC OVERLAY OF CONVENTIONAL AC PAVEMENT (BANKS SCL)

EXISTING PAVEMENT DESIGN

AC SURFACE	8.00 inches	SUBGRADE A-4
GRAN BASE	4.00	
GRAN SUBBASE	0.00	
TOTAL THICKNESS	12.00	

Future design lane ESALs = 1,362,000 (FLEXIBLE ESALs)

DETERMINE SN_f

Vary trial SN_f until computed ESALs equal future design ESALs.

SN_f	M_R, psi	R	Z	S_o	P_1	P_2	ESAL
3.49	4,762	50	0	0.45	4.2	2.5	1,352,984
4.02	4,762	80	0.841	0.45	4.2	2.5	1,363,644
4.32	4,762	90	1.282	0.45	4.2	2.5	1,373,199
4.57	4,762	95	1.645	0.45	4.2	2.5	1,368,562
5.06	4,762	99	2.327	0.45	4.2	2.5	1,359,575
TRIAL			INPUT	INPUT	INPUT	INPUT	

DETERMINE SN_{eff} BY NDT METHOD

Vary trial E_p/M_R until computed D_0 equals actual value.

	ACTUAL			SUBGRADE		TRIAL COMPUTED			
STATION	LOAD, lbs	D_0, mils	D_r, mils	M_R, psi	C FACTOR	E_p/M_R	D_0, mils	E_p, psi	SN_{eff}
AVE	9,000	22.76	4.20	14,286	3	3.82	22.74	54,571	2.05

r = 36 inches
Check r > 0.7 ae = 13.77 inches

DETERMINE SN_{eff} BY CONDITION SURVEY METHOD

LAYER	STR COEF	DRAIN m	SN_{eff}
AC SURFACE	0.22	1.00	1.76
BASE	0.08	1.00	0.32
SUBBASE	0.00	1.00	0.00

SN_{eff} = 2.08

DETERMINE SN_{eff} BY REMAINING LIFE METHOD

Past design lane ESALs = ??? (flexible ESALs)

LAYER	THICK, in	NEW ST CF	SN_o
AC SURFACE	8.00	0.44	3.52
BASE	4.00	0.14	0.56
SUBBASE	0.00	0.00	0
TOTAL	12.00		4.08

SN_o	M_R, psi	Z	S_o	P_1	P_2	$N_{1.5}$	R_L, %	CF	SN_{eff}
4.08	4,762	0	0	4.2	1.5	7,401,770			
		INPUT	INPUT	INPUT	INPUT				

DETERMINE OVERLAY THICKNESS

AC OL structural coefficient = 0.44

DESIGN RELIABILITY	NDT METHOD, in	CONDITION METHOD, in	REM LIFE METHOD, in
50	3.28	3.20	
80	4.48	4.41	
90	5.16	5.09	
95	5.73	5.66	
99	6.84	6.77	

REVISED CHAPTER 5 AASHTO DESIGN GUIDE OVERLAY DESIGN

NW-6 AC OVERLAY OF CONVENTIONAL AC PAVEMENT (SALISBURY JCT)

EXISTING PAVEMENT DESIGN

AC SURFACE	4.00 inches	SUBGRADE A-4, A-6, A-7-6
GRAN BASE	14.00	
GRAN SUBBASE	0.00	
TOTAL THICKNESS	18.00	

Future design lane ESALs = 1,576,000 (FLEXIBLE ESALs)

DETERMINE SN_f

Vary trial SN_f until computed ESALs equal future design ESALs.

SN_f	M_R, psi	R	Z	S_o	P_1	P_2	ESAL
3.59	4,739	50	0	0.45	4.2	2.5	1,591,144
4.11	4,739	80	0.841	0.45	4.2	2.5	1,553,384
4.42	4,739	90	1.282	0.45	4.2	2.5	1,578,705
4.68	4,739	95	1.645	0.45	4.2	2.5	1,589,225
5.17	4,739	99	2.327	0.45	4.2	2.5	1,565,579
TRIAL			INPUT	INPUT	INPUT	INPUT	

DETERMINE SN_{eff} BY NDT METHOD

Vary trial E_p/M_R until computed D_0 equals actual value.

	ACTUAL			SUBGRADE		TRIAL COMPUTED			
STATION	LOAD, lbs	D_0, mils	D_r, mils	M_R, psi	C FACTOR	E_p/M_R	D_0, mils	E_p, psi	SN_{eff}
AVE	9,000	23.57	4.22	14,218	3	2.90	23.62	41,232	2.80

r = 36 inches
Check r > 0.7 ae = 18.44 inches

DETERMINE SN_{eff} BY CONDITION SURVEY METHOD

LAYER	STR COEF	DRAIN m	SN_{eff}
AC SURFACE	0.32	1.00	1.28
BASE	0.10	1.00	1.40
SUBBASE	0.00	1.00	0.00

SN_{eff} = 2.68

DETERMINE SN_{eff} BY REMAINING LIFE METHOD

Past design lane ESALs = ??? (flexible ESALs)

LAYER	THICK, in	NEW ST CF	SN_o
AC SURFACE	4.00	0.44	1.76
BASE	14.00	0.14	1.96
SUBBASE	0.00	0.00	0
TOTAL	18.00		3.72

SN_o	M_R, psi	Z	S_o	P_1	P_2	$N_{1.5}$	R_L, %	CF	SN_{eff}
3.72	4,739	0	0	4.2	1.5	3,679,271			
		INPUT	INPUT	INPUT	INPUT				

DETERMINE OVERLAY THICKNESS

AC OL structural coefficient = 0.44

DESIGN RELIABILITY	NDT METHOD, in	CONDITION METHOD, in	REM LIFE METHOD, in
50	1.80	2.07	
80	2.98	3.25	
90	3.69	3.95	
95	4.28	4.55	
99	5.39	5.66	

REVISED CHAPTER 5 AASHTO DESIGN GUIDE OVERLAY DESIGN

NE-1 AC OVERLAY OF CONVENTIONAL AC PAVEMENT (SR756-01E)

EXISTING PAVEMENT DESIGN

AC SURFACE	3.00 inches	SUBGRADE: CBR = 8
BIT BASE	2.00	M_R EST. AASHTO APP. FF
GRAN SUBBASE	10.00	
TOTAL THICKNESS	15.00	

Future design lane ESALs = 931,327 (flexible ESALs)

DETERMINE SN_f

Vary trial SN_f until computed ESALs equal future design ESALs.

SN_f	M_R, psi	R	Z	S_o	P_1	P_2	ESAL
2.79	8,100	50	0	0.45	4.2	3	947,912
3.30	8,100	80	0.841	0.45	4.2	3	962,019
3.60	8,100	90	1.282	0.45	4.2	3	960,669
3.85	8,100	95	1.645	0.45	4.2	3	942,538
4.37	8,100	99	2.327	0.45	4.2	3	941,446
TRIAL			INPUT	INPUT	INPUT	INPUT	

DETERMINE SN_{eff} BY NDT METHOD

Vary trial E_p/M_R until computed D_0 equals actual value.

	ACTUAL			SUBGRADE		TRIAL COMPUTED			
STATION	LOAD, lbs	D_0, mils	D_r, mils	M_R, psi	C FACTOR	E_p/M_R	D_0, mils	E_p, psi	SN_{eff}
0		0.00	0.00	ERR	3	0.00	ERR	ERR	ERR

r = 0 inches
Check r > 0.7 ae =   inches

DETERMINE SN_{eff} BY CONDITION SURVEY METHOD

LAYER	STR COEF	DRAIN m	SN_{eff}
AC SURFACE	0.30	1.00	0.90
BASE	0.30	1.00	0.60
SUBBASE	0.18	1.00	1.80

SN_{eff} = 3.30

DETERMINE SN_{eff} BY REMAINING LIFE METHOD

Past design lane ESALs = 0 (flexible ESALs)

LAYER	THICK, in	NEW ST CF	SN_o
AC SURFACE	3.00	0	0
BASE	2.00	0.00	0
SUBBASE	10.00	0.00	0
TOTAL	15.00		0.00

SN_o	M_R, psi	Z	S_o	P_1	P_2	$N_{1.5}$	R_L, %	CF	SN_{eff}
0	ERR	0	0.45	4.2	1.5	ERR	ERR	ERR	ERR
		INPUT	INPUT	INPUT	INPUT				

DETERMINE OVERLAY THICKNESS

AC OL structural coefficient = 0.44

DESIGN RELIABILITY	NDT METHOD, in	CONDITION METHOD, in	REM LIFE METHOD, in
50		0.00	
80		0.00	
90		0.68	
95		1.25	
99		2.43	

REVISED CHAPTER 5 AASHTO DESIGN GUIDE OVERLAY DESIGN

NE-2 AC OVERLAY OF CONVENTIONAL AC PAVEMENT (SR239-04M)

EXISTING PAVEMENT DESIGN

AC SURFACE	2.50 inches	SUBGRADE: CBR = 5
GRAN BASE	0.00	MR EST. AASHTO APP. FF
GRAN SUBBASE	8.00	
TOTAL THICKNESS	10.50	

Future design lane ESALs = 574,900 (flexible ESALs)

DETERMINE SN_f

Vary trial SN_f until computed ESALs equal future design ESALs.

SN_f	M_R, psi	R	Z	S_o	P_1	P_2	ESAL
2.81	5,800	50	0	0.45	4.2	2.5	576,370
3.25	5,800	80	0.841	0.45	4.2	2.5	579,932
3.50	5,800	90	1.282	0.45	4.2	2.5	576,358
3.74	5,800	95	1.645	0.45	4.2	2.5	595,324
4.16	5,800	99	2.327	0.45	4.2	2.5	575,180
TRIAL			INPUT	INPUT	INPUT	INPUT	

DETERMINE SN_{eff} BY NDT METHOD

Vary trial E_p/M_R until computed D_0 equals actual value.

	ACTUAL			SUBGRADE		TRIAL COMPUTED			
STATION	LOAD, lbs	D_0, mils	D_r, mils	M_R, psi	C FACTOR	E_p/M_R	D_0, mils	E_p, psi	SN_{eff}
	0	0.00	0.00	ERR	3	0.00	ERR	ERR	ERR

r = 0 inches
Check r > 0.7 ae = inches

DETERMINE SN_{eff} BY CONDITION SURVEY METHOD

LAYER	STR COEF	DRAIN m	SN_{eff}
AC SURFACE	0.35	1.00	0.88
BASE	0.00	1.00	0.00
SUBBASE	0.25	1.00	2.00

SN_{eff} = 2.88

DETERMINE SN_{eff} BY REMAINING LIFE METHOD

Past design lane ESALs = 0 (flexible ESALs)

LAYER	THICK, in	NEW ST CF	SN_o
AC SURFACE	2.50	0	0
BASE	0.00	0.00	0
SUBBASE	8.00	0.00	0
TOTAL	10.50		0.00

SN_o	M_R, psi	Z	S_o	P_1	P_2	$N_{1.5}$	R_L, %	CF	SN_{eff}
0	ERR	0	0.45	4.2	1.5	ERR	ERR	ERR	ERR
		INPUT	INPUT	INPUT	INPUT				

DETERMINE OVERLAY THICKNESS

AC OL structural coefficient = 0.44

DESIGN RELIABILITY	NDT METHOD, in	CONDITION METHOD, in	REM LIFE METHOD, in
50		0.00	
80		0.85	
90		1.42	
95		1.97	
99		2.92	

Appendix N

REVISED CHAPTER 5 AASHTO DESIGN GUIDE OVERLAY DESIGN

NE-3 AC OVERLAY OF CONVENTIONAL AC PAVEMENT (SR26-06M)

EXISTING PAVEMENT DESIGN

AC SURFACE	4.00 inches	SUBGRADE: CBR = 7.5
STONE BASE	8.00	MR EST. AASHTO APP. FF
GRAN SUBBASE	0.00	
TOTAL THICKNESS	12.00	

Future design lane ESALs = 147,816 (flexible ESALs, 10-YEAR DESIGN LIFE)

DETERMINE SN_f

Vary trial SN_f until computed ESALs equal future design ESALs.

SN_f	M_R, psi	R	Z	S_o	P_1	P_2	ESAL
2.00	7,800	50	0	0.45	4.2	2.5	149,952
2.32	7,800	80	0.841	0.45	4.2	2.5	151,551
2.50	7,800	90	1.282	0.45	4.2	2.5	150,256
2.66	7,800	95	1.645	0.45	4.2	2.5	149,807
2.99	7,800	99	2.327	0.45	4.2	2.5	149,450
TRIAL			INPUT	INPUT	INPUT	INPUT	

DETERMINE SN_{eff} BY NDT METHOD

Vary trial E_p/M_R until computed D_0 equals actual value.

	ACTUAL			SUBGRADE		TRIAL COMPUTED			
STATION	LOAD, lbs	D_0, mils	D_r, mils	M_R, psi	C FACTOR	E_p/M_R	D_0, mils	E_p, psi	SN_{eff}
	0	0.00	0.00	ERR	3	0.00	ERR	ERR	ERR

                      r =   0 inches
Check r > 0.7 ae =   inches

DETERMINE SN_{eff} BY CONDITION SURVEY METHOD

LAYER	STR COEF	DRAIN m	SN_{eff}
AC SURFACE	0.30	1.00	1.20
BASE	0.18	1.00	1.44
SUBBASE	0.00	1.00	0.00

SN_{eff} = 2.64

DETERMINE SN_{eff} BY REMAINING LIFE METHOD

Past design lane ESALs = 0 (flexible ESALs)

LAYER	THICK, in	NEW ST CF	SN_o
AC SURFACE	4.00	0	0
BASE	8.00	0.00	0
SUBBASE	0.00	0.00	0
TOTAL	12.00		0.00

SN_o	M_R, psi	Z	S_o	P_1	P_2	$N_{1.5}$	R_L, %	CF	SN_{eff}
0	ERR	0	0.45	4.2	1.5	ERR	ERR	ERR	ERR
		INPUT	INPUT	INPUT	INPUT				

DETERMINE OVERLAY THICKNESS

AC OL structural coefficient = 0.44

DESIGN RELIABILITY	NDT METHOD, in	CONDITION METHOD, in	REM LIFE METHOD, in
50		0.00	
80		0.00	
90		0.00	
95		0.05	
99		0.80	

REVISED CHAPTER 5 AASHTO DESIGN GUIDE OVERLAY DESIGN

NE-4 AC OVERLAY OF CONVENTIONAL AC PAVEMENT (ROUTE 9, 49-104)

EXISTING PAVEMENT DESIGN

AC SURFACE	4.00 inches	SUBGRADE: ?
BIT. BASE	3.00	
CACL2 STAB	4.00	
GRAN SUBBASE	14.50	
TOTAL THICKNESS	25.50	

Future design lane ESALs = 7,040,000 (20-YEAR DESIGN, FLEXIBLE ESALs)

DETERMINE SN_f

Vary trial SN_f until computed ESALs equal future design ESALs.

SN_f	M_R, psi	R	Z	S_o	P_1	P_2	ESAL
3.45	10,000	50	0	0.45	4.2	2.5	7,051,276
3.97	10,000	80	0.841	0.45	4.2	2.5	7,043,437
4.26	10,000	90	1.282	0.45	4.2	2.5	7,009,068
4.51	10,000	95	1.645	0.45	4.2	2.5	7,005,057
5.01	10,000	99	2.327	0.45	4.2	2.5	7,090,364
TRIAL			INPUT	INPUT	INPUT	INPUT	

DETERMINE SN_{eff} BY NDT METHOD

Vary trial E_p/M_R until computed D_0 equals actual value.

	ACTUAL			SUBGRADE		TRIAL COMPUTED			
STATION	LOAD, lbs	D_0, mils	D_r, mils	M_R, psi	C FACTOR	E_p/M_R	D_0, mils	E_p, psi	SN_{eff}
	0	0.00	0.00	ERR	0	0.00	ERR	ERR	ERR
	r =	36 inches							
Check r > 0.7 ae =		inches							

DETERMINE SN_{eff} BY CONDITION SURVEY METHOD

LAYER	STR COEF	DRAIN m	SN_{eff}
AC SURFACE	0.40	1.00	1.60
BIT BASE	0.30	1.00	0.90
SUBBASE	0.14	1.00	0.56
SUBBASE	0.11	1.00	0.44

SN_{eff} = 3.50

(continued on next page)

Appendix N

REVISED CHAPTER 5 AASHTO DESIGN GUIDE OVERLAY DESIGN

NE-4 AC OVERLAY OF CONVENTIONAL AC PAVEMENT (ROUTE 9, 49-104) *(continued)*

DETERMINE SN_{eff} BY REMAINING LIFE METHOD

Past design lane ESALs = 0 (flexible ESALs)

LAYER	THICK, in	NEW ST CF	SN_o
AC SURFACE	0.00	0	0
BASE	0.00	0.00	0
SUBBASE	0.00	0.00	0
TOTAL	0.00		0.00

SN_o	M_R, psi	Z	S_o	P_1	P_2	$N_{1.5}$	R_L, %	CF	SN_{eff}
0	ERR	0	0.45	4.2	1.5	ERR	ERR	ERR	ERR
		INPUT	INPUT	INPUT	INPUT				

DETERMINE OVERLAY THICKNESS

AC OL structural coefficient = 0.44

DESIGN RELIABILITY	NDT METHOD, in	CONDITION METHOD, in	REM LIFE METHOD, in
50		0.00	
80		1.07	
90		1.73	
95		2.30	
99		3.43	

N4.0 AC OVERLAY OF FRACTURED SLAB PCC PAVEMENT

Region-Project	Overlay Type	Existing Pavement	Design ESALs	Design Reliability	Overlay Thickness (in)
MW-3	AC	JRCP	6,700,000 (10 years)	50	4.1
				80	5.5
				90	6.3*
				95	6.9*
				99	8.2*

SHRP LTPP section that was overlaid with 6 and 8 inches of AC after being rubblized.

SW-5	AC	JRCP	9,532,300 (15 years)	50	6.1
				80	7.6
				90	8.4
				95	9.1
				99	10.6

No State design is available. Overlay design is for rubblized JRCP.

MW-4	AC	JRCP	318,000 (20 years)	50	0.0
				80	0.7
				90	1.3
				95	1.7
				99	2.8*

State recommends a 2.75-inch AC overlay after pavement is broken and seated.

SW-6	AC	JPCP	7,370,000 (20 years)	50	2.4
				80	3.9*
				90	4.7*
				95	5.4
				99	6.8

State recommends a 4.2-inch AC overlay plus crack relief fabric after cracking and seating.

SW-7	AC	JPCP	7,370,000 (20 years)	50	1.2
				80	2.6
				90	3.4
				95	4.0*
				99	5.4

State recommends a 4.2-inch AC overlay plus crack relief fabric after cracking and seating.

Region-Project	Overlay Type	Existing Pavement	Design ESALs	Design Reliability	Overlay Thickness (in)
SW-8	AC	JPCP	7,370,000 (20 years)	50	1.7
				80	3.2
				90	4.0*
				95	4.7
				99	6.1

State recommends a 4.2-inch AC overlay plus crack relief fabric after cracking and seating.

Region-Project	Overlay Type	Existing Pavement	Design ESALs	Design Reliability	Overlay Thickness (in)
NE-5	AC	JPCP	329,288	50	0
				80	0.6
				90	1.0
				95	1.3
				99	2.1

State constructed 3.5-inch AC overlay after crack and seating. Subgrade soil has CBR = 15 which results in high estimated resilient modulus (12,000 psi) and thin overlay. No deflection data available.

Summary of Results for Fractured Slab PCC Overlay Designs

1. There are not enough projects to judge the adequacy of the procedure. The limited results show that the required AC overlay thickness of fractured slab PCC appears reasonable for most projects and generally agrees with the State recommendations. A design thickness at 95-percent reliability vs the agency recommendation is given in Figure N7 along with data points from the conventional AC overlays previously shown.
2. The backcalculated subgrade moduli were all divided by 4 (C = 0.25) which is apparently needed to give overlay adequate thickness. One section in the Northeast that had a CBR = 15 (and a corresponding estimated modulus of 12,000 psi), resulted in a very thin overlay requirement. It is believed that the subgrade modulus is too high for this project.
3. The design reliability level is very significant. For these projects, a design reliability level of 90 to 95 percent appears to provide reasonable overlay thicknesses, and in general agrees with agency recommendations.

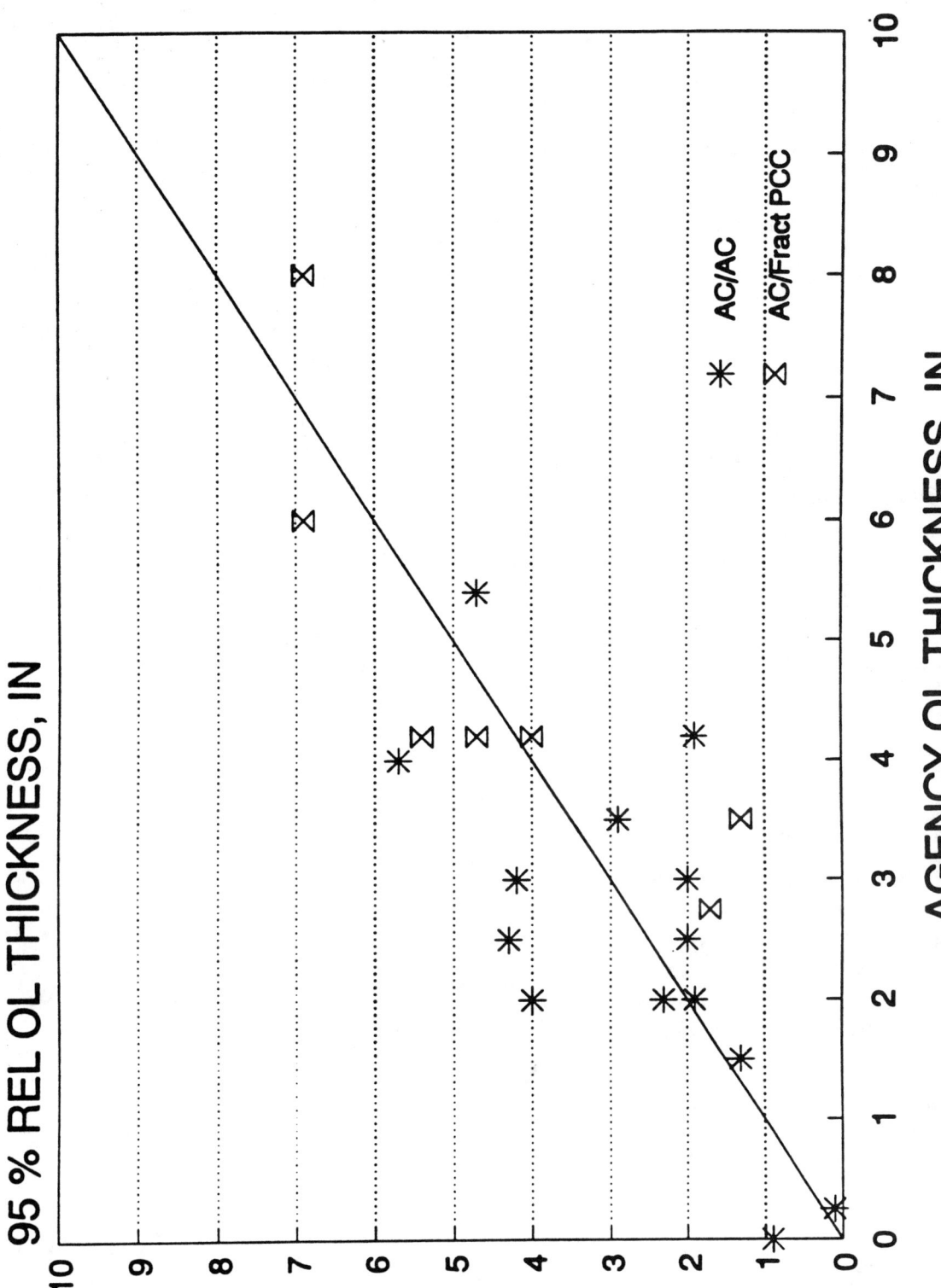

Figure N7. Comparison of Design AC Overlay Thicknesses and Agency AC Overlay Thicknesses for Fractured Pavements (95-percent reliability level)

REVISED CHAPTER 5 AASHTO DESIGN GUIDE OVERLAY DESIGN

MW-3 AC OVERLAY OF RUBBLIZED JRCP PAVEMENT (I57)

EXISTING PAVEMENT DESIGN

RUBBLIZED PCC	10.00 inches	SUBGRADE: A-6
GRAN BASE	6.00	
SUBBASE	0.00	
TOTAL THICKNESS	16.00	

Future design lane ESALs = 6,700,000 (FLEXIBLE ESALs)

DETERMINE SN_f

Vary trial SN_f until computed ESALs equal future design ESALs.

SN_f	M_R, psi	R	Z	S_o	P_1	P_2	ESAL
4.10	5,556	50	0	0.49	4.5	2.5	6,833,081
4.70	5,556	80	0.841	0.49	4.5	2.5	6,839,800
5.05	5,556	90	1.282	0.49	4.5	2.5	6,982,040
5.35	5,556	95	1.645	0.49	4.5	2.5	7,096,596
5.90	5,556	99	2.327	0.49	4.5	2.5	6,895,101
TRIAL			INPUT	INPUT	INPUT	INPUT	

DETERMINE SUBGRADE MR BY NDT METHOD

Vary trial E_p/M_R until computed D_0 equals actual value.

	ACTUAL			SUBGRADE		TRIAL COMPUTED		
STATION	LOAD, lbs	D_0, mils	D_r, mils	M_R, psi	C FACTOR	E_p/M_R	D_0, mils	E_p, psi
1	9,000	4.10	2.70	22,222	4	38.80	4.10	862,222

r = 36 inches
Check r > 0.7 ae = 38.14 inches*

DETERMINE SN_{eff}

LAYER	STR COEF	DRAIN m	SN_{eff}
RUBBLIZED PCC	0.20	1.00	2.00
SUBBASE	0.05	1.00	0.30
SUBBASE	0.00	1.00	0.00

SN_{eff} = 2.30

DETERMINE OVERLAY THICKNESS

AC OL structural coefficient = 0.44

DESIGN RELIABILITY	CONDITION METHOD, in
50	4.09
80	5.45
90	6.25
95	6.93
99	8.18

*Sensors spaced at farther distances were not available, or they would have been used.

Appendix N

REVISED CHAPTER 5 AASHTO DESIGN GUIDE OVERLAY DESIGN

SW-5 AC OVERLAY OF RUBBLIZED CRCP (I610)

EXISTING PAVEMENT DESIGN

RUBBLIZED PCC	10.00 inches	SUBGRADE: ???
GRAN BASE	6.00	
SUBBASE	0.00	
TOTAL THICKNESS	16.00	

Future design lane ESALs = 9,532,300 (FLEXIBLE ESALs)

DETERMINE SN_f

Vary trial SN_f until computed ESALs equal future design ESALs.

SN_f	M_R, psi	R	Z	S_o	P_1	P_2	ESAL
4.70	4,344	50	0	0.49	4.5	2.5	9,983,205
5.35	4,344	80	0.841	0.49	4.5	2.5	9,934,515
5.70	4,344	90	1.282	0.49	4.5	2.5	9,733,044
6.00	4,344	95	1.645	0.49	4.5	2.5	9,573,480
6.65	4,344	99	2.327	0.49	4.5	2.5	9,919,464
TRIAL			INPUT	INPUT	INPUT	INPUT	

DETERMINE SUBGRADE MR BY NDT METHOD

Vary trial E_p/M_R until computed D_0 equals actual value.

	ACTUAL			SUBGRADE		TRIAL COMPUTED		
STATION	LOAD, lbs	D_0, mils	D_r, mils	M_R, psi	C FACTOR	E_p/M_R	D_0, mils	E_p, psi
1	0	0.00	0.00	ERR	0	0.00	ERR	ERR
	r =	36 inches						
Check r > 0.7 ae =		inches						

DETERMINE SN_{eff}

LAYER	STR COEF	DRAIN m	SN_{eff}
RUBBLIZED PCC	0.20	1.00	2.00
SUBBASE	0.00	1.00	0.00
SUBBASE	0.00	1.00	0.00

SN_{eff} = 2.00

DETERMINE OVERLAY THICKNESS

AC OL structural coefficient = 0.44

DESIGN RELIABILITY	CONDITION METHOD, in
50	6.14
80	7.61
90	8.41
95	9.09
99	10.57

REVISED CHAPTER 5 AASHTO DESIGN GUIDE OVERLAY DESIGN

MW-4 AC OVERLAY OF BREAK/SEATED JRCP (JAC-32-12.47)

EXISTING PAVEMENT DESIGN

RUBBLIZED PCC	8.00 inches	SUBGRADE: A-6
SUBBASE	6.00	
SUBBASE	0.00	
TOTAL THICKNESS	14.00	

Future design lane ESALs = 318,000 (FLEXIBLE ESALs)

DETERMINE SN_f

Vary trial SN_f until computed ESALs equal future design ESALs.

SN_f	M_R, psi	R	Z	S_o	P_1	P_2	ESAL
2.70	4,885	50	0	0.49	4.5	2.5	336,217
3.15	4,885	80	0.841	0.49	4.5	2.5	346,855
3.40	4,885	90	1.282	0.49	4.5	2.5	345,152
3.60	4,885	95	1.645	0.49	4.5	2.5	332,752
4.05	4,885	99	2.327	0.49	4.5	2.5	337,800
TRIAL			INPUT	INPUT	INPUT	INPUT	

DETERMINE SUBGRADE MR BY NDT METHOD

Vary trial E_p/M_R until computed D_0 equals actual value.

	ACTUAL			SUBGRADE		TRIAL COMPUTED		
STATION	LOAD, lbs	D_0, mils	D_r, mils	M_R, psi	C FACTOR	E_p/M_R	D_0, mils	E_p, psi
	10,376	5.27	3.54	19,540	4	53.50	5.27	1,045,416

r = 36 inches
Check r > 0.7 ae = 37.16 inches*

DETERMINE SN_{eff}

LAYER	STR COEF	DRAIN m	SN_{eff}
BREAK/SEATED	0.25	1.00	2.00
SUBBASE	0.14	1.00	0.84
SUBBASE	0.00	1.00	0.00

SN_{eff} = 2.84

DETERMINE OVERLAY THICKNESS

AC OL structural coefficient = 0.44

DESIGN RELIABILITY	CONDITION METHOD, in
50	0.00
80	0.70
90	1.27
95	1.73
99	2.75

*Sensors spaced at farther distances were not available, or they would have been used.

Appendix N

REVISED CHAPTER 5 AASHTO DESIGN GUIDE OVERLAY DESIGN

SW-6 AC OVERLAY OF CRACKED/SEATED JPCP (PROJ STN 353)

EXISTING PAVEMENT DESIGN

RUBBLIZED PCC	8.20 inches
C.T. BASE	3.70
SUBBASE	0.00
TOTAL THICKNESS	11.90

Future design lane ESALs = 7,370,000 (2/3 OF 11,000,000 USED AS FLEXIBLE ESALs)

DETERMINE SN_f

Vary trial SN_f until computed ESALs equal future design ESALs.

SN_f	M_R, psi	R	Z	S_o	P_1	P_2	ESAL
4.50	4,350	50	0	0.49	4.5	2.5	7,364,787
5.15	4,350	80	0.841	0.49	4.5	2.5	7,516,147
5.50	4,350	90	1.282	0.49	4.5	2.5	7,452,560
5.80	4,350	95	1.645	0.49	4.5	2.5	7,401,524
6.40	4,350	99	2.327	0.49	4.5	2.5	7,354,079
TRIAL			INPUT	INPUT	INPUT	INPUT	

DETERMINE SUBGRADE MR BY NDT METHOD

Vary trial E_p/M_R until computed D_0 equals actual value.

	ACTUAL			SUBGRADE		TRIAL COMPUTED		
STATION	LOAD, lbs	D_0, mils	D_r, mils	M_R, psi	C FACTOR	E_p/M_R	D_0, mils	E_p, psi
	8,952	6.31	3.43	17,399	4	44.00	6.32	765,574

r = 36 inches
Check r > 0.7 ae = 29.70 inches

DETERMINE SN_{eff}

LAYER	STR COEF	DRAIN m	SN_{eff}
RUBBLIZED PCC	0.35	1.00	2.87
C.T. SUBBASE	0.15	1.00	0.56
SUBBASE	0.00	1.00	0.00

SN_{eff} = 3.43

DETERMINE OVERLAY THICKNESS

AC OL structural coefficient = 0.44

DESIGN RELIABILITY	CONDITION METHOD, in
50	2.44
80	3.92
90	4.72
95	5.40
99	6.76

REVISED CHAPTER 5 AASHTO DESIGN GUIDE OVERLAY DESIGN

SW-7 AC OVERLAY OF CRACK/SEATED JPCP (PROJ 7456)

EXISTING PAVEMENT DESIGN

CRACK/SEATED JPCP	8.20 inches
C.T. BASE	4.80
SUBBASE	0.00
TOTAL THICKNESS	13.00

Future design lane ESALs = 7,370,000 (2/3 OF 11,000,000 USED AS FLEXIBLE ESALs)

DETERMINE SN_f

Vary trial SN_f until computed ESALs equal future design ESALs.

SN_f	M_R, psi	R	Z	S_o	P_1	P_2	ESAL
4.45	4,597	50	0	0.49	4.5	2.5	7,743,986
5.05	4,597	80	0.841	0.49	4.5	2.5	7,401,737
5.40	4,597	90	1.282	0.49	4.5	2.5	7,384,677
5.70	4,597	95	1.645	0.49	4.5	2.5	7,370,609
6.30	4,597	99	2.327	0.49	4.5	2.5	7,390,745
TRIAL			INPUT	INPUT	INPUT	INPUT	

DETERMINE SUBGRADE MR BY NDT METHOD

Vary trial E_p/M_R until computed D_0 equals actual value.

	ACTUAL			SUBGRADE		TRIAL COMPUTED		
STATION	LOAD, lbs	D_0, mils	D_r, mils	M_R, psi	C FACTOR	E_p/M_R	D_0, mils	E_p, psi
	8,496	3.68	3.08	18,390	4	114.00	3.68	2,096,416

r = 36 inches
Check r > 0.7 ae = 44.32 inches*

DETERMINE SN_{eff}

LAYER	STR COEF	DRAIN m	SN_{eff}
CRACK/SEAT JPCP	0.35	1.00	2.87
C.T. SUBBASE	0.22	1.00	1.06
SUBBASE	0.00	1.00	0.00

SN_{eff} = 3.93

DETERMINE OVERLAY THICKNESS

AC OL structural coefficient = 0.44

DESIGN RELIABILITY	CONDITION METHOD, in
50	1.19
80	2.55
90	3.35
95	4.03
99	5.40

*Sensors spaced at farther distances were not available, or they would have been used.

Appendix N

REVISED CHAPTER 5 AASHTO DESIGN GUIDE OVERLAY DESIGN

SW-8 AC OVERLAY OF CRACK/SEATED JPCP (3005, STN 305)

EXISTING PAVEMENT DESIGN

RUBBLIZED PCC	8.20 inches
C.T. BASE	3.70
SUBBASE	0.00
TOTAL THICKNESS	11.90

Future design lane ESALs = 7,370,000 (2/3 OF 11,000,000 USED AS FLEXIBLE ESALs)

DETERMINE SN_f

Vary trial SN_f until computed ESALs equal future design ESALs.

SN_f	M_R, psi	R	Z	S_o	P_1	P_2	ESAL
4.45	4,522	50	0	0.49	4.5	2.5	7,453,483
5.10	4,522	80	0.841	0.49	4.5	2.5	7,656,722
5.45	4,522	90	1.282	0.49	4.5	2.5	7,615,343
5.75	4,522	95	1.645	0.49	4.5	2.5	7,581,902
6.35	4,522	99	2.327	0.49	4.5	2.5	7,567,726
TRIAL			INPUT	INPUT	INPUT	INPUT	

DETERMINE SUBGRADE MR BY NDT METHOD

Vary trial E_p/M_R until computed D_0 equals actual value.

	ACTUAL			SUBGRADE		TRIAL COMPUTED		
STATION	LOAD, lbs	D_0, mils	D_r, mils	M_R, psi	C FACTOR	E_p/M_R	D_0, mils	E_p, psi
	9,144	3.89	3.37	18,089	4	157.00	3.89	2,839,976

r = 36 inches
Check r > 0.7 ae = 45.13 inches*

DETERMINE SN_{eff}

LAYER	STR COEF	DRAIN m	SN_{eff}
CRACK/SEAT JPCP	0.35	1.00	2.87
C.T. SUBBASE	0.22	1.00	0.81
SUBBASE	0.00	1.00	0.00

SN_{eff} = 3.68

DETERMINE OVERLAY THICKNESS

AC OL structural coefficient = 0.44

DESIGN RELIABILITY	CONDITION METHOD, in
50	1.74
80	3.22
90	4.01
95	4.70
99	6.06

*Sensors spaced at farther distances were not available, or they would have been used.

REVISED CHAPTER 5 AASHTO DESIGN GUIDE OVERLAY DESIGN

NE-5 AC OVERLAY OF CRACK/SEAT JPCP (SR611-27M)

EXISTING PAVEMENT DESIGN

CRACK/SEAT JPCP	10.00 inches	SUBGRADE: CBR = 15
SUBBASE	0.00	M_R = 12,000 PSI
SUBBASE	0.00	(AASHTO, APPENDIX FF)
TOTAL THICKNESS	10.00	

Future design lane ESALs = 329,288 (flexible ESALs)

DETERMINE SN_f

Vary trial SN_f until computed ESALs equal future design ESALs.

SN_f	M_R, psi	R	Z	S_o	P_1	P_2	ESAL
1.93	12,000	50	0	0.45	4.2	2.5	330,546
2.25	12,000	80	0.841	0.45	4.2	2.5	342,787
2.42	12,000	90	1.282	0.45	4.2	2.5	335,762
2.59	12,000	95	1.645	0.45	4.2	2.5	346,645
2.90	12,000	99	2.327	0.45	4.2	2.5	337,721
TRIAL			INPUT	INPUT	INPUT	INPUT	

DETERMINE SUBGRADE MR BY NDT METHOD

Vary trial E_p/M_R until computed D_0 equals actual value.

	ACTUAL			SUBGRADE		TRIAL COMPUTED		
STATION	LOAD, lbs	D_0, mils	D_r, mils	M_R, psi	C FACTOR	E_p/M_R	D_0, mils	E_p, psi
0	0.00	0.00		ERR	0	0.00	ERR	ERR
	r =	36 inches						
Check r > 0.7 ae =		inches						

DETERMINE SN_{eff}

LAYER	STR COEF	DRAIN m	SN_{eff}
CRACK/SEAT JPCP	0.20	1.00	2.00
SUBBASE	0.00	1.00	0.00
SUBBASE	0.00	1.00	0.00

SN_{eff} = 2.00

DETERMINE OVERLAY THICKNESS

AC OL structural coefficient = 0.44

DESIGN RELIABILITY	CONDITION METHOD, in
50	0.00
80	0.57
90	0.95
95	1.34
99	2.05

N5.0 AC OVERLAY AND BONDED PCC OVERLAY OF JPCP AND JRCP

Region-Project	Existing Pavement	Design ESALs	Design Reliability	AC Overlay Thickness (in)	Bonded PCC Overlay Thickness (in)
MW-5	JRCP	424,000	50	0	0
			80	0	0
			90*	0	0
			95	0.8	0.4
			99	2.4	1.2

State design procedure indicates no structural overlay is needed for this pavement.

Region-Project	Existing Pavement	Design ESALs	Design Reliability	AC Overlay Thickness (in)	Bonded PCC Overlay Thickness (in)
SW-9	JRCP	17,668,000 (20 years)	50	0	0
			80	1.4	0.6
			90	2.8	1.4
			95*	3.6	1.8
			99	5.5	3.0

Agency recommends a 4-inch AC overlay for a 20-year design.

Region-Project	Existing Pavement	Design ESALs	Design Reliability	AC Overlay Thickness (in)	Bonded PCC Overlay Thickness (in)
SW-10	JRCP	12,800,000 (15 years)	50	0	0
			80	0.3	0.2
			90	1.8	0.8
			95	2.8	1.4
			99	4.5	2.4

Agency recommends a 4-inch AC overlay for a 20-year design.

Region-Project	Existing Pavement	Design ESALs	Design Reliability	AC Overlay Thickness (in)	Bonded PCC Overlay Thickness (in)
SW-11	JPCP	11,000,000	50	3.8	1.9
			80	5.6	3.0
			90	6.5	3.6
			95	7.4	4.2
			99	8.8	5.2

No agency recommendations for conventional AC overlay. Recommendations for AC overlay over crack and seat JPCP was 4.2 inches. No condition data are available for this project.

Region-Project	Existing Pavement	Design ESALs	Design Reliability	AC Overlay Thickness (in)	Bonded PCC Overlay Thickness (in)
SW-12	JPCP	11,000,000	50	3.2	1.6
			80	5.1	2.7
			90	6.0	3.3
			95	6.8	3.8
			99	8.3	4.8

No agency recommendations for conventional AC overlay. Recommendations for AC overlay over crack and seat JPCP was 4.2 inches. No condition data are available for this project.

Region-Project	Existing Pavement	Design ESALs	Design Reliability	AC Overlay Thickness (in)	Bonded PCC Overlay Thickness (in)
SW-13	JPCP	11,000,000	50	4.9	1.5
			80	5.9	2.6
			90	6.6	3.7
			95	8.1	4.7
			99	8.3	4.8

No agency recommendations for conventional AC overlay. Recommendations for AC overlay over crack and seat JPCP was 4.2 inches. No condition data are available for this project.

Region-Project	Existing Pavement	Design ESALs	Design Reliability	AC Overlay Thickness (in)	Bonded PCC Overlay Thickness (in)
SE-1	JPCP	25,500,000	50	1.6	0.8
			80	3.6	1.8
			90	4.5	2.4
			95	5.3	2.8
			99	7.0	4.0

No agency recommendations for this overlay design.

Region-Project	Existing Pavement	Design ESALs	Design Reliability	AC Overlay Thickness (in)	Bonded PCC Overlay Thickness (in)
MW-6	JRCP	22,834,000	50	4.0	2.1
			80	6.1	3.4
			90	7.2	4.1
			95*	8.0	4.7
			99	9.7	5.9

State constructed 4.5-inch bonded PCC overlay. No deflection data are available.

Region-Project	Existing Pavement	Design ESALs	Design Reliability	AC Overlay Thickness (in)	Bonded PCC Overlay Thickness (in)
MW-7	JRCP	10,000,000 (10 years)	50	0	0
			80	1.0	0.5
			90	2.2	1.1
			95*	3.2	1.6
			99	4.6	2.4

State policy design for this pavement is a 3.25-inch AC overlay.

Region-Project	Existing Pavement	Design ESALs	Design Reliability	AC Overlay Thickness (in)	Bonded PCC Overlay Thickness (in)
NW-7	JRCP	80,000,000 (20 years)	50	4.7	2.5
			80	6.2	3.4
			90	7.0	3.9
			95	7.6	4.4
			99	8.9	5.2

Extremely high traffic. The deflection basin used is an average for the project. Existing pavement is in fair to poor condition. Low J factor used to determine D_f, since the State has observed that in this mild climate, JRCP pavements perform much better than AASHTO design equation predicts. State recommended 5-inch AC overlay, based on good performance of 5-inch AC overlay on adjacent section of highway, in service 8 years. It is unlikely, however, that a 5-inch AC overlay could handle 80 million ESALs.

Region-Project	Existing Pavement	Design ESALs	Design Reliability	AC Overlay Thickness (in)	Bonded PCC Overlay Thickness (in)
NW-8	JRCP	20,000,000 (15 years)	50	2.5	1.2
			80	4.1	2.1
			90	4.9	2.6
			95	5.5*	3.0
			99	6.8	3.8

State constructed 6-inch AC overlay in 1976, has carried 20 million ESALs since overlay, current PSI is 3.5. The deflection basin used is average for bare JRCP project of same design. Low J factor used to determine D_f, since the State has observed that in this mild climate, JRCP pavements perform much better than AASHTO design equation predicts.

NE-6	JRCP	12,255,000 (20 years)	50	0.2	0.1
			80	2.2	1.1
			90	3.2*	1.6
			95	4.0	2.1
			99	5.5	3.0

State recommends a 3-inch AC overlay. No deflection data are available.

NE-7	JRCP	16,000,000 (20 years)	50	0	0
			80	1.7	0.8
			90	2.8*	1.4
			95	3.7*	1.9
			99	5.1	2.7

State recommends an AC overlay of at least 3 inches.

NE-8	JRCP	4,650,000 (10 years)	50	0	0
			80	1.9	0.9
			90	2.8*	1.4
			95	3.7*	1.9
			99	5.2	2.8

State design procedure indicates 2.5-inch AC overlay required. State constructed 3.5-inch AC overlay. No deflection data are available. Subgrade resilient modulus was estimated from CBR using AASHTO Guide Appendix FF.

NE-9	JRCP	10,050,000 (10 years)	50	0.4	0.2
			80	2.4	1.2
			90	3.5*	1.8
			95	4.4	2.3
			99	6.0	3.3

State design procedure indicates 3-inch AC overlay required. State constructed 3.5-inch AC overlay. No deflection data are available. Subgrade resilient modulus was estimated from CBR using AASHTO Guide Appendix FF.

Summary of Results for AC Overlay and Bonded PCC Overlay of JPCP and JRCP

1. Overall it appears that the revised AASHTO overlay design procedures produce reasonable conventional AC overlay and bonded PCC overlay thicknesses for jointed PCC pavements that are consistent with State recommendations. For example, no overlay requirement is shown for project MW-5 for a reliability level below 95 percent, which is consistent with the State's assessment that the project does not need an overlay. In another example, the bonded PCC overlay thickness indicated for project MW-6 at the 95-percent reliability level matches very closely the bonded overlay thickness actually constructed by that State. Project MW-7 is another good example: the 3.2-inch overlay requirement indicated at the 95-percent reliability level matches the State's design for 10 years and 10 million ESALs. A survival analysis of overlays in this State has shown that this type of overlay lasts an average of 11.9 years and carries an average of 18 million ESALs. A plot of design AC overlay thickness vs agency specified overlay thickness for these projects is shown in Figure N8.
2. Specific difficulties in AC and bonded PCC overlay thickness design include the sensitivity of the J factor for load transfer and the necessity of imposing practical minimum and maximum values for the PCC elastic modulus, the PCC modulus of rupture, and the effective k-value.
3. The design reliability level is very significant. Most of the projects were Interstate-type highways. A design reliability level of 95 percent appears to be reasonable for AC overlays of JRCP and JPCP.
4. Specific examples of overlays that appear to be too thick are projects SW-11, SW-12, and SW-13. These are located in a State with a very mild climate, which may have a very significant effect on improving overlaid pavement performance and reducing overlay thickness requirements. This could be addressed by using a lower design reliability level, or by using a lower J factor to determine D_f.

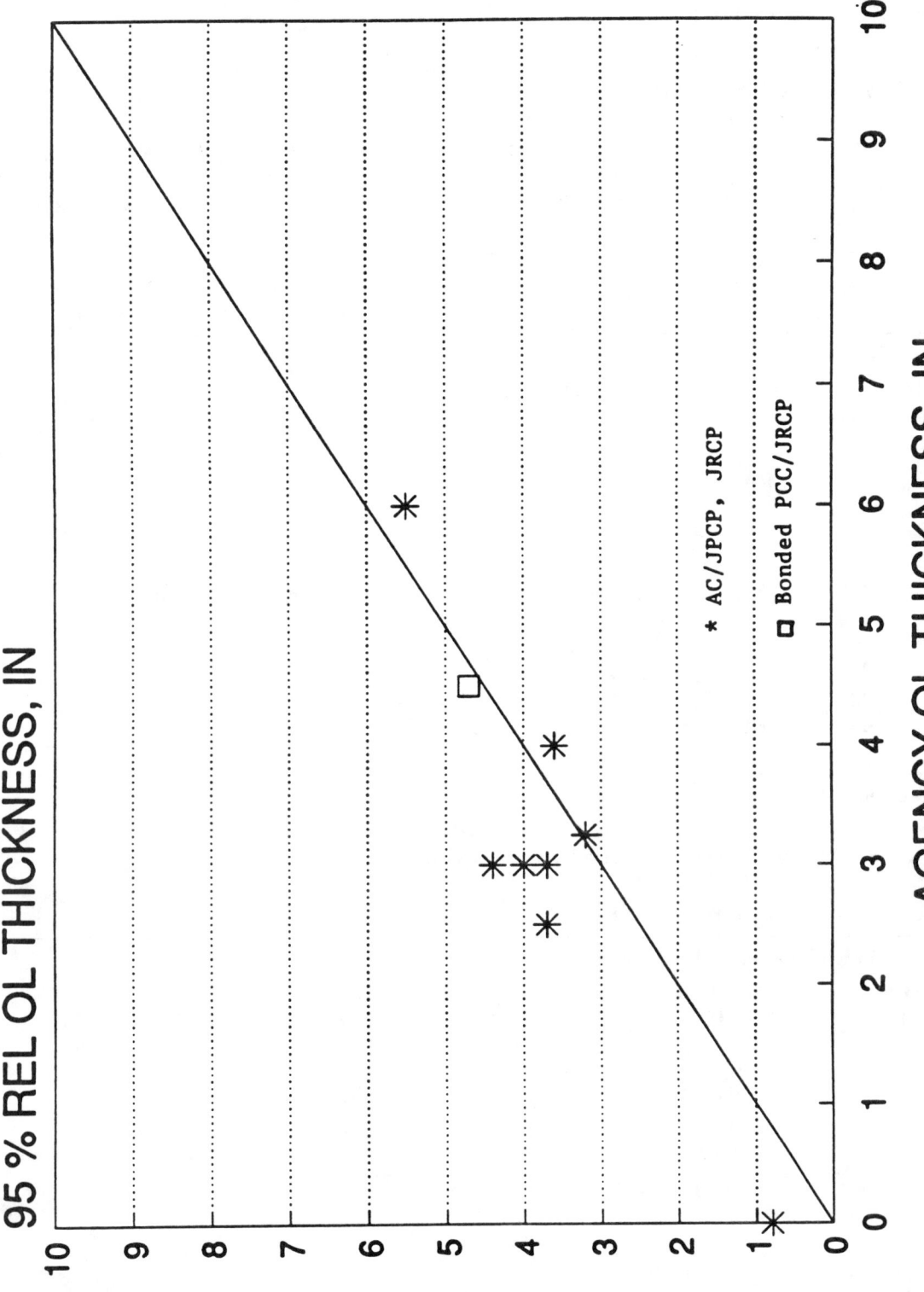

Figure N8. Comparison of AC Overlay Thickness and Bonded PCC Overlay Thickness over JRCP or JPCP (at 95-percent reliability) vs that Recommended by Agencies

REVISED CHAPTER 5 AASHTO DESIGN GUIDE OVERLAY DESIGN

MW-5 AC AND BONDED PCC OL OF EXISTING JRCP (JAC-32-12.47)

EXISTING PAVEMENT DESIGN AND FUTURE TRAFFIC
Slab thickness   8.00 (in)

Future design lane ESALs = 424,000

BACKCALCULATION OF K_{eff} AND E_c

INPUT LOAD (lbs)	INPUT D_0 (mils)	INPUT D_{12} (mils)	INPUT D_{24} (mils)	INPUT D_{36} (mils)	AREA (in)	RADIUS RELSTIFF (in)	K_{dyn} (pci)	SLAB E_c (psi)
10,565	4.76	4.40	3.77	3.18	30.61	36.66	204	8.4E+06
10,376	5.27	4.88	4.21	3.54	30.73	37.31	175	7.7E+06
10,328	5.47	5.11	4.33	3.58	30.64	36.82	172	7.2E+06
10,249	4.52	4.17	3.62	3.03	30.70	37.18	202	8.9E+06
							188	8.1E+06

DETERMINE D_f
Vary trial D_f until computed ESALs equal future design ESALs.

K_{eff} (psi/in)	INPUT J	S_c (psi)	INPUT P_1	INPUT P_2	E_c (psi)	INPUT S_o	INPUT LOS	INPUT C_d
94	4.2	700 ***	4.2	2.5	8.1E+06	0.39	0.00	1.00

TRIAL D_f (in)	R	Z	COMPUTED ESALs (millions)	
6.10	50	0	459,546	***Selected by engineer
6.90	80	0.84	422,939	
7.40	90	1.282	425,110	
7.90	95	1.645	453,572	
8.70	99	2.327	448,616	

DETERMINE D_{eff}
INPUT F_{jc} = 0.98 (10 FAILURES/MI UNREPAIRED)
INPUT F_{fat} = 0.96 (50 MID-SLAB WORKING CRACKS)
INPUT F_{dur} = 1.00

D_{eff} (in) = $F_{jc} * F_{dur} * F_{fat} * D_{exist}$ = 7.53

DETERMINE OVERLAY THICKNESS

RELIABILITY LEVEL	PCC BOL THICK	PCC to AC FACTOR	AC OL THICK
50	0.00	2.46	0.00
80	0.00	2.32	0.00
90	0.00	2.24	0.00
95	0.37	2.17	0.81
99	1.17	2.06	2.41

REVISED CHAPTER 5 AASHTO DESIGN GUIDE OVERLAY DESIGN

SW-9 AC AND BONDED PCC OL OF EXISTING JRCP (I-30)

EXISTING PAVEMENT DESIGN AND FUTURE TRAFFIC

Slab thickness 10.00 (in)

Future design lane ESALs = 17,668,158 (20 YEARS)

BACKCALCULATION OF K_{eff} AND E_c

INPUT LOAD (lbs)	INPUT D_0 (mils)	INPUT D_{12} (mils)	INPUT D_{24} (mils)	INPUT D_{36} (mils)	AREA (in)	RADIUS RELSTIFF (in)	K_{dyn} (pci)	SLAB E_c (psi)
0	0	0	0	0	ERR	ERR	ERR	ERR
							ERR	ERR

DETERMINE D_f

Vary trial D_f until computed ESALs equal future design ESALs.

K_{eff} (psi/in)	INPUT J	S_c (psi)	INPUT P_1	INPUT P_2	E_c (psi)	INPUT S_o	INPUT LOS	INPUT C_d
53	3.2	710	4.5	2.5	5.1E+06	0.39	0.00	1.01
***		***			***			

TRIAL D_f (in)	R	Z	COMPUTED ESALs (millions)
9.30	50	0	17,483,863
10.40	80	0.84	17,272,377
11.10	90	1.282	18,026,295
11.60	95	1.645	17,540,132
12.70	99	2.327	17,638,108

***Selected by engr.

DETERMINE D_{eff}

INPUT F_{jc} = 1.00
INPUT F_{fat} = 0.975
INPUT F_{dur} = 1.00

D_{eff} (in) = $F_{jc} * F_{dur} * F_{fat} * D_{exist}$ = 9.75

DETERMINE OVERLAY THICKNESS

RELIABILITY LEVEL	PCC BOL THICK	PCC to AC FACTOR	AC OL THICK
50	0.00	2.29	0.00
80	0.65	2.13	1.38
90	1.35	2.03	2.75
95	1.85	1.97	3.65
99	2.95	1.86	5.48

REVISED CHAPTER 5 AASHTO DESIGN GUIDE OVERLAY DESIGN

SW-10 AC AND BONDED PCC OL OF EXISTING JRCP (I-30)

EXISTING PAVEMENT DESIGN AND FUTURE TRAFFIC

Slab thickness   10.00 (in)

Future design lane ESALs = 12,801,929 (15 years)

BACKCALCULATION OF K_{eff} AND E_c

INPUT LOAD (lbs)	INPUT D_0 (mils)	INPUT D_{12} (mils)	INPUT D_{24} (mils)	INPUT D_{36} (mils)	AREA (in)	RADIUS RELSTIFF (in)	K_{dyn} (pci)	SLAB E_c (psi)
0	0	0	0	0	ERR	ERR	ERR	ERR
							ERR	ERR

DETERMINE D_f

Vary trial D_f until computed ESALs equal future design ESALs.

K_{eff} (psi/in)	INPUT J	S_c (psi)	INPUT P_1	INPUT P_2	E_c (psi)	INPUT S_o	INPUT LOS	INPUT C_d
53	3.2	710	4.5	2.5	5.1E+06	0.39	0.00	1.01
***		***			***			

TRIAL D_f (in)	R	Z	COMPUTED ESALs (millions)
8.90	50	0	13,112,728
9.90	80	0.84	12,426,728
10.60	90	1.282	13,210,489
11.10	95	1.645	13,011,773
12.10	99	2.327	12,670,888

***Selected by engr.

DETERMINE D_{eff}

INPUT F_{jc} = 1.00

INPUT F_{fat} = 0.975

INPUT F_{dur} = 1.00

D_{eff} (in) = $F_{jc} * F_{dur} * F_{fat} * D_{exist}$ = 9.75

DETERMINE OVERLAY THICKNESS

RELIABILITY LEVEL	PCC BOL THICK	PCC to AC FACTOR	AC OL THICK
50	0.00	2.36	0.00
80	0.15	2.20	0.33
90	0.85	2.10	1.79
95	1.35	2.03	2.75
99	2.35	1.92	4.51

REVISED CHAPTER 5 AASHTO DESIGN GUIDE OVERLAY DESIGN

SW-11 AC AND BONDED PCC OL OF EXISTING JPCP (3005, STN 353)

EXISTING PAVEMENT DESIGN AND FUTURE TRAFFIC

Slab thickness   8.20 (in)

Future design lane ESALs = 11,000,000

BACKCALCULATION OF K_{eff} AND E_c

INPUT LOAD (lbs)	INPUT D_0 (mils)	INPUT D_{12} (mils)	INPUT D_{24} (mils)	INPUT D_{36} (mils)	AREA (in)	RADIUS RELSTIFF (in)	K_{dyn} (pci)	SLAB E_c (psi)
8,952	6.31	5.31	4.36	3.43	27.65	26.04	256	2.5E+06
							256	2.5E+06

DETERMINE D_f

Vary trial D_f until computed ESALs equal future design ESALs.

K_{eff} (psi/in)	INPUT J	S_c (psi)	INPUT P_1	INPUT P_2	E_c (psi)	INPUT S_o	INPUT LOS	INPUT C_d
128	4.0	650	4.5	2.5	3.0E+06	0.35	0.00	1.00
		***			***			

TRIAL D_f (in)	R	Z	COMPUTED ESALs (millions)
9.70	50	0	11,009,829
10.80	80	0.84	11,071,373
11.40	90	1.282	11,005,188
12.00	95	1.645	11,474,284
13.00	99	2.327	11,211,036

***Selected by engr.

DETERMINE D_{eff}

INPUT F_{jc} = 1.00 (REPLACE ALL SLABS WITH CRACKS)
INPUT F_{fat} = 0.95
INPUT F_{dur} = 1.00

D_{eff} (in) = $F_{jc} * F_{dur} * F_{fat} * D_{exist}$ = 7.79

DETERMINE OVERLAY THICKNESS

RELIABILITY LEVEL	PCC BOL THICK	PCC to AC FACTOR	AC OL THICK
50	1.91	1.97	3.76
80	3.01	1.85	5.57
90	3.61	1.80	6.49
95	4.21	1.75	7.38
99	5.21	1.69	8.82

REVISED CHAPTER 5 AASHTO DESIGN GUIDE OVERLAY DESIGN

SW-12 AC AND BONDED PCC OL OF EXISTING JPCP (7456)

EXISTING PAVEMENT DESIGN AND FUTURE TRAFFIC

Slab thickness   8.20 (in)

Future design lane ESALs = 11,000,000

BACKCALCULATION OF K_{eff} AND E_c

INPUT LOAD (lbs)	INPUT D_0 (mils)	INPUT D_{12} (mils)	INPUT D_{24} (mils)	INPUT D_{36} (mils)	AREA (in)	RADIUS RELSTIFF (in)	K_{dyn} (pci)	SLAB E_c (psi)
8,496	3.68	3.08	2.64	2.23	28.29	27.76	367	4.6E+06
							367	4.6E+06

DETERMINE D_f

Vary trial D_f until computed ESALs equal future design ESALs.

K_{eff} (psi/in)	INPUT J	S_c (psi)	INPUT P_1	INPUT P_2	E_c (psi)	INPUT S_o	INPUT LOS	INPUT C_d
184	4.0	689	4.5	2.5	4.6E+06	0.35	0.00	1.00

TRIAL D_f (in)	R	Z	COMPUTED ESALs (millions)
9.40	50	0	10,780,339
10.50	80	0.84	11,039,263
11.10	90	1.282	11,072,656
11.60	95	1.645	11,006,869
12.60	99	2.327	10,932,237

DETERMINE D_{eff}

INPUT F_{jc} = 1.00   (REPAIR ALL DETERIORATED AREAS)
INPUT F_{fat} = 0.95
INPUT F_{dur} = 1.00

D_{eff} (in) = $F_{jc} * F_{dur} * F_{fat} * D_{exist}$ = 7.79

DETERMINE OVERLAY THICKNESS

RELIABILITY LEVEL	PCC BOL THICK	PCC to AC FACTOR	AC OL THICK
50	1.61	2.00	3.22
80	2.71	1.88	5.10
90	3.31	1.82	6.04
95	3.81	1.78	6.79
99	4.81	1.71	8.25

Appendix N

REVISED CHAPTER 5 AASHTO DESIGN GUIDE OVERLAY DESIGN

SW-13 AC AND BONDED PCC OL OF EXISTING JPCP (3005, STN 305)

EXISTING PAVEMENT DESIGN AND FUTURE TRAFFIC
Slab thickness   8.20 (in)

Future design lane ESALs = 11,000,000

BACKCALCULATION OF K_{eff} AND E_c

INPUT LOAD (lbs)	INPUT D_0 (mils)	INPUT D_{12} (mils)	INPUT D_{24} (mils)	INPUT D_{36} (mils)	AREA (in)	RADIUS RELSTIFF (in)	K_{dyn} (pci)	SLAB E_c (psi)
9,144	3.89	3.37	2.85	2.40	28.89	29.62	329	5.4E+06
							329	5.4E+06

DETERMINE D_f
Vary trial D_f until computed ESALs equal future design ESALs.

K_{eff} (psi/in)	INPUT J	S_c (psi)	INPUT P_1	INPUT P_2	E_c (psi)	INPUT S_o	INPUT LOS	INPUT C_d
165	4.0	723	4.5	2.5	5.4E+06	0.35	0.00	1.00

TRIAL D_f (in)	R	Z	COMPUTED ESALs (millions)
9.30	50	0	10,980,171
10.40	80	0.84	11,405,797
11.00	90	1.282	11,517,791
11.50	95	1.645	11,508,298
12.50	99	2.327	11,534,896

DETERMINE D_{eff}
INPUT F_{jc} = 1.00   (REPAIR ALL DETERIORATED AREAS)
INPUT F_{fat} = 0.95
INPUT F_{dur} = 1.00

D_{eff} (in) = $F_{jc} * F_{dur} * F_{fat} * D_{exist}$ = 7.79

DETERMINE OVERLAY THICKNESS

RELIABILITY LEVEL	PCC BOL THICK	PCC to AC FACTOR	AC OL THICK
50	1.51	2.01	3.04
80	2.61	1.89	4.93
90	3.21	1.83	5.88
95	3.71	1.79	6.64
99	4.71	1.72	8.10

REVISED CHAPTER 5 AASHTO DESIGN GUIDE OVERLAY DESIGN

SE-1 AC AND BONDED PCC OL OF EXISTING JPCP (I-10)

EXISTING PAVEMENT DESIGN AND FUTURE TRAFFIC

Slab thickness   9.00 (in)

Future design lane ESALs = 25,500,000 (20 YEARS)

BACKCALCULATION OF K_{eff} AND E_c

INPUT LOAD (lbs)	INPUT D_0 (mils)	INPUT D_{12} (mils)	INPUT D_{24} (mils)	INPUT D_{36} (mils)	AREA (in)	RADIUS RELSTIFF (in)	K_{dyn} (pci)	SLAB E_c (psi)
9,016	1.73	1.45	1.26	0.94	28.06	27.11	867	7.5E+06
9,499	1.61	1.46	1.30	0.98	30.22	34.81	600	1.4E+07
9,177	1.61	1.34	1.26	0.94	28.88	29.60	798	9.9E+06
9,338	1.93	1.57	1.26	1.14	27.14	24.80	959	5.8E+06
							806	9.3E+06

DETERMINE D_f

Vary trial D_f until computed ESALs equal future design ESALs.

K_{eff} (psi/in)	INPUT J	S_c (psi)	INPUT P_1	INPUT P_2	E_c (psi)	INPUT S_o	INPUT LOS	INPUT C_d
403	4.0	895	4.5	2.5	9.3E+06	0.35	0.00	1.00

TRIAL D_f (in)	R	Z	COMPUTED ESALs (millions)
9.30	50	0	25,372,489
10.40	80	0.84	26,051,167
10.90	90	1.282	24,673,324
11.40	95	1.645	24,609,486
12.50	99	2.327	25,928,583

DETERMINE D_{eff}

INPUT F_{jc} = 0.96

INPUT F_{fat} = 0.99

INPUT F_{dur} = 1.00

D_{eff} (in) = $F_{jc} * F_{dur} * F_{fat} * D_{exist}$ = 8.55

DETERMINE OVERLAY THICKNESS

RELIABILITY LEVEL	PCC BOL THICK	PCC to AC FACTOR	AC OL THICK
50	0.75	2.11	1.58
80	1.85	1.97	3.64
90	2.35	1.92	4.50
95	2.85	1.87	5.31
99	3.95	1.77	6.99

Appendix N

REVISED CHAPTER 5 AASHTO DESIGN GUIDE OVERLAY DESIGN

MW-6   AC AND BONDED PCC OL OF EXISTING JRCP (I-80)

EXISTING PAVEMENT DESIGN AND FUTURE TRAFFIC
Slab thickness   10.00 (in)

Future design lane ESALs = 22,834,400 (20 YEARS)

BACKCALCULATION OF K_{eff} AND E_c

INPUT LOAD (lbs)	INPUT D_0 (mils)	INPUT D_{12} (mils)	INPUT D_{24} (mils)	INPUT D_{36} (mils)	AREA (in)	RADIUS RELSTIFF (in)	K_{dyn} (pci)	SLAB E_c (psi)
0	0	0	0	0	ERR	ERR	ERR	ERR
							ERR	ERR

DETERMINE D_f

Vary trial D_f until computed ESALs equal future design ESALs.

K_{eff} (psi/in)	INPUT J	S_c (psi)	INPUT P_1	INPUT P_2	E_c (psi)	INPUT S_o	INPUT LOS	INPUT C_d
155	3.2	640	4.2	2.5	4.2E+06	0.39	0.00	1.00
***		***			***			

TRIAL D_f (in)	R	Z	COMPUTED ESALs (millions)
10.00	50	0	22,173,569
11.30	80	0.84	22,868,477
12.00	90	1.282	22,800,733
12.60	95	1.645	22,700,972
13.80	99	2.327	22,528,954

***Selected by engr.

DETERMINE D_{eff}

INPUT F_{jc} = 0.95

INPUT F_{fat} = 0.95

INPUT F_{dur} = 0.88

D_{eff} (in) = $F_{jc} * F_{dur} * F_{fat} * D_{exist}$ = 7.94

DETERMINE OVERLAY THICKNESS

RELIABILITY LEVEL	PCC BOL THICK	PCC to AC FACTOR	AC OL THICK
50	2.06	1.95	4.01
80	3.36	1.82	6.11
90	4.06	1.76	7.16
95	4.66	1.72	8.03
99	5.86	1.66	9.75

REVISED CHAPTER 5 AASHTO DESIGN GUIDE OVERLAY DESIGN

MW-7 AC AND BONDED PCC OL OF EXISTING JRCP (I-57)

EXISTING PAVEMENT DESIGN AND FUTURE TRAFFIC

Slab thickness   10.00   (in)

Future design lane ESALs = 10,000,000 (10 YEARS)

BACKCALCULATION OF K_{eff} AND E_c

INPUT LOAD (lbs)	INPUT D_0 (mils)	INPUT D_{12} (mils)	INPUT D_{24} (mils)	INPUT D_{36} (mils)	AREA (in)	RADIUS RELSTIFF (in)	K_{dyn} (pci)	SLAB E_c (psi)
11,144	4.39	3.97	3.49	3.01	30.51	36.16	239	4.8E+06
10,864	4.90	4.57	4.18	3.70	31.96	45.36	133	6.6E+06
10,928	4.51	4.09	3.69	3.14	30.88	38.12	206	5.1E+06
10,824	4.55	4.17	3.77	3.30	31.29	40.58	179	5.7E+06
							189	5.6E+06

DETERMINE D_f

Vary trial D_f until computed ESALs equal future design ESALs.

K_{eff} (psi/in)	INPUT J	S_c (psi)	INPUT P_1	INPUT P_2	E_c (psi)	INPUT S_o	INPUT LOS	INPUT C_d
95	3.5	730	4.5	2.5	5.6E+06	0.35	0.00	1.00

TRIAL D_f (in)	R	Z	COMPUTED ESALs (millions)
8.70	50	0	9,973,718
9.70	80	0.84	10,214,587
10.30	90	1.282	10,619,093
10.80	95	1.645	10,851,107
11.60	99	2.327	10,095,415

DETERMINE D_{eff}

INPUT F_{jc} = 0.97 (10 FAILURES/MI UNREPAIRED)
INPUT F_{fat} = 0.95 (50 MID-SLAB WORKING CRACKS)
INPUT F_{dur} = 1.00

D_{eff} (in) = $F_{jc} * F_{dur} * F_{fat} * D_{exist}$ = 9.22

DETERMINE OVERLAY THICKNESS

RELIABILITY LEVEL	PCC BOL THICK	PCC to AC FACTOR	AC OL THICK
50	0.00	2.30	0.00
80	0.48	2.15	1.04
90	1.09	2.07	2.24
95	1.59	2.01	3.18
99	2.39	1.91	4.56

Appendix N

REVISED CHAPTER 5 AASHTO DESIGN GUIDE OVERLAY DESIGN

NW-7 AC AND BONDED PCC OVERLAY OF EXISTING JRCP (N. Albany-N. Jefferson)

EXISTING PAVEMENT DESIGN AND FUTURE TRAFFIC

Slab thickness   8.00 (in)

Future design lane ESALs = 80,000,000 (20 years)

BACKCALCULATION OF K_{eff} AND E_c

INPUT LOAD (lbs)	INPUT D_0 (mils)	INPUT D_{12} (mils)	INPUT D_{24} (mils)	INPUT D_{36} (mils)	AREA (in)	RADIUS RELSTIFF (in)	K_{dyn} (pci)	SLAB E_c (psi)
9,000	5.20	4.50	3.50	2.90	27.81	26.44	302	3.4E+06

DETERMINE D_f

Vary trial D_f until computed ESALs equal future design ESALs.

K_{eff} (psi/in)	INPUT J	INPUT S_c (psi)	INPUT P_1	INPUT P_2	E_c (psi)	INPUT S_o	INPUT LOS	INPUT C_d
151	2.2	636	4.5	2.5	3.4E+06	0.30	0.00	1.00

TRIAL D_f (in)	R	Z	COMPUTED ESALs (millions)
9.70	50	0	80,032,494
10.63	80	0.84	79,968,401
11.15	90	1.282	80,157,839
11.59	95	1.645	80,149,803
12.45	99	2.327	79,857,251

DETERMINE D_{eff}

INPUT F_{jc} = 0.95
INPUT F_{fat} = 0.95
INPUT F_{dur} = 1.00

D_{eff} (in) = $F_{jc} * F_{dur} * F_{fat} * D_{exist}$ = 7.22

DETERMINE OVERLAY THICKNESS

RELIABILITY LEVEL	PCC BOL THICK	PCC to AC FACTOR	AC OL THICK
50	2.48	1.90	4.72
80	3.41	1.82	6.19
90	3.93	1.77	6.97
95	4.37	1.74	7.61
99	5.23	1.69	8.85

REVISED CHAPTER 5 AASHTO DESIGN GUIDE OVERLAY DESIGN

NW-8 AC AND BONDED PCC OVERLAY OF EXISTING JRCP

EXISTING PAVEMENT DESIGN AND FUTURE TRAFFIC

Slab thickness   8.00 (in)

Future design lane ESALs = 20,000,000 (TO PSI = 3.5)

BACKCALCULATION OF K_{eff} AND E_c

INPUT LOAD (lbs)	INPUT D_0 (mils)	INPUT D_{12} (mils)	INPUT D_{24} (mils)	INPUT D_{36} (mils)	AREA (in)	RADIUS RELSTIFF (in)	K_{dyn} (pci)	SLAB E_c (psi)
9,000	5.20	4.50	3.50	2.90	27.81	26.44	302	3.4E+06

DETERMINE D_f

Vary trial D_f until computed ESALs equal future design ESALs.

K_{eff} (psi/in)	INPUT J	S_c (psi)	INPUT P_1	INPUT P_2	E_c (psi)	INPUT S_o	INPUT LOS	INPUT C_d
151	2.2	636	4.5	3.5	3.4E+06	0.30	0.00	1.00

TRIAL D_f (in)	R	Z	COMPUTED ESALs (millions)
8.79	50	0	20,070,512
9.68	80	0.84	19,978,527
10.17	90	1.282	20,004,306
10.58	95	1.645	19,965,651
11.40	99	2.327	20,094,732

DETERMINE D_{eff}

INPUT F_{jc} = 1.00
INPUT F_{fat} = 0.95
INPUT F_{dur} = 1.00

D_{eff} (in) = $F_{jc} * F_{dur} * F_{fat} * D_{exist}$ = 7.60

DETERMINE OVERLAY THICKNESS

RELIABILITY LEVEL	PCC BOL THICK	PCC to AC FACTOR	AC OL THICK
50	1.19	2.05	2.45
80	2.08	1.95	4.05
90	2.57	1.89	4.87
95	2.98	1.85	5.53
99	3.80	1.78	6.78

REVISED CHAPTER 5 AASHTO DESIGN GUIDE OVERLAY DESIGN

NE-6 AC AND BONDED PCC OL OF EXISTING JRCP (ROUTE 9, 60-135)

EXISTING PAVEMENT DESIGN AND FUTURE TRAFFIC

Slab thickness   9.00 (in)

Future design lane ESALs = 12,255,000 (20 YEARS, RIGID ESALS)

BACKCALCULATION OF K_{eff} AND E_c

INPUT LOAD (lbs)	INPUT D_0 (mils)	INPUT D_{12} (mils)	INPUT D_{24} (mils)	INPUT D_{36} (mils)	AREA (in)	RADIUS RELSTIFF (in)	K_{dyn} (pci)	SLAB E_c (psi)
0	0	0	0	0	ERR	ERR	ERR	ERR
							ERR	ERR

DETERMINE D_f

Vary trial D_f until computed ESALs equal future design ESALs.

K_{eff} (psi/in)	INPUT J	S_c (psi)	INPUT P_1	INPUT P_2	E_c (psi)	INPUT S_o	INPUT LOS	INPUT C_d
129	3.5	749	4.5	2.5	6.0E+06	0.35	0.00	1.00

TRIAL D_f (in)	R	Z	COMPUTED ESALs (millions)	
8.80	50	0	12,461,306	SUBGRADE: TILL AND/OR
9.80	80	0.84	12,611,798	ARTIFICIAL FILL,
10.30	90	1.282	12,228,971	GOOD MATERIAL
10.80	95	1.645	12,466,916	K-VALUE, M_R, E ASSUMED
11.70	99	2.327	12,241,576	SAME AS NE-7 MEASURED

DETERMINE D_{eff}

INPUT F_{jc} = 1.00 (REPAIR DETERIORATED JTS/CRACKS)
INPUT F_{fat} = 0.97
INPUT F_{dur} = 1.00

D_{eff} (in) = $F_{jc} * F_{dur} * F_{fat} * D_{exist}$ = 8.73

DETERMINE OVERLAY THICKNESS

RELIABILITY LEVEL	PCC BOL THICK	PCC to AC FACTOR	AC OL THICK
50	0.07	2.21	0.15
80	1.07	2.07	2.22
90	1.57	2.01	3.15
95	2.07	1.95	4.03
99	2.97	1.86	5.51

REVISED CHAPTER 5 AASHTO DESIGN GUIDE OVERLAY DESIGN

NE-7 AC AND BONDED PCC OL OF EXISTING JRCP (ROUTE 2, SHRP 094020)

EXISTING PAVEMENT DESIGN AND FUTURE TRAFFIC

Slab thickness   9.00 (in)

Future design lane ESALs = 16,000,000 (20 YEARS, RIGID ESALS)

BACKCALCULATION OF K_{eff} AND E_c

INPUT LOAD (lbs)	INPUT D_0 (mils)	INPUT D_{12} (mils)	INPUT D_{24} (mils)	INPUT D_{36} (mils)	AREA (in)	RADIUS RELSTIFF (in)	K_{dyn} (pci)	SLAB E_c (psi)
9,504	3.74	3.39	2.84	2.35	29.76	32.80	290	5.4E+06
9,400	3.79	3.47	2.93	2.40	30.06	34.09	263	5.7E+06
9,648	3.99	3.72	3.19	2.65	30.77	37.51	212	6.8E+06
12,848	5.02	4.59	3.89	3.25	30.16	34.50	265	6.0E+06
							258	6.0E+06

DETERMINE D_f

Vary trial D_f until computed ESALs equal future design ESALs.

K_{eff} (psi/in)	INPUT J	S_c (psi)	INPUT P_1	INPUT P_2	E_c (psi)	INPUT S_o	INPUT LOS	INPUT C_d
129	3.5	749	4.5	2.5	6.0E+06	0.35	0.00	1.00

TRIAL D_f (in)	R	Z	COMPUTED ESALs (millions)
8.70	50	0	11,580,501
9.70	80	0.84	11,784,838
10.30	90	1.282	12,212,641
10.80	95	1.645	12,450,060
11.60	99	2.327	11,544,556

DETERMINE D_{eff}

INPUT F_{jc} = 1.00 (REPAIR DETERIORATED JTS/CKS)
INPUT F_{fat} = 0.99
INPUT F_{dur} = 1.00

D_{eff} (in) = $F_{jc} * F_{dur} * F_{fat} * D_{exist}$ = 8.91

DETERMINE OVERLAY THICKNESS

RELIABILITY LEVEL	PCC BOL THICK	PCC to AC FACTOR	AC OL THICK
50	0.00	2.26	0.00
80	0.79	2.11	1.67
90	1.39	2.03	2.82
95	1.89	1.97	3.72
99	2.69	1.88	5.06

REVISED CHAPTER 5 AASHTO DESIGN GUIDE OVERLAY DESIGN

NE-8 AC AND BONDED PCC OL OF EXISTING JRCP (SR119-408, INDIANA COUNTY)

EXISTING PAVEMENT DESIGN AND FUTURE TRAFFIC

Slab thickness   9.00 (in)

Future design lane ESALs = 4,650,000 (RIGID ESALs, 10 YEARS)

BACKCALCULATION OF K_{eff} AND E_c

INPUT LOAD (lbs)	INPUT D_0 (mils)	INPUT D_{12} (mils)	INPUT D_{24} (mils)	INPUT D_{36} (mils)	AREA (in)	RADIUS RELSTIFF (in)	K_{dyn} (pci)	SLAB E_c (psi)
0	0	0	0	0	ERR	ERR	ERR	ERR
							ERR	ERR

DETERMINE D_f

Vary trial D_f until computed ESALs equal future design ESALs.

K_{eff} (psi/in)	INPUT J	S_c (psi)	INPUT P_1	INPUT P_2	E_c (psi)	INPUT S_o	INPUT LOS	INPUT C_d
150	3.2	650	4.2	3.0	4.0E+06	0.35	0.00	1.00

TRIAL D_f (in)	R	Z	COMPUTED ESALs (millions)				
8.10	50	0	4,735,838				
9.10	80	0.84	4,735,538				
9.60	90	1.282	4,604,577				
10.10	95	1.645	4,720,641				
11.00	99	2.327	4,698,339				

CBR = 5 (SUBGRADE)
K-VALUE EST. = 150 PSI/IN

DETERMINE D_{eff}

INPUT F_{jc} = 0.95
INPUT F_{fat} = 0.98
INPUT F_{dur} = 0.98

D_{eff} (in) = $F_{jc} * F_{dur} * F_{fat} * D_{exist}$ = 8.21

DETERMINE OVERLAY THICKNESS

RELIABILITY LEVEL	PCC BOL THICK	PCC to AC FACTOR	AC OL THICK
50	0.00	2.24	0.00
80	0.89	2.09	1.86
90	1.39	2.03	2.82
95	1.89	1.97	3.72
99	2.79	1.87	5.22

REVISED CHAPTER 5 AASHTO DESIGN GUIDE OVERLAY DESIGN

NE-9 AC AND BONDED PCC OL OF EXISTING JRCP (I-80, COLUMBIA COUNTY)

EXISTING PAVEMENT DESIGN AND FUTURE TRAFFIC

Slab thickness   10.00 (in)

Future design lane ESALs = 10,050,000 (RIGID ESALs, 10 YEARS)

BACKCALCULATION OF K_{eff} AND E_c

INPUT LOAD (lbs)	INPUT D_0 (mils)	INPUT D_{12} (mils)	INPUT D_{24} (mils)	INPUT D_{36} (mils)	AREA (in)	RADIUS RELSTIFF (in)	K_{dyn} (pci)	SLAB E_c (psi)
0	0	0	0	0	ERR	ERR	ERR	ERR
							ERR	ERR

DETERMINE D_f

Vary trial D_f until computed ESALs equal future design ESALs.

K_{eff} (psi/in)	INPUT J	S_c (psi)	INPUT P_1	INPUT P_2	E_c	INPUT S_o	INPUT LOS	INPUT C_d
150	3.2	650	4.2	3.0	4.0E+06	0.35	0.00	1.00

TRIAL D_f (in)	R	Z	COMPUTED ESALs (millions)
9.30	50	0	10,643,283
10.30	80	0.84	10,215,852
10.90	90	1.282	10,283,303
11.40	95	1.645	10,258,834
12.40	99	2.327	10,267,810

CBR = 5 (SUBGRADE)
K-VALUE EST. = 150 PSI/IN

DETERMINE D_{eff}

INPUT F_{jc} = 0.95
INPUT F_{fat} = 0.98
INPUT F_{dur} = 0.98

D_{eff} (in) = $F_{jc} * F_{dur} * F_{fat} * D_{exist}$ = 9.12

DETERMINE OVERLAY THICKNESS

RELIABILITY LEVEL	PCC BOL THICK	PCC to AC FACTOR	AC OL THICK
50	0.18	2.20	0.39
80	1.18	2.06	2.42
90	1.78	1.98	3.52
95	2.28	1.93	4.38
99	3.28	1.83	5.99

N6.0 AC OVERLAY AND BONDED PCC OVERLAY OF CRCP

Region-Project	Existing Pavement	Design ESALs	Design Reliability	AC Overlay Thickness (in)	Bonded PCC Overlay Thickness (in)
MW-8	CRCP	10,000,000 (10 years)	50	0.8	0.4
			80	2.4	1.2
			90	3.4	1.7
			95	4.1	2.1
			99*	5.2	2.8

State design procedure indicates 6.2-inch AC overlay is needed. State policy design is 3.25-inch AC overlay.

Region-Project	Existing Pavement	Design ESALs	Design Reliability	AC Overlay Thickness (in)	Bonded PCC Overlay Thickness (in)
MW-9	CRCP	18,000,000 (10 years)	50	2.1	1.0
			80	3.9	2.0
			90	4.8	2.5
			95	5.4	2.9
			99*	6.6	3.7

Pavement has medium- to high-severity "D" cracking. State design procedure indicates 6.3-inch AC overlay is needed. State policy design is 3.25-inch AC overlay.

Region-Project	Existing Pavement	Design ESALs	Design Reliability	AC Overlay Thickness (in)	Bonded PCC Overlay Thickness (in)
MW-10	CRCP	20,000,000 (10 years)	50	0.9	0.4
			80	2.7	1.3
			90	3.6	1.8
			95	4.3	2.2
			99*	5.7	3.1

State design procedure indicates 6.5-inch AC overlay is needed. State policy design is 3.25-inch AC overlay.

Region-Project	Existing Pavement	Design ESALs	Design Reliability	AC Overlay Thickness (in)	Bonded PCC Overlay Thickness (in)
MW-11	CRCP	11,000,000 (10 years)	50	0.6	0.3
			80	2.4	1.2
			90	3.3	1.7
			95	4.0	2.1
			99*	5.2	2.8

State design procedure indicates 6.25-inch AC overlay is needed. State policy design is 3.25-inch AC overlay.

Region-Project	Existing Pavement	Design ESALs	Design Reliability	AC Overlay Thickness (in)	Bonded PCC Overlay Thickness (in)
MW-12	CRCP	21,966,725 (20 years)	50	2.4	1.2
			80	4.3	2.2
			90	5.3	2.9
			95	6.1	3.4
			99*	7.6	4.4

No deflection data are available. State design procedure indicates a 6-inch bonded PCC overlay is needed.

MW-13	CRCP	23,305,980 (20 years)	50	2.8	1.4
			80	4.7	2.5
			90	5.7	3.1
			95	6.5	3.6
			99*	7.9	4.6

No deflection data are available. State design procedure indicates a 5-inch bonded PCC overlay is needed.

MW-14	CRCP	14,066,735 (20 years)	50	1.9	0.9
			80	3.8	1.9
			90	4.8	2.5
			95	5.6	3.0
			99*	6.9	3.9

No deflection data are available. State design procedure indicates a 5-inch bonded PCC overlay is needed.

SE-2	CRCP	57,000,000 (20 years)	50*	3.6	1.8
			80*	5.3	2.8
			90	6.2	3.4
			95	7.0	3.9
			99	8.4	4.9

Extremely high traffic, very soft subgrade (k = 66 psi/inch). State constructed 4.5-inch AC overlay.

SW-14	CRCP	15,405,600 (15 years)	50	0	0
			80	0.9	0.4
			90	2.1	1.0
			95	3.0	1.5
			99	4.6	2.4
			99.9	6.5	3.6

No design recommendation available for 15-year design period.

Region-Project	Existing Pavement	Design ESALs	Design Reliability	AC Overlay Thickness (in)	Bonded PCC Overlay Thickness (in)
SW-15	CRCP	21,726,600 (20 years)	50	0	0
			80	1.9	0.9
			90	3.0	1.5
			95	3.9	2.0
			99	5.6	3.0
			99.9*	7.5	4.3

State constructed 4-inch bonded PCC overlay for a 20-year design period.

Region-Project	Existing Pavement	Design ESALs	Design Reliability	AC Overlay Thickness (in)	Bonded PCC Overlay Thickness (in)
SW-16	CRCP	35,585,400 (30 years)	50	1.1	0.5
			80	3.2	1.6
			90	4.3	2.3
			95	5.2	2.8
			99	6.9	3.9
			99.9	8.8	5.2

No design recommendation available for 30-year design period.

Summary of Results for AC and Bonded PCC Overlay of CRCP

1. Overall, it appears that the revised AASHTO overlay design procedures produce reasonable AC overlay and bonded PCC overlay thicknesses for CRCP consistent with State recommendations, provided different reliability levels are used. For AC overlays, a reliability level of 95 percent produces agency recommendations. For bonded PCC overlays, a reliability of 99 or greater produces agency recommendations. Figure N9 shows the comparison between design overlay thickness and agency recommendations for these levels of reliability.
2. A J factor in the range of 2.2 to 2.6 is needed to produce a reasonable overlay thickness for CRCP. Each agency must determine an appropriate value for J; it appears to vary from State to State.
3. The examples illustrate the importance of condition data and deflection data for overlay design. The condition factor F_{jc}, which indicates the amount of pavement deterioration left unrepaired prior to overlay, has a significant effect on the overlay thickness requirement. Agencies will find that much greater overlay thicknesses are required to meet desired performance lives if overlays are placed without adequate preoverlay repair. Most agencies specified thorough repair for CRCP.
4. The design reliability level is very significant. Most of the projects were Interstate-type highways. A design reliability level of 95 percent appears to be reasonable for AC overlays. Bonded PCC overlays appear to be designed at a 99-percent reliability level.

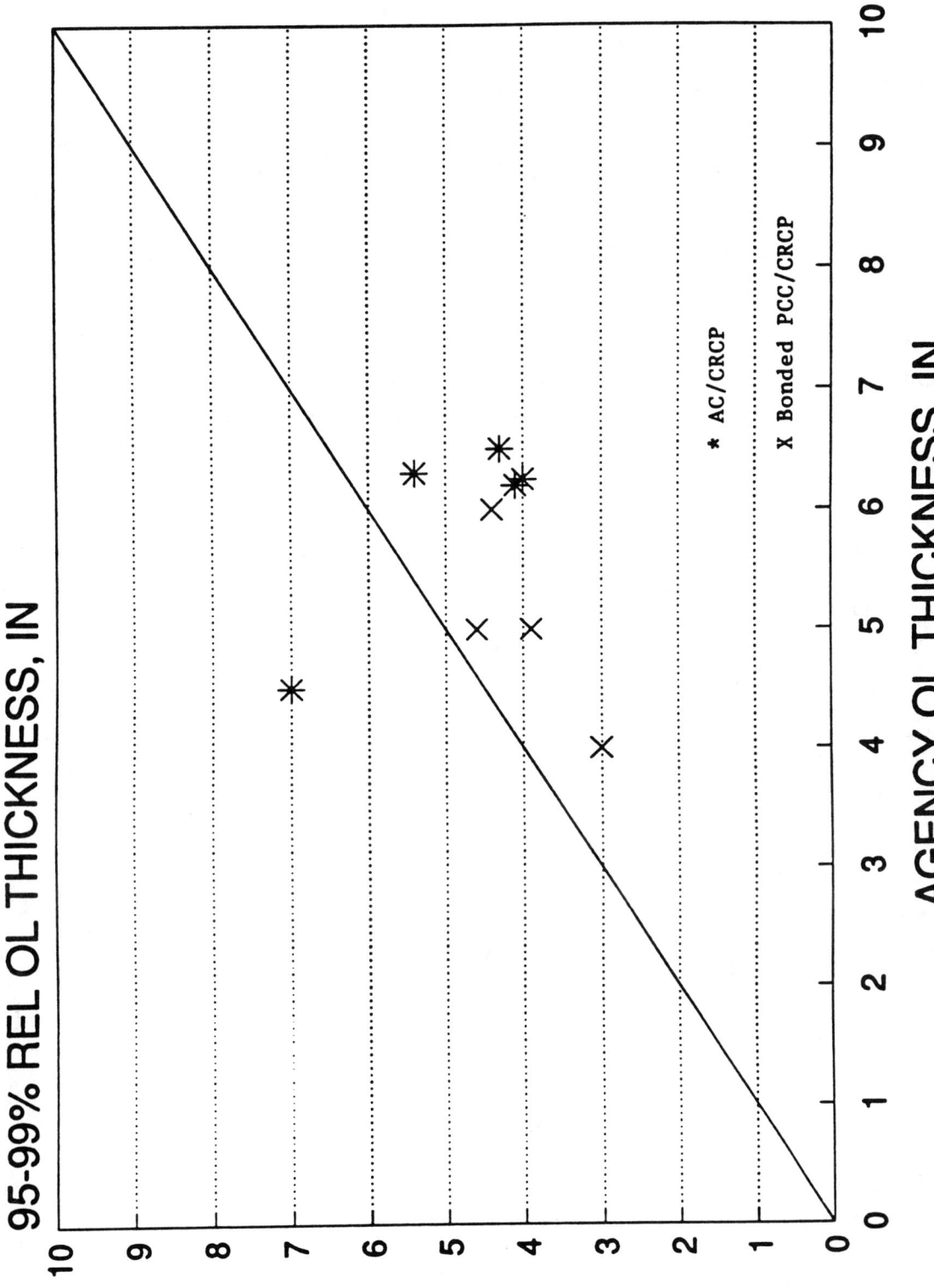

Figure N9. Comparison of AC Overlay Thickness (at 95-percent reliability) and PCC Bonded Overlay Thickness (at 99-percent reliability) vs that Recommended by Agencies

REVISED CHAPTER 5 AASHTO DESIGN GUIDE OVERLAY DESIGN

MW-8 AC AND BONDED PCC OVERLAY OF EXISTING CRCP (I-57)

EXISTING PAVEMENT DESIGN AND FUTURE TRAFFIC
Slab thickness   7.00 (in)

Future design lane ESALs = 10,000,000 (5% ESAL GROWTH RATE, 10 YEARS)

BACKCALCULATION OF K_{eff} AND E_c

INPUT LOAD (lbs)	INPUT D_0 (mils)	INPUT D_{12} (mils)	INPUT D_{24} (mils)	INPUT D_{36} (mils)	AREA (in)	RADIUS RELSTIFF (in)	K_{dyn} (pci)	SLAB E_c (psi)
9,000	6.79	5.96	4.98	4.00	28.87	29.55	186	4.9E+06
9,000	6.01	5.46	4.61	3.82	29.92	33.47	165	7.1E+06
9,000	6.27	5.44	4.42	3.43	28.15	27.38	234	4.5E+06
9,000	7.13	6.94	5.79	4.18	30.94	38.48	105	7.9E+06
							173	6.1E+06

DETERMINE D_f

Vary trial D_f until computed ESALs equal future design ESALs.

K_{eff} (psi/in)	INPUT J	S_c (psi)	INPUT P_1	INPUT P_2	E_c (psi)	INPUT S_o	INPUT LOS	INPUT C_d
86	2.2	753	4.5	2.8	6.1E+06	0.35	0.00	1.00

TRIAL D_f (in)	R	Z	COMPUTED ESALs (millions)
6.90	50	0	10,695,830
7.70	80	0.84	10,204,702
8.20	90	1.282	10,470,333
8.60	95	1.645	10,512,776
9.30	99	2.327	9,997,773

DETERMINE D_{eff} (CONVENTIONAL AC OVERLAY AND BONDED PCC OVERLAY)

INPUT F_{jc} = 0.97
INPUT F_{fat} = 0.96
INPUT F_{dur} = 1.00

D_{eff} (in) = $F_{jc} * F_{dur} * F_{fat} * D_{exist}$ = 6.52

DETERMINE OVERLAY THICKNESS

RELIABILITY LEVEL	PCC BOL THICK	PCC to AC FACTOR	AC OL THICK
50	0.38	2.17	0.83
80	1.18	2.06	2.43
90	1.68	1.99	3.35
95	2.08	1.95	4.05
99	2.78	1.87	5.21

Appendix N

REVISED CHAPTER 5 AASHTO DESIGN GUIDE OVERLAY DESIGN

MW-9 AC AND BONDED PCC OVERLAY OF EXISTING CRCP (I-80)

EXISTING PAVEMENT DESIGN AND FUTURE TRAFFIC

Slab thickness   8.00 (in)

Future design lane ESALs = 18,000,000 (5% ESAL GROWTH RATE, 10 YEARS)

BACKCALCULATION OF K_{eff} AND E_c

INPUT LOAD (lbs)	INPUT D_0 (mils)	INPUT D_{12} (mils)	INPUT D_{24} (mils)	INPUT D_{36} (mils)	AREA (in)	RADIUS RELSTIFF (in)	K_{dyn} (pci)	SLAB E_c (psi)
9,000	4.05	3.60	3.04	2.48	29.35	31.22	280	6.1E+06
9,000	4.16	3.49	2.7	1.91	26.61	23.63	471	3.4E+06
9,000	3.49	3.04	2.59	2.14	29.04	30.12	349	6.6E+06
9,000	5.29	4.84	4.16	3.38	30.25	34.93	172	5.9E+06
							318	5.5E+06

DETERMINE D_f

Vary trial D_f until computed ESALs equal future design ESALs.

K_{eff} (psi/in)	INPUT J	S_c (psi)	INPUT P_1	INPUT P_2	E_c (psi)	INPUT S_o	INPUT LOS	INPUT C_d
159	2.2	727	4.5	2.8	5.5E+06	0.35	0.00	1.00

TRIAL D_f (in)	R	Z	COMPUTED ESALs (millions)
7.40	50	0	17,743,150
8.40	80	0.84	18,815,526
8.90	90	1.282	18,776,535
9.30	95	1.645	18,436,227
10.10	99	2.327	17,988,918

DETERMINE D_{eff} (CONVENTIONAL AC OVERLAY)

INPUT F_{jc} = 0.96

INPUT F_{fat} = 0.98

INPUT F_{dur} = 0.85 ("D" CRACKING)

D_{eff} (in) = $F_{jc} * F_{dur} * F_{fat} * D_{exist}$ = 6.40

DETERMINE OVERLAY THICKNESS

RELIABILITY LEVEL	PCC BOL THICK	PCC to AC FACTOR	AC OL THICK
50	1.00	2.08	2.08
80	2.00	1.96	3.92
90	2.50	1.90	4.76
95	2.90	1.86	5.40
99	3.70	1.79	6.63

REVISED CHAPTER 5 AASHTO DESIGN GUIDE OVERLAY DESIGN

MW-10 AC AND BONDED PCC OVERLAY OF EXISTING CRCP (I-80)

EXISTING PAVEMENT DESIGN AND FUTURE TRAFFIC

Slab thickness   8.00 (in)

Future design lane ESALs = 20,000,000 (5% ESAL GROWTH RATE, 10 YEARS)

BACKCALCULATION OF K_{eff} AND E_c

INPUT LOAD (lbs)	INPUT D_0 (mils)	INPUT D_{12} (mils)	INPUT D_{24} (mils)	INPUT D_{36} (mils)	AREA (in)	RADIUS RELSTIFF (in)	K_{dyn} (pci)	SLAB E_c (psi)
9,000	4.39	3.94	3.26	2.70	29.37	31.30	257	5.7E+06
9,000	4.73	4.27	3.6	2.93	29.68	32.50	222	5.7E+06
9,000	4.95	4.39	3.71	3.04	29.32	31.12	231	5.0E+06
9,000	5.18	4.73	4.05	3.26	30.12	34.32	182	5.8E+06
							223	5.5E+06

DETERMINE D_f

Vary trial D_f until computed ESALs equal future design ESALs.

K_{eff} (psi/in)	INPUT J	S_c (psi)	INPUT P_1	INPUT P_2	E_c (psi)	INPUT S_o	INPUT LOS	INPUT C_d
129	2.2	728	4.5	2.8	5.5E+06	0.35	0.00	1.00

TRIAL D_f (in)	R	Z	COMPUTED ESALs (millions)
7.70	50	0	20,808,480
8.60	80	0.84	20,492,475
9.10	90	1.282	20,429,537
9.50	95	1.645	20,032,921
10.40	99	2.327	20,748,051

DETERMINE D_{eff} (CONVENTIONAL AC OVERLAY)

INPUT F_{jc} = 1.00
INPUT F_{fat} = 0.91
INPUT F_{dur} = 1.00

D_{eff} (in) = $F_{jc} * F_{dur} * F_{fat} * D_{exist}$ = 7.28

DETERMINE OVERLAY THICKNESS

RELIABILITY LEVEL	PCC BOL THICK	PCC to AC FACTOR	AC OL THICK
50	0.42	2.16	0.91
80	1.32	2.04	2.69
90	1.82	1.98	3.60
95	2.22	1.93	4.29
99	3.12	1.84	5.74

REVISED CHAPTER 5 AASHTO DESIGN GUIDE OVERLAY DESIGN

MW-11 AC AND BONDED PCC OVERLAY OF EXISTING CRCP (I-57)

EXISTING PAVEMENT DESIGN AND FUTURE TRAFFIC

Slab thickness   7.00 (in)

Future design lane ESALs = 11,000,000 (5% ESAL GROWTH RATE, 10 YEARS)

BACKCALCULATION OF K_{eff} AND E_c

INPUT LOAD (lbs)	INPUT D_0 (mils)	INPUT D_{12} (mils)	INPUT D_{24} (mils)	INPUT D_{36} (mils)	AREA (in)	RADIUS RELSTIFF (in)	K_{dyn} (pci)	SLAB E_c (psi)
9,000	5.15	4.60	3.89	3.09	29.38	31.34	219	7.2E+06
9,000	5.05	4.42	3.59	2.72	28.27	27.70	284	5.7E+06
9,000	4.92	4.48	3.88	3.08	30.15	34.46	190	9.2E+06
9,000	6.24	5.28	4.29	3.42	27.69	26.14	258	4.1E+06
9,000	4.63	4.12	3.44	2.73	29.13	30.44	257	7.6E+06
							242	6.8E+06

DETERMINE D_f

Vary trial D_f until computed ESALs equal future design ESALs.

K_{eff} (psi/in)	INPUT J	S_c (psi)	INPUT P_1	INPUT P_2	E_c (psi)	INPUT S_o	INPUT LOS	INPUT C_d
121	2.2	782	4.5	2.8	6.8E+06	0.35	0.00	1.00

TRIAL D_f (in)	R	Z	COMPUTED ESALs (millions)
6.70	50	0	11,129,126
7.60	80	0.84	11,449,870
8.10	90	1.282	11,725,647
8.50	95	1.645	11,762,312
9.20	99	2.327	11,178,438

DETERMINE D_{eff}

INPUT F_{jc} = 1.00 (REPAIR ALL FAILURES)
INPUT F_{fat} = 0.97
INPUT F_{dur} = 0.95 ("D" CRACKING)

D_{eff} (in) = $F_{jc} * F_{dur} * F_{fat} * D_{exist}$ = 6.45

DETERMINE OVERLAY THICKNESS

RELIABILITY LEVEL	PCC BOL THICK	PCC to AC FACTOR	AC OL THICK
50	0.25	2.19	0.55
80	1.15	2.06	2.37
90	1.65	2.00	3.29
95	2.05	1.95	4.00
99	2.75	1.88	5.16

REVISED CHAPTER 5 AASHTO DESIGN GUIDE OVERLAY DESIGN

MW-12 AC AND BONDED PCC OL OF EXISTING CRCP (I-80)

EXISTING PAVEMENT DESIGN AND FUTURE TRAFFIC
Slab thickness   8.00 (in)

Future design lane ESALs = 21,966,725

BACKCALCULATION OF K_{eff} AND E_c (ROAD RATER DEVICE)

INPUT LOAD (lbs)	INPUT D_0 (mils)	INPUT D_{12} (mils)	INPUT D_{24} (mils)	INPUT D_{36} (mils)	AREA (in)	RADIUS RELSTIFF (in)	K_{dyn} (pci)	SLAB E_c (psi)
2,000	1.50	1.30	1.30	1.10	31.20	40.01	103	6.0E+06
2,000	3.20	2.90	2.63	2.20	30.86	38.04	53	2.6E+06
2,000	1.40	1.30	1.10	1.00	30.86	38.01	122	5.8E+06
2,000	1.20	1.10	1.00	0.90	31.50	41.95	117	8.3E+06
							99	5.7E+06

DETERMINE D_f
Vary trial D_f until computed ESALs equal future design ESALs.

K_{eff} (psi/in)	INPUT J	S_c (psi)	INPUT P_1	INPUT P_2	E_c (psi)	INPUT S_o	INPUT LOS	INPUT C_d
49	2.5	736	4.2	2.5	5.7E+06	0.39	0.00	1.00

TRIAL D_f (in)	R	Z	COMPUTED ESALs (millions)
8.60	50	0	22,076,633
9.70	80	0.84	22,747,903
10.30	90	1.282	22,842,426
10.80	95	1.645	22,682,553
11.80	99	2.327	22,428,121

DETERMINE D_{eff} (CONVENTIONAL AC OVERLAY)
INPUT F_{jc} = 0.95
INPUT F_{fat} = 0.98
INPUT F_{dur} = 1.00

D_{eff} (in) = $F_{jc} * F_{dur} * F_{fat} * D_{exist}$ = 7.45

DETERMINE OVERLAY THICKNESS

RELIABILITY LEVEL	PCC BOL THICK	PCC to AC FACTOR	AC OL THICK
50	1.15	2.06	2.37
80	2.25	1.93	4.34
90	2.85	1.87	5.32
95	3.35	1.82	6.10
99	4.35	1.74	7.59

Appendix N N-73

REVISED CHAPTER 5 AASHTO DESIGN GUIDE OVERLAY DESIGN

MW-13 AC AND BONDED PCC OL OF EXISTING CRCP (I-280)

EXISTING PAVEMENT DESIGN AND FUTURE TRAFFIC

Slab thickness   8.00 (in)

Future design lane ESALs = 23,305,980

BACKCALCULATION OF K_{eff} AND E_c (ROAD RATER DEVICE)

INPUT LOAD (lbs)	INPUT D_0 (mils)	INPUT D_{12} (mils)	INPUT D_{24} (mils)	INPUT D_{36} (mils)	AREA (in)	RADIUS RELSTIFF (in)	K_{dyn} (pci)	SLAB E_c (psi)
2,000	1.60	1.50	1.40	1.20	32.25	47.84	68	8.1E+06
2,000	1.40	1.20	1.10	0.90	29.57	32.06	171	4.1E+06
2,000	1.20	1.10	1.00	0.90	31.50	41.95	117	8.3E+06
2,000	2.30	2.00	1.70	1.30	28.70	28.99	127	2.1E+06
							121	5.7E+06

DETERMINE D_f

Vary trial D_f until computed ESALs equal future design ESALs.

K_{eff} (psi/in)	INPUT J	S_c (psi)	INPUT P_1	INPUT P_2	E_c (psi)	INPUT S_o	INPUT LOS	INPUT C_d
60	2.5	735	4.2	2.5	5.7E+06	0.39	0.00	1.00

TRIAL D_f (in)	R	Z	COMPUTED ESALs (millions)
8.60	50	0	22,945,510
9.70	80	0.84	23,530,930
10.30	90	1.282	23,578,012
10.80	95	1.645	23,375,539
11.80	99	2.327	23,049,084

DETERMINE D_{eff} (CONVENTIONAL AC OVERLAY)

INPUT F_{jc} = 0.95
INPUT F_{fat} = 0.95
INPUT F_{dur} = 1.00

D_{eff} (in) = $F_{jc} * F_{dur} * F_{fat} * D_{exist}$ = 7.22

DETERMINE OVERLAY THICKNESS

RELIABILITY LEVEL	PCC BOL THICK	PCC to AC FACTOR	AC OL THICK
50	1.38	2.03	2.80
80	2.48	1.90	4.72
90	3.08	1.84	5.68
95	3.58	1.80	6.45
99	4.58	1.73	7.92

REVISED CHAPTER 5 AASHTO DESIGN GUIDE OVERLAY DESIGN

MW-14 AC AND BONDED PCC OL OF EXISTING CRCP (I-35)

EXISTING PAVEMENT DESIGN AND FUTURE TRAFFIC
Slab thickness   8.00 (in)

Future design lane ESALs = 14,056,735

BACKCALCULATION OF K_{eff} AND E_c (ROAD RATER DEVICE)

INPUT LOAD (lbs)	INPUT D_0 (mils)	INPUT D_{12} (mils)	INPUT D_{24} (mils)	INPUT D_{36} (mils)	AREA (in)	RADIUS RELSTIFF (in)	K_{dyn} (pci)	SLAB E_c (psi)
2,000	2.40	2.00	1.60	1.20	27.00	24.48	169	1.4E+06
2,000	1.70	1.50	1.30	1.00	29.29	31.02	150	3.2E+06
2,000	2.00	1.90	1.80	1.50	32.70	52.34	45	7.8E+06
2,000	1.60	1.50	1.40	1.10	31.88	44.69	78	7.1E+06
							111	4.9E+06

DETERMINE D_f
Vary trial D_f until computed ESALs equal future design ESALs.

K_{eff} (psi/in)	INPUT J	S_c (psi)	INPUT P_1	INPUT P_2	E_c (psi)	INPUT S_o	INPUT LOS	INPUT C_d
55	2.5	700	4.2	2.5	4.9E+06	0.39	0.00	1.00

TRIAL D_f (in)	R	Z	COMPUTED ESALs (millions)
8.20	50	0	14,644,036
9.20	80	0.84	14,308,405
9.80	90	1.282	14,559,817
10.30	95	1.645	14,614,977
11.20	99	2.327	13,887,627

DETERMINE D_{eff} (CONVENTIONAL AC OVERLAY)
INPUT F_{jc} = 0.95
INPUT F_{fat} = 0.96
INPUT F_{dur} = 1.00
D_{eff} (in) = $F_{jc} * F_{dur} * F_{fat} * D_{exist}$ = 7.30

DETERMINE OVERLAY THICKNESS

RELIABILITY LEVEL	PCC BOL THICK	PCC to AC FACTOR	AC OL THICK
50	0.90	2.09	1.89
80	1.90	1.97	3.75
90	2.50	1.90	4.76
95	3.00	1.85	5.56
99	3.90	1.78	6.93

REVISED CHAPTER 5 AASHTO DESIGN GUIDE OVERLAY DESIGN

SE-2 AC OVERLAY OF EXISTING CRCP (I-85)

EXISTING PAVEMENT DESIGN AND FUTURE TRAFFIC

Slab thickness   8.00 (in)

Future design lane ESALs = 57,000,000 (20 YEARS)

BACKCALCULATION OF K_{eff} AND E_c

INPUT LOAD (lbs)	INPUT D_0 (mils)	INPUT D_{12} (mils)	INPUT D_{24} (mils)	INPUT D_{36} (mils)	AREA (in)	RADIUS RELSTIFF (in)	K_{dyn} (pci)	SLAB E_c (psi)
9,502	8.08	7.19	5.94	4.67	28.97	29.88	162	3.0E+06
9,621	7.05	6.54	5.56	4.39	30.33	35.32	135	4.8E+06
9,463	5.56	5.17	4.5	3.64	30.80	37.68	148	6.8E+06
9,415	6.63	6.35	5.61	4.77	31.96	45.39	85	8.3E+06
							132	5.7E+06

DETERMINE D_f

Vary trial D_f until computed ESALs equal future design ESALs.

K_{eff} (psi/in)	INPUT J	S_c (psi)	INPUT P_1	INPUT P_2	E_c (psi)	INPUT S_o	INPUT LOS	INPUT C_d
66	2.2	737	4.5	3.0	5.7E+06	0.35	0.00	1.00

TRIAL D_f (in)	R	Z	COMPUTED ESALs (millions)
9.50	50	0	57,706,327
10.50	80	0.84	56,656,066
11.10	90	1.282	57,588,463
11.60	95	1.645	57,839,217
12.60	99	2.327	58,521,832

DETERMINE D_{eff} (CONVENTIONAL AC OVERLAY)

INPUT F_{jc} = 0.99
INPUT F_{fat} = 0.97
INPUT F_{dur} = 1.00

D_{eff} (in) = $F_{jc} * F_{dur} * F_{fat} * D_{exist}$ = 7.68

DETERMINE OVERLAY THICKNESS

RELIABILITY LEVEL	PCC BOL THICK	PCC to AC FACTOR	AC OL THICK
50	1.82	1.98	3.59
80	2.82	1.87	5.27
90	3.42	1.81	6.20
95	3.92	1.77	6.95
99	4.92	1.71	8.40

REVISED CHAPTER 5 AASHTO DESIGN GUIDE OVERLAY DESIGN

SW-14 PCC BONDED AND AC OVERLAY OF EXISTING CRCP (I-610)

EXISTING PAVEMENT DESIGN AND FUTURE TRAFFIC

Slab thickness   8.00 (in)

Future design lane ESALs = 15,405,600 (15 YEARS)

BACKCALCULATION OF K_{eff} AND E_c

INPUT LOAD (lbs)	INPUT D_0 (mils)	INPUT D_{12} (mils)	INPUT D_{24} (mils)	INPUT D_{36} (mils)	AREA (in)	RADIUS RELSTIFF (in)	K_{dyn} (pci)	SLAB E_c (psi)
0	0.00	0.00	0.00	0.00	ERR	ERR	ERR	ERR

DETERMINE D_f

Vary trial D_f until computed ESALs equal future design ESALs.

K_{eff} (psi/in)	INPUT J	S_c (psi)	INPUT P_1	INPUT P_2	E_c (psi)	INPUT S_o	INPUT LOS	INPUT C_d
371	2.4	700	4.5	2.5	4.9E+06	0.39	0.00	0.97
***		***			***			

TRIAL D_f (in)	R	Z	COMPUTED ESALs (millions)					
7.10	50	0	15,478,190		***Selected by Engr.			
8.20	80	0.84	15,781,542					
8.80	90	1.282	15,935,304					
9.30	95	1.645	15,948,597					
10.20	99	2.327	15,145,957					
11.40	99.9	3.09	15,279,993					

DETERMINE D_{eff} (CONVENTIONAL AC OVERLAY)

INPUT F_{jc} = 1.000

INPUT F_{fat} = 0.975

INPUT F_{dur} = 1.000

D_{eff} (in) = $F_{jc} * F_{dur} * F_{fat} * D_{exist}$ = 7.80

DETERMINE OVERLAY THICKNESS

RELIABILITY LEVEL	PCC BOL THICK	PCC to AC FACTOR	AC OL THICK
50	0.00	2.34	0.00
80	0.40	2.16	0.87
90	1.00	2.08	2.08
95	1.50	2.02	3.02
99	2.40	1.91	4.59
99.9	3.60	1.80	6.48

Appendix N N-77

REVISED CHAPTER 5 AASHTO DESIGN GUIDE OVERLAY DESIGN

SW-15 PCC BONDED AND AC OVERLAY OF EXISTING CRCP (I-610)

EXISTING PAVEMENT DESIGN AND FUTURE TRAFFIC

Slab thickness   8.00 (in)

Future design lane ESALs = 21,726,600 (20 YEARS)

BACKCALCULATION OF K_{eff} AND E_c

INPUT LOAD (lbs)	INPUT D_0 (mils)	INPUT D_{12} (mils)	INPUT D_{24} (mils)	INPUT D_{36} (mils)	AREA (in)	RADIUS RELSTIFF (in)	K_{dyn} (pci)	SLAB E_c (psi)
0	0.00	0.00	0.00	0.00	ERR	ERR	ERR	ERR

DETERMINE D_f

Vary trial D_f until computed ESALs equal future design ESALs.

K_{eff} (psi/in)	INPUT J	S_c (psi)	INPUT P_1	INPUT P_2	E_c (psi)	INPUT S_o	INPUT LOS	INPUT C_d
371	2.4	700	4.5	2.5	4.9E+06	0.39	0.00	0.97
***		***			***			

TRIAL D_f (in)	R	Z	COMPUTED ESALs (millions)
7.60	50	0	22,101,395
8.70	80	0.84	22,152,028
9.30	90	1.282	22,094,923
9.80	95	1.645	21,874,239
10.80	99	2.327	21,588,322
12.10	99.9	3.09	22,310,661

***Selected by Engr.

DETERMINE D_{eff} (CONVENTIONAL AC OVERLAY)

INPUT F_{jc} = 1.000
INPUT F_{fat} = 0.975
INPUT F_{dur} = 1.000

D_{eff} (in) = $F_{jc} * F_{dur} * F_{fat} * D_{exist}$ = 7.80

DETERMINE OVERLAY THICKNESS

RELIABILITY LEVEL	PCC BOL THICK	PCC to AC FACTOR	AC OL THICK
50	0.00	2.25	0.00
80	0.90	2.09	1.88
90	1.50	2.02	3.02
95	2.00	1.96	3.91
99	3.00	1.85	5.56
99.9	4.30	1.75	7.51

REVISED CHAPTER 5 AASHTO DESIGN GUIDE OVERLAY DESIGN

SW-16 PCC BONDED AND AC OVERLAY OF EXISTING CRCP (I-610)

EXISTING PAVEMENT DESIGN AND FUTURE TRAFFIC

Slab thickness   8.00 (in)

Future design lane ESALs = 35,585,400 (30 YEARS)

BACKCALCULATION OF K_{eff} AND E_c

INPUT LOAD (lbs)	INPUT D_0 (mils)	INPUT D_{12} (mils)	INPUT D_{24} (mils)	INPUT D_{36} (mils)	AREA (in)	RADIUS RELSTIFF (in)	K_{dyn} (pci)	SLAB E_c (psi)
0	0.00	0.00	0.00	0.00	ERR	ERR	ERR	ERR

DETERMINE D_f

Vary trial D_f until computed ESALs equal future design ESALs.

K_{eff} (psi/in)	INPUT J	S_c (psi)	INPUT P_1	INPUT P_2	E_c (psi)	INPUT S_o	INPUT LOS	INPUT C_d
371	2.4	700	4.5	2.5	4.9E+06	0.39	0.00	0.97
***		***			***			

TRIAL D_f (in)	R	Z	COMPUTED ESALs (millions)
8.30	50	0	35,968,459
9.40	80	0.84	35,003,036
10.05	90	1.282	35,344,543
10.60	95	1.645	35,450,661
11.70	99	2.327	35,737,516
13.00	99.9	3.09	35,373,827

DETERMINE D_{eff} (CONVENTIONAL AC OVERLAY)

INPUT F_{jc} = 1.000
INPUT F_{fat} = 0.97
INPUT F_{dur} = 1.000
D_{eff} (in) = $F_{jc} * F_{dur} * F_{fat} * D_{exist}$ = 7.80

DETERMINE OVERLAY THICKNESS

RELIABILITY LEVEL	PCC BOL THICK	PCC to AC FACTOR	AC OL THICK
50	0.50	2.15	1.07
80	1.60	2.00	3.21
90	2.25	1.93	4.34
95	2.80	1.87	5.24
99	3.90	1.78	6.93
99.9	5.20	1.69	8.81

N7.0 AC OVERLAY OF AC/PCC PAVEMENT

Region-Project	Existing Pavement	Design ESALs	Design Reliability	AC Overlay Thickness (in)
SE-3	AC/JPCP	3,000,000	50	0
			80	0.3
			90	1.3
			95*	2.2
			99	3.7

State feels second AC overlay of about 2.5 inches is reasonable.

SW-17	AC/JRCP	17,668,158 (20 years)	50	0
			80	0.2
			90	1.6
			95	2.6
			99*	4.6

Existing 3-inch AC overlay is poor quality. State plans to mill off completely and replace with 4-inch AC overlay for a 20-year design.

SW-18	AC/JRCP	12,801,929 (15 years)	50	0
			80	0
			90	0.5
			95	1.6
			99	3.6

Existing 3-inch AC overlay is poor quality. State plans to mill off completely and replace with 4-inch AC overlay for a 20-year design.

MW-15	AC/JRCP	10,000,000 (10 years)	50	0
			80	1.0
			90*	2.2
			95	3.3
			99	5.1

State plans to mill 0.5 inch AC from surface, AC patching, and place 2-inch AC overlay. This pavement has serious "D" cracking in the JRCP slab.

MW-16	AC/CRCP	10,000,000 (10 years)	50	4.1
			80*	5.7
			90	6.5
			95	7.2
			99	8.5

Extensive severe "D" cracking and poor quality AC, and more than 80 deteriorated areas per mile which will not be repaired with PCC. State plans to mill off 0.75 inches and place 5-inch second AC overlay.

Summary of Results for AC Overlay of AC/PCC Pavement

1. Overall, it appears that the revised AASHTO overlay design procedures produce reasonable second AC overlay thicknesses that are consistent with State recommendations. The reliability level required to match the State recommendations is variable, however. This is not too surprising since agencies have little performance experience with second overlays.
2. All of the condition factors have a significant effect on overlay thickness, indicating that the amount of pavement deterioration left unrepaired prior to overlay, has a significant effect on the overlay thickness requirement. Some existing AC/PCC pavements are very badly deteriorated due to PCC durability problems.
3. The design reliability level is very significant. A design reliability level of 90 to 95 percent appears to be reasonable for second AC overlays.

Appendix N

REVISED CHAPTER 5 AASHTO DESIGN GUIDE OVERLAY DESIGN

SE-3 AC OVERLAY OF EXISTING AC/JPCP PAVEMENT SR-25

EXISTING PAVEMENT DESIGN AND FUTURE TRAFFIC

AC layer thickness	2.50 (in)
Slab thickness	8.00 (in)
Future design lane ESALs	= 3,000,000

BACKCALCULATION OF K_{eff} AND E_c

AC temp = 78 (deg F)
AC modulus = 533,638 (psi)
AC/PCC = 1 (0 for bonded, 1 for unbonded)

INPUT LOAD (lbs)	INPUT D_0 (mils)	INPUT D_{12} (mils)	INPUT D_{24} (mils)	INPUT D_{36} (mils)	AC AREA (in)	PCC D_0 (mils)	PCC AREA (in)	RADIUS RELSTIFF (in)	K_{dyn} (pci)	SLAB E_c (psi)
9,096	5.54	4.48	3.77	2.91	27.02	5.21	28.36	27.98	273	3.8E+06
9,112	4.84	3.64	3.03	2.32	25.41	4.51	26.84	24.13	422	3.3E+06
9,056	6.59	4.69	3.86	2.95	24.25	6.26	25.22	21.04	395	1.8E+06
9,096	4.49	3.76	3.16	2.49	27.82	4.16	29.57	32.04	262	6.3E+06
									338	3.8E+06

DETERMINE D_f

Vary trial D_f until computed ESALs equal future design ESALs.

K_{eff} (psi/in)	INPUT J	S_c (psi)	INPUT P_1	INPUT P_2	E_c (psi)	INPUT S_o	INPUT LOS	INPUT C_d
169	3.2	654	4.5	2.8	3.8E+06	0.35	0.00	1.00

TRIAL D_f (in)	R	Z	COMPUTED ESALs (millions)
6.93	50	0	3,007,488
7.86	80	0.84	3,005,003
8.36	90	1.282	3,003,153
8.78	95	1.645	2,998,692
9.61	99	2.327	3,008,863

DETERMINE D_{eff}

INPUT F_{jc} = 0.85
INPUT F_{dur} = 1.00
INPUT Fac = 0.94
Thickness of AC to be milled = 0.50 (in)
D_{ac} = Original D_{ac} − milled Dac = 2.00 (in)
$D_{eff} = (F_{jc}*F_{dur}*D_{exist}) + (F_{ac}*D_{ac}/2.0)$ = 7.74 (in)

DETERMINE OVERLAY THICKNESS

RELIABILITY LEVEL	PCC BOL THICK	PCC to AC FACTOR	AC OL THICK
50	0.00	0.00	0.00
80	0.12	2.21	0.26
90	0.62	2.13	1.32
95	1.04	2.07	2.16
99	1.87	1.97	3.69

REVISED CHAPTER 5 AASHTO DESIGN GUIDE OVERLAY DESIGN

SW-17 AC OVERLAY OF EXISTING AC/JRCP (I-30)

EXISTING PAVEMENT DESIGN AND FUTURE TRAFFIC

AC layer thickness   3.00 (in)
Slab thickness   10.00 (in)

Future design lane ESALs = 17,668,158 (20 years)

BACKCALCULATION OF K_{eff} AND E_c

AC temp   =   59 (deg F)
AC modulus = 1,067,303 (psi)
AC/PCC   =   0 (0 for bonded, 1 for unbonded)

INPUT LOAD (lbs)	INPUT D_0 (mils)	INPUT D_{12} (mils)	INPUT D_{24} (mils)	INPUT D_{36} (mils)	AC AREA (in)	PCC D_0 (mils)	PCC AREA (in)	RADIUS RELSTIFF (in)	K_{dyn} (pci)	SLAB E_c (psi)
9,000	2.99	2.60	2.23	1.86	29.12	2.90	29.85	33.16	348	4.9E+06
									348	4.9E+06

DETERMINE D_f

Vary trial D_f until computed ESALs equal future design ESALs.

K_{eff} (psi/in)	INPUT J	S_c (psi)	INPUT P_1	INPUT P_2	E_c (psi)	INPUT S_o	INPUT LOS	INPUT C_d
174	3.2	703	4.5	2.5	4.9E+06	0.39	0.00	1.00

TRIAL D_f (in)	R	Z	COMPUTED ESALs (millions)
8.96	50	0	17,683,315
10.10	80	0.84	17,687,324
10.74	90	1.282	17,681,982
11.29	95	1.645	17,662,473
12.39	99	2.327	17,640,735

DETERMINE D_{eff}

INPUT F_{jc} = 1.00
INPUT F_{dur} = 1.00
INPUT F_{ac} = 0.80

Thickness of AC to be milled   = 3.00 (in)
D_{ac} = Original D_{ac} − milled D_{ac} = 0.00 (in)
$D_{eff} = (F_{jc} * F_{dur} * D_{exist}) + (F_{ac} * D_{ac}/2.0) = 10.00$ (in)

DETERMINE OVERLAY THICKNESS

RELIABILITY LEVEL	PCC BOL THICK	PCC to AC FACTOR	AC OL THICK
50	0.00	0.00	0.00
80	0.10	2.21	0.22
90	0.74	2.12	1.57
95	1.29	2.04	2.63
99	2.39	1.91	4.57

Appendix N N-83

REVISED CHAPTER 5 AASHTO DESIGN GUIDE OVERLAY DESIGN

SW-18 AC OVERLAY OF EXISTING AC/JRCP (I-30)

EXISTING PAVEMENT DESIGN AND FUTURE TRAFFIC

AC layer thickness   3.00 (in)
Slab thickness   10.00 (in)

Future design lane ESALs = 12,801,929 (20 years)

BACKCALCULATION OF K_{eff} AND E_c

AC temp = 59 (deg F)
AC modulus = 1,067,303 (psi)
AC/PCC = 0 (0 for bonded, 1 for unbonded)

INPUT LOAD (lbs)	INPUT D_0 (mils)	INPUT D_{12} (mils)	INPUT D_{24} (mils)	INPUT D_{36} (mils)	AC AREA (in)	PCC D_0 (mils)	PCC AREA (in)	RADIUS RELSTIFF (in)	K_{dyn} (pci)	SLAB E_c (psi)
9,000	2.99	2.60	2.23	1.86	29.12	2.90	29.85	33.16	348	4.9E+06
									348	4.9E+06

DETERMINE D_f

Vary trial D_f until computed ESALs equal future design ESALs.

K_{eff} (psi/in)	INPUT J	S_c (psi)	INPUT P_1	INPUT P_2	E_c (psi)	INPUT S_o	INPUT LOS	INPUT C_d
174	3.2	703	4.5	2.5	4.9E+06	0.39	0.00	1.00

TRIAL D_f (in)	R	Z	COMPUTED ESALs (millions)
8.50	50	0	12,792,393
9.60	80	0.84	12,803,172
10.22	90	1.282	12,839,528
10.75	95	1.645	12,840,425
11.80	99	2.327	12,787,792

DETERMINE D_{eff}

INPUT F_{jc} = 1.00
INPUT F_{dur} = 1.00
INPUT Fac = 0.80

Thickness of AC to be milled = 3.00 (in)
D_{ac} = Original D_{ac} − milled Dac = 0.00 (in)
D_{eff} = ($F_{jc}*F_{dur}*D_{exist}$) + ($F_{ac}*D_{ac}/2.0$) = 10.00 (in)

DETERMINE OVERLAY THICKNESS

RELIABILITY LEVEL	PCC BOL THICK	PCC to AC FACTOR	AC OL THICK
50	0.00	0.00	0.00
80	0.00	0.00	0.00
90	0.22	2.19	0.48
95	0.75	2.11	1.59
99	1.80	1.98	3.56

REVISED CHAPTER 5 AASHTO DESIGN GUIDE OVERLAY DESIGN

MW-15 AC OVERLAY OF EXISTING AC/JRCP (I-74)

EXISTING PAVEMENT DESIGN AND FUTURE TRAFFIC

AC layer thickness   3.00 (in)
Slab thickness   10.00 (in)

Future design lane ESALs = 10,000,000 (20 years)

BACKCALCULATION OF K_{eff} AND E_c

AC temp =   (deg F)
AC modulus = 1,626,000 (psi) from lab tests of cores
AC/PCC = 0 (0 for bonded, 1 for unbonded)

INPUT LOAD (lbs)	INPUT D_0 (mils)	INPUT D_{12} (mils)	INPUT D_{24} (mils)	INPUT D_{36} (mils)	AC AREA (in)	PCC D_0 (mils)	PCC AREA (in)	RADIUS RELSTIFF (in)	K_{dyn} (pci)	SLAB E_c (psi)
9,000	5.19	3.99	3.40	2.79	26.31	5.14	26.49	23.39	389	1.4E+06
9,000	3.82	3.20	2.85	2.38	28.74	3.77	29.02	30.06	324	3.1E+06
9,000	4.05	3.50	3.09	2.65	29.45	4.00	29.72	32.65	259	3.5E+06
9,000	3.84	3.19	2.80	2.41	28.48	3.79	28.76	29.19	341	2.9E+06
									328	2.7E+06

DETERMINE D_f

Vary trial D_f until computed ESALs equal future design ESALs.

K_{eff} (psi/in)	INPUT J	S_c (psi)	INPUT P_1	INPUT P_2	E_c (psi)	INPUT S_o	INPUT LOS	INPUT C_d
164	3.2	606	4.5	2.5	2.7E+06	0.39	0.00	1.00

TRIAL D_f (in)	R	Z	COMPUTED ESALs (millions)
8.59	50	0	10,066,278
9.73	80	0.84	10,036,274
10.37	90	1.282	10,036,705
10.92	95	1.645	10,034,620
12.02	99	2.327	10,048,532

DETERMINE D_{eff}

INPUT F_{jc} = 0.90 (50 unrepaired areas/mile)
INPUT F_{dur} = 0.90 (localized failures from "D" cracking)
INPUT Fac = 0.95 (fair AC mixture)

Thickness of AC to be milled   = 0.50 (in)
D_{ac} = Original D_{ac} − milled D_{ac} = 2.50 (in)
$D_{eff} = (F_{jc} * F_{dur} * D_{exist}) + (F_{ac} * D_{ac}/2.0) = 9.29$ (in)

DETERMINE OVERLAY THICKNESS

RELIABILITY LEVEL	PCC BOL THICK	PCC to AC FACTOR	AC OL THICK
50	0.00	0.00	0.00
80	0.44	2.16	0.95
90	1.08	2.07	2.24
95	1.63	2.00	3.26
99	2.73	1.88	5.13

REVISED CHAPTER 5 AASHTO DESIGN GUIDE OVERLAY DESIGN

MW-16 AC OVERLAY OF EXISTING AC/JRCP (I-74)

EXISTING PAVEMENT DESIGN AND FUTURE TRAFFIC

AC layer thickness   3.00 (in)
Slab thickness   7.00 (in)

Future design lane ESALs = 10,000,000 (20 years)

BACKCALCULATION OF K_{eff} AND E_c

AC temp =   (deg F)
AC modulus = 1,700,000 (psi) from lab tests of cores
AC/PCC = 1 (0 for bonded, 1 for unbonded)

INPUT LOAD (lbs)	INPUT D_0 (mils)	INPUT D_{12} (mils)	INPUT D_{24} (mils)	INPUT D_{36} (mils)	AC AREA (in)	PCC D_0 (mils)	PCC AREA (in)	RADIUS RELSTIFF (in)	K_{dyn} (pci)	SLAB E_c (psi)
9,000	6.67	4.87	4.04	3.26	24.96	6.55	25.31	21.18	370	2.5E+06
9,000	7.86	5.85	4.90	3.87	25.37	7.74	25.66	21.80	296	2.3E+06
9,000	8.42	5.64	4.54	3.60	23.07	8.30	23.32	18.26	390	1.5E+06
9,000	6.86	4.89	4.14	3.31	24.69	6.74	25.02	20.70	376	2.4E+06
									358	2.2E+06

DETERMINE D_f

Vary trial D_f until computed ESALs equal future design ESALs.

K_{eff} (psi/in)	INPUT J	S_c (psi)	INPUT P_1	INPUT P_2	E_c (psi)	INPUT S_o	INPUT LOS	INPUT C_d
179	2.6	583	4.5	2.5	2.2E+06	0.35	0.00	1.00

TRIAL D_f (in)	R	Z	COMPUTED ESALs (millions)
7.55	50	0	10,027,995
8.54	80	0.84	10,049,317
9.08	90	1.282	10,055,760
9.53	95	1.645	10,004,655
10.44	99	2.327	10,056,223

DETERMINE D_{eff}

INPUT F_{jc} = 0.80
INPUT F_{dur} = 0.80
INPUT Fac = 0.85

Thickness of AC to be milled = 0.75 (in)
D_{ac} = Original D_{ac} − milled Dac = 2.25 (in)
$D_{eff} = (F_{jc} * F_{dur} * D_{exist}) + (F_{ac} * D_{ac}/2.0) = 5.44$ (in)

DETERMINE OVERLAY THICKNESS

RELIABILITY LEVEL	PCC BOL THICK	PCC to AC FACTOR	AC OL THICK
50	2.11	1.94	4.11
80	3.10	1.84	5.72
90	3.64	1.80	6.54
95	4.09	1.76	7.21
99	5.00	1.70	8.53

N8.0 UNBONDED PCC OVERLAY OF JPCP, JRCP, AND CRCP

Region-Project	Overlay Type	Existing Pavement	Design ESALs	Design Reliability	Unbonded Overlay Thickness (in)
SW-19	JPCP	JPCP	11,000,000	50	5.4
				80	7.1
				90*	8.0
				95	8.7
				99	10.0

State design procedure indicates an 8-inch unbonded PCC overlay is needed.

SW-20	JPCP	JPCP	11,000,000	50	5.4
				80	7.0
				90*	7.9
				95	8.5
				99	10.0

State design procedure indicates an 8-inch unbonded PCC overlay is needed.

SW-21	JPCP	JPCP	11,000,000	50	5.7
				80	7.3
				90*	8.1
				95	8.8
				99	10.1

State design procedure indicates an 8-inch unbonded PCC overlay is needed.

MW-17	JPCP	JRCP	22,834,400	50	3.4
				80	6.3
				90	7.5
				95	8.4
				99	10.1

No recommendation available from agency.

MW-18	JRCP	CRCP	18,000,000	50	4.9
				80	6.6
				90	7.3
				95	8.0
				99	9.3

No recommendation available from agency.

Region-Project	Overlay Type	Existing Pavement	Design ESALs	Design Reliability	Unbonded Overlay Thickness (in)
SE-4	JPCP	CRCP	57,000,000	50	8.0
				80	9.5
				90	10.4
				95	11.1
				99	12.4

No recommendation available from agency. Extremely high traffic loadings.

SW-22	JPCP	JRCP	17,668,158	50	0
				80	4.5
				90	5.9
				95	6.8
				99	8.5

No recommendation available from agency.

SW-23	CRCP	JRCP	17,668,158	50	0
				80	3.6
				90	5.3
				95	6.3
				99	8.1

No recommendation available from agency.

MW-19	CRCP	CRCP	18,000,000	50	4.9
				80	6.6
				90	7.3
				95	8.0
				99	9.3*

Agency recommends a 9-inch CRCP unbonded overlay.

Summary of Results for Unbonded PCC Overlays

1. Overall, it appears that the revised AASHTO overlay design procedures produce reasonable unbonded PCC overlay thicknesses that are consistent with State recommendations at a reliability level of 95 percent. Figure N10 shows a plot of design thickness vs agency recommendations for the few points available.
2. The unbonded overlay thicknesses were obtained using the original Corps of Engineers equations developed for airfields. An improved design methodology can and should be developed in the future to replace this empirical equation.
3. The design reliability level is very significant. Most of the projects were Interstate-type highways. A design reliability level of 95 to 99 percent appears to be reasonable.

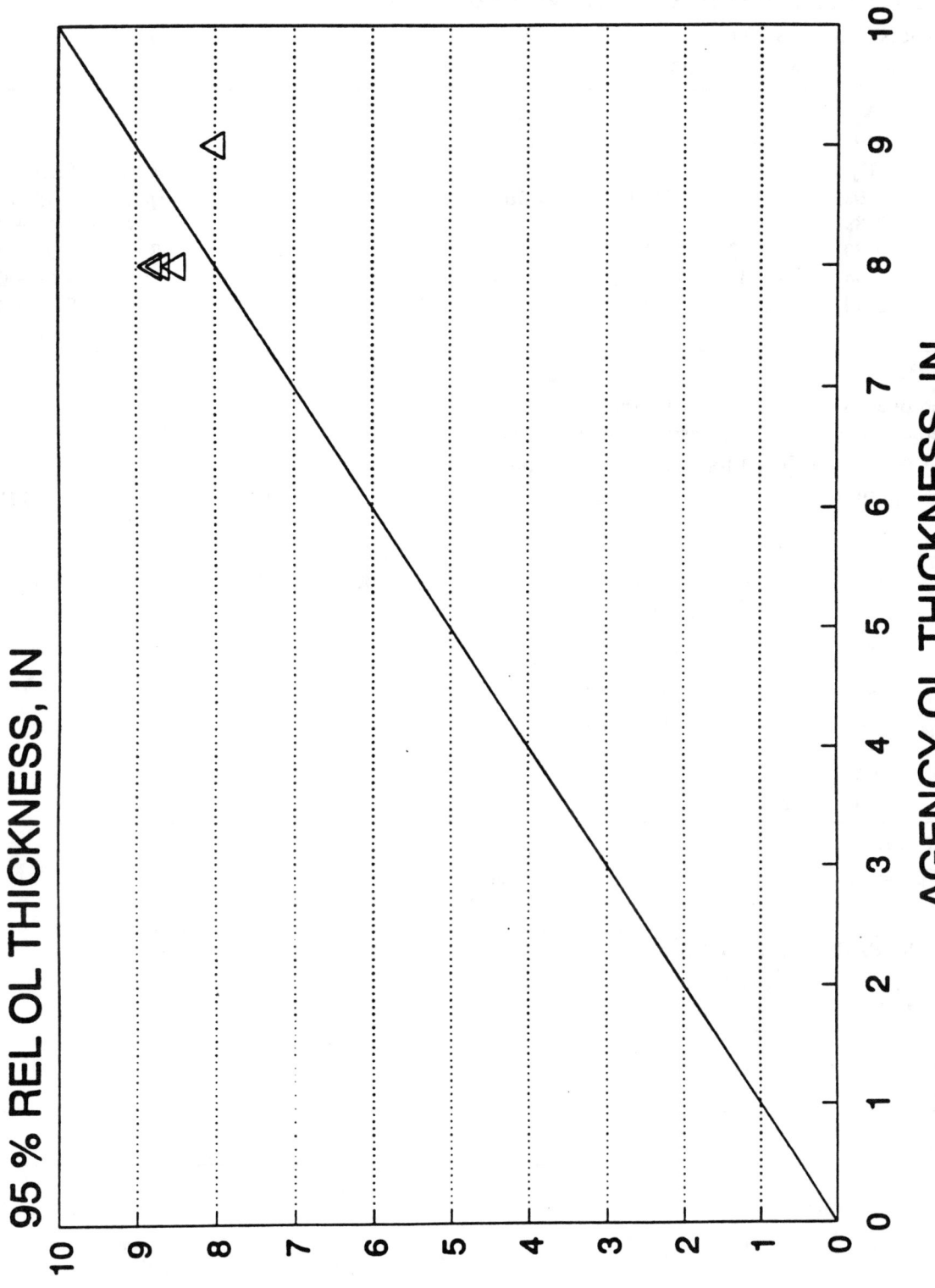

Figure N10. Comparison of Unbonded PCC Overlay Thickness (at 95-percent reliability) vs that Recommended by Agencies

REVISED CHAPTER 5 AASHTO DESIGN GUIDE OVERLAY DESIGN

SW-19   UNBONDED JPCP OVERLAY OF JPCP (Proj. 3005, Stn. 305)

EXISTING PAVEMENT DESIGN AND FUTURE TRAFFIC

Slab thickness 8.20 (in)

Future design lane ESALs = 11,000,000

BACKCALCULATION OF K_{eff}

INPUT LOAD (lbs)	INPUT D_0 (mils)	INPUT D_{12} (mils)	INPUT D_{24} (mils)	INPUT D_{36} (mils)	AREA (in)	RADIUS RELSTIFF (in)	K_{dyn} (pci)	SLAB E_c (psi)
9,144	3.89	3.37	2.85	2.40	28.89	29.62	329	5.4E+06
9,088	3.89	3.33	2.81	2.31	28.50	28.40	355	4.9E+06
9,104	3.94	3.33	2.81	2.36	28.29	27.78	366	4.6E+06
9,128	3.94	3.42	2.85	2.40	28.75	29.17	334	5.1E+06
							346	5.0E+06

DETERMINE D_f

Unbonded overlay modulus of rupture (psi)   =   700

Unbonded overlay modulus of elasticity (psi) = 4,900,000

Vary trial D_f until computed ESALs equal future design ESALs.

K_{eff} (psi/in)	INPUT J	S_c (psi)	INPUT P_1	INPUT P_2	E_c (psi)	INPUT S_o	INPUT LOS	INPUT C_d
173	4.0	700	4.5	2.5	4,900,000	0.35	0.00	1.00

TRIAL D_f (in)	R	Z	COMPUTED ESALs (millions)
9.40	50	0	10,972,879
10.50	80	0.84	11,282,235
11.10	90	1.282	11,337,203
11.60	95	1.645	11,285,326
12.60	99	2.327	11,235,624

DETERMINE D_{eff}

INPUT F_{jcu} = 0.94 (assume 100 deteriorated transverse cracks/mi)

D_{eff} (in) = $F_{jcu} * D_{exist}$ = 7.71

DETERMINE OVERLAY THICKNESS

RELIABILITY LEVEL	UBOL THICK
50	5.38
80	7.13
90	7.99
95	8.67
99	9.97

Appendix N

REVISED CHAPTER 5 AASHTO DESIGN GUIDE OVERLAY DESIGN

SW-20 UNBONDED JPCP OVERLAY OF JPCP (Proj. 7456)

EXISTING PAVEMENT DESIGN AND FUTURE TRAFFIC

Slab thickness   8.20 (in)

Future design lane ESALs = 11,000,000

BACKCALCULATION OF K_{eff}

INPUT LOAD (lbs)	INPUT D_0 (mils)	INPUT D_{12} (mils)	INPUT D_{24} (mils)	INPUT D_{36} (mils)	AREA (in)	RADIUS RELSTIFF (in)	K_{dyn} (pci)	SLAB E_c (psi)
8,496	3.68	3.08	2.64	2.23	28.29	27.76	367	4.6E+06
8,456	3.60	2.99	2.56	2.14	28.07	27.14	390	4.5E+06
8,520	3.64	3.12	2.64	2.23	28.66	28.90	343	5.1E+06
8,472	3.64	3.08	2.60	2.19	28.34	27.90	366	4.7E+06
							367	4.7E+06

DETERMINE D_f

Unbonded overlay modulus of rupture (psi) = 700
Unbonded overlay modulus of elasticity (psi) = 4,900,000

Vary trial D_f until computed ESALs equal future design ESALs.

K_{eff} (psi/in)	INPUT J	S_c (psi)	INPUT P_1	INPUT P_2	E_c (psi)	INPUT S_o	INPUT LOS	INPUT C_d
183	4.0	700	4.5	2.5	4,900,000	0.35	0.00	1.00

TRIAL D_f (in)	R	Z	COMPUTED ESALs (millions)
9.40	50	0	11,172,267
10.40	80	0.84	10,781,303
11.00	90	1.282	10,854,402
11.50	95	1.645	10,821,104
12.60	99	2.327	11,388,209

DETERMINE D_{eff}

INPUT F_{jcu} = 0.94 (assume 100 deteriorated cracks/mi)

D_{eff} (in) = $F_{jcu} * D_{exist}$ = 7.71

DETERMINE OVERLAY THICKNESS

RELIABILITY LEVEL	UBOL THICK
50	5.38
80	6.98
90	7.85
95	8.53
99	9.97

REVISED CHAPTER 5 AASHTO DESIGN GUIDE OVERLAY DESIGN

SW-21 UNBONDED JPCP OVERLAY OF JPCP (Proj. 3005, Stn. 353)

EXISTING PAVEMENT DESIGN AND FUTURE TRAFFIC

Slab thickness   8.20 (in)

Future design lane ESALs = 11,000,000

BACKCALCULATION OF K_{eff}

INPUT LOAD (lbs)	INPUT D_0 (mils)	INPUT D_{12} (mils)	INPUT D_{24} (mils)	INPUT D_{36} (mils)	AREA (in)	RADIUS RELSTIFF (in)	K_{dyn} (pci)	SLAB E_c (psi)
8,952	6.31	5.31	4.36	3.43	27.65	26.04	256	2.5E+06
8,904	6.35	5.31	4.4	3.47	27.63	25.98	254	2.5E+06
8,936	6.27	5.23	4.32	3.43	27.56	25.81	261	2.5E+06
8,984	6.35	5.31	4.36	3.47	27.55	25.79	260	2.4E+06
							257	2.5E+06

DETERMINE D_f

Unbonded overlay modulus of rupture (psi) = 700

Unbonded overlay modulus of elasticity (psi) = 4,900,000

Vary trial D_f until computed ESALs equal future design ESALs.

K_{eff} (psi/in)	INPUT J	S_c (psi)	INPUT P_1	INPUT P_2	E_c (psi)	INPUT S_o	INPUT LOS	INPUT C_d
129	4.0	700	4.5	2.5	4,900,000	0.35	0.00	1.00

TRIAL D_f (in)	R	Z	COMPUTED ESALs (millions)
9.60	50	0	11,515,723
10.60	80	0.84	11,109,475
11.20	90	1.282	11,177,562
11.70	95	1.645	11,135,832
12.70	99	2.327	11,101,416

DETERMINE D_{eff}

INPUT F_{jcu} = 0.94 (assume 100 deteriorated cracks/mi)

D_{eff} (in) = $F_{jcu} * D_{exist}$ = 7.71

DETERMINE OVERLAY THICKNESS

RELIABILITY LEVEL	UBOL THICK
50	5.72
80	7.28
90	8.13
95	8.80
99	10.09

REVISED CHAPTER 5 AASHTO DESIGN GUIDE OVERLAY DESIGN

MW-17 UNBONDED JPCP OVERLAY OF JRCP (I-80)

EXISTING PAVEMENT DESIGN AND FUTURE TRAFFIC

Slab thickness   10.00 (in)

Future design lane ESALs = 22,834,400

BACKCALCULATION OF K_{eff}

INPUT LOAD (lbs)	INPUT D_0 (mils)	INPUT D_{12} (mils)	INPUT D_{24} (mils)	INPUT D_{36} (mils)	AREA (in)	RADIUS RELSTIFF (in)	K_{dyn} (pci)	SLAB E_c (psi)
0	0	0	0	0	ERR	ERR	ERR	ERR
							ERR	ERR

DETERMINE D_f

Unbonded overlay modulus of rupture (psi)   =   640

Unbonded overlay modulus of elasticity (psi) = 4,200,000

Vary trial D_f until computed ESALs equal future design ESALs.

K_{eff} (psi/in)	J	INPUT S_c (psi)	INPUT P_1	INPUT P_2	E_c (psi)	INPUT S_o	INPUT LOS	INPUT C_d
155	3.2	640	4.2	2.5	4,200,000	0.39	0.00	1.00

TRIAL D_f (in)	R	Z	COMPUTED ESALs (millions)
10.00	50	0	22,173,569
11.30	80	0.84	22,868,477
12.00	90	1.282	22,800,733
12.60	95	1.645	22,700,972
13.80	99	2.327	22,528,954

DETERMINE D_{eff}

INPUT F_{jcu} = 0.94 (assuming 100 deteriorated cracks/mi)

D_{eff} (in) = $F_{jcu} * D_{exist}$ = 9.40

DETERMINE OVERLAY THICKNESS

RELIABILITY LEVEL	UBOL THICK
50	3.41
80	6.27
90	7.46
95	8.39
99	10.10

REVISED CHAPTER 5 AASHTO DESIGN GUIDE OVERLAY DESIGN

MW-18 UNBONDED JRCP OVERLAY OF CRCP (I-80)

EXISTING PAVEMENT DESIGN AND FUTURE TRAFFIC

Slab thickness   8.00 (in)

Future design lane ESALs = 18,000,000

BACKCALCULATION OF K_{eff}

INPUT LOAD (lbs)	INPUT D_0 (mils)	INPUT D_{12} (mils)	INPUT D_{24} (mils)	INPUT D_{36} (mils)	AREA (in)	RADIUS RELSTIFF (in)	K_{dyn} (pci)	SLAB E_c (psi)
9,000	4.05	3.60	3.04	2.48	29.35	31.22	280	6.1E+06
							280	6.1E+06

DETERMINE D_f

Unbonded overlay modulus of rupture (psi)   =   700
Unbonded overlay modulus of elasticity (psi) = 4,200,000

Vary trial D_f until computed ESALs equal future design ESALs.

K_{eff} (psi/in)	INPUT J	S_c (psi)	INPUT P_1	INPUT P_2	E_c (psi)	INPUT S_o	INPUT LOS	INPUT C_d
140	3.2	700	4.5	2.5	4,200,000	0.35	0.00	1.00

TRIAL D_f (in)	R	Z	COMPUTED ESALs (millions)
9.10	50	0	18,868,828
10.10	80	0.84	18,552,355
10.60	90	1.282	17,756,688
11.10	95	1.645	17,885,002
12.09	99	2.327	18,092,891

DETERMINE D_{eff}

INPUT F_{jcu} = 0.96 (assume 50 deteriorated areas/mi in existing CRCP)

D_{eff} (in) = $F_{jcu} * D_{exist}$ = 7.68

DETERMINE OVERLAY THICKNESS

RELIABILITY LEVEL	UBOL THICK
50	4.88
80	6.56
90	7.31
95	8.01
99	9.34

Appendix N

REVISED CHAPTER 5 AASHTO DESIGN GUIDE OVERLAY DESIGN

SE-4 UNBONDED JPCP OVERLAY OF CRCP (I-85)

EXISTING PAVEMENT DESIGN AND FUTURE TRAFFIC

Slab thickness 8.00 (in)

Future design lane ESALs = 57,000,000

BACKCALCULATION OF K_{eff}

INPUT LOAD (lbs)	INPUT D_0 (mils)	INPUT D_{12} (mils)	INPUT D_{24} (mils)	INPUT D_{36} (mils)	AREA (in)	RADIUS RELSTIFF (in)	K_{dyn} (pci)	SLAB E_c (psi)
9,502	8.08	7.19	5.94	4.67	28.97	29.88	162	3.0E+06
							162	3.0E+06

DETERMINE D_f

Unbonded overlay modulus of rupture (psi) = 700
Unbonded overlay modulus of elasticity (psi) = 4,200,000

Vary trial D_f until computed ESALs equal future design ESALs.

K_{eff} (psi/in)	INPUT J	S_c (psi)	INPUT P_1	INPUT P_2	E_c (psi)	INPUT S_o	INPUT LOS	INPUT C_d
81	3.2	700	4.5	2.5	4,200,000	0.35	0.00	1.00

TRIAL D_f (in)	R	Z	COMPUTED ESALs (millions)
11.10	50	0	59,227,683
12.20	80	0.84	56,580,476
12.90	90	1.282	57,770,394
13.50	95	1.645	58,671,505
14.60	99	2.327	57,732,613

DETERMINE D_{eff}

INPUT F_{jcu} = 0.96 (assume 50 unrepaired areas/mi in existing CRCP)

D_{eff} (in) = $F_{jcu} * D_{exist}$ = 7.68

DETERMINE OVERLAY THICKNESS

RELIABILITY LEVEL	UBOL THICK
50	8.01
80	9.48
90	10.36
95	11.10
99	12.42

REVISED CHAPTER 5 AASHTO DESIGN GUIDE OVERLAY DESIGN

SW-22 UNBONDED JPCP OVERLAY OF JRCP (I-30)

EXISTING PAVEMENT DESIGN AND FUTURE TRAFFIC

Slab thickness   10.00 (in)

Future design lane ESALs = 17,668,158

BACKCALCULATION OF K_{eff}

INPUT LOAD (lbs)	INPUT D_0 (mils)	INPUT D_{12} (mils)	INPUT D_{24} (mils)	INPUT D_{36} (mils)	AREA (in)	RADIUS RELSTIFF (in)	K_{dyn} (pci)	SLAB E_c (psi)
0	0	0	0	0	ERR	ERR	ERR	ERR
							ERR	ERR

DETERMINE D_f

Unbonded overlay modulus of rupture (psi) = 710
Unbonded overlay modulus of elasticity (psi) = 5,100,000

Vary trial D_f until computed ESALs equal future design ESALs.

K_{eff} (psi/in)	INPUT J	S_c (psi)	INPUT P_1	INPUT P_2	E_c (psi)	INPUT S_o	INPUT LOS	INPUT C_d
53	3.2	710	4.5	2.5	5,100,000	0.39	0.00	1.01

TRIAL D_f (in)	R	Z	COMPUTED ESALs (millions)
9.30	50	0	17,483,863
10.40	80	0.84	17,272,377
11.10	90	1.282	18,026,295
11.60	95	1.645	17,540,132
12.70	99	2.327	17,638,108

DETERMINE D_{eff}

INPUT F_{jcu} = 0.94 (assume 100 unrepaired deteriorated areas/mi)

D_{eff} (in) = $F_{jcu} * D_{exist}$ = 9.40

DETERMINE OVERLAY THICKNESS

RELIABILITY LEVEL	UBOL THICK
50	ERR
80	4.45
90	5.90
95	6.80
99	8.54

Appendix N

REVISED CHAPTER 5 AASHTO DESIGN GUIDE OVERLAY DESIGN

SW-23 UNBONDED CPCP OVERLAY OF JRCP (I-30)

EXISTING PAVEMENT DESIGN AND FUTURE TRAFFIC

Slab thickness   10.00 (in)

Future design lane ESALs = 17,668,158

BACKCALCULATION OF K_{eff}

INPUT LOAD (lbs)	INPUT D_0 (mils)	INPUT D_{12} (mils)	INPUT D_{24} (mils)	INPUT D_{36} (mils)	AREA (in)	RADIUS RELSTIFF (in)	K_{dyn} (pci)	SLAB E_c (psi)
0	0	0	0	0	ERR	ERR	ERR	ERR
							ERR	ERR

DETERMINE D_f

Unbonded overlay modulus of rupture (psi) = 710

Unbonded overlay modulus of elasticity (psi) = 5,100,000

Vary trial D_f until computed ESALs equal future design ESALs.

K_{eff} (psi/in)	INPUT J	S_c (psi)	INPUT P_1	INPUT P_2	E_c (psi)	INPUT S_o	INPUT LOS	INPUT C_d
53	3.2	710	4.5	2.5	5,100,000	0.39	0.00	1.01

TRIAL D_f (in)	R	Z	COMPUTED ESALs (millions)
9.30	50	0	17,483,863
10.40	80	0.84	17,272,377
11.10	90	1.282	18,026,295
11.60	95	1.645	17,540,132
12.70	99	2.327	17,638,108

DETERMINE D_{eff}

INPUT F_{jcu} = 0.98 (assume 25 unrepaired deteriorated areas/mi)

D_{eff} (in) = $F_{jcu} * D_{exist}$ = 9.75

DETERMINE OVERLAY THICKNESS

RELIABILITY LEVEL	UBOL THICK
50	ERR
80	3.62
90	5.31
95	6.28
99	8.14

REVISED CHAPTER 5 AASHTO DESIGN GUIDE OVERLAY DESIGN

MW-19 UNBONDED CRCP OVERLAY OF CRCP (I-80)

EXISTING PAVEMENT DESIGN AND FUTURE TRAFFIC

Slab thickness   8.00 (in)

Future design lane ESALs = 18,000,000

BACKCALCULATION OF K_{eff}

INPUT LOAD (lbs)	INPUT D_0 (mils)	INPUT D_{12} (mils)	INPUT D_{24} (mils)	INPUT D_{36} (mils)	AREA (in)	RADIUS RELSTIFF (in)	K_{dyn} (pci)	SLAB E_c (psi)
9,000	4.05	3.60	3.04	2.48	29.35	31.22	280	6.1E+06
							280	6.1E+06

DETERMINE D_f

Unbonded overlay modulus of rupture (psi) = 700

Unbonded overlay modulus of elasticity (psi) = 4,200,000

Vary trial D_f until computed ESALs equal future design ESALs

K_{eff} (psi/in)	INPUT J	S_c (psi)	INPUT P_1	INPUT P_2	E_c (psi)	INPUT S_o	INPUT LOS	INPUT C_d
140	3.2	700	4.5	2.5	4,200,000	0.35	0.00	1.00

TRIAL D_f (in)	R	Z	COMPUTED ESALs (millions)
9.10	50	0	18,868,828
10.10	80	0.84	18,552,355
10.60	90	1.282	17,756,688
11.10	95	1.645	17,885,002
12.09	99	2.327	18,092,891

DETERMINE D_{eff}

INPUT F_{jc} = 0.96 (assume 50 deteriorated areas/mi in existing CRCP)

D_{eff} (in) = $F_{jcu} * D_{exist}$ = 7.68

DETERMINE OVERLAY THICKNESS

RELIABILITY LEVEL	UBOL THICK
50	4.88
80	6.56
90	7.31
95	8.01
99	9.34

N9.0 JPCP, JRCP, AND CRCP OVERLAY OF AC PAVEMENT

Region-Project	Existing Pavement	Design ESALs	Design Reliability	PCC Overlay Thickness (in)
NW-9	AC	3,600,000	50	5.8
			80	6.8
			90	7.4
			95	7.8
			99	8.7

Effective k-value of 550 psi/inch used in design

NW-10	AC	2,300,000	50	5.5
			80	6.4
			90	6.9
			95	7.4
			99	8.2

Effective k-value of 425 psi/inch used in design

NW-11	AC	4,200,000	50	6.5
			80	7.4
			90	7.9
			95	8.4
			99	9.2

Effective k-value of 390 psi/inch used in design

MW-20	AC	150,000	50	—
			80	—
			90	4.1
			95	4.5
			99	5.3

Effective k-value of 460 psi/inch used in design

SE-5	AC	1,100,000	50	3.8
			80	5.3
			90	5.9
			95	6.4
			99	7.4

Effective k-value of 600 psi/inch used in design. State design method indicates 6.4-inch overlay is needed. Agency constructed experimental sections 6, 7, and 8 inches thick which provides for a range in design reliability from 90 to 99 percent.

Region-Project	Existing Pavement	Design ESALs	Design Reliability	PCC Overlay Thickness (in)
SW-24	AC	11,000,000	50	7.3
			80	8.4
			90	9.0
			95	9.5
			99	10.4

Effective k-value of 650 psi/inch used in design

Region-Project	Existing Pavement	Design ESALs	Design Reliability	PCC Overlay Thickness (in)
SW-25	AC	11,000,000	50	7.7
			80	8.7
			90	9.3
			95	9.8
			99	10.7

Effective k-value of 460 psi/inch used to obtain above thicknesses.

Appendix N

Summary of Results for JPCP, JRCP, and CRCP Overlay of AC Pavement

1. Overall, it appears that the revised AASHTO overlay design procedures provide reasonable JPCP and JRCP overlay thicknesses for AC pavements. One project for which the State designed and constructed three experimental thicknesses showed consistent results.
2. The effective k-value exhibited by an AC pavement as determined by Figure 3.3, Part II, appears to be quite high. No loss of support was applied to the k-value in these examples. However, the sensitivity of PCC overlay thickness to k-value is small. Additional work is greatly needed to investigate effective k-values for PCC overlays of AC pavements, including deflection testing after overlay construction to verify the effective k-value.
3. The design reliability level is very significant. Most of the projects were Interstate-type highways. A design reliability level ranging from 95 to 99 percent appears to be reasonable for most projects.

REVISED CHAPTER 5 AASHTO DESIGN GUIDE OVERLAY DESIGN

NW-9 JPCP OVERLAY OF AC PAVEMENT

EXISTING PAVEMENT DESIGN AND FUTURE TRAFFIC

EXISTING PAVEMENT DESIGN

AC SURFACE	4.25	SUBGRADE: SANDY SILT, SANDY GRAVEL
CR STONE BASE	8.00	
SUBBASE	0.00	
TOTAL THICKNESS	12.25	

Future design lane ESALs = 3,600,000

DETERMINE K_{eff}

Vary E_p/M_r until actual M_R*D_0/P matches computed M_R*D_0/p.

STATION	INPUT LOAD (lbs)	INPUT D_0, in (mils)	INPUT D_r, in (mils)	SUBGRADE M_R (psi)	ACTUAL M_R*D_0/P	TRIAL E_p/M_R	COMPUTED M_R*D_0/E_p	E_p
	9,000	12.80	3.55	16,901	24.04	2.52	24.04	42,592

r = 36

Check r > 0.7 ae = 12.38

Using Figure 3.3, Part II:
  K_{eff} (dynamic) = 1,100 psi/in   INPUT
  K_{eff} (static)   = 550 psi/in

DETERMINE D_f

INPUT

PCC overlay modulus of rupture (psi) =   690 (mean)
PCC overlay modulus of elasticity (psi) = 4,200,000 (mean)

Vary trial D_f until computed ESALs equal future design ESALs.

K_{eff} (pci)	INPUT J	S_c (psi)	INPUT P_1	INPUT P_2	E_c (psi)	INPUT S_o	INPUT LOS	INPUT C_d
550	3.2	690	4.5	2.5	4,200,000	0.35	0.00	1.00

TRIAL D_f (in)	R	Z	COMPUTED ESALs (millions)	Dol (in)
5.80	50	0	3,667,987	5.80
6.80	80	0.84	3,581,992	6.80
7.40	90	1.282	3,720,304	7.40
7.80	95	1.645	3,607,745	7.80
8.70	99	2.327	3,717,080	8.70

Appendix N N-103

REVISED CHAPTER 5 AASHTO DESIGN GUIDE OVERLAY DESIGN

NW-10 JPCP OVERLAY OF AC PAVEMENT (Whitney Hwy 071)

EXISTING PAVEMENT DESIGN AND FUTURE TRAFFIC

EXISTING PAVEMENT DESIGN

AC SURFACE	4.00	SUBGRADE: SANDY SILT, SANDY GRAVEL
CR STONE BASE	14.00	
SUBBASE	0.00	
TOTAL THICKNESS	18.00	

Future design lane ESALs = 2,300,000

DETERMINE K_{eff}

Vary E_p/M_r until actual M_R*D_0/P matches computed M_R*D_0/p.

STATION	LOAD (lbs)	D_0, in (mils)	D_r, in (mils)	SUBGRADE M_R (psi)	ACTUAL M_R*D_0/P	TRIAL E_p/M_R	COMPUTED M_R*D_0/E_p	E_p
AVG	9,000	23.57	4.22	14,218	37.24	1.52	37.22	21,611

r = 36

Check r > 0.7 ae = 15.07

Using Figure 3.3, Part II:
   K_{eff} (dynamic) = 850 psi/in   INPUT
   K_{eff} (static) = 425 psi/in

DETERMINE D_f

                               INPUT
PCC overlay modulus of rupture (psi) = 690 (mean)
PCC overlay modulus of elasticity (psi) = 4,200,000 (mean)

Vary trial D_f until computed ESALs equal future design ESALs.

K_{eff} (pci) INPUT	J	S_c (psi)	P_1 INPUT	P_2 INPUT	E_c (psi)	S_o INPUT	LOS INPUT	C_d INPUT
425	3.2	690	4.5	2.5	4,200,000	0.35	0.00	1.00

TRIAL D_f (in)	R	Z	COMPUTED ESALs (millions)	Dol (in)
5.50	50	0	2,342,259	5.50
6.40	80	0.84	2,260,563	6.40
6.90	90	1.282	2,240,718	6.90
7.40	95	1.645	2,358,520	7.40
8.20	99	2.327	2,339,577	8.20

REVISED CHAPTER 5 AASHTO DESIGN GUIDE OVERLAY DESIGN

NW-11 JPCP OVERLAY OF AC PAVEMENT (Warm Springs Hwy 053)

EXISTING PAVEMENT DESIGN AND FUTURE TRAFFIC

EXISTING PAVEMENT DESIGN

AC SURFACE	5.50	SUBGRADE: SANDY SILT, SANDY GRAVEL
CR STONE BASE	12.00	
SUBBASE	0.00	
TOTAL THICKNESS	17.50	

Future design lane ESALs = 4,200,000

DETERMINE K_{eff}

Vary E_p/M_r until actual M_R*D_0/P matches computed M_R*D_0/p.

STATION	LOAD (lbs)	D_0, in (mils)	D_r, in (mils)	SUBGRADE M_R (psi)	ACTUAL M_R*D_0/P	TRIAL E_p/M_R	COMPUTED M_R*D_0/E_p	E_p
AVG	9,000	24.10	4.45	13,483	36.10	1.71	36.13	23,056

r = 36
Check r > 0.7 ae = 15.22

Using Figure 3.3, Part II:
K_{eff} (dynamic) = 780 psi/in   INPUT
K_{eff} (static) = 390 psi/in

DETERMINE D_f

INPUT
PCC overlay modulus of rupture (psi)   =   690 (mean)
PCC overlay modulus of elasticity (psi) = 4,200,000 (mean)

Vary trial D_f until computed ESALs equal future design ESALs.

K_{eff} (pci)	INPUT J	S_c (psi)	INPUT P_1	INPUT P_2	E_c (psi)	INPUT S_o	INPUT LOS	INPUT C_d
390	3.2	690	4.5	2.5	4,200,000	0.35	0.00	1.00

TRIAL D_f (in)	R	Z	COMPUTED ESALs (millions)	Dol (in)
6.50	50	0	4,495,669	6.50
7.40	80	0.84	4,286,492	7.40
7.90	90	1.282	4,238,316	7.90
8.40	95	1.645	4,433,596	8.40
9.20	99	2.327	4,312,918	9.20

Appendix N

REVISED CHAPTER 5 AASHTO DESIGN GUIDE OVERLAY DESIGN

MW-20 JPCP OVERLAY OF AC PAVEMENT (Newmark Drive)

EXISTING PAVEMENT DESIGN AND FUTURE TRAFFIC

EXISTING PAVEMENT DESIGN

AC SURFACE	1.50	SUBGRADE: SANDY SILT, SANDY GRAVEL
ASPHALT BASE	6.00	
SUBBASE	0.00	
TOTAL THICKNESS	7.50	

Future design lane ESALs = 150,000

DETERMINE K_{eff}

Vary E_p/M_r until actual M_R*D_0/P matches computed M_R*D_0/p.

STATION	LOAD (lbs)	D_0, in (mils)	D_r, in (mils)	SUBGRADE M_R (psi)	ACTUAL M_R*D_0/P	TRIAL E_p/M_R	COMPUTED M_R*D_0/E_p	E_p
0+00	9,000	16.10	6.08	9,868	17.65	37.00	17.72	365,132
	r =	36						
Check r > 0.7 ae =	17.98							
2+00	9,000	18.10	4.37	13,730	27.61	4.45	27.43	61,098
	r =	36						
Check r > 0.7 ae =	9.57							
4+00	9,000	15.10	3.60	16,667	27.96	2.70	27.66	45,000
	r =	36						
Check r > 0.7 ae =	8.40							
6+00	9,000	18.00	5.67	10,582	21.16	18.00	21.24	190,476
	r =	36						
Check r > 0.7 ae =	14.37							
				MEAN = 12,712			MEAN =	165,427

Using Figure 3.3, Part II:
K_{eff} (dynamic) = 920 psi/in   INPUT
K_{eff} (static) = 460 psi/in

DETERMINE D_f

INPUT
PCC overlay modulus of rupture (psi) = 690 (mean)
PCC overlay modulus of elasticity (psi) = 4,200,000 (mean)

Vary trial D_f until computed ESALs equal future design ESALs.

K_{eff} (pci)	J	S_c (psi)	P_1	P_2	E_c (psi)	S_o	LOS	C_d
	INPUT	INPUT	INPUT			INPUT	INPUT	INPUT
460	4.0	690	4.5	2.5	4,200,000	0.35	0.00	1.00

TRIAL D_f (in)	R	Z	COMPUTED ESALs (millions)	Dol (in)
0.00	50	0	0	0.00
0.00	80	0.84	0	0.00
4.10	90	1.282	153,368	4.10
4.50	95	1.645	150,312	4.50
5.30	99	2.327	155,907	5.30

REVISED CHAPTER 5 AASHTO DESIGN GUIDE OVERLAY DESIGN

SE-5 JPCP OVERLAY OF AC PAVEMENT (US 1)

EXISTING PAVEMENT DESIGN AND FUTURE TRAFFIC

EXISTING PAVEMENT DESIGN

AC SURFACE	2.00	SUBGRADE: SAND
CR STONE BASE	8.50	
SUBBASE	12.00	
TOTAL THICKNESS	22.50	

Future design lane ESALs = 1,100,000

DETERMINE K_{eff}

Vary E_p/M_r until actual M_R*D_0/P matches computed M_R*D_0/p.

STATION	LOAD (lbs)	D_0, in (mils)	D_r, in (mils)	SUBGRADE M_R (psi)	ACTUAL M_R*D_0/P	TRIAL E_p/M_R	COMPUTED M_R*D_0/E_p	E_p
	9,000	12.96	1.86	24,604	35.43	0.80	35.63	19,683

r = 47.2

Check r > 0.7 ae = 15.19

Using Figure 3.3, Part II:
  K_{eff} (dynamic) = 1,200 psi/in   INPUT
  K_{eff} (static) =   600 psi/in

DETERMINE D_f

INPUT

PCC overlay modulus of rupture (psi) = 635 (mean)
PCC overlay modulus of elasticity (psi) = 4,000,000 (mean)

Vary trial D_f until computed ESALs equal future design ESALs.

K_{eff} (pci) INPUT	J	S_c (psi) INPUT	P_1 INPUT	P_2 INPUT	E_c (psi)	S_o INPUT	LOS INPUT	C_d INPUT
600	3.2	635	4.2	2.5	4,000,000	0.35	0.00	1.00

TRIAL D_f (in)	R	Z	COMPUTED ESALs (millions)	Dol (in)
3.80	50	0	1,173,786	3.80
5.30	80	0.84	1,127,398	5.30
5.90	90	1.282	1,114,201	5.90
6.40	95	1.645	1,108,802	6.40
7.40	99	2.327	1,162,870	7.40

Appendix N N-107

REVISED CHAPTER 5 AASHTO DESIGN GUIDE OVERLAY DESIGN

SW-24 JPCP OVERLAY OF AC PAVEMENT (Proj. 6044)

EXISTING PAVEMENT DESIGN AND FUTURE TRAFFIC

EXISTING PAVEMENT DESIGN

AC SURFACE	8.00	SUBGRADE: ?
CR STONE BASE	3.00	
SUBBASE	10.40	
TOTAL THICKNESS	21.40	

Future design lane ESALs = 11,000,000

DETERMINE K_{eff}

Vary E_p/M_r until actual M_R*D_0/P matches computed M_R*D_0/p.

STATION	LOAD (lbs)	D_0, in (mils)	D_r, in (mils)	SUBGRADE M_R (psi)	ACTUAL M_R*D_0/P	TRIAL E_p/M_R	COMPUTED M_R*D_0/E_p	E_p
	8,222	7.65	3.25	16,866	15.69	3.40	15.62	57,343

r = 36

Check r > 0.7 ae = 22.90

Using Figure 3.3, Part II:
K_{eff} (dynamic) = 1,300 psi/in   INPUT
K_{eff} (static)   = 650 psi/in

DETERMINE D_f

INPUT
PCC overlay modulus of rupture (psi) = 690 (mean)
PCC overlay modulus of elasticity (psi) = 4,200,000 (mean)

Vary trial D_f until computed ESALs equal future design ESALs.

K_{eff} (pci)	INPUT J	INPUT S_c (psi)	INPUT P_1	INPUT P_2	E_c (psi)	INPUT S_o	INPUT LOS	INPUT C_d
650	3.2	690	4.5	2.5	4,200,000	0.35	0.00	1.00

TRIAL D_f (in)	R	Z	COMPUTED ESALs (millions)	Dol (in)
7.30	50	0	11,020,270	7.30
8.40	80	0.84	11,221,706	8.40
9.00	90	1.282	11,401,365	9.00
9.50	95	1.645	11,503,539	9.50
10.40	99	2.327	11,183,953	10.40

Dowels used in transverse joints due to high traffic.
Edge drains recommended.

REVISED CHAPTER 5 AASHTO DESIGN GUIDE OVERLAY DESIGN

SW-25   JPCP OVERLAY OF AC PAVEMENT (Proj. 0512)

EXISTING PAVEMENT DESIGN AND FUTURE TRAFFIC

EXISTING PAVEMENT DESIGN

AC SURFACE	4.50	SUBGRADE: ?
CR STONE BASE	7.50	
SUBBASE	20.00	
TOTAL THICKNESS	32.00	

Future design lane ESALs = 11,000,000

DETERMINE K_{eff}

Vary E_p/M_r until actual M_R*D_0/P matches computed M_R*D_0/p.

STATION	LOAD (lbs)	D_0, in (mils)	D_r, in (mils)	SUBGRADE M_R (psi)	ACTUAL M_R*D_0/P	TRIAL E_p/M_R	COMPUTED M_R*D_0/E_p	E_p
	9,171	19.29	4.07	15,022	31.60	1.70	31.31	25,538

r = 36

Check r > 0.7 ae = 27.05

Using Figure 3.3, Part II:
K_{eff} (dynamic) = 920 psi/in   INPUT
K_{eff} (static)   = 460 psi/in

DETERMINE D_f

INPUT
PCC overlay modulus of rupture (psi)   =   690 (mean)
PCC overlay modulus of elasticity (psi) = 4,200,000 (mean)

Vary trial D_f until computed ESALs equal future design ESALs.

K_{eff} (pci)	INPUT J	S_c (psi)	INPUT P_1	INPUT P_2	E_c (psi)	INPUT S_o	INPUT LOS	INPUT C_d
460	3.2	690	4.5	2.5	4,200,000	0.35	0.00	1.00

TRIAL D_f (in)	R	Z	COMPUTED ESALs (millions)	Dol (in)
7.70	50	0	11,406,841	7.70
8.70	80	0.84	11,208,695	8.70
9.30	90	1.282	11,497,917	9.30
9.80	95	1.645	11,664,562	9.80
10.70	99	2.327	11,410,746	10.70

INDEX

AASHO ROAD TEST
    Background, ix to x, vii to viii
    In design procedures, II-3 to II-4
    Joint faulting in, II-37
    Limitations of, I-12 to I-13, II-4
    Objectives of, II-3
AGENCY COSTS, I-44
AGGREGATE INTERLOCK, III-62
AGGREGATE-SURFACED ROADS
    Aggregate loss in, II-12
    Low-volume, II-69 to II-77, II-81
    Rutting in, II-12, II-72 to II-77
AGING EFFECTS, I-8, I-13
ANALYSIS PERIOD, I-43, I-46 to I-47
    Definition of, II-6
ASPHALT-AGGREGATE SURFACE
    TREATMENTS, III-71, III-72
ASPHALT CONCRETE. *See also* Flexible
    pavements
    Layer coefficients, I-6, III-104, III-105
    Milling of, III-81
    Overlay over fractured portland cement concrete
        pavement, III-106 to III-113
    Overlays, over asphalt concrete, III-94 to III-106
    Rutting in, III-81
    Temperature effects, I-22, I-27

BASE COURSE
    Bituminous-treated, layer coefficient, II-22, II-24
    Cement-treated, layer coefficient, II-22, II-23
    Crushed stone, layer coefficient value, I-6
    Drainage, I-28
    Flexible pavement
        compaction of, I-17
        definition of, I-17
        Layer coefficient for, I-17
        materials for construction, I-17
        pozzolonic stabilized, I-17
    Granular, layer coefficients, II-17 to II-20
    Seasonal effects, I-27
    Thickness, II-35
BITUMINOUS MIXTURES
    For drainage layer, I-19
    For patching, III-63 to III-64
BONDED CONCRETE OVERLAYS, III-136 to
    III-145

BREAK AND SEAT TECHNIQUE, III-106, III-107
    to III-108

CAPITAL COSTS, I-44 to I-47
CASAGRANDE FLOW EQUATION, I-20
CBR VALUE
    Converting to resilient modulus, I-14
CITY STREETS
    Load equivalency values for, I-13
COEFFICIENT OF PERMEABILITY. *See*
    Permeability
COLD MILLING
    Prior to overlay, III-105, III-135
    Rehabilitation, III-67, III-68, III-81
COMPOSITE PAVEMENTS, I-15
    Load equivalency factors for, I-10
COMPUTERS/COMPUTER PROGRAMS
    For design, II-4
    In mechanistic-empirical design, IV-9
CONCRETE
    Shrinkage, II-28
    Tensile strength, II-28
    Thermal coefficient, II-29
CONDITION SURVEY, I-5, III-28 to III-30. *See
    also* Field data collection
CONTINUOUSLY REINFORCED
    CONCRETE PAVEMENTS. *See also*
    Rigid pavements
    Asphalt concrete over, III-113 to III-125
    Distress survey, III-28, III-30
    Friction factor, II-29
    Full-depth repairs to, III-63, III-114
    Load transfer in, II-26
    Reinforcement steel design in, II-29, II-51 to
        II-62
    Reinforcement variables in design of, II-28 to
        II-II-29
    Subsealing repairs, III-66
CORK EXPANSION JOINT FILLER, I-21
CORNER DEFLECTION ANALYSIS, III-41 to
    III-45
COST EFFECTIVENESS. *See also* Economic
    analyses
    Pavement management strategies and, I-31
CRACK AND SEAT TECHNIQUE, III-106,
    III-107

CRACKS/CRACKING
    Full-depth repair of, III-62 to III-63
    Joint load transfer efficiency, III-70 to III-71
    Process of, II-51
    Reflection crack control in overlay rehabilitation, III-145, III-153
    Reflection crack control in overlays, III-80, III-95, III-108, III-114 to III-115, III-127 to III-128, III-137
    Repairs in asphalt concrete overlay to asphalt concrete pavement, III-94 to III-95
    Sealing techniques for, III-65, III-66, III-72
CRCP. *See* Continuously reinforced concrete pavements
CREEP SPEED DEFLECTION
    Seasonal variation, I-26
CRUSHED STONE
    Layer coefficient value, I-6

DEFLECTION ANALYSIS, III-30 to III-32. *See also* Nondestructive testing
DEFORMED WIRE FABRIC
    Allowable working stress, II-28
    Development in slab, I-21
DESIGN CONSIDERATIONS. *See also* Design requirements
    Basic design equations, I-5 to I-7
    Drainage, I-27 to I-29
    Environmental effects, I-22 to I-27
    Flexible pavement
        basic equations, I-5 to I-6
        materials for construction of, I-16 to I-20
    Freezing index in, I-25
    Frost heave in, I-8
    Initial pavement smoothness in, I-8
    Local experience in, I-5
    In mechanistic-empirical design, IV-8
    In overlays, III-79, III-80 to III-83
    Pavement management systems and, I-31 to I-34, I-35
    Rigid pavement
        basic equations, I-6 to I-7
        materials for construction, I-21 to I-22
    Roadbed soil in, I-13 to I-15
    Shoulder, I-29
    Tie bars, I-22
    Traffic loads, I-10 to I-12
DESIGNED PAVEMENT SECTION, I-53 to I-54
DESIGN PERIOD
    Definition of, I-53
DESIGN REQUIREMENTS
    Aggregate loss, II-12
    Allowable rutting, II-12
    Analysis period in, II-6
    Drainage, II-22 to II-25
    Effective modulus of subgrade reaction, II-16
    Effective roadbed soil resilient modulus, II-12 to II-15
    Environmental variables in, II-10
    Input requirements, II-5, II-7
    Layer coefficients for flexible pavements, II-17 to II-22
    Limitations, II-4
    Load transfer, II-25 to II-27
    Modulus of rupture, II-16 to II-17
    Pavement layer materials characterization, II-16
    Pavement structural characteristics, II-22 to II-27
    Performance criteria in, II-10 to II-12
    Reinforcement variables, II-27 to II-29
    Reliability as variable in, II-9 to II-10
    Scope of, II-3 to II-4
    Time variables in, II-5 to II-6
    Traffic variables in, II-6 to II-9
DESIGN TRIALS, I-34, I-37
    Overlays for, I-34, I-35, I-36
DESTRUCTIVE TESTING, III-45
    Asphalt concrete overlays, III-97, III-101, III-110
    Jointed pavement evaluation, III-120
    Necessity for, III-49
    For structural capacity evaluation, III-88
DIAMOND GRINDING, III-67 to III-68, III-76
DISCOUNT RATE, I-43, I-47 to I-48, I-49
DISTRIBUTION OF LOAD
    in Continuously reinforced pavements, II-26
    Design inputs, II-25
    Directional, I-11
    In jointed pavements, II-25 to II-26
    Joint transfer load analysis, III-32, III-35, III-38 to III-41
    Lane distribution, I-11, II-6 to II-9
    Load transfer coefficient for, II-25
    In tied shoulders, II-26 to II-27
DOWELS
    Full-depth repairs, III-62
    Load-transfer, I-22
    Placement, II-25
    In restoring joint load transfer efficiency, III-71
DRAINAGE
    Climactic zone map, III-26 to III-27
    Flexible pavement design inputs, II-22 to II-25
    Major sources of water infiltration, III-65 to III-66
    Pavement surface, I-28
    Quality levels, II-22
    In rehabilitation, I-28 to I-29
    Rehabilitation survey, III-21 to III-28
    Rigid pavement design inputs, II-25

Roadbed soil considerations, I-15
Subdrainage design in rehabilitation, III-68, III-76 to III-77
Thawing effects and, I-23, I-27
Water entrapment effects, I-27 to I-28
DRAINAGE LAYER, I-28
Casagrande flow equation for, I-20
Flexible pavement
cross section, I-17, I-18
materials for construction, I-17 to I-20
Subbase as, I-16
DURABILITY ADJUSTMENT FACTOR, III-123

ECONOMIC ANALYSES. *See also* LIFE-CYCLE COSTS
Agency costs in, I-44
Analysis period in, I-46 to I-47
Basic concepts in, I-41 to I-42
Discounting in, I-47 to I-48
Equations for, I-49 to I-51
Inflation costs in, I-48 to I-49
Investment costs in, I-44 to I-47
Pavement benefits in, I-46
Pavement evaluation expenditures, III-12
Of rehabilitation, I-44 to I-45
Reliability in, I-63
Residual/salvage value, I-43, I-45
Terminology, I-42 to I-44
Transportation improvement costs in, I-42
User benefits in, valuation of, I-42 to I-44
Valuation methods, I-47
ECONOMIC ANALYSIS
Present worth calculation, I-49 to I-51
ELASTIC MODULUS, II-16. *See also* RESILIENT MODULUS (M_r)
Correlation for portland cement concrete, II-16
ENVIRONMENTAL EFFECTS. *See also* Temperature effects
In deflection testing, III-32
As design variables, II-10
Resilient modulus varying with, I-23, I-24, I-25 to I-27, II-13
Seasonal variation in resilient modulus, II-13
Serviceability and, I-8, I-9, II-10, II-11
EQUIVALENT SINGLE AXLE LOADS (ESALS), I-10
In calculating performance period, I-10 to I-12
In overlay design procedures, III-80 to III-81
In reliability calculations, I-54, I-55
Truck equivalency factors in estimating, I-10
EQUIVALENT UNIFORM ANNUAL COST (BENEFIT), I-43, I-47, I-51
Equation for, I-49

ESALS. *See* EQUIVALENT SINGLE AXLE LOADS (ESALS)
EVALUATION PROCEDURES
For pavement rehabilitation, I-5
EXPANSION JOINTS
In rehabilitation, III-69 to III-70
Sealing materials, I-21, II-50

FIELD DATA COLLECTION. *See also* Nondestructive testing
Asphalt concrete/continuously reinforced concrete, III-129, III-132
Asphalt concrete/jointed pavements, III-129, III-132
Asphalt concrete over portland cement concrete, III-109, III-110
Asphalt concrete pavements, III-96, III-97
Condition survey, III-28 to III-30
Continuously reinforced concrete evaluation, III-147
Destructive testing for, III-45, III-49
Drainage survey, III-21 to III-28
Functional condition, assessment of, III-60
Goals of, III-19
Jointed pavement evaluation, III-117 to III-120, III-138, III-140, III-141 to III-143, III-146 to III-147
Jointed pavement overlays, III-154
Joint load transfer efficiency, III-70 to III-71
Limits of statistical accuracy, III-49 to III-50 to III-57
Major parameters of, III-49
In mechanistic-empirical design, IV-8 to IV-9, IV-10
For nonoverlay rehabilitation, III-59 to III-60
For overlay design, III-83 to III-84
For partial-depth pavement repair, III-64
Pavement response variables, III-19 to III-21
Rehabilitation concepts, III-9 to III-12
Sampling tests, III-45, III-49
For structural capacity evaluation, III-86 to III-88
Unit of analysis in, III-19 to III-21, III-49
FILTER LAYER
Flexible pavement, I-20
FLEXIBLE PAVEMENTS
Asphalt concrete overlays over, III-94 to III-105
Base course, I-17
Basic design equations, I-5 to I-6
Cold milling in rehabilitation of, III-67, III-68
Cross section, I-3, I-4
Distress survey, III-28, III-29
Drainage design inputs, II-22 to II-25
Drainage effects, I-28

Drainage layer, I-17 to I-20
Effective roadbed resilient modulus, II-13 to II-15
Filter material, I-20
Frost heave in, II-33 to II-35
Layered design analysis, II-35 to II-37
Low-volume road design, II-69, II-77 to II-81
Materials for construction, I-16 to I-20
Mechanistic-empirical design procedures for, IV-3
Nonoverlay rehabilitation strategies, III-60, III-61
Prepared roadbed, definition of, I-16
Resilient Modulus for, I-15 to I-16
Roadbed swelling, II-33 to II-35
Selection of layer thickness, II-35
Shoulder design and, I-29
Stabilization materials, I-16
Stage construction of, II-33
Structural capacity of, III-85
Structural capacity survey, III-87
Structural number design nomograph, II-31 to II-32
Subbase course, I-16 to I-17
Surface course, I-20
FOG SEAL, III-72
FREEZING INDEX, I-25
FREEZING-THAWING
Effects of, I-23 to I-27
Seasonal variations, I-23, I-24
FRICTION
Functional evaluation of, III-84
Resistance at pavement-tire interface, I-7
FRICTION FACTORS
Continuously reinforced concrete pavement, II-29
Jointed reinforced concrete pavements, II-28
FROST HEAVE
Cause of, I-23
Design considerations, I-8
Effect on present serviceability index, I-8
In flexible pavement structural design, II-33 to II-35
Predicting frost penetration, I-25
In rigid pavements, II-47 to II-48
Roadbed effects, I-14 to I-15, I-23 to I-27
FULL-DEPTH REPAIRS, III-62 to III-64, III-76, III-114
FUNCTIONAL PERFORMANCE
Definition of, I-7
Structural performance vs., in overlay design, III-81
Surface friction evaluation, III-84
Surface roughness evaluation, III-84 to III-85

GRANULAR LAYERS
Base, coefficients for, II-17 to II-20

Drainage assessment of, III-25, III-26
Subbase, coefficients for, II-20 to II-22
GROUT MIXTURES, III-66 to III-67

HIGHWAY INVESTMENT COST, I-42
HIGHWAY MAINTENANCE COST, I-42, I-44
HIGHWAY USER COSTS, I-42, I-44, I-45 to I-46
HYDROPLANING EFFECTS, III-84

INCREMENTAL COSTS, I-43
INFLATION, ECONOMIC, I-48 to I-49
INITIAL SERVICEABILITY INDEX
Definition of, I-8
In performance criteria, II-10
INTEREST RATE, I-43
INVESTMENT COSTS, I-44 to I-47

J-FACTORS. *See* Distribution of load
JOINTED PAVEMENTS
Asphalt concrete overlay over, III-113 to III-125
Distress survey, III-28, III-29
Full-depth repair of, III-62 to III-63
Load transfer in, II-25 to II-26
Overlay design, III-153 to III-156
Plain, II-27
Reinforced concrete, II-27
friction factor, II-28
slab length, II-27 to II-28
steel reinforcement design nomograph, II-51
steel working stress, II-28
Restoration of joint load transfer, III-70 to III-71
Slab length, design variables in, II-27 to II-28
Subsealing repairs, III-66
JOINT LOAD TRANSFER
Analysis, III-32, III-35, III-38 to III-41
Asphalt concrete/continuously reinforced concrete pavements, III-131 to III-132
Restoration, III-70 to III-71
JOINTS
In bonded concrete overlay, III-143
Cracks as, III-66
Dimensions of, II-49 to II-50
In jointed pavement overlays, III-155
Layout of, II-49
Load transfer efficiency, restoring, III-70 to III-71
Longitudinal, I-22
Nondowelled, J-factor for, II-37
Portland cement concrete overlay, III-82
Pressure relief, III-69 to III-70, III-76
Rigid pavement, structural design, II-48 to II-50
Role of, II-48 to II-49
Sealing of, I-21, II-50, III-65 to III-66, III-76
Slab-void detection, III-32, III-35, III-41 to III-45

Index

Spacing of, II-49
Transverse, I-65 to II-66
  load-transfer devices for, I-22
In unbonded overlays, III-151

JOINT SEALING MATERIALS
Categories of, III-66
In construction joints, II-50
In contraction joints, II-50
Cork expansion joint filler, I-21
In expansion joints, II-50
Liquid, I-21
Preformed elastomeric, I-21

LANE DISTRIBUTION FACTORS, I-11, II-6 to II-9
LAYER COEFFICIENTS
Asphalt concrete surface course, II-17, II-18
Average values, I-6
Bituminous-treated base, II-22, II-24
Cement-treated base, II-22, II-23
Drainage, II-22 to II-25
For flexible pavement base course design, I-17
For flexible pavement subbase design, I-16
Granular base, II-17 to II-20
Granular subbase, II-20 to II-22
For in-service asphalt concrete pavements, III-104, III-105
Resilient modulus test for, II-3, ix
Role of, II-17

LAYERED DESIGN ANALYSIS, II-35 to II-37
LIFE-CYCLE COSTS
Definition of, I-41
In design trials, I-34, I-36
Discounting and, I-47 to I-48, I-49
Inflation in, I-48 to I-49
Mechanistic-empirical design procedures and, IV-7
Of nonoverlay rehabilitation techniques, III-73 to III-74
Pavement management systems and, I-34
Role of, in economic analysis, II-31
In selecting rehabilitation solution, III-15

LOAD EQUIVALENCY FACTORS
Limitations of, I-12 to I-13
Source of, I-10
For urban streets, I-13

LOAD TRANSFER. *See* Distribution of load
LOAD-TRANSFER DEVICES, I-22
LOCAL CONDITIONS
Climactic zones, map of, III-26 to III-27
In design considerations, I-5
In drainage evaluation, III-25

Effective roadbed resilient modulus calculations for, II-13
Freezing index, I-25
regional season length, map of, II-69, II-70, II-71
Seasonal effects, I-27

LONGITUDINAL JOINTS, I-22
LOSS OF SUPPORT
Design inputs, II-27
Role of, II-37

LOW-VOLUME ROADS
Aggregate-surfaced, II-69 to I-77, II-81
Design catalog, II-77 to II-86
Flexible pavement, II-69, II-77 to II-81
Rigid pavement, II-69, II-81

MAINTENANCE COSTS, I-42, I-44
MATERIALS OF CONSTRUCTION
Of flexible pavements, I-16 to I-20
Nonstandard, I-3
Overlay considerations, III-81
Pavement layer, characterization of, II-16
Recycling, I-45
Reinforcement, design variables in, II-27 to II-29
Rigid pavement, I-21 to I-22

MECHANISTIC-EMPIRICAL DESIGN PROCEDURES, x
Background, IV-3 to IV-4
Benefits of, IV-4, IV-10
Design considerations, IV-8
Equipment for, IV-9
Framework for, IV-4 to IV-7
Implementation, IV-7 to IV-10
Input data, IV-8 to IV-9
Testing procedures, IV-10
Training personnel in, IV-9 to IV-10

MODULUS OF RUPTURE
Concrete, II-28
Field data collection for, III-49
Portland cement concrete, II-16 to II-17

MODULUS OF SUBGRADE REACTION, II-16
Estimating, II-37 to II-44
In low-volume road design, II-69
Variables in, II-37

MOTOR VEHICLE RUNNING COSTS, I-42, I-46

NETWORK LEVEL PAVEMENT MANAGEMENT, I-31, I-39
Project feasibility analysis in, I-41
Rehabilitation considerations in, III-12

NONDESTRUCTIVE TESTING
Asphalt concrete pavements, III-101 to III-102
Deflection interpretation in, III-30 to III-32

Evaluating structural capacity, III-32, III-35 to III-38
Joint load transfer analysis, III-32, III-35, III-38 to III-41
Slab-void detection, III-32, III-35, III-41 to III-45
For structural capacity evaluation, III-88
For subgrade resilient modulus, III-91
In thickness deficiency approach, III-4
Types of, III-32, III-35, iii-45

NONOVERLAY REHABILITATION STRATEGIES
Cold milling, III-67, III-68
Development of, III-60 to III-62
Diamond grinding, III-67 to III-68, III-76
Evaluation of pavement condition for, III-59 to III-60
Full-depth repair, III-62 to III-64, III-76
Partial-depth repair, III-64 to III-65, III-76
Performance prediction of, III-73 to III-78
Pressure relief joints, III-69 to III-70, III-76
Restoring joint load transfer efficiency, III-70 to III-71
Subdrainage design considerations in, III-68, III-76 to III-77
Surface treatments, III-71 to III-73, III-76

OPEN-GRADED FRICTION COURSES, III-71, III-72

OVERLAYS. *See also* Rehabilitation
Adjustments to thickness design considerations, III-83
Alternatives to, I-5
Asphalt concrete over asphalt concrete
construction tasks, III-94
feasibility, III-94
preoverlay repair, III-94 to III-95
reflection crack control, III-95
shoulders, III-105
structural number calculations, III-101 to III-104
subdrainage, III-95
surface milling, III-105
thickness design, III-95 to III-105
widening, III-106
Asphalt concrete over asphalt concrete/continuously reinforced concrete
construction tasks, III-125
feasibility, III-125 to III-127
preoverlay repair, III-127
reflection crack control, III-127 to III-128
shoulders, III-135
subdrainage, III-128
surface milling, III-135
thickness design, III-128 to III-135
widening, III-136
Asphalt concrete over asphalt concrete/jointed pavements
construction tasks, III-125
feasibility, III-125 to III-127
preoverlay repair, III-127
reflection crack control, III-127 to III-128
shoulders, III-135
subdrainage, III-128
surface milling, III-135
thickness design, III-128 to III-135
widening, III-136
Asphalt concrete over continuously reinforced concrete
construction tasks, III-113
feasibility, III-113
preoverlay repairs, III-113 to III-114
reflection crack control, III-114 to III-115
shoulders, III-125
subdrainage, III-115
thickness design, III-115 to III-125
widening, III-125
Asphalt concrete over jointed pavements
construction tasks, III-113
feasibility, III-113
preoverlay repair, III-113 to III-114
reflection crack control, III-114 to III-115
shoulders, III-125
subdrainage, III-115
thickness design, III-115 to III-125
widening, III-125
Asphalt concrete over portland cement concrete
construction tasks, III-106 to III-107
feasibility, III-107 to III-108
pavement widening in, III-82 to III-83, III-111
preoverlay repair, III-108
reflection crack control, III-108
shoulders, III-111
subdrainage, III-108
thickness design, III-108 to III-111
Bonded concrete
bonding procedures, III-145
construction tasks, III-136
feasibility, III-136 to III-137
joints, III-143
preoverlay repair, III-137
reflection crack control, III-137
shoulders, III-143
subdrainage, III-137
thickness design, III-137 to III-143
widening, III-145

Design considerations
  existing portland cement concrete slab durability, III-82
  levels of reliability, III-82
  materials, III-81
  milling asphalt concrete surfaces, III-81
  pavement evaluation, III-83 to III-94
  pavement widening, III-82 to III-83
  portland cement concrete overlay joints, III-82
  portland cement concrete overlay reinforcement, III-82
  preoverlay repair, III-80
  recycling existing pavement, III-81
  reflection crack control, III-80, III-108
  resilient modulus in, III-91 to III-94
  rutting in asphaltic concrete, III-81
  shoulders, III-81 to III-82
  structural, vs. functional, overlays, III-81
  subdrainage, III-81
  traffic loadings, III-80 to III-81
Feasibility of, III-79 to III-80
Functional evaluation for, III-84 to III-85
Jointed pavement
  construction tasks, III-153
  feasibility, III-153
  joints, III-155
  preoverlay repair, III-153
  reflection crack control, III-153
  reinforcement, III-155
  shoulders, III-155
  subdrainage, III-153
  thickness design, III-154 to III-155
  widening, III-155
  worksheet, III-156
Mechanistic-empirical design procedures for, IV-4, IV-7
Point-By-Point Approach, III-84
Role of, III-79
Skid-resistance considerations, III-3
Structural analysis methodology, III-3 to III-4
Structural evaluation for, III-85 to III-91
Thickness deficiency concept of, III-4
Types of, III-79
Unbonded
  construction tasks, III-145
  Joints, III-151
  preoverlay repair, III-145
  reflection crack control, III-145
  reinforcement, III-151
  separation interlayer, III-153
  shoulders, III-151
  subdrainage, III-146
  thickness design, III-146 to III-151
Uniform section approach, III-84

PARTIAL-DEPTH PAVEMENT REPAIR, III-64 to III-65, III-76
PATCHING, III-63 to III-64
  Partial-depth, III-64 to III-65
PAVEMENT BENEFITS, I-46
PAVEMENT DESIGN-PERFORMANCE PROCESS, I-56 to I-62
PAVEMENT MANAGEMENT SYSTEM (PMS)
  Definition, I-31
  Design and, I-31 to I-34, I-35
  Guidelines for, I-39
  Mechanistic-empirical design procedures in, IV-4
  Network level in, I-31, I-32, I-39
  Pavement type selection in, I-39
  Project level in, I-31, I-32
  Role of, I-31, I-34, I-36
  State use of, I-34, I-39
PAVEMENT PERFORMANCE
  Definition of, I-56
  Drainage effects, I-28
  Elements of, I-7 to I-8
PERFORMANCE PERIOD
  Definition of, I-10 to I-11, II-5 to II-6
  Maximum, II-6
  Minimum, II-6
PERFORMANCE TRENDS, I-8, I-9
PERMEABILITY
  Of drainage layer materials, I-17, I-19
PMS. See PAVEMENT MANAGEMENT SYSTEM
POROUS CONCRETE LAYERS, I-21
PORTLAND CEMENT CONCRETE
  Asphalt concrete overlay on, III-106 to III-113
  Diamond grinding of, III-67 to III-68, III-76
  Elastic modulus correlation for, II-16
  Frost penetration, calculating, I-25
  Modulus of rupture, II-16 to II-17
  Overlay design considerations, III-82
  Specifications for, I-21
PREPARED ROADBED
  Definition of, I-16
PRESENT SERVICEABILITY INDEX (PSI)
  In definition of pavement performance, I-56
  Environmental effects on, I-8, I-9, I-27
  Minimum, I-44
  Pavement life-cycle and, I-8
  In reliability calculations, I-54, I-55
  Role of, II-10
  Source of, I-7 to I-8
PRESENT VALUE, I-43
PRESENT WORTH
  Calculation for, I-49 to I-51
PRESSURE RELIEF JOINTS, III-69 to III-70, III-76

PRESTRESSED CONCRETE PAVEMENTS
  Fatigue in, II-66
  Magnitude of prestress in, II-66
  Pavement thickness, II-66 to II-67
  Prestress losses, II-67
  Slab length, II-65 to II-66
  Structural design, II-65, II-66 to 67
  Subbase, II-65
  Subgrade restraint, II-67
  Tendon spacing, II-66

RECYCLING, I-45
  Rehabilitation and, III-7, III-81
REHABILITATION. *See also* Nonoverlay
    rehabilitation strategies; Overlays
  Construction considerations, III-7
  Cost analysis in, III-15 to III-16
  Definition of, I-45, III-7
  Drainage considerations in, I-28 to I-29, III-21
  Evaluation procedures, I-5
  Initial capital costs in, I-44 to I-45
  Limitations, III-4
  Maintenance vs., III-7
  Major categories of, III-7
  Major factors in, III-7 to III-8
  Method of, selection process for, III-8 to III-16
  Nonmonetary considerations in, III-15 to III-16
  In performance-based approach, III-3
  Problem definition in, III-9 to III-12
  Project constraints, III-12
  Project-specific decision-making in, III-8 to III-9
  Recycling concepts in, I-45, III-7, III-81
  Slab subsealing, III-41, III-66 to III-67, III-76
  Traffic delay costs in, I-46
REINFORCEMENT. *See also* Jointed pavements,
    reinforced concrete
  Design variables, II-27 to II-29
  In jointed pavement overlays, III-155
  Rigid pavement, structural design, II-51 to II-65
  Steel, in rigid pavement slab, I-21
  Transverse, II-62 to II-65
  In unbonded overlays, III-151
RELIABILITY
  Axle load variables in, I-54
  Compounding of, in stage construction, II-33,
    II-44
  Definition of, I-53
  Definition of pavement condition in, I-54 to I-56
  Design factor, I-56 to I-62
  As design variable, II-9 to II-10
  Factor, I-6 to I-7, I-12, II-9
  Level, I-60 to I-63, II-9
  Overall standard deviations in calculating, I-62

  In overlay design, III-82
  Pavement performance variables in, I-56
  Probability distribution of basic deviations, I-57
    to I-60
  Role of, II-3
  Stage construction alternatives and, I-63
REMAINING LIFE EVALUATION, III-88 to III-91
  Asphalt concrete pavements, III-104 to III-105
  Continuously reinforced concrete pavements,
    III-123 to III-125
  Jointed pavement, III-123 to III-125, III-143
  Portland cement concrete pavements, III-151
RESIDUAL VALUE, I-43, I-45
RESILIENT MODULUS (Mr)
  CBR conversions to, I-14
  Climactic region and, II-71
  Definition of, I-13
  Design value, I-15
  Direct measurement of, ix
  Effective annual, I-27
  Effective roadbed soil, II-12 to II-15
  Laboratory, vs. field, tests for, II-17
  For low-volume roads, II-69, II-71
  Overlay design, III-91 to III-94, III-96 to III-97
  Role of, I-13 to I-14
  R-value conversions to, I-14
  Seasonal variations, I-23, I-24, I-25 to I-27, II-13
RIGID PAVEMENTS. *See also* Continuously
    reinforced concrete pavements; Jointed
    pavements
  Application of rehabilitation procedures to, III-4
  Basic design equations, I-6 to I-7
  Cross section, I-3, I-4
  Diamond grinding in rehabilitation of, III-67 to
    III-68, III-76
  Drainage design inputs, II-25
  Drainage effects, I-28
  Frost heave in, II-47 to II-48
  Joint dimensions, II-49 to II-50
  Joint layout, II-49
  Joint load transfer analysis, III-32, III to 35,
    III-38 to III-41
  Joint sealant dimensions, II-50
  Joint sealing materials, I-21, II-50, III-66
  Joint spacing, II-49
  Joint types, II-48 to II-49
  Load transfer in, I-22, II-25
  Longitudinal joint materials, I-22
  Loss of support in, II-27
  Low-volume road design, II-69, II-81
  Mechanistic-empirical design procedures for,
    IV-3
  Modulus of rupture for, II-16 to II-17

Modulus of subgrade reaction, II-16, II-37 to II-44
Nonoverlay rehabilitation strategies, III-60, III-61
Partial-depth repair, III-64 to III-65
Patching, with bituminous mixtures, III-63 to III-64
Pavement slab, I-21
Portland cement concrete for, I-21
Reinforcing steel in, I-21
Roadbed swelling in, II-47 to II-48
Slab thickness design nomograph, II-44
Slab-void detection, III-32, III-35, III-41 to III-45
Stage construction, II-44 to II-47
Structural capacity, III-85, III-87
Subbase, I-21
Subsealing of, III-66 to III-67, III-76
Tie bars in, I-22
Transverse reinforcement design, II-62 to II-65

ROADBED SOIL(S)
Compaction criteria, I-14
In design equation, I-53 to I-54
Drainage considerations, I-15
Exceptional types of, I-3, I-14 to I-15
Expansive soils, I-14
Freezing-thawing in, I-23 to I-27
Frost effects, I-14 to I-15, I-25, I-26
Highly organic, I-15
Placement considerations, I-14 to I-15
Resilient modulus of, I-13 to I-15, II-12 to II-15, II-16
Resilient soils, I-14
Subbase course design and, I-16
Swelling soils in, I-8, II-33 to II-35, II-47
Thaw-weakening effects, I-25 to I-27

ROAD OILING, III-72

ROUGHNESS
Functional evaluation of, III-84 to III-85
In Present Serviceability Index, I-7 to I-8
User costs and, I-45 to I-46

RUBBERIZED ASPHALT SEAL, III-71 to III-72

RUBBLIZE AND COMPACT TECHNIQUE, III-106 to III-107

RUTTING, I-27
Allowable, II-12
In low-volume aggregate-surfaced road design, II-72 to II-77
In overlay design procedures, III-81

R-VALUE
Converting to resilient modulus, I-14

SAFETY CONSIDERATIONS, I-7
SALVAGE VALUE, I-43, I-45
SAND SEAL, III-72

SANDY GRAVEL
layer coefficient value, I-6

SEASONAL EFFECTS, I-23 to I-27. *See also* Environmental effects
Climactic zones, map of, III-26 to III-27
Effective resilient modulus, I-26, II-13
In modulus of subgrade reaction calculations, II-37, II-44
Regional season length, map of, II-69, II-70, II-71

SEPARATION INTERLAYER, III-153

SERVICEABILITY
Definition of, II-10
In design trials, I-34, I-37
Environmental effects on, I-8, II-10, II-11
Factors in loss of, I-8
Frost-heave effects, I-23
Initial serviceability index, I-8, II-10
Pavement benefits and, I-46
Present serviceability index (PSI), I-7 to I-8, I-9, I-44, I-54, I-55, I-56, II-10
Terminal serviceability index, I-8, II-10

SERVICEABILITY-PERFORMANCE CONCEPT
Basis of, I-7

SHOULDERS, I-22
Definition, I-29
Design criteria, I-29
Overlay design considerations, III-81 to III-82, III-105, III-111, III-125, III-135, III-143, III-151, III-155
Tie bars in, II-26
Tied
definition of, II-26 to II-27
load transfer calculations for, II-26 to II-27

SKID-RESISTANCE
Of overlays, III-3

SLAB LENGTH
Jointed reinforced concrete pavements, II-27 to II-28
Prestressed concrete pavement, I-65 to II-66

SLAB-VOID DETECTION, III-32, III-35, III-41 to III-45

SLURRY SEAL, III-72

SMOOTH WIRE MESH
Development in slab, I-21

SOIL(S). *See* ROADBED SOIL(S)
SOIL SUPPORT NUMBER, II-3
SOIL SUPPORT VALUE
Definitive test for, ix

STABILIZATION
Flexible pavement base, I-16, I-17

STAGE CONSTRUCTION
Compounding of reliability in, II-33, II-44

Cumulative traffic calculations in, II-6
Design considerations, I-12
Flexible pavement, II-33
Reliability and, I-63
Rigid pavement, II-44 to II-47

STEEL REINFORCEMENT
Allowable working stress, II-28, II-53, II-56
Bar/wire diameters, II-29
In CRCP, design variables of, II-28 to II-II-29
Design nomograph for jointed reinforced concrete, II-51, II-52
Design procedure for CRCP, II-51 to II-62
Jointed pavement design variables, II-27 to II-28
Role of, II-51
Thermal coefficient, II-29
Transverse, II-62 to II-65

STEEL WORKING STRESS, II-28

STRESS-ABSORBING MEMBRANE INTERLAYER, III-72, III-95

STRESS STATE
Flexible pavement, I-15 to I-16

STRUCTURAL CAPACITY
Evaluation for overlays, III-85 to III-91
Nondestructive deflection analysis of, III-32, III-35 to III-38
Remaining life evaluation of, III-88 to III-91

STRUCTURAL NUMBER
Asphalt concrete pavements, determination of, III-101 to III-104
Flexible pavement design nomograph, II-31 to II-32
For rigid pavements, III-110 to III-111
Role of, I-6

STRUCTURAL PERFORMANCE
Definition of, I-7
Functional performance vs. in overlay design, III-81

SUBBASE COURSE
Drainage, I-28, III-25, III-26
Flexible pavement
  as drainage layer, I-16
  layer coefficients for, I-16
  materials, I-16 to I-17
  role of, I-16 to I-17
Friction factors, II-28, II-29
Granular, layer coefficient, II-20 to II-22
Prestressed concrete pavement, I-65
Rigid pavement, I-21
Sandy gravel, layer coefficient value, I-6
Seasonal effects, I-27
Subsealing repairs, III-41, III-66 to III-67, III-76
Thickness, II-35

SUBDRAINAGE
Design, III-68, III-76 to III-77
In overlay design procedures, III-81, III-87 to III-88, III-95, III-108, III-115, III-137, III-146, III-153
Survey, III-87 to III-88

SUBGRADE RESTRAINT
In prestressed concrete pavements, II-67

SUBSEALING TECHNIQUES, III-41, III-66 to III-67, III-76

SURFACE COURSE
Asphalt concrete, layer coefficients, I-6, II-17
Drainage, I-28, III-25, III-26
Flexible pavement, I-20
Functional assessment for overlay design, III-84 to III-85
Nonoverlay rehabilitation of, III-71 to III-73, III-76
Thickness, II-35

SURFACE TREATMENTS, III-71 to III-73, III-76

SWELLING SOILS
Effect on pavement serviceability, I-8
In flexible pavement structural design, II-33 to II-35
In rigid pavements, II-47 to II-48

TEMPERATURE EFFECTS, I-22 to I-23. *See also* Freezing-Thawing; Frost heave
Asphalt concrete performance, I-27
Climactic zones, map of, III-26 to III-27
Concrete thermal coefficient, II-29
Design temperature drop, II-29
Freezing index, I-25
Steel thermal coefficient, II-29
Thaw-weakening, I-25 to I-27

TENDONS
In prestressed concrete pavement, II-66

TERMINAL SERVICEABILITY INDEX
Definition of, I-8
Public acceptance in defining, II-10

TERMINOLOGY, I-3, I-4
Economic analysis, I-42 to I-44

THAWING. *See* Freezing-thawing

THAW-WEAKENING EFFECTS, I-25 to I-27

THERMAL COEFFICIENT
Concrete, II-29
Steel, II-29

TIE BARS, I-22
Design procedure, II-62 to II-65
Full-depth repairs, III-62
Placement in tied shoulders, II-26

TIME
Constraints, as design variable, II-5 to II-6
Cumulative ESALS vs., II-8

Index

Environmental serviceability loss vs., II-10, II-11
And expansion joint installation, III-70
Fatigue in prestressed concrete pavement, II-66
Interactive effects of, I-8, I-13
In life-cycle costing, I-47 to I-48
Value of, in economic analysis, I-42, I-43

TIRE INFLATION, I-12

TRAFFIC
Accident costs, I-42, I-46
Delays, cost of, I-46
As design variable, II-6 to II-9
Estimating growth in, I-11 to I-12
Evaluation of, I-10 to I-12
Mixed, converting to ESALs, I-10
Reliability factors in estimating, I-12
Scheduling repairs around, III-63, III-70

TRANSPORTATION IMPROVEMENT COSTS, I-42
TRIDEM AXLES, I-10
TRUCK EQUIVALENCY FACTORS, I-10
Truck weight information for, I-11 to I-12

UNDERCUTTING, III-62
UNIT VALUE OF TIME, I-42
USER BENEFITS, I-42 to I-44, I-46
Indirect, I-50
USER COSTS, I-42, I-44, I-45 to I-46

VALUE OF TRAVEL TIME, I-42, I-46
VEHICLE TRAVEL TIME, I-42, I-46

WELDED WIRE FABRIC
Allowable working stress, II-28